SCIENCE *as a* PROCESS

An Evolutionary Account
of the Social and Conceptual
Development of Science

DAVID L. HULL

The University of Chicago Press
Chicago and London

The University of Chicago Press, Chicago 60637
The University of Chicago Press, Ltd., London

Paperback edition 1990
Printed in the United States of America

07 06 05 04 03 02 01 00 99 98   4 5 6 7 8

Library of Congress Cataloging-in-Publication Data

Hull, David L.
Science as a process: an evolutionary account of the social and conceptual development of science / David L. Hull.
p. cm.—(Science and its conceptual foundations)
Bibliography: p.
Includes index.
ISBN: 0-226-36051-2 (paperback)
1. Science. 2. Science—Philosophy. 3. Science—History.
I. Title. II. Series.
Q158.5.H84 1988
575.016—dc19   88-2743
CIP

This book has been brought to publication with the generous assistance of the Publication Subvention Program of the National Endowment for the Humanities, an independent federal agency which supports such fields as history, philosophy, literature, and languages.

Additional support for the publication of this book has been provided through the generosity of Nancy Gray Puckett.

∞ The paper used in this publication meets the minimum requirements of the American National Standard for Information Sciences—Permanence of Paper for Printed Library Materials, ANSI Z39.48–1992.

To Bella
who still retains
her unreal expectations
of intellectuals

Contents

Cont[illegible]

Illustrations and Tables

Tables

Preface

This book has had three successively more proximate causes. In general I had become increasingly dissatisfied through the years with the logical empiricist analysis of science that had been so popular for over a generation. I did not find the "received view" so much mistaken as too far removed from the ongoing process of science. Too much of science was being left out. But I was no happier with the critics of the logical empiricist analysis of science. Some of these critics, operating within the logical empiricist tradition, presented convincing objections to the received view but offered nothing to put in its place. A second group of critics was bent on debunking science at all costs. Science was the enemy and had to be smashed for the "good of the people." Because science did not possess the ideal characteristics that the "positivists" insisted that it should, knowledge-claims made by scientists had no more warrant than those of magicians, faith healers, and politicians. All is one, and anything goes. Something had to exist, other than the rarefied abstractions of logical empiricist philosophers of science and the wild-eyed proclamations of the more radical critics, but what? That was the problem.

Then, in April 1973, Bert Rowell, the editor of *Systematic Zoology*, sent me a manuscript by Gareth Nelson to referee. In this manuscript, Nelson (1973c) complained of the way that the views of Leon Croizat had been treated through the years by such authorities as G. G. Simpson and Ernst Mayr. I decided that the sort of thing Nelson was investigating with respect to Croizat was the sort of thing I would like to do in philosophy of science. What is the relative importance in science of reason, argument, and evidence on the one hand, and power, prestige, and influence on the other? I thought that answers couched totally in terms of one sort of influence or the other were sure to be wrong and that the interplay between the two was likely to be fascinating. Because I had been involved in the systematics community, I decided to study the fate of a particular group of systematists—the pheneticists, or numerical taxonomists, at the University of Kansas in Lawrence, Kansas. To study these systematists, I had to interact with them—attend meetings, become involved in professional societies, referee papers, etc. While I was studying the pheneticists, I realized that Nelson, without intending to, had given rise to a new "school"—the cladists, or phylogeneticists. I decided to study them as well.

The scientists I have studied investigate such things as fruit flies, fossil fish, and slime molds. As diverse as these organisms are, they have one thing in common: they cannot read the conclusions published about them. My subjects can. One of my goals in this book has been to present a fair and balanced estimate of the influence of various factors, including professional allegiances and alliances, on the course of systematics and evolutionary biology. I obviously have my biases and personal preferences. I also have my own professional goals. Perhaps there is no such thing as objectivity in the abstract, but at least there can be third-party objectivity. For most disputes, I have no strong preference for one position or the other; for some I do. When I do, I say so explicitly and try to present all sides as fairly as possible. One sign that I may have attained this elusive goal of fairness has been that members of all camps find that I have treated their enemies too gently.

I have not even tried to neutralize the effects of my professional goals. I could not, on pain of contradiction. I very much want others to accept my view of the factors that make science function as it does. I did not begin my research with the general view presented in this book fully worked out. As I studied science and scientists, I was forced to change my mind on a variety of topics, but I also did not proceed inductively. There was a complex interplay between my convictions and my experience, the sort of interplay that characterizes all scientific investigations. One conclusion I found inescapable is that the professional relations among scientists influence, at least locally, the content of science. As a result, I spend a good deal of time discussing these relations. I have tried not to indulge in gossip for the sake of gossip without omitting details that are relevant to understanding how science works. I suspect that I have erred on the side of conservatism rather than excess, but others will have to decide this. In consequence of my sustained professional and personal relations with many of the scientists whom I discuss in this book, I have acquired a responsibility to them. I hope I have not compromised it. If I am mistaken on this score, I have total confidence that my lapses will not go unmentioned.

The research for this book was supported by grants from the National Science Foundation (1975–80, 1986) as well as the John Simon Guggenheim Memorial Foundation (1980–81). Without this support I could not have written the book. Numerous friends, students, and colleagues have read the book in its entirety at various stages in its production. I owe special appreciation to Daniel Brooks, David Buller, Stephen Kellert, Edwin Kyle, Ernst Mayr, Joseph Pearson, Robert Richards, Bella Selan, Elliott Sober, and an anonymous referee. Robert Sokal and E. O. Wiley read the entire first half of the book. Steven Farris, Gareth Nelson, and Norman Platnick commented primarily on the parts that dealt with cladistic analysis, while Robin Craw and John Grehan both supplied information about Leon Croizat and suggested improvements in those parts of my manuscript dealing with his theory of panbiogeography. I owe special thanks to David Krantz for supplying the questionnaire he used in his study of the founders of behavioral psychology, as well as to the following people who provided me with photographs: Jon Buskin, George Byers, Gerrit Davidse, Steven Farris, David Kitts, and Gareth Nelson. Joel Cracraft, Vickie Funk, and Pamela Henson also provided very important information, not to mention personal support. Berndt Ehmann prepared the figures. During the fifteen years while I worked on this book, time has marched on, changes have occurred. The most grievous is the untimely death of Donn Rosen. We all have our own memories of our relationship with this complex and engaging man. I appreciate the time and effort that all of these readers generously contributed to making this book more accurate and well balanced.

1

Science, Philosophy of Science, and the Science of Science

The most difficult job for the historian is to develop a double vision, seeing his subjects' choices both as they saw them and as he, the retrospective outsider, sees them, free of the pressures that made them gasp and rage. He does not increase wisdom by laughter at their folly, by indignation at their tyranny, or by sentimental substitutes for ridicule and anger. The historian's scorn, rage, or facile charity are really self-congratulation at bottom.

David Joravsky, *The Lysenko Affair*

The general problem is to show that philosophical conclusions may be supported by historical facts and just how this comes about. Until this is done, the historical approach to philosophy of science is without a conceptually coherent programme.

Ronald Giere, "History and Philosophy of Science: Intimate Relationship or Marriage of Convenience?"

THIS BOOK concerns science. How do scientists choose between alternative views of the world? Through the years, certain evolutionary biologists have come to believe that biological evolution is largely gradual, while others have opted for evolution occurring in relatively discrete steps. At times the gradualists have been in the majority; at other times more saltatory views have prevailed. Beliefs about the role of natural selection have undergone similar vicissitudes, its importance rising and falling in the estimation of evolutionary biologists. What has caused these conceptual changes?

Those of us who study science itself are in anything but agreement about the correct answers to these questions. At one extreme are those who think that, by and large, scientists accept, reject, or ignore new ideas because of the weight of evidence and cogency of arguments. Because these factors are commonly held to be part of science, proponents of this view of science are termed "internalists." At the other extreme are the "externalists," those who believe that a wide variety of other factors are actually operative, factors as diverse as benthic social forces and idiosyncratic personal motives. According to Shapin and Barnes (1977: 60), "people cannot be controlled by ideas"; the only way that social order can be effectively promoted or broken down is through "coercion, the manipulation of rights, or the generation of interests." They go on to argue that what is true of society at large is equally true of science. If other special-interest groups function primarily to promote their own interests, there is no reason to expect scientists to behave differently. For example, externalists commonly maintain that Darwin formulated his theory of biological evolution in terms of individual competition in the struggle for existence because the society in which he lived was so individualistic and competitive. Other scientists were prone to accept his theory because they too were living in precisely the same social context.

This book also concerns the "science" of science. How do those of us who study science choose between alternative views about the nature of science itself? At times commentators on science have viewed scientific change as being largely gradual, while at other times the emphasis has been on scientific revolutions. What has caused these changes in our understanding of science? How much have we been influenced by the internalist trinity of reason, argument, and evidence, and how much by the diverse forces and factors postulated by externalists? Those of us who study science have had a great deal to say about the sorts of factors that actually influence the course of science. Some of us have even gone so far as to state how scientists *should* make their decisions, regardless of past practices. Perhaps scientists think that they can perform crucial experiments and define their theoretical terms operationally, but we know better. Perhaps scientists on occasion are influenced by the factors alleged by the externalists, but, to the extent that they are, they are not behaving scientifically—they are not true scientists. However, we have been much less explicit about ourselves. We have not gone on at any great length about the sorts of factors that actually influence our own decisions, let alone those that *should.*

One possibility is to apply the same conclusions that we have drawn about scientists to ourselves. If scientists make up their minds primarily through reason, argument, and evidence, then perhaps the same can be said of us. The problem here is that the connections between the claims made by students of science, especially philosophers of science, and anything that might count as evidence are not at all clear. Sometimes scientists do what we say they do, sometimes not. And what about our stipulative claims for how scientists should behave? Not only *do* freely falling bodies near the surface of the earth accelerate at 32 feet/second/second, they *must.* Not all empirical claims are merely descriptive. Some also seem to exhibit a kind of necessity. Can a comparable distinction be made for science itself? Do the processes that produce scientific conceptual change possess any intrinsic nomic necessity, or are they merely contingent?

Or it may just be that those of us who study science are influenced by the same sorts of external factors that influence scientists—personal gain, social interests, social forces, and so on. Just as all the time that scientists spend running experiments and gathering data may be as much mythical behavior as athletes crossing themselves before the big game, all the show that some students of science make in gathering data about science may be just so much propaganda. In its most extreme form, this view of the factors that produce changes in the substantive content of science have led to claims that Mendelian genetics is "capitalist" and quantum theory "Jewish." Extrapolated to science itself, this view of the pervasive influence of society on our conceptual systems has given rise to claims that science as it has been practiced since its beginnings in the West is itself "capitalist" or "male."

Self-Reference

If the preceding line of argument is carried to its logical conclusion, it follows that those who are making these assertions must hold the views that they do because of the very same sorts of social causes. Thus, anyone who studies science scientifically is caught up in a self-referential tangle. For example, in discussing the various borders between science and other human activities such as magic, religion, and philosophy, Gieryn (1983: 64) declares, "How science differs from art is not a matter for sociologists to *decide,* but a matter for scientists and artists to negotiate (as sociologists *watch*)." But sociologists are themselves scientists. They are engaged in the very process that they are watching. They are a party to the negotiations. The simpleminded sincerity with which people endlessly repeat the injunction that "one should not generalize" is stupefying. There is a point to be made, but the point is *not* that one should not generalize. In the first place, this claim is itself a generalization. "One shouldn't generalize? But you just did." The point is that one should not generalize facilely or assume that something that is true of a significant proportion of a group is true of everyone who belongs to the group just because they belong to it. "Scientists believe that species evolve primarily via interorganismic competition because these scientists live in highly competitive, individualistic societies? But so do you. Perhaps this is why you think that social forces determine the beliefs that scientists happen to have."

Because the distinction between internalism and externalism is couched in internalist terms and tends to work against externalism, externalists would like to retire it. Built into the notion of "internalism" is the presumption that reliance on reason, argument, and evidence is "internal" to science, while other influences, operative though they may be, are "external" to it. But this is the very point at issue. The main thrust of the externalists is that factors such as class interests are as internal to science as is reliance on evidence. But every author must use some terminology or other, and this choice implicitly commits him or her on a variety of issues. In this book I spend considerable time investigating the social relations that exist among scientists and the effects that these social relations have on the content of science. If rationalism is the view that there are purely rational principles for the evaluation of scientific theories and that science would function optimally if scientists always adhered to these principles, then the position that I advocate in this book is not rationalistic. Instead, to use Giere's terminology (1988: 144), my treatment is "naturalistic" if naturalism is the "view that theories come to be accepted (or not) through natural processes involving both individual judgment and social interaction."

The system of cooperation and competition, secrecy and openness, rewards and punishments that has characterized science from its inception is both social and internal to science itself. The conceptual development of science would not have the characteristics it has without this social system. Quite obviously science is a social process, but it is also "social" in a more significant sense. The objectivity that matters so much in science is not primarily a characteristic of individual scientists but of scientific communities. Scientists rarely refute their own pet hypotheses, especially after they have appeared in print, but that is all right. Their fellow scientists will be happy to expose these hypotheses to severe testing. Science is so structured that scientists must, to further their own research, use the work of other scientists. The better they are at evaluating the work of others when it is relevant to their own research, the more successful they will be. The mechanism that has evolved in science that is responsible for its unbelievable success may not be all that "rational," but it is effective, and it has the same effect that advocates of science as a totally rational enterprise prefer.

Thus, to some extent I find myself in the externalist camp. Perhaps class interests are not as operative in the straightforward fashion that some externalists have claimed, but career interests do matter. However, externalists have not been content to study science and comment on it. Inevitably they have been seduced into trying their hand at philosophy, and just as inevitably they end up espousing some form of social relativism with respect to empirical truth; and here I find myself squarely in the internalist camp. One form of externalism has come to be known as the "strong programme" of the "Edinburgh School." According to Bloor (1976), sociologists should treat knowledge claims made by scientists in the same empirical, causal way that they treat the rest of nature. As a research strategy in the sociology of science, the strong programme has much to recommend it. Scientists really do think that reason, argument, and evidence are the crucial factors in the decisions they make in the course of their research, but time and again in the history of science we see in retrospect that other factors have had significant influence.

The scientific literature on sexual dimorphism in the human species makes depressing reading. If such biases have influenced scientists in the past, they might well still be influencing them today. If sociologists go to science assuming that scientists believe what they believe primarily because of the conclusions forced on them ineluctably by nature, they are liable to miss those instances in which other factors are also operative. Thus, Collins's (1981a: 218) assertion that sociologists of science, in their investigations, "must treat the natural world as though it in no way constrains what is believed to be" can serve as a useful antidote to our usual prejudices. But how about sociologists of science themselves? Advocates of the strong programme urge extensive empirical investigations of the actual practice of science. But to what end if the natural world in no way constrains our beliefs?

As Laudan (1982) has pointed out, there seems to be something inherently self-contradictory about the externalist penchant for presenting data to show how irrelevant data actually are. Collins (1981c: 217) has addressed this difficulty head-on, suggesting that social scientists "treat the social world as real, and as something about which we can have sound data, whereas we should treat the natural world as something problematic—a social construct rather than as something real." I must admit that I share to some extent the perverse delight Collins takes in turning the tables on natural scientists, who frequently consider social scientists as "scientists" only out of professional courtesy. Still, Collins's position founders on the self-referential gambit.

When the Congress of the United States passed a civil rights bill a few years back, it explicitly exempted itself. No one in the United States can refuse to hire or promote anyone because of race, religion, or country of origin—except Congress. National governments can enact any law they see fit, no matter how apparently unfair or hypocritical it might be, but, as Collins is well aware, social scientists have no such legislative powers. If social scientists can have "sound data," then so can physicists. Collins (1982a, 1982b) responds by distinguishing the practice of social scientists as scientists and the meta-level understanding of this practice. When actually engaged in their research, social scientists should act as if the natural world constrains their own beliefs, but periodically they should also step back from their labors and turn their attention to the warrant that their own knowledge-claims have. When they do, they are forced to admit that the warrant for their knowledge-claims is no different from that of any other scientist. According to Collins (1982b: 142), sociologists of science believe what they believe because of the same social forces that are operative on everyone else:

I believe that questions of sociological methodology and questions of the construction of natural scientific knowledge can be kept separate. The findings of the sociology of scientific knowledge *need* only inform sociology when the attempt is made to justify certain sociological methods by reference to canonical versions of scientific method. Here the sociology of scientific knowledge may have a salutary effect.

Collins's position with respect to sociology and the sociology of knowledge is no more contradictory than Popper's (1934, 1959) well-known views on the relation of science and the philosophy of science. Popper became famous for distinguishing between science and pseudoscience by means of his principle of falsifiability. In order to count as being genuinely scientific, a theory must imply something about the empirical world that can be tested. It must be open to refutation by observation. Because scientific theories, according to Popper, must be universal in form, only falsifiability is possible. Statements of the form "All A's are B's" cannot be verified because they apply to indefinitely many instances; but one unambiguous exception can falsify a universal statement. The immediate facile response to Popper's principle of falsifiability is, "Yes, but is it falsifiable?" And the immediate response to this question is that Popper never purported to be setting out a scientific theory. Instead, his principle of falsifiability is part of a metaphysical research program. Whether in the case of Collins or Popper, there is nothing inherently contradictory about evaluating scientific and metaphysical claims by different standards. Even though one might insist on data for scientific claims, nothing like data might exist for metaphysical claims. One might maintain that they are a priori, or analytic, or intuitive.

As implausible as the position Collins (1982a, 1982b) terms "special relativism" may be, it is not strictly contradictory. Instead, it is a bit of heuristic, albeit marginally schizophrenic, advice. According to Collins, sociologists do better work if they pretend that their investigations afford them "sound data," when actually no such thing exists. Even though I accept the distinction between an activity and the study of that activity, I am unable to sustain such a pretense. I cannot spend the hours necessary to study refereeing practices in scientific journals without believing that these investigations actually "constrain" what I come to believe about refereeing practices in scientific journals. But more than this, my empirical investigations *have* forced views on me that I was disinclined to accept. For example, I thought that older scientists would be less prone than younger scientists to adopt new ideas. I was sure that referees' professional and conceptual allegiances would have marked effects on their recommendations for publication. I would not be in the least surprised to discover that my predispositions on these issues were strongly influenced by all sorts of "external" factors, but something caused me to change my mind on both counts. The most likely candidate for such a cause is the research I did on these issues. If evidence can influence me, it can influence others as well.

Weasel Words in Science and Philosophy

In the past, reflexivity, or the self-referential gambit as I have termed it, has been used as a weapon to stun an opponent. For example, certain sociobiologists have argued that all human behavior is ultimately genetically selfish. If so, then their own behavior as scientists must also be genetically selfish. If all that sociobiologists are doing in their research program is attempting to increase their own genetic inclusive fitness, then why should the rest of us help them by accepting the views that they are promoting? The self-referential gambit can be used as a weapon, but it need not be. It can also be used as a tool to help one probe one's own unreflective beliefs. Time and again in my own research, as I reached a conclusion about the scientists I was studying, I stopped to ask myself, "How about me? Is this conclusion equally true of me?" The frequency with which the answer to this question turned out to be affirmative served both to puncture any nascent pretensions I might have had and to suggest that possibly my conclusions had some substance. If I am engaged in the same activity as my subjects, then anything that is true of them had better be true of me. The self-referential gambit has been used so effectively against the relativist inclinations of externalists that they have become highly sensitized to it. Anyone who introduces it is immediately marked as an enemy. In taking this position, externalists have denied themselves one of the most powerful investigatory tools at their disposal.

For example, I noticed that key words in the scientific disputes I was studying had a certain "plasticity" to them. They expanded and contracted in systematic ways to fulfill the needs of the scientists using them. When a group that I term "pheneticists" criticized their opponents, they tended to treat "phenetic characters" as referring to theory-free, direct observations. When pheneticists were developing their own views, "phenetic characters" included codons, entities that are neither theory-free nor observable in any straightforward way. The terms "cladistic" and "cladogram" have had a similar history. Sometimes they are used in a generic sense to refer to any nested pattern, sometimes they refer to nested patterns more closely connected to phylogeny. The term "individual" has been used in the same ambiguous way in the dispute over the nature of biological species. Are they classes or individuals? Unfortunately, the term "individual" has been used in a generic sense to refer to any spatiotemporally localized entities and in a specific sense as being equivalent to "organism," and species are clearly not organisms.

Yet another ambiguity constantly crops up in our discussions of scientific theories. Are they hypotheses or facts? Can they be "proved"? Do scientists have the right to say that they "know" anything? While interviewing the scientists engaged in the controversies under investigation, I asked, "Do you think that science is provisional, that scientists have to be willing to reexamine any view that they hold if necessary?" All the scientists whom I interviewed responded affirmatively. Later, I asked, "Could evolutionary theory be false?" To this question I received three different answers. Most responded quite promptly that, no, it could not be false. Several opponents of the consensus then current responded that not only could it be false but also it was false. A very few smiled and asked me to clarify my question. "Yes, any scientific theory *could* be false in the abstract, but given the current state of knowledge, the basic axioms of evolutionary theory are likely to continue to stand up to investigation."

Philosophers tend to object to such conceptual plasticity. So do scientists—when this plasticity works against them. Otherwise they do not mind it at all. In fact, they get irritated when some pedant points it out. From the perspective of finished knowledge, systematic ambiguity is a fault to be decried and immediately eliminated. From the perspective of knowledge acquisition as a temporal process being carried on by fallible human beings whose careers have an inevitable temporal limit, it may be an evil, but it is at the very least a necessary evil. In fact, I find that such equivocation in science is not in the least evil but a powerful method of conceptual improvement. Often, I was forced to conclude that the standards dictated by philosophers of science, if taken literally, would destroy the very mechanisms that produce the characteristics of science that philosophers value so highly.

In science, "weasel words" serve an important positive function. They buy time while the scientists develop their positions. It would help, one might think, if scientists waited until they had their views fully developed before they publish, but this is not how the process of knowledge development in science works. Science is a conversation with nature, but it is also a conversation with other scientists. Not until scientists publish their views and discover the reactions of other scientists can they possibly appreciate what they have actually said. No matter how much one might write and rewrite one's work in anticipation of possible responses, it is impossible to avoid all possible misunderstandings, and not all such misunderstandings are plainly "misunderstandings." Frequently scientists do not know what they intended to say until they discover what it is that other scientists have taken them to be saying. Scientists show great facility in retrospective meaning-change.

In general, too much emphasis in communication is put on the intentions of the sender. In the ongoing process of science, intentions receive very rough treatment. What the receiver thinks the sender intended is what matters. Perhaps Mendel did not intend his so-called laws to be incompatible with evolutionary theory. No matter, that is how they were construed. Perhaps Sewall Wright did not intend his shifting-balance theory to be incompatible with more traditional selectionist views on evolution, but that was how it was interpreted. Intellectual justice in matters such as these is important to historians but to almost no one else, scientists included. Was Darwin really a cladist? As anachronistic as this question might seem, it and others like it play important roles in science, and historical accuracy makes precious little difference. Anyone who doubts the preceding claims can check their accuracy by following any scientific dispute over a period of several years.

If these observations apply to scientists in their disputes, then they should apply with equal force to those of us attempting to understand science. As one might expect, I find it much easier to notice and acknowledge the positive roles that vagueness and equivocation play in other people's disciplines than I do in my own. The way in which some externalists use the term "interests" exasperates me no end. Sometimes interests are indistinguishable from psychological states. Some scientists really are interested in winning a Nobel Prize. Science is also organized so that the professional interests of scientists are promoted. Here "interests" is being used in a sociological sense. And these professional interests frequently gradate into broader social interests of a more subterranean sort (Geertz 1973). Add to this list, intellectual interests (Morrell and Thackray 1981) and cognitive interests such as "prediction" (Shapin 1979a), and the externalist position differs from that of the internalist only in emphasis (for a critical evaluation of the role of "interests" in our understanding of science, see Woolgar 1981).

Precisely the same remarks apply with equal force to such terms as "theory" in the writings of more internalist philosophers of science. According to one widely held view, theoretical terms in science such as "pion," "gene," and "Id" are quite obviously theory-laden. They get much of their meaning from the theories in which they function. When a new term is invented to serve as a theoretical term, the connection between this term and its theory is obvious. We are likely to identify such neologisms as "allelomorph" as theoretical terms because they are so strange. "Allelomorph" had no antecedent meaning prior to its being invented by early Mendelians. These connections become less clear when terms are borrowed from one area of discourse for use in another. As theoretical terms are transferred from theory to theory, they trail clouds of previous connotations. For example, the term "gene" was coined initially in the context of the dispute over Mendelian genetics and only later introduced into evolutionary biology and appropriated by molecular biologists. The situation is even more problematic when the terms are borrowed from that protean, heterogeneous assemblage termed "ordinary usage." The borrowing of "altruistic" from moral discourse by sociobiologists to modify "gene" is a good case in point.

Philosophers of science have not been content to claim that theoretical terms are theory-laden. They extend this position to observation terms as well. Not only are such theoretical terms as "gene" and "species" theory-laden, but so are such apparently nontheoretical terms as "animal," "male," "dorsal," and "wing." Perhaps it might seem a matter of simple observation that one side of an organism is dorsal and the other ventral, but the amount of effort involved in making such decisions and the extent of the disputes that have broken out over them attest to the theoretical nature of this predicate. And as everyone knows, not all wings are actually "wings" in the evolutionary sense of this term.

In this discussion, I have been playing fast and loose with the term "theory." The theories to which I have alluded have ranged from evolutionary theory and particular hypotheses about phylogenetic descent to such informal theories as the theory of the organism. But such systematic equivocation over the term "theory" is necessary if the global claim about the theory-ladenness of all scientific terms is to be sustained. I must admit that equivocation over the term "theory" in the internalist literature was for me much more difficult to notice than was the comparable equivocation over the term "interests" in the externalist literature, and, once I noticed it, it bothered me much less. To complicate matters even further, scientists themselves have addressed the issue of the role of theories in concept formation. Some scientists think that, in the early stages of a scientific investigation, scientists can and should stick to the facts and nothing but the facts. Others argue that theories can, should, and even must be operative in scientific studies from beginning to end. Hence, I am forced to take sides on one of the issues that I purport to describe in the disinterested and impartial fashion required of those of us who study science. Science may be distinguishable from philosophy, but workers who are officially scientists not infrequently address philosophical issues. Much less frequently, professional philosophers engage in scientific research.

An Empirical Account of Empirical Science

Because I myself make recourse to empirical data in my own research, I am hardly in a position to dismiss it as a cover for my class interests. If everyone has class interests, then I most surely have class interests. The question is the determinate influence class interests have on my beliefs about science. When we are caught up in an empirical study, all sorts of factors that are not narrowly connected to the subject under investigation are likely to influence our preceptions—not to mention conceptions—but we do have the ability to stand back and view ourselves in relation to our subject matter (Nagel 1980, Collins 1982a, 1982b). When we do, other factors not narrowly connected with this investigation might also "intrude." We must then stand back from this activity and view ourselves as part of it, and so on. How many times we can repeat this process is problematic. Most of us are capable of at least one meta-perspective, but I suspect few of us are capable of sustaining a meta-meta-perspective very successfully or for very long.

In any case, others have somehow been able to "transcend" such external factors as class interests—Marx, for example. Such transcendence is not easy to come by, but if others can do it, so can I. One good way to transcend such things as class interests is to pay attention to empirical data. When the men studying honeybees first noticed one large bee in a hive among all the others, they immediately dubbed it the king bee, indicating at least two social biases. Without changing their own gender, however, these scientists soon came to realize that this bee was female and promptly termed it the queen bee. Recently, critics of sociobiology have objected to the remnants of the monarchical bias in this appellation. With the possible exception of Queen Victoria, human queens have very few characteristics in common with the "queens" in such eusocial insects as bees and termites.

I have no intention of playing down the insidious effects that the connotations of terms can have on the ways that we think. Language is the all-too transparent bottle that constrains Wittgenstein's hapless flies. For example, scientists usually distinguish between sexual and asexual forms of reproduction, the former being "normal" and the latter somehow suspect. In reaction, the authors in a recent book on the evolution of clonal organisms (Jackson, Buss, and Cook 1986) consistently refer to clonal and aclonal organisms. Initially the usage is jarring—sexual organisms being called "aclonal"? But the effects are subtly pervasive. Growth becomes indistinguishable from reproduction. The notion of an organism becomes blurred. Before long, the evolutionary process begins to look strikingly different.

As constraining as language is, we *can* become self-conscious about its influence. When we do, we can begin to notice these constraints and selectively question them. We cannot suspend all of language all at once, but we can examine parts of it serially in ways that are partially independent of language. The same can be said for social interests. We cannot cease having them, but we can become suspicious when our own social interests turn out to coincide consistently with the greater good. Sometimes what is good for General Motors *may* be good for the country, but not always. If you just happen to be born all the best things that a person can be—American, white, male, middle class—then perhaps you are merely reading your own status into your evaluation of the world. All of us, scientists included, are influenced in our beliefs by a wide range of considerations. Among these are reason, argument, and evidence. It is also a fact of nature that sometimes some of us can step back and view ourselves in the context of our actions and evaluate them from a different standpoint. This different standpoint is not Archimedean. It is not outside everything. But it is also more than just "different." The preceding are empirical claims and can be tested in the same ways that all empirical claims are tested. One of the goals of this book is to put them to the test.

Externalists are confronted by the apparent hypocrisy of urging sociologists to pay attention to data while maintaining that such internalist factors play at best a minor role in the process of knowledge acquisition. Presenting data to show the irrelevance of data seems a bit contradictory. One response to this apparent contradiction is extremely cynical. Because most of the intellectuals who study science suffer from the illusion that data influence the views that they hold, externalists are forced to kowtow to this mistaken belief and to present data. A less cynical interpretation of the externalists' penchant for using data to show the irrelevance of data is that they are merely adopting the tactics of their opponents to show that, even by the opponents' own standards, their views are mistaken.

Internalists confront just the opposite problem. They maintain that recourse to reason, argument, and evidence is crucial in science. In fact, they tend to define "science" in these terms. Anyone who is paying attention to these factors is on this account engaged in science, and anyone who does not, regardless of academic degrees and professional positions, is not. Granted that all scientific theories are underdetermined by anything that might be called evidence, it does not follow from this that evidence plays no role in science whatsoever. How are these pronouncements to be evaluated? Presenting data to show the relevance of data looks a bit circular. Internalists are able to avoid the charge of circularity by the same means that Collins was able to avoid allegations of self-contradiction—by distinguishing between science and meta-science.

One way for scientists to decide between two competing scientific theories is to find data that distinguish between the two. Although two different theories cannot be brought into conflict in the strict, precise way that certain analyses of scientific theories as ideal conceptual entities require, real theories are sufficiently blunt that they can be forced into a rough sort of confrontation. Although lots of leeway existed, nineteenth-century scientists were able to settle on data that made acceptance of both thermodynamics and Darwin's theory of evolution difficult. According to the new science of thermodynamics, the earth could not possibly be older than 20 to 200 million years. According to such uniformitarian geologists as Darwin, the wearing down of a single geological stratum could take as long as 300 million years. The mechanisms Darwin postulated for the evolution of species were compatible with the geological, not the thermodynamic, time scale. Hence, either thermodynamics or uniformitarian geology and evolutionary theory were wrong—or maybe both were.

Enough disagreement exists about the ways in which scientists choose between conflicting scientific theories. How are those of us who study science to choose between different theories about the nature of science? In the nineteenth century, philosopher-scientists such as John Herschel (1830), William Whewell (1840), and John Stuart Mill (1843) set out conflicting views on proper scientific method. We now look back with a sigh of relief that scientists took the pronouncements of these authorities with a grain or two of salt. Today such internalists as Laudan (1981a) and Roll-Hanson (1983, 1985) are doing battle with such externalists as Collins (1981a, 1981b, 1981c, 1982a, 1985) and Barnes (1977, 1981, 1985a; see also Hollis and Lukes 1982 and Gellner 1985). Who is right? How are we to decide?

Even if one agrees that class interests are the crucial factors in scientific change, it is still possible to disagree about which class interests were actually operative in any particular episode (e.g., Cantor 1975a, 1975b; Shapin 1975, 1979a, 1979b; Collins, Pinch, and Shapin 1984). The larger question remains: which sorts of factors influence the decisions which scientists make? For one, Barnes (1985b: 760, 762) now acknowledges that a realist orientation is both "ubiquitous and probably essential in the natural sciences." The realism that he opposes involves the "existence of universals or of essences or of permanent unchanging entities of any kind." The realism that he now advocates is that on occasion "speech points to something beyond itself." Nothing else is required.

Differences of opinion also exist among internalists. Even if reason, argument, and evidence are the crucial factors in leading scientists to make the decisions that they do, there is still considerable room for disagreement about which of these factors were actually operative. Traditional histories of science are made up almost exclusively of such disputes. Disagreements among internalists also exist on more general issues. For example, Laudan (1977, 1984) is as unhappy with traditional realist philosophers of science as relativists are. Empirical truth plays no more central a role in Laudan's view of science than it does in Collins's view, but for quite different reasons and with different effects (see also van Fraassen 1980, Cartwright 1983, Fine 1986a). Among those who study science, there is as much intragroup conflict as intergroup conflict, possibly more.

be "justified" in the sense that generations of epistemologists have attempted to justify them. The reason that epistemologists have not been able to justify knowledge-claims in their sense of "justify" is that no such justification exists. They want the impossible.

But even on a more realistic notion of "justify," some epistemologists have construed their task in such a way as to make it impossible. They view knowledge acquisition in terms of an individual subject who confronts objects of knowledge in total isolation from other knowing subjects. On this view, languages mirror the world, but the "world" is the sensations of the isolated knowing subject. Because each subject has access only to his or her own sensations, languages that refer to these sensations are necessarily private. This epistemological stance places obstacles in the path of knowledge acquisition that are all but insurmountable. It assumes that languages are not only descriptive but also descriptive of necessarily private objects of knowledge. Philosophers have long recognized that languages function primarily as devices for communication and only secondarily for description. Because we can use languages to communicate, we can also use them to describe (Wittgenstein 1953).

Whenever the same organismal structure must perform two functions, conflicts are likely to arise. This observation holds for languages. The communicative and descriptive functions of languge sometimes come into conflict. Similar observations hold for science. When one expands one's perspective from the scientist as an isolated knowing subject to include other scientists with different interests, communication becomes more difficult but genuine testing becomes a real possibility. On the one hand, scientists go to great lengths to salvage the views to which they are publicly committed. That way scientific hypotheses get an honest run for their money. On the other hand, scientists holding conflicting views test these hypotheses as rigorously as possible. Scientific polemics are not very efficient, but they are highly effective. I realize that philosophers will find these observations irrelevant to their interests (Passmore 1983), but that cannot be helped.

In this book I do not attempt to answer any traditional problems in epistemology. I see no reason why everyone must do epistemology all the time. Instead I set out a general analysis of selection processes that is equally applicable to biological, social, and conceptual development. Selection processes have several features that are worth pointing out, but one of them is not that they guarantee the generation of perfectly adapted organisms or infallibly true statements. I am not interested in justifying selection processes but in seeing how they work and why their products have the characteristics they do. I do not know what it might mean to claim that a theory of biological evolution "justifies" that evolution. Similarly, I am not sure what it might mean for a theory of social or conceptual evolution to "justify" that evolution. I do talk about the effects of certain practices in science. Some methods have proved more successful than others, given certain goals. If urging the use of successful methods amounts to epistemology, then my concerns are "epistemological," but only in the most anemic sense of this term. The most I am prepared to claim for the mechanism I present is instrumental epistemic merit (Firth 1981). The only necessity I claim for any of my assertions is nomic necessity, and it is more than enough. The danger in setting out such a general analysis of selection processes, one that is equally applicable to biological, social, and conceptual change, is that it can become so general that it is empirically vacuous. I think that I have succeeded in avoiding that danger. The views I set out about the mechanisms of biological and scientific change may be mistaken, but they are not devoid of empirical content.

Some things cannot be expressed in a natural language. The relations that exist between the genetic material and ordinary phenotypic traits is one of them. Many people have very peculiar beliefs about the ways in which ordinary phenotypic traits are transmitted and developed. On the basis of these half-articulated, highly erroneous views, they conclude that behavioral traits cannot possibly have any genetic basis. The arguments which they present against the literal inheritance of behavioral predispositions, if cogent, would count against the inheritance of ordinary phenotypic traits such as hair color, body height, and blood type. The "transmission of traits" is shorthand for a much more complicated expression. Literally, only genes are transmitted. Genes organized into genomes, interacting through time with successive environments produce phenotypes, which in turn interact with their more inclusive environments to influence the process. The most that genes code for is reaction norms.

Once one grasps the minimum conceptual tools necessary to understand ordinary biological inheritance, it is impossible not to realize that certain behaviors are just as genetically transmitted as are some ordinary phenotypic traits. No, a trait need not be impervious to environmental influences for it to be, in a significant sense, genetically transmitted. No, genetically transmitted traits need not be universally distributed in a species, etc. But, having said this much, I ignore any influence that genes and changes in gene frequencies might or might not have on conceptual change in science.

Sociocultural change is influenced by two sorts of heredity—genetic and cultural. This dual inheritance is one reason why cultural change is so difficult to model (Boyd and Richerson 1985). Such cultural entities as practices and artifacts and such conceptual entities as beliefs are *traits* from the perspective of genetic transmission; they are analogs to *genes* from the perspective of cultural transmission. To switch from one perspective to the other in attempting to provide a dual-inheritance model for sociocultural change is a formidable task. In this book I limit myself to cultural inheritance, chiefly because I am concerned with change in our scientific conceptions about the world, conceptions that have changed too quickly for alterations in gene frequencies to have played any role. I do not investigate the gradual evolution of our ability to know scientifically but the changes that have occurred since. Every explanation must take something for granted. I simply assume that human beings in general and scientists in particular are extremely curious about the world in which they live, but do not attempt to explain it. This is simply the sort of species that we are.

My use of the phrase "social development" and the attention I pay to the professional relations among scientists are likely to signal externalism or social determinism of some sort, as if I thought that the content of scientific theories is determined (or at least strongly influenced) by the character of the wider society in which the scientists themselves live. As the early discussion in this chapter indicates, such a conclusion would be erroneous. Science quite obviously proceeds in a "social context"; it could not possibly exist in total isolation from the rest of society. It is also true that science itself exhibits social organization. Scientists form social groups. They cooperate with each other, compete, build on each other's work, sometimes give credit where it is due. Studying the world in which we live is a social process, but from this platitude it does *not* follow that our knowledge of the world is socially determined. It might be, but it need not be.

That science is a social institution functioning in various societies is neither a controversial nor an especially interesting assertion. That scientists form social groups as small and tightly knit as research teams or as large and loosely structured as those that grow up around a particular subject matter is not especially controversial, but one of the messages of this book is that this "demic" structure of science is *extremely* interesting. Several of the most important features of science flow from it. For example, Boyd and Richerson (1985) list several factors that affect the pace of evolution. One of them is the size of the effective population. Evolution occurs most slowly in large panmictic populations, most quickly in small populations largely isolated from the rest of their species. If science is a selection process, then conceptual change should occur most rapidly when the scientific community at large is subdivided into "demes."

Scientists cooperate with each other in both a metaphorical and a literal sense, and the two are interconnected. Speaking from personal experience, Ziman (1968: 134) comments that occasionally "one finds oneself rather emotionally engaged in a controversy upon a scientific matter; but this, after all, is one's job, and one must learn to advocate powerfully, but to concede defeat gracefully, without making a personal issue of it. In a sense, a well-fought controversy between two spirited champions is a form of cooperation." Sometimes scientists disagree amicably; sometimes not. As Hallam (1983: 51) notes, the controversy in geology between catastrophists and uniformitarians in the 1830s "never descended to the depths of acrimony that had characterized the neptunist-vulcanist/plutonist controversy." Scientists cooperate with each other in a metaphorical sense when they engage in disputes, whether amicable or not, but they also cooperate in a literal sense when they form research groups.

The animosities that arise during the course of science are complicated by the allegiances and alliances scientists form. A common response among scientists is that if you attack mine, you attack me. The other side of the coin is that if you support mine, you also support me. To some extent, scientists evaluate substantive views on the basis of their source. Periodically in the history of science, national origin, sex, race, religious affiliation, and the like have influenced the spread of scientific ideas, but such influences have been extremely variable. The effects of affiliation of scientists with particular research groups have been constant and widespread.

Conceptual Lineages

In my exposition I use the terminology of the "new" philosophy of science—terms such as "theory-laden" and "research program." I avoid using such expressions as "interests" and the "negotiation of facts." I realize that in choosing terminology that has grown out of the internalist camp I am identifying myself with one side of the internalist/externalist dispute. In doing so, I automatically tend to alienate one important audience, but I see no alternative. Just as there is no neutral observation language in which scientific theories can be compared, there is no neutral metalanguage in which opposing views about science can be evaluated. My choice of terminology also has the virtue of accurately reflecting my own historical development and intellectual legacy. My early philosophical training was in logical empiricism or, more accurately, in internal criticisms of the logical empiricist analysis of science. I have learned a lot about science from externalist sociologists of science but always in reaction to it. I strongly suspect, however, that once the current "spirited" dispute over internalism and externalism has run its course, neither program will remain unchanged.

My choice of terminology has a second virtue as well. It is likely to avoid alienating one of the chief audiences for which I have written this book—scientists. Certainly I want the people who study science to read and appreciate what I have to say, but I would judge the book a failure if scientists themselves got nothing out of it. Convincing scientists that scientific theories are always "grossly underdetermined by experience" is difficult enough without arguing that in their laboratories they are "negotiating facts," even though the assertive content of these two claims may not be all that different.

Finally, I pay considerable attention to the conceptual development of science in the way traditional internalist students of science always do, but my treatment of concepts is radically different. Traditionally, intellectual historians have treated concepts the way that ideal morphologists treat the phenotypic characteristics of organisms. According to ideal morphologists, traits are traits. Certain basic patterns (or bauplans) exist in nature. Particular organisms exhibit these bauplans to a greater or lesser degree. But most important, according to some ideal morphologists, these bauplans can be recognized and organized in total isolation from any knowledge of "time, space, or cause" (Borgmeier 1957: 53). We can and should develop our classifications independently of any theories about the processes that have produced the patterns which we reflect in our classifications. Organisms come and go but, according to ideal morphologists, these patterns are forever.

Although the perspective of ideal morphology is not totally incompatible with species evolving, it does recast evolution in a peculiar way. Successive generations of organisms can change. In doing so they might wander from one bauplan to another, but the bauplans themselves remain unchanged during such transitions. Although any one species might not be exemplified now and again, species are eternal and immutable. Some day dinosaurs might once again come into being. According to ideal morphologists, biological taxa and their traits are defined structurally. They exist in space and time, but they are not historical entities. Taxa such as Mammalia need not be monophyletic. An organism need not be part of any particular phylogenetic lineage in order to be a mammal. Similarly, according to ideal morphologists, characters such as wings need not be evolutionary homologies. A wing can count as a wing regardless of its genesis.

Although this way of viewing organisms might seem counterintuitive to anyone brought up in the Darwinian tradition, a comparable view of concepts is liable to appear perfectly acceptable, in fact so obvious that only a lunatic would question it. However, in this book I treat concepts as historical entities in the same way that evolutionary biologists treat taxa and traits. I organize term-tokens into lineages, not into classes of similar term-types. In the past, biologists have treated species in the same way philosophers have treated concepts—as classes of similar entities regardless of their genesis. Not until the work of Ghiselin (1969, 1974b) were biologists forced to recognize the conflict between this traditional way of treating species and the requirements of the evolutionary perspective. A consistent application of what Mayr (1963, 1969) has termed "population thinking" requires that species be treated as lineages, spatiotemporally localized particulars, individuals. Hence, if conceptual change is to be viewed from an evolutionary perspective, concepts must be treated in the same way. In order to count as the "same concept," two term-tokens must be part of the same conceptual lineage. Population thinking must be applied to thinking itself.

This change in perspective is radical, so radical that some readers are liable to dismiss it out of hand. Terms are not important. Concepts are. Some scientists can use the same term and mean different things, while other scientists can use different terms and mean the same thing. The gene concept is the gene concept, regardless of space and time, regardless of conceptual replication sequences. All those who have the same views about genes are using the gene concept in the same way, accidents of terminology and history to one side. As a result, several scientists can make the "same" discovery independently. Just as wings have evolved several times independently, Mendel's laws were discovered independently by several workers. If "Mendel's laws" or "the gene concept" are treated as conceptual evolutionary homologies, then independent discoveries are ruled out by definition. If the discoveries are genuinely independent, then they are not the "same" discoveries.

One might be willing to treat wings as evolutionary homologies, refusing to consider the wings of birds and insects as the same structure, while drawing the line at concepts. But at the very least, these two different notions of sameness must be kept distinct. Just as species cannot be treated simultaneously as historical entities and as eternal and immutable natural kinds, neither can concepts. Nothing but confusion can result from treating something as being both spatiotemporally restricted and spatiotemporally unrestricted. More strongly, if an evolutionary account of conceptual change is going to have any chance of succeeding, the basic units of both evolutionary biology and conceptual evolution must be viewed as the same sort of thing—either as spatiotemporally unrestricted classes or as spatiotemporally connected lineages. In my account, I opt for the second alternative. The things which are evolving as a result of selection, whether species or concepts, must be treated as historical entities.

Histories of ideas as ideal types have one virtue: they are relatively easy to write. All one must do is to argue for essential similarities between concepts. For example, anyone who organized plants and animals into anything like a Great Chain of Being belongs in a history of this concept. Showing actual descent is much more difficult. As difficult as it may be, it is also absolutely essential for an evolutionary account of conceptual change. Conceptual ideal morphologies will not do. An evolutionary account of concepts must go even further. I cannot content myself with just the history of concepts. I must also specify the mechanisms involved in this change, and one of these mechanisms requires heredity. Selection occurs between tokens, not types. To be sure, each token must be of some type or other. If conceptual change is to be viewed as a selection process, then one term-token may well have been selected over another because of the term-type which it instantiates, but ancestor-descendant connections are also necessary. In term-token lineages, a token of one type can give rise to a token of another type, a proposition can evolve into its contradictory. Any account of conceptual change that ignores conceptual lineages cannot count as being genuinely "evolutionary."

The Structure of the Book

In the first half of this book, I set out a series of interconnected historical narratives. I begin each chapter in this first half with a short discussion of a general issue or two, just enough for the reader to understand why I have structured the narratives as I have. Although these narratives may appear to be simple descriptions, they are anything but. They are structured as they are because of my general theory about the scientific process. Thus, I cannot claim complete objectivity in the stories that I tell. Case studies in science are no more theory-neutral than are the experiments run by empirical scientists. But at the very least, on *some* issues my biases are different from those of the scientists whom I have studied. With respect to these issues, I can at least claim third-party objectivity. On those issues that bear directly on my own ideas about the nature of science, I am engaged in the same activity as my subjects and can claim no greater independence of outlook than they have.

In the second half I present a general analysis of selection processes as well as a social mechanism which, I argue, explains the most important features of science as a process. Normally books such as mine are structured around the general argument with examples drawn from science interspersed in strategic places to illustrate the main themes of the general analysis. Such a structure was not open to me for two reasons. First, one of my main contentions is that science is a selection process. As a result, the evidence for my perspective cannot be presented episodically. It must be provided in the form of connected lineages. In order to understand why scientists challenge certain tenets and take others for granted, one must understand the history of these convictions. The detailing of connected lineages takes space and time. Any attempt to "intersperse" these lengthy narratives within a general argument would totally obscure the structure of the argument. For instance, by the time the reader has followed the recurrent emergence and successive defeat of "idealist" views of biological classification, he or she is very likely to have forgotten the point of this extended example.

The second reason why I do not follow the usual organization of books about the nature of science is that I intend to use at least *some* of my illustrations as evidence, and the requirements for evidence are a good deal higher than for illustrations. Once again, by the time the reader has slogged his or her way through the details I present to distinguish between alternative hypotheses about the nature of science, he or she is likely to have lost the general structure of the argument. Thus, in the first half of this book, I present the interconnected narratives that I use later to both illustrate and test my general views about science; then I refer briefly to these narratives in the second half. Although this organization is far from ideal, it is the best I could do.

In the past, students of science have introduced episodes in the history of science primarily to illustrate their views. As illustrations, these episodes need not be all that detailed or even accurate. In fact myths will do. For example, inductivist philosophers were once fond of illustrating their views by reference to Kepler plotting the orbit of Mars, Galileo dropping balls from the Leaning Tower of Pisa, the role that Darwin's finches played in the genesis and development of his theory of evolution, and Einstein's putative operationism. Externalists are also fond of noting that Marx wrote to Darwin to ask permission to dedicate *Das Kapital* to him and that John D. Rockefeller justified the rapacious character of American capitalism by reference to the survival of the fittest. That all of these stories are myths should surprise no one (see Hanson [1958] for Kepler and Galileo, Sulloway [1982] for Darwin, Popper [1975] for Einstein, Colp [1982] for Marx, and R. Wilson [1967] for Rockefeller).[1]

Myths serve a variety of functions in both science and philosophy of science, but one function they cannot serve is that of evidence. If Kepler did not formulate his laws on the basis of a massive accumulation of data, then his practice cannot be used to support an inductive view of science. Claims about the causes of a particular event in history are extremely difficult to test, regardless of whether the causes alleged are internal or external. It is certainly no easier to decide whether the competitive character of Darwin's theory stemmed primarily from the data he had gathered about selection processes or largely as a result of his personal experiences in Victorian culture.

One of the tenets of the "new" philosophy of science is that there are no "raw data," no theory-neutral atomic facts; but not until I attempted to test views about the nature of science did I realize how malleable data can be. Evidence is incredibly difficult to obtain and usually it turns out to be indeterminate. Even when one thinks that at last the evidence gathered impinges on a proposition under investigation, there are always loopholes, alternative interpretations, and so on. Fine (1986b: 152–3) reminds us that the "application of science involves an enormous amount of plain old trial and error; hence, it always entails an enormous amount of error." But more than error, it entails indeterminacy.

Although data about science are to some extent malleable, they are not totally plastic. Perhaps students of science should not be able to use episodes in science to choose between alternatives analyses of science, but they do. In the philosophical rewriting of Aesop's fable, the hare should never be able to catch the tortoise, but in real life hares easily hop past plodding tortoises. Hence, something is wrong. In this book I present evidence to show the effect that evidence can have on changing people's beliefs. I also apply this same line of reasoning to myself. I am willing to abandon my beliefs about the nature of science if enough evidence points in another direction—but not without a fight and plenty of finagling. I do not expect my behavior to differ materially from the behavior of the scientists whom I have studied.

In the first half of this book, I follow the interconnected paths of evolutionary biology and systematics, starting with Darwin for evolutionary biology and Aristotle for systematics. I cover the early history of these two disciplines quite briefly, too briefly

1. The Rockefeller story is a good example of the tenaciousness of such myths. For instance, Hubbard (1979: 52) attributes "the survival of the fittest" remark to John D. Rockefeller and cites Hofstadter (1955: 45), who cites Ghent (1902: 29). Ghent properly attributes the quotation to John D. Rockefeller, Jr., but gives no citation. Even though R. Wilson (1967: 97) pointed out this minor error twenty years ago, it continues to be perpetuated.

for them to count as evidence. As I approach the present, the narrative slows down appreciably and becomes more substantive. To some extent, my choice of evolutionary biology and systematics is a matter of contingency: I happen to know most about these areas of science. However, I had a more important reason for setting out the series of controversies that have occurred through the years in these two disciplines. In order to understand my evolutionary account of science, the reader has to understand several basic issues in evolutionary biology. I use the substance of my early discussions of evolutionary biology and systematics later in the development of my more general views about the nature of science. For example, an understanding of genetic inclusive fitness is helpful in understanding the notion of conceptual inclusive fitness. Thus, the stories I tell in the first half of this book serve triple duty, illustrating my views about the nature of the scientific process, serving as evidence, and embodying this very same process.

I have also included a discussion of sociobiology, in part because the goals of this research program are easily confused with my own and in part because it is supposed to be a paradigm example of the influence of broad social factors on the content of science. Sociobiologists propose to extend the principles of evolutionary biology to include the behavior and social organization of human beings. Scientists are human beings. Hence, if sociobiologists can explain territoriality and incest avoidance in terms of changes in gene frequencies, they should also be able to explain citation patterns among scientists in these same terms. I do not propose to extend a gene-based biological theory of evolution to include conceptual development in science. Instead, I provide a general analysis of selection processes which is intended to apply equally to both biological and conceptual change. Neither biological nor conceptual selection is more fundamental than the other.

If the nature of the societies in which scientists live strongly influences both the hypotheses which they generate to explain the natural world as well as the reception of those hypotheses by other scientists, then one would expect this influence to be most clear-cut in those areas of science that bear most directly on human beings. Hence, if the externalists are right about science, sociobiology should be their best case. Just as a direct connection existed between Victorian England and the character of Darwin's theory, there should be an equally direct connection between the societies in which sociobiologists have been raised and their theories about the biological nature of the human species. If no correlations of this sort can be discovered, then the externalist research program is in trouble.

The final research program I chronicle concerns the work of Leon Croizat in biogeography. From Hooker and Wallace to the present, biogeography is one of the most important areas in evolutionary biology. If species evolve from preexisting species, then the geographic distribution of species should tell us something about the actual course that phylogeny has taken. In addition, geographic distribution should also serve as evidence for and against various hypotheses about the evolutionary process; e.g., very similar species living in very different habitats, as well as different species living in the same environment, should bear on the relation between environments and the character of the species that inhabit them. But I had two additional reasons for paying special attention to Croizat and his intellectual descendants. One is important: the connection that Gareth Nelson and Donn Rosen forged between their preferred principles of systematics and Croizat's theory of biogeography has influenced the course of cladistic analysis. The second is idiosyncratic: the problems that Nelson and Rosen confronted instigated my own research.

One thing that impressed me and that is sure to strike the reader is how much narrative is required for so little message, but this feature of the stories that I tell is itself one of the main messages of my book. When scientists first opt one way or the other on important issues, the causal circumstances that are relevant to these decisions are extremely particularized. Only an intensive and extensive investigation of these circumstances can explain why science took the course it did. Because these causal situations are so particularized and the requirements for evidence so stringent, rarely can episodes in the history of science serve as evidence for or against particular views about the nature of science. The necessary evidence too often is missing. For example, Rudwick (1985) recently studied a controversy among a dozen or so geologists in the early years of the nineteenth century over the existence of the Devonian period. In spite of intensive study of the historical records of a sort likely to be repeated only rarely by other historians, time and again Rudwick was unable to discover what he needed to know about his protagonists. Hence, episodes in the history of science are liable to be of only marginal aid in deciding disputes about the nature of science.

According to one prevalent view of biological evolution, the primary locus of evolutionary change is very restricted. If biological evolution is to be understood, the specific interactions between small groups of organisms in their particular environments must be studied. A broader focus misses everything of interest. A parallel observation holds for science. In science a small number of scientists are disproportionately important, and these scientists make their decisions in extremely particularized contexts. If one wants to understand the course of science, it must be studied just as minutely as evolutionary biologists study changes in gene frequencies in local populations. Such studies are not easy. Just reading published reports is not good enough. Even the sort of historical inquiries for which historians of science are justly proud are also not good enough. Like it or not, the sorts of inquiries that hold out the greatest hope of distinguishing between alternative views about the nature of science are those concerned with present-day science. Only in such circumstances are the relevant data available. That is why I turned to recent disputes in biological systematics, in particular the fates of phenetics (or numerical taxonomy) and cladistics (or phylogenetic systematics).

Scientific Communities

The empirical world is a large subject. Scientists, in order to make some progress, have consistently narrowed their focus. For example, as long as biologists studied hereditary phenomena at large, little progress was made. Different biologists settled on different phenomena, hoping that they might be the key to understanding heredity, again with little success. As it turned out, Mendelian three-to-one ratios in hereditary transmission turned out to be the point of entry for unraveling hereditary phenomena. Science itself is a fairly large subject. If scientific practice is any guide, progress can be made in understanding it only by narrowing one's focus. Hence, I have narrowed the scope of my investigations to certain restricted sorts of phenomena, phenomena that may or may not turn out to be an appropriate point of entry for understanding the scientific enterprise. I have settled on the small groups that scientists form as they pursue particular avenues of research.

For example, hamadryas baboons exist in increasingly more inclusive groups and these groups perform different functions. The smallest groups of baboons, averaging twelve members, are primarily concerned with mating and care of the young. When feeding, these baboons form groups composed of between forty and fifty individuals. The troops of 125 to 750 that come together at night serve primarily for protection against predators. Such neat stratification and division of labor is rare in organisms at large. It is equally rare among scientists. However, scientists do form socially defined groups that perform different functions. I treat these groups as genuinely social groups, defined by their interactions. In the past, facile reference has been made to "the scientific community" or "physicists" as social groups. No such groups exist. They are worse than useless fictions; they are highly misleading fictions.

As Blau (1978) has shown, even so narrowly defined a subject matter as theoretical high-energy particle physics does not serve to denote a genuine research group, even though scientists working in this area in the United States at the time numbered under five hundred. Instead Blau found that roughly half of the scientists working in this area were organized into small "cliques" or research teams, while the rest worked primarily alone. These isolates attended an occasional professional meeting and read the relevant literature, but that was about all. Blau also discovered that about half of the cliques were organized more loosely into a federation or "invisible college." The scientists working in cliques that were part of this invisible college were much more productive than those who were not. In science cliques are the largest units of hereditary transmission, while more inclusive units function both in the development of research programs and in their defense (see also Crane 1972, Griffin and Mullins 1972).

In order to understand science, it is absolutely essential that we not view scientists as isolated knowing subjects confronting their experiences. Individual workers can make some headway during the course of their lifetimes in understanding the empirical world, but this sort of understanding lacks several characteristics essential to science. Because each of these investigators must start anew each generation, such an activity cannot be transgenerationally cumulative the way that science is. Because the only person available to test the beliefs of such isolated investigators is the investigator himself, individual bias is extremely difficult to eliminate. One of the strengths of science is that it does not require that scientists be unbiased, only that different scientists have different biases.

None of the preceding requires that scientists be organized differentially into research groups, any more than natural selection necessitates the evolution of species or the subdivision of species into partially isolated demes. In fact, it took a long time for species to evolve. However, once organisms were able to exchange genetic material rather than just simply to pass it on in replication, a higher level of organization evolved that opened up new possibilities as well as new problems for the constituent organisms. In this book I argue that one of the fundamental mechanisms in the conceptual development of science is conceptual inclusive fitness. Scientists can pass on replicates of their ideas *as their ideas* directly to later generations of scientists, but they also can cooperate with their contemporaries in promoting their collective goals. Initially, in the history of science, scientists worked in relative isolation from their contemporaries. They built on past work but did not band together to pursue joint research. Rather rapidly, however, the demic structure of science materialized and continues to characterize science to the present. As the years have gone by, it has grown increasingly difficult for scientists to work in social isolation from their contemporaries. However, just as the advent of sexual reproduction introduced a new set of partially conflicting goals into biological evolution, as organisms began to cooperate with their genetic competitors, the formation of research groups in science introduced a comparable set of partially conflicting goals into science, as scientists had to cooperate closely with their conceptual competitors.

Although Darwin (1859: 20) was concerned to argue that all species evolve, he found it desirable to begin his *Origin of Species* with a single species—the domestic pigeon. "Believing that it is always best to study some special group, I have, after deliberation, taken up domestic pigeons." Darwin settled on pigeons, Mendel on garden peas, and modern geneticists on the fruit fly. After deliberation, I too have taken up a special group to study. My fruit flies are systematic biologists; not all systematic biologists, but two small research groups—the pheneticists and the cladists. I follow the course of evolutionary biology from Darwin to the present primarily to illustrate how conceptual selection occurs in science. As intensively as the Darwinian revolution has been studied, our knowledge of it is still not sufficiently detailed to count as very firm evidence. But at least readers are liable to have heard of it and be interested in it. I intend my study of the pheneticists and cladists (obscure though they may be) to serve primarily as evidence for my views about the nature of science. Plenty of relevant evidence is still missing, equivocal, or downright mushy, but the recent history of systematic biology is likely to present as ample a documentation of science as students of science are going to get.

One problem with small research groups such as the pheneticists and cladists is that, from the human perspective, they are relatively ephemeral. By the time that one notices that such a group has developed, it is already well on its way to extinction. In this respect, those of us attempting to study scientific development are presented with problems just opposite to those that confront evolutionary biologists. From the human perspective, biological evolution is much too slow. The human species is liable to become extinct before evolutionary biologists are able to study firsthand the evolution of very many species. I was fortunate to happen upon the pheneticists during the early years of their formation, although at the time I had no intention of studying the course of their development as an instance of a scientific research group. That came later. I was even more fortunate in connection with the group that has come to be termed the cladists. It materialized before my very eyes as I was studying the pheneticists, although the realization that a new group was emerging was necessarily retrospective.

These two groups have much to recommend them as subjects for investigation. For the crucial periods in their development, they were geographically localized: the pheneticists at the University of Kansas in Lawrence, Kansas, and the cladists at the American Museum of Natural History in New York City. In addition, systematic biologists use the work of many other scientists, and other scientists in turn use their work. In an age of specialization, they are among the few biologists who still study the "whole organism" and its interactions with the environment. Although taxonomy is not one of the currently "hot" areas of science, it is centrally located. It also helps that those scientists engaged in the disputes under investigation had, until quite recently, only one journal in which to publish—*Systematic Zoology.* The history of these two groups can be traced with only minimal distortion by studying the development of this one journal (Hull 1983a, 1985a).

It also is fortunate for my purposes that one of these groups seems to have "lost," while the other is succeeding beyond anyone's early expectations, save its own. In the past, students of science have tended to concentrate on those groups of scientists who either won or at least lost in big ways, but such a biased perspective is liable to be as misleading as the study of biological evolution only through those species that have succeeded in surviving to the present. The vast majority of species that have ever existed are now extinct, and only a very few ever became especially prevalent. Similarly, the vast majority of research groups arise, publish for a while, and then go extinct without leaving a ripple on the surface of science. Anyone who ignores these research groups—the vast majority—is ignoring most of science. Success may be more interesting than failure, but failure is no less a part of science than success. Any theory about science that does not account for the activities of the vast majority of scientists has to be inadequate.

The Essence of Science

In genetics, progress was determined as much by the choice of an appropriate research organism as by the selection of a tractable problem. Mendel did not simply study three-to-one ratios but three-to-one ratios in the garden pea. For this purpose the common garden pea could not have been better. When an infestation of pests forced Mendel to change to another species and he adopted Carl von Naegeli's suggestion that he switch to a common composite in the group *Hieracium,* Mendel's budding career rapidly came to a close. Geneticists only recently have succeeded in unraveling the genetics of this plant. De Vries and Johannsen were just as unfortunate. The balanced heterozygous chromosome rings of *Oenothera* led de Vries to think that the "mutants" which he observed were the founders of new species, while the self-fertilizing, near homozygous beans of the genus *Phaseolus* misled Johannsen to conclude that natural selection could be only a minor force in evolution (Mayr 1982: 731). Of course, the scientists just mentioned did not simply "happen" upon their research organisms. They sought them out for specific reasons. However, these organisms also happened to have unexpected characteristics which either frustrated or facilitated the research conducted on them.

One inevitable criticism of my research is that I have picked abnormal, atypical areas of science to investigate. Some critics may complain that anything I might find about successive groups of evolutionary biologists and biological systematists may well be true of them, but that I cannot reason from these monstrous cases to science itself. Real science, or mature science, or, to be blunt, physics is likely to exhibit quite a different structure. This response to inductive inference is quite common and not always misplaced. Mendel has become a paradigm example of an unappreciated precursor. What if Darwin had read Mendel's paper? He might well have. If he did, his conclusion was likely to be the same as that drawn by others at the time. "What a curious pattern of inheritance garden peas happen to have." Even at the turn of the century, the main objection to Mendel's laws was not that they were inapplicable to garden peas but that they were not generalizable to other species.

The features of science that I have discovered in the groups that I have studied may turn out not to be generalizable. All I can say in response to this objection is that we will have to wait to see as other groups are studied in sufficient detail. Thus far the data look good. Certainly, Rudwick (1985) found among nineteenth-century British geologists exactly the mix of cooperation and competition, camaraderie and animosity, as well as the allegiances and alliances that I detail in this book. Raup (1986) tells a very similar story for the present-day astrophysicists, chemists, geologists, and paleontologists involved in the Nemesis affair (see also Wade 1981, Bliss 1982, Provine 1986, and Giere 1988). In connection with the particular groups I studied, Dawkins (1986: 275) characterizes taxonomy as the "most rancorously ill-tempered of biological fields" and finds pattern cladists especially nasty (see also Ridley 1986). I find this conclusion puzzling, coming as it does from someone who was so caught up in the controversy over sociobiology. I for one have not found the recent history of sociobiology any less nasty than anything that has taken place in taxonomy.

However, I have a more fundamental reason for rejecting the atypicality objection, a reason internal to my own account of science. Prior to Darwin, with only a very few exceptions that were as noteworthy as they were rare, nearly everyone viewed species as having eternal, immutable essences. Each species had to be characterized by a set of properties which *all* (but abnormal) members of that species possessed, and *only* members of that species possessed. Most people, including most philosophers and some evolutionary biologists, still view species in this way. A comparable view about such things as research groups, research programs, theories, and even science itself is just as mistaken and even more insidious. Science as a historical entity no more has an essence than do particular scientific theories or research programs. The sorts of activities that are part of science at any one time are extremely heterogeneous, and they change through time. Once systematics was the queen of the sciences; now it is a handmaiden at best. At one time evolutionary biology did not exist. It blossomed under the influence of Darwin, went into an eclipse at the turn of the century, became rejuvenated after World War II, and once again is now becoming the focus of intense activity by some of the most brilliant scientists working today. All these boundaries and evaluations pose serious problems for anyone who thinks they can make easy judgments about "normal" versus "abnormal" science. The nature of science is constantly under negotiation (in the literal sense of this term), and the chief currency in these negotiations is the success of particular research programs (Rudwick 1972, Cohen 1981, Laudan 1981b, Hull 1983b).

Evolutionary biologists have studied the evolutionary process by studying a variety of species without any special concern about how "typical" or "atypical" they might be. If species evolve, evolutionary theory had better be able to handle them. At the very least, I fail to see in what sense garden peas and domestic pigeons are any more (or less) typical species than any other. Similarly, I fail to see in what sense high-energy physics is any more typical a science than systematic biology. Each has some characteristics that are fairly common in science. Each has a few characteristics that are rare. When asked whether he thought that systematists are especially atypical scientists, Humphrey Greenwood of the British Museum (Natural History) responded, "If I knew what a typical scientist was, I'd answer your question. I can no more tell you what a typical scientist is than I can tell you what a typical fish is."

After reading about the infighting and personal vendettas that have occupied so much of the time of the scientists whom I have studied, the reader is likely to conclude that these scientists are really not behaving the way that scientists should behave. To the contrary, I argue not only that these scientists are behaving the way that all innovative scientists behave but also that this sort of behavior actually facilitates scientific development. One of the chief messages of this book is that factionalism, social cohesion, and professional interests need not frustrate the traditional goals of knowledge-acquisition. Affection frequently binds the members of research groups together, but devotion to a "higher" goal allows them at times to treat each other with a steeliness that the Borgias would have admired. Science does have the characteristics usually attributed to it by internalists but not for the reasons they usually give. Scientists do cooperate with each other extensively, but it is a form of cooperation that does not differ materially from mutual exploitation.

Although scientists themselves are frequently uneasy about using terms like "true" and "objective," all the time that they spend running experiments and making extensive and careful observations is inexplicable on the assumption that knowledge is in any significant sense socially determined. Scientists are not infallible. Science is not a process by which we go from no knowledge to some knowledge, or from some knowledge to total knowledge. Rather it is a process by which scientists go from some knowledge to more knowledge. The important feature of science is not that it *always* produces increased knowledge but that *sometimes* it does. Science is not a perfect machine for grinding out true claims about the world in which we live, but it is the best of all the imperfect machines developed to date. Scientists are not objective all the time, but given the social organization of science a little objectivity can go a long way. Nor is the process necessarily complete. Better methods may wait just over the horizon. They may already be here, and we have not even noticed them yet.

But isn't science merely the way that Western and Westernized societies establish their beliefs? Science *is* one of the major ways that people in Western and Westernized societies today establish their beliefs, but it is neither the only way nor merely the way that they do so. What science has in its favor is that it beats all other ways hollow. There is no contest. Totalitarian nations of every political stripe can safely suppress artistic expression, but they find it very difficult to suppress freedom of inquiry in science, that is, if they want to remain in power. Although life in a society without artistic productions might not be worth living, it is possible. It is no longer even possible without science. Success in the knowledge game is hardly an incidental feature of *Homo sapiens*. It is our chief adaptation. It is the only thing in the struggle for existence that we do better than any other species.

Science is not the only important human endeavor. There are many others. Many areas of human endeavor exhibit some of the same characteristics of science. Science is not the only scholarly activity. Philosophers, literary critics, and biblical scholars build on past work and give each other credit. However, no other intellectual activity exhibits in precisely the same way all the same elements that characterize science. In particular, in no other area of human endeavor is the notion of "evidence" so clear and direct. In fact, the domain of science is determined primarily by means of the availability of evidence. To the extent that evidence in a very literal sense can be obtained in a particular endeavor, that endeavor counts as science.

In paying so little attention to institutions other than science, I do not mean to denigrate them. No one can investigate everything. Even in the context of science, I pay little attention to the psychological makeup of scientists, concentrating instead on their social organization. For the questions that I am addressing, the psychology of the individual actors is less important than the structure of the communities in which they function. Many scientists really are interested in truth for its own sake. Many are concerned with giving credit where credit is due. Some even want to help humanity. However, I have not found these psychological characteristics especially useful in understanding the general characteristics of the scientific enterprise. Even scientists who lack them can function quite effectively in science. Nor do I spend much time on the sources of creativity. The "wow feeling" is as important in science as is orgasm in human procreation. In their absence, the frequency of both sorts of activities would no doubt decline. Without innovation, scientific change would cease, but I have come up with very little of any interest to say on the subject. Instead I concentrate on the selection process that follows upon innovation (for science as a cognitive process, see Giere 1988).

The Essence of Philosophy

One need not read very far in this book to discover that I do not treat many traditional issues in the philosophy of science. Most books on philosophy of science begin with a long, detailed description of past attempts by philosophers to set out the nature of science, showing precisely where those attempts went wrong. Such descriptions and criticisms are not only the strongest part of these expositions but also the easiest to write. By now the outlines of inductivism, deductivism, hypothetico-deductivism, falsificationism, etc. are well known, almost as well known as are their shortcomings (see Lakatos and Musgrave [1970] and Suppe [1977]). I do not rehearse either here. As I see it, the chief weakness of the logical empiricist analysis of science has been the emphasis of its advocates on inference to the near total exclusion of everything else about science, especially its temporal and social dimensions. Logical empiricists were as aware as anyone else that science as an ongoing process is both temporal and social, but these aspects were thought to be irrelevant to a "philosophical" analysis of science. Philosophers must limit themselves to more modest activities, e.g., to the analysis of rationality as such, the abstract relation between theories and data, and the like. I suspect that one reason for philosophers ignoring science as it actually takes place is that they fear that it will prove to be too complicated and haphazard to allow for any general conclusions. One purpose of this book is to show that this fear is mistaken.

Perhaps science as it is actually practiced is more complicated than textbook expositions, but it is not so complicated as to be unmanageable.

Another "unphilosophical" feature of this book is that it includes no homespun, science-fiction, or deliberately silly examples. Traditional philosophers, especially in the analytic tradition, are frequently suspicious of those philosophers of science who spend so much time on detailed case studies, suspecting that their primary function is to lend a spurious air of "dignity and significance" to otherwise prosaic philosophical points (Bromberger 1962). I must admit that sometimes the extensive detail introduced in philosophical works serves only to disguise the absence of any real philosophical content, but if philosophical claims about science are to be *testable*, the only tests that count are those drawn from science. One traditionally tests the limits of concepts by introducing contrary-to-fact conditionals. "You say that species are spatiotemporal particulars? Well, what would you say if on Alpha Centauri . . . ?" The truth of empirical claims, including laws of nature, are tested by what in fact does happen. According to one present-day version of evolutionary theory, speciation can occur in sexual organisms only in the presence of an appropriate isolating mechanism. Such claims involve physical, not conceptual, possibility. It is certainly possible to imagine speciation occurring under other conditions, but even so, such occurrences may be physically impossible.

The point I wish to make is that issues of conceptual and physical possibility, as problematic as they may be on a host of counts, are too closely intertwined to be treated separately (see Kuhn 1977). The counterexamples used to test the limits of a scientific concept must be scientifically informed. Perhaps ordinary people in ordinary contexts can imagine centaurs—sort of. But biologists are likely to have problems. Centaurs have two upper torsos. Do they have two hearts or one? Two pairs of lungs or one? If they have one heart, where is it and how is it connected to the lungs? If they have two hearts, how are they connected to each other and the rest of the circulatory system? Does the fecal material of the human torso empty into the stomach of the horse torso? In the context of mythology, such questions are out of place. In the context of biological science, they are not. They are exactly the sorts of questions which a biologist *must* ask. Any analysis of scientific concepts must be in the context of the relevant science. Ordinary concepts, as important as they are to ordinary people and ordinary-language philosophers, are irrelevant.

Even if one accepts the independence of scientific discourse, as a technical discourse, from that amorphous hodgepodge termed ordinary language, the status of philosophical claims about science is more problematic. In retrospect, one can distinguish between object languages that refer to the empirical world (science) and metalanguages that refer to these languages (philosophy of science). Officially, philosophy of science is about science, not part of it. Certainly the discourse of philosophers of science is just as technical as that of scientists. Philosophical claims about science should no more be rejected because of conflicts with ordinary discourse than should scientific claims because of parallel conflicts.

In practice, however, science and philosophy of science are more than closely connected. They interpenetrate each other. In particular, issues of conceptual possibility in metascientific contexts are too closely intertwined with issues of metaphysical possibility for unconstrained science-fiction examples to be relevant. Whatever else philosophy of science is, it is not a subdiscipline of ordinary-language philosophy. Rorty (1980: 729) expresses the same exasperation I feel with the analytic tradition in philosophy:

> No sooner does one discover the categories of the pure understanding for a Newtonian age than somebody draws up another list that would do nicely for an Aristotelian or an Einsteinian one. No sooner does one draw up a categorical imperative for Christians than somebody draws up one which works for cannibals. No sooner does one develop an evolutionary epistemology which explains why our science is so good than somebody writes a science-fiction story about bug-eyed and monstrous evolutionary epistemologists praising bug-eyed and monstrous scientists for the survival value of their monstrous theories. The reason this game is so easy to play is that none of these philosophical theories have to do much hard work. The real work has been done by the scientists who developed the explanatory theories by patience and genius, or the societies which developed the moralities and institutions in struggle and pain.

If philosophical theories about science are to be taken seriously, then they cannot remain mere gestures containing nothing but indefinitely renewable promissory notes. Some "hard work" must be done. If such hard work is incompatible with a theory counting as a genuine "philosophical" theory, then so be it, but I do not think that is the case. Rorty (1980: 729) argues that relativism "only seems to refer to a disturbing view, worthy of being refuted, if it concerns *real* theories, not just philosophical theories." But I see no reason why philosophical theories cannot be *real* theories. Richter (1972: 5) remarks that scientists "want freedom from control by society, but only in order that they may submit more fully to 'control' by nature." Such control by nature is not easy to come by; the route to it is more tortuous and studded with more perils than the road to the Emerald City, but it is a trip worth taking.

Truth and Responsibility

T. H. Huxley deplored the term "scientist" as an unacceptable barbarism, but he, more than anyone else at the time, exemplified the ideal of the Victorian scientist. Huxley's guiding maxim was that, "in matters of the intellect, follow your reason as far as it will take you, without regard to any other consideration." In today's climate, we are likely to smile indulgently. Truth is a human good, but it is not the only human good, possibly not even the most important. What is the value of knowledge if it destroys us? Knowledge is power, and power is dangerous. But in Huxley's day scientists were also reminded of their broader responsibilities. For example, *The Times* of London chastized Darwin for publishing *The Descent of Man* (1872). With Paris in flames, Darwin's behavior was "more than unscientific—it is reckless" (Desmond 1982: 182). A quarter of a century later, Wilhelm Johannsen (1896: 94) found Weismann's conception of the germ plasm "morally dangerous."

Once again, modern readers are likely to smile indulgently. Many of the fears of earlier generations seem almost quaint, but comparable fears are being expressed today with equal fervor concerning issues as diverse as standardized performance tests, the study of statistical differences between races of human beings, physiological differences between men and women, gene splicing, and the flouridation of water. As in Darwin's day, critics of views that they take to be dangerous are not content to argue merely that these conclusions are dangerous but that they are "unscientific" as well.

Throughout history, scientists have been urged to suppress their views about nature for the sake of the public welfare. Authorities feared that society would be destroyed if it became widely known that the sun is at the center of the solar system, that species evolve, that Lamarckian inheritance does not occur, that genes influence mental as well as ordinary phenotypic traits, and so on. I do not question the sincerity of those who would sacrifice our best estimates of the truth for our best estimates of the public good. Changing ideas about the world in which we live have influenced the way in which we perceive ourselves as well as the societies in which we live. As quaint as the fears expressed in Darwin's day may appear to intellectuals today, evolutionary theory has destroyed the faith of numerous people from Darwin's day to the present. Thus far, however, those who have urged the suppression of new views for the "good of the people" have underestimated the ability of both societies and individual people to survive successive challenges to their conceptions of the world and how it works.

To make matters worse, the very labeling of certain areas of knowledge as "dangerous" carries some weight of social responsibility. In the past, all the wrong people have attempted to proscribe all the wrong views for all the wrong reasons. Given such a long history of past abuses, the onus is on those who would suppress apparent truth for apparent good to present a very solid case to back up their proscriptions. Even so, it remains the case that knowledge can be and often has been misused. Those people who are most committed to following their reason as far as it will lead in other matters of the intellect should not abandon this maxim when attempting to follow out the social consequences of their own research.

Although I do not think that academics have the great impact on society that they apparently think they do, on occasion ideas about society have changed society, and not always in especially desirable ways. In this book I present a picture of scientists that some are likely not to find all that flattering when compared to the traditional stereotype of scientists working selflessly for the good of humankind. Some readers may object to my views because they are mistaken, "unscientific." Science in general and individual scientists in particular exhibit the very characteristics that I claim are little more than romantic hypocrisy. Others are just as likely to find my views about science dangerous. If the public at large is exposed to the portrait of science presented in this book, people may become disillusioned with science and resist supporting it with public funds. As false and self-serving as the traditional picture of science may be, it is necessary if science is to persist. To paraphrase a well-known response to Darwin's theory, "Scientists want credit for their contributions? My dear, let us hope that it isn't true! But if it is true, let us hope that it doesn't become widely known!"

All societies, especially pluralist societies, can persist only by maintaining a certain degree of tacit hypocrisy. Hypocrisy, so it seems, is an essential social lubricant. Without it, society would grind to a halt. Sometimes people are vaguely aware that they are joining in this tacit hypocrisy; just as often they are not. An example might bring home how important hypocrisy is in society. One of the most cherished beliefs in American society is that the Supreme Court does not make new laws but only "interprets" the law in the light of the Constitution. The picture that comes to mind is of the justices in their black robes poring over the Constitution late into the night, word by word, trying to discover a word or phrase that past justices have overlooked or misinterpreted. The Constitution is the rock upon which the republic is built. If so, the Constitution is made of Silly Putty.

"Interpret" has become a technical term in legal philosophy. The Constitution can be changed officially by amending it, but it is also modified just as radically through the years by successive "interpretations." The tacit understanding is that these latter changes are acceptable only if they are gradual and made under a verbal smoke screen. The justices of the Supreme Court can change the Constitution, but only so long as they never admit openly that this is what they are doing (see Brown 1986 and Meese 1986).

Perhaps the general public needs the illusion that the laws of the land are based on rock rather than on the shifting views of nine human beings, no matter how learned and well-intentioned; but I cannot see the advantage of those studying legal institutions (or the justices themselves, for that matter) joining in such fictions. Perhaps legal scholars, out of social responsibility, should publish their potentially destructive views in technical journals written in sufficiently arcane jargon to ensure that those who are incapable of facing the truth are not disillusioned and, hence, society is not brought down. Perhaps the general public also needs the illusion that scientists are dispassionate, disinterested arbiters of scientific truth, unsullied by any of the baser motives that characterize the rest of us. Perhaps the only justification that taxpayers are willing to accept for supporting science is the prospect of practical gain, although I doubt it (see Norman 1983).

As necessary as hypocrisy may be for other social institutions, I think that science can survive a critical examination without being destroyed in the process. In fact, once all the layers of hypocrisy have been stripped away, science comes through largely unscathed. No amount of debunking can detract from the fact that scientists do precisely what they claim to do. The effects of science on human beings have not been unmixed. There have been trade-offs. But the growth of our understanding of the empirical world has been spectacular. Perhaps those of us who are studying science should publish our subversive views in prose so dense that only a few other devotees are liable to slog through it, but students of science (not to mention scientists themselves) can accept the traditional caricature of science only at some peril.

All of the preceding concerns science as a social institution. What about individual scientists? I am afraid that no one will read the first half of this book without periodic gasps of dismay. Perhaps scientists are not disembodied intellects, but there are limits. The political infighting, the name-calling, the parody and ridicule, the arrogance, elitism, and use of raw power are likely to strike some readers as distasteful. This response calls for two comments. First, some of the virtues which scientists fail to exemplify are not and never have been part of the ethos of science. Neither humility nor egalitarianism has ever characterized scientists, and no one has ever given any good reasons why they should. Second, scientists behave no worse in these respects than do members of other professions. The behavior of the scientists whose careers I chronicle may not look very good when compared to some Platonic ideal scientist, but it looks very good indeed when compared to the behavior of doctors, politicians, or bankers.

I argue an even stronger thesis: some of the behavior that appears to be the most improper actually facilitates the manifest goals of science. Mitroff (1974: 591) remarks that the "problem is how objective knowledge results in science not despite bias and commitment but because of them." Although objective knowledge through bias and commitment sounds as paradoxical as bombs for peace, I agree that the existence and ultimate rationality of science can be explained in terms of bias, jealousy, and irrationality. As it turns out, the least productive scientists tend to behave the most admirably, while those who make the greatest contributions just as frequently behave the most deplorably. You pays your money; you gets your choice.

2

Up from Darwin

Evolution is a process of Variation and Heredity. The older writers, though they had some vague idea that it must be so, did not study Variation and Heredity. Darwin did, and so begot not a theory, but a science.

William Bateson, *Heredity and Variation in Modern Lights*

WHEN A novelist such as Tolstoy wants to introduce his cast of characters, all he has to do is stage a ball. Although a soirée at which Charles Darwin chats with Gregor Mendel and T. H. Morgan about topics of mutual interest is pleasant to contemplate, no such devices are open to historians. To the best of their ability, they must tell history the way that it happened, but such a dictum is easier to enunciate than to follow. Philosophers of history worry endlessly about the bias introduced into historical narratives by the interests and experiences of the historian, but historians inevitably look at the past from the only perspective available to them—the present.

Retrospective bias in history is a vice, but it has a source that is somewhat less exciting and even more recalcitrant than the interests and experiences of historians. Given the basic structure of both the written and spoken word, historical narratives lend themselves much more easily to linear chains than to diverging trees or interlocking networks. Even though an individual life may be experienced as one damned thing after another, human history forms a web. Pick an event, any event, and numerous sequences of past events can be found converging on it, while a symmetrical tree of consequences fans out from it. Events have as many effects as causal antecedents. Pick another event and perform a causal analysis of it. Another pair of symmetrical causal trees will be discovered. Superimpose enough of these trees on one another, and the resulting interlocking network becomes a hopelessly complicated maze. The task that confronts historians is how to describe this maze in a comprehensible way within the linear constraints imposed by the narrative form.

Historians are not the only ones faced with such problems. We commonly hear of family trees, but such "trees" are actually networks. They can be made to look like trees only by ignoring the ancestry of one of the marriage partners through successive generations. Children fan out from their parents, but with each marriage two genealogies become interconnected. Sexism may explain why family trees in Western cultures are almost always traced through the male partner, but it does not explain our penchant for causal chains over trees and trees over networks. Some people can visualize all three equally well, but not all three are equally easy to describe.

A similar story can be told for paleontologists, as they attempt to reconstruct phylogenetic trees. Genealogy is the stuff of phylogeny. It is what produces phylogenetic sequences, but not until the genealogical web is rent does it count as phylogeny. As it is usually portrayed, the phylogenetic tree consists of species that split and diverge. In actual fact, merging also occurs—rarely in animals, quite commonly in plants. Even if one limits oneself to a section of the phylogenetic tree that contains no hybrid species, it is not easy to describe in a comprehensible way. The simplest thing to do is pick an extant species and trace its ancestry back through time. In the process, additional lineages join in, but they are largely ignored. Then one can turn around and retell the story in the reverse order, showing how an ancient lineage eventuated in an extant species such as the modern domestic horse. The result is that the past seems to home in on the present with ineluctable finality. It is this sort of phylogenetic story that makes *Homo sapiens* appear to be the ultimate product of three and a half billion years of biological evolution. If the data were available, paleontologists could trace the human lineage back to the first glob of glub. Conversely, this same lineage could be traced without interruption from the origin of life to the human species. Of course, one can do the same thing for any species, including the German cockroach. It is as much an ultimate product of three and half billion years of evolution as is *Homo sapiens*.

Professional paleontologists do not distort phylogeny in this way, but the tracing of lineages in branching trees lends itself to just this sort of abuse. Unfortunately, one of the most common forms of human history amounts to tracing lineages. Historians begin by identifying a later development, such as the Modern Synthesis in evolutionary biology. They then proceed to trace its origins back through time. Avenues of research that led elsewhere or nowhere are at most mentioned. Once a natural starting point is reached, such as Darwin, the story is retold in the opposite direction. Tracing lineages in this fashion may not seem to be all that pernicious a practice, but it tends to obscure too many significant connections. Phylogenetic trees are actually phylogenetic bushes containing numerous branches that lead nowhere. But few of us have the patience to follow up numerous leads that turn out to be dead ends; we much prefer pruned trees.

If scientific development were as treelike as phylogeny, tracing lineages would be difficult enough, but science is commonly portrayed as containing considerable interlineage borrowing. As its name might imply, the Modern Synthesis is supposed to be an instance of a "synthesis." The intent of the founders of the Modern Synthesis was to bring together the best in population genetics, systematics, paleontology, and so on into one grand theory. If any episode in the history of science should reveal the difficulties in describing the web of scientific development, it should be the Modern Synthesis. As we shall see, the interconnections are not as extensive as one might expect. Recall that phylogenetic trees have more interconnections than they are usually portrayed as having. Reticulation characterizes both sorts of development.

No totally satisfactory solution exists for the historian's dilemma. My solution in this book is to follow one strand in the interconnected web of events that led from Darwin's various versions of evolutionary theory to the Modern Synthesis—the "main" strand. Then I backtrack to follow another strand. Eventually, after enough backtracking, the outlines of the actual course of events begin to emerge. However, the influence of my initial choice of the primary lineage is impossible to neutralize. Like it or not, other lineages will seem more like offshoots and extraneous pathways than they actually are. The reader can test my claims about our intellectual predilections for chains over trees and trees over webs by noting his or her irritation at each interruption of the narrative. "Why don't I get on with it instead of backtracking to pick up other sequences of events?" But the complexity of scientific development precludes any other solution. Even the simplest story in the history of science has as many characters as a Russian novel, and there is no way to trace all their fates simultaneously. James Joyce notwithstanding, narrative fugues are all but incomprehensible.

I decided to begin my narrative with Darwin and the birth of evolutionary biology for two reasons. First, if science proceeds by means of a mechanism anything like the one operative in biological evolution, then it must be viewed as consisting of lineages, not isolated episodes. The contours of a particular scientific research program are determined as much by the sequences of competitors that it confronts as by its own internal dynamic. Although the history of evolutionary theory from Darwin to the present has been told many times before, it has not been structured consistently along lines required by my theory. Second, many of the issues which arise in the development of evolutionary biology are central to the evolutionary account of science that I present in the second half of the book. Because anyone reading the first half of this book will already be familiar with the points at issue, I can characterize more succinctly my theory about the way that science evolves.

The Darwinians

In very short order, Captain Fitz-Roy (1805–65) decided that the man who had been appointed surgeon-naturalist to the Beagle, Robert McKormick (1800–1890), was no fit companion for a gentleman on such a long voyage. He sent out inquiries for a naturalist-companion. These inquiries soon reached the Reverend John Stevens Henslow (1796–1861), the kindly professor of botany at Cambridge University. He could not even consider such an undertaking because of family obligations, but perhaps his brother-in-law, Leonard Jenyns (1800–1893), might be interested. However, Jenyns was married as well and did not want to absent himself for several years on an arduous and dangerous trip. Finally, Henslow approached his favorite student, a bachelor, Charles Robert Darwin (1809–82).

When Captain Fitz-Roy first met Darwin, he was put off by the shape of his nose. According to the principles of phrenology, it implied a lack of energy and determination, two characteristics necessary for any long sea voyage. Despite his nose, Darwin eventually succeeded in convincing Fitz-Roy that he would be an acceptable companion. For his own part, Darwin hoped to realize his dream of taking a voyage of discovery of the sort celebrated by Alexander von Humboldt (1769–1859) and in the process make a name for himself among the geologists and naturalists of the world. After two deflating false starts, the *Beagle* finally succeeded in setting sail on December 27, 1831.

The voyage turned out to be even more arduous than Darwin had anticipated. The quarters that he shared with Captain Fitz-Roy were extremely cramped, and Darwin was seasick much of the time. But Darwin doggedly pursued his course of geological observations and natural-history collecting. The crowding was alleviated somewhat at Rio de Janeiro in April of 1832 when McKormick left the *Beagle* to return to England. In September Darwin stumbled across his first fossil, the head of a giant armadillo-like creature. A month later he received a copy of volume 2 of Charles Lyell's (1797–1875) *Principles of Geology.* Darwin had used volume 1 as his chief guide in his geological observations. In volume 2 Lyell confronted the species problem, primarily in the context of refuting Lamarck's transmutation theory. Lyell viewed extinction as entirely a natural process. When the conditions of existence changed in a particular locality, the species that could not survive these changes went extinct. Lyell was less explicit about the manner in which new species came into existence, but if species regularly go extinct and the number of species through time stays roughly the same, then new species had to arise somehow.

Lyell was sure that new species did not come into being the way that Lamarck claimed, but he was none too clear about how they actually did arise, although conditions of existence had something to do with it. In fact, Lyell (1830:v.1,123) thought that if the appropriate conditions ever reappeared, formerly extinct groups might also reappear:

> Then might those genera of animals return, of which the memorials are preserved in the ancient rocks of our continents. The huge iguanadon might reappear in the woods, and the ichthyosaur in the sea, while the pterodactyle might flit again through umbrageous groves of ferns.

In retrospect, the most important period in Darwin's voyage was the month that he spent in the fall of 1835 exploring the Galapagos Islands off the coast of Ecuador. The great tortoises, the marine iguanas, the mockingbirds, the many small finches were all fascinating, but at the time Darwin saw nothing special about the inhabitants of these piles of volcanic rock. In fact, not until he had circumnavigated the globe and was on his way home did Darwin register in his journal any doubts about the fixity of species. Less than three months after he began to think seriously about the species problem, Darwin arrived back in England, on October 2, 1836. He had been gone almost five years. When his father saw him again, he remarked that the very shape of his head had changed. Perhaps he would amount to something after all.

Darwin spent most of the next six years in London. They proved to be among the most productive in his long life. He began work on the specimens he had collected and stored with Henslow at Cambridge. He presented papers before the Geological Society of London, started work on the story of his journey, and began a notebook on the species question. He was convinced that species change through the course of time, but he did not know how. At first Darwin toyed with the idea that numerous living particles (or monads) are constantly generated spontaneously from inorganic material;

perhaps these simple forms of life became more complex as they changed in reaction to the direct effects of the environment. Perhaps species are like organisms and undergo life cycles. In the ensuing months Darwin would try out a variety of hypotheses, producing one tentative theory after another, until, in 1838, he stumbled upon Thomas Malthus's *Essay on Population* (1798).

A necessary effect of limited resources and an ever-increasing population is, as Malthus saw it, misery and death. In the human species, Malthus recommended sexual abstinence as a remedy, especially among the lower classes. Darwin reasoned that the struggle for existence, instead of keeping species within preestablished boundaries, might cause them to change through time as the less favored races were weeded out. In the fullness of time, one species might be transmuted into another the way that alchemists were forever trying to transmute baser elements into gold. Finally, Darwin had come up with a mechanism that he thought might be adequate to account for both the extinction and the generation of species—variation and the selection of better-adapted forms. As Darwin recalled, his reading of Malthus produced one of those ecstatic feelings that fix themselves with such clarity in the minds of scientists. Even so, Darwin scholars have shown that the psychological processes that led to this experience of instantaneous elation were quite protracted.

During his stay in London, Darwin kept quiet about his heretical views. Instead of proclaiming his discovery far and wide, he continued his geological work on the formation of coral reefs and the possible causes of the shelves or "roads" that run parallel to each other around the sides of certain glens in Scotland. During this period, Lyell was Darwin's most important intellectual and professional mentor. Not only did Lyell impress his own "uniformitarian" way of thinking on Darwin, he also introduced Darwin into scientific society. From Lyell Darwin learned what it was to be a "scientist," a term that had only recently been coined by the great William Whewell (1794–1866).[1] During this same period, Darwin married his cousin, Emma Wedgwood (1808–96), and made the acquaintance of the man who was to be his closest friend throughout his life, the botanist, J. D. Hooker (1817–1911). Hooker was about to depart on his own voyage of discovery accompanied by the ubiquitous Robert McKormick. Hooker and McKormick seem to have gotten along, but just barely.

In 1839 Darwin wrote a fourteen-page draft of his views on species and in 1842 expanded it to thirty-five pages just prior to moving to Down, a small village near London. For the rest of his life, Darwin was to live in this rural setting, leaving it only rarely for very short periods, usually to take some sort of cure for a recurrent illness. In 1842 the government added an "e" to the name of the village to distinguish it from the county in Ireland, but Darwin never adopted the newfangled spelling for his own

1. Whewell (1834) coined the term "scientist" in a review of a book by Mary Somerville (1780–1872) on the nature of the physical sciences. Whewell noted that the French had the term *savant,* while the Germans had *Naturforscher,* and concluded that the English deserved a term as well. He reasoned that if people who do art are termed "artists," then people who do science might well be termed "scientists." T. H. Huxley for one was appalled by the barbarous neologism. That the term "scientist" was introduced in a review of a book by a woman and that Huxley was to become the paradigmatic Victorian scientist says something about the vagaries of history.

Down House. In 1844 Darwin expanded his 1842 essay on species to 230 pages, and two years later began work on his beloved—and hated—barnacles. He also began, ever so gingerly, to broach the subject of species to his fellow scientists. Some he told of his doubts about the immutability of species. A few he even trusted with his proposed mechanism for species change—natural selection. Among his earliest confidants about species were two friends from his Cambridge days—Jenyns and yet another cousin, William Darwin Fox (1805–80). But Hooker was the first to learn of both his heretical view on the transmutation of species and his mechanism for its accomplishment.

In 1855 Darwin noticed a paper entitled "On the law which has regulated the Introduction of New Species" by an obscure naturalist, a Mr. Alfred Russel Wallace (1823–1913), but all Wallace claimed was that new species are created in close proximity to the species that they resemble most closely. As Darwin read this paper, Wallace was not suggesting evolution but the creation of similar species in close proximity to each other. Lyell and Edward Blyth (1810–73) thought otherwise. Both men brought the paper to Darwin's attention, and in April 1856 Lyell urged him to publish lest he be forestalled. Darwin refused to rush into print. He had seen what had happened to earlier authors who in the absence of adequate inductive foundations had published theories urging the transmutation of species—Lamarck (1744–1829) in his *Philosophie zoologique* (1809) and the infamous *Vestiges of the Natural History of Creation* (1844) published anonymously by the Scottish publisher Robert Chambers (1802–83). Darwin was not about to have his brainchild savaged in the same way. However, he did begin to work in earnest on his species book and became somewhat more open about his views on species, meeting with Huxley, Hooker, and T. V. Wollaston (1822–78) to discuss his ideas. One can hardly claim that thereafter Darwin's heretical views were a secret.

A quarter of a century after Darwin had set off on the *Beagle,* a young naturalist lay shivering in the tropical heat as he suffered through another bout of fever. Like Darwin before him, Wallace was inspired by Humboldt's *Personal Narrative of Travels in South America* (1818), but he had a second inspiration as well, Darwin's *Voyage of the Beagle* (1839). Wallace first traveled to the Amazon and, when the results of these explorations failed to make him a name, then to the Malay archipelago. Darwin was still a creationist during his voyage. Wallace was more fortunate. Prior to departing on his voyages, he had already been convinced by Chambers that species are mutable. Thus Wallace could study nature with evolution in mind. During his voyage, he already knew what he wanted—evidence for the transmutation of species and a mechanism adequate to bring it about. Darwin had to reconstruct his experiences, frequently on the basis of abysmally inappropriate and incomplete records. It took ornithologists decades to straighten out the mess that he had made of his bird collection (Sulloway 1982).

Wallace was able to dismiss Lyell's contention that similar conditions of existence might bring forth similar species. "The Islands of Baly and Lombock, for instance, though of nearly the same size, of the same soil, aspect, elevation and climate, and within sight of each other, yet differ considerably in their productions, and, in fact, belong to two quite distinct zoological provinces, of which they form the extreme limits" (Wallace 1857: 1514). Although the strait separating the two islands is no wider than ten miles, Baly's fauna is clearly Asian while that of Lombock is Austrailian (Brooks 1984: 138). Furthermore, in the sometimes lush, sometimes fetid jungles of the Malay archipelago, Wallace saw the constant battle for survival. As he lay shivering in his crude hut, suddenly he recalled a book that he had read thirteen years earlier—Malthus's *Essay on Population.* It had the same effect on Wallace that it had had twenty years earlier on Darwin—a tremendous flash of insight and elation.

As soon as he was well enough, Wallace wrote up the results of his brainstorm. But where to send the paper? His earlier paper had gone totally unnoticed, at least so Wallace thought. How could he get his fellow naturalists to pay attention this time? With a working-class background, he knew no one of any influence. A year earlier he had exchanged letters with one well-established naturalist, Charles Darwin, and Darwin had mentioned that he too was working on the species question. Who better than Darwin? Darwin it was. On March 9, 1858, the mail boat pulled out from Ternate in the Moluccas carrying Wallace's bolt from the blue.

When Darwin read Wallace's manuscript, he realized with a sinking heart that Lyell's prediction that he would be forestalled had come true with a vengeance. What was he to do? If he simply forwarded Wallace's paper to some journal for publication and then waited a decent period before publishing something of his own, Wallace would get credit as the originator of the theory of natural selection. As a dedicated scientist, Darwin was not supposed to be concerned with matters of priority. Truth is truth no matter who gets credit for it. But he could not pretend, even to himself, that he did not want credit for the idea of natural selection and for all the work he had done over the previous twenty years to develop and expand his theory. It is one thing to think that possibly natural selection might cause species to evolve; it is quite another thing to explain sexual dimorphism, neuter insects, gaps in the fossil record, the evolution of intersterility between species, and on and on. Of course, Darwin could have tucked Wallace's manuscript away and quietly published something of his own. Wallace was far away, mail took a minimum of six months, and he would probably never suspect what had happened.

Darwin appealed to Lyell for advice. After all, it was from Lyell that he had assimilated the scientific ethic. Lyell suggested that both Wallace's manuscript and a couple of short pieces by Darwin be read at a meeting of the Linnean Society of London and then published in its journal. In the midst of these negotiations, two of Darwin's children fell sick, and the younger, Charles Waring, died. Darwin was so distraught that he washed his hands of the whole affair. Whatever Lyell decided to do was fine with him. On July 1, 1858, Lyell and Hooker presented the Darwin-Wallace papers at the Linnean Society. Their presentation caused hardly a ripple. A month later, on August 20, 1858, the papers appeared in the *Journal of the Linnean Society.* Still there was no reaction from the scientific community—two passing comments in a couple of journals and that was all. In fact, in his presidential address to the Linnean Society, Thomas Bell (1792–1880) remarked that 1858 had not been a very exciting year in science. It had not "been marked by any of those striking discoveries which at once revolutionize, so to speak, the department of science on which they bear" (Bell 1860: vol.4, viii–ix).

His grief notwithstanding, Darwin set to work abstracting a brief explanation of his theory from the partially completed *Natural Selection.* Eight months later he sent a draft of this abstract to his publisher and shortly thereafter it appeared. With the publication of Darwin's *On the Origin of Species by Means of Natural Selection, or the Preservation of Favoured Races in the Struggle for Life* (1859), one can safely say that all hell broke loose. Darwin did respond to his critics, primarily in successive drafts of the *Origin of Species.* However, he spent most of his time at Down House grinding out book after book: one presenting pangenesis, his theory of heredity; two extending his views to include human beings; six major works on plants; and finally his most popular book, *The Formation of Vegetable Mould, through the Action of Worms, with Observations on Their Habits* (1881).

Wallace did not return to England until 1862. By then the controversy over "Darwinism" was well under way. Except for a paper on the origin of human races and the antiquity of man published in 1864, Wallace did not publish extensively on evolution, not until his *Contributions to the Theory of Natural Selection* (Wallace 1870). By then evolutionism of one sort or another was fairly well established. Although the letters Darwin and Wallace exchanged on such topics as the origin of intersterility and the effect of geographic isolation on speciation indicate a deep understanding on Wallace's part, Wallace did not have the impact on the development of evolutionary biology that he might have had. Instead his main influence was in biogeography.

The controversy between the Darwinians and their opponents was extremely bitter, in part because its resolution went far beyond the species question. It had ramifications for our basic conception of science itself. The most common objection to Darwin's theory was that it was not sufficiently "scientific." This was precisely the conclusion that Darwin had urged with respect to the views of his opponents. The *Origin of Species* was directed primarily against two solutions to the species question: supernatural special creation and reverent silence. Just as physicists such as John Herschel (1792–1871) had eliminated any reference to God's direct intervention in the functioning of the solar system and geologists such as Lyell had expelled miracles from the formation of the earth, Darwin argued that explaining the origin of species in terms of the miraculous flashing together of elemental atoms into living creatures no longer had a place in science. And, in reaction to Lyell, Darwin maintained that reverent silence also was no longer acceptable. Scientists could no longer shun the species question. They had to declare themselves—miracles or science.

Darwin chose creationists as his main opponents in the *Origin* because he knew how to confront them. Species cannot simultaneously arise over millions of years by means of one species evolving into another and by instantaneous miracles. Although Darwin was willing to contradict his mentor, albeit ever so indirectly, he carefully avoided confronting another alternative to his theory—a view which for want of a better term I call "idealism." In the *Origin* Darwin (1859: 434) limits his discussion of "Morphology," which he states is the "very soul" of natural history, to less than a half dozen pages, characterizing references to the "unity of type" as being "interesting."

The term had the same equivocal connotations in the 19th century that it still has today.[2]

Idealism became influential in science in Germany with the work of Goethe (1749–1832) and Lorenz Oken (1779–1851) and in France with Etienne Geoffroy Saint-Hilaire (1772–1844). Although these Continental idealists differed among each other as much as any scientists do, they were partial to explaining natural phenomena in terms of timeless general patterns or "archetypes." For example, from the diversity exhibited by plants, Goethe abstracted one archetype or *Urbild* for all plants, the *Urpflanze,* several *Baupläne* for animals, and the *Urgestein* for all minerals (Baldridge 1984). However, his most detailed discussion concerned the *Urpflanze.* According to Goethe, all the various parts of a plant are modifications of a single structure—the leaf. But Goethe also thought that his *Urpflanze* might exist exemplified in nature in all its simplicity and generality in the form of an actual plant (Wetzels 1985), and granite was the *Urgestein.*

E. Geoffroy Saint-Hilaire in turn explained all the various parts of the skeleton of vertebrates as modifications of a single structure—the vertebra. For a while he even tried to assimilate all animals to a single archetype, a view which finally goaded the powerful Cuvier to defend his own preference for four *embranchements.* On occasion idealists suggested processes that might be responsible for the patterns that exist in nature. For example, Oken claimed that life results from a "polarizing force" that produces unity and repetition and an "organizing principle" that is responsible for diversity and adaptation. However, they also maintained that these timeless patterns were in and of themselves explanatory. No additional explanations in terms of processes were necessary. All parts of a plant are "transformations" of the leaf in the same sense that a circle is a "transformed" ellipse.

Explanations in terms of timeless relations and polarizing forces never gained much of a following in Great Britain. In the mid-seventeenth century, Platonism of sorts became popular at Cambridge but then rapidly died out. In the nineteenth century a series of proponents tried to import some form of idealism to the island fortress. Chief among them were P. M. Roget (1779–1869) of thesaurus fame, the philosopher William Whewell, and the biologists William Sharp Macleay (1792–1865), Richard Owen (1804–92), W. B. Carpenter (1813–85), and Edward Forbes (1815–54). For example, Forbes (1854) noticed that the distribution of the fossil corals he was studying formed a figure eight, with its constriction between the Permian and Triassic epochs (fig. 2.1).

2. The reactions of many intellectual historians to my referring to a pattern of thought that has periodically surfaced in Western thought as "idealism" is likely to exemplify two of the principles that I set out in the second half of this book. First, these authorities are sure to object that what I refer to as "idealism" was an extremely heterogeneous assemblage of views. I could not agree more, but heterogeneity of content cannot preclude the use of a general term without the total elimination of general terms from discourse; heterogeneity is the rule, not the exception. Idealism is no more a hodgepodge than Darwinism, Christianity, postmodernism, or any of the terms which these critics use to express their criticisms. Of course, each of us knows what *the* meaning of any term is: it is the meaning which is most familiar to us. Our idiolect is common usage. Second, among those authors who are willing to apply general terms to intellectual movements, each has his or her own preferred term. Anyone who uses some other term is chastized for using the "wrong" term. I have no strong preference for one of the dozen or so terms which have been used to refer to what I term "idealism," but I cannot use all of them simultaneously.

The four-starred corals were most prevalent in the Paleozoic and then became increasingly rare till they disappeared in the Permian to be replaced by six-starred corals, which proceeded to become increasingly prevalent thereafter. Forbes generalized this pattern to all organisms and argued that, because of the symmetry around the Permian-Triassic break, the conventional Cenozoic and Mesozoic eras should be collapsed into a single era, the Neozoic. Although Forbes (1854:432) believed that this pattern itself was explanatory, he also provided a second explanation of it in terms of a "manifestation of force of development at opposite poles of an ideal sphere" (fig. 2.2). Although Forbes's figure eight extended through time, polarity did not. It was timeless. According to Forbes, time is merely an "attribute with which man's mind invests creation."

This view of the nature of science conflicted with the more "empirical" conceptions of Herschel and Mill, not to mention Darwin and Wallace. Darwin found Forbes's claims about polarity "absolutely unintelligible" (Rehbock 1983: 104). Darwin was even blunter in a letter which he wrote to Hooker on July 2, 1854: " 'Polarity' makes me sick—it is like 'magnetism' turning a table" (F. Darwin 1903: v. 1, 77). Wallace was so incensed by Forbes's paper that he wrote his own 1855 species paper as a genuinely scientific alternative to Forbes's "ideal absurdity." However, before Wallace's paper appeared, Forbes died. Questions about what might have happened in history cannot be answered with much in the way of justification, but one wonders how different the history of Darwinism might have been had Forbes not died so prematurely. Wallace and Forbes might have engaged in a dispute over the proper way of explaining species' distributions. If so, then the novice Wallace would have had to confront the powerful Forbes over the origin of species without natural selection in hand. More

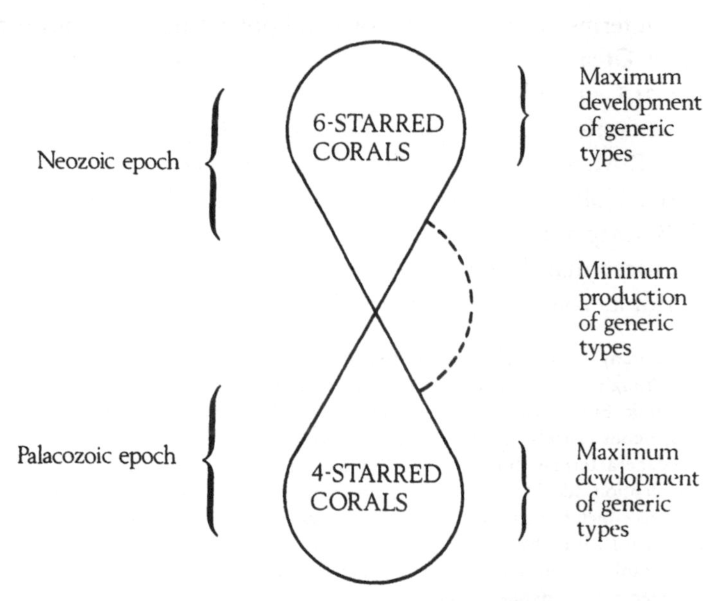

Figure 2.1. Polarity of generic types. (From Edward Forbes, "On the Manifestation of Polarity in the Distribution of Organized Beings in Time," *Proceedings of the Royal Institution of Great Britain* 1 [1854]: 428–33.)

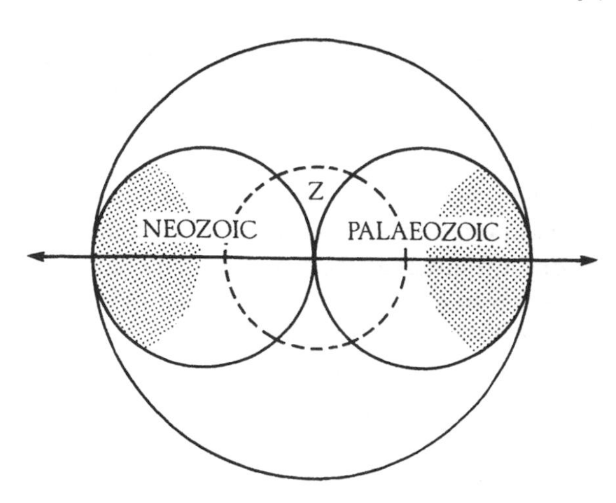

Figure 2.2. Polarity in time. (From Edward Forbes, "On the Manifestation of Polarity in the Distribution of Organized Beings in Time," *Proceedings of the Royal Institution of Great Britain* 1 [1854]: 428–33.)

than this he would have had to explain why idealist explanations are not really scientific explanations at all. Evolution and idealism would have met head-on over the species question. In an ironic turn of fate, Forbes's chief protégé at the time was none other than T. H. Huxley. In 1855 Huxley's sharp beak and talons might well have been used to defend idealism against evolution. Wallace would not have had a chance.

As things turned out, Darwin was able to avoid outright confrontation with idealists over idealism. In fact, the relation between this particular view of science and the evolution of species was far from clear at the time and has become no clearer since. Two of Darwin's most influential critics were idealists of sorts—Owen and Louis Agassiz (1807–73). In the first edition of the *Origin,* Darwin (1859: 435) concluded that Owen's explanations in terms of archetypes was "interesting," his "unity of type" one of the great "laws" in biology. Later, after Owen came out strongly against his theory, Darwin added, "but this is not a scientific explanation." Owen's explanations in terms of the conflict between the Platonic *eidos* and a general polarizing force were even less acceptable. Agassiz's explanations were worse still. Darwin complained that Agassiz's " 'categories of thought', 'prophetic types' & his views on classification are to me merely empty words. To others they seem full of meaning" (de Beer 1958: 113). Among the others were Huxley and Carpenter. These two early converts to evolution saw no incompatibility between ancestors and archetypes.

Huxley is an especially interesting case. Ernst Haeckel, one of Darwin's earliest converts in Germany, claimed that all life must have evolved from a very simple life form, which he named Monera. In the process of laying the transatlantic cable, the engineers dredged from the depths of the ocean a transparent gelatinous substance which Huxley thought was an excellent candidate for the *Urschleim* of the idealists and named it *Bathybius Haecklii* in honor of Hackel. As it turns out, this mysterious substance was only an organic precipitate. Although Huxley saw nothing unscientific about reference to *Urschleim,* he had no patience with explanations in terms of polarizing forces and the like. After dismissing Agassiz's theological explanations, Huxley (1893: v.1, 549) remarked with respect to Owen, "Neither did it help me to be told by an eminent anatomist that species had succeeded one another in time in virtue of a 'continually operative creational law.' "

The prevalence of quotes around so many words used by idealists when they were being discussed by Darwin indicates a profound difference in outlook. Darwin did not find idealist explanations just mistaken; he found them incomprehensible. Descriptions of generalized types were at best descriptions, at worst misleading. Ancestors were no more "generalized" than their progeny. According to Darwin, explanations of homologies and the unity of type in terms of ancestors were genuinely scientific; explanations in terms of archetypes were not. Darwin was not content to recognize timeless patterns in nature. He wanted to explain them, and explanations for Darwin had to be couched in terms of processes, not patterns. In one of his earliest letters to Huxley, Darwin (1903:1:73) distinguished between "types" as Huxley used them in his anatomical studies and as Owen and Agassiz used them. But what was Huxley's use? According to Huxley (1853:50), the "plan" for a particular group is the "conception of a form embodying the most general propositions that can be affirmed respecting" the organisms belonging to that group, "standing in the same relation to them as the diagram to a geometrical theorem, and like it, at once, imaginary and true."

As di Gregorio (1984:123) put the contrast, "It is as if in observing the world, what Huxley saw was the *order* of nature, whereas what Darwin saw was its *process*." Although di Gregorio's comment is appropriate for Huxley, it is not for those idealists who added explanations in terms of such things as polarizing forces. According to more empiricist Victorian scientists, polarizing forces represented processes, albeit the wrong sort of processes. In general, evolution seemed to make reference to archetypes superfluous. The unity of type alone was no explanation. Instead it was in need of explanation, and that explanation was in terms of history, not pure form. According to evolutionists, descent with modification was the *cause* of the group-within-group pattern so obvious in the living world. The pattern was an *effect*, and scientists explain in terms of *causes*, not effects. In the struggle for existence, flesh-and-blood ancestors beat ethereal archetypes hollow (Rudwick 1972, Bowler 1976, Winsor 1976, Desmond 1982, di Gregorio 1982, Rehbock 1983, and Richards 1987; for a contemporary evaluation, see Hopkins 1860).

Darwinism was as much a triumph of one view of science over its competitors as the triumph of one scientific theory over its alternatives. Because today we are all Darwin's intellectual heirs, we are likely to agree with him and his allies about the unintelligibility of idealistic explanations, so much so that it is difficult to explain the idealist alternatives without implicitly denigrating them. Even direct quotations take on an air of ridicule. According to idealists, nature exhibits a sort of cystalline structure that never changes. One explains phenomena by referring them to their appropriate compartment in this crystalline lattice. Although viewing the world in this way does not preclude the acceptance of organic evolution, a certain dissonance exists between the two. As eternal and immutable as this lattice might be, nothing keeps organisms from wandering from compartment to compartment. Nor must all compartments be

A hundred years from now, future students of science may well look back at present-day disputes with the same amused tolerance with which we view similar controversies in the nineteenth century.

Social and Conceptual Development in Science: An Evolutionary Account

Kuhn (1962) ended his highly influential book, *The Structure of Scientific Revolutions,* with a chapter arguing that science progresses through revolutions. Science has progressed from primitive beginnings but not toward any particular goal; surely it consists in nothing so crude as a Popperian successive approximation to the truth. Instead, Kuhn (1962: 171–72) alludes to an analogy with biological evolution. The resolution of revolutions in science is a matter of

> selection by conflict within the scientific community of the fittest way to practice future science. The net result of a sequence of such revolutionary selections, separated by periods of normal research, is the wonderfully adapted set of instruments we call modern scientific knowledge. Successive stages in that development process are marked by an increase in articulation and specialization. And the entire process may have occurred, as we now suppose biological evolution did, without benefit of a set goal, a permanent fixed scientific truth, of which each stage in the development of scientific knowledge is a better exemplar.

In later publications, Kuhn (1970a, 1977) apologized for the considerable confusion that his poetic style had caused commentators. If he were to write his famous book over again, he would have expanded on both the community structure of science and the evolutionary allusion. Shortly thereafter Toulmin (1972) also distinguished between what he termed intellectual disciplines and professions and emphasized that both biological and conceptual evolution are special instances of a more general selection process. In the case of genes, the process is natural selection. In the case of concepts, the process is more appropriately termed "rational selection," a phrase coined by Rescher (1977). Popper (1972) in turn spawned an active metaphysical research program commonly referred to as evolutionary epistemology (Campbell 1974; for additional references, see Bradie 1986 and Plotkin 1987).

To put the goal of this book succinctly, it is to present an evolutionary account of the interrelationships between social and conceptual development in science. Of course, words in general connote much more than their literal meanings, and the words in the preceding description are certainly no exception. For example, "evolutionary account" is likely to be read as implying some sort of evolutionary epistemology, as if I intended to justify the content of science, the methods of science, or both, by reference to the successful evolution of the human species. In this book I have no intention of deriving an epistemic "ought" from a descriptive "is." Putnam (1982: 6) once remarked that what is "wrong with evolutionary epistemology is not that the scientific facts are wrong, but that they don't answer any of the philosophical questions." According to a very common understanding of philosophy, Wittgenstein (1922: 77) was quite right that "Darwin's theory has no more to do with philosophy than any other hypothesis in natural science." However, I draw just the opposite conclusion. The main problem with past work in evolutionary epistemology is not that it is evolutionary but that it is epistemology. As far as I can see, neither the content nor the methods of science can occupied at all times. An idealist might well accept evolution. In the beginning, the crystalline lattice was created, and through time organisms have evolved from compartment to compartment. As Desmond (1982:192) remarks with respect to the work of H. G. Seeley (1839–1909), idealists saw science as being concerned with the eternal and immutable, not the temporary and variable: "Seeley assumed that the whole point of doing science was to reveal the enduring Thought, the transcendent reality, not chronicle shadows flitting across the globe."

As Winsor (1985:252) has observed, "To put ourselves in the shoes of a thoughtful naturalist of the pre-*Origin* days requires an imaginative power greater than most of us can muster. To do so means to surrender not only our knowledge, but our materialist world view. We flatter ourselves that we can comprehend the pious natural theologians, the speciesmongers, and the Humboldtians, but the subtle mind of Edward Forbes escapes the grasp of most of us, as it puzzled some of his own contemporaries." Darwin triumphed over the idealists, not so much by confronting them head-on in print but by ignoring them; but, as we shall see, his victory has not proved to be permanent. The contrast between pattern and process and the relative priority of the two recur throughout the continuing debate over evolution, right up to the present. (For further discussion of the nineteenth-century conflict, see Rudwick 1972, Bowler 1976, 1983, Winsor 1976, Gould 1977a, Desmond 1982, Hull 1983b, Rehbock 1983, Baldridge 1984, Rainger 1985, Appel 1987.)

Darwin dismissed both supernatural explanations and explanations in terms of ideal types as not being truly scientific, but many criticisms of his theory could not be dealt with in this way. Some criticisms were genuinely scientific, even according to Darwin's conception of science. Chief among the critics who mixed supernatural and naturalistic criticisms of Darwin's theory, especially its reliance on natural selection, was St. George Jackson Mivart (1827–1900). At an early age Mivart had shunned the religion of his family for Catholicism and was shunned in turn by the Darwinians when he tried to ally himself with them. But several of his objections were scientific. Mivart agreed that species evolved but maintained that Darwin's theory was inadequate to explain this evolution. Instead he insisted that some sort of internal program directed the development of species and that speciation had to occur quite suddenly in the space of a single generation. In his autobiography, Darwin (1958:125) remarked the he had always been treated fairly by his scientific critics. The single exception was Mr. Mivart, who "has acted towards me 'like a pettifogger,' or as Huxley has said, 'like an Old Bailey lawyer.' "

The Darwinian and anti-Darwinian camps were divided on substantive issues, but personal animosities served to widen the gap. For example, after the publication of the *Origin*, Owen both attacked Darwin's theory and claimed priority for the idea of organic evolution. However, the idealistic language in which he couched his views so obscured them that few of his contemporaries read him as advocating evolution. As if the conceptual gap between Darwin and Owen was not great enough, Hooker and Huxley's early advocacy of Darwin's theory all but guaranteed Owen's opposition for professional reasons. Owen had grown to hate the Hookers of Kew, possibly because of the rivalry between Kew Gardens and his own beloved British Museum (Natural

History) (Turrell 1963:90). Although Owen had helped Huxley early in the younger man's career, the two men had come to despise each other as they found themselves on opposite sides of several controversies. When the *Origin* appeared, Huxley and Owen were engaged in a dispute over issues in paleontology and comparative anatomy. Huxley transmuted their dispute into a disagreement over evolution. Once Hooker and Huxley had come out in favor of Darwin, it would have been impossible for Owen to join the group, even if he had been inclined to join with other scientists in furthering his scientific pursuits. One characteristic of anti-Darwinian idealists at the time was that they were not capable of cooperating to mount an organized attack on evolutionary theory. Instead they took individual potshots.

The same cannot be said for a second class of critics, who could not be dismissed lightly—physicists and mathematicians, especially William Thomson (1824–1907), later Lord Kelvin, P. C. Tait (1831–1901), and Fleeming Jenkin (1833–1907). These men argued that, on the basis of the most up-to-date principles of physics, the earth could not possibly be old enough to allow for the evolution of species. For example, by calculating the heat loss of a globe the size of the earth, they concluded that the earth was between 24 million and 100 million years old, while Darwin estimated that the denudation of a single geological stratum in Great Britain, the Weald, probably took about 300 million years. If these physical theories were true, Darwin's theory had to be false. Even in the face of such apparently clear refutations of his theory, Darwin persisted. Some way would be found to reconcile biology, geology, and physics. Darwin was right, but the reconciliation was a half-century in coming.

By and large the Darwinians presented a united front to their enemies, but on occasion there was dissension in the ranks. The publication of Lyell's *The Geological Evidences of the Antiquity of Man* in 1863 was just one such occasion. In his book Lyell dealt with three topics: glaciers, biological evolution, and the antiquity of man. Although Lyell was primarily a geologist, he was concerned with biological evolution because of its implications for his uniformitarian view of natural processes and with the antiquity of man for moral reasons. His book is best remembered for Darwin's disappointment at Lyell's failure to come out in favor of evolution, something which Lyell would not do for another five years and then only grudgingly. However, Lyell did change his mind on the antiquity of man. Too much evidence had accumulated for him to deny it.

Lyell was not a comparative anatomist. Hence he wrote to Huxley for advice in treating the Huxley-Owen dispute over primate brains. On reading Lyell's book, Owen immediately recognized Huxley's hand and wrote an angry rebuttal. Huxley was not in the least upset about Lyell using his work to do in Owen and, in the ensuing quarrel, Huxley gleefully aided Lyell behind the scenes. Other allies such as George Rolleston (1829–81) and William Flower (1831–99) also joined in against Owen. However, Owen was not the only one unhappy with Lyell's book. So were some of the men who had done much of the original investigation on the antiquity of man, several of whom were allies of Darwin, in particular Hugh Falconer (1808–65) and John Lubbock (1834–1913). Lyell described the evidence for the antiquity of man that others had first uncovered almost as if he himself had made the discoveries. The ensuing controversies over inadequate acknowledgment among allies and former friends were even more acrimonious than the dispute with Owen. Darwin, Hooker, and Huxley were caught in the middle. No matter which way they went, they would lose professional friends and allies, and the cause of evolutionary theory would suffer. Huxley in particular did what he could to ameliorate the hard feelings but without any great success. As Hooker remarked in 1865, the dispute was only aggravated by Mrs. Lyell's refusal to invite George Busk (1807–86) and his wife to her parties, apparently because Mrs. Busk was socially unacceptable. Because Busk was a friend of Lubbock and Huxley, this social slight caused considerable discomfort. That the Owens were also not invited caused no problem at all (for further discussion of the strains that Lyell's behavior caused in the Darwinian camp, see Bynum 1984, Grayson 1985, Richards 1987).

In spite of opposition from so many quarters, "Darwinism" flourished. Scientists and even large segments of the general public admitted that species change through time; it was admitted that species "evolve," to use the word that was to become prevalent. As Charles Kingsley (1890:253) observed in 1863, "The state of the scientific mind is most curious; Darwin is conquering everywhere, and rushing in like a flood by the mere force of truth and fact." The role of truth and fact in the success of Darwinism must remain for the moment a moot question, but no one can deny that Darwinism was successful. It spread from Great Britain to Germany, the United States, Russia, the Netherlands, and even Spain, but not to France.[3] The odd fact is that the "Darwinism" that became so widespread had very little in common with anything that Darwin might have said. Darwin saw evolution as gradual, perhaps in some sense progressive, but not directed, and natural selection as the chief though not the only force in determining the path that evolutionary development might take. The version of "Darwinism" that rushed in like a flood was saltative, directed, clearly progressive, and natural selection played at best a minor role.

Of all the evolutionists, the Germans were most enthusiastic. Haeckel embarked on a frenzy of phylogeny reconstruction and popularized the slogan that ontogeny recapitulates phylogeny. According to Reif (1986:110), the Darwinians were so successful in Germany in the first decades after the publication of the *Origin of Species* that they all but "wiped out the memory and influence of the Naturphilosphen and idealists." Soon thereafter another German, August Weismann (1834–1914), forced to abandon his microscopical work because of failing eyesight, turned to an examination of the hereditary process. Although Darwin assumed a strong principle of heredity in his theory of evolution, his own suggested mechanism for hereditary transmission was met with embarrassed silence by his contemporaries. According to Darwin's theory of pangenesis, each part of the body produces gemmules characteristic of that part. These

3. Of all the nations with a strong scientific tradition in the second half of the nineteenth century, France was most immune to the charms of Darwinism. It remains so to the present. For example, Boesiger (1980: 309) remarked that in 1974 France remains a "kind of living fossil in the rejection of modern evolutionary theories: about 95 percent of all biologists and philosophers are more or less opposed to Darwinism." Numerous hypotheses have been suggested for the reaction of French biologists to Darwinism and for the back seat which they have taken in biology in most of its areas ever since (Conry 1974, Stebbins 1974, Farley 1974, Limoges 1980, Buican 1982, 1984, Mayr 1982, Cahn 1984, and Corsi and Weindling 1985).

gemmules find their way to the sex organs and are passed on in reproduction. The characters that are transmitted depend on the number, strength, and affinity of the gemmules received by the offspring from its parents. Although the provisional hypothesis of pangenesis allowed for Lamarckian inheritance, the inheritance of acquired characteristics always played a subsidiary role in Darwin's theory. In a series of books and papers, Weismann (1883, 1885) attacked the foundations of "Lamarckism," a collection of views that had even less to do with Lamarck than "Darwinism" had to do with Darwin. Weismann argued that the hereditary material (the germplasm) remains unaffected by changes in the body of an organism (its somatoplasm). Since only the germplasm is transmitted in reproduction, Lamarckian inheritance is impossible. In addition to arguing for a barrier between the germplasm and somatoplasm, Weismann claimed that natural selection as a mechanism for evolutionary change is totally sufficient. He would have none of Darwin's subsidiary hypotheses.

At the time, nearly everyone assumed that acquired characters can be inherited. Wallace was an exception. Thus, when Weismann presented his heretical views at a meeting of the British Association for the Advancement of Science in Manchester in 1885, Wallace was pleased. Lamarckian evolutionists such as Herbert Spencer (1820–1903) were livid. At the time, George John Romanes (1848–94) had declared himself Darwin's heir apparent and had set himself the goal of defending Darwin's pluralist view of evolution against such extremists as Weismann. Needless to say, Romanes was dismayed by Weismann's hard line on soft inheritance. Dismay turned to alarm when a whole group of young evolutionists in Great Britain took up the Weismannian cause. Romanes found these young zealots too extreme. In placing so much emphasis on selection, they were, he feared, endangering the entire theory. In order to distinguish these zealots from true Darwinians such as himself, Romanes coined the derogatory term "neo-Darwinian" to imply an extreme form of Darwinism. The term stuck but not as the name of an aberrant offshoot of Darwinism. Instead it came to denote the genuine intellectual descendants of Darwin, the trunk leading from Darwin to the present, much to the dismay of such Darwinians as Haeckel (1909:140), who concluded that it was "quite improper" to describe Weismann's hypothetical system of ids and idants as "Neodarwinism."

Later evolutionists were also uneasy about Darwin's conclusions concerning the gradual nature of evolution. In order for this question to be addressed cogently, however, evolutionary biology had to become more quantitative. Although Darwin made reference to the potential for a "geometric" increase in numbers of organisms, he presented his theory in a singularly nonmathematical form at a time when equations had become the hallmark of genuine science. Darwin's first cousin Francis Galton (1822–1911) set out to rectify this weakness. He was joined in this enterprise by Karl Pearson (1857–1936) and W. F. R. Weldon (1860–1906). The result was the formation of a committee sponsored by the Royal Society to conduct statistical inquiries into the measurable characteristics of plants and animals. From the first there were problems. Galton agreed with Pearson and Weldon that statistical methods were appropriate for studying both evolution and hereditary variation but disagreed with their main conclusion. Darwin believed that hereditary variations were usually quite small—a slight change in color or shape. As a result evolution had to occur in small, imperceptible steps. The net effect was that, at a slight distance, evolution looked gradual.

Galton thought that hereditary changes were quite abrupt and large, such that a new species could appear in the space of a single generation. The model he chose to illustrate his view of evolution was a faceted spheroid tumbling from one facet to another. Any intermediate position was unstable. Pearson and Weldon disagreed. As in the case of Darwin before them, they adopted a blending theory of inheritance. Importing a term from mathematics, they argued that variation and hence evolution are "continuous." By this, they did not mean that variation from parent to offspring was "continuous," a notion that makes no sense, but that the slight variations that occur in species can be arrayed on a normal curve so that both extremes of variation can be connected by intermediates. Because Darwin had opted for gradual evolution, Pearson and Weldon considered themselves to be defenders of true Darwinism. Hence, anyone who opposed continuous variation opposed Darwinism.

Pearson also had decided views about the nature of science. In his influential *The Grammar of Science* (1892), he argued that all we have access to are our own sense impressions. Thus, the brain is like a central telephone exchange, processing sense impressions. As a result Pearson objected to the postulation of unobservable entities as actual material bodies. Even causal claims were too metaphysical for Pearson if they implied more than the constant conjunction of sense impressions. Because he thought of himself as replacing the crude materialism of many early physicists with a "sound idealism," Pearson was more than a little surprised when Darwin's harrier Mivart (1892) attacked his philosophy as being essentially materialistic.

In 1894 a former student and good friend of Weldon, William Bateson (1861–1926), published a book entitled *Materials for the Study of Variation, Treated with Especial Regard to Discontinuity in the Origin of Species.* As the title implies, Bateson was championing a saltative form of evolution. This was precisely the aspect of Bateson's book that Weldon (1894) attacked in his review. Weldon argued that Bateson would do better to adopt the statistical methods advocated by Weldon's biometrical school. Bateson responded by attacking Weldon in a series of letters to Galton. The disagreement was partly semantic, turning on two different senses of "discontinuous." Parents could produce an offspring which was significantly different from them. For example, they could both be quite tall and one of their offspring quite short. The change would be abrupt and large. However, both extremes might lie within the normal range of variation of the species. It would be hereditarily discontinuous and populationally continuous.

Galton was in a quandary. On the one hand, he approved of the quantitative turn that Pearson and Weldon had given the study of inheritance, but he did not share their preference for continuous variation. On the other hand, he agreed with Bateson's partiality to discontinuous variation, but he did not join in Bateson's antipathy to quantitative methods in biology. Galton's solution was to increase the membership of his committee to include several critics of traditional Darwinian theory, including Bateson. The committee rapidly divided into two warring camps. Eventually, in 1900, Galton, Pearson, and Weldon resigned from the committee, hoping that as a result it would cease to exist, but Bateson managed to salvage it. In the midst of these disputes, Mendelian genetics was launched (for further discussion, see Provine 1971, Mayr 1982, Bowler 1983, MacKenzie and Barnes 1979, Roll-Hansen 1980).

The Divorce of Genetics from Embryology

Not all the activity in evolutionary biology was confined to Germany and Great Britain. In the Netherlands Hugo de Vries (1848–1935) developed two interconnected theories—his grand mutation theory and its subsidiary theory of heredity, intracellular pangenesis (de Vries 1889, 1901). De Vries named his theory of hereditary transmission "intracellular pangenesis" to emphasize the continuity of his views with those of the great Darwin. In response to Darwin's theory of pangenesis, Galton (1871) reasoned that blood transfusions should have an effect on an organism's progeny. After all, if gemmules travel from all parts of the body to the sex organs, the blood is the most likely conduit. Galton found that blood transfusions had no effect on hereditary transmission in the rabbits he studied. To Galton's dismay, Darwin did not treat his findings as clearly refuting the theory of pangenesis. Instead Darwin suggested that his gemmules must find their way to the sex organs by some means other than the blood. De Vries (1889) circumvented this problem by restricting his "pangens" to the cells that produce them. They remain "intracellular." However, as Darwin had before him, de Vries thought that the number of pangens influencing the production of a trait is quite variable.

De Vries searched for a species of plant that exemplified his mutation theory. He found it in *Oenothera lamarckiana,* the evening primrose. One day he noticed two strikingly different plants in a field of primroses. When he grew them, each bred true, leading him to consider them the first members of a new species. De Vries (1901) reasoned that there are two sorts of hereditary variation, small and large. The fluctuating variation that one observes in the process of normal reproduction is a function of small variations and is fundamentally irrelevant to the origin of new species. Species do not arise through an accumulation of small variations, as Darwin thought, but saltatively in the space of a single generation through the appearance of a marked change or "mutation." Selection played a role in weeding out the new species that were poorly adapted in the struggle for existence but it did not play a role in their construction. The lasting effect on de Vries of his early observations on the evening primrose is evident in his continued promotion of his mutation theory, even though he was unable to discover any other species that produced mutations as the evening primrose did. His solution was that, at any one time, only a very few species are undergoing mutable periods. Most are in periods of immutability. As it turns out, De Vries's mutations are not "mutations" in the modern sense of this term but result from the peculiar nature of the genus *Oenothera*.

Although de Vries's mutation theory along with its subsidiary theory of hereditary transmission were of major importance during the first decade of the century, in retrospect we are more interested in only one small part of his work—his three-to-one ratios. In the final years of the nineteenth century, de Vries conducted a series of experiments in which he crossed breeding stocks that exhibited contrasting states of the same character; e.g., the seedlings of some plants are hairy, those of others are not. De Vries crossed such plants to see what distributions were produced in the next generation, then crossed these plants, and so on. At first he discovered 2:1 ratios in the transmission of these contrasting characters. If he crossed even numbers of hairy and smooth seedlings, the ratio in the next generation was twice the number of hairy seedlings to smooth. In his next series of crosses, he obtained 4:1 ratios, and then, finally, the magic number that was to become the golden mean of genetics—3:1.

In 1900, de Vries published three papers in rapid succession in which he not only set out 3:1 ratios but also presented their most likely explanation. Each pair of contrasting characters must be controlled by a single pair of pangens. In such cases, hereditary particles come in pairs, not variable numbers. At reproduction, each parent gets to pass on only one of each of its pairs of pangens. The door was unlocked.

The details of the publication of de Vries's papers in 1900 would not be of any great significance if it were not for the extreme, though highly selective, importance scientists place on priority. According to the conventional mores of science, one scientist can work twenty years developing a theory; another may spend only a couple of weeks on his theory. If the second scientist publishes first, then officially this scientist is supposed to receive all the credit for the contribution. Of course, the system does not always work this way. For example, Darwin is endlessly castigated for not behaving more honorably toward Wallace. What he actually should have done upon receiving Wallace's paper was to write to Wallace asking permission to have it published, wait the year necessary for Wallace to receive the letter and for Wallace's reply to get back to England, and then publish Wallace's paper. After that, he could properly publish his own work.

I find this scenario of proper scientific etiquette absurd in the extreme, but even if Darwin had been foolish enough to behave so irresponsibly, the subsequent history of Darwinism would have been little altered. Darwin still would have received most of the credit. But, one might justifiably object, historians are not supposed to be able to reach such contrary-to-fact conclusions. In this case, however, the conclusion is unexceptional because, after Darwin and Wallace published, an obscure author established priority over both men. In 1831 Patrick Matthew (1790–1874) had published a clear statement of natural selection in an appendix to a book on naval timber and arboriculture. If Wallace deserved priority over Darwin because of his 1858 paper, then Matthew surely deserved priority over both men because of his 1831 paper. All sorts of reasons can be given for ignoring Matthew's claim to priority for natural selection: he published his views in an out-of-the-way place, he did not develop them in sufficient detail, he did not continue to develop his theory in later publications, and so on. The problem with these excuses is that they apply to numerous unappreciated precursors to whom we have awarded retrospective credit.

Another variation of the priority principle occurred in 1900 when de Vries published his three papers on 3:1 ratios, two in French, one in German. Within weeks of the appearance of his first paper (de Vries 1900a), he received preprints of papers by two young workers, Carl Correns (1864–1935) and Erich Tschermak (1871–1962), each claiming that they too had discovered 3:1 ratios and had come up with the same explanation as de Vries had. Of greater importance, they also pointed out that an obscure monk, Gregor Mendel, had anticipated all of them by thirty-five years. In his paper published in German, de Vries (1900b) acknowledged that he too had noticed Mendel's work, but only after he had formulated his own laws of heredity. Perhaps he had been anticipated by Mendel, but he had not derived his own ideas from those of Mendel. It was an original discovery on his part. Although de Vries's (1900c) second paper in French appeared before the German version, it was written after it and included a reference to the papers by Correns (1900) and Tschermak (1900).

Historians have subjected this particular episode to intense investigation, showing that none of the so-called rediscoverers actually discovered 3:1 ratios and Mendel's explanation for them in total independence from Mendel. Correns claims to have thought of the likely explanation for his 3:1 ratios in October 1899, several weeks before he read Mendel's paper, but he made no mention of either in a paper in December 1899 and did not get around to writing up his results until he received de Vries's reprint—five months after his flash of discovery. Tschermak is even a less likely candidate for a rediscoverer of Mendel's laws because he still did not see the connection between 3:1 ratios and Mendel's explanation, even after reading Mendel's paper.

De Vries poses a more difficult case. We know that in July 1899 he had not paid any special attention to 3:1 ratios. Nine months later, in March 1900, he did. In the interim he had received a copy of Mendel's paper from a friend in Delft, and Correns had published a paper in which he referred to Mendel's work on peas. Possibly de Vries first saw the importance of 3:1 ratios from reading Mendel, possibly not. However, as far as de Vries was concerned, priority for 3:1 ratios was subsidiary to the fate of his total research program. The ratios were only part of his theory of intracellular pangenesis, and it in turn was only part of his larger mutation theory. More than this, he thought 3:1 ratios and the paired pangens that produce them are exceptions to the general rules of inheritance, not the norm (Kottler 1979, Olby 1966, 1979, 1985, Brannigan 1979, Campbell 1980, Darden 1985, Meijer 1985, MacRoberts 1985).

The fate of scientists' work does not lie totally in their own hands. De Vries could not keep other scientists from fixing on 3:1 ratios and, as de Vries saw it, from blowing their significance far out of proportion, and that is precisely what happened. Upon receiving a reprint from de Vries, Bateson looked up Mendel's original paper and read it on a train to London, where he was scheduled to give a paper to the Royal Horticultural Society. Instead of presenting his prepared address, he proclaimed the birth of a new science and declared its basic principles to be in total opposition to the Darwinian theory of continuous evolution. Mendel's constantly differentiating characters were precisely the discontinuities that he had been claiming for years as the material basis of evolution (Bateson 1894). According to Bateson, one could accept Mendel's laws or Darwin's theory, but not both. At the turn of the century, Bateson viewed the science that he termed "genetics" as being incompatible with Darwinian evolution, just as Kelvin had viewed thermodynamics forty years earlier. As Bateson saw it, there was no contest. Darwinism had to go.

Pearson and Weldon saw the conflict in the same terms as Bateson but came to the opposite conclusion. Mendelism must be mistaken. E. B. Poulton (1856–1943) reacted just as negatively to what he took to be Bateson's narrowness, dogmatism, prejudice, and contemptuous deprecation of research about which he was regretably ill-informed, not to mention Bateson's exaggerated estimation of the importance of his own work. Because Bateson so completely dominated Mendelism in Great Britain and his views departed so significantly from those of Mendel and American Mendelians, Poulton (1908:xv) urged that British Mendelians should be more correctly termed "Batesonians." Of all the controversies that have erupted in biology over the years, the skirmishes between the biometricians (or Darwinians) and the Mendelians were among the most acrimonious. Conceptual confusions were only exacerbated by personal animosities (Provine 1971, MacKenzie and Barnes 1979, Roll-Hansen 1980, 1983, Cock 1973, 1983).

The dispute between the biometricians and the Mendelians was part of a larger dispute that partially cut across it, a long-running debate between naturalists and experimentalists. Naturalists at the time studied organisms in two ways, in their natural habitats as well as in museum collections. Experimentalists studied relatively few sorts of organisms reared in the artificial environments of the laboratory. When experimentalists published findings widely at variance with the experience of the naturalists, these naturalists quite naturally concluded that something was wrong. For example, they saw significant geographic variation with respect to numerous slight changes in the makeup of organisms. The idea that single large changes could be the major, let alone the sole, cause of speciation seemed incredible to them. They concluded that mutations are inconsequential aberrations, more a function of the artificial conditions in which the research organisms were being reared than anything else. Experimentalists in turn dismissed the work of naturalists as too impressionistic to take seriously. If biology was to make any headway, controlled experiments were necessary. As much as naturalists and experimentalists disagreed on a variety of issues, by and large they were in agreement on one point: natural selection was of only minor importance in the evolutionary process. Similarly, as much as Bateson might disagree with Pearson and Weldon about the value of Mendelian genetics, he agreed with them that it was unscientific to postulate the existence of genes as material bodies. They were merely calculation devices. Genuine scientists did not postulate unobservable theoretical entities.

In the early years of Mendelian genetics, the emphasis was on the difference between large and small variations. In 1903 a Danish biologist, William Johannsen (1837–1927), published a book in which he argued that a different distinction was actually more important—in modern terms, the distinction between genotype and phenotype. Some of the observable variation in nature is due to changes in the hereditary material, the rest to environmental differences. One way to distinguish between the two is to establish "pure lines" by intensive inbreeding. Once a pure line has been established, any subsequent phenotypic variation in it has to be the result of the variable influence of the environment. In one sense Johannsen was fortunate in choosing to study beans of the genus *Phaseolus* because genetically they are extremely homozygous. Hence, it did not take much inbreeding to establish pure lines. But as a result Johannsen was led to go along with the majority opinion that natural selection does not play a very important role in evolution (Roll-Hansen 1978, Mayr 1982).

Weldon and Pearson were not impressed with Johannsen's work. They found his statistical methods too crude and refused to adopt his distinction between what we now term genotype and phenotype because it seemed to hinge on the real existence of genes, even though Johannsen himself later warned (1909) against conceiving of genes as material bodies characterized by morphological structure. Instead genes should be considered merely as useful fictions. Traits appear in successive generations in various ratios, and that is that. Anything else is irresponsible speculation. Organisms exist but not genes. How then can genotypes be contrasted sensibly with phenotypes? Johannsen does not say. At the very least, Johannsen's genotype-phenotype distinction was a far cry from the distinction to which it gave rise.

A solution to the conflict over blending versus particulate inheritance was also suggested quite early in the controversy. According to both Darwin and de Vries, hereditary patterns are a function of the transmission of material bodies of some sort, gemmules and pangens respectively. In this sense, their theories of heredity are "particulate." But they also believed that in most cases the net effect of all these particles is a blending of the parental characters. Hence, both Darwin and de Vries set out particulate theories of blending inheritance. In this respect, they were not especially unusual. Most of those working on inheritance in the nineteenth century, to the extent that they were willing to entertain the existence of unobserved entities at all, thought of the material basis of heredity as being particulate. They disagreed about the manner in which these particles are transmitted and their effects on visible characters. The postulation of hereditary particles did not preclude biologists from considering inheritance to be phenotypically blending.

Almost immediately upon the birth of Mendelian genetics, both Bateson and the mathematician G. U. Yule (1871–1951) realized that the blending of traits can be explained wholly within the principles of Mendelian genetics. Even if each pair of alleles at a single locus is constantly differentiating, the effects of several such pairs might appear to be relatively continuous. Add environmental variation, and continuous variation is guaranteed (Bateson 1901, Yule 1902). But no one was listening.

Part of the problem was conceptual. Nearly everyone treated hereditary transmission as if it had a material basis and as if this material basis consisted of particles, but no one was supposed to mention these particles. Reference to unobservable particles was not properly "scientific." Instead everyone had to couch their discussions totally in terms of the things that are observable—traits. As a result, early geneticists consistently confused genes with traits. If for every trait a single gene existed, and vice versa, this confusion might not have been too serious, but just the opposite is the case. No matter how one subdivides the genetic material into genes and the phenotype into traits, no simple mapping between the two can be constructed. Most genes influence several traits, and any one trait is influenced by numerous genes. Similarly, genes function serially in causal chains so that some genes appear to influence other genes rather than traits. Nor were matters helped any by the pugnacious character of some of the participants, especially Bateson. The perspectives of the two groups were so different and the interpersonal relations between them so bitter that mutual understanding, let alone agreement, soon became impossible.

Across the Atlantic Ocean, Thomas Hunt Morgan (1866–1945) grew impatient with all this wrangling and set about doing something to end it. Like so many of the scientists involved in these disputes, Morgan was trained as an embryologist and was highly suspicious of idle speculation. Anything that could not be put to experimental test was not science. Even though he had never read the *Origin of Species,* Morgan agreed with Bateson that Darwin's theory was sorely deficient. From what he had heard, it was teleological, and the principle of the survival of the fittest was a tautology (Weinstein 1980:437). Morgan was no less suspicious of Mendelian genetics. To him it smacked of preformationism in embryology, as if organisms existed preformed in their genetic makeup. But he was attracted to de Vries's mutation theory. According to Morgan (1919:273), de Vries's belief that species consist of very many "small species" that "are made up of many common genes and differ in a relatively small number of genes" is "so novel that it has not yet received the recognition which we may expect that it will obtain in the future when relationship by common descent will be recognized as of minor importance as compared with relationship due to community of genes."

Morgan decided to try to find some de Vriesian mutations, to see if they obeyed Mendel's laws. Most early work in Mendelian genetics had been performed on plants. As his research organism, Morgan settled on *Drosophila melanogaster,* a species related to the common fruit fly. These tiny insects breed quickly and can be raised cheaply. Although Morgan had no way of knowing it at the time, they also have only four pairs of chromosomes, a nice manageable number, and extremely large chromosomes in their salivary glands. Initially, Morgan had no success. It took him two years before he found any mutants in his flies, and then in rapid succession he discovered four, including one male fly with white eyes instead of the normal red. This single white-eyed fly was to change the course of both genetics and evolutionary biology. To help in his research, Morgan gathered around him a group of students who were to become famous in their own right, including A. H Sturtevant (1891–1970), C. B. Bridges (1889–1938), and later H. J Muller (1890–1967).

In the early years of genetics, another American embryologist joined in the crusade against the Darwinian orthodoxy. William Castle (1867–1962) of Harvard University initially accepted Bateson's identification of Mendelian genetics with discontinuous variations—large, discrete changes. Rather rapidly, however, he decided that both Bateson and his biometrician opponents were mistaken. The variations that are operative in evolution are neither continuous, as the biometricians claimed, nor large and discrete, as Bateson and de Vries insisted, but small and discrete. Darwin had been right all along. Of greater importance, Castle attacked the most central tenet of Mendelian genetics—the purity of the gametes. In the simplest Mendelian crosses, one character state was thought to be dominant to the other. If a pure dominant is crossed with a pure recessive, all the offspring are dominant in appearance. However, if their offspring are crossed, the recessive trait reappears. When it reappears, it should remain unaffected by its temporary submergence. Wrinkled peas remain just as wrinkled and smooth peas just as smooth no matter how often they might be crossed.

For those who believed in genes as material bodies, this meant that two different alleles do not materially combine or mix when they reside together in the same individual. They come out as pure as they go in. Darwin had argued for a "blending" theory of inheritance primarily because he believed that the number of hereditary particles (his gemmules) is both large and highly variable. As a result, all possible gradations of a character can occur. Darwin also thought that on occasion gemmules themselves might fuse. Castle suggested a similar hypothesis. Characters appear to blend because alleles tend to contaminate each other. Through successive crosses, smooth peas might become increasingly wrinkled because the alleles that influence the surface of the peas can "leak" into each other. Of course, those scientists who refused to talk about genes as material bodies could not even discuss Castle's hypothesis cogently. If genes do not exist, then they can neither blend nor remain pure.

The Mendelians responded by accepting the hypothesis suggested by Bateson (1901) and Yule (1902) that numerous pairs of alleles at different loci can affect the same trait, producing phenotypic blending with no blending of genes. In fact, they argued that some genes (epistatic genes) are not correlated with characters directly but only indirectly via other genes. Some genes influence genes. Castle objected vehemently. This endless multiplication of genes looked very ad hoc to him, like the piling of epicycle on epicycle in ancient astronomy. Castle stimulated the members of the Morgan school to put their views to vigorous experimental test. Their opposition, in turn, made Castle even more adamant. However, the evidence against Castle's contamination theory became increasingly convincing. In 1911 Castle publicly capitulated, only to reassert his old view a year later. But finally, in 1919, Castle admitted that the gradations he had studied were due to modifier genes, not modified genes (Provine 1985a, 1986).

One of the great achievements of the Morgan group was the discovery of "linkage." According to Mendel's second law, traits are inherited independently of each other. If two pairs of constantly differentiating characters are followed together through successive generations, they should assort independently. The Morgan group discovered that some characters do assort independently of each other, while others tend to be transmitted together; they are "linked." The most obvious explanation for linkage was that genes are material bodies residing in or on chromosomes, especially when the number of linkage groups and chromosome pairs in the species being examined turns out to be the same. Morgan resisted. He had been able to make the progress he did by divorcing transmission studies from embryology and concentrating on the former. Furthermore, he insisted that the postulation of unobservable entities was "unscientific." Chromosomes could be observed during certain stages in a cell's development, but no one had seen individual genes. Morgan retained these views even though his close friend and valued colleague, E. B. Wilson (1856–1939), was a strong proponent of the chromosome theory. But the coupling of eye color in *Drosophila* with the chromosome that determined gender was too much for him. By 1911 Morgan's white-eyed male forced him to identify linkage groups with chromosome pairs.

The course of Mendelian genetics in Great Britain was no less influenced by the preferences of individual workers than it was in the United States. Like Morgan, Bateson opposed both continuous evolution and chromosomes as carriers of hereditary factors, but he was much more resistant to changing his mind than Morgan. Not until Bateson visited Morgan's laboratory in 1921 and saw for himself the observational evidence marshalled by the Morgan group in favor of the chromosome theory did he capitulate, and then neither totally nor without occasional vacillation. Part of the problem was that Bateson did not have a very deep understanding of the theory he was championing. For example, E. B. Ford (1980:340) recalls spending an afternoon with Bateson, trying to explain the binomial theorem to him, with little success. For Bateson, the essence of Mendelism was discontinuous variation, not 3:1 ratios or their more complex descendants. German biologists continued in the tradition of developmental genetics, concentrating on departures from the regularities of Mendelian genetics (Harwood 1984, 1985). With the exception of L. Cuénot, France contributed very little to the emerging science of genetics. Although Russian biologists began to turn to genetics after the Revolution, their work had little immediate impact on the development of Mendelian genetics. As a result, Mendelian genetics was almost exclusively a product of English-speaking scientists.

At the turn of the century, nearly all biologists thought that the next burst of achievement in biology was going to be in experimental embryology. Ford recalls that this opinion was still strong in the 1920s. We are still waiting (Thomson 1985). Most of the founders of Mendelian genetics, as well as their critics, were trained in embryology. In order to make progress in what was to become genetics, these early workers had to abandon many of the assumptions, standards, and goals of their early training. The questions that early geneticists asked were not the questions that embryologists of the day wanted answered, but they were the questions that early geneticists were able to answer. By divorcing themselves from embryology, geneticists were able to make progress. These early geneticists were also opposed to material interpretations of the gene, but the development of their field forced them, however grudgingly, to abandon this metaphysical preference. They also found themselves in opposition to what they understood as Darwinian evolution. Evolutionary biologists had long awaited an adequate theory of heredity to supplement their theory of organic evolution. When one at long last arrived, it was presented as if it were in total opposition to Darwinian evolution. It took a while, but eventually biologists were able to uncover their mistake. No sooner had geneticists served final papers on embryology than the founders of what came to be known as the "synthetic theory of evolution" set about wedding genetics to evolutionary biology (Olby 1966, 1985, Mayr and Provine 1980, Mayr 1982, Bowler 1983, Harwood 1985).

The Wedding of Genetics to Evolutionary Biology

The first steps in the Modern Synthesis were taken by an unlikely triumvirate: R. A. Fisher (1890–1962), J. B. S. Haldane (1892–1964), and Sewall Wright (1889–1988).

Fisher was a brilliant, arrogant, irascible visionary who helped found both modern statistics and mathematical genetics. He also was an enthusiastic supporter of eugenics. From the beginning Fisher had trouble getting his work published. Part of the problem was that he was trying to join in happy matrimony two groups of scientists who despised each other for both personal and professional reasons. Fisher's papers were too mathematical for the Mendelians and too genetical for the biometricians. Even after his papers appeared in print, few people were able to read them with any understanding, let alone sympathy. Another problem was that Fisher did not quite understand what made other human beings tick. He blithely sent a paper attacking Pearson's work to the journal that Pearson edited. Needless to say, Pearson rejected it. Fisher quarreled with everyone—his neighbors, the butcher, university administrators. For example, he caused such a fuss at Cambridge over the plan to erect a new library building close to his favorite pear tree that the officials were forced to change the proposed site. With the exception of E. B. Ford (b. 1901) and Darwin's son, Major Leonard Darwin (1850–1943), Fisher sooner or later fell out with every one of his professional colleagues. Yet he never could understand why, during the Second World War, scientists of lesser stature were called upon to do work of national importance while he was not.

Fisher exhibited no greater sensitivity in his personal life. In order to do his bit to put right the imbalance of births in the upper and lower classes, he proceeded to have eight children. He also had decided ideas about how his wife should raise their children. For instance, he would not allow her to comfort her babies when they cried because he was sure that this would only increase the frequency with which they cried. Fisher's wife raised her eight children under extremely primitive conditions, carrying water from a well, firing a wood-burning stove, and helping her husband care for his research animals. Because of Fisher's extremely poor eyesight, she often read to him. Once, when she pleaded weariness, Fisher flew into a rage. The older he got, the more frequent such tantrums became. Yet, he was truly hurt when at last his wife decided that twenty-five years of abuse were enough and left him. He died in Australia, where his funeral was held without benefit of family (Box 1978).

Haldane was in the best tradition of a British upper-class eccentric—a mathematical biologist, a popularizer, a brilliant conversationalist, and an ardent Communist. His father, J. S. Haldane, a famous biologist in his own right, encouraged his son to get to know farmers and miners, while Haldane's mother rejected Christianity, regarding it as a religion fit only for servants. Haldane fell in love with a young reporter, Charlotte Burghes, who happened already to be married. In 1925 they went through the charade necessary at the time for her to obtain a divorce. They checked into a hotel in the presence of a detective. Cambridge used the ensuing scandal as an excuse to get rid of a professor who had become an impossible irritant. Haldane appealed his removal and won. In 1933 he left Cambridge anyway for a position at University College, London, where his relations with the administration were no smoother. As Medawar (1982: 267) remembers Haldane, he was no less difficult than Fisher. He was so "ignorant of anything to do with administration that he did not even know how to call the authorities' attention to the contempt in which he held them. When he burst into terrible anger about grievances, it was over the heads of minor functionaries and clerks. The cleaners were terrified of him, and the electricians were said to have demanded danger money for working in his room."

No sooner did Haldane arrive at University College, than he attracted the attention of a young geneticist, Helen Spurway. As in the case of Charlotte, Helen shared Haldane's commitment to the Communist cause. At the time it was not unusual for idealistic intellectuals to be attracted to communism. Haldane continued to champion the Communist cause even when the magnitude of the atrocities being committed under Stalin could not be denied, and his first wife had defected from the party. In 1948 Charlotte divorced Haldane so that he could marry Helen. Eventually, Haldane did resign from the Communist party because of its repeated attacks on Haldane's own profession of genetics that had become commonplace once Stalin embraced Lysenkoism. As did Fisher, Haldane spent his last years in India, where he seemed to find some degree of peace. Like Fisher, he cared about people in the abstract but could not get along with them for very long in the flesh (Clark 1968).

Wright, a graduate of Lombard College in Galesburg, Illinois, could not have been more different from his British counterparts. He was as gentle and unassuming as they were confident and demanding. Although he could be firm in his convictions, he did not throw tantrums or terrify janitors. The lives of people such as Wright provide very few striking anecdotes. He worked, married, had children, taught, raised his guinea pigs, and published. About the only excitement in his life was his famous feud with Fisher. It began amicably enough when Wright (1929) criticized Fisher's theory of the evolution of dominance. Some traits are dominant to others. How could such a relation evolve? The issues raised in this exchange turned out to run deep, signaling fundamental differences in the conceptions of the two men about the evolutionary process. As these differences surfaced, their exchanges became increasingly polemical until, by the late 1940s, their feud had reached legendary proportions (Provine 1985b, 1986).

Fisher, Haldane, and Wright are invariably cited as initiating the resurgence of evolutionary biology in the twentieth century after its long eclipse. These mathematically inclined workers claimed to show the theoretical limits of the evolutionary process—what could and could not happen given certain assumptions. For example, evolutionary biologists at the time were convinced that very small selection pressures could not have much of an impact on evolution. Fisher (1930) argued that these nonmathematical intuitions were seriously in error. He also showed how dominance might evolve, how selection could maintain two or more alleles in a population, and what the direct connection was between genetic variance in fitness in a population and the rate of increase in fitness—his famous fundamental theorem of natural selection.

One of Fisher's most frequent simplifying assumptions was that populations in nature are large enough to be treated as if they were infinite. His paradigm of evolution was the mass selection of slight variations. Wright (1929, 1931) took just the opposite tack. According to Wright, a much more common state of affairs in nature is the subdivision of species into partially isolated demes. Within demes a new mutant might have a chance of becoming established even if it were not especially adaptive. As Wright remembers his own work, he was proposing a shifting-balance theory of evolution, a process that takes place in three phases: random drift, selection within a deme, and then interdemic selection. Because the results of drift can at times be different from those implied by selection, Wright was popularly interpreted as presenting an alternative theory in opposition to Darwinian selection, a theory in which random drift is a major determinant of the direction of evolution.

Wright also ran into trouble with those scientists who continued to hold a Pearsonian view of science. One of Wright's (1921) contributions was a mathematical technique designed to distinguish between genuine causal chains and mere correlations, a technique which he appropriately dubbed "path coefficients." One of Pearson's disciples, Henry E. Niles, found Wright's contrast between causation and correlation to be sheer nonsense. As far as Niles (1923:259) was concerned, causation is nothing more than correlation based on sufficient experience. "The combination of knowledge of correlations with knowledge of causal relations" means "merely a combination of knowledge of correlations with knowledge of other correlations."

In his work, Haldane (1932) not only emphasized the role of chromosomes in evolution but also supported Fisher's conclusions about the influence of slight selective advantages and raised the issue of "altruism." How can a behavior or trait that helps another organism evolve if it is detrimental to the organism that exhibits it? Such behaviors and traits seem fairly common. Haldane's answer was that altruism can evolve if the recipient of the benefit is closely related to the benefactor. One can pass on one's genes directly through one's progeny or indirectly through one's collateral relatives. Haldane is supposed to have remarked, half seriously, that one should be prepared to give one's life for more than two brothers, four half-brothers, eight first cousins, etc. (see Hamilton 1964).

As different as Fisher, Haldane, and Wright were, they had one thing in common. They had a knack for devising novel mathematical techniques to deal with biological phenomena. They did not simply apply proven mathematical methods; they invented new ones. The trouble was that very few of their contemporaries, especially among biologists, were able to understand their work, let alone appreciate it. Fisher (1930) and Haldane (1932) did publish short books on the evolutionary process designed to be understood by biologists at large, but neither sold very well. Wright restricted himself to publishing long, very technical papers. Not until 1968 did he get around to publishing a synopsis of his views, and then it eventually ran to four volumes (Wright 1968–1978).

In actual fact, the more technical aspects of mathematical population genetics had little impact on evolutionary biologists, especially those working in the field. Evolutionary biologists were not sure how much the conclusions reached by these mathematicians flowed from empirical aspects of the evolutionary process and how much came from the simplifying assumptions introduced to make natural phenomena more mathematically tractable. Lewontin (1980:65) has remarked that this situation remains unaltered to the present. "If you ask me in a broader context how much field population genetics has depended directly for its direction on theoretical work, the answer is very little."

The preceding comments apply only to genetics in the West. The story was quite different elsewhere. In Russia, genetics had a late start. The first course in genetics was not offered there until 1913. However, after the revolution the Bolshevik government set up three large research institutes in biology. By 1921 Iurii Filipchenko (1882–1930) at the University of Leningrad and Nikolai Vavilov (1887–1943) at the Institute of Applied Botany, also located in Leningrad, were the two leading geneticists in Russia. Unfortunately, they had been trained in the West and brought with them all the beliefs about the incompatibility between genetics and a Darwinian theory of evolution that frustrated progress in the West. The story was quite different with Nikolai Koltsov (1872–1940) and his Institute of Experimental Biology in Moscow. He was not especially interested in either genetics or evolutionary theory, but these were the only promising areas of biology that were within the financial means of his institute. Fruit flies were cheap. In 1921 he hired a butterfly systematist and naturalist, Sergei Chetverikov (1880–1959), to develop a research group in experimental population genetics. The first order of business for the members of this group was to work their way painfully through *The Mechanisms of Mendelian Heredity,* produced by the Morgan group (Morgan et al. 1915). They learned English as they learned genetics. In 1922 Muller dropped off several cultures of *Drosophila* during a short visit to Koltsov's institute. By 1925 the Chetverikov group began to study four species of *Drosophila* found in the environs of Moscow. The result was a series of papers on the genetics of natural populations that equaled anything produced by Anglo-American workers (Chetverikov 1961).

The Chetverikov group was able to do so much so quickly because they benefited from the Russian tradition of field work in natural history and because they had so little to unlearn. Because they gained their genetics from a textbook rather than from the primary literature, they were unaware of all the difficulties inherent in any attempt to synthesize Mendelian genetics and Darwinian evolution. For them no synthesis was necessary because the two areas of study were never estranged. At a time when only a dozen or so workers were developing population genetics in the West, three schools of population genetics were flourishing in Russia. Unfortunately, all the support that the Soviet government had given to science in the early years began to turn into suppression. In 1929 Filipchenko was forced to resign his post, and a year later he died. In the same year Chetverikov was arrested, imprisoned, and eventually exiled, and his research group was disbanded. Koltsov gathered together another group, but what began as a general suppression of political dissent continued as a more focused suppression of genetics. In a dispute between the agronomist T. D. Lysenko (1898–1976) and established biologists, the Russian government took the side of Lysenko. Vavilov was stripped of his administrative duties and sent to prison in 1941, dying there in 1943. By the time the Lysenko affair had run its course, both genetics and population genetics were all but finished in Russia. Whatever influence Chetverikov was to have on future developments, it was going to have to be indirect (Joravsky 1970, Adams 1980, Weiner 1985).

To understand the interrelations between mathematical population genetics and more traditional evolutionary biology among Anglo-American biologists, a second triumvirate must be introduced—Theodosius Dobzhansky (1900–1975), George Gaylord Simpson (1902–84), and Ernst Mayr (b. 1904).

Dobzhansky was born in Nemirov, Russia, at the turn of the century. In spite of the pressures of World War I, the Russian Revolution, and the deaths of both his parents, by 1921 he had completed the requirements for an undergraduate degree at the University of Kiev. Dobzhansky began his professional career teaching biology at the Polytechnic Institute of Kiev. He immediately began to study populations of ladybird beetles (ladybugs) in nature and, in 1923, the genetics of Muller's *Drosophila* in the laboratory. From his work in the field, Dobzhansky concluded that the variation within and between geographic races was basically of the same sort. In the laboratory, he discovered that changes in genes at a single locus could influence several characters (pleiotropy). He would have liked to perform comparable genetic studies on ladybugs, but the genetics of this insect turned out to be much too complicated. As a result of these studies, however, he was invited to join Filipchenko's Department of Genetics at the University of Leningrad. During this time, Dobzhansky periodically visited the Chetverikov group in Moscow but did not appreciate what they were doing. Then in

Theodosius Dobzhansky. (Photo by Vadim Pavlovsky, reproduced by permission of Columbia University Press.)

1927 he had the good fortune to be awarded support from the Rockefeller Foundation to visit the Morgan group at Columbia University.

Dobzhansky arrived at Morgan's fly room in December of 1927. He was amazed to discover how small, dirty, and crowded the cradle of Mendelian genetics was. The accommodations that he was to share made the institutes back in Russia look luxurious. Sturtevant took Dobzhansky under his wing and introduced him to the work the Morgan group was doing in pursuing further implications of Mendelian genetics. The physical surroundings were greatly improved a year later when the Morgan group moved to the California Institute of Technology in Pasadena, California. When Dobzhansky's support from the Rockefeller Foundation ran out in 1929, Sturtevant convinced Morgan to make Dobzhansky an assistant professor. Until 1936 Sturtevant and Dobzhansky worked in close consort, but by then their friendship had begun to cool. In 1940 Dobzhansky left Cal Tech and returned to Columbia University, this time as a full professor.

The breakdown in relations between Sturtevant and Dobzhansky had several causes. Soon after joining the Morgan group, Dobzhansky developed a none too high opinion of Morgan's intellect. Dobzhansky found the man too literal, crude, and philistine in his thinking. As part of the banter that went on in the laboratory, Dobzhansky got in the habit of making disparaging remarks about Morgan. Sturtevant made no protest. Then, in 1933 Morgan won the Nobel Prize for establishing the very principles of genetics he had set out to refute. Morgan quietly divided the prize money among his own grandchildren and the children of Sturtevant and Bridges. The next time Dobzhansky made a cutting remark about Morgan, Sturtevant blew up at him. Dobzhansky was equally irritated by his colleagues. In general, members of the Morgan group had a very low opinion of anything that smacked of religion. It was all ignorant bigotry, an attitude that bruised Dobzhansky's own religious feelings. Certainly Dobzhansky did not find his colleagues very cosmopolitan.

The increasing tension between Dobzhansky and Sturtevant also had a more narrowly scientific source. Not until 1932, when Dobzhansky published a paper on the species concept, did his work begin to have any connection to evolution. In this paper Dobzhansky produced a first rough outline of what was to develop into the "biological species concept." According to this concept, gene flow and its interruption are primary in the speciation process. If diagnostic characters accompany species formation, so much the better, but reproductive isolation is prior to diagnostic characters. Sturtevant took a more traditional taxonomic view of species. Any two groups that are morphologically indistinguishable should be considered the same species regardless of reproductive isolation.

One source of the Morgan group's success was the complementary abilities of the various members. The same can be said for Dobzhansky and Sturtevant in their collaboration. Dobzhansky worked very quickly and published his results as soon as possible without exercising excessive care in designing his experiments or in handling his data. Sturtevant worked slowly, taking great pains with his experiments and the data they generated. Only when he was sure did he publish. As a result, he published only a small proportion of the data that he generated. According to several scientists who knew the two geneticists, one reason Sturtevant ceased working with Dobzhansky was that he found too much of his colleague's work "sloppy" and feared that it would not hold up well. Sturtevant did not want to endanger his own reputation by continued association (Provine 1981:49–50).

This unhappy state of affairs was brought home to Dobzhansky in 1936 when J. T. Patterson at the University of Texas in Austin offered him the position which H. J. Muller had resigned so that he could stay on in Russia. Morgan made a counteroffer, but after considerable thought Dobzhansky decided to accept the position at Texas. Sturtevant expressed sadness at Dobzhansky's leaving Cal Tech. Dobzhansky reconsidered, sending Patterson a telegram that he had decided to decline the offer. When Dobzhansky told Sturtevant that he was staying at Cal Tech after all, Sturtevant's face fell. Dobzhansky sent yet another telegram to Patterson informing him that he had changed his mind again and would happily accept his offer at Texas. Patterson never replied to this telegram, and Dobzhansky was forced to remain at Cal Tech.

By 1937 Dobzhansky's interest in evolution had flowered in his *Genetics and the Origin of Species*. This book, more than any other, was the work that initiated the Modern Synthesis. In looking back over his rupture with Sturtevant, Dobzhansky suggested that another reason Sturtevant turned against him was "plain jealousy" over the success of this book. The breakup, Dobzhansky concluded, "did vastly greater harm to him. Since then, more than a quarter of a century, he did very little in the way of research" (Provine 1981:31–2). Yet another reason that Dobzhansky was willing to go his own way was that a replacement for Sturtevant was in the wings—Sewall Wright. Initially Dobzhansky had needed Sturtevant to teach him transmission genetics, but as Dobzhansky began to move from genetics back to his first love, evolution, Sturtevant was of less use. In 1932 Dobzhansky had heard Wright deliver a paper at the Sixth International Congress of Genetics in Ithaca, New York. Wright had been put in the position of presenting his views on evolution in the space of an hour or so. The sort of mathematical presentation that Wright would have preferred was impossible. Hence, Wright resorted to a pictorial model of an adaptive landscape with organisms clustered atop adaptive peaks separated by maladaptive valleys. Dobzhansky was impressed. Here was a mathematician who understood evolution. Dobzhansky himself had very little training in or facility for mathematics. If he was going to make any contributions to population genetics, he was going to need some help. Wright was ideal.

However, it was not until the fall of 1936 that the two men were able to set aside a couple of weeks when they could work together. After this meeting, Dobzhansky sat down to begin serious work on a book that would put some empirical flesh on the mathematical bones of Wright's general conception of the evolutionary process. Wright also needed Dobzhansky. In Wright's dispute with Fisher, Ford had supplied the data about natural populations that Fisher had needed. Although Wright had studied guinea pigs for years, he had little access to detailed knowledge of populations in nature. This Dobzhansky was willing to supply. For almost ten years Dobzhansky worked in close collaboration with Wright, until the press of Wright's own independent research forced Wright to call the collaboration to a halt in 1945. The parting was completely amicable.

One important lesson that Dobzhansky learned from Wright was to be "certain *before* gathering any data about the theoretical relation of the data to theory" (Provine 1986:398).

Although Wright himself had very little direct impact on field naturalists, he had a major influence through Dobzhansky's *Genetics and the Origin of Species* (1937). The same can be said for the Chetverikov group in Moscow. Although Dobzhansky had not appreciated their work when he was in Russia, he came to understand its significance while he was writing his book. For more than a generation, successive editions of Dobzhansky's book formed the bible of evolutionary biology. It struck both mathematical biologists and field naturalists alike as being basically right-headed. Not only did Dobzhansky seem to understand the techniques devised by such mathematicians as Fisher, Haldane, and Wright, he also applied them to real cases in nature. Here at last was a geneticist who made sense to naturalists. Dobzhansky's book also had the effect of saddling Wright with the "Wright Effect," the view that the primary mechanism of evolution is random drift in small populations, a view that in later years Wright would disown.

Dobzhansky's proposed synthesis was not without its critics. One of the most influential was Richard Goldschmidt (1878–1958). Goldschmidt was one of those people who have a knack for being in the wrong place at the wrong time as well as for rejecting a view just as it becomes widely accepted. At the outbreak of World War I, he was visiting Harvard's Bussey Institution, where he made himself unpopular by siding with his native Germany. Eventually Goldschmidt was interned as an enemy alien. He never forgot being marched the length of lower Manhattan in handcuffs, escorted by two armed soldiers, or the sweltering summer he spent under the corrugated tin roofs of Fort Oglethorpe in Georgia. Twenty-seven years later, when Japan bombed Pearl Harbor, Goldschmidt was once again in the United States but this time as an émigré. As an enemy alien, he was restricted in his movements but allowed to continue work. The net effect was that Goldschmidt, as a Jew, was spared a second and probably lethal stay in an internment camp.[4]

Goldschmidt's opposition to the Modern Synthesis was especially damaging because there was very little in the way of genetic data at the time about natural populations, and Goldschmidt's work on the gypsy moth was one of the two major studies. The other was by F. B. Sumner (1932) on deer mice. The problem was that both men opposed key elements in the synthesis. Sumner not only opposed Mendelian genetics but also rejected natural selection in favor of neo-Lamarckism. Goldschmidt was hardly opposed to Mendelian genetics. He utilized the principles of Mendelian genetics in his own work and castigated those geneticists who refused to acknowledge a material basis for heredity, but like so many biologists of his day, he was trained in embryology. As an embryologist, he found the work of the Mendelians simplistic. Perhaps genes could be treated as independent particles in transmission studies but not in the context of development. In the production of organisms, Goldschmidt insisted, genomes function

4. Goldschmidt was not just Jewish. His family was used by the Nazis as an exemplar of the evil perpetrated by Jews on the German people through the centuries; for further information, see Goldschmidt (1960), Piternick (1980), and Gould (1982a).

as integrated wholes, or "fields." Genes cannot be altered piecemeal without affecting the entire functional system.

Goldschmidt was no happier with extrapolations from Mendelian genetics to the evolutionary process. He agreed with Bateson that the speciation process was "utterly mysterious." Of one thing he was sure, the subspecies recognized by taxonomists were not incipient species. His own tentative hypothesis was that new species arise in the space of a single generation by means of a "systematic mutation," a minor change that totally reorganized the genomic field. Goldschmidt (1933) dubbed these newly mutated individuals "hopeful monsters," an appellation that he would grow to regret. Natural selection might weed out nonviable monsters, but it played no role in their construction (see also Goldschmidt 1940, 1950).

As is so often the case, Goldschmidt was not being totally fair to his opponents. Right from the start, Fisher, Haldane, and Wright realized that genes interact, and they included in their earliest works ways of treating such interactions. Wright for one was especially interested in physiological genetics. However, these discussions were among the most difficult in their writings as well as being among the most inconclusive. All in all, it is truly amazing how much progress could be made in understanding both the local transmission of characters and short-term evolutionary change by treating genes as if they were independent, isolated particles—beans in a bag. The chief effect of Goldschmidt's *The Material Basis of Evolution* (1940) was to rouse Ernst Mayr to write, in white-hot indignation, his *Systematics and the Origin of Species* (1942).

Mayr benefited from the sort of thorough education in natural science for which German universities were justly famous. Officially he was enrolled as a medical student, but his true love was ornithology. During the years when Nazism was beginning to develop in Germany, Mayr was studying geographic variation in birds in New Guinea and the Solomon Islands. When he returned to Germany, political conditions looked so unsettled that he decided a career in science there would be impossible. Although he himself was not Jewish, he immigrated to the United States in 1930 and joined the staff of the American Museum of Natural History in New York in January 1931. In 1953 Mayr joined the staff of the Museum of Comparative Zoology at Harvard University, serving as director from 1961 to 1970. He remains on the staff of the museum to this day as professor emeritus.

When Mayr joined the staff at Harvard he was amazed to discover that no one had taught a course on evolution there for the past twenty-five years. He set about rectifying that omission. Mayr agreed with Goldschmidt that many geneticists and population geneticists treated genes in an overly simplistic way, a tendency he derided as "bean-bag genetics," but he was no more impressed by Goldschmidt's suggested mechanism for speciation, which he dismissed as speciation by means of "hopeless monsters." The term caught on; thereafter, critics of Goldschmidt invariably dismissed his theory with this epithet.

For his part, Mayr agreed with Dobzhansky that speciation is a populational affair. Biological species are internally quite heterogeneous, breaking up into geographic races under the influence of natural selection. According to Mayr, these geographic races are the raw material for speciation. As a special case, Mayr suggested his "Founder Principle." According to Mayr, speciation can occur quite rapidly when a very few organisms become isolated from the main body of their species and succeed in becoming established as an isolated colony. Such population "bottlenecks" usually result in extinction, but when they do not, speciation can be speeded up considerably. Although Mayr's views on speciation overlapped those of Wright to some extent, they were quite different from Goldschmidt's hopeful monsters. Hopeful monsters result from changes in single organisms, while Mayr's "hopeful populations" result from reduced numbers of unchanged individuals. The organisms are no different; only the makeup of the population is different.

Ernst Mayr. (Photo by M. K. Kelly. Courtesy of Harvard University Press.)

Mayr was—and still is—an outgoing man who is very much at home at professional meetings and conferences, where he vigorously and tirelessly defends his views. He makes no distinctions between august experts and hesitant graduate students. He is as willing to spend time setting one straight as the other. The vigor and definiteness with which Mayr addresses all comers is at times intimidating, so much so that one day Stephen Jay Gould became annoyed with the way Mayr appeared to be asserting his authority over a younger colleague in a dispute about some point in evolutionary systematics. But then it struck Gould (1984:257) that this is Mayr's essence: "He remains so in love with his subject, so enthusiastic about its promise and intellectual content, that he couldn't hold back. He was arguing with all the verve of a graduate student because, by God, he remains one himself in heart, energy, and commitment." Mayr might tell you in no uncertain terms that you are wrong and where you are wrong, but he will never walk away in an indignant huff.

George Gaylord Simpson. (Photo by Bachrach Photographers, supplied by David Kitts.)

The third member of the American triumvirate, G. G. Simpson, was a native of the United States. Although he was born in Chicago, he was raised on the edge of Denver, when Denver was still part of the Old West. He received his degree in geology from Yale University in 1926 but declined a position at the university when the faculty, at the urging of Margaret Mead, inquired into his marital problems. Instead he went to the American Museum, where he eventually became chairman of the Department of Paleontology. One goal of Mayr's *Systematics and the Origin of Species* was to show that the basic principles of biological systematics were totally in accord with the newly emerging synthetic theory. Simpson proposed to do the same for paleontology in his *Tempo and Mode in Evolution* (1944). Prior to Simpson's book, the most that could be said for paleontologists and geneticists was that they tolerated each other. Thus, in the preface, Simpson (1944:xv–xvi) was led to remark, "As a paleontologist, I confess to inadequate knowledge of genetics, and I have not met a geneticist who has demonstrated much grasp of my subject; but at least we have come to realize that we do have problems in common and to hope that difficulties encountered in each separate type of research may be resolved or alleviated by the discoveries of the other." Simpson had one advantage, however. He was sufficiently adept mathematically that he was able to read the works of Fisher, Haldane, and Wright with more understanding than most.

If Simpson had a bête noire in his *Tempo and Mode in Evolution* (1944), it was as it had been for Mayr,Goldschmidt. Simpson argued that the tempo of evolution might be irregular but that in general evolutionary change is gradual and slow. Here Simpson was arguing for a minority view in paleontology. Although idealism suffered an eclipse among German-speaking paleontologists after the *Origin,* it underwent a resurgence in the 1920s and 1930s. The leading ideal morphologist at the time was O. H. Schindewolf (1896–1971). According to Schindewolf, the gaps in the fossil record are real. The steps between all taxa at all levels are necessarily abrupt. A reptile could no more evolve gradually into a bird than a triangle could evolve gradually into a rectangle: "The first bird hatched from a reptile's egg" (1936:59). Thus, Simpson had to refute Schindewolf's views. Because both orthogenesis and neo-Lamarckism were popular among paleontologists, he was forced to argue against these theories as well. Paleontologists thought they could discern trends in the fossil record, trends toward increased size, complexity, and the like, and postulated non-Darwinian mechanisms to explain them. Frequently the environment figured in these mechanisms as a directive force. Simpson argued, regardless of the opinions of all these authorities, that the fossil record is totally compatible with the synthetic theory of evolution.

Because Simpson's chief opponents among paleontologists were advocates of the principles of ideal morphology, he took special exception to a paper published by Rainer Zangerl (1948) in *Evolution* arguing that these principles are prior to any inferences about phylogeny. In fact, he was so angry that he never spoke to the man again. Simpson was no more pleased by a paper published a decade later by a philosopher, Marjorie Grene (1958), who compared his views with those of Schindewolf and concluded that Schindewolf's principles were epistemologically preferable because Schindewolf acknowledged the necessary role of types in science. Simpson himself did not respond to Grene, but several younger workers did, e.g., Bock and von Wahlert (1963), Van Valen (1963) and Carter (1963). Niles Eldredge, who was an undergraduate at the time, also noticed Grene's paper, but it did not have quite the same negative impact on him that it had on the others.

Simpson was an extremely private man. He had very few close professional friends or students. Bobb Schaeffer of the American Museum was a lifelong friend, and David Kitts, a vertebrate paleontologist turned philosopher, was the student whom he remembered with greatest affection. His second wife, Anne Roe, was all the company he needed. He always hated to lecture and disliked professional meetings even more. In later life he was able to avoid both as much as possible by using as an excuse his leg that had been crippled when a massive tree fell on him in the midst of the Amazon. The trip out by canoe, boat, and eventually plane to a hospital in New York took eight days. Thereafter, escalator rides at convention hotels were too much for his leg, though he somehow could manage mountain trails in the Himalayas. In later life Simpson was harder to meet in person than the Great Oz himself.

While he was recuperating from his accident, an acting chairman took over his administrative duties in the Department of Paleontology at the American Museum. Once Simpson's health had returned, the director of the museum gave him the choice of assuming his duties as chairman again or continuing as a full-time researcher. Simpson interpreted the director as telling him that no more work was expected of him. All he had to do until he retired was to clock in every day and go home if he liked. Simpson resigned, instead of accepting what he took to be a humiliating situation, and in 1959 accepted a position at Harvard's Museum of Comparative Zoology. Eight years later he left Harvard for the University of Arizona under equally unpleasant circumstances. As Gould (1985:229), one of his colleagues at the time, remarked in his obituary notice, Simpson "left Harvard for a complex mix of reasons, including poor health, impending retirement and general dissatisfaction (born of our stodginess and his irascibility), and spent the rest of his career in Tucson." Gould (1985:232) concluded that Simpson, throughout his life but more particularly near the end, was trapped in a Catch-22 of his own construction:

> He couldn't tolerate a toady, for his intellectual honesty was too great and his perceptions too acute to misidentify this genre. But neither could he bear disagreement, however gently expressed. He took offense easily, placing the worst possible interpretation on any event that displeased him. This bitterness seemed so inconsistent with the grandeur of his prose, the generosity of his vision, the warm and expansive humanism of his general writings.

With some justification one might add yet a third triumvirate in the formation of the Modern Synthesis—Julian Huxley (1887–1975), C. D. Darlington (1903–81), and E. B. Ford (b. 1901). After all, Ford published his *Mendelism and Evolution* in 1931, Darlington produced a highly imaginative work on the evolution of genetic systems in 1939, and the term "Modern Synthesis" came from Huxley's *Evolution: the Modern Synthesis* (1942). One might wonder why, in the face of the early work in both Russia and Great Britain, the leading student of the Modern Synthesis, William Provine (1980:354), concludes that the synthesis occurred primarily in the United States. In both cases, the answers are as externalist as an explanation can be. In Russia, political oppression brought an end to work in population genetics. Few outsiders were able to read the papers published in Russian, and Russian geneticists published very few papers in Western languages. The major conduits for Russian advances were N. W. Timofeeff-Ressovsky in Germany, Haldane in Great Britain, and Dobzhansky in the United States. As remarkable as the Russian achievements were, considering the short period during which they were produced, they are overshadowed by the Anglo-American contributions. In Great Britain it was World War II that brought research in population genetics to a halt. Great Britain entered the war in 1939, while the United States held off until 1941. Great Britain was devastated, while the United States remained untouched. After the war, scientists in the United States were able to pick up their research again almost immediately, with the considerable financial support of the National Science Foundation, while the British had to rebuild their country and start over again almost from scratch. What was true for Great Britain was even truer for Germany and Russia.

By 1947 at a meeting held at Princeton, New Jersey, it became clear that an amazing level of agreement had developed about the evolutionary process. Because of political upheavals and World War II, scientists located in the United States were largely responsible for this consensus. Dobzhansky, a Russian émigré, Simpson, a fiery American whose red hair turned white in later years, and Mayr, a quintessential German, who in his eighties still stands ramrod straight, submerged their differences sufficiently to integrate the disparate elements of genetics, paleontology, systematics, and evolutionary biology into a single, coherent theory of evolution. In the face of considerable opposition, proponents of the synthetic theory presented a united front. However, the amount of agreement and the degree to which the founders of the Modern Synthesis actually worked together should not be exaggerated. For example, when Cedric A. B. Smith (1984:299) got to know such British luminaries as Fisher, Haldane, and Lancelot Hogben after the war, it became immediately clear to him that "relations between them were alas far from ideally cordial and that each in his own way could sometimes be inexplicably prickly, hindering both close intellectual contact and scientific advance."

Nor did the founders of the Modern Synthesis read each other's works as closely as one might expect. Mayr (1980b:421) acknowledged that he knew nothing of Fisher's work until he read Dobzhansky and that he remained unaware of Haldane until 1947. In his 1942 book, Mayr mentions Ford and Simpson but not Darlington. As one might expect in a book on systematics, Mayr does refer to Julian Huxley's *The New Systematics* (1940). Simpson in turn pays ample attention to Fisher, Haldane, Wright, and Dobzhansky but hardly mentions the work of Huxley, Ford, Darlington, and Mayr. Simpson joined the staff of the American Museum in 1926; Mayr in 1931. Mayr moved to the Museum of Comparative Zoology at Harvard in 1953. In 1959 Simpson also moved to Harvard. On the face of it, one might expect that having two founders of the synthetic theory of evolution working at the same institutions would prove to be an especially happy accident, but Simpson and Mayr rarely spoke to each other during these years on professional subjects, even though they and their wives were social friends. Early in their friendship, Mayr attempted to discuss issues of mutual interest with Simpson, but Simpson discouraged such conversations. Eventually Mayr ceased trying. Mayr did give Simpson the last chapter of his 1942 book to read prior to publication, but Mayr had read none of Simpson's manuscript before Simpson sent it to the publisher. Simpson had completed his book prior to the war, although it did not appear until 1944 (Mayr and Provine 1980, Mayr 1982).

Nor was the synthesis as seamless as it was made to appear. Propaganda to one side, the conceptual gap between Mendelian geneticists working in the laboratory and naturalists working in the field was as large as the gap that saltationists claimed separated biological species. It has not been totally bridged to this day. Mayr opted for a nondimensional species concept, species at a moment in time. Simpson argued for temporally extended, evolutionary species. Additional points of disagreement among the founders of the synthetic theory of evolution could be enumerated almost indefinitely, so many that an occasional voice has been raised questioning the justification of terming the synthetic theory "synthetic." In spite of all the continuing differences among the founders of the synthetic theory, Mayr (1980a:40) insists that there is "every justification to designate this process as a synthesis." The crucial significance of the synthesis was the "fusion of the widely diverging conceptual frameworks of experimentalists and naturalists into a single one."

Conclusion

In this chapter I have traced a few strands in the development of evolutionary theory from Darwin to the Modern Synthesis. Even though I have paid more attention than usual to actual causal connections rather than just to inferential relations, my narrative has been unduly conceptual and "whiggish." I have not attempted to trace every twig in the tangled causal bush. Instead I have pruned it extensively to reveal what I take to be a few major channels of development. In subsequent chapters I fill out the story somewhat more fully, but I never tell it totally as it happened. The amount of research necessary for such an undertaking is prohibitive, and besides no one would be willing to read the results if I were to write the full account.

I have also paid no attention to Darwin's precursors. If by "precursors," one means earlier scientists who held views similar to those that Darwin was to publish later, I have no apologies to make. If I am forced to omit reference to numerous scientists who actually influenced Darwin, I am not about to include discussions of those scientists who had little or no effect on the development of evolutionary biology. Unappreciated precursors are of antiquarian interest only. With the exception of Lamarck, none of the scientists who influenced Darwin were advocates of anything that might be considered evolution in a significant sense. Similar observations hold for Mendel. If the fundamentals of Mendelian genetics were formulated and published without any influence from Mendel's early papers, then Mendel warrants an occasional footnote in any chronicle of the development of the science that bears his name; nothing more. However, if current scholarship is correct, Mendel did influence the development of Mendelian genetics. Hence, his work deserves at least some of the attention it receives.

An occasional passing comment notwithstanding, I have also not paid sufficient attention in this chapter to the effects that the professional relations among scientists have on science. The alliances which Darwin forged with Lyell, Hooker, Huxley and others influenced both the development and the reception of his theory. That Owen despised the Hookers as well as Huxley exaggerated the gulf between the evolutionists and "idealists." As the quotation marks indicate, idealists in Great Britain were singularly unable to form any sort of a united front to promote their common interests. Darwin and Kelvin succeeded where Owen failed, and this success contributed to their influence.

One message of this chapter is that historical contingencies influence the course of science. If Romanes had been more successful, the Weismannians might have been viewed as anti-Darwinians instead as descendants of Darwin. Although the version of evolution favored by de Vries was a good deal more saltative than any of the versions held by Darwin, de Vries portrayed himself as continuing in Darwin's footsteps. It was Bateson who insisted that Mendelian genetics was incompatible with Darwinism. As a result, Mendelian genetics and the versions of evolutionary theory set out by the biometricians were developed in opposition to each other. Historical contingencies produced quite different effects in Russia. At the very least, such historical contingencies have significant short-term effects on the conceptual development of science.

A second message of this chapter is that our understanding of the nature of science goes hand in hand with our understanding of nature. The dispute over Darwinism was as much a disagreement over the nature of genuine science as over the existence of evolution. Because Darwin's theory triumphed, the view of science implicit in it gained support; conversely, because Darwin's theory exemplified a particular view of science which had gained considerable currency at the time, Darwin's theory appeared more acceptable. Similar observations hold for Mendelian genetics. Certain biologists at the turn of the century argued that genuine scientists should limit themselves to discovering which characteristics tend to be passed on from generation to generation to form linkage groups. The postulation of genes residing on chromosomes to explain linkage was totally unscientific. Others insisted that genuine empirical science must provide mechanisms capable of producing such regularities. Abstract patterns are not good enough. Central to this recurrent dispute are two quite different senses of "empirical." Is empirical science "empirical" because it is limited to observable phenomena or because it specifies causal mechanisms?

The development of the synthetic theory of evolution did not cease in 1947. It continues to the present in ongoing controversies over the relative importance in the evolutionary process of natural selection, neutral mutations, stochastic processes, structural constraints, saltative forms of evolution, and so on. One of the most insistent calls has been to reintroduce embryological development into evolution (Thomson 1985). Perhaps at the turn of the century the time was not ripe, and genetics had to be divorced from embryology before it could be wed to evolutionary theory; but many present-day evolutionary biologists are beginning once again to urge a reconciliation. Some are arguing for a hierarchical view of evolution, one in which processes can operate simultaneously at a variety of levels of organization.

Some evolutionary biologists view all these hypotheses as enclosed within the all-encompassing arms of the synthetic theory of evolution (Stebbins and Ayala 1981). Others disagree. But this story must be postponed while I backtrack to trace the early history of biological systematics, or natural history, as it was termed for most of its career. Most of Darwin's "precursors" belong in this narrative, but more importantly, many of the issues central to the conception of science that I develop in this book are pursued most relentlessly in this literature.

In his massive *The Growth of Biological Thought* (1982), Mayr argues that too much attention is paid to the merger of genetics with evolutionary biology in the development of the Modern Synthesis and not enough credit given to the contributions of systematists and naturalists. Evolutionary biology emerged from natural history, not genetics, and the development of the two have been intertwined ever since. In my discussion thus far I have perpetuated the biased perception of which Mayr complains. It is now time to rectify this imbalance, but this time, instead of going back a hundred and fifty years to Darwin, I must begin in ancient Greece with the intellectual father of us all—Aristotle.

3

Up from Aristotle

Systematic work would be easy were it not for this confounded variation, which, however, is pleasant to me as a speculatist, though odious to me as a systematist.

Charles Darwin, *The Life and Letters of Charles Darwin*

AFTER NUMEROUS false starts, science limped into existence. With the power of hindsight, we can recognize here and there early investigators who were on occasion engaged in activities that count as "scientific," but these early protoscientists were like hermits working in isolation from each other. One hermit would solve this problem, another that, and each would live and die in ignorance of the other. Throughout human history, people have tried to unravel the mysteries of nature, but they did not make much headway because each had to start from scratch. Eventually early scientists became aware of the accomplishments of their predecessors and immediate contemporaries. They were able to use each other's work, to build on it, to criticize it, improve on it, ignore it, sometimes parody it. As Richter (1972: 2) notes, "After many centuries of irregular growth and decline, science came to acquire a self-reinforcing capacity which has been reflected in the massive, sustained growth of science from the seventeenth century to our own time."

Historians are divided over the justification for calling anyone a "scientist" before modern science became firmly established in the seventeenth century. Even the term "modern science" is frequently treated as questionable. If it is not modern, then it is not science. One source of this reticence is a serious misunderstanding of language. The justification for declining to use certain terms with respect to the past is that these terms had yet to be coined at the time. For example, Lamarck is usually credited with inventing the term "biologie" in 1802, although two other men had introduced the term somewhat earlier. Not until three decades later did William Whewell coin the term "scientist." Hence, it follows that most figures in the history of biology were not actually biologists, nor were most figures in the history of science scientists. Once presented this blatantly, the terminological convention being suggested by purists is so patently silly that it hardly warrants refutation. Every author must write in the language of his own time and place. I am writing for my contemporaries, and they speak contemporary English. It does not help me in the least if a word exists in some other language or at some other time if my readers are unaware of it. Even though words denoting scientists existed in French and German, Whewell still had to introduce a term into English. For this same reason, Lamarck's introduction of the term "biologie" into the French language is irrelevant to the existence of this subdivision of science in other scientific communities. Prior to 1802, French-speaking people could not refer to Buffon (1707–88) as contributing to "biologie" because the term had not been invented yet. Afterwards they could, and in doing so they were in no way sinning against the canons of acceptable historiography—pedantry notwithstanding.

Even so, the pedants have a point. A society quite obviously influences the character of the language used by its members and, somewhat more subtly, the language affects the society. All one has to do is to ponder the connotations of the English words for men who have never married and women who have never married to realize something about the society in which the terms "bachelor" and "spinster" are used. Today many people are engaged in activities that we term "scientific." A subclass of these people are engaged in the narrower pursuit of "biology." Many of these activities do not characterize the work of early scientists and biologists. Terming Buffon a "biologist" or even a "scientist" might lead the present-day reader to attribute inappropriate attitudes and activities to the man. Today anyone engaged in producing an encyclopedic work of the sort that made Buffon famous would be considered only marginally a scientist. Similarly, scientists felt perfectly free to refer to God in the midst of a scientific treatise well into the nineteenth century. Sometimes these references were *pro forma*, but sometimes God played an important role in the theory. Today no editor of a scientific journal would allow such a reference to stand. In some societies, references to political leaders have taken the place of appeals to the deity. Scientists elsewhere pass over these references in pained silence.

In short, modern science has evolved and continues to evolve. The only science that we know anything about is Western science. After it emerged in the Western world, it spread to other societies, extinguishing in certain areas alternative indigenous undertakings that might well have warranted the name "science." We know the forms that modern science has taken so far. It arose in Christian, individualistic, competitive, elitist societies. Not surprisingly, it too was Christian, individualistic, competitive, and elitist. We now take the Christian part to be an accident of history. Knowledge acquisition in science is not inherently Christian. Which other characteristics are accidental and which essential? Might not science in the future be very different from science today? Perhaps science in the future will turn out to be more genuinely cooperative and truly egalitarian than it has been thus far. Science from its inception has been extremely competitive, but perhaps it need not be. Or perhaps the science of the future will put virtue above truth. But if science changes too much from our current conceptions of it, what right have we to continue to term it science?

The tension in the preceding paragraph has two sources. One concerns the character of the entities being referred to. Does "science" refer to a particular sort of activity regardless of when and where it is conducted, or is it a tradition with a beginning in time and a continued development through time? On the first interpretation, anyone anywhere who engages in enough of the appropriate activities has the right to be termed a "scientist." Although some of Aristotle's ways of reasoning were a far cry from modern science, enough of the things he did were similar enough for us to consider him a scientist rather than as a painter or a ballet dancer. In this same sense, science was carried on in ancient China and the Arab world after the early demise of science in Greece and Rome. In the future, people who term themselves scientists may take seriously all sorts of considerations that we now dismiss as unscientific. If they depart too far from contemporary practice, then they will not really be scientists, no matter what they might think.

Historians, however, deal in traditions, periods, epochs, and the like. They treat some terms as referring to particular times and places. In order to count as Baroque, a building must have been built at a particular time and place. It also helps if it is built in a particular style. However, no matter how similar a building might be to a Baroque building, it cannot count as Baroque if it was built at the wrong time or in the wrong place. Later structures can be built in the Baroque style, but they cannot actually be Baroque. In short, the Baroque period is a historical entity. To deflect the inevitable objections from ordinary language, let me explicitly acknowledge that the term "Baroque" is used in a variety of senses in ordinary English, at times referring to a general style, at times to a historical period, at times to both, or possibly to neither. A distinction can be important even it is not consistently reflected in a particular natural language. I have nothing against natural languages, only the use of ordinary discourse to reject technical usage, whether scientific or philosophic.

Part of the reticence that historians have with respect to extending terms such as "science" back in time is that Plato and Aristotle were not engaged in the same sort of activity as Newton and Galileo or Watson and Crick. Here "science" is being used as a general term referring to any activity that fits the definition of "science." However, it is quite difficult to define "science" in such a way that the activities of all these men fall under this definition. Part of the reticence also stems from the possible lack of the continuous development of "science" from the ancient Greeks to the present. As the story goes, science began with the ancient Greeks, sort of, was passed on to the Romans and a few workers in the Arab world, then reintroduced into Europe in the Renaissance when modern science was at long last born. In this usage, "science" refers to a tradition. If the Dark Ages had really been as dark as they are sometimes portrayed, then science would not form a continuous tradition, and the term "science" should not be applied to the work of the ancient Greeks and modern scientists.

The issue is similarity versus descent. Which should take precedence in a discussion of science? If the two always went together, it would not matter, but they do not. Some terms such as "Baroque" are tied down to particular times and places. Most are not. What sort of term is "science"? Historians feel the pull of both usages. Although a continuous tradition might well have connected Aristotle to J. D. Watson, these two men have so little in common that it seems strange to refer to both of them by the same name. They may both be "men" in precisely the same sense, but they are hardly "scientists" in anything like the same sense.

Although the distinction between general terms, terms which refer to entities, processes, and properties that can exist at any time and any place in the universe, and terms whose referents are limited to particular places and times may not seem all that important, it is absolutely crucial in the way that we conceptualize the world. From the beginning, one of the chief goals of science, possibly *the* chief goal of science, has been to discover classes of phenomena that are lawfully related—classes commonly termed natural kinds.[1] If natural laws are to be spatiotemporally unrestricted, then it follows that natural kinds must themselves be spatiotemporally unrestricted. Anything anywhere in the universe that has mass must be related in the way specified by Newton's law of universal gravitation. If not, the law is false. In order to test a natural law, the terms in it must refer to something. For instance, Sol, Earth, and Earth's moon have masses and must attract each other in the way specified. Science requires natural kinds—classes whose names figure in laws of nature. It also requires natural individuals—entities referred to by these general terms. Science requires "planet" and "Mars," "geological period" and "Cambrian," and "nation-state" and "France."

The status of the terms listed above is relatively unproblematic. Few are likely to insist that, to count as a planet, a heavenly body must revolve around the sun; any star will do. Conversely, no one is likely to argue that anything with the appropriate characteristics can count as "Mars" regardless of its connections to the body called by that name in our solar system. Mars is clearly a proper name and denotes a particular individual. However, not all terms fall so easily into one category or the other. For example, does "Tiffany lamp" refer to a lamp of a certain style or a lamp actually built in the studio of L. C. Tiffany? Ordinary language provides no answer to this question. The phrase is ambiguous. To those who would reject the distinction that I am drawing as being of no practical significance, all I can say is that they should compare the prices of Tiffany lamps and Tiffany-style lamps.

Before I return to the term "science," a brief discussion of this same distinction with respect to one other sort of term might prove helpful. More than helpful, the recognition that such terms as "Carnivora" and "*Dodo ineptus*" have been used ambiguously from at least the time of Darwin has led to a major advance in our understanding of the evolutionary process. Do such terms refer to chunks of the genealogical nexus or to ecological types? Must all organisms that belong to Carnivora be carnivorous, and conversely do all carnivorous organisms belong to Carnivora? As these terms are used in biology, the answer to both questions is no. A few organisms that belong to Carnivora rarely, if ever, eat meat, and millions of species that subsist primarily on meat are not part of this chunk of the genealogical nexus. Similarly, the dodo was characterized by a cluster of characteristics. It is now unfortunately extinct. Can it evolve again? If a group of organisms were to evolve exhibiting this same cluster of characteristics, would these organisms count as dodos? They would belong to the same type but not to the same lineage.

Should species be treated as classes of similar organisms, regardless of descent, or should they be treated as spatiotemporally continuous lineages? If the two ways of grouping organisms into species always produced the same result, a decision on this

1. Although John Venn (1889: 83) claims that John Stuart Mill was the first to introduce "natural kind" as a technical term in Mill's *A System of Logic* (1843), the idea has been implicit in the work of nearly all early philosophers from Plato and Aristotle to the present. Not all ways of grouping entities into kinds are equally "natural."

issue would not be all that crucial, but they do not. Species that are extremely similar to each other frequently evolve in various parts of the phylogenetic tree. If similarity alone were a sufficient basis for individuating species, all these different species might well be considered a single species. But for the purposes of evolutionary theory, each twig in the phylogenetic tree is a distinct species. Species are individuated primarily by means of their location in the tree. Similarity is of secondary importance. Highly diverse groups of organisms count as a single species if they exchange genes frequently enough. Conversely, groups of organisms that are all but indistinguishable count as distinct species if they do not exchange genetic material. One of the main messages of this book is that species, if they are to play the roles assigned to them in evolutionary theory, must be treated as historical entities. *Dodo ineptus* is conceptually the same sort of thing as the Baroque period. Both are gone and can never return. Extinction *is* forever.

One thing follows from treating species as historical entities: they are not natural kinds. They cannot function in any natural laws. They are natural individuals, they may instantiate a natural kind, but they themselves are not natural kinds. Spatiotemporally restricted entities cannot be spatiotemporally unrestricted. Historians treat human history historically, tracing traditions through time. If one chooses, one can subdivide such lineages into sequential segments and name them. For example, the Baroque period developed into the Rococo, but no one can expect to find laws of these historical entities qua particular historical entities. One might hope to find lawlike regularities about kinds of episodes in architecture, e.g., florid styles of architecture. If so, then the Rococo period had better exemplify this putative law, because if any period of architecture counts as florid, the Rococo period does. Similarly, one might be able to develop process theories about kinds of events in biological evolution, e.g., runaway positive sexual selection that produces male secondary sexual characteristics no less excessive than the ornamental filigree of Rococo architecture. If the positive feedback explanation of highly developed secondary sex characteristics is adequate, it had better apply to the tails of peacocks and male widow birds.

Exactly parallel observations hold for "science." Philosophers of science take it as one of their tasks to analyze terms such as explanation, law, and science. In doing so, they provide lists of defining characteristics. Whether one treats these definitions in the classic sense as providing necessary and sufficient conditions or as cluster concepts, they map out a place in conceptual space. Any activity anywhere and at any time which exhibits the appropriate characteristics counts as science. Gellner (1985: 104–105), for one, insists that "*in our society,* the concept of the 'scientific' is precisely of this kind." In his mind, there is "no shadow of doubt but that in fact, discussions concerning what is and is not 'scientific' are carried out in this utterly Platonistic, normative and non- conventionalist spirit. These are debates about whether something is really, really scientific." We may not know what precisely science is, but "we do know that it is important and that we cannot tinker with it at will." The issue of realism to one side, there is nothing wrong with philosophers providing general definitions of metascientific terms if they go on to *do* something with them. In science the chief role of such general terms is to function in scientific laws. What analogous function are philosophical analyses of metascientific terms to perform? If none, then I see no point to all this effort (Sober 1984b).

Similarly, one can develop a perfectly consistent notion of species as Aristotelian natural kinds. As a result, biologists would have at their disposal millions upon millions of natural kinds. However, as natural kinds, these Aristotelian species cannot "evolve." Aristotelian species are eternal and immutable. As fuzzy as the boundaries between Aristotelian species in conceptual space may be, they are stable. The "species" that evolve are temporary and mutable. They come and go, and once gone cannot return. Sometimes their boundaries in conceptual space are fuzzy, but their important boundaries occur in physical space. Species have ranges. Although Aristotelian species cannot fulfill the role of evolutionary species in evolutionary biology, they might do something else. They might function in some other natural regularities. The trouble is that they do not. If, to the contrary, species are treated as historical entities (natural individuals), then their names cannot function as terms in the statement of natural laws, but at least they might function as instances of regularities in the evolutionary process.

By now I hope the general distinction is obvious: it is between Tiffany lamps and Tiffany-style lamps, between Rococo architecture and florid architecture, between Cambrian and basalt, between Carnivora and carnivores. The first member of each pair is a historical entity located in space and time; the second is timeless. The first member of each pair may be an instance of a natural kind. If so, then it is a natural individual functioning in a natural process. Because the second member of each pair is spatiotemporally unrestricted, it is at least a candidate for a natural kind. It can function in a law of nature. In biological evolution, the natural individuals have always seemed to be patent—genes, organisms and species. The big challenge has been to discover appropriate natural kinds—classes that are lawfully related. In scientific development, neither the natural individuals nor the natural kinds have seemed all that apparent. However, if there is going to be a point in defining something like "science" as a natural kind, then it had better be put to some purpose. Providing lists of defining criteria for words as we happen to use them at the moment is not entirely pointless. After all, we do keep updating telephone books. But something more is required of scientists, not to mention philosophers.

The general question is how we treat conceptual entities: as historical entities individuated in terms of their insertion in history or as general notions which are spatiotemporally unrestricted. We cannot have it both ways. Science cannot be treated both as a historical entity and as a natural kind. If it is treated as a historical entity, then it can evolve but its name cannot function in the statement of any lawlike regularity. The most it can do is exemplify such a regularity. If, to the contrary, science is treated as being a natural kind, then we might well discover regularities about science as a process. If we move one level up, parallel observations hold for the term "science" itself. It can be construed as a lineage of ancestor-descendant instances (term-tokens) or as a class of identical instances (a term-type). If concepts are treated as lineages, then they can evolve, but their names cannot function in the statement of any lawful regularity.

One reason "evolutionary epistemology" seems so counterintuitive to most philosophers and continues to be stuck in a programmatic stage is that these two ways of individuating terms have not been kept sufficiently distinct. In the past, philosophers have glimpsed these alternatives but have failed to pursue the distinction in all its ramifications. The classic example, cited several times by Toulmin in his *Human Understanding* (1972), occurs in Kierkegaard's discussion of the concept of irony. According to Kierkegaard (1841: 47):

> Concepts, like individuals, have their histories and are just as incapable of withstanding the ravages of time as are individuals. But in and through all this they retain a kind of homesickness for the scenes of their childhood.

As might be expected, Kierkegaard does not stop here but goes on to add:

> As philosophy cannot be indifferent to the subsequent history of this concept [irony], so neither can it content itself with the history of its origin, though it be ever so complete and interesting a history as such. Philosophy always requires something more, requires the eternal, the true, in contrast to which even the fullest existence as such is but a happy moment.

No blanket decision needs to be made at this stage on terms as such. However, I must choose one way or the other on terms such as "science" and "biology." Much of the confusion over terms such as these arises from their being used sometimes as if they were confined to activities performed at particular times and places, and at other times in a spatiotemporally unrestricted sense. For most purposes, it does not matter what usage we choose as long as we stick to it. Wavering unconsciously back and forth between these two usages is guaranteed to generate confusion of a sort likely to be anything but constructive. In this book, I intend to use "science" in a general sense. Science is the sort of thing that can reemerge when the appropriate conditions are met. Western science, science as we know it, is one instance of science in this general sense. However, it is only *instances* of science that can evolve, and science as we know it might well evolve into nonscience. If I were concerned with human understanding in the broad sense and with the competition among various ways of knowing, then I would have to treat science as a historical entity in competition with religion, humanistic activities, political orthodoxy, and so on. The point of this discussion should become clearer in the second half of this book when I turn to an account of conceptual change in science as a selection process.

The Old Systematics

To the extent that members of the general public have heard of taxonomy, they are liable to be shocked by references to it as the "most lovable of the sciences" (Lindroth 1983: 1). As most people view taxonomists, they are more librarians than scientists and just about as lovable. But in its heyday, biological systematics was the queen of the sciences, rivaling physics. Originally no distinction existed between systematics and natural history. Collectors and classifiers were the ones who had sufficient knowledge to appreciate the true diversity of life. They were the ones who fanned out across the face of the globe to discover exotic new species to stock the cabinets of Europe. Granted, a collector had to pay the price of his adventures. When he returned, he had to curate his collections. Because collecting is so much more fun than curating, crates began to pile up in the basements and attics of museums. Rumor has it that some of the *Beagle* specimens Darwin sent home to England have yet to be unpacked. Eventually, as collections grew, a cadre of workers arose who spent most of their time organizing and cataloguing the specimens collected by others. These are the people who spring to mind when one hears the term "taxonomist" today, but as we shall see, the world of the museum taxonomist is not quite the safe backwater that it might seem. Perhaps the seminar rooms of the American Museum are not as perilous as Wallace's upper Amazon, but they come close. And certainly the life of the naturalist/systematist throughout most of the history of this science was anything but sedentary.

Aristotle has been designated the father of many areas of human endeavor, but this appellation is most clearly warranted in the case of natural history. In his works he not only gathered together everything that was believed about plants and animals in his day but also organized these beliefs in accordance with his general theory about the nature of the universe. The two most fundamental questions in systematics are the nature of the basic units of classification (traditionally termed "species") and the rules according to which these basic units are grouped into increasingly inclusive classes (or "taxa" as they are now called). Aristotle says quite a bit about his species concept. It is absolutely central to his entire system of thought, but he does not present a global, coherent classification of living creatures. Instead, he sets out bits and pieces of a classification that are not always mutually compatible. Later workers, however, extracted a hierarchical classification from the works of Aristotle that became enshrined as the Great Chain of Being.

According to Aristotle, all nature can be subdivided into natural kinds that are, with appropriate provisions, eternal, immutable, and discrete. For example, living organisms are of two sorts—plants and animals. Although Aristotle's works on plants have been lost, we do know that he took over the distinction common at the time between trees, shrubs, and herbs. He subdivided animals into those that have red blood and give birth to their young alive and those that do not. He further subdivided each of these groups until finally he reached the lowest level of the hierarchy—the species. Arrayed at the base of the great taxonomic pyramid are such species as the horse, human beings, and cephalopods. It is this unbroken array of species that forms the Great Chain of Being stretching from the human species through hairy quadrupeds, reptiles, fish, worms, and sea cucumbers, to corals, petunias, and mosses. Because we now believe that species evolve, the temptation is to read some form of evolution into the work of everyone from Aristotle and Dioscorides to Linnaeus and Richard Owen. But, according to Aristotle and generations of Aristotelians, all species are as eternal and immutable as are the physical elements. No one today expects lead to evolve into gold; Aristotle had no higher expectations of fish evolving into frogs. To be sure, an organism might change its species, just as a sample of lead might be transmuted into a sample of gold, but the species themselves remain unchanged in the process. Species are at least potentially eternal (for various views on the subject, see Atran 1985, Lennox 1985, and Granger 1987).

Aristotle was aware that organisms that belong to different species tend to be intersterile, an occasional fertile hybrid notwithstanding. He also thought that species could be distinguished by means of defining characters. Hairy quadrupeds are distinguished from other animals in part because they have four legs. Rare monsters aside, all hairy quadrupeds have four legs. Conversely, not everything with four legs belongs in this group, but once hair is added the group that we now know as mammals can be distinguished from all other quadrupeds. Later Aristotelians developed Aristotle's system further on the basis of hints in the works of the master. For instance, hairy quadrupeds can be subdivided according to the sorts of legs that they have, then further by the number and kinds of toes, and so on, until the species level is reached. No further subdivision is then possible. Even though characters can be found that distinguish males and females, Aristotle held that these differences are accidental, not essential. He believed that in general intersterility and diagnostic character criteria go together. Aristotle did not view nature as perfect. Natural phenomena need not occur with perfect regularity, only "always or for the most part." He also did not think the boundaries in conceptual space between species are perfectly discrete; they are usually a matter of "the more and the less." Even so, the general drift of Aristotle's worldview is patent. The differences that exist within species are not just variations; they are deviations (Lennox 1980, 1982, 1985; Sober 1980, Atran 1985; Granger 1985, 1987).

Aristotle was not all that interested in the identification of specimens. Instead, the hints that he gave of a possible global classification of animals reveal that he viewed classification as a function of his basic theory. His most fundamental theoretical notions are two pairs of polar principles—the hot and the cold, the dry and the moist. At bottom Aristotle explained everything in terms of some combination of these polar principles, much as we explain natural phenomena in terms of mass and acceleration. For example, corresponding to the four possible combinations of the polar principles are Aristotle's four humors, four elements, four seasons, and so on. Blood is a function of the hot and the moist. As a consequence, animals without blood tend to be cold and dry. Corresponding to this decrease in vital heat are other changes as well; for example, in mode of reproduction. Blooded animals give birth to their young alive without the interposition of an egg. Animals lower down on the scale produce eggs, but they retain them within their bodies and give birth to their young alive. Others still lower down lay true eggs, and so on. The manner in which animals digest (concoct) their food also corresponds to the preceding order as well.

As peculiar and at times mistaken as the details of the system derived from Aristotle's writings might seem to us, it does not differ in kind from comparable systems today. We now believe that the characteristics a chemical compound exhibits are a function of the number, kind, and arrangement of the elements that make it up. We now think that there are over a hundred elements. Aristotle believed that there were only four—fire, air, earth, and water. But the explanatory strategy was not all that different. For our purposes, however, the important feature of Aristotle's system is that his principles of classification are part and parcel of his views on how nature works.

People die and, once dead, remain in their graves. Ideas recorded on stone, papyrus, or parchment have the ability to lie dormant, like recessive genes, to burst forth again much later. Although the connections between the ancient Greeks and modern science detoured through the Muslim world and became somewhat attenuated in the process, a continuous tradition existed in at least some areas of science, in particular optics and astronomy. Other subjects started up again, primarily through the rediscovery of documents. Although scientists came to decry the leaden influence of past knowledge on the development of science, it served to reawaken Europe. Might not science as we know it have developed in the absence of the rediscovery of the works of Aristotle and Plato? Perhaps. We will never know. Of the many ways in which science might have originated, it happened to grow out of the rediscovery of the systems of thought developed by the ancient Greeks.

The first real pinnacle in natural history occurred in the work of Carolus Linnaeus (1707–78). Linnaeus consciously based his system of classification and nomenclature on his understanding of Aristotle. The major difference between Linnaeus and Aristotle is that Linnaeus was a Christian and Aristotle was not. For Aristotle, species stretch infinitely into the past and just as infinitely into the future. Horses always were and always will be. As a good Christian, Linnaeus reserved eternal existence for God. For Linnaeus species are in no sense eternal. The original progenitors of all species were created by God in the Garden of Eden, and eventually all species will go extinct at the Second Coming. However, within the confines of Creation and Judgment Day, Linnaeus maintained species are as Aristotle said they are.

Aristotle was interested in finding some order among the natural kinds that populated the world, but he showed little concern for their identification. Not only was Linnaeus more concerned than Aristotle had been in classifying plants and animals, he also wanted to produce classifications that would help in identifying particular specimens. In both respects, Linnaeus was extraordinarily successful. He brought order to the naming of organic kinds not only in his native Sweden but in the rest of the Western world. In flowering plants his system was "sexual" in the sense that he used the number and arrangements of the stamens and pistils as his first criteria in subdividing plants. He even extended his system of classification to include minerals. Although Linnaeus had some reservations about species existing in minerals, he did develop a theory of "sexuality" for them. In minerals, the fundamental principle of water generated two offspring—"a saline male and a terrene female" (Turton 1802: vol. 7:3).

Although Linnaeus became renowned in his day for his views on classification, later in life he drastically altered his views. He came to believe that plants are made of two substances—the medulla and the cortex. In the beginning, God created a very few elemental species, each with its own medulla and cortex. Organisms belonging to these elemental species then proceeded to hybridize, producing a whole array of secondary species. The characteristics of these secondary species depended on the particular combination of medulla and cortex in the parents. Hybrids from the same two elemental species might be different, depending on which species bequeathed its medulla and which its cortex. These secondary species proceeded to hybridize to produce tertiary species, and so on until all possible combinations had been produced. Linnaeus thought that animals and minerals could be explained in this same way.

Because of Linnaeus's later ideas, some authors have suggested that he developed an evolutionary view of the world. It all depends on what one means by "evolution." Linnaeus did think that species appear sequentially through time, but according to a single, predetermined plan built into the constitution of the elemental species. Once the elemental species had been created, all possible combinations were fixed. Even though this arrangement filled out through the course of time, the structure of the living world in Linnaeus's system is eternal and immutable. To use a present-day example, God might well have created all the basic physical particles in the beginning. In doing so, he simultaneously structured the physical world so that only a certain prescribed set of molecules could ever come into existence. Whether all molecules that could exist do exist, and the question of the order in which they are formed, are quite separate issues. In this view of nature, the structure of the world is eternal; the filling out of this structure is a temporal process but one with a single, predetermined terminus. If this is "evolution," it is evolution in a sense very different from that employed by Darwin and later evolutionists.[2]

In his later years Linnaeus suggested that the living world could be likened to a map on which each elemental species is represented as the capital of a province. Successive hybridizations produce interconnected cities until the map is filled out. As a result, species cannot be ordered serially in a Great Chain of Being. One species may be related to a particular array of species on one side, another on a second side, and so on. Linnaeus also did not say how these elemental species are supposed to hybridize. He thought he had observed the production of new species by hybridization between species, but these species were quite similar to each other. The elemental species that were supposed to produce secondary species by hybridization were as different as any species could be. It is as if a willow were to hybridize successfully with a petunia, and the petunia in turn with a moss.

As fascinating as Linnaeus's more mature views are, they had no influence whatsoever in his own day or since. We would not know that they existed if historians had not dug them up in their inquisitive rummaging about in old manuscripts. An occasional scholar knew of their existence, but they had no impact on the continuing development of biology. The Linnaeus who influenced generations of systematists was the Linnaeus who believed in the special creation of all species in the Garden of Eden, a spot that Linnaeus suggested must have been a mountainous island near the equator. As the oceans receded, the resident plants and animals spread out over the globe. One message that collectors received from Linnaeus was that God created only a finite number of species. Young explorers had better go out and discover theirs before they were all taken. The voyages of exploration that took place during the next two centuries brought

2. I have no strong preferences with respect to the "proper" use of the term "evolution" and its cognates. To some, it means merely change. Hence, stars evolve. Others prefer a much narrower usage which excludes the evolution of partially programmed change that occurs in embryological development, excluding as a result the most common use of the term "evolution" in Darwin's day. But I have no intention of speaking Victorian English. "Evolution," as it is commonly used today, refers to indefinite change through time in which the entities that are evolving themselves change. I have no brief against those who prefer other usages, just so long as they warn me.

systematics to the height of scientific prestige. They also served to undermine the foundations of the system of thought that instigated them (Hull 1985d).

Linnaeus's chief opponent throughout his life was the renowned Georges-Louis Leclerc, comte de Buffon (1707–1788). From the age of twenty-five, when he settled at Montbard in France, Buffon contributed to many areas of science. Although he translated Newton's *Fluxions* (1740), he was known chiefly for his multivolume *Histoire naturelle, générale et particulière*. Whereas Linnaeus classified organisms, arranging them in hierarchies by means of succinct descriptions, Buffon treated species seriatim in their order of importance to man. As in the case of Linnaeus, Buffon changed his mind during the course of his research. He began by claiming that only organisms exist and that all groupings of organisms into species and the like are entirely arbitrary, but eventually he came to believe that species were in some sense real and explained their existence in terms of a unique *moule intérieure* for each. Man, the dog, and the willow tree exist as separate natural kinds, but classes such as vertebrates, mammals, and carnivores are entirely products of the human mind. Species could be grouped together in an indefinite number of ways. No single preferable classification existed.

Buffon heaped ridicule on Linnaeus and the "*nomenclateurs*." In return, some of Buffon's contemporaries doubted that he was a genuine scientist. Perhaps his observations were original and his prose both beautiful and uplifting, but Buffon claimed to shun the very sort of activity that they took to be the hallmark of genuine science—systematization. It is certainly true that Buffon was an encyclopedist, but he had a second area of concern as well—an explanation of the living world in terms of organic molecules and internal molds. As Darwin was to discover many years later, Buffon had devised a system of heredity not all that different from his own theory of pangenesis. According to Buffon, millions of organic molecules permeate all nature. What passes for spontaneous generation is the coming together and proliferation of these molecules to form living creatures. Just as different elements result from different combinations of organic molecules, different species of plants and animals result from various combinations of organic molecules. The notion of *moule intérieure* was intended to be comparable to Newton's concept of gravitational force. Once formed, a species might also degenerate into another species. For example, through time the domestic horse might degenerate into the ass, or the human species into various species of ape. In general, each family of animals might have resulted by means of degeneration from an original progenitor. Scholars continue to debate the extent to which Buffon accepted, rejected, or was just toying with these ideas. They are also of several minds as to whether or not these musings count as "evolution" (Sloan 1973, 1987).

Buffon's greatest contribution to "evolutionism" was unintentional. It occurred when he hired a young botanist to accompany his son on a tour of northern Europe. This young botanist was Jean-Baptiste Pierre Antoine de Monet Lamarck (1744–1829). Lamarck began his career as a systematist and ended it in the same way, but in between he published works on physics, meteorology, and transformism. The image we have of Lamarck is one that has been distorted by defenders and detractors alike. Lamarck was not ridiculed in his day because he believed in the inheritance of acquired characteristics, a view that was widely accepted at the time. His innovation was to suggest that a combination of the inheritance of acquired characteristics and an innate, mechanical drive to perfection could cause more complex organisms to arise through successive generations. In this process, the drive to perfection produced by vital fluids coursing through living organisms was the more fundamental force. Nor was a belief in some sort of evolution all that unusual in Lamarck's day. For instance, of the four leading zoologists at the Museum of Natural History in Paris, only the powerful Cuvier (1769–1832) adhered strictly to the immutability of species. In addition to Lamarck, B. G. E. de la ville Lacépède (1756–1825) and Etienne Geoffroy Saint-Hilaire (1772–1844) held views on species that might be considered evolutionary.

Lamarck's reputation as an idle speculator was due to his bizarre and frequently old-fashioned views on subjects in which he had little professional competence. Any scientist who issues daily weather reports from his apartment is courting ill-repute. When Napoleon rebuked Lamarck for damaging the reputation of French science, it was not for his *Philosophie zoologique* (1809) but for his ventures into meteorology. In his later life Lamarck isolated himself from his fellow scientists. He was not in the least interested in cultivating protégés. For example, in spite of persistent attempts, A. P. de Candolle (1778–1841) failed to get close to the man. As Henri de Blainville recalled, Lamarck became so totally wrapped up in his own ideas that no one could have a rational discussion with him (Appel 1987).

In addition, Lamarck's transmutation theory is not very "evolutionary." Although Lamarck's system does allow for organisms to move up the tree of life to which they belong, species themselves do not "evolve" in the sense that they do in Darwin's theory. According to Lamarck, the simplest organisms are produced by spontaneous generation and then proceed to become modified by the environments in which they find themselves, propelled by an inner hydraulic impetus. Once an organism embarks on a particular sort of development, the possible paths that its descendants can take are highly constrained. Organisms can occasionally degenerate, but the drive to perfection guarantees that in the main they will become progressively more complex. It is as if creases exist in the environment to channel successive generations up alternative trees of life. Particular environments might disappear in one part of the globe only to reappear elsewhere. Whenever an appropriate environment appears, the appropriate organisms move up to fill it. Perhaps one segment of a branch in one tree of life might not be occupied for a while, but there are organisms immediately below it being impelled upward. Sooner or later, the gap must be filled.

According to Lamarck's particular brand of transformism, species in an important sense are immutable. Organisms through successive generations can proceed from one series to another, and that is all. Species themselves do not change. There is also a sense in which species are eternal. Lamarck maintained that the basic structure of nature is eternal. Only the happenstance of exemplification is contingent. According to the present-day view, environmentalists are extremely concerned about extinction because once a species goes extinct, it is lost forever. Short of science-fiction remedies, we will never get it back. As far as Lamarck was concerned, extinction was not all that serious; perhaps sperm whales might go extinct, either terminally or by evolving into some other species, but they would re-evolve.

Invertebrate Animals
I. Infusorians
II. Polyps
III. Radiarians
IV. Worms
V. Insects
VI. Arachnids
VII. Crustaceans
VIII. Annelids
IX. Cirrhipedes
X. Molluscs
Vertebrate Animals
XI. Fishes
XII. Reptiles
XIII. Birds
XIV. Mammals

Figure 3.1. The arrangement and classification of animals according to the order most in conformity with the order of nature. (From J. B. Lamarck, *Zoological Philosophy* [1809], pp. 131–33.)

Lamarck represented his conception of the order in nature in the form of both a traditional classification and a treelike diagram. The classification is linear, ranging from infusorians to mammals (fig. 3.1). His diagram, however, is branching (fig. 3.2). Lamarck's classification gives no hint of the branching character of his diagram, and the branching in his diagram concerns only large groups of organisms, not species giving rise to species. How these branchings could be translated into classifications, Lamarck does not say.

Lamarck's chief adversary was the younger and much more adept Léopold Chrétien Frédéric Dagobert Cuvier (1769–1832). Cuvier preferred to be called simply "Georges Cuvier" although it was not his legal name. He was born in the duchy of Württemberg, of French-speaking parents, and received his training in Germany. Not until the beginning of 1795 was Cuvier able to make it to Paris, then the center of European culture. In the aftermath of the Revolution and the ensuing Terror, the established system of patronage in science was laid waste. Scientists who were too closely connected to royalty had to flee or to go into hiding to save their lives. As a result, opportunities for advancement opened up for young scientists who happened to have the appropriate backgrounds (Outram 1984). One young scientist who initially rose rapidly in the French scientific community was Geoffroy Saint-Hilaire (1772–1844). In 1792, at the age of twenty, Geoffroy was admitted to the Institut de France (Académie des Sciences after 1815). Both Lamarck and Geoffroy were instrumental in bringing Cuvier to Paris from Normandy—against the advice of several friends, it must be added. Within a year of arriving in Paris, Cuvier was also admitted to the Institute. By 1803 he was made permanent secretary of the First Class.

At first Cuvier was on good terms with both Lamarck and Geoffroy. He even coauthored papers with Geoffroy. The turning point in their relationship occurred

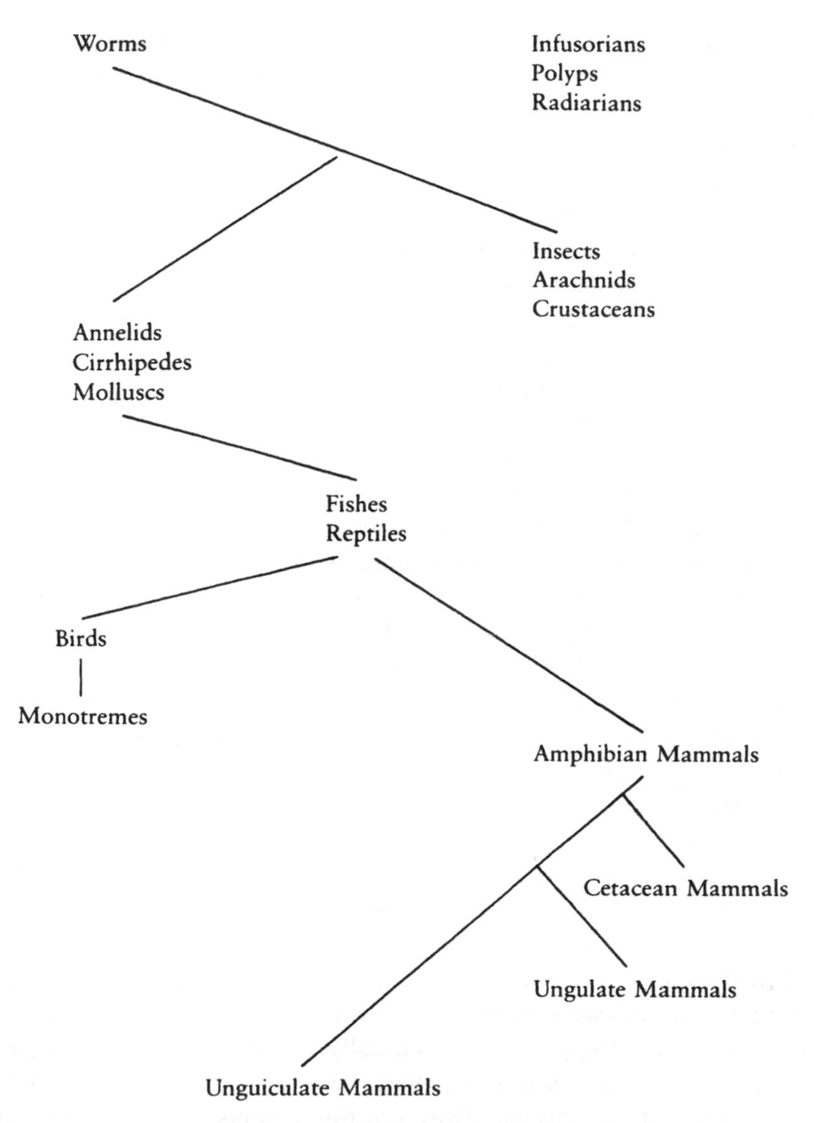

Figure 3.2. Lamarck's table showing the origin of the various animals. (From J. B. Lamarck, *Zoological Philosophy* [1809], pp. 179.)

when Geoffroy joined Napoleon and his troops in their conquest of Egypt. Cuvier remained in Paris, where he solidified his power. When his military campaign went badly, Napoleon slipped out of Egypt, leaving his plague-ridden troops and camp followers behind. Geoffroy was one of those abandoned in Egypt. As his eyesight began to fail, he wrote Cuvier letters pleading for help in getting back to France. There is no indication that Cuvier answered his friend's letters, let alone interceded on his behalf. However, Geoffroy eventually did obtain his freedom and returned to Paris, where Cuvier was in firm control of a large segment of French science. Even so, he declined to help Geoffroy regain a position in the Parisian scientific community. Needless to say, it did not take long for Cuvier and Geoffroy to become professional enemies. To be sure, many substantive issues divided the two men, but their subsequent feuds were made only more bitter by what Geoffroy perceived as less than generous treatment at the hands of a man whom he had befriended (Appel 1987).

Cuvier was first and foremost a comparative anatomist. According to Cuvier, each organism is a delicately balanced, organized system. If any significant variation occurs in its parts, it perishes. As he put it (1812: 95), "Every organized creature forms a whole, a unique closed system, whose parts mutually correspond to one another and concur toward the same definite action by a reciprocal reaction." It was his belief in the correlation of parts that allowed Cuvier to reconstruct extinct species from their incomplete fossil remains and gave rise to the story of a student who burst into Cuvier's bedchamber in the middle of the night, dressed as Satan, exclaiming, "I am the devil, and I am going to eat you!" Cuvier quietly sneered, "Horns, cloven hooves, a tail: herbivore. You can't eat me." The story is surely apocryphal, not because the application of Cuvier's principles is mistaken, but because no one, least of all a student, would dare disturb the great Cuvier in his bedchamber.

Cuvier believed that organisms as functional systems are narrowly constrained by their conditions of existence. Change those conditions, and the organism will die. The fossils that Cuvier and his workers discovered in the Paris basin appeared quite suddenly in the strata they occupied. Cuvier concluded that these species had been abruptly extinguished by a series of limited though major catastrophes. Because no new species were created to take their place, the earth was gradually being depleted of its original stock of species. What appeared to be new species actually were migrants from other parts of the earth. For the time, Cuvier's catastrophism was quite a radical position and was one reason why Lamarck was driven to his peculiar version of transformism. According to Cuvier, species go extinct and no new ones are created to take their place. According to Lamarck, genuinely new species are neither created nor go extinct, at least not permanently. Another difference between Lamarck and Cuvier was that Lamarck divided animals into vertebrates and invertebrates, while Cuvier subdivided them into four great *embranchements*—Vertebrata, Mollusca, Articulata, and Radiata.

In the history of science, Copernicus is usually honored as having initiated the first great scientific revolution in astronomy when he claimed that the sun, not the earth, occupies the center of the universe. However, this was just about the only alteration that Copernicus made in the trappings of Ptolemaic astronomy. For example, he continued to utilize Ptolemaic circular motion and its ensuing epicycles. It was Kepler who abandoned circular motion for ellipses. Given the relative importance of these two modifications, Hanson (1961) demoted Copernicus's contribution to a "disturbance" and elevated Kepler's modification to the status of a true "revolution." Shortly thereafter, Foucault (1966) passed a similar judgment on the relative importance of the contributions made by Cuvier and Darwin. According to Foucault, the real revolution in biology occurred not when Darwin argued for the transformation of species but when Cuvier broke up the Great Chain of Being into four distinct *embranchements*. Cuvier's revolution preceded Darwin's minor disturbance.

Of course, Cuvier was not the first to break up the Great Chain of Being anymore than Darwin was the first to suggest that species evolve, or Copernicus was the first to place the sun in the middle of the universe. In the history of science, no one is ever the first to do anything. But Cuvier was the first to obtain a wide following for the practice. Changing from circles to ellipses or from one chain to four *embranchements* may seem pale in comparison to moving the earth from the center of the universe or convincing biologists that species change through time, but the tacit assumptions that Kepler and Cuvier overturned were among the most fundamental in Western thought. However, though Hanson and Foucault have a point, scientists then as now are somewhat more impressed with empirical than with formal achievements.

Throughout Lamarck's life, Cuvier combated what he took to be the older man's irresponsible speculations, first by ignoring them, then by objecting to them in print and denigrating them in private. In Cuvier's most widely read publication, the *Discours préliminaire sur les révolutions du globe,* he repeatedly attacked Lamarck. However, it was in his infamous *Eloge* that Cuvier did his greatest damage to his predecessor's reputation as a scientist. According to Cuvier, not only was Lamarck mistaken with respect to his scientific views he also had strayed from the path of true science. He was not a mistaken scientist; he was no scientist at all. Cuvier used the same tactic with respect to Geoffroy. According to Cuvier, science is the collection of positive facts, not idle speculation about possible causes. In this connection Geoffroy seemed almost to go out of his way to help his enemy portray him as irresponsibly speculative.

While he was trapped in Egypt, Geoffroy formulated all sorts of theories on topics such as the nature of light, but his friends convinced him not to publish them but instead to stick to his area of expertise—comparative anatomy. In 1818 he published his famous *Philosophie anatomique.* Although reviewers of Geoffroy's book complained of what they took to be the gratuitous and speculative physiological explanations he gave for the patterns he discovered, they were generally impressed. However, while Geoffroy was examining some Jurassic fossil reptiles which he expected to be quite similar to the Mesozoic fossils of *Plesiosaurus* (a sort of long-necked swimming dinosaur), he was amazed to discover that they appeared even more similar to living crocodilians. In 1833 he published a paper suggesting that changes in the environment had acted directly on the embryos of these earlier reptiles, releasing a potential already present in them to take on the form of present-day crocodiles. But more importantly, Geoffroy began to urge that all animals could be assimilated to a single plan or archetype, rather than to Cuvier's four *embranchements.*

In 1830 the disagreements between Geoffroy and Cuvier erupted into a public confrontation before the members of the Academy of Science. With the bias of hindsight, the famous debate between Geoffroy Saint-Hilaire and Cuvier has been portrayed as centering on evolution, Geoffroy arguing for it, Cuvier against; actually it concerned the number of archetypes necessary to capture the patterns exhibited by animals. Just as Goethe had proposed a single structural plan for all plants, Geoffroy thought that he perceived a single plan exemplified in all animals, both living and extinct. According to Geoffroy, the vertebrae in vertebrates are homologous to the integuments of insects. Cuvier and Geoffroy also disagreed about the relative priority of structure and function. Cuvier thought that function determines structure, while Geoffroy insisted that structure determines function. That the substance of this famous dispute seems unimportant to us today merely indicates how different our view of science is from the one within which Cuvier and Geoffroy were operating.

In the conduct of their dispute both Cuvier and Geoffroy behaved badly, conduct that the popular press followed with great interest. By the time that it was over, professional naturalists found themselves on Cuvier's side. The general public tended to side with Geoffroy. It was at this time that Cuvier wrote his infamous *Eloge* of Lamarck. Just as Huxley used his review of Chambers's *Vestiges of the Natural History of Creation* to attack Owen, Cuvier's *Eloge* was directed as much against Geoffroy as against Lamarck. Although Cuvier died in 1832, before he was able to read his biographical memoir before the French Academy of Science, Geoffroy lived on for another dozen years. In later life he became an increasing embarrassment to his colleagues as he resurrected some of the theories which he had conceived while he was in Egypt and as he insisted on his privilege as a member of the Academy to read these interminable papers to largely empty seats (Appel 1987).

Cuvier used two chief weapons in his arguments with Lamarck and Geoffroy. One was to portray himself as a hardheaded empiricist who began with the massive accumulation of positive facts. Any general hypothesis he might enunciate grew out of these facts. He was the true scientist, while Lamarck and Geoffroy were idle speculators. Cuvier also attempted to discredit Geoffroy's views by identifying them with the brand of science so popular in Germany at the time, the sort epitomized in the work of Goethe (1749–1832) and Lorenz Oken (1779–1851). Just as Geoffroy explained the basic structure of all animals in terms of the transformation of the primordial vertebra, Goethe explained the organization of plants by means of the primordial leaf. Although such views gained wide popularity in Germany, they did not travel well, either to France or to Great Britain. (For a discussion of how the brands of ideal morphology of Goethe, Oken, and Owen differed from each other, see Brady 1987.)

In the debate between Cuvier and Geoffroy, neither the existence nor the explanatory adequacy of archetypes were issues in dispute. Their disagreement was over the number and particular makeup of the archetypes needed. The story was quite different in Great Britain. The issue was the legitimacy of archetypes as explanatory devices in science in the first place. British scientists were led to ask in what sense reference to archetypes could explain anything. While the debates between Geoffroy and Cuvier were going on, a young British comparative anatomist was studying in Paris—Richard Owen. As Owen was to discover when he tried to import ideas such as these into Great Britain, British scientists were more than a little skeptical. They found his comparative anatomy first-rate but passed over what they took to be his more metaphysical musings in silence. Most British systematists proceeded in a more empirical fashion to see what patterns emerged, no matter how variable and unsystematic these patterns might turn out to be. These systematists had general views that influenced their work. For example, they were not satisfied until they found one or more characters that uniquely defined their taxa. But these general views remained largely tacit. The only systematist who was able to rouse much interest in any sort of an elaborate system of classification in Great Britain was William Sharp Macleay (1792–1865).

Macleay began by noticing certain parallels between two subgroups of species in a genus of beetle (*Scarabaeus*). If the species in these two subgroups are ordered into two series, striking parallels can be discerned between corresponding members of the series. These parallels become even more striking if the first and last members of each series are brought together so that each subgroup forms a circle. These two circles then intersect at a single point. Here at last was the clue needed to reveal the true order of nature. Past classifications had failed because they attempted to represent in a linear fashion relationships that are actually circular. Perhaps planets do not travel in perfect circles, but plants and animals can be classified in nested sets of interlocking circles. Macleay subdivided Animalia into five circles arranged in a circle (fig. 3.3). On one side, Vertebrata meets Mollusca; on the other side it joins Annulosa (insects, spiders, etc.). At each point of intersection, Macleay placed what he termed "osculant" groups. For example, cephalopods are the osculant group between Vertebrata and Mollusca,

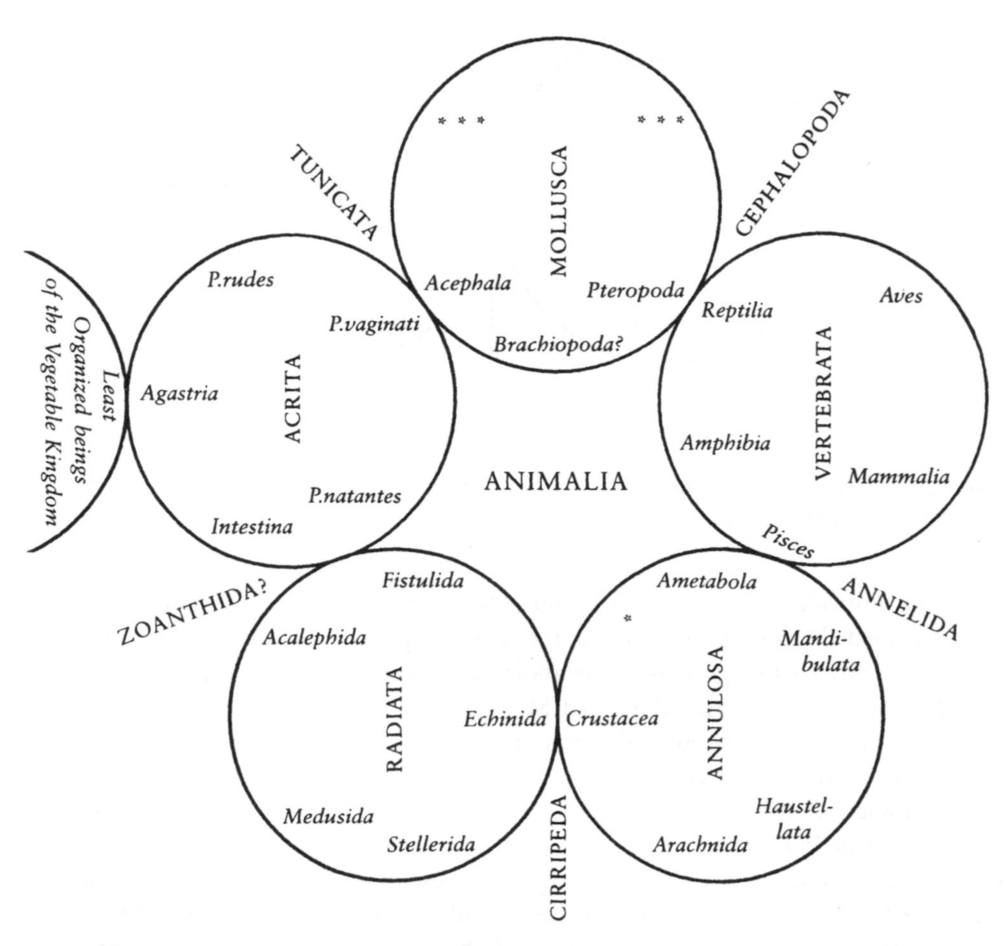

Figure 3.3. Macleay's classification of Animalia, Annulosa lower right. (From W. S. Macleay, *Horae Entomologicae* [1819], 1: 318.)

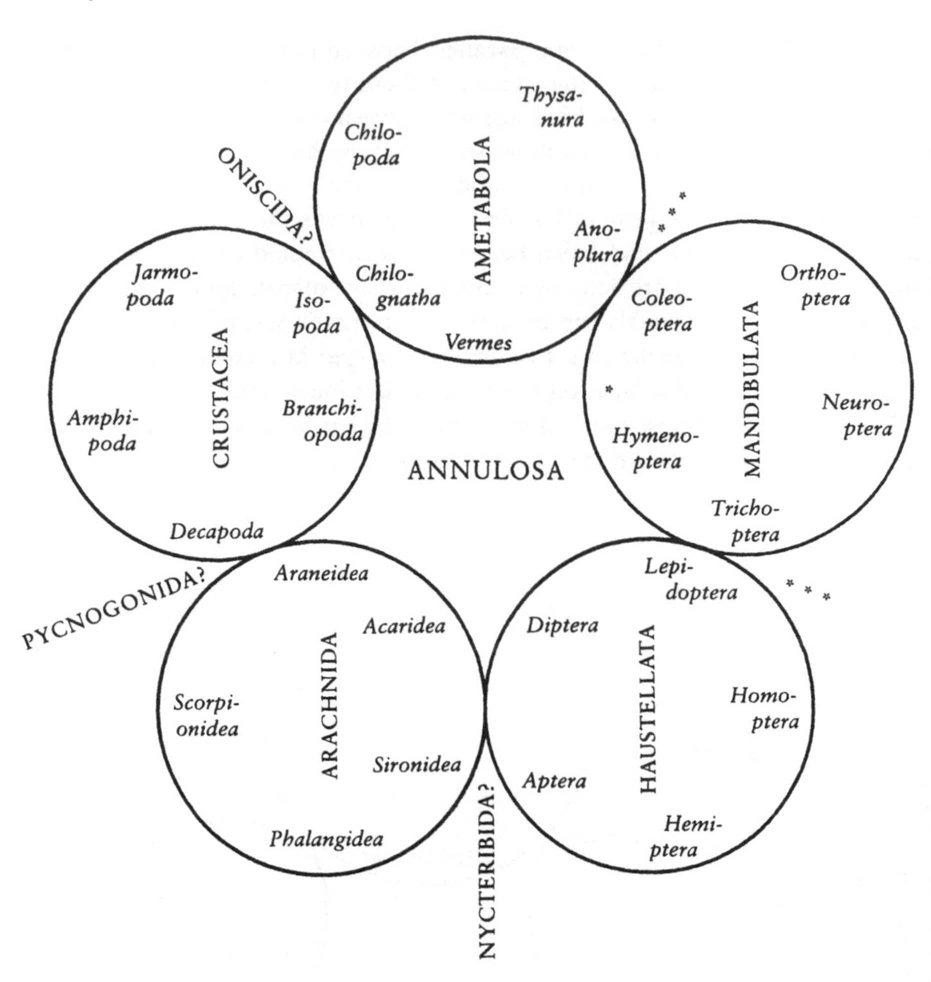

Figure 3.4. Macleay's classification of Annulosa, Mandibulata upper right. (From W. S. Macleay, *Horae Entomologicae* [1819], 1: 390.)

while the annelids are the osculant group between Vertebrata and Annulosa. Macleay then repeated the process for each of the five groups of animals, e.g., he subdivided the Annulosa into five groups—Ametabola, Crustacea, Arachnida, Haustellata, and Mandibulata (fig. 3.4). He subdivided Mandibulata in turn into five smaller groups, and so on (fig. 3.5). Appropriately his system came to be known as "Quinarianism."

Soon after Macleay (1819, 1821) set out his views, a group of naturalists formed the Zoological Club at the Linnean Society to discuss Macleay's system, and in 1825 the *Zoological Journal* began to appear, in large part to foster Quinarianism. For a while at least, it looked as if a new, progressive research program was well underway, but in 1826 Macleay left for Cuba and eventually settled in New South Wales. During this period a fire destroyed most copies of the second volume of his major work. Thereafter, his system was known chiefly through William Swainson's (1835) more popular exposition. The impact of any contribution to science is always difficult to

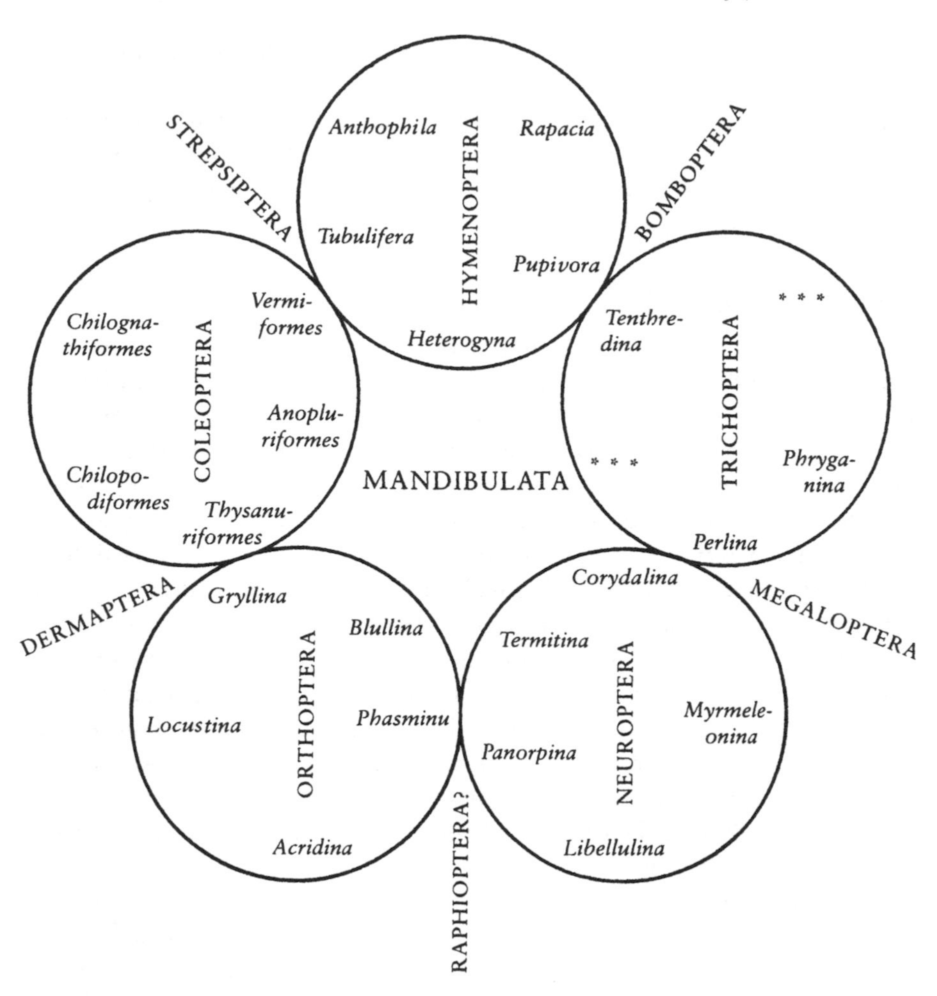

Figure 3.5. Macleay's classification of Mandibulata. (From W. S. Macleay, *Horae Entomologicae* [1819], 1: 439.)

guage. Quinarianism was never dominant among British systematists, but it also was not dismissed out of hand. As fate would have it, Darwin's barnacles formed one of Macleay's osculating groups between Radiata and Annulosa (fig. 3.3), but at the time Darwin's chief concern was the species question. What mechanism could produce such geometrically symmetrical patterns? Certainly not natural selection (Winsor 1976, Mayr 1982).

The young Huxley was much taken with Macleay when the two men met in Sydney in 1847. Huxley was extremely impressed when he discovered that certain parallels he had noticed between hydrozoan and anthozoan polyps fit perfectly into Macleay's system. In general, however, Macleay's disciples were no match for the big guns arrayed against them. At the 1843 meetings of the British Association for the Advancement of Science, some of the leading naturalists in Great Britain roundly condemned Quinarianism as pseudoscience. Nature did not fit neatly into geometric patterns. With an unerring instinct for lost causes, Chambers nonetheless built Quinarianism into his *Vestiges of the Natural History of Creation* (1844). Chambers's endorsement only added to the poor reputation of Quinarianism. Later, in a letter to Owen about the fate of his newly published *Origin of Species,* Darwin half-seriously related his fear that his own theory might suffer the same fate as Macleay's Quinarian system. If Owen had had his way, it would have (Winsor 1976, Rehbock 1983).

The Revolution That Failed

By the time Darwin came on the scene, taxonomy had lost much of its early status. The order latent in the living world seemed as apparent as ever, lying just beneath the surface of appearances. The trick was to find just the right combination of characters to make everything fall into place. A few systematists devised elaborate explanatory schemes from which natural classifications were to flow. They were not very successful. As a result, other systematists condemned such a priori theorizing. They argued that systematists should extract systematic patterns inductively, letting facts speak for themselves. Most systematists did classify in a highly empirical way, studying their specimens, discerning bits and pieces of the divine plan. However, no system materialized. No two systematists working on the same organisms seemed to be able to discern the same pattern. Successive generations of systematists churned their taxa as regularly as stockbrokers today churn the portfolios of their clients, and with much the same effect. One result of adequate scientific methods is, supposedly, that different workers can sooner or later come to agree with each other about nature. This result continued to elude systematists, regardless of their methodological proclivities. Darwin proposed to put an end to all this wrangling.

The two most fundamental problems in systematics are the species problem and the ordering of species into natural classifications. Darwin thought that he had solved both problems. If he was right and species evolve, then systematists were mistaken in attempting to find some way to distinguish unequivocally between species and varieties, because no such unequivocal distinction exists. Varieties are merely incipient species. Not all varieties become species, but all species at one time were varieties. Furthermore, if speciation is as gradual as Darwin thought, then at any one time all stages of the speciation process should be discernible. Hence, the difficulties that systematists continued to have in distinguishing species and varieties were only to be expected. If species must be eternal and immutable to be "real," Darwin concluded that they are not real. If the boundaries between species must be sharp for species to be "real," Darwin concluded that species are not real. If varieties must be clearly distinguishable from species for species to be "real," then Darwin concluded once again that species are not real. But species are not nothing either. After all, they are the things that evolve. Darwin was never able to see his way clearly on the species question. Sometimes he was inclined to think that failure to breed successfully was the relevant consideration, sometimes the presence of differentiating characters was what counted (Kottler 1978). If the two criteria always covaried, there would be no problem, but they do not.

In *Origin of Species* Darwin (1859: 484) predicted that when his views were generally admitted, we can dimly foresee that there will be a considerable revolution in natural history. Systematists will be able to pursue their labours as at present; but they will not be incessantly haunted by the shadowy doubt whether this or that form be in essence a species. This I feel sure, and I speak after experience, will be no slight relief. The endless disputes whether or not some fifty species of British brambles are true species will cease.

Darwin could not have been more mistaken. Systematists did not skip a beat in their arguments over either the species question in general or the status of particular species. If anything, the acceptance of evolution intensified debates over the nature of species, debates that continue down to the present. Present-day adherents of one solution or other of the species problem believe that the debate is over. Their arguments are decisive. Some of these disputants interpret their solution as the one implicit in Darwin. Hence, on this score, the Darwinian revolution in systematics succeeded, but even these authors admit that success had to await their own efforts. Darwin won but not right away. Others insist that their views about species are incompatible with those of Darwin. Hence, Darwin was mistaken.

Darwin also thought (1859: 420) his theory would lay to rest disagreements over the right away to arrange species in classifications:

community of descent is the hidden bond which naturalists have been unconsciously seeking, and not some unknown plan of creation, or the enunciation of general propositions, and the mere putting together and separation of objects more or less alike.

For Darwin, the bond naturalists had been seeking *was* hidden. Extant species are similar to each other because they are descended from common ancestral species, but the past is past. Phylogeny must be reconstructed. On the basis of current records, paleontologists must infer past phylogenetic relations. The chief form of data about the past is character distributions, but Darwin insisted that these character distributions involve more than just degrees of similarity and dissimilarity. Some similarities are true homologies, others are deceptive analogies. One task of the systematist is to distinguish between the two.

Darwin's solution to the problem of classification was not accepted as widely and as enthusiastically as Darwin had hoped it would be. The chief sticking point was the reconstruction of phylogeny. Disputes over the true phylogeny of a group were no less endless than disputes over degrees of nonevolutionary affinity. Huxley (1874: 101) for one was willing to reject phylogeny as the proper basis for classification:

Darwin, by laying a novel and solid foundation for the theory of Evolution, introduced a new element into Taxonomy. If species, like an individual, is the product of a process of development, its mode of evolution must be taken into account in determining its likeness or unlikeness to other species; and thus "phylogeny" becomes not less important than embryogeny to the taxonomist. But while the logical value of phylogeny must be fully admitted, it is to be recollected that, in the present state of science, absolutely nothing is positively known respecting the phylogeny of any of the larger groups of animals. Valuable and important as phylogenetic speculations are, as guides to, and suggestions of, investigation, they are pure hypotheses incapable of any objective test; and there is no little danger of introducing confusion into science by mixing up such hypotheses with Taxonomy, which should be a precise and logical arrangement of verifiable facts.

For Huxley, that a particular character accurately described a group of organisms and that two groups shared the same character are "verifiable facts." As such, they could be used to construct a classification. The additional claim that these characters are evolutionary homologies depends on reconstructing the past, and this endeavor is too speculative for Huxley. Although Darwin (1859: 357) was no doubt right that "each species has proceeded from a single birthplace," these birthplaces were not knowable with sufficient certainty to allow such considerations to enter into something as important as taxonomy.

This was how the situation remained for decades after Darwin's great revolution. Some systematists claimed that their classifications were "phylogenetic," others not, but no one was very clear about what this claim entailed. Darwin himself had his doubts. For example, in a letter to Hooker dated December 23, 1859, Darwin (F. Darwin 1899, 2:42) remarked with respect to the taxonomic system of the French naturalist, Charles Naudin (1815–99), "His simile of tree and classification is like mine (and others), but he cannot, I think, have reflected much on the subject, otherwise he would see that genealogy by itself does not give classification." A century later taxonomists in their feuds over the proper way to classify plants and animals are absolutely sure that in this and other quotations Darwin was endorsing their particular view of classification. Darwin was really a cladist or an advocate of evolutionary systematics. I, for one, find his scattered comments somewhat cryptic. Even so, in remarking that genealogy by itself does not give classification, he could not have been more prescient.

The relationship between a branching phylogenetic tree and the successive subdivisions of a hierarchical classification could not seem more patent. Yet, it is not. Whether Darwin noticed the seductive though deceptive similarity in form between phylogeny and classifications is not clear, but he does discuss formal problems of classification. Although this discussion is somewhat technical, it must be introduced here because the deceptive similarity between the overall form of phylogeny and classification has emerged as a central issue in later disputes and because it is likely to deceive the reader as well. So much attention was paid to difficulties in reconstructing phylogenies that no one noticed the difficulties inherent in representing phylogeny in a classification (assuming that one already has a reasonably good phylogeny in hand).

In the *Origin of Species* Darwin presented no phylogenetic reconstructions. He had enough to do to convince his fellow scientists that species evolve without getting himself embroiled with specialists about particular relationships within specific groups. But he did present a hypothetical phylogeny to give a general idea of the overall structure of phylogenetic development (fig. 3.6). Darwin began by postulating eleven genera in the Silurian epoch (A through L). Seven of these genera rapidly went extinct (B, C, D, G, H, K, L), but species in the other four genera (A, E, F, I) succeeded in leaving descendants down to the present. Two of these (A and I) proliferated extensively, while the other two (E and F) remained largely unchanged. In his diagram, Darwin indicates 58 species, represented by lowercase letters with superscripts. Many other species no doubt existed, but these are the only ones that have left traces. How should these species be classified? Darwin (1859: 422) answers as follows:

This natural arrangement is shown, as far as is possible, in the diagram, but in much too simple a manner. If a branching diagram had not been used, and only the names of the groups had

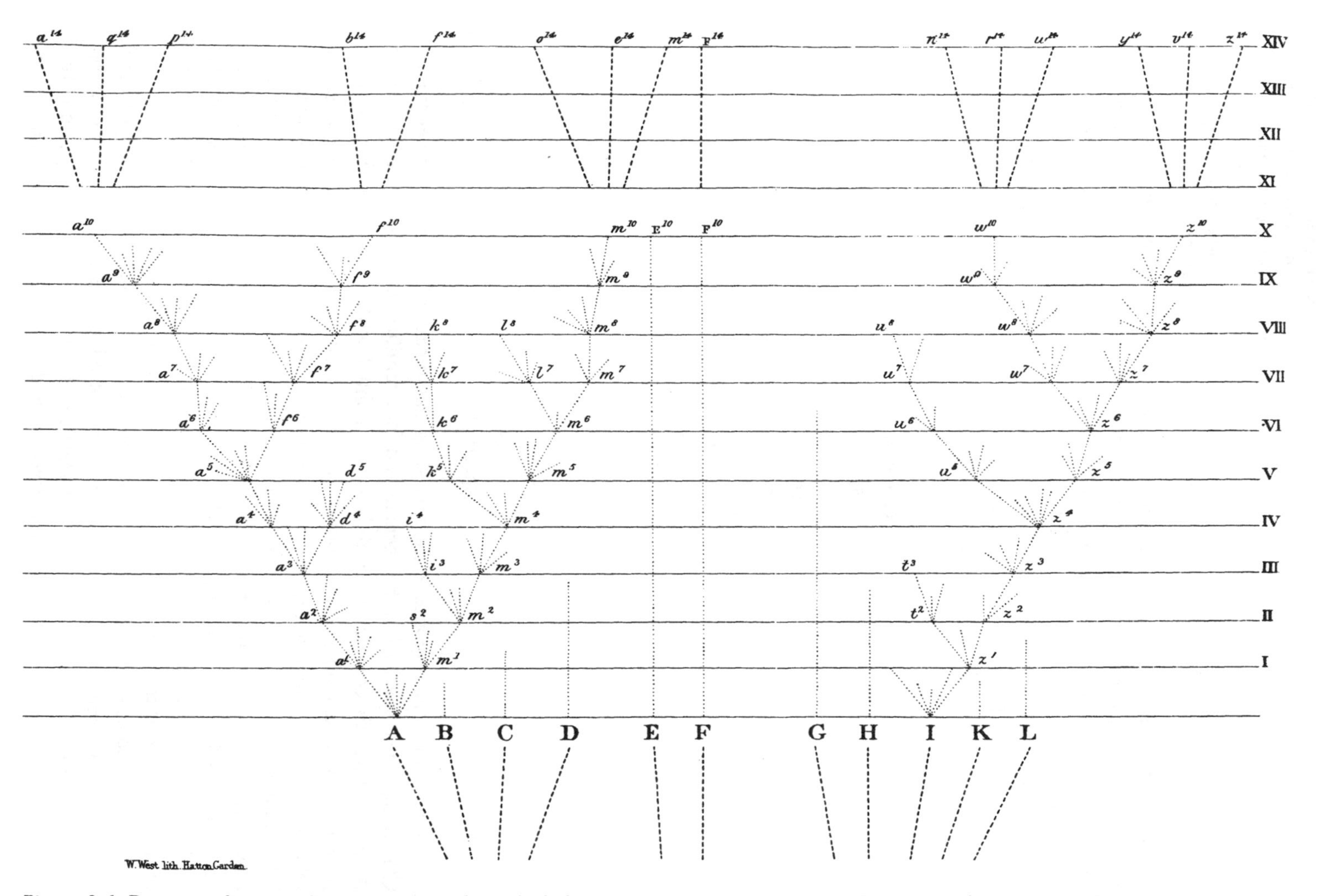

Figure 3.6. Darwin's diagram depicting a hypothetical phylogeny. (From C. Darwin, *On the Origin of Species* [1859], p. 117.)

been written in a linear series, it would have been less possible to have given a natural arrangement; and it is notoriously not possible to represent in a series, on a flat surface, the affinities which we discover in nature amongst the beings of the same group. Thus, on the view which I hold, the natural system is genealogical in its arrangement, like a pedigree; but the degrees of modification which the different groups have undergone, have been expressed by ranking them under different so-called genera, sub-families, families, sections, orders, and classes.

Because of a dispute that broke out a century later among systematists over the proper way to classify organisms, Darwin's reference in the preceding quotation to "degrees of modification" has become an important bone of contention. However, my concern here lies in the ways in which the relations exhibited in Darwin's hypothetical genealogy can be expressed unequivocally in a hierarchical classification. Limiting ourselves for the moment just to extant species, i.e., those arrayed along the top of Darwin's diagram with the superscript 10, they might be classified as follows:

Family 1
  Genus A
    species a^{10}
    species f^{10}
  Genus M
    species m^{10}
Family 2
  Genus E
    species E^{10}
Family 3
  Genus F
    species F^{10}
Family 4
  Genus I
    species w^{10}
    species z^{10}

The most obvious response to the preceding classification is that it is too elaborate. Classifying seven species into five genera and four families might seem a bit excessive. As justified as this complaint certainly is, the point I wish to make is that even this overly elaborate classification contains very little information about the phylogenetic relationships of these seven species. Less elaborate classifications would contain less information. The preceding classification indicates that species a^{10} and f^{10} are more closely related to each other than either is to species m^{10} and that these three species are more closely related to each other than to any of the other extant species indicated. It also indicates that species w^{10} and z^{10} are more closely related to each other than they are to any of the other extant species indicated. And that is all.

In the preceding classification, all the species treated are contemporaneous with each other. What happens when the ancestor-descendant relation is included? For example, in the first fork on the left-hand side of Darwin's hypothetical genealogy, an extinct species a^5 goes extinct as it gives rise to two descendant species, a^6 and f^6. How is this relationship to be represented in a classification? Darwin never says. The following two possibilities suggest themselves:

(i)	(ii)
Genus A	Family I
species a^5	Genus A
species a^6	species a^5
species f^6	Genus B
	species a^6
	species f^6

Because of the presence of superscripts, one might easily infer that in classification (i) above, a^5 is ancestral to a^6 and f^6, but these superscripts are an artifact of Darwin's mode of representation. In real classifications, only species names include any information about their taxonomic relationship. For example, from the names *Ursus arctos* and *Ursus americanus* alone, one can infer that these two species of bear are closely related. They both belong to the same genus. But above the species level, no such hints exist. Certain endings imply category level; that is all. Names such as Hemichordata, Thallophyta, and Nematoda imply nothing about either taxonomic or phylogenetic relations. And even the names of species imply nothing beyond their own genus. For example, from the name *Thalarctus maritimus,* no one would be led to suspect that this species of bear, the polar bear, is thought to have evolved from a species in the genus *Ursus.* This feature of taxonomic names might be remedied by adopting the convention that the first name in any series of species-names be considered the common ancestor. Without such a convention, classification (i) indicates nothing about ancestry.

The situation in classification (ii) is different. It implies that something is special about species a^5. It is placed by itself separate from the other two species listed. One possibility is that it is the common ancestor of the other two species. However, if this way of arranging taxonomic names is adopted to indicate common ancestry, no way remains for indicating degree of relationship among contemporaneous species. The same mode of representation cannot serve two functions without introducing ambiguity. Thus, by using indentation one can indicate common ancestry *or* phylogenetic relationships among contemporaneous species, but not both. But systematists who viewed themselves as advocates of phylogenetic classifications wanted to indicate both. Although diagrams representing phylogenetic trees and the arrangements of taxonomic names in classifications might appear to be related to each other in a systematic and straightforward way, this appearance is extremely deceptive. I am not claiming that either Darwin or his contemporaries noticed this problem or discussed it in any but the most tangential way. To the contrary, I have spent so much time investigating this relationship because no one at the time or for many years subsequently noticed the problems inherent in any attempt to make classifications "genealogical."

What then was the fate of Darwin's revolution in systematics? Mayr (1982: 212) states that theoretical "discussions of evolutionary classification in the ensuing century have consisted of little more than footnotes to Darwin. None of Darwin's rules or principles has been refuted and none of any special consequence has since been added." But Mayr (1982: 220) also notes that the "splendid start which Darwin had made in developing a theory and methodology of macrotaxonomy was largely ignored in the post-Darwinian period." If enunciating a view that later workers take to be correct is enough, then the Darwinian revolution in systematics succeeded even though Darwin's theory and methodology were largely ignored in the post-Darwinian period. If getting one's fellow scientists to adopt one's views is also necessary, then Darwin failed miserably until quite recently. Not until a century later did systematists expose these issues to intensive investigation, and the results have been mixed. Some present-day systematists claim to have decisively refuted Darwinian principles of classification. Others claim to have established them just as decisively, but these workers disagree about the precise nature of these principles. It is to this topic that I now turn. But Darwin was successful in one respect: he killed off archetypes. Because we are all, like it or not, descendants of Darwin, a world composed of eternal, immutable archetypes existing in timeless relations to one another is all but incomprehensible to us. Those who attempt to replace ancestors with archetypes have their work cut out for them.

The New Systematics

Systematics and evolutionary biology overlap in two places: the species concept and the relation between phylogeny and classification. If the basic units of classification (taxonomic species) are to coincide with some basic units of the evolutionary process (evolutionary species), then theoretical disputes about the nature of the evolutionary process have clear implications for systematics. Similarly, if biological classifications are somehow to reflect or represent phylogeny, then problems of phylogeny reconstruction are of equal interest to the taxonomist. Advocates of the "New Systematics" concentrated mainly on the former question. As Julian Huxley (1940: 2) expressed his conviction in his contribution to *The New Systematics,* "Fundamentally, the problem of systematics . . . is that of detecting evolution at work." But in retrospect, very little that could be counted as "new" actually appeared in the pages of Huxley's prolegomenon. With the exception of a philosophical paper by the botanist J. S. L. Gilmour, there was nothing much new in the New Systematics.

Both Dobzhansky (1937) and Mayr (1942) emphasized in their seminal works the importance of interbreeding and reproductive isolation in the formation of species. All populations bound together by extensive interbreeding belong unproblematically to the same species, but not all populations between which interbreeding is absent or greatly reduced belong in different species. It is enough that they have the potential to interbreed even if at the moment that potential is not being realized. As Mayr (1942: 120) phrased his famous biological species concept:

> Species are groups of actually or potentially interbreeding natural populations, which are reproductively isolated from other such groups.

Mayr's biological species concept engendered both theoretical and practical objections. The first theoretical objection is that interbreeding is not an all-or-nothing affair. For example, about 30,000 years ago, glaciers from the Alps and Scandinavia divided

Europe in two. Species that had been spread across Europe were consequently subdivided into two geographically isolated populations. When the glaciers receded thousands of years later, these pairs of populations came into contact again. In the interim all possible degrees of reproductive isolation had developed. Some pairs of populations had become totally incapable of interbreeding. They had clearly become separate species. Other pairs merged together with no difficulty. They had originally been a single species and remained one. But in some cases, reproductive isolation was only partial. For example, when the descendants of what had once been a single species of crow met, they mated with no problem but strangely enough did not merge (introgress). Instead they formed a hybrid zone running down the center of Europe, the hooded crow on one side, the carrion crow on the other. This hybrid zone has remained roughly where it is for thousands of years and has never expanded in width to more than 75 to 100 kilometers. Even though reproductive isolation is incomplete, the gene flow between these two groups of crows appears not to be significant. They are evolving in relative independence of each other. Hence, they should be considered separate species. Although decisions can be made in particular cases, the general problem remains how much interbreeding is enough.

Life would be simpler if reproductive isolation, once attained, was never lost, but in plants at least life is not so simple. Species not only split but also merge. According to recent estimates, 47 percent of angiosperms and 95 percent of pteridophytes (ferns) are allopolyploids (Grant 1985). In plants, pollen from one species commonly fertilizes ova from another species. Normally nothing comes of such matings because the chromosomes are so different that they cannot pair at meiosis. However, sometimes the chromosomes in these hybrids double in preparation for mitosis, but mitosis does not occur. Now each chromosome has a mate—the other half of its former self. Mitosis can now take place. The result is a genuine species of hybrid origin. Such hybrid species have a single, unique origin, but they stem from two ancestral species instead of one. Estimates of the proportion of plant species in general that are of hybrid origin run as high as 30 or 40 percent.

A final theoretical problem for delimiting species by means of interbreeding and reproductive isolation concerns asexual organisms. Millions of organisms exist that are capable of exchanging genetic material at most parasexually and then only rarely (Jackson, Buss, and Cook 1986). Blue-green algae are a good case in point. Do they form species? Dobzhansky (1937: 316–21) was willing to bite the bullet and admit that they do not. Like hybrid swarms, asexual organisms belong to no species whatsoever. They evolve, but they do so without forming species. To those who want to recognize species of asexual organisms at least for taxonomic purposes, Dobzhansky (1951: 275) responded that nothing is saved by this stratagem except the word "species." Although later authors modified Dobzhansky's conclusions with respect to asexual species, this is how matters stood in the early years of the new systematics.

The most pervasive criticisms of the biological species concept, however, were not theoretical but pragmatic. Assuming that one knows how much interbreeding is enough to maintain unity within a species, actual interbreeding can be tested only in populations of extant species and only when they come into proximity with each other. In cases of geographically isolated populations of extant species, the most one can do is transport individuals from one population to another to see what happens. For earlier populations of extant species and all extinct species, neither actual gene flow nor the potentiality for it can be tested. Instead, inferences must be made from morphological gaps. All sides readily admitted that morphological gaps do not always coincide with reproductive isolation. Both polytypic and sibling species are common among extant species. There is no reason to expect that they were any less common in the past. Hence, in such cases, inferences from morphology to reproductive isolation will lead to error. As theoretically appropriate as the biological species might be, its critics complained that it was not sufficiently "operational." The question that remained was the extent to which scientific terms, especially those that are central to scientific theories, must be operational. If all terms in science must be operationally defined, then the biological species concept is in trouble. However, if terms that are not totally operational are nevertheless scientifically acceptable, then possibly the biological species concept belongs to this class (Hull 1968).

The major voice raised in dissent to Huxley's proposed New Systematics was that of J. S. L. Gilmour (1940). Gilmour began his contribution to Huxley's volume by noting that a new research program termed "logical positivism" had recently arisen among physicists, philosophers, and mathematicians. In agreement with the new program, Gilmour argued that the fundamental principles of classification cannot be formulated in isolation from an adequate epistemological theory of how scientists obtain knowledge of what is commonly termed the "external world." Accordingly he distinguished between sense-data and the concepts that we construct to clip these sense-data together. Sense-data are objective and unalterable. The "clips" are of our own making. Thus, Gilmour (1940: 467) concluded that all classifications should be regarded as "rational concepts constructed by the classifier to clip together certain sense-data experienced by him." Natural classifications group individuals that have a large number of attributes in common; artificial classifications group individuals that have only a small number of attributes in common.

But what of gene flow and phylogeny? According to Gilmour (1940: 469), "A species is a group of individuals which, in the sum total of their attributes, resemble each other to a degree usually accepted as specific, the exact degree being ultimately determined by the more or less arbitrary judgment of taxonomists." Organisms that look the same are the same, regardless of the possibility of phylogenetic convergence and divergence. In justifying this position, Gilmour distinguished between two senses of "phylogenetic relationship." One is the group concept which concerns the relationships between groups; e.g. group A is more closely related to group B than either is to any other group, if they share a more recent group ancestor. The other is a lineage concept, which concerns the actual genealogical relationships between organisms. Gilmour (1940: 470) argued that decisions about how closely related two groups are cannot serve as an independent criterion for constructing classifications, because "we must make our groups *before* constructing our phylogeny." Any attempt to construct phylogeny by means of correlating attributes that uses this phylogeny as an explanation of the correlation of these very same attributes is simply "arguing in a circle." Gilmour did not think that such circular reasoning infected the lineage concept. He thought that the fossil record afforded criteria in addition to attributes for constructing lineages. However, even though Gilmour believed that a phylogenetic classification based on the lineage concept is logically impeccable, it is only a special-purpose classification. "A natural classification is that grouping which endeavours to utilize *all* the attributes of the individuals under consideration, and is hence useful for a very wide range of purposes" (Gilmour 1940: 472).

The other biologist influenced at the time by logical positivism was J. H. Woodger. One day in 1928 while browsing through the science library of University College, London, Woodger happened upon a copy of the *Principia Mathematica* (2d ed., 1925) by A. N. Whitehead and Bertrand Russell. On the train ride home that night, he began to read it and was thrilled to discover the sort of precise language that he could use to transmute traditional biology into a genuine science. He spent the rest of his life turning, as he put it, the Boole-Frege searchlight on the slovenly language so common in biology. One of his chief goals was to eliminate from biological discourse any reference to such mysterious things as properties, relations, sets, and classes. Woodger realized that practicing biologists saw nothing especially problematic about the claim that a particular garden pea has yellow cotyledons, but he did. He wanted to dump such unnecessary baggage. Woodger's (1952: 195) solution was to treat names "not as names of sets but of the objects which are said to be the members of sets."

Because of the esoteric notation that Woodger adopted, he had very little impact on biology. As with *Finnegans Wake,* everyone seems to have heard of Woodger's *The Axiomatic Method in Biology* (1937), but very few have ever read it, and the few who have have not come away very impressed (see Ruse 1975 and Roll-Hansen 1984 for a negative assessment, Williams 1970 and Balzer and Dawe 1986 for a more positive response). The only issue discussed by Woodger that struck a responsive chord at the time concerned the relation between phylogeny and classification. Woodger (1945) began by distinguishing between time-stretches and time-slices in the lives of particular organisms. The life cycle of a particular organism is a sequence of time-slices integrated into a time-stretch, initiated by fertilization and terminated by death. Homologies can then be established by setting up the maximal number of one-to-one correspondences between the parts of particular organisms. Taxonomic groups can be recognized by finding which correspondences have maximum taxonomic distribution. The result is a bauplan. These bauplans are what determine groups (taxa). (The term "bauplan" is a technical one borrowed by Woodger from the German and Anglicized.)

The obvious feature of the preceding discussion is that neither evolution nor phylogeny is mentioned. Woodger (1945: 109) objected to reserving the term "homology" for correspondences between two organisms that share a common ancestor. Defining "homology" in terms of common ancestry is "putting the cart before the horse, because descent from a common ancestor is something assumed, not observed. It belongs to theory, whereas morphological correspondence is observed." On the assumption that similarity in bauplans implies community of descent, ancestor-descendant relations can be inferred, but they are implications of morphology, not vice versa. Evolutionary theory can explain the results of morphological research, but in the order of knowledge purely formal morphological correspondence is primary. To be logical, systematists must admit that they discover the taxonomic relations *first* and only *then* postulate phylogenetic relations.

Most advocates of the New Systematics considered the only impediment to representing phylogeny in classification to be lack of evidence. As Dendy (1924: 241) saw it, if only our knowledge of phylogeny were complete, "we should then doubtless see at once that the taxonomic tree and the phylogenetic tree are, after all, one and the same thing." Julian Huxley (1940: 19–20), like his illustrious grandfather before him, was a bit more skeptical. Although he admitted that "there are certain cases where taxonomy does *not* have a phylogenetic basis," he still thought that, whenever possible, classifications should be considered to have a "phylogenetic background." Mayr (1942: 103) was even more pessimistic, not because of the epistemological problems raised by Gilmour and Woodger, not because of scarcity of data, but because "*no system of nomenclature and no hierarchy of systematic categories* is able to represent adequately the complicated set of interrelationships and divergences in nature" (italics in the original).

There is a sense of the term "sound" in which a tree that falls in a forest makes no sound unless someone hears it. No one at the time seems to have heard Mayr's warning. It tolls forebodingly only in retrospect. No one else noticed the difficulties in representing phylogeny in a hierarchical classification because of the apparent similarity in form between the two. The leading spokesmen of the systematic community declared themselves part of the New Systematics, and that was good enough.

One of the most familiar ways to classify scientists is according to their subjects of inquiry. There are physicists, biologists, and social scientists. Among physicists there are those who study solid-state physics, others who study heat transfer, and so on. Nowhere is this practice carried to such extremes as in systematics. Taxonomic subdivisions are mirrored in the structure of the discipline. Just as there are beetles, fish, and spiders, there are coleopterists, ichthyologists, and arachnologists. In most cases, each of these taxonomically defined groups of scientists has its own society and journal. As a result, communication among systematists is frequently minimal.

The dozen or so systematists who met for breakfast at the 1947 meeting of the American Association for the Advancement of Science were unhappy with the isolation of practicing systematists from each other as well as from the scientific community at large. They also were alarmed by the low esteem in which taxonomic studies were held. In order to bring systematics into the mainstream of science and to obtain their fair share of financial support from the National Science Foundation, they had to organize. They also had to be careful not to be swallowed up in some more inclusive group such as the Society for the Study of Evolution, which had been formed the year before. About the only concerns that systematists shared were their general methods of classification and the need to name their taxa. A journal devoted to nomenclature and the principles of systematics might be able to bridge the taxonomic boundaries that separated systematists.

In 1947 the Society of Systematic Zoology was formed, and in 1952 the Society began to publish its journal, *Systematic Zoology*. R. E. Blackwelder supervised the publication of the first issue, until the president of the society, Alexander Petrunkevitch, an arachnologist from Yale University, was able to find an editor. He settled on John L. Brooks at his own university. In 1953 Ernst Mayr, in collaboration with E. Gorton Linsley and Robert L. Usinger, published the first modern text in systematics, *Methods and Principles of Systematics* (for a list of the annual meetings of the Society of Systematic Zoology, presidents of the society, and editors of the journal, see, respectively, appendices A, B, and C).

Neither Huxley's call for a New Systematics nor the founding of The Society of Systematic Zoology was enough to revitalize the discipline. During its first half-dozen years, less than 20 percent of the pages of *Systematic Zoology* were devoted to discussions of taxonomic theory. Most of the authors of papers on systematic philosophy considered themselves to be "evolutionary" in their outlook, in the tradition of the New Systematics, but were not at all clear as to what this meant. The chief exceptions were R. E. Blackwelder and Alan Boyden of the United States, R. S. Bigelow from Canada, and Thomas Borgmeier, a German living in Brazil. Blackwelder at the time was secretary-treasurer of the Society of Systematic Zoology. In a paper with Boyden, Blackwelder (1952: 31) declared that the "grand object of classification everywhere is the same. It is to group the objects of study in accordance with their essential natures." Borgmeier (1957: 53) agreed, contending that, as the science of order, "systematics is a pure science of relation, unconcerned with time, space, or cause."

These systematists agreed that evolution had occurred and that the distribution of organisms so apparent in the living world is the result of evolution, but they saw no reason to allow such considerations to intrude into taxonomy and found numerous reasons to exclude them. Phylogeny is too difficult to reconstruct, the little fossil evidence available is too spotty, our understanding of the evolutionary process is highly contentious, and characters are basic anyway. If the only data that a systematist usually has available are characters, if everyone always begins with characters, then why not just stick with characters and abandon idle speculation about phylogeny and gene flow, at least in conjunction with classification?

These attacks on the New Systematics elicited little in the way of response. As Blackwelder (1959: 204) complained, anyone who questioned the belief, then so prevalent, that classifications should be based on phylogeny or at least be consistent with it was treated as a "crank not to be taken seriously." Little by little, however, an active research program began to develop. In 1961 G. G. Simpson published his *Principles of Animal Taxonomy.* Mayr's *Principles of Systematic Zoology* followed in 1969. In these two books, evolutionary systematics reached its high-water mark. Although Simpson acknowledged difficulties with Mayr's genetic (biological) definition of "species," he concluded that it was appropriate for time-slices of sexual species but that a more inclusive definition was needed to integrate successive time-slices defined by gene flow into temporally extended lineages. An adequate definition of the species category also had to accommodate asexual organisms. To these ends, Simpson (1961: 153) presented his evolutionary species concept:

> An evolutionary species is a lineage (an ancestral-descendant sequence of populations) evolving separately from others and with its own unitary evolutionary role and tendencies.

According to evolutionary biologists at the time, gene flow is very important in maintaining the cohesiveness of gene pools. Interbreeding promotes unity of evolutionary role. However, in the absence of mating, asexual organisms also seemed capable of having their own unitary evolutionary roles and tendencies. Simpson (1961: 163) concluded, "the evolution of uniparental and biparental populations is different in many important ways. That does not alter the fact that both form species and, by appropriate definition, the same kind of species."

Simpson also addressed the question of the relation between phylogeny and classification. The rallying cry for evolutionary systematics was that higher taxa have to be "monophyletic." At the time monophyly required at least that all members of a higher taxon be descended from a single, immediately ancestral species. Hence, hybrid species posed serious problems. Although they might have a single origin, that origin was in two immediately ancestral species. Hybrids aside, Simpson thought the principle of monophyly that was prevalent at the time was "impractical," and he suggested a weaker notion. According to Simpson (1961: 124), a higher taxon counts as monophyletic if it can be traced back through one or more lineages to a single, immediately ancestral taxon of its own or lower rank. Thus, a genus counts as monophyletic for Simpson if it resulted from two or more species just so long as these species are placed in the same immediately ancestral genus.

The net effect of Simpson's weakening of the requirement of monophyly for higher taxa is that, given a phylogeny, numerous alternative classifications are equally legitimate. According to Simpson (1961: 112), a classification cannot and should not try to *express* phylogeny. In a strict sense, it is not even *based on* phylogeny. The most that one can expect of a biological classification is that it be *consistent* with phylogeny, that it not *contradict* the classifier's views on the phylogeny of the group. Systematics as Simpson construed it was both a science and an art. No cut-and-dried rules exist for the construction of classifications. Instead a systematist has to balance a variety of considerations (for more detailed discussions, see Hull 1964, Wiley 1981b).

In his *Principles of Systematic Zoology* (1969), Mayr expanded upon his earlier position on the nature of species as temporal time-slices of evolving lineages, what he termed the "nondimensional" species concept. According to Mayr (1969: 26), a species is a reproductive community because a "multitude of devices ensures intraspecific reproduction in all organisms," an ecological unit that "interacts as a unit with other species with which it shares the environment," and a genetic unit "consisting of a large, intercommunicating gene pool, whereas the individual is merely a temporary vessel holding a small portion of the contents of the gene pool for a short period of time." In his restatement of his biological species concept, Mayr omitted his earlier reference to "potential interbreeding," because everything that he had said in the past in terms of two populations being potentially interbreeding could be said just as well in terms of their not being reproductively isolated. Reference to "potential interbreeding" was thus redundant. According to Mayr's (1969: 26) reformulated definition:

Species are groups of interbreeding natural populations that are reproductively isolated from other such groups.

Mayr noted that, by definition, asexual organisms do not form populations; as a result, the biological species concept does not apply to them. He went on to acknowledge that groups of asexual organisms do tend to occupy their own ecological niches and perform their own evolutionary roles. In this sense, they form species. Mayr also acknowledged that species as evolving systems extend vertically through time. However, he did not explicitly endorse Simpson's (1961) evolutionary species concept or Simpson's characterization of the biological species concept as a special case of his own evolutionary species concept.

With respect to higher taxa, Mayr emphasized that adaptation is also part of the evolutionary process. Lineages periodically invade new adaptive zones and proliferate. These adaptive grades should be recognized in classifications. As a result, he endorsed Simpson's weakened definition of "monophyly," agreeing that phylogeny is not based on classification nor is classification based on phylogeny. Even so, Mayr (1969: 78) concluded that there is "no excuse for abandoning the evolutionary approach to classification because no man-made, hierarchical system of categories is capable of expressing precisely all the known or inferred facts of evolution."

Summary

In this chapter I have traced all too briefly the history of natural history and systematics from Aristotle to Darwin and from there to the New Synthesis. For Aristotle, species are first and foremost secondary substances. As such, species are as eternal, immutable, and conceptually distinct as triangles and gold. They are part of the fundamental makeup of the universe. Species could no more come and go for Aristotle than gravity could come and go for Newton. Numerous individual organisms might well present borderline cases between two or more species, but the borders in conceptual space remained distinct in the face of such failures. Species as they exist at any one time might well have geographic ranges, but the borders which they form are not the borders that count for Aristotelian species.

The historically efficacious Linnaeus held a basically Aristotelian view of species. He differed from Aristotle only in placing the origin of species in the Garden of Eden and their demise at the Second Coming. In addition, Linnaeus developed a sophisticated theory about the genesis of species that was starkly at variance with Aristotle. On this theory, God created a few primary species in the beginning. Once created, they defined the contours of the living world. In time these primary species hybridized to produce additional species, which hybridized, and so on until they filled out all the slots in God's great map. Linnaeus believed in evolution of a sort—saltative evolution without contingency.

Lamarck and Geoffroy Saint-Hilaire had two things in common: they advocated a form of biological evolution in their later years and neither was a match for Cuvier. According to Lamarck, very simple life forms are generated spontaneously at the base of several trees of life and are infused with a tendency to progress up their respective trees. The contours of these trees are determined, according to Lamarck, by environmental factors. As long as water collects in rivers, pools, and oceans, organisms are presented with the water-air interface and must react to it. For Lamarck, evolution is gradual, though rapid, because heritable changes occur in the space of a single lifetime. The general shape of Lamarck's trees are inherent in nature. However, which path an organism might take at a particular branch point is largely a matter of historical contingency.

Geoffroy viewed organic change through the eyes of an idealist—embryonic change in the context of a world of eternal, immutable forms. The great debate with Cuvier was more a matter of the number of archetypes needed to account for the patterns discernible in the living world than a matter of transmutation. Cuvier succeeded in defeating both men by portraying himself as a true scientist and playing off the tendency of his opponents to present theories which, at the time, seemed highly speculative.

In this chapter I have also argued that systematics and a Darwinian view of biological evolution were not synthesized by Darwin or later Darwinians up to and including the early years of the New Synthesis. Present-day systematists are likely to disagree. My explanation of this false consciousness rests on two considerations: the strong appeal of Darwin as a patron saint and the misleading similarity in form between hierarchical classifications and phylogentic trees. Darwin was so great he *must* have included at least implicitly all subsequent developments in his occasional pronouncements, cryptic as they might seem to the unconvinced. Similarly, Darwin *must* have wrought a major revolution in systematics because he altered its subject matter so profoundly. Even though my discussion in this chapter may well be insufficient to cast doubt on these convictions, later discussions should prove to be more convincing. If Darwin clearly saw the problems inherent in establishing a determinate relationship between phylogeny and classifications, then we would not have needed Mayr, Simpson, and Hennig. But we did.

Also in this chapter I have detailed the sporadic appeal of "idealist" views of nature in the English-speaking world, especially in the work of Woodger, Gilmour, Blackwelder, Boyden, and Borgmeier. Although their opponents dismissed them as "typologists," they viewed themselves as hard-headed "empiricists." They were convinced that sooner or later other systematists would come to see how much damage idle speculation about phylogeny and the evolutionary process was doing to systematics. It is to this topic that I now turn.

4

A Clash of Doctrines

Numerical taxonomy uses statistical methods to form groups whereas traditional taxonomy only uses them to discriminate more precisely between groups already perceived. If it becomes increasingly apparent that there is a fundamental divergence here, let us remember Whitehead's dictum, that a clash of doctrines is not a disaster—it is an opportunity.

B. L. Burtt, "Adanson and Modern Taxonomy"

Statements such as these [by Janvier (1984) criticizing evolutionary systematists] may be technical violations of the Marquis of Queensbury Rules, but they are mild compared to the invective common in contemporary systematics, in which there is more bitter infighting than in any field of contemporary science. Some day someone will write the history of this infighting; perhaps only those who were there will believe it.

J. Felsenstein, "Waiting for Post-Neo-Darwinism"

IN THE early 1960s the philosophy of science was ripe for a revolution. The philosophical research program usually termed "logical empiricism" had lost its early vitality. While it had not quite begun to degenerate, it certainly had not been making much progress. Thomas Kuhn's *The Structure of Scientific Revolutions* (1962) provided the spark that ignited the imaginations of historians, philosophers, and scientists alike. Each group found something in Kuhn's work to excite them. If professional philosophers had had a say in the matter, they would not have selected Kuhn's book to function as a catalyst to reinvigorate their discipline. With hardly an exception, professional philosophers were irritated if not appalled by the book. It was too casual in its treatment of key issues. Sometimes Kuhn seemed to be saying one thing; sometimes another. One problem was his highly ambiguous use of his central notion of a "paradigm." Sometimes it referred to global conceptual systems; sometimes to one element in such systems—exemplars, "concrete puzzle-solutions which, employed as models or examples, can replace explicit rules as a basis for the solution of the remaining puzzles of normal science."

One of the main themes in my own view of science is that the actual impact that a scientific work has is at least partially independent of its "inherent worth," assuming that the latter notion makes sense. For example, anyone reading Erwin Schrödinger's *What Is Life?* (1947) today is likely to find it trivial, but time and again the founders of the revolution that took place in molecular biology recall that it fired them to join in this emerging research program. Similar remarks hold for other seminal works. Some, such as Dobzhansky's *Genetics and the Origin of Species* (1937), hold up very well. Others, such as Jacques Loeb's *The Mechanistic Conception of Life* (1912), do not. In any case, philosophers of science must admit that Kuhn's book, for all its weaknesses, was efficacious. For good or ill, it was the seminal work in the emergence of the "new" philosophy of science.

Looking back on the reception of his brainchild, Kuhn (1970a) himself was dismayed. All the wrong people seemed attracted to his book for all the wrong reasons. Social scientists in particular read him as advocating a relativist view of scientific knowledge, as if truth were nothing more than what the scientists in power say it is. He intended no such conclusion. He also faulted philosophers for ignoring his most important innovation—conceptual exemplars. The role of concrete problem solutions is to allow scientists to move rationally from one conceptual scheme to another in the absence of any explicitly formulated deductive connections. After all, if two systems are genuinely different, one cannot be a deductive consequence of another. However, rationality involves more than just making deductive inferences. Kuhn (1970a:176) also remarked that if he had his book to write over again he would begin with a discussion of the community structure of science, because a one-to-one isomorphism exists between paradigms and scientific communities. "A paradigm is what the members of a scientific community share, *and,* conversely, a scientific community consists of men who share a paradigm."

Unfortunately, Kuhn's examples of scientific communities include such "groups" as physicists. If there is one thing that physicists are not, it is a "community." That is, if the relevant social relations are traced, the resulting social groups do not even roughly correspond to those scientists working on problems in physics. As I remarked in Chapter 1, even so narrowly defined a specialty as high-energy particle physics cannot be characterized as a community in a social sense. Such things as Crane's (1972) invisible colleges and Blau's (1978) cliques are the operative social groups in science. For example, in the middle of the nineteenth century, the Darwinians formed a clique centered in London (Cannon 1978, Manier 1978, Greene 1981, Rudwick 1982, Hull 1985b, Mishler 1987). They not only supported each other in their scientific endeavors but actually met on occasion to plan strategy. Similar groups sprang up elsewhere. All of these cliques taken together formed the "Darwinians" in a more global sense. One of the equivocations that plagues commentaries on science is that between local research groups and their more global counterparts, especially when they are called by the same name. One contention of the present work is that the small research groups that periodically crop up are the most important focus of rapid, though usually abortive, change in science. They are the locus of innovation and initial evaluation. The groups that are more global are operative only when it comes to general acceptance. And such local groups tend to be very local. For example, during a short period, Edge and Mulkay (1976) discovered four different cliques in astronomy at Cambridge University alone.

The members of these separate cliques went to the seminars presented by members of other cliques and exchanged information, but they did not actually work together (see also Griffin and Mullins 1972 and Cambrosio and Keating 1985).

Research groups as social groups are as difficult to treat accurately as are biological populations. The tendency to read more homogeneity into a group than actually exists is all but impossible to avoid, especially when the scientists working together to promote a particular conceptual system exaggerate how extensively they agree with each other. This appearance of agreement is accentuated by the use of similar terminology, a similarity that is sometimes more apparent than real. Conversely, scientists who find themselves in rival camps go to the opposite extreme and exaggerate how much they disagree, an exaggeration that is further enhanced by differences in terminology. These differences are frequently only superficial, but, as in life in general, superficial appearances frequently have substantial effects.

Because the simplifications introduced by scientists about their own relationships and degrees of agreement make the job of those of us who study science much easier, we tend to go along with these fabrications. However, when one investigates with any care either scientific conceptual systems or their associated scientific communities, the correspondence between the two is far from perfect, Kuhn's conjecture notwithstanding. If a paradigm is defined in terms of a neat set of propositions, then no sociologically defined scientific community can be found associated with it. Inevitably, one or more members of such a group will differ on important issues with others in the group. Although the members of scientific research groups frequently go to great lengths to hide their disagreements both from each other and from outsiders, they still exist. The preceding observations apply even to the smallest, most highly integrated research groups (or cliques). Such diversity only increases when the groups under investigation are larger and more amorphous. In short, cooperation in science does not require total agreement, nor does agreement necessarily imply extensive cooperation. Pick any two members of a research group and they will agree with each other on all but a few of the issues relevant to their research. However, it does not follow from this sort of pairwise agreement that there is one set (or even cluster) of propositions about which all members of the group agree. The conceptual systems associated with research groups exhibit a multimodal distribution.

Perhaps an accurate portrayal of conceptual systems requires that they be treated as multimodal clusters, but how in the world can such variegated systems be individuated, picked out? One answer is to take up Kuhn's suggestion and substitute exemplars for paradigms in his equation: "An exemplar is what the members of a scientific community share, *and,* conversely, a scientific community consists of men who share an exemplar." Although this equation is not totally accurate, it comes close. One way to individuate a research group is to pick a member, any member, and trace his or her scientifically relevant social relations. If a genuine research group exists, its contours will materialize. Central members work better in this connection than peripheral members, but such distinctions cannot be made in advance of delineating the group; and in tightly knit groups, it makes no difference. One way of delineating conceptual systems of the highly particularized sorts that function in science is to pick a particular conceptual element, any element, and trace its actual conceptual connections. If a genuine conceptual system exists, its contours should materialize. In the context of conceptual systems, the appeal of "central" elements (exemplars) is even greater, but more peripheral elements can also fulfill this function. On my view exemplars need not be especially exemplary (Hull 1983c).

The greatest danger in treating scientific change is the ease with which spurious groups can be manufactured simply by classing together disparate individuals who happen to hold similar views regardless of their actual connections. For example, several of Darwin's more idealistically inclined contemporaries were among his most effective critics, but they were unable to form any sort of a cooperative endeavor. When H. G. Seeley (1839–1909), one of the strongest advocates of idealism in Great Britain, agreed with Huxley that pterodactyls might not be reptiles but warm-blooded vertebrates closely related to birds, Owen reacted to ally and enemy with equal hostility. When Mivart was attacked by the Darwinians for his conversion to idealism, he approached Owen for support and received no encouragement. As Desmond (1982: 142) emphasizes, the Darwinians formed a social network; the Platonists (in my terminology idealists) did not and suffered accordingly. Each attacked Darwinism in isolation and was met with an organized response. Huxley had his mafia; Owen did not. Even so, it is all too easy to refer to Darwin's idealist opponents as if they formed a genuine group.

Conversely, just because no one at the time recognized a group of scientists and gave it a name, it does not follow that no such group existed. For example, a group of physicists and mathematicians formed around Lord Kelvin to oppose Darwin. One reason that the Kelvin conspiracy was so effective was that the Darwinians got hit from several sides by mutually supportive critiques (e.g., Bowen 1860, Haughton 1860, Hopkins 1860, Jenkin 1867, Tait 1869). In the dispute at the turn of the century between the neo-Darwinians and neo-Lamarckians, neither side formed much in the way of effective social groups. It was more a matter of individuals with opposing views taking pot shots at each other (Bowler 1983).

In the ongoing process of science, the inherent worth of ideas is far from irrelevant, but it is also far from sufficient. If it were, there would be no unappreciated precursors, yet the history of science is littered with them. Being "right" is not enough. Scientists must convert their fellow scientists as well. To move up one level, one reason why internalist commentators on science have been forced to deal with epistemological relativism once again is the rise of the Edinburgh School. As the name implies, it is a "school." Although the members of this school differ with each other on a host of counts, they support each other in their intellectual endeavors. As a result they have been extremely effective in forcing defenders of more conventional views to meet their attacks.

Most histories of science are conceptual histories. They detail the discoveries that scientists make, the problems they had, their insights, and the evidence that they presented to substantiate their claims. That most histories of science are histories of ideas is no surprise. Scientists are interested primarily in content. They are simultaneously engaged in numerous diverse activities, but all the while they keep their attention doggedly riveted on their subject matter. Everything else, including teaching, administration, and even their private lives, is a distraction. The content of science is also fascinating in its own right. It is a rare investigator who can study science without becoming intrigued by the issues and ideas. Good conceptual histories are difficult to write for a variety of reasons. One reason is that historians must understand the work of the scientists under investigation so that the genesis of these ideas can be traced. Because scientists seldom concern themselves with the future needs of historians, they keep only haphazard records. As a result, historians are not always able to find the data necessary to tell as complete a story as they would like.

The greatest danger in writing conceptual histories is to read into the work of past scientists connections that the scientists themselves did not see and to gloss over apparently incomprehensible lines of reasoning; in short, to make the work of past scientists more "logical" than it actually was. A related danger is to substitute "logical" connections for actual connections to produce a sort of "ideal morphology of ideas"; to view the history of ideas in terms of intellectual archetypes. Just because it would have been "only natural" for a scientist to draw a particular conclusion, it does not follow that he or she did. The temptation is to pass off "rational reconstructions" as histories because such ideal morphologies of ideas are so much easier to write than histories which trace actual causal influences.

No sooner does a historian decide to write a conceptual history as it actually took place than he or she is forced to confront scientists and their historical relations. Some commentators have claimed that Mayr's ideas on the nature of biological species "flow naturally" from the work of Sewall Wright. Mayr understandably objects. First, he fails to see the logical connection, but more importantly he had formed his own ideas before reading Wright. If there are connections between the views of Wright and Mayr, they are not direct. Perhaps the principles of Mendelian genetics and Darwin's theory of evolution are in some abstract sense perfectly compatible, but given the understanding of the protagonists at the time, the two were treated as being incompatible for over a decade. The conceptual systems that are efficacious in science are extremely heterogeneous, but the interests of everyone concerned conspire to disguise this heterogeneity. Philosophers tend to be interested in the abstract inferential relations of their rational reconstructions. The conceptual idiosyncrasies of the scientists themselves are worth an occasional aside, but that is all. Historians are committed in principle to detailing all the causal factors that actually influenced the views which the scientists under study held and the decisions that they made. The trouble is that these causal influences are extremely difficult to uncover, and the resulting story tends to be so complicated and detailed that few are willing or able to read it.

Textbooks written by scientists are infamous for the degree to which they disguise the differences of opinion that characterize scientists. They are designed to indoctrinate students, not to present an accurate portrayal of the diversity of opinion that characterizes the scientific community at any one time. Besides, students want to memorize the truth, not a range of opinions. If asked, scientists are likely to acknowledge the same admiration for conceptual variation as Gerard Manley Hopkins expressed for nature's pied beauty. Without variation, conceptual change would be impossible. However, when it comes to their own case, they are just as likely to draw the line. *Their* view was *the* view. Scientists in the midst of scientific controversies do not obscure heterogeneity by abstracting some ideal pattern but by substituting their particular version for all the variety that actually existed. The net effect, however, is the same: the obliteration of conceptual heterogeneity.

To summarize, conceptual histories require actual causal connections between conceptual tokens. The ultimate subjects of genuine conceptual histories are conceptual lineages. Periodically, individual scientists elaborate a well-articulated conceptual system of the logical, timeless sort that so attracts conceptual idealists, but the only things that ever get passed on are tokens. These causal connections may sometimes be via single scientists, or sequences of single scientists, but periodically social groups materialize to develop and promote one of these elaborations. Just as single scientists and particular bright new ideas blink on and off in the history of science, conceptual systems and groups of scientists proliferate and then die out. Sometimes, very rarely, a particular system and its associated group proliferate extensively.

In this chapter I chart the rise of two scientific research groups: the pheneticists while they were based at the University of Kansas in Lawrence, Kansas, and the cladists while they were centered at the American Museum of Natural History in New York City. Neither Robert Sokal at Kansas nor Gareth Nelson at the American Museum intended to found research groups. They were interested in the *issues*. But nevertheless they were the social foci around which these groups formed. Others made important contributions to these research programs, some might argue more important than those made by Sokal and Nelson.[1] But social roles are at least partially independent of substantive contribution. Both of these research groups were sufficiently successful so as to be able to send out founder populations elsewhere, resulting in the formation of more inclusive "communities" of the same name.

The literature on the recent history of taxonomy is permeated by a systematic equivocation between two senses of "pheneticist" and "cladist." Sometimes these terms refer to the original research groups centered for a decade or so in Kansas and New York respectively; sometimes they refer to anyone who happens to be aligned with these original groups. Because I think that the existence of small research groups is so important in the conceptual development of science, I find such equivocation to be extremely misleading (see also Mishler 1987). To those who are dismayed by all the attention that I pay to two groups of scientists who appear in the larger picture to be at best obscure, all I can recommend is consideration of the role of fruit flies in the history of genetics. Geneticists did not have to pick extremely "important" species to

1. If citations are any indication, Farris has been easily as important in the development of cladistics as Nelson. Of the twenty-eight most frequently cited papers published in *Systematic Zoology* between 1961 and 1983, Farris either authored or coauthored three of them to Nelson's one (see Appendix D). These citations are by cladists and noncladists alike and are from journals not devoted primarily to taxonomic methods and philosophy. Of the most frequently cited authors in the three early books on cladistic analysis by cladists, Nelson comes in fourth and Farris fifth (see Appendix E). However, for individuating cladists, either man serves as well. Nelson and Farris's most immediate colleagues were different, but once the group is expanded to a dozen or so people, the membership, as determined by either focus, is nearly identical.

study in order to penetrate the workings of hereditary transmission. Similar observations hold for understanding the workings of science.

Numerical Phenetics

Just when the evolutionary systematics of the New Synthesis was becoming well established, an increasing number of voices were being raised in opposition. In a series of papers, A. J. Cain traced the methods that systematists from Aristotle, John Ray, Linnaeus, Cuvier, Lamarck, and Augustus P. de Candolle to Darwin had used to construct their classifications. As Cain (1962: 1) explicitly stated, his purpose for studying the history of systematics was "to cast light on its present state." The lesson that Cain learned from the history of taxonomic methodology was the superiority of the empirical, a posteriori approach to the deductive, a priori method that afflicted most taxonomists from Aristotle to the present. As Cain characterized the a priori method, a systematist reasoned from certain first principles to the sorts of characters that are essential in classification. If movement is essential to animality, then classifications according to mode of locomotion should be most natural for animals. Too often, Cain complained, systematists then reasoned from the constancy of a character in a group to its importance, closing the vicious circle.

Cain found the first principles used by such systematists as Cuvier to be totally without foundation and viewed the attempts of these great figures to support their preferred principles by showing how they accorded with observed distributions equally fallacious. He himself opted for theory-free, inductive methods. "We can only proceed empirically, simply finding out what subjects exist and what are their attributes, not deducing them from known principles and axioms" (Cain 1959a: 146). The only advocate of this empirical approach that Cain could find in the history of taxonomy was the French botanist, Michel Adanson (1727–1806), whose career spanned the later years of Linnaeus and Buffon and the early years of Lamarck. Quoting from de Candolle's (1813) description, Cain (1959a: 205) characterized Adanson's method as classifying according to the overall similarity of organisms using numerous, equally weighted characters. According to Cain, Adanson set up sixty-five separate classifications of the same group of plants on the basis of different organs. He concluded that those plants that were found side by side in the greatest number of these single-character classifications must be most closely related to each other. De Candolle rejected Adanson's method because Adanson ignored the functional importance of certain organs over others. Cain recommended its adoption for exactly the same reason. (For more accurate descriptions of Adanson's method, see Stafleu 1963 and Burtt 1966).

Cain (1959a: 210) attributed the eventual downfall of a priori methods of classification to the success of Darwin's theory of evolution, but "no sooner does one *a priori* method fall than another is put in its place." After Darwin, those characters that indicate common ancestry (evolutionary homologies) became more important in classification than those that obscure it (convergences). The problem became how to distinguish between these two sorts of similarities. Cain concluded that the best a systematist has to go on are "probabilities," with really no way to evaluate these probabilities. Hence, he recommended classifying according to overall similarity. These are the truly "natural" classifications. Even so, because Darwin's deductions did not rest solely on previous classifications, as Cain claimed the first principles of Darwin's predecessors had, he was not arguing in a circle.

Cain was not content with merely criticizing the inadequacies of early a priori methods. In collaboration with G. A. Harrison, he presented explicit methods for estimating both overall similarity and ancestor-descendant relations. Building on Woodger's (1945) earlier work, Cain and Harrison (1958: 86, 96) argued that estimations of "true" affinity in a phylogenetic sense are "both logically and historically posterior to that of affinity as overall similarity." Hence, overall similarity must be ascertained first, before any attempts to infer ancestor-descendant relations. Because of the nature of our notion of similarity, any attempt to express degree of divergence is "hopeless and can at best produce only an approximation of very uncertain value, perhaps materially misleading." However, in their second paper, Cain and Harrison (1960) note that, once we allow information in addition to comparative data to inform our inferences, certain conclusions about evolutionary relationships become more probable than others. These additional sorts of data include geological age and distribution as well as analyses of function, ecology, genetics, and intraspecific variation. Hence, phyletic weighting had some justification, but the weighting must be an all-or-nothing affair. Although the confidence which a systematist may have in the validity of a particular character might vary continuously, this character must be included or excluded—either it is indicative of actual descent or it is not.

Cain (1959b: 241) concluded his survey of post-Linnean taxonomy with the following prophesy:

> I think that we are about to see a considerable revision of the whole basis of taxonomic theory, especially the relation of phylogenetic to natural classifications, a clear separation of phylogenetic weighting from recognition of the co-variation of characters, and the development of methods for making comparisons, whether phylogenetic or natural, more precise.

Predictions about the future course of science are notoriously faulty. In this case, Cain was more than a little prescient. The occasion for this revolution in systematics was a snowstorm in Kansas.

Although few are likely to expect Lawrence, Kansas, to be the center of anything save possibly the continental United States, it is the home of one of the most active research museums in the country, the Snow Museum of Natural History. At the time, the leading member of its staff and the biggest bee man in the world was (and continues to be) C. D. Michener. Michener is a quiet, gentle man of the sort usually associated with the Old South. He does not enjoy participating in raucous polemics and expresses himself, especially with respect to his more radical views, as judiciously as possible. He is quiet but in scientific combat also highly effective, some might say deadly. Early in his career, Michener came to see one of his roles in science as supporting bright young workers who showed unusual promise. He recognized such promise in a tall Austrian immigrant who joined the staff at Kansas in 1951, Robert R. Sokal. When the Nazis took over his homeland, Sokal and his family had to flee Vienna. The only country that would take them was China. While in China, Sokal earned a bachelor's degree from St. John's University in Shanghai and married the daughter of a well-placed Chinese family. After the war, he and his wife, Julie, moved to the University of Chicago. Even though Sokal's scholarship barely covered tuition, as aliens they had no permission to work outside the university. To survive, Sokal worked at the library, and he and his wife took care of a paraplegic professor, becoming in effect servants in someone else's home.

At Chicago Sokal began by studying traditional evolutionary ecology under A. E. Emerson (1897–1976), the biggest termite man in the world. At the time Sokal had no interest whatsoever in mathematics, even though his laboratory mate, Clyde Stroud, tried to get him to share his own enthusiasm for applying quantitative techniques to biological phenomena. Sokal resisted his friend's enthusiasn for well over two years until he finally gave in and agreed to collaborate on a paper. Before they could begin their project, Stroud died of cancer. After Stroud's death, Emerson extracted a paper from Stroud's dissertation that applied factor analysis to the classification of termites and arranged to have it published in *Systematic Zoology*. Stroud's (1953) paper was one of the earliest numerical studies to be published in the journal. It went unnoticed, but as a result of Stroud's influence Sokal's work became increasingly mathematical, so much so that his dissertation was actually supervised by Sewall Wright. Because the University of Chicago had a rule that no student could be supported for more than four years, Sokal had to find a position elsewhere and in 1959 signed on as a research assistant to Michener at Kansas to study DDT resistance. Seven years later, the Sokals became American citizens.

At weekly luncheon meetings devoted to taxonomic method, Sokal raised repeated objections to the intuitive methods used by practicing taxonomists. More objective classifications could be produced by explicitly formulated statistical methods. After weeks of such exchanges, one of the graduate students dared Sokal to put up or shut up. If it could be done, he should do it. Michener offered to supply the data. The prize would be a six-pack of beer (Sokal 1985). Several weeks later Sokal returned with a phylogeny generated by a computer program, a phylogeny that differed only slightly from the one that Michener had produced by means of his own more intuitive methods. One result of the lively discussions over liverwurst sandwiches was that Michener and Sokal decided to collaborate on a paper applying quantitative methods to Michener's bee data. They quickly decided that one and the same classification cannot depict *both* the order of branching in a phylogenetic tree *and* degrees of overall similarity. Michener still preferred the more flexible principles of classification suggested by Mayr, Linsley, and Usinger (1953), even if the application of these principles was somewhat subjective. Sokal to the contrary was disposed to methods that are more uniform and objective. He went to graduate school to become a scientist, not an artist. Instead of attempting to include both descent and divergence in a single classification, Sokal thought that systematists should settle ‹ n a single relation. It is better to express one relation clearly than several obscurely. He himself preferred overall similarity. Classifications should unequivocally represent the resemblances exhibited by organisms without regard to their descent. Descent could then be indicated in an accompanying diagram.

Their residual disagreements notwithstanding, Michener and Sokal completed a paper (Michener and Sokal 1957). They did not even consider sending it to *Systematic Zoology.* It was too important to bury in such a minor journal. Instead they submitted it to *Evolution.* The skepticism that Michener as an experienced taxonomist had initially shown to the project was the most common response to the paper from other practicing taxonomists. One systematist volunteered that he hoped they would never succeed in making taxonomic judgments sufficiently quantitative so that a computer could make them because, if they did, it would take all the fun out of systematics. Drudges could collect data and feed it into computers. Another colleague was less genteel. He found the next paper produced by Sokal and Michener (1958) the "biggest mound of bullshit" he had ever seen.

Although Sokal had experienced considerable give-and-take at the luncheon seminars over his radical views, he confronted his first public opposition when he presented a paper at the Missouri Botanical Gardens in 1957. Robert Inger of the Field Museum of Natural History in Chicago objected strongly, not to the use of quantitative techniques in taxonomy, but to Sokal's refusal to weight characters differentially. As far as Sokal was concerned, all characters used in a taxonomic study must be given equal weight. To Inger, assigning greater weight to more important characters was the heart of taxonomic procedure. As the two men contended from opposite sides of the room, the discussion became somewhat heated. Sokal was especially dismayed when Inger lumped him with Blackwelder, Boyden, and Borgmeier as "typologists." After the exchange was over, Sokal suspected that he had not convinced Inger totally, but he did think that he must have had some effect. Thus, he was dismayed when he read a paper published later in *Evolution* by Inger (1958) in which Inger reiterated all the same objections he had raised at the Missouri meeting. Once again he argued that for a classification to be genuinely scientific it had to be connected intimately to the evolutionary process. Pattern without process is not science. Sokal disagreed vehemently. To be genuinely scientific, recognition of overall similarities must precede speculations about process. Any other procedure is viciously circular. Sokal's (1959) reply was so heated that the editor of *Evolution,* Everett Olson, did not want to publish it; Sokal insisted, however, and Olson's successor, I. Michael Lerner, published it. If Michener had been in the country at the time, he might have counseled moderation, but he was not.

Sokal did not convince Inger at St. Louis, but the exchange did have one important effect. It revealed to Sokal not only that he was able to stand on his feet and defend himself but also that he enjoyed doing it. He was no longer the insecure young immigrant. He was his own man. It also redirected his research interests toward taxonomy. Even though Sokal had held a scholarship at the Field Museum in Chicago for a short time, he had really never been much interested in systematics, in part because he lacked the intuitive ability to see patterns in complex relationships. But money was money. However, once in a conversation with Herbert Ross (1908–1978), a major figure in the field who was visiting the museum, Sokal revealed his true attitude. When Ross asked Sokal if he was interested in taxonomy, the young graduate student blurted out, "Of course not," an admission that Ross did not forget. As long as taxonomic method remained largely intuitive, Sokal would have to look elsewhere for research topics, but if quantitative methods could be developed to discover these patterns, even those people who lacked intuitive ability could practice systematics. However, it was what Sokal took to be the blind rejection by traditional taxonomists of his efforts that determined the direction of his future research. He would show those "sonsabitches."

At fifteen Paul Ehrlich volunteered to help Michener work on butterflies while Michener was at the American Museum in New York. However, when Ehrlich got to college, he found the temptations of "booze and broads" so irresistible that his academic record was poor, so poor that he would never have gotten into graduate school if Michener had not interceded for him. At first Ehrlich thought that Sokal's ideas on classification were crazy, but when Sokal was able to reproduce Michener's phylogeny of bees solely on the basis of a list of characters given him by Michener, Ehrlich was converted and, like so many converts, became more Catholic than the pope. Although Sokal was willing to ignore phylogeny in the initial stages of classification, he still retained the usual attitude toward the central role of the biological species concept. Ehrlich (1961b) saw no reason to draw the line at species. If all higher taxa are to be grouped according to character distributions with no consideration of phylogenetic descent, Ehrlich saw no reason to treat species differently. They too should be defined solely in terms of character distributions without any attention to reproductive habits. Actual interbreeding was usually unknown, and potential interbreeding was unknowable. It took a while, but eventually Ehrlich convinced Sokal to treat all taxa as Operational Taxonomic Units (OTUs for short).[2]

If anything, Ehrlich was more combative than Sokal, both in print and in public debate. In a paper published in 1961, Ehrlich (1961a) made what he knew would be unpopular predictions for systematics in 1970: electronic data-processing equipment would be the systematist's most important tool, nomenclature would be deemphasized, and traditional taxonomic monographs would largely be replaced by computer print-outs of data matrices. At the St. Louis meeting, when one taxonomist asked indignantly, "You mean to tell me that taxonomists can be replaced by computers?" Ehrlich responded, "No, some of you can be replaced by an abacus." Thereafter, Ehrlich did not consider the give-and-take after a paper truly successful unless he brought at least one taxonomist to the point of tears. When he was hired years later at Stanford University, he put his own preachings into practice by getting rid of its huge collection of butterflies and moths, donating it to the California Academy of Sciences.

At the National Institute for Medical Research in London, Peter Sneath, a medical doctor who was doing research on microbial systematics, was coming to conclusions very similar to those of Sokal and the Kansas group. According to Sneath (1957b), classifications should be ideally based on numerous, equally weighted unit characters,

2. Because of the similarity of the views expressed by pheneticists to those set out a generation earlier by logical positivists, one might suspect some connection between the writings of these philosophers and the philosophical predilections of the pheneticists. There was a connection, but it was indirect, via Gilmour (1940). Prior to developing their views on systematics, none of the founders of phenetic taxonomy was aware of the logical positivists or their views. Although Rudolph Carnap was on the faculty while Sokal attended the University of Chicago, Sokal had no contact with the man.

clustered by some quantitative algorithm. Natural classifications should reflect overall similarity, not phylogeny. Unlike Sokal, Sneath was aware that others before him had urged similar views; e.g., Bather (1927), Gilmour (1940), and Turrill (1942). Sneath was even aware of Adanson's prescient views two centuries earlier. In tribute, Sneath suggested that the sort of taxonomy that he envisaged should be named "Adansonian." When Sneath stumbled upon a report of a paper that Sokal had given at a regional meeting, he was delighted. Here was a man who thought the way that he did. He immediately wrote Sokal, and the two men exchanged reprints, the accepted greeting ritual in science.

In 1959 Sneath joined Joshua Lederberg at the University of Wisconsin, just in time to move with him from Wisconsin to Stanford. Sneath loaded his family and possessions in an old car and started the long trek. One good thing about the trip was that he could stop along the way to see Sokal. What did Sneath expect of Kansas? Did he anticipate the great flat sweep of Dorothy's gray prairie, cracks running through sun-baked sod and all? He arrived in a snowstorm. One of his children had been sick in the car. He and his family were stranded. The two men finally met—a tall, urbane Austrian, perfectly comfortable at all times in a three-piece suit, and an equally tall but slightly stooped Englishman wearing National Health Insurance glasses. During the next three days of intense conversation, Sokal and Sneath discovered that their abilities and interests meshed nicely. Soon thereafter Sokal accepted a National Science Foundation postdoctoral fellowship at University College, London, and several months later Sneath also returned to England. The two men used this opportunity to begin work on a paper in which they intended to reorient the foundations of their discipline. The paper rapidly expanded into a book. When Sokal and Sneath started their book, very little had been published on quantitative methods for making taxonomic judgments, perhaps a dozen papers in all. By the time that their *Principles of Numerical Taxonomy* appeared in 1963, the field had expanded considerably. The chief outlet for this activity in the United States was *Systematic Zoology.*

When Libbie Henrietta Hyman (1888–1969) became president of the Society for Systematic Zoology in 1959, she discovered that the original editor, John L. Brooks, had fallen a year behind in publishing the society's journal. Hyman promptly took a train to New Haven, marched into Brooks's office, and informed him that he was through as editor. She then proceeded to install herself as editor *pro tem.* A year later, the Council of the Society of Systematic Zoology officially installed her as editor. Journals such as *Evolution, Nature,* and *Science* were willing to publish an occasional paper on taxonomic theory, but these journals were unlikely to publish the numerous works necessary for a research program in taxonomy to get off the ground. *Systematic Zoology* was the ideal outlet. Right from the beginning, Ehrlich had troubles with Hyman. She returned the first manuscript that he submitted to the journal without sending the paper out to be reviewed, recommending that the next time he considered submitting a manuscript for publication, he let Michener see it first. In fact Ehrlich had shown Michener his paper before sending it to Hyman. In 1961, however, Hyman did publish several of the papers presented at a symposium on the philosophical basis of systematics that had been held at the Davis, California, meeting of the Pacific Section of the Society of Systematic Zoology, including two papers by Ehrlich (1961a, 1961b).

By this time another "slow starter" had joined the Kansas group—James Rohlf. Although Rohlf had come to Kansas with no background in computers, he turned out to have a natural bent for programming. Together Rohlf and Sokal wrote a paper on factor analysis and submitted it to Hyman. Hyman rejected it as inappropriate for the journal, once again without sending it out for review. Sokal complained, and Hyman relented, but in her 1961 editor's report to the Council of the Society of Systematic Zoology she included the following comment:

One article was rejected because written in incredibly bad English and another because too mathematical. Inquiries among subscribers indicate that the journal has had enough for the present of articles about numerical taxonomy. However, further opinions on this matter are desired. An article of this nature is scheduled for the March, 1962, issue but that will be the last of this nature for the present.

In the folklore of numerical taxonomy, Hyman's comment has come down as, "One paper with numbers is enough." True to her word, Hyman rejected the next manuscript that Rohlf submitted to the journal, a manuscript in which he showed that numerical classifications of larval and adult mosquitoes turn out to be quite different from each other. Again Sokal applied pressure and Hyman once again relented, publishing not only Rohlf's (1963) paper but also a long paper by Michener (1963) in which, for the first time, he came out in favor of classifying without regard to phylogeny. Sokal and Michener decided that battles could not be fought manuscript by manuscript. Hyman had to go. When the Kansas group first began having trouble with Hyman, Blackwelder was president of the society. Blackwelder was second only to Waldo Schmitt (1887–1977) in founding and sustaining the Society of Systematic Zoology. He brought out the first number of the journal and was secretary-treasurer from 1952 until 1959. Both Schmitt and Blackwelder thought that systematists should get on with the task of classifying their specimens and stop wasting time talking about evolution and phylogeny. By the time that the dispute over Hyman's editorship came to a head, Simpson was president-elect. Sokal and Michener appealed to Simpson for help. Although G. G. Simpson was one of the strongest proponents of the taxonomic philosophy that Sokal and Michener were attacking, he was in no way opposed to the use of mathematics in biology. He had in fact coauthored with his wife, Anne Roe, the first text on quantitative methods in biology (Simpson and Roe 1939).

As president-elect of the society in 1962, Simpson circulated his personal observations on the state of *Systematic Zoology* to the officers of the society and the members of the editorial board of the journal. Simpson was surprised to discover that, four years earlier, a questionnaire concerning just these issues had been circulated. In light of the results of that questionnaire and Simpson's discontent, the then current president, C. W. Sabrosky (b. 1910), placed the topic of the future of the society and its journal on the agenda of the 1962 meeting of the council in Philadelphia. In her editor's report, Hyman objected to Simpson's having used the word "bias" with respect to her editorial policy in his original observations, but she acknowledged that a couple of areas had received too much attention in recent years—the subspecies question and numerical taxonomy! Hyman concluded her report by stating that she was willing to continue as editor but felt that the "journal should be in the hands of someone younger, more vigorous, and more aggressive than I."

Once Simpson was president, he took Hyman at her word and replaced her as editor with George Byers at the University of Kansas. Kansas was a natural location for the new editor because the firm that published the journal, Allen Press, was located in Lawrence. That the University of Kansas was also the home of numerical taxonomy was also not an irrelevant consideration. However, Simpson's replacement of Hyman was only part of a larger effort by Simpson to rejuvenate the society. In the early years, systematists had consciously set up their own society independent from the Society for the Study of Evolution because they had interests at least partly distinct from those of evolutionary biologists, and they did not want to get lost in this larger group. Simpson and Mayr now proposed to redirect the goals of the society toward increased attention to evolution as the only perspective that could serve to integrate taxonomic studies. When Waldo Schmitt saw the report of the policy committee presented at the 1963 meeting in Cleveland, he fired off a six-page letter to Simpson objecting, but his protests were futile. The old guard had been effectively replaced.

The new guard, however, did not consist of numerical taxonomists but advocates of Simpson-Mayr evolutionary systematics. During the next ten years, while the journal resided in Lawrence, the only president of the society who had any allegiance to the Kansas group was Michener. Instead, the society was led by a series of critics of numerical taxonomy, including Mayr, Inger, and Ross (see Appendix B). The conflicts that arose in the Society of Systematic Zoology were not between advocates of the Simpson-Mayr school and numerical taxonomists but between evolutionary systematists and the deposed advocates of more traditional methods of systematics—the old guard. For example, Blackwelder maintained a collection of books on taxonomy which he brought to various meetings and set up as a systematic zoology book exhibit. He augmented this collection by writing to publishers requesting free copies. Those who attended the exhibit and contributed books to it were under the impression that it was an official organ of the Society of Systematic Zoology. Successive presidents of the society requested Blackwelder to make the private character of this traveling book exhibit plain. Finally, as partial as Blackwelder was to his book exhibit and as valuable as he thought it was, he abandoned the practice.

The journal was quite another matter. Even though the three men who edited the journal while it resided at Kansas were not especially strong advocates of numerical taxonomy, they were located at the center of this new school. During this period, *Systematic Zoology* was strongly identified with numerical taxonomy, almost as if it were an arm of this movement. In their early papers, Michener, Sokal, and Rohlf were measured in their criticisms of traditional taxonomic methods and goals. The role of banderillero was willingly adopted by Ehrlich as he happily skewered sacred bull after sacred bull. Later, after receiving more than their share of hostile criticism, other members of the movement began to elevate the polemical tone of their publications. For example, Sokal and Sneath (1963: 101) complained that Simpson had rejected their empirical approach to taxonomy "out of hand." They agreed that typology is unacceptable when "contaminated with idealistic, metaphysical concepts," but saw nothing wrong with it when it "represents an empirical summation of the information available on a given taxon without phylogenetic value judgment of these characters"

(Sokal and Sneath 1963: 266), a view not that different from the one expressed by Darwin on Huxley's use of archetypes. Sokal and Sneath also complained of Simpson's failure to answer their charge of vicious circularity.

Sokal and Sneath (1963) were clear about their goal for biological classifications. They proposed to ascertain the overall similarity among organisms by using numerous, equally weighted unit characters. Whatever methods are used they must be objective, explicit, quantitative, and repeatable. One problem they confronted was the apparently indefinite number of ways that organisms can be subdivided into characters, and the absence of any reason to prefer one subdivision to any other. At the genetic level, however, all codons are the same size and occur in nice linear sequences. It would seem that codons are the ideal unit characters for numerical analysis. The problem with codons is that they are not easy to ascertain. They may be in some sense "phenetic," but they are far from "observable."

In response to this problem, Sokal and Sneath set out several hypotheses, which they called by the terms nexus, nonspecificity, factor asymptote, and matches asymptote. According to the nexus hypothesis, the relation between genes and characters is many-many; i.e., each gene influences many characters, and each character is affected by many genes. According to the nonspecificity hypothesis, the various ways that systematists divide characters into kinds are unrelated to the genetic basis of these characters. For example, characters drawn from one region of an organism's body are no better measures of its overall genetic makeup than characters drawn from any other region. Similarly, separate classifications constructed for different stages in the life cycle of an organism or for different castes and sexes should be roughly congruent.

Sokal and Sneath reasoned that if the preceding hypotheses are approximately true, then phenetic classifications should converge asymptotically as more characters are used. The factor-asymptote hypothesis consisted of three subsidiary hypotheses: random samples of phenetic characters are accurate reflections of random samples of genes; the number of characters used to produce a classification is positively correlated with the information content of that classification; and, as the number of characters is increased, the information added by each additional character decreases. Finally, according to the matches-asymptote hypothesis, "as the number of characters sampled increases, the value of the similarity coefficient becomes more stable; eventually a further increase in the number of characters is not warranted by the corresponding mild decrease in the width of the confidence band of the coefficient" (Sokal and Sneath 1963: 114). Although Sokal and Sneath were aware that none of the correlations they postulated are perfect, they thought such imperfections could be neutralized by taking enough characters into account. Sooner or later, all classifications should converge on one another. As in the case of the ideal morphologists before them, the pheneticists were convinced that out there in nature there is something appropriately termed "overall similarity."

Herbert Ross, the man whom Sokal had shocked a decade earlier by admitting he was really not interested in systematics, reviewed Sokal and Sneath's book for *Systematic Zoology.* Although Ross and Sokal had become friends in the interim, Ross (1964: 108) concluded his largely negative review with his "considered opinion that numerical taxonomy is an excursion into futility." Simpson (1964) in his review published in *Science* emphasized that efforts to make biological classification more quantitative had been around for a long time and decried the "fanatical fervor" of the group led by Sokal and Sneath. Mayr (1965) followed up these two short reviews with a long paper in *Systematic Zoology*. He began by objecting to Sokal and Sneath's appropriating the term "numerical taxonomy." He himself had nothing against the introduction of either quantitative methods or computers into taxonomy. In fact, he welcomed both. What he objected to was the highly empirical, overly inductive, and antitheoretical philosophy that Sokal and Sneath promulgated under the guise of making taxonomy truly scientific. Mayr proposed instead to adopt the terminology introduced by Cain and Harrison (1960) and to term the principles of classification advocated by Sokal and Sneath "numerical phenetics." Mayr's chief objection was the refusal of pheneticists to weight their characters differentially. In response to Ehrlich's (1961a) predictions about the future of taxonomy, Mayr (1965: 94) made a prediction of his own: the "future development of taxonomy will see the incorporation of the most useful taxometric methods into evolutionary taxonomy but a rejection of the backward ideology underlying phenetics."

Although the big guns had come out against them, Sokal and Sneath were buoyed by the increased number of papers appearing in *Systematic Zoology* and elsewhere advocating what they continued to call numerical taxonomy, Mayr's admonitions notwithstanding. They had the right to feel confident. While *Systematic Zoology* was housed at the University of Kansas, it more than doubled in size, from 200 pages a year to 500 pages. This increase in size was accomplished even though acceptance rates for papers submitted had dropped from an average of 90 percent under Brooks and Hyman to 70 or 80 percent under the Kansas editors. During this same period, the circulation of the journal increased from 1,900 to almost 2,600. After a fierce struggle, Sokal thought that at long last his research program was well established, a force to be reckoned with. Thus, he was dumbfounded when he saw the program for the upcoming First International Conference of Systematic Biology to be held in June of 1967 at the University of Michigan in Ann Arbor. Papers were by invitation only, and no one from Sokal's group had been scheduled to present a paper. He and Rohlf were reduced to being discussants of papers on topics that they themselves had pioneered. They complained, but at that late date there was nothing they could do but attend the meeting in their subordinate roles (see Charles Sibley's Foreword to the proceedings of this conference, Sibley 1969).

No sooner had Sokal and Rohlf settled into their seats in the first row of the hall at the Ann Arbor conference than their worst fears began to be realized. After an introductory paper on the history of systematics by Frans Stafleu, a very nervous young comparative anatomist, Michael Ghiselin (1969), not only objected sharply to the philosophy of systematics favored by the pheneticists but also ridiculed Gilmour's (1940) paper on which it was based. The notion of overall similarity was a metaphysical delusion. In his comments on Ghiselin's paper, a young philosopher, David Hull, seconded Ghiselin's criticisms and concluded by quoting P. W. Bridgman's (1959) lament that in his operationalism he had created a Frankenstein monster that had gotten away from him. Hull (1969) concluded that if the original author of the highly empirical views that Sokal and Sneath were championing had his doubts, perhaps taxonomists should be wary.

When T. H. Huxley was just beginning his career, his mentor Edward Forbes suggested that he go to the 1851 meeting of the British Association for the Advancement of Science at Ipswich and make himself "notorious." He did. Over a century later, an intense, dark-haired graduate student at the University of Michigan, Steven Farris, received similar advice from one of his professors, Arnold Kluge. No sooner had the official commentator on Ghiselin's paper sat down than Farris rose from the audience to challenge Ghiselin. "Do you believe," Farris (1969a: 65) asked, "that there is such a thing as an order-quantifiable, or order-specifiable—metric or pseudometric—overall similarity over some specified and finite set of characteristics of organisms?" Ghiselin mumbled an answer but was furious. During the next three days, the proceedings were enlivened by Farris grabbing the microphone to ask equally audacious and at times downright impolite questions. With a little help from another professor at Michigan, Herbert Wagner, Farris even succeeded in presenting an unscheduled paper.

Sokal, Rohlf, and their students were not quite so successful. At several sessions they tried to direct the discussion to numerical taxonomy, but it kept being deflected elsewhere. After ten long years of publishing scores of papers and a book, of attending meetings and presenting papers, of arguing and cajoling, numerical taxonomists were still on the outside looking in. Sokal decided that it was time to organize. On his return to Kansas, he began to plan a meeting to be held at Lawrence five months later, in November. Just as there would be something seriously wrong with a newspaper headline declaring that World War I had just broken out, no one at the time called the meeting in Lawrence, Kansas, "Numerical Taxonomy One," but that was what it was eventually termed, or in the sort of symbolism that numerical taxonomists prefer, NT-1. Of the people invited, only thirty-three were able to attend on such short notice, and no formal papers were presented. The participants simply discussed issues of mutual interest. The atmosphere was of comradeship and high expectations. As Sneath (1968: 92) concluded his report of the meeting, "It was certainly refreshing to attend a conference on numerical taxonomy at which the best way to perform it was discussed, rather than whether it was permissible at all." (For additional meetings of the Numerical Taxonomy Group, see Appendix A.)

Although Michener was not much older than Sokal, he served as Sokal's model for what a scientist should be. From Michener, Sokal acquired critical standards as well as an understanding of the amount of diligence and perseverance necessary to succeed in science. Even though Michener was initially skeptical of Sokal's more radical views, he was supportive and sympathetic. He had a good feel for what Sokal was after, and he had a massive store of data for Sokal to utilize, not to mention a secure place in the taxonomic community. He knew the ropes of science and taught them to Sokal. He was well placed in the profession, with numerous professional contacts and positions, not the least of which was membership in the National Academy of Science. For a while, he even came to adopt Sokal's general position on the goals of biological classification. In their initial enthusiasm, Michener and Sokal began to make plans to establish an international center for taxonomy at Kansas, but gradually Michener began to cool. His style was not one of confrontation and polemics. Michener was always in control, but his control was exercised largely by indirection. He saw no reason to alienate all the most influential people in the field. Although he joined with Sokal, Sneath, Rohlf, and J. H. Camin in drafting a response to Mayr's (1965) blast at numerical phenetics, he declined to include his name among its authors (Sokal et al. 1965). Even his old doubts about the fundamental principles of numerical taxonomy began to reassert themselves.

For his part Sokal was beginning to grow discontented with his lot at Kansas. Plans for an international center for systematics were going nowhere, and strangely enough for someone who was so opposed to letting evolutionary considerations enter into classification, Sokal thought that the coming area in biology was going to be population biology. If Kansas was not to become an international center for taxonomy, at least the program could be restructured to climb on the population-biology bandwagon. Most of Sokal's colleagues did not share his enthusiasm. In 1968 Sokal accepted a position at the State University of New York at Stony Brook, although he did not actually join the staff until 1969. Before Sokal agreed to the position, he insisted that Rohlf be hired as well, and when he was consulted about Stony Brook hiring Steve Farris, fresh from graduate school, Sokal was enthusiastic. The administration at Stony Brook was intent on developing a high-powered department, and they were well on their way toward doing so.

Back in Great Britain, things had hardly stood still while numerical taxonomy was developing in the United States. In 1963 two botanists, P. H. Davis and V. H. Heywood, brought out their *Principles of Angiosperm Taxonomy,* in which they opted for the phenetic position. The Systematics Association, which had sponsored the conference a quarter of a century earlier that had resulted in Julian Huxley's *The New Systematics* (1940) sponsored in 1964 another meeting on the topic suggested by Heywood—phenetic and phylogenetic classification. In his introduction, Heywood (1964: 1) prophesied that taxonomy is "poised on the edge of a far greater revolution than that promised (but not fully achieved) by the New Systematics of the 1940's." In that same year Sneath helped found the Classification Society and its journal, *The Classification Society Bulletin.* This society was not limited to biologists but was expressly organized to include anyone interested in classification, from anthropologists and soil scientists to librarians and computer specialists. The bulletin, however, appeared only once a year and rarely exceeded fifty pages. In 1967 a bright young philosopher arrived on the British scene, Nicholas Jardine. He began by publishing a paper in a philosophy journal on the concept of homology in biology. Building on the work of Woodger (1945), Jardine (1967) developed a notion of common structure which was independent of any phylogenetic considerations. Sneath predicted that Jardine was destined to make major contributions to the science of systematics. In 1971 Jardine published with R. Sibson a book entitled *Mathematical Taxonomy.*

Near the end of 1971, Sokal discovered that history was about to repeat itself. The First International Congress of Systematics and Evolutionary Biology, to be held in Boulder, Colorado, in 1973, was being organized, and even though Sneath was on the international committee, once again numerical taxonomists were being excluded. Although a session was planned on the computer revolution in systematics, it was being organized by James Peters of the United States National Museum. In the call for papers which Peters sent out, he stated emphatically that there was to be no discussion of the controversy over numerical taxonomy. Sokal fired off irate letters to the chairman of the steering committee, Hans Stafleu, as well as to the presidents of the two sponsoring societies, the International Association for Plant Taxonomy and the Society of Systematic Zoology. Quite rapidly the organizers of the congress agreed that a session on numerical taxonomy was desirable and settled on a young botanist from the University of Notre Dame to organize it, Theodore Crovello. Of all the numerical taxonomists, Crovello was the most acceptable because he had actually published some traditional taxonomic work and had served on the systematics panel of the National Science Foundation. He was a known quantity.

Unaware that these decisions had already been made, Sokal arranged to meet with the program chairman of the congress, Paul D. Hurd, Jr., in December of 1971. Sokal and two colleagues used this meeting to urge a session on taxonomic philosopohy as well as to complain of the proposed sessions on evolutionary biology. If no one else could be found, Sokal would volunteer to organize the symposium on contemporary systematic philosophy. Once again the organizers of the congress rapidly agreed that a session on taxonomic philosophy was called for but decided that Sokal was too closely identified with one faction to organize the session. Sneath was on the international committtee for the congress, but he too would be viewed as too partisan. Instead, they settled on another member of the international committee, a philosopher who had responded to Ghiselin's paper at the Ann Arbor meeting. Hull was a professional philosopher and not committed to any of the schools of systematics.

For speakers at the session on numerical taxonomy, Crovello chose himself, Sokal, Sneath, Rohlf, Farris, Jardine, Peter Raven, and M. M. Goodman. By this time, three quite distinct schools of systematics had emerged: evolutionary systematics, supported by Simpson and Mayr; phenetics, favored by the numerical taxonomists; and a school that Mayr (1965) had dubbed "cladistics" (more of cladistics shortly). Hull chose a Canadian entomologist, G. C. D. Griffiths, to introduce the symposium. Although Griffiths considered himself a cladist, he learned his principles from George Ball at the University of Alberta and was not a party to the uproar over cladistics south of the Canadian border. Sokal represented the pheneticists, Walter Bock presented the views of traditional evolutionary systematics, and Gareth Nelson of the American Museum argued for the principles of the emerging cladistic school. The panel of discussants included Heywood, Michener, a young pheneticist named W. W. Moss, and Peter Ashlock, recently hired by Michener at Kansas.

If attendance is any sign, the session on contemporary systematic philosophy was a great success. The large hall was so full that latecomers had to sit on the steps leading down to the stage. Griffiths summarized the basic principles of the three leading schools of taxonomy. Bock argued that traditional evolutionary systematics provides the best approach to classification, according to Karl Popper's demarcation principle, while Nelson set out in startling detail the difficulties that confront anyone who seriously tries to represent explicitly various sorts of phylogenetic relationships in a hierarchical classification. As Sokal rose to speak, his wife whispered to the person next to her, "I hope they listen to him this time." Surprisingly, Sokal's paper did not concern taxonomic philosophy, phenetic or otherwise. Instead Sokal summarized the great diversity in modes of speciation to be found in the living world. Although the papers themselves were subdued in tone, the discussion afterward was not. Tempers flared. Nelson took special offense at Michener's supporting a point by reference to what a majority of taxonomists believed. Nelson objected that science is not a matter of majority rule. Several members of the audience ventured that they had heard all this before. As the afternoon wore on, Mayr got up from his place in the audiencee, strode to the stage, and suggested to the moderator that it was time to bring the proceedings to a close.

In 1973 Sokal and Sneath published a second edition of their book, this time with the order of authors reversed and the title shortened simply to *Numerical Taxonomy.* In their original book, Sokal and Sneath were able to cite only about two dozen papers in which the principles of systematics they favored were advocated or developed. By 1973 the bibliography in Sneath and Sokal boasted almost 2,000 titles, the vast majority of which either supported their taxonomic philosophy or simply took it for granted. The major change in this second edition was that Sneath and Sokal had decided that their early emphasis on genes was misplaced. Because only a small proportion of an organism's genes are functioning at any one time and some never become functional, Sneath and Sokal (1973: 96) concluded that the nexus hypothesis "has lost some of its relevance to phenetic taxonomy." Furthermore, the congruence between classifications based on different sorts of characters had turned out to be even less close than they had at first expected. As a result, their earlier factor-asymptote hypothesis was of little utility and the matches-asymptote hypothesis not much better. Even though Sneath and Sokal (1973: 107) admitted that possibly "phenetic similarity is not a single quantity but a shifting concept depending on the method of measurement as well as the character base," they were not willing to go along with Ehrlich and reject totally the notion of overall similarity (Ehrlich and Ehrlich 1967). Instead they suggested a principle of inertia:

> As more and more characters are added, it takes an increasingly large number of characters with quite different phenetic information to alter appreciably a given estimate of phenetic similarity. Thus, while classifications of the same OTU's based on different sets of characters might start out as different constellations in phenetic hyperspace, they would eventually converge toward the same general region, though they would not necessarily be identical as more characters were added to the system.

In the hubbub over numerical taxonomy, hardly anyone noticed scurrying around in the underbrush the ancestors of the next dominant group in systematics, as inconspicuous and active as the progenitors of the mammals had been in the age of the Dinosaurs.

Cladistic Analysis

Science is supposed to be international, and to some extent it is, but language differences can form very real barriers. In 1950 an East German entomologist, Willi Hennig (1913–

76), published a formidable treatise entitled *Grundzüge einer Theorie der Phylogenetischen Systematik*. In this work, Hennig took seriously the claim that phylogenetic classifications are to represent phylogeny. As numerous systematists had noted before him, there is no way that all the details of phylogenetic development can be represented in so simple a system as a traditional hierarchic classification. Instead of opting for a vague reflection of several factors in a classification, Hennig settled on one—the sister-group relation. Two taxa, B and C, are sister groups if they are more closely related to each other than to any third taxon, A. The evidence for this relationship is the presence of characters that B and C exhibit but A lacks. The sister-group relationship is collateral, not ancestor-descendant. B and C must share a more recent common ancestor with each other than either does with A, but none of the taxa mentioned in the statement of a sister-group relation are claimed to be ancestral to any other (Hennig 1966: 74). In the epigraph introducing his *Grundzüge,* Hennig agreed with Francis Bacon that truth emerges more readily from error than from confusion. Perhaps his phylogenetic systematics might prove to be mistaken, but no one was going to be able to accuse it of being confused.

Initially, Hennig had little impact on taxonomic disputes among English-speaking systematists. The first mention of Hennig in the pages of *Systematic Zoology* was by G. C. Steyskal (1953: 41), who merely translated two passages from Hennig asserting that the goal of systematics is to provide a universal reference system for biology. In 1959, in response to Bigelow's (1956, 1958) objections to the intrusion of phylogeny into classification, Sergius G. Kiriakoff (1959: 118), an invertebrate systematist working in Belgium, brought Hennig's system to the attention of the readers of the journal, complaining that it is a "pity that modern phylogenetic systematics seems quite unknown in the United States." In a footnote to his major work on systematics, Simpson (1961: 71) apologized for not paying greater attention to Hennig's ideas, but he had to admit that he had not read Hennig's book until after he had finished his own.

Kiriakoff (1962, 1963) continued to defend phylogenetic systematics in subsequent papers but now against the criticisms of the neo-Adansonians (i.e., the pheneticists). As a result of Kiriakoff's papers and Simpson's footnote, Sokal obtained a copy of Hennig's *Grundzüge* and worked his way through it. Even though German was Sokal's native language, he found Hennig's prose very tough going, but by the time that he had struggled through the book he was more than impressed by the sophistication of Hennig's views. For example, Hennig distinguished clearly between two senses of "similarity"—in Sokal's preferred terminology, between phenetic similarity and phylogenetic relationship. Hennig then went on to distinguish between two senses of "phylogenetic relationship"—his own preferred sister-group relationships and ancestor-descendant relationships. According to Hennig, a truly phylogenetic classification does not indicate overall similarity, or ancestor-descendant relationships, but only sister-group relationships. One might think that species A gave rise to species B and C, but such information cannot be included in a phylogenetic classification.

Hennig steadfastly maintained that systematics had to be based on some all-inclusive explanatory principle and that the only candidate for this principle is phylogeny. In their book Sokal and Sneath (1963: 265) conceded that phylogeny is "indeed an all-explanatory principle" but objected that it cannot be used in "classificatory procedures, since we mostly do not know (and in many cases cannot know) its true course." Sokal and Sneath were not impressed by the various procedures that Hennig set out for inferring phylogeny. They found them too liable to produce error. Nor were they impressed by Hennig's method of "reciprocal illumination," according to which scientists reason from one sort of evidence to another, each time correcting past errors and expanding the scope of their hypotheses. To Sokal and Sneath, reciprocal illumination looked too much like reasoning in a circle. Their goal was to develop clear, logical, straight-line methods for constructing classifications—no circles, no spirals.

Soon after the appearance of the book he wrote with Sneath, Sokal met Hennig at the Twelfth International Congress of Entomology in London in August of 1964. Sokal hoped to discuss systematic philosophy with Hennig. Although Hennig spoke very little English, Sokal was fluent in German. However, Hennig was a very shy and self-effacing man. It also turned out that he was unaware of the work that was being done in the English-speaking world on quantifying taxonomic decisions. Sokal had read Hennig, but Hennig had not read Sokal. The meeting produced very little reciprocal illumination. Hennig's (1965: 100) only response was to dismiss Sokal's repeated contention that phenetic classifications are more fundamental than phylogenetic classifications. However, Sokal and Sneath (1963) did convince Kiriakoff (1965: 63) that phylogenetic classifications are usually not possible "owing to lack of necessary information."

Language was also no barrier for Mayr in reading Hennig's book. Although Mayr's 1965 paper was billed as a critique of numerical phenetics, he devoted at least as much attention to Hennig's system. In order to distinguish the principles of systematics which he himself preferred from those of Hennig on the one hand and those of Sokal and Sneath on the other, Mayr decided that each system had to be given a clear and unequivocal name. Sokal and Sneath had settled on "numerical taxonomy" over "neo-Adansonian," while Hennig termed his system "phylogenetics." In the early years of systematics, taxonomists insisted that the names of biological taxa accurately describe them. If a plant originally termed "alba" turned out to exist in a variety of other colors as well, the name had to be changed. However, the instability introduced into classifications by this practice forced systematists to abandon it. Certainly, Mayr was not about to reject the name "Carnivora" just because pandas are largely herbivorous. Even so, Mayr was not willing to extend this same principle to the names of taxonomic schools. They had to be accurate, and neither "numerical taxonomy" nor "phylogenetic systematics" accurately distinguished the relevent groups. After all, nothing prevented advocates of any of the major schools from using numerical techniques, and the Mayr-Simpson school had as much right to be termed "phylogenetic" as did the system advocated by Hennig. Splitting was certainly a phylogenetic relation, but then so was subsequent divergence.

Mayr (1965) proposed to term the Sokal-Sneath school "phenetic" because all that its members wanted to represent in their classifications were appearances—the overall similarities exhibited by organisms regardless of descent. He found the word "cladistic," formed on the Greek word for "branch," appropriate for Hennig's system because all Hennig wanted to represent was branching sequences. In Mayr's terminological conventions, it became possible to distinguish, for instance, numerical phenetics from numerical cladistics. Both are part of numerical taxonomy, but numerical cladistics is at variance with the basic principles of numerical phenetics. For his part, Mayr thought that his own "evolutionary" classifications should reflect both order of branching (cladistics) and various degrees of divergence (phenetics).

Although his opponents were not initially very happy with the names that Mayr had chosen for them, there was very little they could do about it, especially when others began to adopt them. Numerical taxonomists still refer to themselves as numerical taxonomists and treat any papers on quantitative techniques in systematics as contributions to their own research program; but others do not. Outsiders identify the school started by Sokal and Sneath more with their phenetic philosophy than with their advocacy of numerical techniques. Initially cladists despised the term "cladists," but they reluctantly adopted it. More recently, certain cladists have resurrected "phylogenetics" to distinguish themselves from what has come to be termed "pattern cladism," but more of this later. At the very least, Mayr's paper forced Kiriakoff to realize that Hennig's system was quite different from the one urged by Mayr and Simpson. Kiriakoff (1963: 93) concluded sadly that "cladists are but a rather rare subspecies of taxonomists." That state of affairs was not to last for long.

In 1965 Hennig published a short summary of his views in English; a year later, he published *Phylogenetic Systematics* (1966), a translation of an extensive revision of his

Willi Hennig, Robert L. Usinger, and Robert R. Sokal, at the Twelfth International Congress of Entomology, London, August 1964. (Photo by George W. Byers.)

Grundzüge. When D. Dwight Davis of the Field Museum contacted Hennig in 1960 about translating his German work, Hennig was in the midst of revising it. Hennig sent this revised manuscript to Davis. Although Davis was fluent in German, he found that he was having great trouble with Hennig's prose. He appealed to one of his colleagues at the museum to help him, Rainer Zangerl, the Swiss-born paleontologist who had so angered Simpson years before by claiming that the science of morphology was not only independent of any evolutionary considerations but actually prior to them. However, these two men did not just translate the manuscript. They also heavily edited it, eliminating what they took to be repetitive passages, simplifying Hennig's Teutonic sentences, and clarifying his ideas. In this midst of this undertaking, Davis died, and Zangerl had to carry on alone. As fate would have it, both men who translated Hennig's manuscript were trained in the very philosophical tradition that Hennig attacked in his book—idealistic morphology. Both Zangerl (1948) and Davis (1949) emphasized the necessary role that the hierarchies of morphological types of neoclassical morphology play in phylogeny construction, a position that Hennig himself adamantly opposed. Only a German scholar studying the relevant manuscripts can say how much the idealist presumptions of Davis and Zangerl influenced their translation. Hennig himself was unable to help much in the project because he was in the midst of fleeing from East to West Germany. An accurate translation of his manuscript was the least of his worries.

In his *Phylogenetic Systematics* Hennig treated three main topics: the species question, inferring phylogenetic relationships, and translating these relationships into classifications. Hennig accepted the gene-pool notion of species so central to the synthetic theory of evolution and the New Systematics. Although he recognized that the limits of particular species are generally inferred indirectly from morphology, he distinguished between the theoretical factors included in the definition of species category (e.g., gene flow) and the evidence that one uses to decide that the theoretically significant criteria have been met (e.g., character distributions). For Hennig, as for Simpson and Mayr, species are the basic units of the evolutionary process. They are the things that evolve as a result of mutation and selection. They are also the basic units in classification. Evolution and classification intersect at the species category. Finally, Hennig insisted that the characters used to infer both species status and phylogenetic relationships must be evolutionary homologies.

The only peculiar feature of Hennig's discussion of the evolutionary process is that he considered organisms and species to be basically the same sort of thing. Just as an organism is an integrated sequence of momentary stages (semaphoronts) in its life cycle, a species is an integrated sequence of parent-offspring (tokogenetic) relations. As a result, both organisms and species are spatiotemporally localized systems. Not until the species level is reached does the genealogical network become hierarchical in structure and the principles for inferring phylogeny and constructing classifications come into play. For Hennig, inferring phylogeny is like reconstructing a vase from its parts, not discerning classes of similar objects. If the phylogenetic tree is a connected whole, then Hennig (1966: 93) concluded that the characters used to infer phylogeny must themselves be connected and not just abstract similarities:

> Different characters that are to be regarded as transformation stages of the same original character are generally called homologous. "Transformation" naturally refers to real historical processes of evolution, and not to the possibility of formally deriving characters from one another in the sense of idealistic morphology.

In geometry, ellipses can be made to form atemporal, abstract transformation series as the distance between their foci is increased. Hennig's transformation stages of characters are real changes through time.

One important way to view a scientific research program is by taking note of its opponents, and Hennig's chief enemies in Germany were ideal morphologists. In the late nineteenth century, Darwinism successfully replaced idealism in Germany, but in the 1920s and 1930s it underwent a resurgence. Goethe became the patron saint of this new wave of ideal morphology, and O. H. Schindewolf (1897–1971) its most successful expositor (Grene 1958, Reif 1986). Because ideal morphology had made so little headway in Great Britain in Darwin's day, Darwin was able to ignore this alternative. Hennig was in no such position. German ideal morphologists were his most powerful opponents. Just as Darwin's *Origin of Species* was one long argument against creationism, Hennig's *Grundzüge* was directed primarily against the priority claimed by German idealists for morphology over phylogeny. This same antipathy to "typology" characterized evolutionary systematists in the English-speaking world. That is why Zangerl's (1948) defense of typological methods of comparative anatomy so angered Simpson. When such pheneticists as Sokal (1962) attempted to rehabilitate "typology" as anything but a pejorative term, they were singularly unsuccessful.

Although Hennig found no fault with the synthetic theory of evolution, he was not especially concerned with the evolutionary process in his book. He was more interested in phylogeny reconstruction and the relation of these inferred relationships to classifications. Although he placed much less emphasis on fossils in reconstructing phylogenies than most paleontologists did, the methods he described for reconstructing phylogeny were not especially new. Rather his chief contribution was the clarity with which he set out his principles and the emphasis which he placed on them. According to Hennig, the fundamental relation in phylogenetic systematics is the sister-group relation or, as he usually termed it, the phylogenetic relationship. This relationship can be represented in two ways—either as a hierarchical classification or as a phylogenetic diagram of the sort later to be termed a "cladogram" (see fig. 4.1).

The important feature of (a) and (b) in figure 4.1 is that they express precisely the same information. They are isomorphic. They state that B and C are more closely

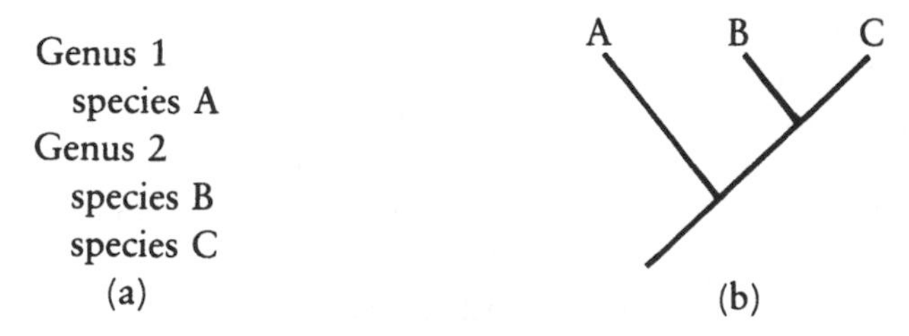

Figure 4.1. Classifications and cladograms.

related to each other than either is to A—and that is all. Nothing more should be read into either the classification or cladogram. Because the diagram looks something like a phylogenetic tree, it is easy to think of it as a highly stylized tree (fig. 4.2), but it is not. In a phylogenetic tree, each line represents a species, the splitting of one line into two represents a stem species splitting into two daughter species, and so on. In both (a) and (b) above none of the letters represent stem species. B and C share a stem species that neither share with C, but that stem species is included in neither the classification nor the diagram.

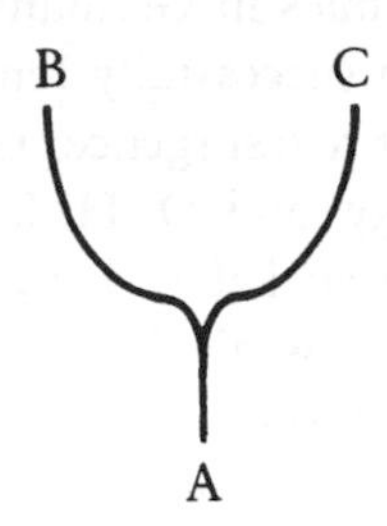

Figure 4.2. A phylogenetic tree.

The confusion engendered by ambiguities in representation was further abetted by the terminology which Hennig chose to distinguish them—phylogeny diagrams versus phylogeny trees, a distinction that often got lost in a complex line of argument. One of the chief sources of the misunderstandings that plagued the introduction of Hennig's work into the English-speaking world was the confusion of cladograms and trees. Both are branching diagrams, but they represent very different relations. Only one aspect of a tree appears in a cladogram—sister-group relations. Furthermore, for Hennig cladograms were only temporary tools to be used to construct classifications and then discarded. Phylogenetic trees have a more lasting value.

According to Hennig, phylogenetic classifications and phylogenetic diagrams (cladograms) must be strictly isomorphic. The same can be said for characteristics—true phylogenetic characteristics. Once the characters used to construct classifications and cladograms have been properly individuated, they too must form perfectly nested sets. The reason that pheneticists kept coming up with polythetic taxa is that the characters that they used were a hodgepodge. The distinction was clear. Once characters have been arranged properly in Hennigian transformation series, they must nest perfectly. Actual practice was much more difficult. Characters that superficially seem to be the same, frequently are not. The result is conflicting sets of characters: one set implies one cladogram, another set implies a second cladogram.

But the part of Hennig's system that caused the most consternation were the conventions he devised to allow for the unequivocal representation of sister-group relations in classifications. For example, he maintained that when an ancestral species speciates, it must be considered extinct. As a result, species are lineages that connect speciation events. A single lineage might persist while budding off successive daughter species without changing any of its own characteristics. Even so, for the purpose of phylogenetic classifications, this persistent lineage should be considered successive species (Hennig

1966: 212). Conversely, a single lineage must not be subdivided into successive chronospecies no matter how much the lineage might change through time. Only when splitting occurs should new species be recognized. In addition, phylogenetic systematists must proceed dichotomously in the production of phylogenetic diagrams and classifications.

These and other conventions raised howls of protest primarily because they were interpreted as descriptions of the evolutionary process, as if Hennig were claiming that speciation is always dichotomous. Although Hennig occasionally justified the preceding conventions by reference to the empirical world, he took them to be first and foremost methodological conventions necessary to facilitate unequivocal representation in classifications of sister-group relations. At first glance, the relations between Hennig's conventions and the difficulties in constructing cladistic classifications are far from apparent. However, anyone who sits down with paper and pencil and works out the various outcomes that result from the modification of any of Hennig's prinicples soon realizes that he did not choose them capriciously. Because these lines of reasoning tend to be extremely involved, I can treat only one here.

The only relation represented in cladograms and their isomorphic classifications are sister-group relations. If a species is considered to go extinct when speciation occurs, then an ancestor species cannot co-exist with its descendant species. No two contemporaneous species can be related by the ancestor-descendant relation. The only relation possible is sister-group. Problems arise when we try to include species at different time horizons in the same classification, as Hennig (1965, 1969) thought that we can and must. What if by chance we happen on the remains of a species that actually was the immediate common ancestor of two extant species (fig. 4.2)? Because it appears earlier, it might be a common ancestor, but is it? What character distributions might lead a systematist to recognize it as a common ancestor? Some of the characters of A might have been transformed as part of it evolved into B; others may have been transformed as another part of it evolved into C; or one and the same characters may have evolved differently into the two descendant species.

If we chart these transformations, the end result is that character distributions do not nest. Some imply one set of sister-group relations; some imply another. If this were the only situation that gave rise to character conflicts, then whenever such conflicts arose, we might safely assume that one of the species under investigation actually was the common ancestor of the other two, but unfortunately numerous different explanations can be given for this state of affairs. The first, and most likely, is that some of the characters being used are not genuine characters. They have been misidentified or have been arranged inappropriately in transformation series. Another is ignorance. Not enough characters have been studied. But mistakes and ignorance to one side, other events in phylogenetic development can produce these same sorts of character distributions. For example, A may have resulted from hybridization between B and C, or these three species may have arisen simultaneously from a single unknown ancestral species. By and large, well-chosen characters should nest. When they do, the most likely explanation is that the groups under study are sister-groups. When, after extensive investigation, characters do not nest, "phylogenetic systematics is up against the limits of the solubility of its problems" (Hennig 1966: 211). The discovery of nesting transformation series is necessary for Hennig's phylogenetics; whether it is also sufficient is another and more controversial question.

Assuming that one thinks that a particular extinct species was the actual stem species of two or more descendant species, one must find some way to classify this stem species so that the actual relations are expressed unambiguously. By simple inspection of a classification or cladogram, one should be able to tell which species are sister-groups and which ancestral. As long as all the species in a classification are contemporaneous, this problem never arises, but as soon as one attempts to include species at more than one time horizon in the same classification it does. As far as I can see, Hennig never solved this problem (see Hennig 1969 for his most detailed discussion).

Although Hennig himself does not refer explicitly to trichotomous cladograms, implicit in his discussion is that any time systematists are forced to include anything other than a dichotomy in their classifications, they are introducing ambiguity (fig. 4.3). Systematists may be mistaken about the relations they express in their dichotomous classifications and cladograms, but if they are not, they are unequivocally representing one and only one relation. Multiple branchings can stand for too many things. The trichotomous classification and cladogram illustrated in figure 4.3 can stand for too many different phenomena: an ancestral species giving rise to two descendant species, two species generating a hybrid species by introgression or allopolyploidy, not to mention three descendant species arising from a single ancestral species.

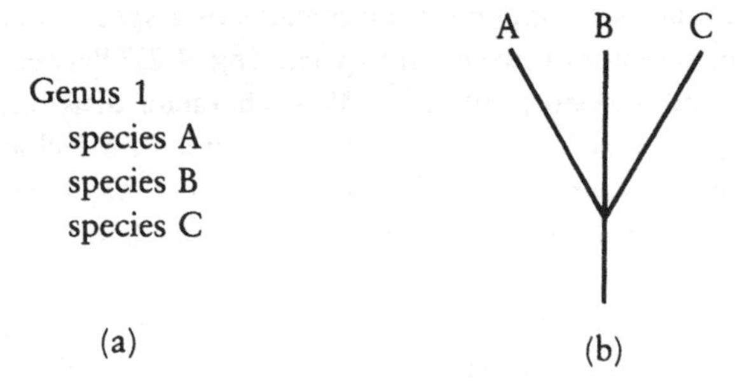

Figure 4.3. A trichotomous classification and cladogram.

In short, as arbitrary as Hennig's methodological conventions may have seemed to many, on a little reflection the reasoning behind them is clear. If cladistic classifications are to be kept as unambiguous as possible, then multiple branchings must be kept at a minimum. The methodological conventions that Hennig so carefully formulated are designed to do just that. Far from being idiosyncratic ideas about the evolutionary process, they are well thought out methodological conventions introduced to fulfill a specific purpose. If they are adopted, trichotomous cladograms are kept at a minimum. They must be kept at a minimum because they are inherently ambiguous, and Hennig wants at all costs to keep his system free from ambiguities. He prefers possible errors to guaranteed confusion. However, few of Hennig's readers saw the issues this way. He was consistently read as claiming that speciation is always dichotomous, that one daughter-species always diverges more from the mother-species than does the other daughter-species, and so on.

Another feature of Hennig's phylogenetic systematics that caused considerable controversy was his definition of "monophyly." Simpson's definition of this important term is quite weak and entirely retrospective. According to Simpson (1961), all it takes for a higher taxon to be monophyletic is for it to arise from a single immediately ancestral taxon of its own or lower rank. For example, in figure 4.4a, genus B counts as being monophyletic for Simpson even though it is derived from two ancestral species, because both of these species belong to the same immediately ancestral genus. Mayr's (1969) preferred definition of "monophyly" is also entirely retrospective but stronger than Simpson's definition, because for Mayr all higher taxa must arise from a single, immediately ancestral species. Higher taxa are not good enough. According to Mayr, genus B in figure 4.4a is polyphyletic, while both genus A and genus B in figure 4.4b are monophyletic because each stems from a single, immediately ancestral species.

Hennig's definition of "monophyly" is both prospective and retrospective. Not only must all higher taxa stem from a single, immediately ancestral species, but in addition *all* the species that arise from a single stem species must be included in the same higher taxon. For Hennig, genus B in figure 4.4a is polyphyletic because it arises from two immediately ancestral species, and genus A is paraphyletic because not all the species arising from its stem species are included in the same taxon with it. Only at the next category level are these two genera grouped together in a single taxon. In figure 4.4b, genus B, when taken by itself, is monophyletic, but genus A is paraphyletic. Not all the descendants of its stem species are included in the same higher taxon with it at the appropriate level. For Hennig, genus A and genus B in figure 4.4c are monophyletic because they fulfill both of his requirements.

As much as one might object to Hennig's definition of "monophyly" on a variety of counts, it can be applied consistently to species in a single time horizon. Problems arise when one attempts to include stem species with the species to which they gave

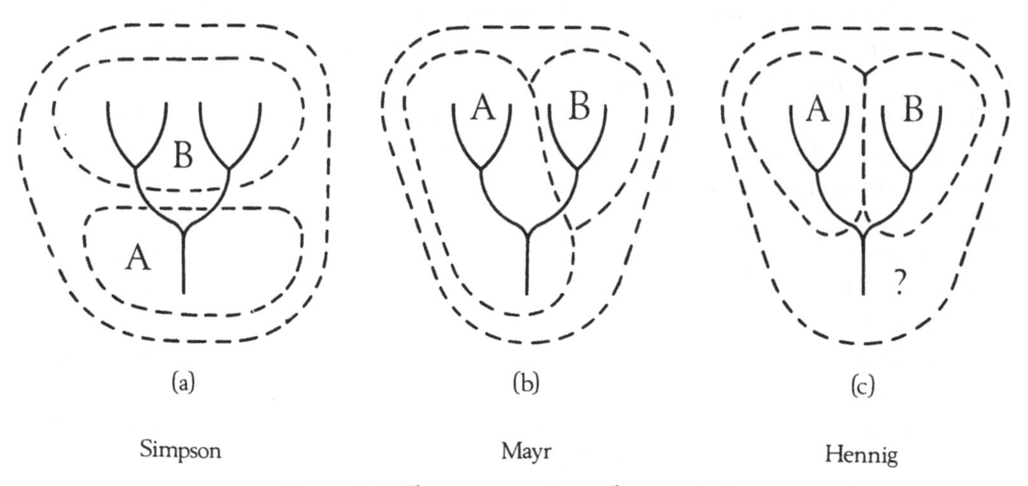

Figure 4.4. Three conceptions of monophyly.

rise. For example, in figure 4.4c, is the stem species which gave rise to genus A and Genus B to be included in both descendant genera or in no genus whatsoever? One way out of this problem is to refuse to recognize any species as being ancestral. Some species *have* to be ancestral, but epistemology can come to the rescue. One can claim that even though some species might be ancestral, there is no way of ever knowing which are which. Hence, problems about classifying stem species unambiguously are only in-principle problems.

The net result of adopting Hennig's definition of "monophyly" is that all higher taxa are so structured that they are complete branches of the phylogenetic tree (i.e., clades). Initially, two points were at issue: whether all taxa should be monophyletic in Hennig's sense or in one of the senses preferred by Simpson and Mayr, and which definition of "monophyly" is the "correct" one. To the outsider neither issue might seem important, but to those involved both were equally momentous. Later, as Hennig's descendants exposed his system to intense analysis, the problem of how to classify ancestors and descendants together so that the resulting taxa are monophyletic (in Hennig's sense) came to the fore.

Everyone from pheneticists to evolutionists tirelessly repeats the slogan that classifications are systems for information storage and retrieval. Hennig took the slogan seriously. Numerous workers before and after him have discussed methods by which phylogenetic relations can be inferred, but he went on to discuss how this information is to be stored in a classification in such a manner that it can be retrieved. If all taxa are "monophyletic" in Hennig's sense, then they will all be clades. As a result cladistic classifications in their purist form turn out to be extremely asymmetrical, resulting in cladograms characterized as "Hennig's comb": A and everything else, B and everything else, C and everything else, and so on. As asymmetrical as strictly cladistic classifications may be, at least all the information that goes into the construction of cladistic classification can be retrieved from it. Phenetic classifications could have this same virtue. If all taxa are phenetic clusters and the principles that are used to form these clusters are stated explicitly, then the information stored in phenetic classifications can be retrieved. Phenetic classifications could have this virtue but they need not, as an exchange at the Boulder Congress between Vernon Heywood (a pheneticist) and Joel Cracraft (a cladist) clearly indicated.

To those assembled on the podium, Cracraft (1973: 399) complained that everyone claimed to put information into a classification but no one was saying very much about how this information was to be gotten out. "If I have a hierarchy here, I may think that I've used this information to construct it, but how do we read it out?"

Heywood (1973: 398–99) responded, "The answer is, of course, that you can't read it out." A hierarchical classification "will *contain* a vast amount of information which it will not *express*, and I think that's the difference."

Cracraft was not satisfied. "I agree that you cannot read it out. Then why pretend that you've put it in?"

Heywood (1973: 399) tried again. "Perhaps you've misunderstood what I said. It can be *retrieved* from a classification but it will not be *expressed* by the classification."

"No," Cracraft objected, "can we *retrieve* it? That's what I'm asking." The exchange went nowhere, as Cracraft and Heywood continued to talk past each other. Cladists and pheneticists alike lodged the same complaint against the evolutionary systematists' habit of interspersing grades and clades in their classifications in such a way that no one can tell which taxa are which. The general problem is how to feed information into the construction of a classification in such a way that it can be retrieved. What does it mean to say that information can be retrieved from a classification when that classification does not express it?

Which sorts of taxa are to be termed "monophyletic" may seem a much less important question. What's in a word? However, the literature on the "proper" definition of "monophyly" was almost as polemical and acrid as the debate over whether or not Darwin was really a cladist. Soon after Sokal and Rohlf left the University of Kansas, Peter Ashlock joined the staff. In his first paper, Ashlock (1971) showed that the systems of terminology preferred by both sides of the dispute were intertranslatable. Anything that anyone wants to say in one set of terms can be said just as easily in the other set. For his part, Ashlock preferred to use "monophyly" in Mayr's sense and to term Hennig's concept "holyphyly." Hennig's descendants would have none of it.

On one thing both sides agreed: there is one and only one "proper" use of the relevant terms—their usage—and their opponents were misusing these terms in an especially pernicious way. The dispute continues undiminished to the present. According to the parties to this feud, anyone who does not acknowledge that terms have proper uses that must be adhered to does not understand language, and anyone who does not see that the usage which the speakers prefers is the one true usage is bigoted, blind, or both.

Terminological decisions always pose problems. When scientists come up with new ideas, they have to decide whether to coin a new term for this new idea or to borrow and transform an old one. No matter what the choice, objections will be raised. Scientists are no more fond of neologisms than are members of the general public. For instance, numerous objections were raised to the introduction into English of such terms as autapomorphy, synapomorphy, and plesiomorphy. Why all this biobabble? But cladists found the closest correlative terms from evolutionary systematics had inappropriate connotations, connotations that were likely to be carried over into their own usage and to cause all sorts of confusion, the very sorts of confusion that their retention of "monophyly" for Hennig's concept generated.

Although scientists not involved in a particular terminological squabble are likely to dismiss it as being "merely semantic," more than conceptual clarity is involved in such disputes. By and large it is much easier to say what one wants to say in one's own terminology than in the terminology preferred by one's opponents, even when on occasion there are some fairly precise mappings (e.g., the equivalence of Hennigian monophyly and Ashlock's holophyly). Getting one's preferred terminology adopted strongly biases the game in one's favor. But more than this, terms are out in front for all to see. Meanings are a good deal less conspicuous. If the terminology of a particular research program is adopted, subsequent generations will assume that this research program succeeded, no matter how much the meanings of these terms may have been transmuted in the process. For example, if cladistic terminology becomes standard, then subsequent generations will assume that the cladists won even if the system that eventually becomes standard turns out to be indistinguishable from the system that Hennig set out to refute. Conceptual lineages, at a fundamental level, are much more important than terminological lineages, but they are also a good deal more difficult to distinguish.[3]

In the same year that Hennig's *Phylogenetic Systematics* appeared, Lars Brundin published a monograph on the transantarctic relationships of chironomid midges in which he championed Hennig's system of phylogenetic analysis. The distribution of species around the South Pole has fascinated biogeographers since the time of J. D. Hooker, Darwin's closest friend. Hooker used the striking resemblances between disjoint populations of the species inhabiting the land masses circling Antarctica to support Darwin's theory of evolution. In a series of books, Hooker explained the biogeography of the antarctic region by means of migrations via Antarctica across land connections. Darwin agreed with Hooker that the biogeography of the antarctic region supported his theory of evolution but that the distributions of plants and animals were best explained by means of long-distance dispersal over intervening stretches of ocean. During the next century, the dispersalist views of Darwin prevailed, culminating in the work of W. D. Matthew (1871–1930), G. G. Simpson, and P. J. Darlington. As these men interpreted the data, older, more primitive species have been crowded south by the expansion of more vigorous species from the north. As a result these primitive species have piled up at the southern tips of Africa, South America, and Asia where a few succeeded in dispersing to various island refuges and to Australia.

Brundin (1966) rehabilitated Hooker's view that Antarctica is a center of evolution. Brundin maintained that the mistake earlier biogeographers made was to misanalyze the phylogenetic relationships among the species under investigation. To do biogeography, one has to adopt Hennig's method of phylogenetic analysis, especially his deviation rule. According to Hennig's (1966: 169) biogeographic method, sister-groups frequently differ in their geographic distribution and, when they do, they replace each other geographically. He termed the phenomenon of sister-groups replacing each other in space "vicariance." In passing, Hennig (1966: 207) remarked that, in the process of speciation, one of the two daughter species "tends to deviate more strongly than the other from the common stem species (or from the common original condition)."

3. Several readers of this manuscript, including Mayr, found it extremely unphilosophical for a philosopher not to come down in favor of correct usage, but this particular terminological conviction runs contrary to one of the main theses of this book. Like species, terminology evolves, and this evolution is not always neat and logical. Endless confusion has been caused by Darwin's continuing to call species "species" when this term had always referred to static entities, by his calling the transmutation of species "evolution" when at the time this term usually referred to embryological development, and so on. If only de Vries's term "mutation" had not caught on, life would have been simpler. I for one find Hennig's polysyllabic terminology irritating. After years, the "heterobathmy of synapomorphy" still does not trip lightly off my tongue. But no matter: the tide of terminological change can no more be arrested by protests of those of us who are more terminologically conservative than can the flow of molten lava from an erupting volcano be stayed by prayer. Like it or not, "unique" is coming to mean "rare" and adverbs are going extinct.

To the extent that sister groups replace each other in space and characters form transformation series, the two phenomena together should give an accurate picture of the biogeography of the species through time.

The first really negative comments on the work of Hennig and Brundin by anyone closely associated with the pheneticists were by the Australian entomologist, Donald Colless. Colless (1967b: 294) termed the basic error committed by the Hennigians the "phylogenetic fallacy," the belief that, in "reconstructing phylogenies, we can employ something more than the observed attributes of individual specimens, plus some concept of 'overall resemblance' and some concept of an 'attribute' of a set or class of such specimens." According to Colless (1967a: 17–18), all attempts to reconstruct phylogeny ultimately rest on a phenetic classification, and a phenetic classification is "not a hypothesis, but a datum, a convenient summary of observed facts." As Colless (1967a: 6–7) expressed himself more fully:

> Phenetic taxonomy, in the sense employed here, is an empirical procedure that demonstrates and describes a particular form of classification (non-overlapping hierarchical), which we recognize as applicable to entities (not necessarily biological) of the material world; and it does that by reference only to the observed properties of such entities, without any reference to inferences that may be drawn *a posteriori* from the patterns displayed. Such a classification can, and, to be strictly phenetic, *must,* provide nothing more than a summary of observed facts.

According to the preceding description, phenetic classifications are nothing but summaries of observed facts. Colless (1967b: 289) did not deny that we have available a "body of reasonably credible phylogenies," but he insisted that many taxonomists have an "erroneous view of the process by which such phylogenies are inferred." To be specific, because all methods of phylogeny reconstruction depend on observations, at bottom they are all phenetic. Anyone who bases his or her work on observations automatically grants priority to phenetics. Hence, all scientists are really pheneticists, whether they realize it or not. As in the case of Sokal and Sneath's' original book, Colless concluded that codons are the ultimate approximation to unit attributes. Given Colless's emphasis on observation and the fact that no one has ever observed a codon, his selection of codons as the closest approximation possible to unit phenetic attributes seems paradoxical. Apparently, Colless was not bothered by inferences to very small things, such as codons, or to very large things, such as black holes, because they all exist in the here and now. He reserved his skepticism only for those things that occur in the past. Finally, Colless (1967b: 292) concluded that Hennig's system is "simply an intuitive, prototypical form of statistico-phenetic taxonomy."

Phylogeneticists and evolutionists alike objected to Colless's characterization of Hennig's system. Dieter Schlee (1969: 133), for example, complained that Hennig's principles are not "phenetic" because cladists use only a small subset of the characters exhibited by their specimens. They are not "statistical" because the number of shared-derived characters (synapomorphies) is of no importance—one will do as well as twenty in grouping two taxa together. Nor is Hennig's system "intuitive" because Hennig spelled out his principles precisely and explicitly (see also Bock 1968 and Ghiselin 1969b). Colless (1969) responded that Hennig's principles are "phenetic" because they depend on the existence of higher-level phylogenies and these are ultimately phenetic.

They are "statistical" in two senses. First, they are empirical and, hence, subject to the laws of chance, and second, true synapomorphies are distinguished from convergences by means of the greater number of congruences. Finally, Colless (1969: 142) insisted that Hennig's system was "intuitive" because higher-level phylogenies are assumed "without overt recourse to a formal sysem of inferences."

While this skirmish between certain pheneticists and advocates of Hennig's system was brewing, a young ichthyologist from Chicago, Gareth Nelson, was visiting the Swedish Museum of Natural History in Stockholm. While thumbing through recent acquisitions in the library of the museum, he came across Brundin's (1966) chironomid midge monograph. He read through the introductory chapters, then read them twice more. Brundin had codified the method that he himself had been using all along. When Nelson shared his discovery with a friend at the museum, he was informed that Brundin was currently working at the museum. To Nelson's deferential questions, Brundin replied clearly and politely. Nelson was even more intrigued.

In November of 1966 Nelson journeyed to London to be interviewed for a position at the American Museum by Donn Rosen. Rosen was visiting the British Museum (Natural History) for a couple of weeks. After the interview, Nelson returned to Stockholm where he urged the members of the organizing committee of an upcoming Nobel Symposium to include Brundin as a speaker. In the spring of 1967 he traveled to Copenhagen and discussed Brundin's ideas with Niels Bonde at the Zoological Museum. Returning to London and the British Museum, he broached the subject of phylogenetics to some of the members of the museum. Because of the flammable spirits used to preserve specimens, smokers had to congregate outside the building under the colonnade. Included in this group were Nelson, Humphrey Greenwood, and Colin Patterson. Although the lanky youngster from the South Side of Chicago was anything but imposing in his white socks, shirt open at the collar, and brown-and-tan-checked sport coat, Nelson was relentless in setting out the logic of Brundin's system. Patterson soon realized that the acceptance of Brundin's system meant all but abandoning his own lifework. All his ancestor-descendant phylogenetic sequences were worthless. The suave Englishman with his resonant voice was convinced by the persistent young visitor from the States.

As it turned out, Patterson, Greenwood, and Bonde had all been invited to attend the Nobel Conference. They were joined by Bobb Schaeffer from the American Museum. Brundin shocked the paleontologists present by arguing that classifications can be produced only for a single time-level. Integrating these sequential classifications into a single all-inclusive classification is impossible. In general, however, those present were not very impressed by Brundin's presentation. Schaeffer for one could not get by the thick accent. When the British contingent returned to London, Nelson discovered that he was being offered a position at the American Museum, and in the fall of 1967 he joined the staff in New York. No sooner did he settle into his new position than he began to press Rosen about phylogenetic systematics. Rosen had a long association with Mayr, in part because he had married Mayr's favorite research assistant. In later life, Rosen discovered that whenever Mayr became too cross with him, he could return the older man to good spirits by bringing up the topic of his former research assistant. Nelson's persistence eventually got to Rosen, and he retorted sharply that he did not have time for such abstract debates. Mayr was writing a book on the subject, and they would both be wise to wait for it to appear. Nelson snapped back, "I'm not about to let Mayr do my thinking for me!" Rosen was brought up short. Perhaps he shouldn't let Mayr do his thinking for him either.

The ensuing heated discussions between Nelson and Rosen struck many at the museum as unseemly. As Nelson and Rosen argued their way up and down the corridors of the museum, other members of the staff ducked into their offices and closed their doors. Eventually Rosen became convinced. Nelson's interaction with Schaeffer was much more peaceful but no less successful. As Nelson expanded his horizons from the Department of Ichthyology to the museum at large, some of the staff became irritated with this upstart babbling incoherently about synapomorphies and making snide remarks about Simpson.

Nelson's views on fossils were the main bone of contention. Exactly what these views were at the time is not easy to determine. In his *Phylogenetic Systematics,* Hennig admitted that fossils are not much help in reconstructing lineages. At most, the presence of a fossil in a particular stratum implies that the taxon to which the species belonged existed during the period that the stratum was laid down. It does not indicate how long the group to which it belonged had already existed or how long it would persist. Given the scanty nature of the fossil record and the relatively few characters provided by fossils, in comparison to extant species, inferences about extinct groups tend to be very uncertain. Hennig, though, did not totally reject such inferences; fossils have characters, and these characters can be used to infer phylogenetic relationships.

Hennig (1950: 44, 1966: 28) was well aware that absolute certainty is not possible in science. Because organisms live on the same time-scale as their scientific observers, ontogenetic as well as short-term genealogical relationships can be observed. A systematist can watch an organism go through its life cycle from conception to death. He or she can also monitor mating and the production of offspring through successive generations, just as long as these generations are shorter than our own. Inferences are involved in such matters, but the proportion of inference to observation in the reconstruction of phylogenies is significantly greater. Even so, fossils are useful, e.g., in deciding between conflicting character distributions. However, for Hennig, the most fundamental problem continued to revolve around integrating ancestral species with their progeny into cladistic classifications, when cladistic classifications are supposed to represent only sister-group relations.

Although Brundin was somewhat more pessimistic than Hennig about the role of fossils in reconstructing phylogeny, he did not reject it entirely (Brundin 1966: 28, 1968: 481, 1972: 116). In his first paper on historical biogeography, Nelson (1969) presented a basically Hennigian view of fossils, but in a second paper he objected to the belief he found so prevalent among paleontologists, that fossils present "evolution frozen into the rocks," just waiting to be read off by paleontologists. Regardless of what paleontologists might think, they have no "ultimate authority." They must use the same phylogenetic methods as everyone else. Extinct species must be treated the same way that all species must be treated—as sister-groups. Common ancestors are "unknowable." According to Nelson (1972c: 368),

> Hennigian relationships embody the principles that all common ancestral species are necessarily hypothetical, and that ancestral species, although they may be reconstructed (e.g., Fitch 1971), will forever remain unknown and unknowable in a directly empirical sense (e.g., in the sense that species are "known" by way of inference from observation of study material).

In place of common ancestors, Nelson recommended the substitution of "hypothetical common ancestors" or "morphotypes." As morphotypes, these hypothetical common ancestors could no more be real ancestors than could the archetypes of ideal morphologists.[4]

When Nelson set out the preceding views at staff seminars in the coffee room of the American Museum, the paleontologists present were furious. It certainly sounded as if Nelson was saying that their lifework was worthless. First, numerical taxonomists wanted to take all the fun out of systematics by turning over taxonomic decisions to computers. Now, Nelson wanted to dump fossil groups as ancestors. Some of the paleontologists in the audience stomped out of his seminars, others engaged in shouting matches with Nelson, while one went so far as to complain to the director of the museum, Thomas Nicholson. Even though Schaeffer found Nelson's views "interesting," he thought that they could be presented more diplomatically. However, Rosen was on his side, and his arguments convinced several graduate students and junior members of the staff, e.g., Niles Eldredge, Joel Cracraft, Eugene Gaffney, and somewhat later Ed Wiley.

Eldredge studied paleontology at the American Museum of Natural History under Roger Batten and Norman D. Newell, who like so many of the curators at the museum held a joint appointment at Columbia University. At the time, the controversy over Hennig was only just getting started. Eldredge's response was that all these new ideas were very threatening and he would do well to postpone dealing with them until he finished his dissertation and was safely settled in a position.

In 1969 Eldredge joined the staff at the American Museum as an assistant curator and subsequently attended a three-week seminar on systematics in the summer of 1970. It was there that he was exposed at length to Nelson and his threatening ideas. Eldredge raised all the usual objections to Hennig's system, but Nelson responded patiently and

4. The extent to which Nelson right from the start was setting out a system at variance with the patron saint of cladistics can be gauged from the following quotation taken from Hennig (1965: 99):

> In morphological systems, the "beginner" which belongs to each group is a formal idealistic standard ("Archetype") whose connections with the other members of the group are likewise purely formal and idealistic. But, in a phylogenetic system, the "beginner" to which each group formation relates is a real reproductive community which has at some time in the past really existed as the ancestral species of the group in question, independently of the mind which conceives it, and which is linked by genealogical connections with the other members of the group and only with these.

Nelson's response to Hennig is that ancestral species may well really exist, but there is no way to decide which species are actually ancestral to which. Hence, he is justified in treating all species in the same way—as sister-groups. Hence, species appear *only* at the termini of cladograms, *never* at branching points. If anything, branching-points represent "morphotypes."

with only a hint of exasperation. He had heard these objections many times before. As Nelson drew cladograms on napkins in pizza parlors near the museum, Eldredge became convinced. The initial result was a paper he wrote with his advisor, Bobb Schaeffer, and Max Hecht on phylogeny and paleontology (Schaeffer, Hecht, and Eldredge 1972). This was the more diplomatic presentation of Nelson's views that Schaeffer had thought was needed. The publication of this paper had two effects. It added stature to views that previously had been dismissed as the product of youthful enthusiasm; Schaeffer and Hecht were both established scientists. It also caused Simpson considerable distress to see his best friend associate himself publicly with views that Simpson found worse than mistaken.

Cracraft's conversion to cladistic analysis was similar to Eldredge's. Cracraft went to the American Museum to study avian comparative anatomy under Walter Bock at Columbia University. At first he was content with the views of Simpson and Mayr that he was being taught, but he found Bock as a teacher somewhat too rigid and authoritarian. Nor was he much impressed by the writings of Brundin and Hennig. Brundin spent too much time attacking others, and Hennig was all but unreadable. Cracraft also found many of the separate tenets of phylogenetic systematics unreasonable, e.g., the substitution of hypothetical ancestors for actual species at real nodes. This convention seemed utterly ridiculous to him because ancestors had to have existed. Too much of phylogenetic systematics seemed contrary to common sense. Once again Nelson patiently set out his ideas, and eventually the internal logic of his arguments convinced Cracraft, much to Bock's consternation.

In 1971 Rosen brought up Nelson for early tenure, but he was turned down. However, at the end of the normal period of five years, Nelson received tenure. Even so, several of the curators complained that Nelson had gotten into areas beyond his competence. He should, they thought, spend more time curating fishes and less time telling others what they can and cannot do. Although Nelson's livelihood was now secure, the controversy over cladistics had only just begun. In 1965 when Brundin was putting the finishing touches on his biogeography of the antarctic region, P. J. Darlington came out with his *Biogeography of the Southern End of the World* (1965), in which he urged conclusions in direct opposition to those of Brundin. Brundin argued that Antarctica was a center of evolution, while Darlington insisted that it was not. According to Darlington, in earlier times Africa and South America may once have formed a single continent and the other continents had once been in closer proximity to Antarctica before they drifted northward, but he rejected a more extensive role for continental drift. Changes in climate were the primary factor in explaining biogeographic distributions. Although Brundin was too far along in publishing his own book to include an extensive refutation of Darlington's conclusions, he was able to insert a section in which he complained of Darlington's high-handed attitude, as if Darlington had a "special predisposition to divine truth," and of his failure to utilize Hennig's biogeographic method.

It took a while, but eventually Darlington (1970) responded to Brundin's criticisms in a lead article in *Systematic Zoology*. In his rebuttal Darlington listed samples of Hennig's rules and ridiculed each in turn, e.g., splitting, deviation, and dichotomy.

Although Darlington noted that Hennig claimed these rules to be fundamentally methodological, he was able to quote places in which Hennig justified his methodological rules by reference to the evolutionary process. For example, in one place, Hennig (1966: 211) justified his rule of dichotomy by stating, "A priori it is very improbable that a stem species actually disintegrates into several daughter species at once." But Darlington (1970: 4) reserved his special scorn for Brundin. While Hennig was tentative in the application of his overly simple idealizations, Brundin applies them "uncritically and sweepingly" and "berates persons," including Darlington, who questioned them. Darlington (1970: 17) complained in particular of Brundin's tone of "self-assured superiority," an attitude that he found unfortunately characteristic of the cladists.

In reaction to Darlington's paper, both Nelson and Brundin wrote responses defending phylogenetic systematics. Nelson submitted his paper to *Systematic Zoology* and it appeared within the year. In it Nelson (1971a) complained that Darlington had totally misunderstood Hennig and Brundin. Of the five secondary principles that Darlington criticized, neither Hennig nor Brundin ever held three of them. Of the two that Darlington got right, the rule of deviation was present in the work of Hennig and Brundin but was not essential. Only the principle of dichotomy was both present and essential. Nelson defended the principle of dichotomy in terms of maximal information content. Multiple branching cladograms are always questionable because they rest only on negative evidence—no characters that resolve them further have yet to be found.

Because English is not Brundin's first language, he sent his paper to Nelson to read and criticize. Nelson at the time was on the editorial board of *Systematic Zoology.* Nelson gave Brundin some advice on how to improve his English and suggested that he tone down certain passages that might be interpreted as being too personal. Both Brundin and Nelson then wrote to Albert J. Rowell, who had taken over as editor of *Systematic Zoology* from Richard F. Johnston, to warn him that Brundin was going to submit a paper in response to Darlington. When it arrived, Rowell had it reviewed and subsequently returned the manuscript along with the referee's report suggesting that Brundin cut the paper by about a third and then resubmit it. Although several comments in the referee's report irritated Brundin, he did rewrite, cutting the paper nearly in half. This time Rowell accepted the paper for publication. During this same period Rowell was considering a second paper submitted earlier by Brundin defending phylogenetic systematics. This one he rejected outright.

Early in January of 1972 Nelson wrote to Brundin to ask him what had happened to his papers. Brundin filled him in on Rowell's reluctant acceptance of his response to Darlington and his rejection of the other paper. Nelson was irate. He sent a copy of Brundin's letter along with a letter of his own to the president of the Society of Systematic Zoology, Norman D. Newell. Cracraft added a letter of his own to Newell. Newell in turn wrote to Rowell about the complaints raised by Brundin, Nelson, and Cracraft, including a copy of Brundin's letter. Upon reflection, Rowell still thought he had behaved properly, noting that he had agreed to publish Brundin's response to Darlington even though he was still unhappy with its personal nature. In his letter to Newell, Rowell revealed the identity of the original referee, asking that it be kept confidential. He also sent a copy of this letter to Nelson. In his response to Rowell,

Nelson detailed his own involvement in the Brundin affair, complaining that Darlington seemed to have much easier access to the pages of *Systematic Zoology* than did Brundin. He also objected to one of the comments made by the eminent systematist who had refereed the paper. For those who might have been puzzled by the vehemence with which Nelson objected to what seemed to be a casual comment made by Michener at the Boulder Congress in August of 1973, it should be noted that Michener was the referee of Brundin's paper.[5]

Nelson's involvement in the Brundin affair had an unexpected effect. As Rowell remembers it, he suggested to Nelson that if he wanted more influence on the editorial policy of the journal, then perhaps he himself should become editor. Rowell had intended to serve as editor for three years, and in 1972 he was already well into his second year. After some discussion, the next president of the society, Herbert Ross, approached Nelson about becoming editor of the journal. Nelson agreed but on the condition that Niles Eldredge be appointed coeditor. Under this arrangement, Nelson had full editorial responsibility. Eldredge would take over only when Nelson was out of the country or when conflicts of interest arose. After an absence of ten years, *Systematic Zoology* returned to the American Museum of Natural History, the home base of Libby Henrietta Hyman (by then Hyman had been dead for four years). It was also at this time that Nelson, Eldredge, and Cracraft decided to coauthor a book tentatively titled the *Principles of Comparative Biology* and in the spring of 1973 signed a contract with Columbia University Press.

In this same year two young men joined the staff of the American Museum, Norman Platnick as an assistant curator in the Department of Entomology and Ed Wiley as a scientific assistant and graduate worker in the Department of Ichthyology. In 1964 when Platnick was twelve, he enrolled at Concord College in Athens, West Virginia, Farris's hometown. While at Concord, Platnick was attracted romantically to another student, Nancy Price, who needless to say was several years his senior. In order to work with her, Platnick would have been happy to change his area of study to hers, but Nancy's specialty was millipedes. The easiest way to ferret out these creatures in nature is by their distinctive odor. Unfortunately Platnick's olfactory sense was deficient and he had to settle for the group closest to millipedes that did not require a highly developed sense of smell—spiders. When Platnick revealed his intentions to his fellow student, she suggested that they wait to marry at least until he was of legal age; sixteen seemed a bit young. In 1968 Platnick became a zoology graduate student at Michigan State University in East Lansing. Finally, in 1970 Nancy relented, and she joined him in his move to Harvard University. After receiving his Ph.D. from Harvard, Platnick joined the staff of the American Museum.

While at Southwest Texas State University in San Marcos, Ed Wiley became interested in fish taxonomy through the influence of W. K. Davis. He also read Sokal and Sneath (1963), but his initial enthusiasm for pheneticism was dampened by Davis's conviction that organisms can never be reduced to numbers. Although Davis encouraged him to go on to graduate school prior to his graduation in 1966, Wiley took a job as a fisheries

5. The preceding narrative has been constructed from interviews with Brundin, Nelson, and Rowell.

biologist with the Texas Parks and Wildlife Department. When it became clear that he would be drafted, he joined the air force in 1967. Upon release from active duty, he went to graduate school at the only university he thought would take him, Sam Houston University in Huntsville, Texas. His work on killifish convinced him that phenetics would not work, because in his phenograms males and females of the same species frequently fell into separate OTUs.

For his graduate seminar, Wiley read Hennig (1966) and contrasted his views with those of the pheneticists as well as those of the evolutionary school of Simpson and Mayr. Wiley liked Hennig's principles for reconstructing phylogeny but was not convinced that cladistic relations should be the sole basis for classification. After obtaining his master's degree, he worked as a project assistant for the Trinity River Environmental Impact Study. On the advice of Berry Hinderstein, a young herpetologist from New York, Wiley applied to Donn Rosen to become an assistant and graduate worker in ichthyology at the American Museum. In 1976 he received his Ph.D. While Platnick and Wiley were at the American Museum, they became ardent advocates of cladistic analysis, so much so that Platnick replaced Rosen as Nelson's chief confidant and sounding board.

As if Nelson had not already taken on enough in championing Hennig's principles of cladistic analysis in the English-speaking world, promoting Brundin's biogeographic extensions of Hennig, and assuming the editorship of a major scientific journal, he picked this time to retrieve a theory of historical biogeography that leading biogeographers of the day had consigned to the trash heap of science—the panbiogeography of Leon Croizat (1894–1982). Croizat was born of French parents in Turin, Italy, near the French border (see photograph). When Croizat was six, his parents separated, and by the time Croizat's father died in 1915, the family was destitute. For the next forty years, Croizat was never more than poor, sometimes close to starvation. While he was in the Italian army from 1914 to 1919, he married and began a family. After the war he quickly earned a law degree and went to work in a textile mill owned by a friend, but the rise of fascism soon forced him to emigrate to the United States. In 1923 he landed penniless in New York City with a wife and two children. For six years he made money any way that he could, including selling his own watercolors. When the stock market crashed, the market for original works of art crashed with it, and Croizat moved to France, but the life of an unknown artist in Paris turned out to be even harder than it had been in New York, and Croizat returned to New York.

Finally, he obtained a job identifying plants for a topographical survey of New York parks. In the Great Depression the federal government not only paid artists to cover the walls of public buildings with murals of rolling farmlands and muscular steelworkers, but also commissioned such make-work scientific projects as a flora of the Bronx Park. While engaged in this undertaking, Croizat met E. D. Merrill (1876–1956). When Merrill became director of the Arnold Arboretum at Harvard University, he hired Croizat as a technical assistant. While at Harvard Croizat submitted a paper to the journal published by the arboretum. Generally such in-house journals automatically publish anything submitted by one of their own, but Croizat was only a technical assistant, and in his paper he sharply criticized the work of an extremely influential botanist, T. A. Sprague at Kew Gardens. When the journal rejected Croizat's paper, he published it elsewhere. In spite of this flap, Merrill kept Croizat on, but when Merrill was replaced as director of the arboretum Croizat was dismissed just a few months shy of tenure.

Once again Croizat emigrated, this time to Venezuela, where he held a number of academic posts in botany between 1947 and 1952. Then he divorced his first wife and married a woman with as remarkable a history as his own, Señora Catalina. During World War II she too had arrived in Venezuela as a refugee, but from Hungary. When her ship was sunk in the Mediterranean, she lost all her possessions. She finally arrived in Caracas with no money, possessions, or family. Before she was done, she owned the most successful landscaping firm in Caracas (further details of Croizat's life can be

Leon Croizat. Caracas, Venezuela, August 1974. (Photo by Jon Baskin.)

found in Craw 1984). After marrying Catalina, Croizat devoted the next thirty years to research and published a stream of papers and books numbering over ten thousand pages. He was able to publish so extensively, in part, because he published much of his work at his own expense. Croizat's biogeographic method was to plot the distributions of plants and animals to see where these distributions consistently overlapped, terming these areas of congruence "standard tracks." To Croizat's surprise, one of the most general tracks marched right across the Atlantic Ocean between Africa and South America. That many chance dispersals seemed very unlikely. Croizat argued that the discovery of standard tracks was the primary tool of biogeography and that equally general causes had to be found for these pervasive phenomena. The leftovers could then be explained in terms of chance dispersal. The converse line of reasoning struck Croizat as totally wrong-headed.

Such authorities as Simpson and Mayr consciously avoided public reference to Croizat's work, fearing to bring attention to views that they found mistaken, but Simpson did correspond with Croizat for a short time in the spring of 1959. Simpson happened upon Croizat's *Panbiogeography* (1958) and was taken aback by the tone of Croizat's extensive criticisms of his own work. Simpson wrote to Croizat to register his dismay at all the unprovoked "emotional venom" that Croizat had lavished on him. Croizat responded that he was sorry that Simpson took his criticisms personally; he himself viewed their disagreements as a "fight in the realms of ideas," a fight between Croizat and the "New York School of Zoogeography," represented by Simpson, Mayr, Darlington, and their master, W. D. Matthew. The next day, Croizat sent off a second letter, this one a good deal more irate over the superior tone which he perceived in Simpson's letter of March 31, 1959:

> Your having introduced into our exchanges such marked enmity as "emotional venom without provocation" (I being evidently volatile, poisonous, quarrelsome; you sedate, sweet, self-contained by definition) has made me recall, eventually, that in the U.S.A. (a country of which I have right complete understanding) historical attitudes are all too often taken when matters of nationality, opinion, and the like enter the stage. It seldom is so that among you those who are *objectively* challenged in point of ideas fail to react in a wholly impersonal manner, and so in a spirit of aroused, quite righteous self-indignation. You are the lilies, the rest are blackberries.[6]

Simpson responded to these two letters, reiterating his objections to the highly personal, vituperative tone of Croizat's criticisms and mentioning how inappropriate the name "New York School of Zoogeography" was. Most of the men listed had very little connection to New York. What they did share was a Darwinian approach to biogeography. Croizat replied with a single-spaced, three-page letter in which he alternated between extreme hostility over the censorship which he perceived had afflicted his own attempts to publish and gracious comments about the generous tone of Simpson's letters. Simpson responded politely but in a way which signaled that he considered their correspondence concluded.

6. The preceding quotations are taken from copies of the correspondence between Croizat and Simpson kindly supplied to me by Robin Craw and Gareth Nelson.

Simpson continued to omit reference to Croizat's work, and in this matter he was joined by Mayr. According to Mayr, "Neither Simpson nor anyone else has affected my treatment of Croizat, but only his totally unscientific style and methodology. Time is too short to argue with such authors and one cannot simply refer to Croizat without a detailed analysis. I am prepared to be criticized for this, but any scientist has to make the decision where to draw the line" (Nelson 1977: 452). Croizat's work, however, did not go totally unnoticed in print. For example, in his response to Bigelow's (1958)

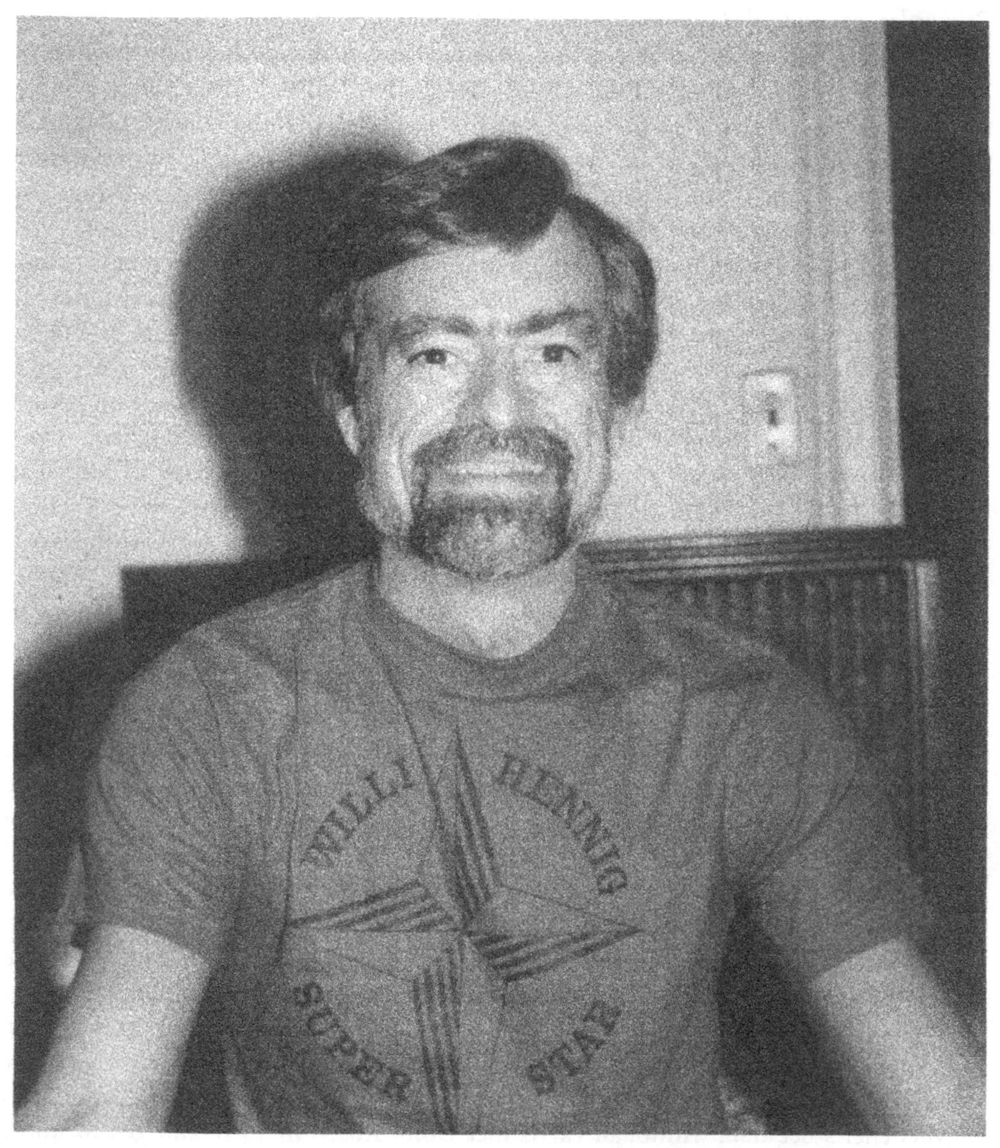

James S. Farris. Manhattan, New York, February 1988. (Photo by James M. Carpenter.)

objections to phylogenetic systematics, Kiriakoff (1959) praised Croizat's panbiogeography, and in his original critique Brundin (1966: 61) chastised Darlington for ignoring Croizat's "blazing sermon" (for data concerning Croizat's influence, see Schmid 1986). The conspiracy of silence from high authorities and the praise from Brundin were all Nelson needed to set out on another crusade; but before the story of vicariance biogeography can be pursued, I must backtrack one final time to J. S. Farris as another source of cladism.

When Steve Farris was twelve, he read a book on fossil reptiles and became an instant paleontology freak. At the University of Massachusetts, he read Simpson's *Principles of Animal Taxonomy* (1961) but could find no one interested in either taxonomy or phylogeny reconstruction. At the University of Michigan, his official advisor was the ichthyologist Robert Rush Miller, but Miller did not know what to make of Farris, who seemed to spend most of his time auditing courses in the Department of Mathematics. At one juncture, Miller was led to observe, "I don't know what Steve is, but he is not an ichthyologist." Farris wanted to reconstruct phylogenies, but no one seemed able to tell him how to do it. No one stated their principles very clearly, and no two paleontologists seemed to use the same principles. To make matters worse, any one paleontologist seemed to use different principles at different times. To use Simpson's term, there was too much "art" in the traditional methods of paleontology. While Farris was in graduate school, J. H. Camin and Sokal (1965: 312) published a method for deducing branching sequences in phylogeny based on "parsimony," or the "least number of evolutionary steps for the character studied." On a visit to Kansas, Farris tried to see Sokal, but Sokal was too busy. Back at Michigan Farris discovered that Herb Wagner had already published a method for estimating phyletic relationships (Wagner 1961), and Wagner was happy to work with Farris. Later, when Farris developed his own methods, he named his constructions "Wagner trees," not "Sokal trees." Arnold Kluge was also interested in the development of quantitative methods for reconstructing phylogenies.

While the people at Michigan were preparing for the 1967 International Conference on Systematic Biology, Farris (1966, 1967a) was publishing his first papers. In the second of these papers, Farris (1967a) distinguished between patristic and cladistic relationships and presented methods for determining both. Because these methods were as operational and mathematical as any proposed by the numerical taxonomists, he saw no reason why they should object to them. In fact, Sokal was sufficiently impressed by Farris's early papers and his performance at the Ann Arbor international conference that he invited him to the first numerical taxonomy conference. Shortly thereafter both men ended up at the State University of New York at Stony Brook. Throughout these early years at Stony Brook, Farris continued to work on numerical methods for inferring phylogenetic trees. The paper that impressed practicing taxonomists most was written with Kluge, a paper in which Farris applied his methods to Kluge's frog data (Kluge and Farris 1969) in the way that Sokal, a generation earlier, had applied his methods to Michener's bee data. Although Farris met Donn Rosen in 1967 and got to know Nelson soon thereafter, he did not immediately become part of the burgeoning group of cladists at the American Museum. His reaction to reading Hennig was just as delayed.

He did not acknowledge the difference between Mayr's retrospective definition of monophyly and Hennig's retrospective and prospective definition until 1970.

During this same period, Farris adopted some of the jargon introduced by the pheneticists (e.g., terming taxa OTUs) and wrote a couple of papers in which he addressed some of the issues within numerical phenetics. For example, Sokal and Sneath (1963) had devised the cophenetic correlation coefficient to compare various clustering techniques. Farris (1969a) examined the properties of this coefficient. Later, in a review article, Farris (1971) investigated the connection between Sokal and Sneath's hypotheses of nonspecificity and congruence. Although he remarked on the anomaly of pheneticists founding their nonevolutionary principles of classification on a hypothesis that is clearly evolutionary in character, Farris (1969b: 383) did not openly attack his more "orthodoxly pheneticist colleagues."

From reading Farris's early papers, one would never guess that he was to become the Black Knight of the cladistic movement, doing battle with the evil pheneticists. Pheneticists, to the contrary, came to view him more in the role of Darth Vader (Felsenstein 1986). To be sure, his early exchanges with Ghiselin at the Ann Arbor meeting and later in the pages of *Systematic Zoology* were a bit sharp (Farris 1967b, Ghiselin 1967), but in general Farris's early publications were not especially polemical. In graduate school, Farris struck Wagner as a "nice warm bear." As far as Sokal was concerned, Farris was a *numerical* cladist and his work a contribution to numerical taxonomy. Of the twenty-one significant achievements in numerical taxonomy that Sneath and Sokal (1973) listed in the second edition of their book, four were in numerical cladistics. But increasingly Farris identified himself with the cladists at the American Museum and considered his work to be contributing to numerical *cladistics* and not to numerical taxonomy. At the same time, the relationship between Farris and Sokal began to degenerate. At a time when the art of scientific polemics frequently degenerated to the level of "na-na-na, so's your old man," Farris approached T. H. Huxley in the delight that he took in intellectual assassination. The acidity of his commentaries on his opponents approached that of aqua regia, one part dark rumbling to three parts flashing wit. Just as certain strains of bacteria can live off sulfur, Farris appeared to thrive on vitriol.

Conclusion

One of the basic premises of this book is that in conceptual change causal connections are essential. Only those concepts that are transmitted can influence the development of science, and the primary vehicles for this transmission are scientists and their published works. Although a particular conclusion might follow inferentially from a particular set of views, that conclusion is irrelevant to science as a process if no one drew that inference. In short, after-the-fact rational reconstructions are irrelevant to science. (Whether they are also irrelevant to our understanding of science is another matter.) In this minimal sense, science is social. Science would not be very cumulative if succeeding generations of scientists did not build upon or at least play off the work of earlier scientists.

But science is social in a second sense as well. During periods of rapid change in science, scientists form small groups to generate, develop, and disseminate new views. Bright new ideas frequently occur to scientists working in relative isolation, but the development of these ideas benefits from both the support and the contributions of other scientists. Dissemination is necessarily a social process. Research groups are not necessary in science, but they are both prevalent and quite effective. In this chapter I have traced the initial stages in the development of two such groups—the numerical pheneticists at Kansas and the cladists at the American Museum of Natural History in New York. Most research groups quietly disappear without anyone noticing that they ever existed. These two have had a lasting impact.

Initially, the formation of both groups was an exhilarating affair. The founders were embarked on a crusade, perhaps not a great crusade, but a crusade nonetheless. They intended to make systematics truly scientific. The interests and abilities of the members of these groups influenced the course of development that their research programs took, but so did the actions of their opponents. Was the equal weighting of characters the most important tenet of numerical phenetics? It was the tenet that evolutionary systematists chose to combat. As a result, any retreat on this issue became magnified. Was dichotomy so central to cladistics? Certainly the opponents of cladistic analysis attempted to saddle the cladists with this interpretation.

In general, the opponents of emerging research programs influence the content of those programs. They also serve to reinforce the internal cohesiveness of the associated research groups. As long as a particular research program is under attack and in danger of being defeated, the advocates of this program tend to present a united front. During such periods in science, those involved tend to decry all the factionalism, name-calling, and polemics, but as disconcerting as all this uproar is to those of us who prefer quieter times, it does serve a purpose. Simpson and Mayr noticed how deceptive the relation between phylogeny and classification is and even remarked on it, but no one noticed. The pheneticists and cladists forced systematists at large to recognize the problem. Both groups were ambitious. They raised their sights. Systematics could do more than it had in the past.

Another feature of the episodes I have described in this chapter is the need for scientists to find outlets for their views. They must be able to publish their papers in journals and present them at conferences and congresses. In the first instance, this meant gaining some control of the chief outlet for papers dealing with systematic philosophy—*Systematic Zoology.* Because old-boy networks are a good deal more resistant to invasion, Sokal formed his own informal group that met yearly to present papers and to discuss issues of mutual interest. Later the cladists were to do the same.

Already in this chapter, some signs have cropped up indicating how difficult continued cooperation among scientists in a research group can get. Within the numerical taxonomists at large, numerous partially independent research groups formed. For all of them, the use of quantitative techniques continued to be important, phenetic philosophy much less so. Numerical taxonomists were willing to produce computer programs designed to reconstruct phylogenetic trees, just so long as the techniques were objective, repeatable, and numerical. Nelson was much more skeptical about designating extinct species as ancestral than either Hennig or Brundin. Certainly, his blunt dismissal of ancestor-descendant relations was the chief cause of hostile feelings among many of his contemporaries. His advocacy of Croizatian biogeography also posed problems. It was one more reason for his opponents to reject cladistics. It also raised problems for some of his fellow cladists. Not everyone who had come to adopt the principles of cladistic analysis appreciated the charms of vicariance biogeography.

5

Systematists at War

Taxonomists have always had the reputation of being difficult. Intransigence may be rooted in the necessity of defending prolonged self-immersion in a taxon that others find a total bore; it is frustrating to have one's life work greeted with a yawn. Numerical taxonomists have proved to be just as prickly as conventional taxonomists, possibly more so because some of the brightest people in systematics are involved in the current taxonomic battles. The political maneuvering and character assassination that characterize certain taxonomists today may not be atypical for science; they certainly provide a fine example of its seamier side. If Feyerabend is correct, it may be even a requirement of human nature that scientific progress occur in this manner.

W. W. Moss, "Taxa, Taxonomists, and Taxonomy."

PEOPLE IN any age and in any society tend to read their own local mores and values into all other ages and cultures. Such biases are easier to perceive in other cultures and other ages than in our own. For example, in the first sociological study of scientists in the English-speaking world, Francis Galton (1874: 111) observed, "My bias has always been in favor of men of science, believing them to be especially manly, honest, and truthful, and the results of this inquiry have confirmed this bias." Today we might well agree with Galton that honesty and truthfulness are relevant to science but wonder about manliness. Most scientists today continue to be men, but we fail to see any special role in science for peculiarly male virtues. To this day scientists are also extolled for being humble, but humility is a Christian, not a scientific, virtue. No one has shown that humble scientists do better work than their more arrogant colleagues.

Not only do we tend to read the mores of the society in which we live into all other societies, but we also tend to assume that virtues extolled for society at large should apply equally to all segments of society. In Western societies today efficiency is a virtue. One periodically hears rumbles of discontent about how much money is wasted in science through the support of unproductive scientists. If 95 percent of all citations are to works published by 5 percent of the practicing scientists, why waste so much money supporting all those third-raters? Would not science be improved if we made it more efficient? Perhaps it is only mindless romanticism, but the call for efficiency in the production of great works of art is likely to strike most of us as worse than wrong-headed. The same may be true for the more creative aspects of science. It may just be possible that creative processes of all sorts are inherently inefficient and that attempts to make them too efficient are likely to destroy them. Some support is lent this hypothesis by the character of biological evolution in the creation of new species and the incredible array of adaptations exhibited by organisms. The vast majority of germ cells produced never unite to form zygotes, the vast majority of zygotes never reach sexual maturity, and so on. Of the thousands of sexual adults released from a termite hill, only one or two ever succeed in founding a new colony. The God of the Galapagos is anything but the Protestant God of waste not, want not.

However, just because manliness, humility, and efficiency are not inherent in science, it does not follow that nothing is. One of the main messages of this book is that both cooperation and competition, once properly understood, are central to science. Science works as well as it does because of the interplay between cooperation and competition. By definition, a society cannot be a society without at least a minimal degree of cooperation. As a result, in most societies, cooperation is a virtue. Even in those societies in which cooperation is not especially extolled, it must nevertheless be practiced. Hence, we are likely to admire the institutionalized cooperation that occurs in science. Scientists really do use each other's work and to an amazing degree give credit where credit is due. (References to the large literature on the topic of cooperation and competition in science can be found in chapters 9 and 10, where it is discussed at great length.)

Attitudes toward competition tend to vary from society to society. In more capitalistic countries, competition at the individual level is extolled, while in people's republics it is disparaged. In both, however, we are constantly urged to submerge individual benefit to the good of society at large. Inducements to promote one's own good are rarely needed. From its inception science has been elitist and competitive. Those people who see nothing wrong with elitism and competition in society at large are not likely to complain about these characteristics of science. However, even those who would prefer societies to be more genuinely egalitarian and cooperative need not insist that all social institutions share equally in these characteristics. Just as science may not be inherently masculine or Christian, it may not be inherently egalitarian either. If one important goal of science is to provide understanding of the world in which we live, this goal may best be realized by an institution structured along lines different from those characterizing society as a whole. In even the most egalitarian societies, science may have to remain as ineluctably elitist as ballet. Both are extremely high-resolution activities.

Scientists both cooperate and compete with each other. If human beings are inherently competitive, then outbreaks of cooperation require special explanations. If we are naturally cooperative, then instances of competition need explaining. To the extent that human history is any guide, people are capable of both extensive cooperation and sustained competition. Both sorts of behavior are equally natural and stand in equal need of explanation, depending on the circumstances. I see no reason to exempt scientists from the same principles that apply to all other people. The issue is not the inherent virtue or vice of cooperation versus competition. Both can be overdone. Too much cooperation is as likely to impede scientific development as too much competition. The issue is the appropriate mix of the two. In science as in biological evolution, an appropriate mix of cooperation and competition is likely to turn out to be most adaptive.

In this chapter, I examine the periodic warfare that has broken out among competing groups of systematists. Most of the readers of this book are likely to be intellectuals, and intellectuals tend not to enjoy barroom brawls. Quite naturally we read our preferences into all segments of society. Just because we do not enjoy squaring off after a football game to beat on each other, we insist that laws be passed prohibiting such behavior in others. We are equally adamant that our prohibitions against street fights do not stem merely from our own personal preferences but from eternal and immutable moral first principles. As vicious as the verbal exchanges between intellectuals frequently are, we rarely resort to physical violence, and when we do, it is rarely effective. We limit our aggression to verbal abuse. Repeated complaints about polemics notwithstanding, we do not urge laws prohibiting it.

We are likely to reason just as automatically in the case of science. However, I think that the function of personal animosity in science is still an open question. It might after all serve a variety of purposes. For instance, one thing that is true of science is that it takes time. Any scientist who is not willing to put in the hours formerly reserved for factory workers in Victorian England is not likely to succeed. Strong motivations are needed to induce such protracted labors. Scientists acknowledge that among their motivations are natural curiosity, the love of truth, and the desire to help humanity, but other inducements exist as well, and one of them is to "get that son of a bitch." Time and again, the scientists whom I have been studying have told stories of confrontations with other scientists that roused them from routine work to massive effort. No matter what the cost, they were going to get even. Although those of us who enjoy peace and quiet tend to decry the vicious polemics that frequently erupt in the midst of scientific controversies, these interpersonal animosities may well serve a positive function in the ongoing process of science. If nothing else, they frequently rouse scientists to near superhuman effort.

Recently a few historians of science have objected to the use of the language of warfare in the characterization of science. Just as "king bee" tells us more about the society in which the scientist who coined the term lived than it does about bees, all our talk about scientists doing battle with each other may simply be an unjustified extrapolation from one area of society to another. For example, Moore (1979) has argued that the history of Darwinism was as much a tale of cooperation between religion and science as of disputation. He could find no correlation between scientists' religious views or allegiances and which way they opted on evolution. Darwin himself found it "absurd to doubt that a man can be an ardent Theist and an Evolutionist" (de Beer 1958: 88). Asa Gray, for one, was both. Certainly historians have tended to exaggerate the hostility between science and religion. However, I do not see that historians have exaggerated the intramural battles that frequently characterize science. If anything, they have tended to play down this aspect of science as little more than occasional lapses in proper scientific conduct to be ignored or explained away. Regardless of whether one approves of hostile confrontation, the language of the battlefield is as appropriate for science as is the language of camaraderie and cooperation. The exuberance and good will that characterize the intragroup behavior of research groups in science most people are likely to find appealing. The intergroup bickering that accompanies it is likely to strike most of us unfavorably. When what was formerly a single group composed of good friends breaks up into warring factions, we are likely to be appalled. But friendship and enmity are equally part of the professional relationships of scientists. Would we really want scientists to behave as "gentlemen" if we discovered that behaving in this way would bring science to a halt?

For those of us who value science, the answer is obvious, but for many the pursuit of knowledge is only one good out of many and possibly far from the preeminent good. Even though they might well value science, they hold other goods in higher regard. When scientific truth conflicts with the good of humankind, truth must be sacrificed. It would be nice if truth and virtue always coincided, but they do not. Certainly misunderstanding the world in which we live is always a danger, but one need not introduce science-fiction examples to show that sometimes increased knowledge can do great harm, whatever one's definition of "harm." These extreme cases aside, scientists tend to fear calls to sacrifice truth to virtue because standards of virtue are so much more variable and problematic than those of truth. As the continuing appeal of epistemological relativism clearly shows, standards of empirical truth are elusive enough, but distinguishing good from evil is vastly more difficult. Even though one might be willing in principle to sacrifice truth to goodness when the two conflict, the history of past conflicts might make one less willing in practice. Time and again, our understanding of the world in which we live has been sacrificed to "goods" which in retrospect we do not find in the least good. In the past, commentators on science have tended to treat such altercations as science gone wrong—as unfortunate occasional anomalies—but anomalies in the science of science are easily as important as they are in science itself. Morgan might have dismissed his white-eyed male fruit fly as an inexplicable anomaly. If he had, the history of genetics would have been much different. Students of science are well advised to cherish their exceptions.

Although Hennig and Brundin were far from unknown in the English-speaking world, it was primarily through the efforts of Gareth Nelson at the American Museum of Natural History that cladistics became a full-fledged movement. The irony that the original home of G. G. Simpson and Ernst Mayr should become a hotbed of opposition to the views of classification that they did so much to foster was lost on no one, least of all Simpson and Mayr. Earlier Simpson (1965) had complained of how "rigid" and "fanatical" the Kansas group was. If anything, charges of "fanaticism" and "fervor of conversion" were even more commonly leveled at the American Museum group and their "New York Rules" of conduct than they had been a decade earlier in connection with the Kansas group. In this book I concentrate on these two groups of scientists. Others outside these two groups also contributed to the development of systematics during this period. For example, David Rogers and Taffee Tanimoto produced several early works in numerical taxonomy (e.g., Rogers and Tanimoto 1960), but both men dropped out of the picture quite early. Even so, one of Rogers's students, George

Estabrook, found his way to Wagner at the University of Michigan. In the early years of cladistics in the English-speaking world, Roy Crowson looked as if he were going to play a major role, but he did not.

Several workers who did make important contributions to these early discussions eventually turned their attention elsewhere. For example, the young man of whom Sneath had such high hopes, Nicholas Jardine, ceased work in systematics and began publishing in the history of science. As it turned out, he had never been very interested in systematics. Quite early, Paul Ehrlich decided that all the really important work had been done in numerical taxonomy and turned his attention increasingly to population biology as well as becoming involved in the population-control movement. Research groups are not liable to survive in the face of the defection of their best and brightest. Cladistics has faced defection of a different sort. Several of the best young cladists, instead of turning their attention to other areas of science, have struck out on their own in the field of systematics. By all indications, the cladists have speciated.

Because I pay so much attention to the *groups* involved, isolated individuals tend to get shunted to the side. Most systematists belonged to none of these groups. Several of the systematists who contributed to the relevant literature were not socially part of any of the groups involved. And through time the character of these groups has itself changed. When the Kansas group arose, most systematists thought of themselves as traditional evolutionary systematists, but no socially defined group existed, only a few well-placed individuals willing to defend this position. By the time the cladists came on the scene, the Kansas group as a tightly-knit group of workers no longer existed. Even though people interested in numerical taxonomy continued to meet each year, they never formed an official society. The fervor for making classifications theory-neutral, at least in the formative stages, became decreasingly prevalent in their writings. They did, however, continue to share an allegiance to the task of making biology more quantitative (e.g., see the mix of papers in the proceedings of the NATO Advanced Study Institute on Numerical taxonomy, edited by J. Felsenstein, 1983). The cladists as a group have been so successful in promulgating at least the basics of their method that a high percentage of systematists today consider themselves "cladists." They are not part of the cladists as a social group, but they are "cladists" in the sense that they utilize cladistic methodology.

In this chapter I describe the disputes that recurrently broke out between pheneticists, cladists, and evolutionary systematists, especially with respect to the editorial practices of three successive editors of *Systematic Zoology*—Gareth Nelson, Randall T. Schuh, and J. D. Smith. *Systematic Zoology* is so important because at the time it was the primary outlet for papers on taxonomic method and philosophy. I concentrate on the period after 1974 because of the availability of data. Nelson and Schuh deposited nearly a complete record of their editorship in the archives of the Smithsonian Institution. It is this record that permits the documentation of this period in the history of the journal. Previous disputes, although somewhat rarer, were no less acrimonious. I have already discussed how unhappy members of the Kansas group were with the editorial policies of Hyman. Byers had a nasty exchange with an author about such matters as punctuation. Johnston became embroiled with Blackwelder for a while. As I noted earlier,

Nelson objected to some of Rowell's editorial decisions, not only his handling of the two manuscripts submitted by Brundin but also Rowell's failure to send him a manuscript critical of cladistics by Stephen Jay Gould (1973) before it was published. However, of the various disputes over the actions of editors prior to 1974, only one made it to the council meetings of the Society of Systematic Zoology—Hyman's decision that one paper with numbers was enough. In later years council meetings took on the character of trials, as motions were introduced to censor first one side and then the other.

One message of this book is that the altercations that arise in science from its competitive aspects are as intrinsic to it as is the camaraderie that results from its more cooperative side. In the early years of both numerical taxonomy and cladistics, the members in each group freely cooperated with one another. Even the intergroup polemics had a joyful aspect to them as the members of these groups banded together to fight the forces of darkness. In this chapter I concentrate on the altercations that arose because of the ways in which certain manuscripts submitted to *Systematic Zoology* were handled. In concentrating on these altercations, I am liable to give the impression that they were the rule rather than the exception. However, during the period under investigation, roughly fifty papers appeared each year in *Systematic Zoology.* Most of these manuscripts worked their way through the editorial process without incident. In most cases even the papers that did not ultimately appear caused no commotion. To help put the relatively few papers that caused trouble in their proper perspective, I provide summary data.

It also happens to be the case that all the editors for whom I have summary data (chiefly Nelson and Schuh) happen to be strongly committed to the principles of cladistic analysis. Was there some intrinsic connection between cladism as a taxonomic philosophy and a penchant for polemics, or were the heightened polemics a function of the personalities involved? Several influential taxonomists at the time had strong convictions about proper scientific conduct, convictions that struck several of the leading cladists as being long out-of-date and counterproductive. Truth is more important than politeness. They suspected that appeals to proper scientific conduct were little more than cover established by those in power to frustrate the introduction of new ideas. According to several of the leading cladists, science should be a free give-and-take of views honestly held—sensitive egos be hanged! Historical contingencies may also have played a role. By the time that the cladists came along, the pheneticists had made some headway, but their position was far from secure. Just when the pheneticists thought they could consolidate their gains and let the natural course of events decide which philosophy of systematics would prevail, they were challenged by a group of scientists even more zealous than they had been. The controversy became three-cornered—advocates of the Simpson-Mayr school, pheneticists, and cladists. The most acrimonious disagreements, however, occurred between the pheneticists and cladists. Although evolutionary systematists hardly sat on the sidelines, they were not as involved as their role as defenders of the received view might have led one to expect. Any explanation of the extreme hostility that arose between pheneticists and cladists must surely include reference to the intense animosity that developed between Sokal and Farris.

Agonistic Displays

Editors of professional journals have considerable power. Some editors run their journals as absolute monarchs, making all the editorial decisions themselves. Others have the manuscripts submitted to their journals read by members of an editorial board. The most common practice is, however, for an editor to send each manuscript to two or three referees, who read it and make recommendations. They may suggest outright acceptance, acceptance but only after suggested changes, or rejection. Recommendations for rejection can stem from the referees' conviction that the manuscript is so bad it is not salvageable or that it is not appropriate for the journal. The editor then reads the referees' reports and decides whether or not the manuscript should be rejected, returned to the author for revision, or accepted as is. Although editors tend to go along with the recommendations of their referees, on occasion they override them, sometimes rejecting a manuscript that received a clean bill of health, sometimes accepting one that all the referees recommended be rejected. In doing so, editors run some risk. However, editors usually do not have to overrule their referees very often because they can influence the results of the refereeing process by their choice of referees. An editor soon learns that certain referees have very stringent standards. They rarely think a paper is good enough to publish. Others are very lenient. Almost anything strikes them as publishable. Also editors are well aware of the preferences of their referees. Some like quantitative papers; others do not. Some insist that even the most theoretical papers contain masses of data; others are not so particular. And so on.

In the early years of *Systematic Zoology*, the editors made most editorial decisions themselves. The use of referees was sporadic and acceptance rates were high, hovering around 90 percent (table 5.1). However, in 1963 a policy committee headed by Bobb Schaeffer recommended to the council of the society that all papers accepted for publication in the journal be refereed by at least one competent judge, and that papers be rejected only after being refereed by two or three authorities in the area. During the first twenty-two years of its existence, the editorial structure of *Systematic Zoology* remained much the same, consisting of an editor, an editorial board, and an occasional assistant or associate editor. In practice the members of the editorial board played no special role other than perhaps refereeing more papers than other experts in the field.

Table 5.1. Acceptance rates in *Systematic Zoology*

Editor	*Acceptance Rates, Percents*
Brooks (1952–57)	90
Hyman (1958–63)	90
Byers (1964–66)	77
Johnston (1967–70)	72
Rowell (1971–73)	80
Nelson (1974–76)	67
Schuh (1977–79)	59
Smith (1980–82)	42*
Schnell (1983–85)	56

*Figure is only for Schnell's first three years as editor.

Assistant and associate editors were frequently graduate students hired to take care of the more laborious tasks. In 1974 when Nelson took over as editor, he made significant changes. First he introduced a book review editor, Michael Ghiselin. Ghiselin received books from presses, decided which warranted review in *Systematic Zoology*, and chose the reviewers. Then, half way through his first year, Nelson added two contributing editors, Stephen Gould and David Hull. In 1975 he added Sokal to the other two contributing editors.

But the most significant change occurred at the end of 1975, when on consent of the Council of the Society of Systematic Zoology, Nelson introduced four associate editors to take over the initial stages of the refereeing process. The associate editors received all manuscripts, sent them for review, and then made a preliminary judgment. They were empowered to reject a manuscript outright, but acceptance was contingent on the editor's approval. Of course, any author whose manuscript was rejected could appeal directly to the editor or to the council of the society for that matter. In 1977 because Ghiselin was in ill health, Nelson replaced him with Platnick as book review editor. When Schuh took over in 1977, he dropped the largely honorific post of contributing editor, but otherwise he retained the editorial structure that Nelson devised. When Schnell became editor in 1983, he ceased using associate editors and returned to the original practice of the editor taking care of the entire review process. He did retain the post of book review editor and replaced Platnick with W. Wayne Moss. During this same period, the acceptance rates for the journal dropped steadily until they reach a low of 42 percent under J. D. Smith. As figure 5.1 shows, acceptance rates dropped sharply when the journal moved to Kansas, but this change was more than offset by a doubling of the size of the journal. Although the percentage of papers accepted dropped, the absolute number of papers published increased dramatically. When the journal moved to the American Museum, acceptance rates dropped sharply again but this time without any increase in the number of pages published.[1]

To anyone who witnessed the events, the controversy over phenetics seemed to have hit its peak in biological systematics when Richard F. Johnston was editor of the journal, while the polemics over cladistics reached its peak during J. D. Smith's tenure as editor. My personal impression is that the controversy over cladistics was somewhat more heated than was the controversy a decade earlier over phenetics. Such things as degrees of polemics and their effects are difficult to estimate, let alone measure. One indirect indication of degree of polemics in a journal is the number of pages devoted to the "Points of View" section. Learned journals frequently set aside a section for shorter, more polemical papers, papers that frequently take exception to something said earlier in the pages of that journal. Contributions to the Points of View section are refereed the same way that ordinary papers are, but the authors are usually allowed somewhat greater leeway in the tone of their contributions than in standard papers. In the Points

1. The figures for acceptance rates for manuscripts submitted to *Systematic Zoology* are somewhat impressionistic because each editor calculated his or her figures differently. In addition, several editors passed on numerous manuscripts to their successor, while others did not. For example, on Schnell's mode of calculation, he published 56 percent of the manuscripts submitted to him. When the method that Smith used for his own tenure in office is used, the result is 42 percent.

of View section, authors are allowed to take each other on somewhat more directly and bluntly than in conventional articles. If number of pages devoted to Points of View in *Systematic Zoology* is any indication, a spurt of polemics occurred in 1969 during the editorship of Johnston, another in 1975 when Nelson was editor, followed by a steady rise under Schuh and then Smith, reaching a peak in 1980 (fig. 5.1). In 1980 the number of pages devoted to standard articles dropped to under 50 percent.

As figures 5.1 and 5.2 indicate, a correlation exists between the amount of polemics that has characterized the pages of *Systematic Zoology* and membership in the Society of Systematic Zoology as well as subscriptions to the journal. Although correlations do not guarantee causation, controversy seems to have been good for the society, at least as far as numbers are concerned. Peaks in circulation and membership correlate nicely with periods of increased confrontation and controversy.

Being the editor of a journal has its advantages. Chief among them is the ability to influence the course of development of a discipline. But being an editor also has its disadvantages as far as one's own career is concerned. First of all, the time required to edit a journal necessarily interferes with one's own research. Secondly, the role of editor is calculated to gain no friends but generate considerable ill will. When editors accept a manuscript, they get little credit. The manuscript was accepted on its merits. If, on the contrary, they reject a paper, the author is sure to be upset and likely to blame the editor. A third problem with being an editor of a journal is that editors tend not to publish in a journal while they are editing it. The likelihood that others will charge conflict of interest is too high. This problem was especially acute with respect

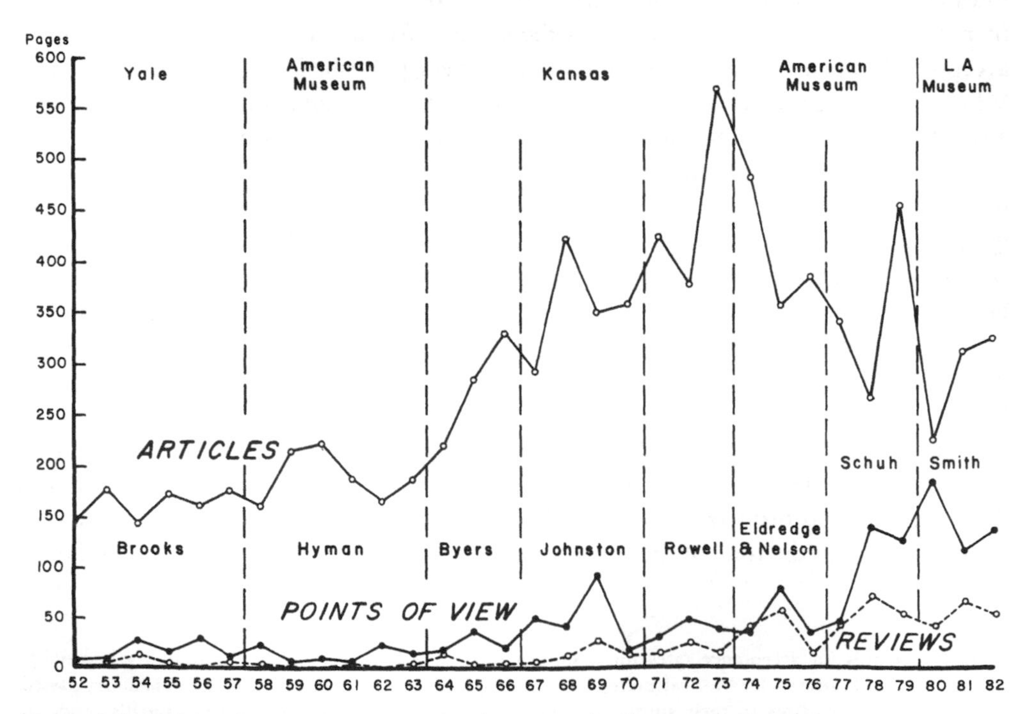

Figure 5.1. Articles, points of view, and book reviews in *Systematic Zoology.*

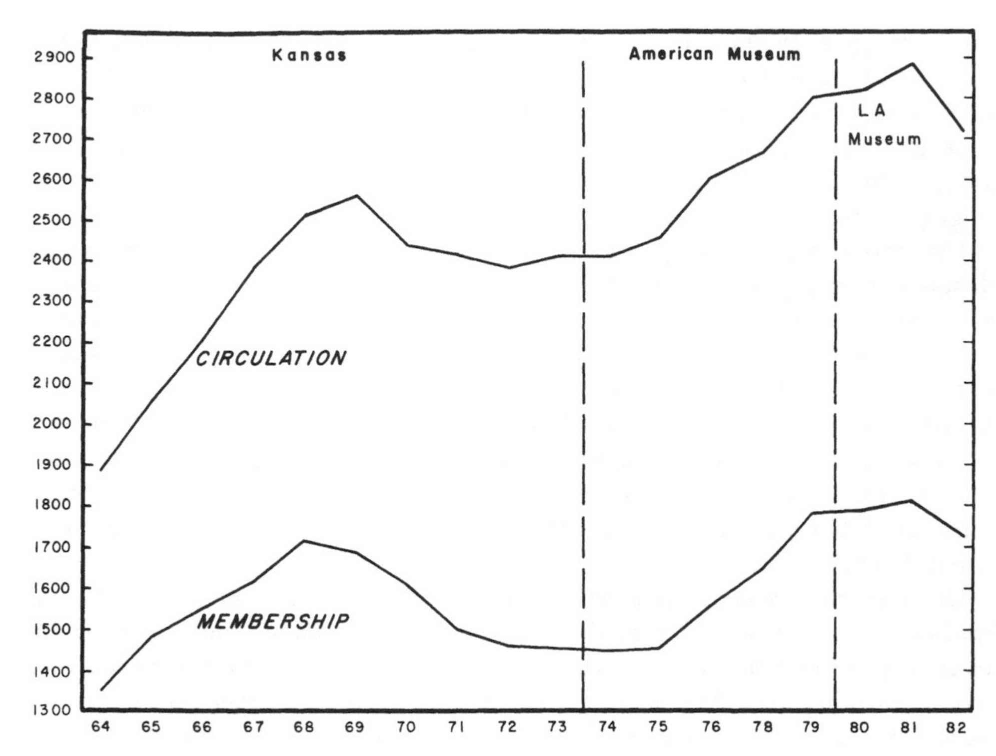

Figure 5.2. Circulation of *Systematic Zoology* and membership in the Society of Systematic Zoology.

to *Systematic Zoology* because for years it was almost the only outlet for papers on systematics philosophy; *Systematic Botany* began to appear in 1976, the *Journal of Classification* in 1984, and *Cladistics* in 1985. During the first twenty-two years of *Systematic Zoology,* the only editors to publish papers in the journal while they were editor were Hyman and Johnston. Hyman (1958, 1959) published two very short pieces amounting to less than one page on the proper use of the term "nema," while Johnston (1969) published a substantial numerical study of character variation and adaptation in European sparrows. During this same period, several editors published short book reviews, all of them favorable: Brooks (2), Byers (1), Johnston (2) and Rowell (1).

When Nelson became editor of *Systematic Zoology,* he was well aware of the difficulties associated with an editor publishing papers in his own journal. However, cladistics and Croizatian biogeography were just then coming to a head, and there were very few places other than *Systematic Zoology* which were liable to publish papers on these topics. He wrote to several associates to ask for advice about how to proceed. He was advised to publish in the journal but only with extreme care. During his first year as editor of the journal, Nelson published two short points of view and coauthored a long paper on vicariance biogeography with Croizat and Rosen. In 1974 Mayr published a long paper in a German journal arguing against the principles of cladistic analysis and claiming that Darwin was the founding father of evolutionary systematics.

One of Nelson's (1974a) short pieces was a reply to Mayr (1974) arguing that Darwin was actually a precursor to cladistics. Rosen (1974) also replied to Mayr (1974), but his response concerned the superiority of Hennig's principles of classification to the more eclectic principles that Mayr espoused. In his second point of view, Nelson (1974b) corrected his own earlier position on proper methods of inference in historical biogeography (Nelson 1969), rejecting Hennig's Progression Rule.[2]

The story of the Croizat paper is much more complicated. In 1973 Croizat sent Nelson a manuscript that attacked Darwin and his views on biogeography in terms of centers of origin and defended Croizat's own theory of panbiogeography. Many of the experts in the field whom Nelson first approached to review Croizat's manuscript for *Systematic Zoology* refused. As he accumulated refusals to referee the manuscript, he started sending it out to increasingly large numbers of potential referees. By the time he was done, he had obtained nineteen reviews of Croizat's manuscript along with ten refusals. Anyone who has had any experience with the refereeing practices of professional journals knows that Nelson's obtaining so many reviews for a manuscript is extremely unusual.

Of the nineteen referees for Croizat's manuscript, only one found it ready for publication. Most thought that it needed major or minor revision. Four urged outright rejection (see Column A in table 5.2). One of the commonest complaints concerned Croizat's writing style. Although one referee found Croizat's paper "clear and comprehensible" and another thought that it was "well-written," most strongly disagreed. As might be expected from Croizat's cosmopolitan background, he had some facility in several languages, including French (the language of his parents), Italian (the language of his birthplace), English, Spanish, as well as Portuguese, Latin, Russian, German, and to some extent Greek. Although Croizat's flair for languages is intimidating, the

Table 5.2. Summary recommendations for six manuscripts which received multiple reviews in *Systematic Zoology*.

	Biogeography			Cladistics		
Recommendation	*A*	*B*	*C*	*D*	*E*	*F*
Ready	1	4	5	3	1	15
Minor modification	6	4	5	4	3	6
Major modification	9	0	8	3	8	1
Reject	4	6	0	3*	8	2*
No rating	0	3	0	0	1	0
Totals	20	17	18	13	21	24

*Referees recommended alternative actions—rejection or major modification.

2. The referees' reports for the replies to Mayr by Nelson (1974a) and Rosen (1974) are the only ones during this period that are not present in the archives of the Society for Systematic Zoology housed at the Smithsonian Institution Archives. The only other manuscripts for which no records exist are Nelson (1974b) and all the papers that appeared in Nelson's final issue, the proceedings of a symposium on biogeography. The contents of the society's archives have yet to be catalogued. I appreciate the aid provided by Pamela Henson in helping me to work my way through this mass of paper.

manuscripts that he submitted for publication frequently departed significantly from standard academic prose. They were filled with numerous noncolloquial expressions and bizarre syntactic constructions. By all indications he submitted first drafts, typed as fast as his fingers could strike the keys. He also tended to express his exasperation with the work of others in overly direct ways. In any field informal conventions exist with respect to the terms one should and should not use in criticizing one's colleagues. For example, saying that a particular view is "interesting" carries a mildly critical connotation; saying that it is "irresponsible garbage" is too strong. Saying that an author "claims" something implies doubt; saying that he "shows" something indicates assent.

Most referees also complained of the excessive space devoted to criticizing mistaken views and insufficient attention to the explication of Croizat's own preferred views. Several took exception to what they took to be mistaken and unfair criticisms of Darwin. For his time, Darwin's views on biogeography were quite reasonable. Croizat also went on at great length about the methodological shortcomings of those whose views he opposed. Their work was too "deductive." He preferred to keep theories out of science and stick to the facts. Several referees objected to the exclusion of theories and hypotheses from science and pointed out that Croizat's method was no less open to the charge of "deductive reasoning." However, even some of the most critical reviewers indicated a long-standing curiosity about Croizat's ideas. They would welcome a concise statement of Croizat's theory of panbiogeography, but the manuscript that Nelson had sent them to review was anything but that.

Most of those who declined to referee Croizat's manuscript either gave no explanation for their refusal or begged off because of too much work. A few gave more particular reasons. One disqualified himself because he had to "admit to a certain prejudice against Dr. Croizat's work," but it was Simpson's comment that especially rankled Nelson. According to Simpson, "Study of Croizat's voluminous work has convinced me that he is a member of the lunatic fringe. Therefore I could not make an unbiased review & it would be unfair for me to act as referee" (Nelson 1977: 451). That summer Simpson presented a paper at the Wenner-Gren Foundation's castle in Austria on phylogeny and classification. One of those who attended this conference at Burg Wartenstein, Malcom McKenna, taped the presentations and, upon his return to the American Museum, let several people, including Nelson, listen to Simpson's remarks. According to Simpson:

> Many people—some people anyway—have said that if you start applying Hennigian taxonomy to a group of organisms, you are likely to wind up with nonsense. And the reply—the only reply that I have seen (a general reply, a reply on a matter of principle rather than attacking an example)—has been that the system cannot generate nonsense because everything that it generates is consistent with its premises. Well, now, that sounds like a marvelous argument; but you know, if your premises are wrong and are idiosyncractic, you're insane if you argue this way. That is known as paranoia in psychiatric terms. So I think literally this system is a paranoic system. Sure, it's perfectly logical if you accept its premises, but its premises are wrong and they are idiosyncratic. That is certainly a very violent, strong criticism to make and it does require some backing up. You may say that I have invested my life in a different system of taxonomy and so, of course, I'm going to think that anybody who doesn't agree with me is paranoid. That may or may not be true. But even if it's true, I don't believe it.

Nelson wrote to the director of research of the Wenner-Gren Foundation to obtain permission to use this quotation in a book that he was writing. After some delay, she responded that he would have to ask Simpson, which he promptly did. Simpson responded that on no account could Nelson quote these comments. "That was a closed conference. Discussion was among friends. My remarks were not intended for publication and were at times facetious, as was understood in that context. The context in your proposed book will certainly not be similar, and I have good reason to think that it will not be friendly." In his reply, Nelson set out the details of the taping and admonished Simpson for his "unwillingness to be candid before the world, for your habit of saying one thing in public and another in private, and in seeking protection in the guise of privacy." In his autobiography, Simpson (1978a: 217, 271) recalled this incident, complaining that some facetious remarks on phylogeny and classification which he had made at a conference in Austria caused him to be "insanely denunciated when an unofficial taping of them was played at my former institution in New York." A Hennigian in the audience had "what our ancestors called a conniption fit."[3]

Nelson returned Croizat's manuscript to him to rewrite in the light of the referees' comments, which he promptly did. Nelson sent this new version of the manuscript to nine of the original referees. Three of the referees who had originally recommended outright rejection saw no reason to change their minds. According to one of these referees, "In general, my criticism of this paper is that it deals too much in straw men and dead horses and too little in either analysis or synthesis of our currently expanded knowledge. The style is as usual with this author polemical and devoid of balance" (Smithsonian Archives). The other six who received the manuscript to reevaluate all had recommended major modification. Three found the revised manuscript still in need of minor modification, while three thought that it was ready for publication.

In an effort to gain a larger audience and more serious consideration for Croizat's views, Nelson wrote to Croizat to suggest that he and Rosen use Croizat's revised manuscript as the basis for a joint paper authored by all three of them. Authorities would find it easy to continue to ignore Croizat, writing in his highly personal style from his isolated home in Venezuela. They would find it much harder to ignore a paper published jointly with two workers located at one of the most prestigious museums in the world, one the editor of *Systematic Zoology,* the other president-elect of the society that published the journal. Although Croizat was taken aback by the suggestion, he consented. Nelson and Rosen made three major changes in Croizat's manuscript. They added a long introductory section explaining the notion of "generalized tracks," numerous lengthy footnotes, and a discussion of Hennig's principles of classification. Only one review of this manuscript survives, and this referee, though closely associated with Nelson, suggested major modification. He especially complained of the excessive

3. The taping took place in July of 1974. Due to delays in corresponding with the officials of the Wenner-Gren Foundation, the exchange between Nelson and Simpson did not occur until May of 1976. Copies of his correspondence with G. G. Simpson were supplied me by Gareth Nelson.

number and length of the footnotes. Nonetheless, the joint paper appeared, footnotes and all (Croizat, Nelson, and Rosen 1974).[4]

Nelson published no other papers in the journal during the rest of his tenure as editor. A half dozen years later, J. D. Smith, with a coauthor, published a paper while he was editor that applied the principles of cladistic analysis to a group of bats (Hood and Smith 1982), but that was the end of editors publishing papers of their own while editing the journal. Book reviews are quite another matter. Because book reviews are not refereed and rarely occasion much in the way of editorial changes, authors of book reviews are put in a peculiar position. They are free to say fairly much what they please about the work of another author without the usual constraints imposed by the refereeing process. Early editors of the journal published an occasional book review: Brooks (2), Byers (1), Johnston (2), and Rowell (1). In each case, these reviews were favorable.

The situation changed somewhat after Nelson introduced the post of book review editor. The book review editor, not the editor in chief, selected reviewers for the books submitted to the journal. While Ghiselin was review editor, he reviewed seven books, all of them favorably. Nelson also published book reviews while Ghiselin was review editor, increasing the number of his reviews when Ghiselin became ill. Before his term as editor was over, Nelson had published twenty-eight book reviews for a total of twenty-nine pages printed in reduced type. In particular, he reviewed seventeen books in a single number of the journal in 1975. Most of the books he reviewed concerned biogeography and plate tectonics. In general he decried how long it takes for scientists to respond to scientific revolutions. Most biogeographers continued to work on the old model of centers of creation followed by dispersal, ignoring the implications of continental drift for this traditional view. And everyone was ignoring the monumental work of Croizat! However, his overall estimations of these books were evenly divided among positive, negative, and mixed.

Eventually Platnick took over as book review editor. During his six years as review editor, he reviewed seventeen books, ten of them with Eugene Gaffney on the philosophy of Sir Karl Popper (Platnick and Gaffney 1977, 1978). Of the reviews authored by Platnick himself, three were favorable, three unfavorable, and one very critical. While Schuh was editor, he published three reviews, while Smith published none. When Schnell became editor, he replaced Platnick with Wayne W. Moss. During his first three years as book review editor, Moss published no book reviews, while Schnell reviewed a book on graph theory quite favorably. A common refrain in the book reviews written by cladists was that the author of the book under review was ignoring Croizat, cladism, or both.

The publication of articles, points of view, and book reviews in a journal by those in a position to make editorial decisions periodically elicited objections. One reader from New Zealand wrote to Gould, who was one of the three contributing editors at

4. Even though the referees found the manuscript by Croizat, Nelson, and Rosen (1974) poorly written, a citation analysis conducted in 1984 lists it as one of the most frequently cited papers to appear in *Systematic Zoology*, the seventeenth most cited paper (see Appendix D).

the time, to complain of Nelson's coming "very close to abusing his editorial privilege" in connection with the responses he published to Mayr's (1974) critique of cladism—responses by Rosen (1974) and himself (Nelson 1974a):

> Rosen and Nelson's replies to Mayr's critique of phylogenetic systematics, however right or wrong Mayr may be, appear to have been published with haste that is possible only to those on the inside of the system. It seems to me that such rapid publication would not have been available to me, or to them in any other journal. . . . Similarly, it seems to me that Nelson's book reviews are little more than an insult to the writers and a further opportunity to champion the cause of Croizat—they are not in any sense of the word book *reviews* (Smithsonian Archives).

This author also wrote a letter directly to Nelson complaining of what he took to be Nelson's excesses, a letter which was, if anything, even blunter than the letter to Gould.

Publication by members of the editorial staff of a journal in that journal was one problem. The number of referees that Nelson consulted in several cases was another. Receiving two or three referees' reports for a manuscript is usual, but in a half dozen cases Nelson sent manuscripts to multiple referees. The first instance was Croizat's paper. The number of referees grew in part because of Nelson's initial difficulty in getting anyone to referee the manuscript. The second instance was a paper submitted by a Canadian paleontologist. Early in 1974 Nelson presented a paper on cladistics and biogeography at a Canadian university where he got into a heated exchange with a local paleontologist. The paleontologist became so incensed that he walked out of the room. Immediately thereafter he submitted two short papers to Rowell criticizing Nelson, thinking that Rowell was still editor of *Systematic Zoology.* Rowell forwarded the two manuscripts to Nelson to edit. The shorter paper of the two concerned Darwin's allegiance to the principles of cladistic analysis. Nelson responded as editor, mentioning that he would follow the recommendations of the referees. He also included additional comments, not in his role as editor, but as an interested party, appealing to "friendliness" and the virtues of "candor." In less than two weeks, Nelson was able to return this short paper to its author along with four referees' reports. All four recommended publication after some revision. The author wrote Nelson a note informing him that he would rewrite and resubmit.

The second, longer paper by the Canadian paleontologist was highly critical of Nelson's (1973c) defense of Croizat. Once again, Nelson attempted to act in two capacities at the same time, as an editor and as an author under siege. As Nelson remarked in a five-page letter, "Your manuscript strikes me as being nothing more than a personal attack on me." Several of the referees agreed. By the time that Nelson was done, he had accumulated reports from seventeen referees (see column B in table 5.2). Four referees judged the manuscript ready for publication, four more suggested publication after minor changes, six opted for outright rejection, and three gave no summary recommendations. Thus, in slightly over a month after submission, the author received seventeen strongly divided referees' reports. This time the author did not respond. A year later, when Nelson had heard indirectly that this author had no intention of resubmitting either paper, he wrote to him expressing his regret. Then in 1976 Nelson appointed this author to the editorial board of the journal.

Late in 1975 the New Zealand biogeographer who had written to Gould and Nelson to complain of Nelson's possible abuse of his editorial privilege submitted a paper critical of Croizat's views on biogeography to Nelson. Nelson obtained eighteen referees' reports for this manuscript (see column C in table 5.2). Although nearly all the referees found this paper too long and polemical, none recommended outright rejection. Five found it ready for publication, five suggested minor modification, while eight recommended major modification. However, two of the recommendations for immediate publication warrant some comment. Croizat urged Nelson to publish the paper so that he could "execute" the author, while Rosen was anxious to see the paper in print so that he could "blast him out of the water." Gould found the paper much too long, poorly written, and polemical but voted for publication because others such as Nelson got to publish such papers, why not this author? Nelson responded with some heat to Gould's poorly informed criticisms of his editorship of *Systematic Zoology.* Gould answered, "You manage to hurt more people's feelings than almost anyone I know, your invariably defensive letters have the unerring character of rendering yourself blameless for everything" (Smithsonian Archives).

Three months after receiving the manuscript, Nelson returned it to the author with the eighteen referee reports. Once again, Nelson asked the author to take account of the reviewers' criticisms and suggestions but agreed to publish the paper regardless. Although the author was surprised by so many reviews, he wrote to Nelson that he would rewrite and resubmit. Six months went by, and Nelson heard nothing from the author, but he did become aware of rumors to the effect that "editorial bias" was holding up publication of the paper. Nelson wrote to the author complaining of such "malicious gossip." The author responded that he had told no one that his paper had been rejected by *Systematic Zoology.* To the contrary, he had told the two people to whom he had written on the matter that the paper had been accepted subject to modification. The author then turned to Nelson's handling of the journal, asking rhetorically, "But can you not see that the sort of reaction you are getting is an inevitable result of your involvement in *Systematic Zoology* as both editor and the author of controversial and very strongly worded and critical papers?" He concluded his letter with some well-intentioned advice for Nelson:

> I once had the cheek to write to you that I thought you were operating on the basis that the more often you say something and the louder you say it, the more likely people are to believe you. You hotly rejected such a view, but I still wonder whether there is some truth to it. Don't cut yourself off from those who disagree with you by writing angry, abusive letters. You'll end up isolating yourself in the way that Croizat has, and I can imagine that he is a very lonely old man. The blame doesn't all belong to him, but in my view much of it does.

By the time that the New Zealand biogeographer finished revising his manuscript, Schuh had taken over as editor of the journal. In accordance with the editorial policy of the journal, the author sent his revised manuscript to one of the associate editors, Hull, to edit; he also sent a covering letter describing the manuscript's tortured history. Schuh in turn sent Hull copies of all past correspondence over the manuscript and the earlier referee reports. Hull sent the revised manuscript to two referees with a letter explaining the earlier reviews. These referees still found the paper too long, poorly written, and polemical but one suggested that under the circumstances it should be published with only a few stylistic, editorial changes. Hull passed on this recommendation to Schuh and reported the results to the author. Three years after its initial submission, the paper appeared in *Systematic Zoology*. However, most of the delay resulted from the author taking so long to resubmit the revised manuscript.

Three of the papers submitted to Nelson that received numerous reviews concerned Croizatian biogeography. The other three concerned Hennig's principles of classification. In 1975 two second-generation cladists submitted a paper about the role of ancestor-descendant relationships in reconstructing phylogeny. Once again Nelson sent the manuscript to numerous referees, and once again they were strongly divided on the merits of the work under review, three recommending acceptance, four minor modification, three major modification, and three outright rejection (see column D in table 5.2). While Nelson was editing this manuscript, he received a manuscript defending the role of ancestor-descendant relationships in phylogeny reconstruction by a paleontologist who held a joint appointment at the City University of New York as well as the American Museum of Natural History, Nelson's own institution. Nelson sent this manuscript to fourteen reviewers. In this case, there was strong consensus among the referees, most recommending major modification or outright rejection. They found the manuscript too polemical and poorly argued. The only referee who voted for publication thought that it was a terrible paper but no worse than several other papers that Nelson had already published.

Upon receiving fourteen reviews, most of them negative, the author objected somewhat testily. Nelson responded that, if he did not like how his manuscript was being handled, he should complain to the Council of the Society of Systematic Zoology. Nelson then turned the manuscript over to his coeditor, Niles Eldredge, to finish editing. By the time that the refereeing process was over, this manuscript had accumulated twenty-one referees' reports (see column E in table 5.2). A majority of the referees continued to be dissatisfied with the manuscript. Even so, the author persisted. In 1976, when Schuh began to take over the editing of the journal, he was willed these two problem cases, one arguing against the traditional role of ancestors in phylogeny reconstruction, the other arguing for it. Schuh's decision was to publish the two manuscripts together. The readers of the journal could compare the two and decide the merits of the case for themselves.

As much commotion as the preceding five manuscripts caused, it was nothing compared to the controversy that resulted from a manuscript submitted by Sokal defending Mayr against the attacks by Rosen (1974) and Nelson (1974a) and complaining about Nelson's handling of manuscripts submitted to *Systematic Zoology*. Nelson received the manuscript on April 8, 1975, and a week later wrote Sokal a three-page response. With respect to his editorship, Nelson remarked that:

> it may appear to you that I take some partisan stance in favor of "cladism" or some other "ism," with the purpose of converting the world to some point of view in order to form some kind of "school" or tradition. And it may appear to you that I am using the editorship of *Systematic Zoology* somehow unfairly to further that purpose. *I know otherwise*. My efforts are aimed at

transcending the partisanship of "schools." Of course, you are free to view this statement as a conceit, but I wished to state it anyway (Smithsonian Archives).

However, with respect to his manuscript, Nelson assured Sokal that he would treat it according to the standard editorial practices of the journal. In fact, he would be willing to publish the original manuscript if Sokal found the reviewers' comments to be of no help.

In slightly more than a month, Nelson returned Sokal's manuscript with twenty-four reviews plus extensive comments from Rosen (see column F in table 5.2). In his letter, Nelson charged Sokal to respond as "constructively as possible to the reviewers' comments." Although Sokal was dismayed at receiving so many reviews, he could not complain about their summary judgments. A majority of the referees found the original manuscript ready for publication. More than one referee commented on the irony of one of Mayr's most persistent critics coming to his defense. It took Sokal a while, but he finally returned a revised version of his manuscript along with twelve pages of detailed responses to the referees' comments. Nelson wrote back that he really had not intended for Sokal to respond so extensively and apologized for the misunderstanding. Sokal replied that he had put the matter out of his head and that it had done no great harm. Sokal's defense of Mayr against his cladist critics appeared late in 1975.

In his original manuscript, Sokal had questioned the propriety of Nelson's publishing a polemical paper in a journal which he himself was editing. In the published version, Sokal did not mention this issue and relegated Nelson's (1974a) claim that Darwin was a cladist to a single paragraph at the end, dismissing the issue as being beside the point. Most of the paper dealt with Rosen's (1974) failure to address several of the objections raised by Mayr (1974) to cladistic analysis. In his revised paper, Sokal added a comment about the numerous referees' reports that he had received—twenty-eight in all. (The discrepancy between table 5.2 and Sokal's count results in part from Sokal considering two letters which Nelson included as reviews.)

As unusual as Nelson's sending these six manuscripts to so many referees was, it does provide insight into the intellectual climate of the time. Because Croizatian biogeography and cladistic analysis were such controversial topics and Nelson was their chief advocate in the English-speaking world, one can understand an author suspecting that the treatment of his manuscript was to some extent biased. Two papers critical of Croizat did receive multiple reviews but so did the paper by Croizat himself. In fact, it was the first paper that Nelson sent to multiple reviewers. Two papers critical of the principles of cladistic analysis also were sent to numerous referees, but once again so was a paper defending Hennig's methodology. Although the ratio is two to one, Nelson was not totally biased in selecting papers for multiple reviews. And, as Sokal remarked in his response to Nelson, receiving so many reviews is not all bad. Although Sokal would not like to be confronted with so many reviews every time he submitted a paper for publication, "it probably did me some good to think myself through the variety of criticisms that were offered." (In this chapter I have discussed the role of possible bias with respect to editors of professional journals. I return to these papers in chapter 9, where I investigate the effect of possible bias on the part of referees. For an informative study of the way that individual papers make their way through the refereeing process, see Myers 1985.)

In the middle of 1976, Nelson passed on the editorship of *Systematic Zoology* to a young entomologist at the American Museum, Randall T. Schuh. In connection with transferring editorial responsibility of the journal, Nelson, in a letter to the associate editors, made the following observation about his own past editorial policies:

> While I have been editor I have operated on the assumption that my job was to produce as much review comment useful to the author as I could. I never felt it was my job to sit in judgment of manuscripts. I did feel that for a manuscript to be published, someone had to support its publication. Someone almost always did, except for a few cases of manuscripts that were unintelligible. Sometimes a manuscript went the review rounds more than once. Almost always, however, I found that authors responded constructively to criticism from reviewers. I never required an author to accept a reviewer's opinion, but I often had a manuscript reviewed enough so that, if there was a problem, it would be obvious to the author. I have found that system to work very well. With one exception, who shall remain nameless, no author refused to respond to the generality expressed by reviewers.

Schuh finished off several manuscripts that were left over from Nelson's tenure and brought the practice of numerous reviews to a halt. However, he retained the general editorial structure which Nelson had initiated. In general, Schuh was a sterner editor than Nelson had been. During his tenure, Nelson rejected only thirty-two papers outright, ten of them because they were inappropriate for the journal. He returned most papers for revision, either minor or major. Of these, twenty-four were either withdrawn or simply disappeared. As a result, Nelson published 67 percent of the manuscripts he received. He would have published more if the authors had bothered to rewrite and resubmit. Schuh was more likely simply to reject a paper without encouraging the author to rewrite and resubmit. The most common reason that Schuh gave for rejection was that the paper was inappropriate for the journal, and the most common reason for this decision was that the paper was simply a description of data with no implications for wider issues.[5]

From the beginning, the founders of *Systematic Zoology* wanted the journal to be primarily an outlet for papers on taxonomic philosophy and on other topics only to the extent that they bore on this one. Such goals were easier to declare than to realize. As table 5.3 shows, during the first twenty-two years of the journal's existence, 46 percent of its pages were devoted primarily to the presentation of data, while only 24 percent of the pages were devoted to discussing general taxonomic views. The figure for description rose to 62 percent under Nelson and that for philosophy dropped to 10 percent. The low number of papers discussing general taxonomic philosophy during Nelson's tenure as editor was not due to authors submitting papers and getting them rejected. They simply were not submitting very many papers on taxonomic philosophy to the journal. Nor were systematists publishing papers on these topics elsewhere.

5. In his editor's reports for 1974–76, Nelson lists sixty-seven manuscripts as being rejected or withdrawn (thirty-four rejected; thirty-three withdrawn). The discrepancies between Nelson's figures and those which I obtained are probably due to Nelson's counting the same manuscript more than once from year to year. In his editor's reports for 1977–79, Schuh lists 109 manuscripts as rejected or withdrawn. He does not provide separate figures for manuscripts which were rejected and those which were returned to be rewritten and never resubmitted. On my count, he rejected seventy-three and returned thirty-six for revision.

Table 5.3. Percentage of pages during tenure of each editor devoted largely to presentation of data and descriptions versus general discussions of taxonomic philosophy.

Editor	*% Description*	*% Philosophy*
Brooks	50	17
Hyman	58	22
Byers	44	24
Johnston	50	30
Rowell	55	17
Nelson	62	10
Schuh	35	32
Smith	28	33
Schnell	33	34

When Schuh became editor, he rectified the imbalance between descriptive and theoretical papers. The decision was conscious. Schuh thought that too many descriptive papers were appearing in the journal, especially numerical studies of various groups. Under Schuh acceptance rates continued to drop, down to 59 percent. Under Smith, who continued Schuh's policy, the drop continued, down to 42 percent. The increase to 56 percent under Schnell's editorship results in part from differences in modes of calculating acceptance rates.

Although acceptance rates are not strictly comparable from editor to editor because of differing methods for computing them, a trend toward a decrease in the percentage of manuscripts accepted for publication is clear (table 5.1). Under such conditions one might expect author dissatisfaction to increase. During the editorships of Schuh, Smith, and Schnell, the papers devoted to taxonomic philosophy began to approach the prevalence that the founders of the journal had anticipated.

Because Schuh did not become as personally involved with the authors who submitted papers as Nelson had, his tenure in office tended to be somewhat less tumultuous. He wrote short, business-like rejection letters, and that was that. If an author complained, Schuh was firm. However, two papers generated considerable controversy. Nelson bequeathed Schuh one of these papers—a manuscript by Mary Mickevich and Michael Johnson using both phenetic and cladistic methods to test congruence; i.e., the production of very similar classifications given different sorts of data. The other problem case Schuh left to Smith was a paper by a mathematician, M. F. Janowitz, criticizing some of Farris's work. In 1971 Mickevich transferred from Boston University to become a graduate student and teaching assistant in biology at Stony Brook. Although Mickevich came to Stony Brook with a high opinion of Farris's work, initially she took a strong dislike to him. He struck her as being too abrasive and opinionated, but eventually Mickevich became convinced that his computer programs for estimating phylogenetic relationships among organisms were vastly superior to the techniques being developed by other members of the department. With support from his National Science Foundation grant, Farris took her on as an assistant, and he became chairman of her dissertation committee. Eventually, when they married, Farris turned over the chairmanship of the committee to Rohlf. Other local members of the committee included Sokal and G. C. Williams. Donn Rosen from the American Museum was the outside member.

By this time the relations between Sokal and Farris had begun to degenerate. The final breakdown came over the program for Numerical Taxonomy 8, to be held in Oeiras, Portugal, in August of 1974 under the sponsorship of the Advanced Study Institutes Program of the North Atlantic Treaty Organization. George Estabrook from the University of Michigan was made program chairman for the meeting. Estabrook invited Farris to present a paper. Farris in turn urged Estabrook to invite Mickevich as well. When Estabrook refused, Farris approached Sokal to intercede, but Sokal declined to enter into the dispute. The program for the meeting was "Estabrook's baby." As a result Farris himself refused to attend the meeting, but he was more than happy to review the proceedings of the conference later. In his review, Farris (1977a: 229) complained of what he took to be a systematic equivocation in the pheneticists' use of the term "phenetic." Sometimes they used it to mean little more than "empirical," making "phenetic" taxonomy the "only defensible approach to classification."

Early in his career, Farris criticized the numerical methods presented by pheneticists, but he did so politely. After the run-in over the Portugal meeting, his critiques became somewhat more pointed. His usual technique was to show that cladistic methods are superior to phenetic methods on the very evaluative criteria that the pheneticists themselves were proposing. For instance, Farris (1971) argued that cladistic classifications are in principle more congruent than phenetic classifications, i.e., that given different sorts of data, the classifications produced by cladistic methods tend to be similar to each other, while those produced by phenetic methods fluctuate erratically. In 1975 Mickevich collaborated with Michael Johnson to test for congruence between Johnson's morphological and allozyme data for five species of fish of the genus *Menidia,* using both phenetic techniques and Farris's minimum-length Wagner trees. Using phenetic techniques they found no congruence between the two sets of data and near perfect congruence when cladistic methods were used. Farris's (1971) arguments were in-principle formal arguments concerning hypothetical data sets. Mickevich and Johnson supplemented Farris's formal arguments with actual data. The threat to phenetics was clear.

When Mickevich and Johnson submitted their manuscript to Nelson, he sent it out to two referees, neither of them especially aligned with any "school" of taxonomy. They promptly returned the manuscript with their recommendations to publish after major rewriting. One referee complained of the unavailability of the relevant data and of the disparaging remarks that Mickevich and Johnson made about his own earlier work. Johnson had yet to publish either his biochemical or his anatomical data. The second referee complained of the characters being classified as either present or absent instead of in terms of their relative frequencies. A population was counted as exhibiting a trait regardless of the percentage of individuals in the population that possessed it, whether 1 percent or 100 percent. However, as much as this referee disagreed with the major conclusions in the manuscript, he recommended publication because it would rouse interest and controversy. A revised version of the paper appeared in the fall of 1976.

No sooner did Mickevich finish her first paper than she started work on a more ambitious project extracted from her dissertation, a paper in which she processed five different data sets using five phenetic and three cladistic methods. She found that all the phenetic methods were extremely unstable. Of the cladistic methods, Farris's Wagner method was the most stable. Once again cladistics was shown to be superior to phenetics on the pheneticists' own criteria. Schuh received Mickevich's paper in the middle of February, 1977, and sent it to two referees, both second-generation pheneticists. Both referees found the paper very good but in need of major modification. Schuh returned the manuscript to Mickevich for revision.

Back in Australia, Colless was beginning to get suspicious of the paper published by Mickevich and Johnson (1976). By then Johnson's (1975) biochemical data had appeared, and Colless found several discrepancies between the data sets presented in the two papers. In addition, he could not get the results claimed by Mickevich and Johnson. In the summer of 1977 Colless wrote to both Johnson and Mickevich about the problems that he was having. By then Johnson had taken a position in Australia himself. Johnson replied to Colless, indicating the errors in the published data that he had discovered. When Mickevich did not respond, Colless wrote up a short manuscript which he sent to Johnson and Mickevich in February of 1978. In this paper he pointed out the problems that he was having with the data base used by Mickevich and Johnson (1976) and argued that their conclusions did not follow quite as automatically as they claimed. He concluded that, for the data supplied, "there are no grounds for choosing a Wagner Tree rather than a phenogram as the basis for classification."

This time Mickevich responded, setting out all the errors that she had found in the published data sets, promising to publish corrections, and expanding on the methods that she had used to process her data sets. Colless reworked his paper using the corrected data sets and sent a copy to Mickevich for her comments. The problems that Mickevich was having with Colless over her original paper with Johnson could not have come at a worse time. Both Rohlf and Sokal were expressing serious reservations about her dissertation, in particular with the methods that she was using. Mickevich and Farris took the objections raised by Rohlf and Sokal to stem more from unhappiness with her antiphenetic conclusions than from any weakness in her methods. It also did not help that Sokal and Farris by this time were no longer on speaking terms. Because of the problems that she was having with Colless and her dissertation committee, and because of the death of her father, Mickevich had not been able to rework the paper she had submitted to Schuh the year before. Finally, in April of 1978 she returned a revised version of her manuscript and it appeared in August of that year.

The appearance of Mickevich's (1978) paper was the last straw as far as three members of her committee were concerned. They resigned in protest over her publishing the substance of her dissertation without giving them a chance to improve or approve it. Farris took Rohlf's place as director, Rosen stayed on, and Walter Fitch of the University of Wisconsin at Madison was brought in from the outside to serve as an independent judge. Fitch was the ideal person to resolve this very difficult situation.

He was an associate editor of *Systematic Zoology* and worked in the same area as Farris and Mickevich—computer methods for reconstructing phylogenetic trees. He was on good terms with the people on both sides of the dispute. He considered Farris to be his most valued professional friend, and he also attended numerical taxonomy meetings. In the end, the reconstituted committee awarded Mickevich her doctorate in 1978.

While all this was going on, M. F. Janowitz, a mathematician from the University of Massachusetts, submitted a paper to Schuh critical of Farris's (1977b) comparison of phenetic and phylogenetic methods. Several years earlier Janowitz had spent a sabbatical at the State University of New York. During this year Farris and Janowitz got along well, so well that Farris suggested to Malcolm McKenna, who was program chairman of the Society of Systematic Zoology, that he invite Janowitz to present a paper at the 1977 meetings of the society in Toronto. In this paper, Janowitz objected to Farris's replication of certain characters, a practice which he claimed amounted to weighting certain characters more heavily than others. After Janowitz delivered his paper, instead of responding in his usual acerbic manner, Farris said nothing. In fact, the paper engendered no response at all. When Janowitz submitted his paper to *Systematic Zoology,* he informed Schuh that Sokal, Rohlf, and Estabrook had all read the paper and thought that it was important enough to publish immediately. Because the paper dealt with Farris's work, Janowitz sent him a copy. Schuh also sent Farris a copy. In this case, Farris did not consider himself a referee. Insead, he wrote a response to Janowitz which he submitted to Schuh. Schuh published Janowitz's paper as it had been submitted, along with Farris's response (Janowitz 1979, Farris 1979). Janowitz's paper was short—two and a half pages. Farris's response ran to over fourteen pages. Needless to say, Farris was critical. Among the more important faults that he found with the paper, Farris (1979: 205) pointed out several arithmetic errors, remarking that, in point of fact, "most of Janowitz's calculations are incorrect."

The appearance of Mickevich (1978) had a second effect as well. It roused Colless to submit his revised critique of Mickevich and Johnson (1976) to Schuh for publication. Schuh sent the manuscript to Farris to referee. As was his habit, Farris wrote extensive comments on Colless's manuscript and promptly returned it to Schuh. By this time, Schuh's tenure as editor of *Systematic Zoology* was drawing to a close. During most of the journal's history, the editorship was passed along in an amazingly informal fashion. The officers of the society made a few inquiries, a couple of letters or telephone calls were exchanged, and a new editor was selected. Although the council of the society had to approve of the new editor, the vote was largly pro forma. For example, when Petrunkevitch was president, he merely picked a junior member of his own university, John L. Brooks, to become the first editor of the journal. When Hyman became president of the society, she installed herself as editor pro tem because of her displeasure with the way that Brooks was letting the journal fall behind in publication. Later the council confirmed her editorship.

The transfer of the journal from the American Museum to Kansas was accomplished somewhat more deliberately as part of a larger program to reorient the society away from more traditional taxonomic concerns, but once at Kansas the editorship was passed from Byers to Johnston and then to Rowell in a relatively informal fashion. The return of the journal to the American Museum was again a more deliberate, official decision, but the transfer from Nelson to Schuh was largely an in-house affair. In each case, of course, the consent of the Council of the Society of Systematic Zoology was required. As Schuh's tenure in office began to draw to a close, the president of the society was Richard J. Johnston, himself one of the Kansas editors. Johnston decided that care had to be taken in selecting a replacement for Schuh. Hence, he appointed Rosen as past-president of the society to organize a nominating committee to pick a slate of candidates for the position. The nominating committee eventually presented the council with a slate of six candidates—three were well-known cladists (Cracraft, Farris, and Platnick), one was an evolutionary systematist (Ashlock, from Kansas), one was a pheneticist (Gary Schnell), and one was relatively unknown (J. D. Smith). In the past Smith had published largely descriptive, numerical studies. The members of the nominating committee were aware of Smith's allegiances, but the members of the council were not. They opted for the unknown quantity.

Soon after J. D. Smith was elected editor of *Systematic Zoology,* Numerical Taxonomy 13 was held at Harvard University. The wrangling at this meeting between the pheneticists and cladists proved to be even more acrimonious than usual. After so much heated debate, the participants were looking forward to the banquet at a Greek restaurant in Cambridge, especially since the after-dinner talk was to be given by the new editor of the journal. Smith began his address, "There has been some discussion here at this meeting about whether I am a pheneticist or a cladist, and I want to set the record straight. I am a born-again cladist." The pheneticists were stunned, the cladists jubilant, and then the belly dancer came on.

One paper in progress that Smith inherited from Schuh was Colless's critique of Mickevich and Johnson (1976). On the basis of Farris's negative comments, Smith rejected the paper and returned it to Colless. When Colless received Smith's rejection letter, he was a bit upset because he had no doubts about the identity of the referee. He found it a somewhat questionable editorial practice to have a paper critical of an author refereed by that author's husband. Colless returned the manuscript to Smith, requesting a second reviewer who was likely to be less "partisan" than the first. Smith responded that he would be happy to consider a revised version of the paper once Colless had taken into account the forty-odd objections raised by the original referee. He also suggested that the tone of the revised manuscript be less "testy." Colless returned a slightly revised text along with an item-by-item rebuttal of Farris's objections. He also sent copies of his correspondence to Johnston, Rohlf, and several other friends in the United States.

At this time, Smith was becoming involved in an even stickier controversy. Now that Rohlf and Sokal were no longer members of Mickevich's dissertation committee, they felt free to rebut Mickevich's published work. They submitted a joint manuscript to Fitch as the appropriate associate editor. Fitch sent a copy of this manuscript to Mickevich in order for her to write a preliminary response for Rohlf and Sokal to see. That way unnecessary confusions and disagreements could be eliminated prior to publication. Although this process can work when the authors involved are not too strongly divided, Mickevich did not like the idea at all. She preferred to have Rohlf and Sokal publish their piece without any help from her. She would then respond to them in her own good time. She appealed to Nelson and Smith. Nelson responded as a former editor of the journal that, as far as he knew, no editorial policy required that she present her objections to Rohlf and Sokal before they published. In reaction to a dispute between an author and one of his associate editors, Smith took over the editing of the paper himself. Just as Janowitz (1979) published in ignorance of Farris's (1979) rejoinder, Rohlf and Sokal would have to publish without any advance notice of the faults that Mickevich was likely to find with their paper.

Bridling Aggression

The hegemony of English as the standard language in science since the Second World War has led scientific journals published in other languages to include at the end of each article a short resumé in English. In 1978 Croizat published in the bulletin of the Venezuelan Academy of Science a discussion of Søren Løvtrup's (1977) defense of Hennig's system of classification. Croizat's paper in Spanish was fifty-five pages long; the English resumé ran to thirty pages. Croizat's resumé had three main messages: Hennig had stolen his principles of classification from D. Rosa (1918); Hennig's principles, especially the principle of dichotomy, were "ludicrous"; and his biogeography was even worse—"pure farce." As might be expected, Hennig had never referred to Croizat's panbiogeography. When Croizat (1978: 117) wrote to Hennig to ask why he had not, Hennig "blandly answered that he had lacked the time to consult it." For his part, Croizat (1978: 124) would stand on his own record:

> . . . over 10,000 printed pages on the subject of dispersal, which allow me to laugh and titter when faced by, for instance, the "zoogeography" of Hennig, poor soul . . . , and the efforts of uncounted ultra-modern "biologists" *ignorant of evolution in factual terms of space and time.* Why do they speak so loud when they positively do not know the scores of life? Why they do not wish to familiarize themselves—like Hennig, for one—with these scores before flying off on the wings of theory, riding "dichotomous cladograms" and similar broom-sticks?

Nor did Mayr get off unscathed. Croizat (1978: 125) denounced him in a footnote for his "aggressive, crass ignorance" of biogeography in claiming that he knew of no authentic case of transatlantic dispersal among birds.

Before leaving the American Museum for Harvard, Mayr started a study group to foster the cause of systematics. This group was still active twenty years later, but things had changed. The main topic had become cladistics, and as Nelson and Rosen (1981: xii) remarked, the meetings had become somewhat "unruly" because they were "governed only by the absence of formality, which has come to be known as the 'New York Rules' of discourse." In May of 1979, the Systematics Discussion Group sponsored a symposium on biogeography to be held at the museum. Because Virginia Ferris in the past had proved so skillful in handling her more rambunctious colleagues, she was selected as moderator. The discussions after the papers turned out to be so "informal" that they offended some of those present. Others found the impassioned give-and-take exciting.

In response to a publicity release from the American Museum that referred to the "fury and emotion" of two opposing groups of biogeographers who were going to do battle at the conference, a reporter from *The New Yorker* showed up. After the first day's session, he joined the participants at a cheese and wine reception in the Museum's Hall of Ocean Life. There he discussed all the furor with Rosen, "a tall, slightly stooping man of fifty, who was wearing subdued brown tweeds," Nelson, a "big man in his early forties who was wearing a vivid checked jacket," and Cracraft, who was "standing in a blue blazer in front of a walrus diorama" (*New Yorker* Sept. 10, 1979: 38). Rosen gave credit to Nelson for popularizing vicariance biogeography, while Nelson was quoted as giving credit to a "wealthy Venezuelan amateur botanist named Leon Croizat." Since everyone whom the reporter met seemed to be receptive to the vicariance theory, he kept asking where he could find a dispersalist opponent. Each time he was directed to Karl Koopman, a "calm-looking middle-aged man sitting quietly at a table beneath the flukes of the great blue whale." Although Croizat was unable to attend the conference, he did send a paper which was read at the conclusion of the meeting.[6]

When Croizat read the reference to himself as a wealthy Venezuelan amateur botanist, he fired off a letter to the editors of the magazine asking for a rectification. He was neither wealthy nor an amateur. In fact, he was not even a "Venezuelan." They declined. At this time Croizat was corresponding with a young biogeographer from New Zealand, Robin Craw. Craw was also intent on gaining more serious attention for Croizat's views. As Croizat was to repeat frequently during the next half-dozen years, he found Craw to be his most intelligent interpreter. No matter how often others misunderstood his ideas, Craw got them just right. In February of 1980 Croizat wrote to Craw to complain of Nelson and Rosen's characterization of him as a wealthy Venezuelan amateur botanist. However, he said nothing of his indignation to Nelson and Rosen, and he heard nothing from either man for several months to come.

Croizat broke his silence when he read Ferris's review of the New York symposium in *Systematic Zoology*. In the review, Ferris (1980: 73) noted that Croizat's paper was read near the end of the three-day affair: "if the paper had been read earlier, I suspect many of the discussions would have taken a different turn." When Croizat read the review, he was once again furious. He interpreted the placement of his paper as a means of precluding discussion of his work, and, to make matters worse, Nelson and Rosen continued to confuse their vicariance biogeography with his own panbiogeography. Croizat immediately sent a manuscript to Platnick, the review editor, as well as to one of the associate editors. In this paper he emphasized the differences between his views on biogeography and those of Nelson and Rosen, objecting to the confusion that had been sown by rewriting his earlier paper and tagging their own names onto the results. Within the month, Croizat received letters from Platnick, Ferris, Nelson, and Rosen apologizing for any unintentional slight that he may have perceived. Nelson

6. Lawrence Abele (1982: 80), in his review of Nelson and Rosen (1981), described the proceedings as follows:

> A symposium was held at the American Museum in May, 1979, with a critique of vicariance biogeography as the stated goal. Forty-one participants contributed to a volume containing the results of the symposium. A review of the symposium was published by V. R. Ferris (1980), who moderated and maintained some order among participants who, as Nelson and Rosen tell us in the preface, usually operate under "New York Rules." These "Rules," based on my own and others' observations at various symposia, are apparently a combination of rudeness, arrogance and indifference to the perspectives of others.

explained that the placement of Croizat's paper at the end of the symposium was not intended as a slight and that the characterization to which Croizat took such exception was the creation of the reporter from *The New Yorker.* Nelson went on to remind Croizat of how much his own career had suffered by his championing Croizat's views of biogeography. Nelson also wrote to Craw at this time, asking him to drop Croizat a line to express his own point of view on the unfortunate situation that had materialized. Craw decided against it.

In a letter to Nelson dated November 2, 1980, Croizat explained why he felt slighted at being termed an amateur botanist and complained of Hennig's work as a "monumental piece of wreckage on the path of constructive scientific thinking." He was especially dismayed that the American Museum of Natural History had rushed to bestow a gold medal on Hennig. He also reminded Nelson that "at the start of our relations, I clearly told you I did not expect you to jeopardize your chances for the purpose of shielding me." But, as a sign of his good will, Croizat renewed his invitation to Nelson and Rosen to visit him and his wife if ever they happened to be in Venezuela. A couple of weeks later he wrote a similar letter to Rosen, thanking him in particular for his "neatness of expression and felicity of feelings." As a result Croizat withdrew the manuscript in which he pointed out the mischief engendered by their joint paper (Croizat, Nelson, and Rosen 1974).

Traditionally the numerical taxonomy conferences were held separately from the meetings of the Society for Systematic Zoology. In 1976 at the tenth anniversary conference, Richard Johnston suggested that the next numerical taxonomy conference meet in Toronto, Canada, with the Society of Systematic Zoology. Although the three sessions organized by Malcolm McKenna for the Toronto meetings of the Society of Systematic Zoology contained papers by numerical taxonomists, including the paper by Janowitz, Numerical Taxonomy 11 was actually held at the University of Wisconsin in Madison and hosted by Walter Fitch. By 1978 Farris was program chairman of both groups. Even so, they met separately once again, the Society of Systematic Zoology in Richmond, Virginia, and NT-12 in Stony Brook, New York. Then, in January of 1979, Toby Schuh wrote to Johnston, the current president of the Society of Systematic Zoology, to urge that the two groups meet together in 1979 in Tampa, Florida. Johnston replied that the only previous attempt to include sessions on numerical taxonomy at the meetings of the Society of Systematic Zoology had not been very successful. "Relatively few NT'ers showed up at Toronto, and they did not enjoy the meeting." One cause of this disenchantment was the preference of numerical pheneticists for appropriate professional behavior at meetings, a preference which some at Toronto found an "odious hangover of the 19th-century":

> Specifically to the point of our concern, there is considerable disparity in basic thinking between numerical pheneticists and others in SSZ, especially the phylogenetic systematists ("cladists"). NT'ers are more numerically and less verbally oriented, and cladists are, nearly, the reverse. And, there is a difference in—how should I say it?—the vigour and accompanying conviction of infallibility with which cladists and NT'ers present their positions. The NT'ers resent manipulated confrontations and being forced to abandon a point at issue because they are less verbally facile or lack the chutzpah for invective. The points at issue will of course be settled not by verbal vigor but by how closely they mirror reality; that always takes more than 4 days. (Smithsonian Archives)

Even so, Johnston wrote to William Fink, the convener of the next meeting of the Numerical Taxonomy Group, to see if he would be interested in organizing a session at the Tampa meeting of the Society of Systematic Zoology. If so, he should contact Farris as program chairman for SSZ (letter dated January 31, 1979, in Smithsonian Archives). The interweaving of informal alliances and professional positions can be seen by the fact that the president of the Society of Systematic Zoology was allied with the numerical pheneticists, while the convener of the next numerical taxonomy conference and the program chairman of the Society of Systematic Zoology were both cladists. Nevertheless, Numerical Taxonomy 13 met at the Museum of Comparative Zoology at Harvard University in October of 1979. This was the meeting at which Smith declared himself a "born-again cladist." The exchanges a dozen or so years earlier between the pheneticists and the defenders of more traditional methods of systematics had been at times sharp, but they were nothing like the altercations at Harvard between the pheneticists and cladists. Discussions after some of the papers turned into shouting matches. The cladists themselves were growing weary of such exchanges. Over a meal at the Wursthaus in Cambridge, Farris, Mickevich, Vicki Funk, Dan Brooks, and Ron Brady toyed with the idea of having a meeting where cladists could discuss problems of mutual interest rather than waste time defending their basic principles. Sokal had de facto control of the numerical taxonomy meetings, and very few systematists actually attended the annual meetings of the Society of Systematic Zoology. The next day the cladists in attendance decided that it was about time to start their own society.

Soon after the Harvard meeting on November 12, 1979, C. D. Michener sent an open letter to the editor and associate editors of *Systematic Zoology* objecting to the "divisiveness, intolerance, and downright unpleasantness"of some of the papers appearing in the journal (Smithsonian Archives). As an example, he cited Farris's (1979) response to Janowitz (1979). Two days later, Johnston, as president of the society, also sent an open letter to the officers of the society in which he seconded Michener's concerns and called for "bridling the aggression in our journal." He concluded by noting that he had "yet to be persuaded that my attitude on this matter is old fashioned or that it should be superseded by contemporary values emphasizing plain talk and getting to the core of things without bothering with manners." He placed an item on the agenda of the council meeting to be held at the end of December in Tampa, Florida, to discuss editorial policy.

The letters from the two men at Kansas elicited rapid responses from several people at the American Museum. On November 24, 1979, Nelson wrote to Johnston to say that he saw nothing wrong with any member of the society, Michener included, advising the editor about the conduct of his office. "As R. F. Johnston, yes, you too could suggest that future editors suppress criticism that you find, according to your own personal standards, unnecessarily vitriolic and aggressive. But as the Society's President? No, I think not, and I think your memo is ill advised." In his letter of November 26, 1979, Rosen interpreted Michener's complaints as merely statements of his own personal "standards of good taste." For his part, he was delighted with the journal's "lively and robust style." Johnston responded to both letters on December 4, 1979. To Nelson he admitted that his earlier memo was "deliberately ugly" in an attempt to show Nelson that tone can interfere with communication. He reassured Rosen in turn that he did not want to stifle controversy but to tone down the polemics so that the medium would not drown out the message. In his response on December 7, 1979, Nelson gave no indication that he was in the least placated. Rosen and Wiley responded in a lengthy memorandum dated December 12, 1979. They pointed out that membership in the society and subscriptions to the journal had never been higher, implying that controversy was good for both.

Several months earlier, on January 2, 1979, Max Hecht from Queen's College of New York and an old adversary of Donn Rosen, had written to Johnston to warn him that Smith was a "closet cladist" and to complain about the "arrogance of the cladist fundamentalists." Upon receipt of Johnston's memorandum of November 14, 1979, Hecht responded on November 27, 1979, to reiterate his earlier objections and to append published comments by a French disciple of Hennig, Claude Dupuis (1979), deploring the plethora of polemical articles appearing in *Systematic Zoology* by North American cladists. Depuis felt that they were giving all cladists a bad name. He also objected that too much substantive material was appearing in book reviews, where it might be overlooked. However, Hecht also registered some surprise at Michener's singling out Farris's point of view for special mention. He did not find it especially objectionable. On December 13, 1979, Hecht also sent an open letter to the officers of the society repeating his charges and complaining that the "level of vitriol" in the journal could not be excused as merely "lively and robust style." (All of this correspondence can be found in the archives of the Society of Systematic Zoology at the Smithsonian Institution.)

Upon receiving Johnston's memorandum, Sokal replied on December 10, 1979, with a long letter of his own in which he detailed his objections to the editorial policies of Nelson and Schuh. In particular he objected to Nelson's publishing so extensively in the journal while he was editor. Too often his book reviews were lightly disguised polemics against his enemies. Sokal also mentioned the twenty-eight reviews he had received of his paper that defended Mayr against his cladist critics. However, his major charge concerned the treatment that the paper by Janowitz had received. Sokal saw nothing wrong with Schuh's sending the Janowitz paper to Farris, because it was a criticism of Farris's work; but he also should have sent it to a referee who was less intimately connected with the dispute. That way any straightforward errors, once discovered, could have been corrected prior to publication. Sokal viewed the publication of a paper with known errors to be a "serious breach of professional ethics." Finally, he complained of the biased and polemical character of the book review section under Platnick.

Upon receipt of a copy of Sokal's letter, Mayr wrote to Johnston on December 26, 1979, in support of Sokal, his long-time adversary. As far as the principles of systematics were concerned, he disagreed with pheneticists and cladists alike. This issue did not concern the substantive content of science but proper conduct. On this score, Mayr found himself allied with Michener, Johnston, and Sokal against Schuh, Smith, Nelson, Rosen, and Farris. Fitch as the associate editor who had initially handled the Janowitz paper felt caught in the middle. In response to all the hubbub over editorial policy, he wrote two letters to Smith on December 26, 1979. In the first, he registered gentle disapproval of the way in which the Janowitz paper had been handled. Authors should be informed of any straightforward errors in their papers and be given the opportunity to correct them. He also raised the issue of a paper by Rohlf and Sokal that Smith had taken out of his hands. Fitch warned of the potential danger of the appearance of bias in this matter and urged that in the future nothing be done to give the impression to authors that they were being "bushwacked or sand-bagged." Fitch sent copies of his second letter to Michener, Johnston, and Farris. In it he observed that the quality of *Systematic Zoology* would have been significantly improved if Janowitz and Farris had eliminated some of their disagreements and misunderstandings prior to publication. Farris for one was not pleased to see Fitch lining up with his enemies. At the time, Farris was fond of quoting Eldridge Cleaver's assertion that you are either part of the solution or part of the problem.

Just at this time Johnston received Colless's packet of correspondence. On December 14, 1979, Johnston wrote Schuh reminding him that he was still editor of the journal and responsible for its contents. He also complained to Schuh of his failure to publish corrections of the errors in the Mickevich and Johnson paper and warned him that Colless was likely to bring charges in connection with the handling of his paper. Schuh responded on December 18, 1979, that he was not responsible for the errors in the Mickevich and Johnson paper because he was not editor when it was accepted and that he had published corrections as soon as he received them (Mickevich and Johnson 1977, 1979). Although Farris was "by his own admission biased in this particular case," Schuh justified using him as a referee for his wife's paper because he was "undeniably competent in the subject area." When Schuh had transmitted the manuscript and review to Smith, he had not intended for this single review to stand as a rejection of Colless's manuscript. With respect to the transfer of editorship, he had completed volume 28 during the early months of 1979 and saw no reason not to list Smith as editor, starting with the second issue, so that prospective authors could send Smith new manuscripts directly instead of detouring them through him.

All of the preceding took place within the space of a single month. It was in this climate that the 1979 council meeting of the Society of Systematic Zoology was called to order. Council meetings of professional societies tend to be ritualized, pallid affairs. Reports are presented, a few motions are passed unanimously, the outgoing officers are thanked for their splendid service, and everyone retires to the bar. This meeting would prove to be an exception to the rule. Of the fifteen councillors eligible to attend the meeting, only two showed up—Virginia Ferris and David Hull. Three councillors who could not attend the meeting designated other members of the society as their proxies. Hecht sent Walter Bock as his representative, William Fink sent George Lauder, and Wiley sent a graduate student, Sadie Coats. Five of the officers were present: Johnston (president), Everett Olson (president-elect), A. H. Clark (secretary), Farris (program chairman), and Schuh (editor). The past president, Rosen, did not attend but sent Nelson as his proxy. Later, Nelson and Coats ended up marrying each other. Thus, of the ten people eligible to vote, six were officials of the society; the other four were proxies. Neither Michener nor Sokal attended the meeting.

As it turned out, none of the specific issues that had been raised with Johnston were discussed. Instead, the only topic that was mentioned was the "aggressive" style of some of the papers in the journal. Ferris for one had gone back to read Farris's paper and did not find it especially polemical. Johnston agreed that Michener's choice was not an especially good one, but he felt that the tone of the papers published in the journal was nevertheless a genuine issue. Coats then read an open letter from Wiley and Rosen urging no restriction on theoretical disputes or "vivid" language in the journal. They noted that the very people who were now complaining about the journal had made ample use of its pages in previous years to have their say. Now that it was the cladists' turn, suddenly the tone of the journal had become too strident.

The consensus of those present was that a "vigorous style" was healthy for the journal. Coats moved that the council express appreciation to Nelson and Schuh as past editors for the excellent work that they had done. The motion passed. Nelson proposed that a committee be formed to prepare a questionnaire concerning the journal to be sent to the members of the society. Then, just as the meeting was about to adjourn, Farris introduced a motion to censure Michener for singling out Farris's paper as an especially egregious example of unnecessary vitriol in the journal. Everyone who spoke to the motion opposed it, including Nelson, and, in the absence of a second, it died. Farris was furious at being abandoned by his friends. As a result, the cladists' victory celebration after the meeting was somewhat less joyous than it might have been.

The flap at the Tampa meeting notwithstanding, the business of the society and its journal had to go on. On January 10, 1980, Johnston wrote to both Sokal and Mayr apologizing for not handling the meeting better and trying to put as good a face on the outcome as he could. At least the issue had been raised and discussed. He concluded his letter to Mayr with the observation that systematics "will survive, as you perfectly well know, and be the better for having treated the young Turks with reasonable civility. I look for a synthetic reapprochement within 5 or 6 years. Meanwhile I guess we grin and bear it."

As far as the editorial controversies then in progress are concerned, Smith wrote to Colless sending him everything that he had been able to obtain from Mickevich and Johnson relevant to their original paper. Even one of the published corrections itself turned out to contain an error. From Smith's letter, Colless gathered that the only way his paper would see print was if he could somehow come up with Mickevich and Johnson's "true data set," but by now Mickevich had provided three distinct data sets. Colless objected that it was up to Mickevich and Johnson to recover the original data set, not him. Johnson then discovered that he had sent Colless erroneous data in his original letter. Finally, Colless revised his manuscript one last time, and Smith published it (Colless 1980). A year later, Mickevich in collaboration with Farris published a response (Mickevich and Farris 1981).

Smith had inherited the Colless imbroglio, but new problems were brewing. In his first number, Smith published a paper putting the issue of congruence once again to the test, this time in a paper by Schuh, the recent editor, and John T. Polhemus. Once again cladistic methods turned out to be superior to phenetic methods, even using the criteria advocated by the pheneticists themselves. In that same issue, Rohlf and Sokal's

(1980) critique of Mickevich appeared. In it they complained that she investigated only one sense in which various sorts of classification could be said to be "stable" and then in a way biased toward her own preferred technique—classifications based on unrooted Wagner trees. Although in her response Mickevich (1980) pointed out what she took to be errors in the paper by Rohlf and Sokal, none were of the simple arithmetic sort that had produced so much consternation in connection with the Janowitz paper. Finally, in the December issue of that year, Janowitz (1980) and Farris (1980) had another go at each other.[7]

In late September, Rohlf and Sokal sent Fitch, as the appropriate associate editor, a long paper to edit for *Systematic Zoology* in which they discussed both Farris's criticisms of their work as well as those of Schuh and Polhemus (1980). Fitch sent the paper to two referees, but soon he received a letter from Smith relieving him of his duties as associate editor. Smith observed that Fitch would no doubt want to finish editing those manuscripts on which he was currently working but all future manuscripts should be forwarded to the office of the managing editor. When Nelson instituted the system of associate editors at the end of 1975, he selected people with a variety of backgrounds and allegiances to open up the refereeing process. The first four associate editors were Richard Eyde, Walter Fitch, David Hull, and Ronald Nussbaum. Schuh retained this same panel of associate editors. When Smith took over in 1979, he retained Fitch and Hull but replaced the other two members with Daniel R. Brooks, Charles Mitter, and Richard Pimentel. At the time, Brooks was a post-doctoral fellow at the National Zoological Park in Washington, D.C. and Mitter was an assistant professor at Fordham University. Although Pimentel had begun publishing papers on quantitative techniques in systematics before even Sokal had entered the field, he had always kept his distance from the Kansas group. Instead he formed allegiances with the cladists.

In 1980 systematic zoologists had become so factionalized that they met in three separate groups: the Hennig Society in the middle of October, the Numerical Taxonomy Conference two weeks later, and the Society of Systematic Zoology at the end of December. After the idea of the cladists organizing a meeting of their own was raised at the Harvard Numerical Taxonomy Conference, it languished until Farris decided that not only a separate meeting but also a separate society would be a good idea. Farris telephoned a variety of people to see what they thought. Nelson and Rosen strongly opposed the formation of a separate society. They preferred to work within the Society of Systematic Zoology. But others were enthusiastic. In particular, Vicki Funk viewed the formation of a separate society for cladists as a "new era in science.

7. Eventually both Sokal and Rohlf (1981) and Colless (1981) published critiques of Schuh and Polhemus (1980), followed by replies from Schuh and Farris (1981). The only conclusion which Rohlf and Sokal (1981: 483) felt was justified at the time was that none of the studies published to date comparing phenetic and cladistic methods had been carried out appropriately and that the only one that had been, a dissertation by one of their graduate students, "yielded complex results and does not uniformly support any one school of taxonomy." Later, Sokal (1985: 15, 16) presented a more determinate conclusion: "Thus, for small character samples (the situation typical of many cladistic analyses), a phenetic method would give a better estimation of the cladogram than a cladistic method (if the results from the Caminalcules can be generalized). With respect to minimum-length trees, there does not seem to have been any study fitting characters to phenograms and so estimating the length of a phenogram viewed as a phylogenetic tree."

Our Society would be free of all the political infighting that characterized SSZ and NT. We would be dedicated to having an open forum for systematics and science." In particular she hoped to see more of the free-wheeling give-and-take which she had witnessed at the biogeography symposium in New York. Farris persuaded Ed Wiley to organize Hennig I (Personal correspondence).

The first meeting of the Hennig Society was held appropriately enough in Lawrence, Kansas, the site of the first Numerical Taxonomy Conference thirteen years earlier. Even the high spirits of the participants were reminiscent of early numerical conferences. The exchanges at the sessions were basically good-humored. Even when Mary Mickevich, while chairing a session on phylogenetic reconstruction, held her husband to his prescribed time limits, he restrained himself. The only sign of his frustration at being cut off was the snapping of the pointer he happened to be holding at the time. Fox from the University of Alberta was invited to attend the meeting to debate Nelson on paleontological challenges to cladism, but he refused to attend. Instead a younger colleague, Bruce Naylor, defended the role of paleontological evidence in phylogeny reconstruction. Nelson's response was perfunctory at best. He observed that some people look at these issues one way; some another. He was beginning to tire of these endless disputes.

In an exchange after one of the papers, Hull raised a point that turned on a confusion between the atemporal way in which pheneticists individuated characters and Hennig's

Gerrit Davidse, Daniel Brooks, Peter Crane, Joel Cracraft, Donn Rosen, Victoria Funk, Edward Wiley, and Walter Fitch, at the Thirtieth Annual Systematics Symposium, Missouri Botanical Garden, St. Louis, Missouri, October 1983. (Photo courtesy of Missouri Botanical Garden.)

Donn Rosen, Gareth Nelson, and Norman Platnick at the American Museum of Natural History, New York, 1983. (Photo by Gareth Nelson.)

transformation series. Farris brought him up short with a good-natured quip. Later Hull alluded to this same distinction in his banquet address, which was entitled "Games Scientists Play." During his remarks about the social organization of science, he noted what he took to be early signs of a rift in the ranks of the cladists between those who wanted to maintain a close connection between classification and phylogeny and those who did not. In an allusion to the remonstrance he had received earlier from Farris, he termed this latter group "transformed cladists." Just as a character can become modified successively through time in biological evolution, scientists can also gradually change their minds. In the question-and-answer period after the address, Dan Brooks asked if the speaker had any advice for the cladists to insure their success. Hull responded, "Stay socially cohesive, terminologically rigid, conceptually open, and make room for the next generation."

Because the first meeting of the Hennig Society was held at the University of Kansas, several of the Kansas people who were not themselves cladists attended some of the functions. When Nelson joined a group at the smoker that included Michener, Michener quietly withdrew. At a second social gathering held at the Allen Press, Farris descended on Peter Ashlock, the systematist who had been hired to carry on the tradition in systematic philosophy begun so many years before by Michener and Sokal at Kansas. Ashlock (1971) had recommended using the term "monophyly" for Mayr's conception and terming Hennig's more inclusive notion "holophyly." Because of Ashlock's early support of Mayr's views, Mayr had asked him to join him in revising his *Principles of*

Systematic Zoology (1969). Nelson and Platnick had obviously done their homework. They came primed with historical references designed to show that the term "monophyly" went back to Haeckel and that Haeckel had used it in Hennig's sense. They celebrated their defeat of Ashlock by raising a toast to the great Haeckel. Ashlock, of course, was in no position at the time to contest the interpretation of Haeckel that he had been given. If he had, he would have discovered at least one puzzling fact: Haeckel accepted the taxon Reptilia, the paradigm example of a non-monophyletic group in Hennig's sense.

At this social, Nelson and Platnick also indicated that they were toying with the idea of rehabilitating Aristotelian principles of systematics. Aristotle had been right all along. As Sneath had remarked previously about NT-1, Schuh (1981: 81) in his summary comments suggested that the first meeting of the Willi Hennig Society was so successful because it was "by and large uncluttered by irrelevant commentary." The participants did not have to argue for the general principles of cladistics but could explore their ramifications. (For other meetings of the Willi Hennig Society, see Appendix A.)

James Rodman (1982) in his review of the Proceedings of the First Meeting of the Willi Hennig Society (Funk and Brooks 1981b) saw things from a different perspective:

> Atop Mt. Oread in October of 1980, seventy of the faithful gathered to found a society in honor of the late Dr. Willi Hennig, entomologist and pioneering methodologist of systematics. Thirteen of the 22 formal presentations are assembled here under the catchall title of Advances in Cladistics, providing "a cross-section of current primary research in phylogenetic systematics by botanists and zoologists." Scoffers and skeptics were carefully excluded; "proponents of old dogma," judged barren of ideas these last several years, were not allowed to profane the dichotomous rites. With the doubters and dissenters banished, the level of self-congratulation was high, the range of uncontested assertion wide, and the prospect of imagined future accomplishments, handsome. From the tone of these proceedings we may presume the victors celebrated "this most remarkable revolution in systematics and comparative biology."

At this same time, the new president of the Society of Systematic Zoology, Olson, began to receive letters warning him that, at the next meeting of the council of the society, the opposition to the current editorial policies of the journal was not going to be caught unprepared. Sokal was the first to write. In his letter of October 29, 1980, he set out five major charges: the failure of Schuh to let Janowitz see Farris's response prior to publication and the refusal of Smith to let Rohlf and Sokal see Mickevich's responses; instances in which Smith sent out manuscripts for additional reviews after an associate editor had recommended that they be accepted for publication; the lack of experience and qualifications of the two junior associate editors appointed by Smith, and the impending departure of Fitch, the only associate editor combining both impartiality and a knowledge of numerical taxonomic techniques; the blatant bias of the book review section; and the need to clarify the roles of the managing editor and associate editors.

Within a week Nelson had come into possession of a copy of Sokal's letter and wrote to Olson on November 5, 1980, to defend the past editorial policies of the journal under cladist editors. On November 6, 1980, Nelson wrote to Sokal apologizing for the way in which he had handled Sokal's manuscript defending Mayr, and followed this with a letter on November 8, 1980, to Johnston, reminding him of the skirmish that he had had with Blackwelder when he himself was editor of *Systematic Zoology.* As Nelson saw it, the real point at issue was Mary Mickevich. Because she had the "bad luck of finding the 'wrong' results in her doctoral research," Sokal and Rohlf obstructed her progress toward her degree. She published her dissertation, hoping that it would force them to approve it. Instead they resigned from her committee and attacked her paper in print. Sneath, Mayr, and Michener all wrote to Olson to second Sokal's objections, as did several authors who felt that manuscripts they had submitted to the journal had not been treated fairly. Included in this latter group was the Canadian paleontologist whose paper on Croizatian biogeography had received a dozen or so reviews. He had submitted a couple of manuscripts to the journal in the interim, and both had been rejected. Colless also wrote a letter detailing his problems with *Systematic Zoology.* Between 1967 and 1974 when the journal was at Kansas, he had submitted thirteen manuscripts and all thirteen had eventually been published. Of the seven manuscripts he had submitted since the journal left Kansas, only two had been accepted. One of these manuscripts, the critique of Mickevich and Johnson, took two years to work its way through the editorial process. Finally, Fitch wrote to Olson to explain the circumstances under which his term of office as associate editor had been terminated.

The 1980 meeting of the Society of Systematic Zoology was held in Seattle, Washington. Four officers of the society attended the council meeting—Olson (president), James Slater (president-elect), Farris (program chairman), and Smith (editor). Johnston, as past president, sent Sokal to represent him. Of the fifteen councillors, four showed up and six sent proxies. Included among these proxies were Dan Brooks (for Virginia Ferris) and Donn Rosen (for Wiley). The meeting began quietly enough—reports, an increase in dues, etc. Farris then moved that Robert Baker, a cladist from Texas Tech University, be approved as associate editor to replace Fitch. Instead of passing unanimously, as such motions usually do, it carried ten in favor and four opposed. When Smith presented names for the new editorial board, Sokal moved that Smith's motion be tabled for the time being. Sokal's motion passed eight to five.

Olson then distributed copies of the results of the poll that had been taken the year before of the members of the society. There had been 418 responses, less than one-quarter of the membership of the society. Of those who did respond, 62 percent thought that the society was serving them satisfactorily, while 24 percent thought that it was doing so at least partially. The question relative to the journal was, unfortunately, ambiguous:

> *Systematic Zoology* is devoted to publication of results of research projects, serves as an open forum for points of view on systematics, is a medium for review of pertinent books. Do you agree that these prescribed aims are appropriate?

Of those who responded, 90 percent answered yes, but from written comments it was apparent that some read the question as referring to the *aims* of the journal, others as referring to *execution.* Specific complaints, however, were in the minority. The most frequent suggestions for improvement were the publication of more substantive, empirical papers (18 percent), reduce the Points of View section (7 percent), reduce polemics (7 percent), more stringent editing of English to reduce wordiness (4.5 percent), broaden review base (4 percent), and decrease editorial bias (3 percent). In general, 61 percent of the responses were favorable and 39 percent unfavorable.

Olson then announced the formation of a committee to revise the constitution of the society and to examine the editorial structure of the journal. If the distribution of the results of the questionnaire and the announcement of the committee were designed to defuse the threatened confrontation, the strategy failed. Sokal rose to present his objections to the editorial practices of the journal since it had been taken over by the cladists. First, in the Points of View section, communication between the protagonists tended to be asymmetrical. In the three cases that he knew anything about, cladists got to see the comments of their critics prior to writing up their responses, while the noncladists did not. Second, Sokal complained of instances in which an associate editor had forwarded a paper to the managing editor with favorable reviews and his own recommendations to publish, and then the managing editor had sent it out for additional reviews. Sokal did not question the editor's right to send manuscripts out for additional reviews, but all four cases that had been brought to his attention were papers critical of the editor's own preferred methodology. Third, authors frequently had to wait ten or eleven months before receiving reviews.

At this point Smith broke in to explain his reasons in particular cases for slow responses, and then Sokal continued to item four. Although an editor might rightly send a manuscript that he finds clearly inappropriate for the journal to only a single referee, Sokal was aware of one instance in which a paper had been rejected on the basis of a single review written by a person who, with only slight exaggeration, might be termed a mortal enemy of the author. The instance was, of course, Farris's refereeing of Colless's critique of Mickevich and Johnson. In several cases, the managing editor had overridden the recommendations of an associate editor. When one associate editor had objected to this treatment, Smith had dismissed him. The associate editor at issue was Walter Fitch. Sokal also thought that it was unwise to appoint very junior people to be associate editors, especially when their sole qualification seemed to be their preference for cladistic philosophy. Sokal found this subject especially painful because one of these new associate editors was in the room (Brooks) and the other (Mitter) was a graduate of his own department. These young systematists should be publishing in the journal, not editing it.

Sokal continued by noting that, at the start of his tenure as editor, Smith had declared himself a "born-again cladist" and had several times declared in front of witnesses that once he was editor he would accept no more phenetic papers. Furthermore, the book review editor had let his patent biases influence his selection of reviewers. Of the fifty most recent reviews, thirty-seven were written by cladists. In the few cases in which reviews of major books were written by noncladists, a cladist always presented a counterreview. In many cases, authors had used book reviews to rebut arguments by other people and to expand at great length on their own views. Finally, Sokal complained of how page charges were handled. The editor should not know prior to editorial decisions which authors were in a position to pay page charges. Smith interjected that this was the way that things were already done.

As Ferris's proxy, Brooks summarized her views on the matter. In the main she found the past refereeing practices of the journal acceptable. Brooks also took this opportunity to pass out copies of his vita to rebut Sokal's charge that he was unqualified to be an associate editor. Although he was admittedly just beginning his career, he had already published sixty-four papers in systematics and evolutionary biology. Rosen then rose to defend recent editorial practices in the journal. Sokal interrupted to ask how many papers by cladists had been sent out to be reviewed a second time after they had been accepted by an associate editor. Smith could think of none. In his own defense, Smith responded that he followed the recommendations of his associate editors in 90 percent of the cases, and his statement about rejecting phenetic papers was made only in jest.

Rosen continued by defending Nelson for sending Sokal's manuscript to twenty-eight referees. It was intended to show Sokal that his views were not as widely shared as he thought they were. Cracraft and Farris interjected at this point that, even though they were cladists, manuscripts of theirs had also been subjected to multiple reviews. (During Nelson's tenure in office, Cracraft had received seven reviews of a paper on biogeography and ended up publishing it elsewhere. Later Platnick received seven reviews and Nelson six reviews of manuscripts they submitted to the journal. No record exists of the manuscript mentioned by Farris.)

Farris rose to rebut several of Sokal's charges. Years before, when Johnston was editor of the journal, Ghiselin wrote a paper highly critical of Farris, and Johnston did not give him any advance notice of the paper's appearance. Farris also took exception to what he termed the libelous and malicious statements that Sokal had made about Brooks and Mitter.

Olson interrupted the heated exchange between Farris and Sokal by interjecting, "Let's try to bring this to a close."

Rosen moved to thank the editor for the fine job that he was doing under extraordinarily difficult circumstances.

Sokal protested. "What happened to my recommendation?"

His protests were brushed aside as Rosen's motion was brought to a vote. It passed seven in favor, five opposed. Rosen had used an apparently perfunctory motion to thank an editor for a job well done as a means of registering the council's approval or disapproval of the recent editorial policies of the journal.

In response to complaints about particular instances of impropriety with respect to individual manuscripts, Olson reminded the councillors that a committee had already been established to look into the charges.

After a few additional exchanges, Farris moved that "Dr. Sokal's letter that stated that in effect that the managing editor, two associate editors and a review editor were unsuitable for their positions was unwarranted and that the Council express its confidence in such editorial personnel" (Council Minutes 1981, *Systematic Zoology* 30: 223). Rosen seconded the motion.

Once again, Sokal and Farris began to wrangle, and once again Olson interrupted. "Let's not get this down to a personal vendetta between the two of you."

Under his breath, Farris interjected, "What do you think this whole thing is?"

Olson snapped back, "I know what this is, and I don't like it."

Farris's motion was finally brought to a vote and passed, but just barely—six in favor, five opposed, and three abstentions. Farris then moved to remove Smith's list of editorial board members from the table. The motion lost: five in favor, seven opposed, and one abstention. The council had not censured the editor, but it had sent a message.

Although the meeting was over, its effects continued. On January 5, 1981, Rosen wrote to Sokal, characterizing his behavior as a "shameless display of petulance." He was especially incensed by Sokal's "tasteless" accusations about Mitter, Brooks, and Smith and took them to be motivated by the growing unpopularity of phenetics. On January 6, 1981, Pimentel wrote to Olson objecting to Sokal's implications about his deficiencies as an associate editor. He was perfectly capable of editing papers in numerical taxonomy. However, with respect to Pimentel's request that his letter be appended to the minutes of the meeting, both Olson and Slater declined. Sokal in turn was shown the letter that Nelson had written to Johnston on November 8, 1980, in which Nelson attributed Sokal's discontent to an unfortunate student-teacher relationship. On January 14, 1981, Sokal wrote to Nelson to inform him that he was totally mistaken about the facts of the matter and the implications he had drawn from them. If Mickevich consented, he would be happy to have a full airing of the situation. Nelson replied to Sokal on January 23, 1981, objecting to his allegations. Olson continued to get letters from readers of the journal, including one who concluded that Farris must have "lost his bearings."

The following year, the committee on editorial policy and procedure looked into the various charges of impropriety that had been lodged against recent editors of the journal and presented their findings in November 1981. They concluded that some of the complaints made by specific authors "are clearly the results of misunderstandings of the current editorial policy and procedures, others are the results of editorial abuses." They also suggested that the instructions to authors wanting to publish in the journal should include the following warning: "Abusive language and statements that are intended to be insulting to another individual, rather than enlightening to the reader, will not be published." They also proposed a detailed editorial structure for the journal, including an editorial board that had the power to overrule decisions made by the editor. When this memorandum was presented at the 1981 meetings of the society at Dallas, Texas, the council rejected the suggested editorial structure as restricting the actions of the editor too severely. Most of the members of the council thought that no one would agree to be editor under such guidelines. However, a proposed constitutional amendment, introducing a means for appealing editorial decisions, received enough votes to be placed on the next ballot to be sent to the membership for ratification. It passed.

Although Croizat claimed that he no longer harbored any hard feelings over the earlier misunderstandings about Hennig and vicariance biogeography, the appearance of Nelson and Platnick's *Systematics and Biogeography* in 1981 was too much. When Sadie Coats, Nelson's wife, wrote to Croizat to ask permission to visit him while they were in Venezuela collecting, Croizat wrote to inform her that he did not care to receive

Dr. Gareth Nelson. "You are welcome in my home at any time and for how long you please, but he will not be admitted." After rehashing his past grievances, he turned to Nelson's most recent offense. He makes "light work of my aversion to Hennig, and trumpets out the *false* claim that he, Gareth, can mince together Hennig, Croizat and Popper to cook thus a tasty ball." He also objected strongly to Nelson and Platnick's (1981: 539) paper claiming that no progress had been made on the problem of "systems of regions" in the hundred years since Wallace, a claim that overlooked his own work totally.

As unlikely as it may seem, the relations between Croizat and Nelson were to degenerate even further. In 1981 the proceedings of the 1979 symposium on vicariance biogeography appeared (Nelson and Rosen 1981). When Croizat read a review of this volume by Brian Rosen (1982), he was led to write a manuscript similar to the earlier manuscript that he had withdrawn, complaining of Nelson and Rosen rewriting the paper that appeared as Croizat, Nelson, and Rosen (1974). Because he did not know who the current editor of *Systematic Zoology* was, he sent the manuscript to Nelson and asked him to forward it, which Nelson promptly did. During this period Croizat kept up his correspondence with Craw in New Zealand, suggesting that the latter "trounce" Nelson and Platnick (1981). In response to Craw's reticence, Croizat reassured his young ally in a letter dated April 28, 1982, "You NEVER stand under a compulsion of accepting my viewpoint, acting accordingly against your better judgment. By no means. . . . I am glad having found in you a reliable soul in defense, but that's all. You are *perfectly right* in handling Nelson with silk gloves for the purpose of getting out of him the greatest possible help, furnishing photocopies, opening outlets for publication etc." In response to a complaint by Craw about Croizat's not referring to the young man's own work, Croizat cited his ill health and the hostility that might be directed at the young man if he were to boast of him as an ally. Later, in a letter of July 12, 1982, Croizat repeated this reason for not referring to Craw's work: "mind you: if I do not publicly manifest my opinion of you and your priceless assistance, the only reason is that I do not care to inform the American public that Leon Croizat and Robin Craw are indeed interchangeable, thus attiring on you the sum of hatred weighing on my shoulders."

In May of 1982 Craw wrote to Croizat to tell him that his paper denouncing Nelson and Rosen was going to be rejected by Smith. Just as Rowell had used the impending appearance of a paper defending Hennig and Brundin against Darlington's (1970) attack, to justify rejection of Brundin's manuscript, Craw warned Croizat that Smith was going to reject Croizat's manuscript because it was too long and overlapped too extensively with a paper of his own that Smith had already accepted for publication. A month later Croizat received Smith's rejection letter, dated June 14, 1982, which concluded with the following paragraph:

> While I find most of this overly long paper unacceptable for publication in *Systematic Zoology*, I would be interested in a *shorter*, point-of-view regarding your view of the difference between panbiogeography and vicariance. If you mean something different than those who would treat these as synonyms, I think that would be a worthwhile paper. I think that most people know that you have little use for Hennig and I would not expect this paper to be overly burdened

with that. I am in the process of accepting a paper by Robin Craw entitled, "Phylogenetics, areas, geology and the biogeography of Croizat: A radical view." You are cited in the acknowledgments as having read the paper so you will know its nature. It seems to be addressing many of the same issues that you present in this paper that I am returning. I look forward to seeing this shorter paper.

Croizat rewrote his paper. It appeared, followed immediately by Craw's paper on the same subject (Craw 1982). In his paper, Croizat (1982: 296) dismissed Hennig's phylogenetics as "monumental verbiage." Hence, he deeply resented that his "lifework, *Panbiogeography,* has been dragged in with Hennigism to the very extent of publicly losing its identity under the improper designation of 'Vicariance Biogeography.' " To Simpson and Mayr, Croizat added Nelson and Patterson as "pontificating" authorities. He explicitly exempted Rosen from his condemnation. Craw promptly wrote Croizat to warn him that his positive references to Rosen put his own work in danger. Croizat acknowledged in a letter dated September 3, 1982, that Craw was right to place Rosen with the Nelsonians and added, "I have used Rosen 'politically' in my support without being at all his dupe or Nelson's."[8]

When Craw mentioned to Croizat that Nelson continued to write to him in a most friendly manner, Croizat warned Craw that Nelson perceived young authors such as Craw as a "nemesis," and he "tries in every possible way to open some crack between you and me that will allow him a point of advantage" (October 8, 1982). Croizat agreed with Craw's own assessment that Nelson continued to humor Craw "because he is scared of my obvious capacity for trenchant critique." Craw sent two more letters but they were never read by Croizat. He died of a heart attack on November 30, 1982. However, he left a manuscript responding to criticisms of his work by Mayr which Craw arranged to have published posthumously. In this manuscript Croizat responded in no uncertain terms to his treatment by such authorities as Mayr. He also used this manuscript to rap pattern cladists on the knuckles. According to Croizat (1984: 59), a "process is always more important than any of its byproducts. . . . It is glaringly evident that *she or he who does not understand the process will be in difficulty when trying to sort out its byproducts.*"

Conclusion

In this chapter, by following a related series of disputes and rivalries in some detail, I have attempted to document how pervasive cooperation and competition are in science. If nothing else I have shown how difficult it is for scientists to maintain alliances when their professional interests do not always coincide. Although Farris and Sokal were in the same department and both wanted to apply numerical techniques to systematics, they eventually became bitter opponents. Although Nelson and Rosen were very effective in publicizing Croizat's work in biogeography, Croizat rebelled when they tried to advance their own views in connection with his work. Panbiogeography did not need any improvement or supplementation. Fitch found how difficult it was to stay on good terms with warring factions. In such disputes, one must choose. The cladists themselves began to feel some tension as the size of their group increased. If anyone

8. Robin Craw kindly supplied me with copies of his correspondence with Croizat.

had been paying attention, they might have noticed the first signs of speciation in the response to the manuscript submitted to *Systematic Zoology* by the two second-generation cladists.

Perhaps the sorts of agonistic displays I have chronicled are anomalies, but if so, they are anomalies that need explaining. I have not attempted to explain them in this chapter or to show what impact they had on the development of the substantive issues under dispute. Certainly those involved thought that the editorial practices of *Systematic Zoology* were increasingly biased by the allegiances of successive editors of the journal and increasingly so. On the surface, at least, the appearance of bias is patent. I have not explored the effect that this apparent bias may or may not have had on the content of the journal. I deal with this topic in chapters 9 and 10.

However, several tentative conclusions can be drawn. First, editors must conduct themselves with utmost caution and conservatism if they are to avoid the appearance of bias. The role of judge and involved participant are mutually incompatible. Authors find it impossible to believe that anyone who is strongly opposed to the ideas expressed in their manuscript can nevertheless edit it in a disinterested fashion. Writing style also matters. Although many authors think that the expository requirements of standard English are a plot, referees tend to react negatively to formless papers that wander from topic to topic in an incoherent way. Some even prefer verbs that agree with their subjects. Whatever scientists are learning in universities, it is not how to write clearly and effectively. By and large, the authors who receive suggestions for improving their expository style ignore them. First drafts are good enough. Referees also tend not to share the strong preference of authors for polemics. They consistently pleaded with authors to tone down their invective.

If the preceding episodes are any indication, the control of a journal by the society that publishes it is very indirect. By the time the elected officials of a society come to realize that problems exist, it is too late to do anything about them. If nothing else, formal societies force members of different informal research groups to interact, no matter how painful these interactions may be. Most scientists are not very adept parliamentarians. They do not take their roles as councillors and officers of professional societies very seriously. They rarely show up for meetings, and when they do, they are not prepared. Thus, anyone who wants to take the time to plan ahead can control the actions of a professional society quite easily, at least in the short run. But in the midst of all the maneuvering and raw emotions, it is equally apparent that evidence does matter. When Mickevich produced data sets that she claimed showed the superiority of Farris's cladistic methods over all others, she was not ignored. Her data were exposed to intense scrutiny. This attention, however, did not come from other cladists but from their opponents.

6

Down with Darwinism—Long Live Darwinism

Why all this silly rigamarole of sex? Why this gavotte of chromosomes? Why all these useless males, this striving and wasteful bloodshed, these grotesque horns, colors, . . . and why, in the end, novels like *Cancer Ward,* about love?

William Hamilton, *The Quarterly Review of Biology*

Christian doctrines have of course changed and developed over two thousand years of their history, and the present time is perhaps a particularly confusing one, when there are wide disagreements about just what the essential doctrines are.

Leslie Stevenson, *Seven Theories of Human Nature*

DURING THE eclipse of Darwinism at the turn of the century, numerous authors argued that area after area of biology was incompatible with Darwinian versions of evolutionary theory. Many of these claims relied on simplistic parodies of the nature of a truly Darwinian view of evolution. According to these parodies, Darwinians and neo-Darwinians alike maintained that evolution is always as continuous as a mathematical continuum and that every variation that spreads through a population does so under the constant action of natural selection. Variations arise, according to this parody, totally by chance, under no constraints whatsoever. Lamarckian, orthogenic, saltative alternatives were urged instead. In reaction, a series of evolutionary biologists showed that the alleged inconsistencies were neither as great nor as numerous as these critics claimed.

Julian Huxley (1942) termed the emerging theory "synthetic" in order to counter persistent claims that experimental genetics, paleontology, and so on were incompatible with Darwinism. In retrospect, Huxley's appellation was hardly based on past accomplishments but was a combination of a public-relations ploy and a hope for the future. Numerous conflicts continued to exist among the founders of the synthetic theory. For example, Fisher and Haldane differed about the significance of Fisher's explanation of Mendelian dominance in terms of the action of natural selection on multiple genes. Haldane agreed that selection could produce dominance but constructed his own model involving multiple alleles at the same locus rather than genes at different loci. Fisher saw Haldane's theory as merely a supplement to his own, while Haldane insisted that it was a genuine alternative (Box 1978: 233). Wright emphasized the role of partially isolated demes in the process of phyletic evolution, while Fisher argued against it, an argument that became increasingly bitter. Simpson and Mayr disagreed just as profoundly over that nature of species, each finding the other's preferred concept fundamentally "unbiological" (Mayr and Provine 1980: 463). As Templeton and Giddings (1981: 770) observe:

> the Modern Synthesis is often treated as if it were a single, unified view of evolution, yet as is evident to anyone who has read and contrasted the works of Fisher, Haldane, and Wright (three of the principal contributors to the Modern Synthesis from the population genetics viewpoint), there never was a single evolutionary theory.

Mayr (1980a: 41–42) admits that some areas of disagreement over details remained among the founders of the Modern Synthesis, e.g., disagreement over the heterogeneity of gene pools, the causes of this heterogeneity, and the relative importance of populous, contiguous populations versus small peripherally isolated populations; but he insists that in retrospect the "basic structure of the theory arrived at in the 1940s seems in no way changed regardless how these and other questions are answered. For me the synthesis was completed in principle in the 1940s." In particular Mayr cites an international conference on genetics, paleontology, and evolution that took place in Princeton on January 2–4, 1947, as the event that marks the formal completion of the Modern Synthesis.

To be sure, in the 1940s the architects of the Modern Synthesis seemed tacitly to have agreed to play down their differences and to emphasize their areas of agreement. They agreed to disagree, but not often and not in popular presentations.[1] As the years went by, the appearance of a united front became increasingly difficult to maintain as evolutionary biologists grew increasingly dissatisfied with the "coarse-grained" agreement over basics and began to insist on a finer-grained analysis. Perhaps Lamarckism and orthogenesis had been laid to rest, but there were too many areas of continued disagreement for evolutionary biologists to remain content for long. The synthesis continued.

Some of the authors involved in subsequent disputes in evolutionary biology agreed with Mayr that, as significant as the changes being suggested might be, the result was still part of the Darwinian research program, once that program was properly understood (Stebbins and Ayala 1981); others contended that, "if Mayr's characterization of the synthetic theory is accurate, then that theory is effectively dead despite its persistence as textbook orthodoxy" (Gould 1980: 120). The modern synthesis achieved a "spurious unity by submerging the concept of levels and opting for an extrapolationist vision that reduces all macroevolution to microevolution extended" (Gould 1982a: xxxix). Elsewhere, Gould (1982b: 382) argued that the synthetic theory is not so much

1. D. Dwight Davis afforded one major exception to the unanimity claimed by Mayr for those at the 1947 Princeton conference. He came out squarely in favor of a typological view of comparative anatomy. When asked if the founders of the synthetic theory had consciously agreed not to criticize each other's work too vehemently, Simpson laughed and said, "It is not quite fair to say that. We do differ." And the example he gave was his differences with Mayr over the appropriate definition of "species."

wrong as limited. Such basic concepts of the synthetic theory as mutation, adaptation, and natural selection need to be expanded to include all levels of the hierarchy of life. The Modern Synthesis is "incomplete, not incorrect." For his part, Eldredge (1985) views the Modern Synthesis as having lain unfinished for the past forty years.

Another characteristic of the Modern Synthesis was that embryological development remained a black box. It had to. No one knew enough about embryological development in the 1940s for it to make any contribution to the synthesis (Hamburger 1980, Churchill 1980). Later, when Løvtrup (1973) urged the introduction of epigenetics into evolutionary biology, he considered his views to be clearly in conflict with the synthetic theory. However, when Rachootin and Thomson (1980: 181) made similar proposals, they hoped that their suggestions would be viewed as a "natural offshoot from the synthetic stock." When King and Jukes (1969) supported Kimura's (1968) neutral theory of evolution, they declared it "non-Darwinian." Later, Stebbins and Ayala (1981: 967) concluded that the selectionist and neutralist "views of molecular evolution are competing hypotheses within the framework of the synthetic theory of evolution." Gould (1982b: 382) responded that "Stebbins and Ayala have tried to win an argument by redefinition. The essence of the modern synthesis must be its Darwinian core. If most evolutionary change is neutral, the synthesis is severely compromised."

Part of the problem in deciding what actually counts as Darwinian and what as anti- or non-Darwinian evolution is that scientists are engaged in the ongoing process of jockeying for recognition in science. Some scientists exaggerate their differences with the received view to emphasize how original their contributions are, while others exaggerate the similarities between their views and those of contemporary Darwinians in order to throw the mantle of the great Darwin around their own shoulders. Their opponents then attempt to unmask these exaggerations.

For example, two issues have been involved in the recent dispute over punctuational views of speciation: how abrupt punctuational changes are and the role of natural selection in the process. Gould (1977c: 24) argues that Darwin was mistaken in thinking that gradualism is essential to his theory and Goldschmidt was just as mistaken in thinking that his theory of hopeful monsters was non-Darwinian. "For Goldschmidt, too, failed to heed Huxley's warning that the essence of Darwinism—the control of evolution by natural selection—does not require a belief in gradual change." If gradualism is not essential to Darwinism, then neither the punctuational views of Eldredge and Gould (1972) nor the saltational views of Goldschmidt (1940) are non- or anti-Darwinian. Andrew Huxley (1981: ii) agrees. Punctuationalism versus gradualism is merely a "debate within the Darwinian framework" (see also Stebbins and Ayala 1981). When it comes to the role of neutral mutations, Gould thinks his views are original and anti-Darwinian. Stebbins and Ayala (1981: 967) disagree. No matter the role that neutral mutations play in evolution, the resulting theory is to be counted Darwinian.

From the beginning of their careers, scientists are presented with a dilemma. They can make their work look as conventional as possible—just one more brick in the great edifice of science—or as novel and controversial as possible—declaring the foundation of a whole new theory or possibly even a whole new science. On the first strategy, their work is likely to be incorporated effortlessly into the greater body of scientific knowledge. If so, then they will get some credit, but not much. On the second strategy, the work is likely to be greeted with silence. If the author is especially lucky, perhaps an authority can be smoked out to attack these radical new views. However, if on the outside chance that these new views become accepted, the author receives considerable credit. The choice is between a safe strategy with minor payoff versus a very dangerous strategy that promises great rewards. From my own reading of the recent history of science, I see no strong correlation between my own estimates of the novelty of an idea and which strategy an author adopts.

But career interests are not the only factors that influence how the conceptual development of science is itself conceptualized. Scientists tend to view scientific theories and research programs as being timeless and immutable, just as many early biologists viewed species. Each theory has its own essence, a set of propositions that all, and only, the adherents of this theory accept. Everyone agrees, friend and foe alike, that Darwinism has an essence. They disagree only about the precise nature of this essence. Lewontin (1974: 4) agrees with Mayr (1963) that, for Darwin, "evolution was the conversion of the variation among individuals within an interbreeding group into variation between groups in space and time. Such a theory of evolution necessarily takes the variation between individuals as of the essence." Gould (1976: 28) argues that the "essence of Darwinism lies in its claim that natural selection creates the fit. Variation is ubiquitous and random in direction. It supplies the raw material only." Michod (1986: 290) contends that the "essence of Darwin's theory, and the 'adaptationist program,' is that fitness depends primarily on adaptedness. In other words, the adaptational position is that the actual rate of increase of a type depends primarily on its functional, adaptive characteristics." Instead of arguing endlessly about the one true essence of Darwinism, one might entertain the hypothesis that scientific theories have no essences. The essence of scientific theories, as is the case for historical entities in general, is to have no essence.

All of the preceding has concerned the synthetic theory construed fairly narrowly as a scientific theory dealing with a particular class of phenomena, but "Darwinism" has also been interpreted more globally as a worldview or disciplinary matrix. Just as biological phenomena are organized hierarchically, so are conceptual phenomena. For example, Lewontin (1977: 4) insists that the "essential nature of the Darwinian revolution was neither the introduction of evolutionism as a world view (since historically that was not the case) nor the emphasis on natural selection as the main motive force (since empirically that may not be the case), but the replacement of a metaphysical view of variation among organisms by a materialist view." The historian Greene (1981: 130–31) also adopts a global perspective of Darwinism but disagrees with Lewontin about the appropriate elements of this worldview. According to Greene, they include a belief in nature as a law-bound system of matter in motion, organic evolution, change by means of competition, Lockean epistemology, and sensationalist psychology. Who was the first to combine all these elements of Darwinism into a single all-embracing synthesis? Herbert Spencer, that is who.

The preceding has concerned developments only in the English-speaking world. What of developments in the rest of the world? In 1972 a group of historians gathered to compare the reception of Darwinism around the world (Glick 1974). One commentator after another presented strikingly different stories, not only with respect to the reception of Darwinism but also with respect to the "Darwinism" that was being received. At the time, I summarized my own reactions:

> As the papers in this volume clearly show, Darwinism was many things to many people. It was rank materialism, an atheistic attack on the Christian faith, unadulterated positivism, a death blow to teleology. Simultaneously it was irresponsible speculation, an outrage against positivistic science, a rebirth of teleology, proof of the beneficent hand of God, a Christian plot to subvert the Muslim faith. It was also an intellectual weapon to use against entrenched aristocracies, a justification for laissez-faire economic policies, an excuse for the powerful to subjugate the weak, and a foundation for Marxian economic theory (Hull 1974: 388).

Leeds (1974: 439) in this same volume agreed with my estimate of the multifarious elements that marched under the banner of Darwinism:

> It is important to note that the [Darwinian] model is not made explicit in most of the papers in this book; the writers assume not only that it is known to the readers but also that it was not merely known but *held* by the writers they discuss. This is demonstrably false for a number of cases, e.g., the Mexican one discussed by Roberto Mareno—the Mexicans *said* they were Darwinian, but any close analysis of what they were actually assuming seems clearly strongly Lamarckian (see the quotations given in Mareno's article). What appears to me striking is how few of the figures discussed in these pages—with the exception of a small number of the Spanish, the Germans, and the English—held a Darwinian view at all. Mostly they assimilated a phrase or an aspect of Darwin's expression of his thought to their own understanding and thought, then, that they were Darwinians. The most striking case is that of the Russians, discussed in James Allen Rogers's paper, in which *not one* of the protagonists of his drama is remotely near the Darwinian model.

Whether Darwinism is treated narrowly as a scientific theory or more globally as a worldview, considerable variation can be observed. Whether we limit ourselves just to British Darwinians in Darwin's day or open up our inquiry to include later Darwinians, not to mention Darwinians around the world, the fact remains that "Darwinism" has been used to apply to a motley assemblage of beliefs and "Darwinians" to an even more motley group of people. Even if we limit ourselves just to Darwin, several generations of Darwin scholars have showed that at any one time in his conceptual development Darwin toyed with a variety of mechanisms, settling strongly on one or another only for periods in his life. At one time he might think that geographic isolation is necessary for speciation; at other times not. Darwin was not above changing his mind. When more recent scholarship is taken into account, the multiplicity only increases with respect to both Darwin and the so-called Darwinians (e.g., Bowler 1983, Buican 1982, 1984, Conry 1974, Farley 1974, Groeben 1982, Kelly 1981, Oldroyd 1984, Shimao 1981, Pancaldi 1983, Pusey 1984, Harwood 1985, and the thirty-one papers in Kohn 1985).

From the beginning, the founders of the synthetic theory were concerned to make the Modern Synthesis appear as complete and coherent as possible. Several of the inconsistencies claimed by the opponents of Darwinian versions of evolutionary theory did turn out to be spurious. Mendelian genetics was fused to a Darwinian theory of microevolutionary processes, but the most that Mayr (1980a: 1) could claim for macroevolutionary processes and speciation was that they "can be explained in a manner that is consistent with the known genetic mechanisms." Provine (1980: 387) found no stronger connection between Mendelian genetics and paleontology. "All that Simpson

(1944) could argue was that Mendelian heredity was consistent with what was known of the fossil record." Hence, the only relation that the various elements of a theory need have to each other is precisely the same relation that evolutionary systematists claim for biological classifications and phylogeny—consistency. Consistency is certainly a necessary first step in bringing about a synthesis, but much more is required. Perhaps a strict deductive hierarchy of laws may be asking for too much, but mere consistency is too weak. At least some sort of mutual support is necessary.

In the history of science, I think that there are facts of the matter. Either Darwin read Lamarck early in his career or he did not. Either Darwin received Wallace's paper on natural selection in time to use it to rework his own principle of divergence or he did not. Either Wright intended his drift hypothesis to imply nonadaptive differentiation at the species level or he did not. We may never have sufficient evidence to decide these questions with reasonable certainty, but they do have answers. But questions about the nature of Darwinism are not of this sort. We can discover whether or not two scientists held extremely similar views, we can find out if they read each other's work or collaborated with each other, and so on, but there is no way on the basis of this knowledge to predict the contours of "Darwinism." Sometimes views which, by all rights, should have been treated as mutually supportive were not, while views that were all but contradictory were assimilated into the same research program. The events that matter in science are not the ones that should have happened but the ones that did.

Deciding which elements belong to a particular theory or research program as it changes through time is entirely a retrospective exercise. In this exercise, the agents involved may claim one thing; historians may well decide that the facts indicate otherwise (e.g., see Provine's 1985b, 1986 disagreement with Wright about his shifting-balance theory). Such disagreements may not be reconcilable, and they need not be. Scientists do care whether other scientists interpret them correctly. They go to great lengths to explain their views, to clear up unfortunate misunderstandings, but such "explanations" frequently take on the creative aspect of the "interpretations" of the Constitution handed down by the Supreme Court of the United States. However, what matters most in science is not what a particular scientist intended to say but what others took him to be saying. Wright's contemporaries interpreted Wright's shifting-balance theory in much the way that Provine portrays it. Intellectual justice to one side, Wright's intentions at the time are largely of interest only to him and his intellectual biographers. For instance, one might ask, why, if Wright is right about his own intentions, did his contemporaries consistently misread him? Similarly, Hennig (1966) intended his principles to be methodological, but initially his critics read them as being empirical. Only later, through sustained efforts by his disciples, did Hennig's intentions become realized.

Species are relatively easy to handle if they are treated as sharply delineated, static entities. All organisms and only those organisms that possess a certain set of traits belong to that species. Mating, parentage, and descent are irrelevant. If two birds belonging to a particular species give birth to an offspring that lacks one of the essential traits for that species, it belongs to a different species. If in turn that aberrant individual gives birth to young with the appropriate traits, they once again belong to the species in question even if one of their parents does not. Conceptualizing species as being characterized by means of clusters of traits helps to avoid some of these counterintuitive consequences but not all of them. Only when these clusters are allowed to change through time do species become genuinely evolutionary lineages, and what determines the order of generations is descent. With respect to evolving species, descent tends to covary with character distributions, but when the two are not coincident, descent must take priority to character distribution if organisms are to be ordered into lineages. As a result, the recognition of species through time is necessarily retrospective. Which of several peripheral isolates will establish itself as a new species, only time will tell.

As counterintuitive as this view is with respect to species, it is even more counterintuitive with respect to conceptual systems and research groups. Mayr, the father of the biological-species concept and of population thinking in evolutionary biology, is uneasy with this perspective when applied to science itself. According to Mayr (1983: 507), "Darwinism in the 1860's was what Darwin believed, and the most suitable type specimen for a Darwinian was Darwin himself. No one who was a saltationist and rejected natural selection qualifies for the designation 'Darwinian.'" As Mayr acknowledges, one consequence of this decision is that neither Huxley nor Lyell count as Darwinians. Even though Wallace disagreed with Darwin over several fundamental issues, including the role of natural selection in the speciation process and its application to man (Kottler 1985), he does count as a Darwinian. If biological species can be extremely heterogeneous at any one time, I fail to see why conceptual systems and research groups cannot be characterized by a parallel heterogeneity. If scientific communities are defined by social criteria, if a Darwinian is "someone who identified with Darwin, but not necessarily someone who accepted all of Darwin's ideas" (Ruse 1979a: 203), the end result is that the Darwinians are conceptually heterogeneous. If, to the contrary, groups of scientists are delineated by means of essential agreement or even large overlap in agreement, the resulting groups are hodgepodges. Darwin finds himself in bed with Richard Owen, and Bob Sokal with Steve Farris.

Although Mayr (1983) acknowledges conceptual heterogeneity at any one time in conceptual systems, he thinks that scientific tenets can be divided into those that are fundamental and those that are peripheral. In order for the term "Darwinian" to apply "appropriately" to one of Darwin's contemporaries in 1860, that person must agree with Darwin on these fundamentals. In short, Mayr accepts the essentialist position for conceptual systems at any one time. However, unlike traditional essentialists, he is willing to accept change through time. Just as species can change gradually through time, so can scientific theories. Thus, Mayr is willing to refer to Darwin, Romanes, Weismann, Fisher, Dobzhansky, and himself as "Darwinians" (or "neo-Darwinians") just so long as one keeps in mind that this term has different definitions for each of these biologists. Who will be the new Darwinians? No one can tell. It may turn out that evolutionary biologists promoting views all but indistinguishable from those of Darwin will prevail, or radically different views may come to predominate. The former may come to be termed non-Darwinian; the latter Darwinian. Stranger things have happened in the history of science. For example, some geologists thought of themselves as having accepted the system of James Hutton (1726–97) even though they rejected the core claim of his system—that the consolidation of calcareous rocks was caused by heat. Similarly, other geologists thought of themselves as disciples of Alfred Werner

(1866–1919) even though they abandoned one of the two means by which he classified geological formations. Werner classified formations by means of age and mode of formation, while his descendants dropped mode of formation and concentrated on age (Laudan 1987). The disciples of Jacques Loeb (1859–1924) abandoned every one of his basic beliefs.

The general position which I am urging in this book is that conceptual systems at any one time are quite heterogeneous, almost as heterogeneous as species. This conceptual heterogeneity only increases if one follows a conceptual system through time. More importantly, heterogeneity is not accidental or incidental to evolving entities; if conceptual systems are to evolve in anything like the way species do, it is essential. Whether one thinks, as Darwin did, that varieties are incipient species or one opts for peripheral isolates as the precursors to new species, variation is essential. Darwinism no more has an essence than does Copernicanism (Curd 1984), sociobiology (Barash 1979, Dawkins 1978, Ruse 1979b, Vetta 1981, Caplan 1984), or Christianity, for that matter, protestations by their advocates and opponents notwithstanding. Or to put the same point differently, conceptual systems can be made to have essences only in retrospect if one exercises maximum ingenuity and minimum historical accuracy (for more detailed discussions of Darwinism as an historical entity, see Hull 1985b and Richards 1987).

In this chapter I discuss some of the issues that have taxed evolutionary biologists during the past thirty years or so. I do not pay much attention to which views are truly Darwinian and which not, or to which authors are part of a Darwinian research program and which not. I am interested in the issues in their own right because I intend to use these ideas in the second half of this book to set out the mechanisms that give science as a selection process the character it has. Because I do not spend sufficient time in discovering which ideas actually gave rise to other ideas and which authors actually collaborated with other authors, the episodes I discuss cannot serve very well as evidence for my views. They are meant to be primarily illustrative and informative. They indicate the various mechanisms which biologists have suggested are operative in biological evolution. In the second half of this book, I argue that very similar mechanisms are operative in the conceptual development of science as well. The evidence for my claims must come from those episodes in which I have access to the necessary data, primarily from the disputes among the pheneticists and cladists in systematic biology. I must also warn the reader that in the following discussions I cannot claim even third-party objectivity. Too many of the issues that I discuss bear too directly on my own research.

The Synthesis Continues

The first major schism to appear among the ranks of the neo-Darwinians concerned the connection between natural selection and the structure of natural populations. According to Muller's (1950) "classical" theory, the vast majority of mutations are deleterious and, hence, should be rapidly eliminated from populations. As a result, populations in nature should be genetically homogeneous. At every locus, a "wild-type" allele should predominate; alternative alleles should be exceedingly rare. Whatever genetic variation occurs in a species should be in conjunction with environmental variation in its range. One problem with this view is that such workers as Chetverikov (1926) had found considerable genetic variability in natural populations, and it did not seem to be correlated in obvious ways with environmental variation. Building on the work of Lerner (1950), Dobzhansky (1951) suggested another possibility—balanced polymorphisms maintained by heterozygote superiority.

According to this hypothesis, a homozygote (an organism possessing two instances of the same allele at the same locus) may be selectively less fit than heterozygotic individuals (organisms possessing different alleles at the relevant locus). Hence, even though the homozygotes are steadily eliminated each generation, the surviving heterozygotes are just as steadily producing new homozygotes in the next generation. The result is the retention in a population of both alleles. Although Ford (1940) was especially instrumental in pointing out such balanced polymorphisms in populations of butterflies, the most well-known example is sickle cell anemia in certain human populations. The sickling allele produces red blood cells that are less efficient in transporting oxygen than nonsickling alleles. Individuals who possess two sickling alleles usually abort spontaneously before birth or die in childhood. Homozygotes for the nonsickling allele produce red blood cells that transport oxygen adequately, but these cells are also susceptible to malarial infections. People who are heterozygotic at this locus are slightly anemic, but they are also resistant to one form of malaria. As a result, in those areas in which this form of malaria is common, populations continue to be polymorphic because of the superiority of the heterozygote. Although this solution to the problem is far from ideal—evolution does not guarantee the best of all possible worlds—it has proven to be good enough. Better anemia than malaria.

Until the perfection of a technique termed gel electrophoresis, no one was aware of exactly how genetically heterogeneous natural populations actually are. Even though this technique tends to underestimate genetic heterogeneity, Hubby and Lewontin (1966) and Harris (1966) discovered the sort of extensive genetic heterogeneity needed by the balancing theory. At a third of the loci tested, two or more alleles exist even within single populations, and on the average any individual is likely to be heterozygotic at 10 percent of its loci. In fact, so much heterogeneity showed up that the balancing theory was presented with an embarrassment of riches, even more heterogeneity than it could plausibly explain. Other mechanisms had to be at work as well, perhaps density dependent selection.

At this juncture a third alternative arose—neutralism (Kimura 1968, King and Jukes 1969, Kimura and Ohta 1971). According to this theory, both the classical and balanced theories as presented are mistaken about the power of selection. Most mutations go unnoticed under the myopic gaze of natural selection. They are selectively neutral. The neutrality thesis does not imply that these genes have no physiological effect, only that alternative alleles produce proteins that are, for all intents and purposes, equally good at fulfilling the needs of the organism. Thus, no selective advantage is required to explain genetic heterogeneity. Advocates of the "non-Darwinian" or neutralist theory acknowledged that those mutations with significant adaptive effects are selected in the way championed by generations of Darwinians, but the neutralist theory is "non-

Darwinian" in the sense that a majority of mutations are adaptively equivalent and become prevalent by means of drift. According to a recent statement by Kimura (1985: 88) the neutralist theory can be summarized as follows:

> From time to time, the position of the optimum shifts due to change of environment, and the species tracks such a change rapidly by altering its mean. Most of the time, however, stabilizing selection predominates. Under this selection, neutral evolution through random fixation of mutant alleles occurs extensively, transforming all genes (including those of living fossils) profoundly at the molecular level. Thus we see that neutral molecular evolution is an inevitable process under stabilizing phenotypic selection.

During this same period, a parallel controversy was going on in evolutionary biology concerning the levels at which selection can occur. In their private correspondence, Darwin and Wallace carried on very sophisticated arguments over several very subtle problems in evolutionary biology (Kottler 1985). For example, Wallace thought that intersterility between species could be selected for and hence was an adaptation. As much as Darwin would have liked to agree with Wallace, he was forced to conclude that intersterility is only an effect of selection occurring for other traits. Darwin also worried about the way in which highly specialized structures could evolve in neuter insects. Because they did not reproduce themselves, selection could not act on them directly. How then could the oversized jaws of soldier ants evolve through selection? Darwin (1859: 242) took the evolution of adaptations in neuter insects to be the "most serious special difficulty which my theory has encountered." His solution was to posit selection at the level of the family.

During the early years of Darwinism, such difficult questions received very little attention. The Darwinians were chiefly concerned to convince their fellow scientists as well as the lay public that evolution actually occurred. Nor did these problems receive much attention during the eclipse of Darwinism at the turn of the century (Bowler 1983), but with the rise of expectations that followed upon the Modern Synthesis evolutionary biologists turned to them once again, this time en masse. One of the early catalysts for this renaissance in evolutionary biology was Wynne-Edwards's *Animal Dispersion in Relation to Social Behavior* (1962). Ever since Darwin, biologists on occasion made passing reference to some trait or other being "for the good of the species." For example, the heavy fur of a female polar bear can be explained in terms of her own personal survival, but her mammary glands present a problem. They aid her not at all in coping with her environment and may actually be detrimental. However, any female polar bear born without mammary glands is likely to leave very few offspring to pass on this mutation. Hence, the prevalence of properties such as mammary glands in mammals might be explained in terms of the perpetuation of the species at the cost of some lowering of the fitness of the individual organism. Certainly organism lineages must persist if evolution is to occur, but the persistence of lineages is not quite the same thing as the selection of traits possessed by individual organisms in order to enhance the survival of species.

Building on the work of several of his predecessors, Wynne-Edwards (1962) pushed the notion of group selection to its logical conclusion. He argued that species can maintain population densities below the carrying capacities of their environments by means of group selection. Hence, even though the individual organisms of these populations would have lower fitness, population-level fitness would be enhanced. In general, Wynne-Edwards argued that selection on individual organisms can produce adaptations in a group if they help that group, even though they might lower the fitness of individual organisms. G. C. Williams (1966) and David Lack (1966) took up the challenge, arguing that group selection is all but impossible because a population of "altruists" is extremely vulnerable to invasion by organisms that are not disposed to hold their reproduction in check. Williams's general principle is that adaptations must be explained by selection occurring at the lowest level possible—preferably at the level of alternative alleles at a single locus. Selection might occur at higher levels of organization, but the necessary conditions at these higher levels become increasingly rare as one proceeds from genes to chromosomes and entire genomes, to organisms, and finally to groups. Although group selection is not physically impossible, it should be at best very rare.

Hamilton (1964) further contributed to the demotion of group selection by distinguishing between kin selection and the selection of kinship groups. Adaptations that in the past had been explained in terms of the selection of kinship groups could be explained just as well in terms of selection taking place at the genetic level. If the goal of biological evolution is to pass on replicates of one's genes, the more the better, then this goal can be accomplished either directly by passing on replicates of one's own genes to one's own offspring or else indirectly by contributing to the perpetuation of replicates of these very same genes in one's immediate kin. Hamilton's theory of kin selection involves selection occurring solely at the genetic level. It explained helping behavior by means of an organism increasing the representation of its genes in later generations by restricting its altruistic acts to close kin or possibly to nonkin who carry the relevant genes. It did not require the existence of kinship groups or competition between these groups. It did not resort to the selection of kinship groups, just kin.

Although Trivers (1971, 1974) was hardly the first to do so, he devised the notion of "parental investment" as the quantity that is maximized in the interrelations between parents and their offspring at just the right time, and his notion caught on. Because parental investment can be extremely asymmetrical, conflicts of interest can arise. If organisms are viewed as unconsciously attempting to increase their inclusive fitness with a minimum of investment, numerous predictions can be made about the sorts of adaptations one should expect in particular circumstances. For example, in species in which females can rear their young without the help of males, males should attempt to mate with as many females as possible. Strategies become more complicated when both parents must invest in their offspring in order for them to survive. Although their work has received less attention, other authors contributed to this early literature (e.g., Alexander 1971, 1974, Ghiselin 1974a, Michener and Brothers 1974).

The preceding way of viewing the evolutionary process was generalized using the techniques of game theory. Although Lewontin (1961) first suggested that a game theoretic approach might prove helpful in evolutionary biology, John Maynard Smith (1972, 1974, 1978, 1984) developed this way of viewing evolution most extensively. In particular, Maynard Smith introduced game theory to analyze the evolution of phenotypic traits, including behaviors, when the fitness of the trait depends on its frequency in the population. In human beings, game theory is useful, according to Maynard Smith, only when individual interests are to some extent independent of social norms. In particular, Maynard Smith developed the notion of an evolutionary stable strategy. In the context of behavior, a strategy is evolutionarily stable if it cannot be bettered just so long as almost everyone else in the population is doing it. A strategy is not evolutionarily stable when the introduction of a different strategy can upset the balance. For example, in certain situations the production of an equal number of males and females is the most stable strategy, in others more females, in others changing ratios of males to females, and so on.

Just as Wynne-Edwards pushed group selection to its logical conclusion, Richard Dawkins in *The Selfish Gene* (1976) pushed gene selection to its limits. Dawkins insisted that the arguments which such authors as G. C. Williams had presented to show that group selection can be at most extremely rare, apply with almost equal force to individual organisms. According to Dawkins, in sexually reproducing organisms only short segments of the genetic material have what it takes to be selected. Organisms are simply survival machines constructed by genes to aid them in their single-minded quest for replication. As Dawkins (1976: 36) put this point, "In sexually reproducing species, the individual is too large and too temporary a genetic unit to qualify as a significant unit of natural selection. A group of individuals is an even larger unit. Genetically speaking, individuals and groups are like clouds in the sky or dust-storms in the desert. They are temporary aggregates or federations." Rarely has a book written for a popular audience roused so much consternation among professionals. It was one thing for Williams to argue for the near impossibility of group selection. Group selection had always seemed somewhat suspect. But dethroning the organism as the primary focus of selection was quite a different matter.

Not all challenges to the synthetic theory stemmed from conflicts within evolutionary biology itself. Paleontologists have also been roused to question its basic premises. Many of Darwin's fellow Darwinians were not as partial to gradual evolution as he was. They tended to prefer more saltative forms of evolution. Although the synthetic theory is not unrelievedly gradualistic, it is closer to Darwin on this score than most versions of evolutionary theory that had come in between. Everyone acknowledges that new species can arise in the space of a single generation by polyploidy, but more traditional Darwinians tend to dismiss this mechanism as a curiosity. After all, it occurs primarily in plants. The question remains, how gradualistic is speciation in other circumstances? According to Mayr's founder principle, new species can arise in the space of a few generations, but even so the phenomenon is gradualistic at the populational level. Simpson's quantum evolution is even more gradualistic. These forms of rapid evolution notwithstanding, the general impression that the literature on the Modern Synthesis gives is that of gradualism—new species arising either through the subdivision of a single species into two or more daughter species (speciation) or through the gradual transformation of a single lineage (phyletic evolution). This emphasis on gradualism stems from the evolutionary mechanisms inherent in the Modern Synthesis, but it also has been exaggerated by the fact that the chief opponents of this synthesis among paleontologists such as Schindewolf (1936) advocated saltationism of the most extreme sort. At other times in other places, the synthetic theory might not have been so gradualistic.

Unfortunately, when paleontologists scoured geological strata for examples of gradual evolution, they were not very successful. Darwin took refuge in the incompleteness of the fossil record, but as the years went by and paleontologists reconstructed numerous fossil sequences that were about as complete as anyone could reasonably expect to have, this explanation began to wear a bit thin. One highly prevalent pattern that materialized was the geologically abrupt appearance of a species and its continued existence largely unchanged until it went extinct. Eldredge and Gould (1972) caused quite a stir by focusing attention on this pattern and claiming that it accurately reflected phylogenetic development. Species actually arise as abruptly as they appear to and remain largely unchanged thereafter, an evolutionary pattern that they termed "punctuated equilibrium."

In their early paper, Eldredge and Gould (1972) were concerned primarily to argue for the prevalence of this pattern and to point out that, on the basis of the synthetic theory, one would not expect it. They also suggested a mechanism. Building on Mayr (1954, 1963), they argued that speciation always (or almost always) occurs by means of the isolation of a relatively few organisms at the periphery of a species (allopatric speciation). Because species are so genetically heterogeneous, a few organisms, no matter which ones happen to become isolated, cannot possibly retain all of this heterogeneity. Most rare alleles will be lost, but on occasion a rare allele might come to predominate in the space of a single generation just by the luck of the draw. If these peripheral isolates stay small for a few generations, drift is likely to magnify these differences even further. Of course, small size also increases the likelihood of extinction. Most peripheral isolates either merge back with the parent species or go extinct. Only rarely is one likely to become established as a new species in its own right. One major difference between Mayr's original exposition and that of Eldredge and Gould (1972) is that Mayr emphasized the selective aspects of this process while Eldredge and Gould deemphasized them.

Speciation by means of peripheral isolates explained the punctuational part of the pattern emphasized by Eldredge and Gould. A comparable explanation for the equilibrium part proved more difficult. According to the principles of the synthetic theory, continued stasis in the face of radically altered environments had to be explained in terms of selection pressures neatly balancing each other. For example, roughly 25 percent of human beings can taste very small concentrations of phenyl thiocarbamide. When Fisher tested eight chimpanzees in the Edinburgh Zoo, he discovered that two of them were also able to taste extremely low concentrations of this chemical as well, and he was led to wonder what selective balance in favor of the heterozygote could have maintained the relative frequencies of these alleles for the millions of generations that separate the anthropoid and hominid stocks (Box 1978: 372). This is selectionism with a vengeance.

Not all architects of the Modern Synthesis thought that stasis could be explained so exclusively in terms of selection. Other mechanisms must also be involved. Mayr

(1963) devoted an entire chapter to the problem. Once again Eldredge and Gould (1972: 114) found themselves in close agreement with Mayr, suggesting that species, like organisms, are homeostatic systems, "amazingly well-buffered to resist change and maintain stability in the face of disturbing influences." However, the coherence of species is "not maintained by interaction among its members (gene flow)," but is a "historical consequence of the species origin as a peripheral isolated population that acquired its own homeostatic system." The stability of species is an "inherent property of individual development and the genetic structure of populations." Thus, from a variety of perspectives, numerous workers came to emphasize the cohesive character of biological species.

A Metaphysical Interlude

In Solzhenitsyn's *Cancer Ward* (1969), the imprisoned Kostoglotov heard Beethoven's four muffled chords of fate thunder above his head first when he read that the personnel of the Supreme Court of the Soviet Union had been completely changed and again when he read that Malenkov had resigned. These same chords sounded for Hamilton (1975) when he read first Ghiselin's *The Economy of Nature and the Evolution of Sex* (1974a) and then Williams's *Sex and Evolution* (1975). Hamilton was impressed with these books because they brought the "full searching gaze of science" on the most outstanding problem in biology—sex. In this book, Ghiselin did something else as well. He turned his gaze on a fundamental problem of even longer standing—the metaphysical status of species.

Throughout the history of Western thought, species have been a constant source of controversy. One of the most common views was, and continues to be, that biological species are the same sort of thing as geometric figures and physical elements.[2] On this view, the horse, triangularity, and gold are universals, secondary substances, or classes, depending on one's preferred philosophical system and terminology. Early theories of "evolution" did not threaten this common conception very directly. They implied that an organism might change its species or that certain species might cease to be exem-

2. As I read the literature on the species concept from Aristotle to the synthesis of the 1940s, the vast majority of naturalists who concerned themselves with the species problem approached it in the context of species being universals of some sort, most frequently classes or kinds. The most likely exception to this rule seems to have been Buffon (Sloan 1987). How about Darwin? How did Darwin view species? Early in the development of his theory of organic evolution Darwin seems to have viewed species very much as historical entities with life cycles like organisms (Hodge 1983). For how long and how consistently Darwin held this position is difficult to decide.

Later evolutionary biologists can be found who hint that species cannot possibly be classes or sets; others who explicitly hold this position. However, I disagree with Mayr (1987: 152) when he states, "Anyone reviewing this literature is puzzled why the virtually universal rejection of the concept of the species as a class by evolutionary biologists was so completely ignored by philosophers." But numerous biologists can be found who explicitly advocate treating species as classes, e.g., Bigelow, Blackwelder, Bormeier, Davis, Gilmour, Gregg, Heywood, Sneath, Sokal, Woodger, and Zangerl to name but a few. Mayr might counter that the preceding biologists are not *evolutionary* biologists. However, Simpson surely counts as an evolutionary biologist, and in an interview that I conducted with him at his home in Tucson, he explicitly rejected treating species as anything but classes. I think that biologists have not been quite as clear-sighted as Mayr claims nor philosophers so blind (see Hull 1987 and other papers in volume 2, number 1 of *Biology and Philosophy*).

plified, but not that species themselves change. One reason why Darwin's theory was so threatening was that it implied that species themselves evolve. The only way that new species can come into existence is from preexisting ancestral species, and once a species becomes extinct that species can never come into existence again. Extinction is forever. On the traditional view, one species evolving into another was as incomprehensible as one element evolving into another. Alchemists kept trying to transmute samples of baser metals into gold, but no one thought that in doing so these elements themselves were being altered. The physical elements, as is the case with all natural kinds, are eternal and immutable.

Periodically, authors noted the problem that Darwin's theory posed for traditional conceptions of species as classes or secondary substances. If species are classes, then they should be definable in terms of sets of characteristics that only and all the organisms belonging to that species possess; but if species evolve as gradually as Darwin claimed, then such definitions are necessarily impossible. Any definitional lines drawn to subdivide a gradually evolving lineage into successive species are inherently arbitrary. In his contribution to the *New Systematics,* Gilmour (1940: 467) noted that certain taxonomists talk as if the names of species are not assigned "on account of the possession of certain characters by the individual concerned, but in the same arbitrary way as the christening of a baby," but he dismissed the idea as not worth discussing. Similarly, two systematists, who read an early draft of a paper by Gregg (1950: 425) on the reality of species advanced the notion that "species are composed of organisms just as organisms are composed of cells: according to this argument a species is just as much a concrete, spatiotemporal thing as is an individual organism, though it is of a less integrated, more spatiotemporally scattered sort." One of these readers was Mayr. Gregg rejected the idea as having unacceptable consequences. If species are classes or sets, then statements such as "the limits of species are blurred" and reference to the "borderlines between species" must be interpreted as metaphorical. But if species are viewed as concrete individuals, then they could literally have ranges and boundaries. That a species could literally have a range, Gregg found unacceptable.

Shortly thereafter, Woodger (1952: 19) noted the same contrast between species as concrete entities and species as abstract entities:

> In the Linnean system of classification of animals and plants a species was a set or class, in fact it originally meant a smallest named class in the system. But a class or set is an abstract entity and thus has neither beginning nor end in time. We cannot, therefore, speak of the origin of species if we are conceiving species in the Linnean manner. The doctrine of evolution is not something that can be grafted, so to speak, onto the Linnean system of classification. The species of Darwin and the species of Linnaeus are not at all the same thing—the former are concrete entities with a beginning in time and the latter are abstract and timeless.

Woodger (1952: 21) used the preceding considerations to bolster his view that classification and phylogeny should be kept separate. "The taxonomic system and the evolutionary phylogenetic scheme are quite different things doing quite different jobs and only confusion will result from identifying or mixing them." As was the case with so many of Woodger's insights, this one went unnoticed.

When confronted with this same problem, several founders of the Modern Synthesis opted for the opposite conclusion. For example, in the opening pages of his *Genetics and the Origin of Species,* Dobzhansky (1937: 6) noted that a species is "not merely a group and a category of classification. It is also a supraindividual biological entity, which, in principle, can be arrived at regardless of the possession of common morphological characteristics." Through the years, Mayr also made similar comments. But it was Ghiselin who finally forced evolutionary biologists to realize the implications for evolutionary theory of construing species in this way. Just as the name "Gargantua" denotes a particular organism from conception to death, "*Gorilla gorilla*" denotes a particular segment of the phylogenetic tree. Because natural laws have been traditionally construed as being spatiotemporally unrestricted, it follows that the names of spatiotemporally restricted entities such as organisms and species cannot function in them. There can no more be a law of *Homo sapiens* than there can be a law of Napoleon. However, both sorts of entities can be instances of natural kinds, which do function in genuine laws of nature. For example, *Homo sapiens* is a cosmopolitan species and Napoleon was a dominant alpha male. Laws governing cosmopolitan species and alpha males are at least feasible (Hull 1976, 1978).

Of course, very few evolutionary biologists took note of Ghiselin's metaphysical musings, and most who did dismissed them as of no importance to their own work. Eldredge and Gould were an exception. They saw the compatibility of their views on species with Ghiselin's nonstandard construal of species. They were soon joined by Stanley (1975, 1979). According to Stanley, the prevalence of sex can be explained as a species-level adaptation that permits sexual species to speciate much more rapidly than asexual "species." In fact, Stanley found himself agreeing with Dobzhansky and Mayr that asexual organisms do not even form species, at least not species in the same sense as sexual organisms do. If gene flow is one of the primary mechanisms for the maintenance of the cohesion of the gene pool and asexual organisms rarely if ever exchange genetic material, then they do not form species. Just as not all organisms form kinship groups, not all organisms form species. Species did not come into existence until sexual reproduction succeeded in evolving. For the first half of life here on Earth, no species existed—just organisms (Schopf, Haugh, Molnar, and Satterthwait 1973).

As long as genes and organisms are viewed as spatiotemporally restricted individuals and species as spatiotemporally unrestricted classes, selection operating on species in the same sense in which it operates on genes and organisms seems very problematic. Only if candidates for selection are construed as the same sort of thing can they possibly perform the same functions in natural processes. Being an individual is not a sufficient condition for an entity functioning as a unit of selection (Sober 1984a). After all, entities as diverse as Mars and the Empire State Building are individuals, and they do not function in selection processes, but as I argue at length in chapter 11, being an individual *is* necessary. Only individuals have what it takes to be selected.

Ghiselin's reinterpretation of species finally brought the reductionist character of gene selection into full relief. Everyone acknowledges the hierarchical structure of the living world. Genes exist in chromosomes, chromosomes in cells, cells in organisms,

organisms (sometimes) in kinship groups, populations, and species. Everyone recognizes the relationship between the early members of this progression to be part/whole. Cells are part of organisms. According to Ghiselin, the relationship between the later members of this progression is also part/whole. Organisms are part of their kinship groups, populations, and even species. Construing species in this way is important because classes are readily reduced to their members. The reduction of organized systems to their constituent parts is a much more problematic process. In his book Wynne-Edwards (1963: 623) argued that some groups are really not "groups." They are entities in their own right with properties of their own. As he developed his ideas,

> it soon became apparent that the greatest benefits of sociality arise from its capacity to override the advantage of the individual members in the interests of the survival of the group as a whole. The kind of adaptations that make this possible, as explained more fully here, belong to and characterize social groups as entities, rather than their members individually. This in turn seems to entail that natural selection has occurred between social groups as evolutionary units in their own right, favoring the more efficient variants among social systems wherever they have appeared, and furthering their progressive development and adaptation.

Traditional adaptations characterize individual organisms. Individual organisms have fur, bear their young alive, and sing mating songs. But some properties characterize groups as such and not their members severally. For example, species exhibit certain gene frequencies, balanced polymorphisms, and frequency-dependent selection. Some species are subdivided into numerous, partially isolated demes, while others have highly convoluted peripheries, ideal for speciation. The question is whether these characteristics can be treated cogently as species-level adaptations and, if so, whether these characteristics can in turn be reduced to characteristics of individual organisms and possibly even genes. As Williams (1966: 108) put the contrast, it is between a "population of adapted insects and an adapted population of insects." For example, he argues that the fleetness of a herd is totally a function of the fleetness of individual deer in the herd. Only if the herd were a well-organized whole could it have adaptations of its own. "Such individual specialization in a collective function would justify recognizing the herd as an adaptively organized entity. Unlike individual fleetness, such group-related adaptations would require something more than the natural selection of alternative alleles as an explanation" (Williams 1966: 17).

Williams (1966: 19) concluded that herds do not have such group-related adaptations. In fact, "adaptations need almost never be recognized at any level above that of a pair of parents and associated offspring." His general principle is that the fitness of a group can be treated as a single summation of the fitnesses of its constituent organisms. Williams (1966: 56–57) extended this same argument to include organisms:

> Obviously it is unrealistic to believe that a gene actually exists in its own world with no complications other than abstract selection coefficients and mutation rates. The unity of the genotype and the functional subordination of the individual genes to each other and to their surroundings would seem, at first sight, to invalidate the one-locus model of natural selection. Actually these considerations do not bear on the basic postulates of the theory. No matter how functionally dependent a gene may be, and no matter how complicated its interactions with other genes and environmental factors, it must always be true that a given gene substitution will

have an arithmetic mean effect on fitness in any population. One allele can always be regarded as having a certain selection coefficient to another at the same locus at any given point in time. Such coefficients are numbers that can be treated algebraically, and conclusions inferred for one locus can be iterated over all loci. Adaptations can thus be attributed to the effect of selection acting independently at each locus.

At the very least, Williams (1985) is claiming that coefficients of fitness of alleles at single loci are good enough for what has come to be called the "bookkeeping aspect" of selection. Hoping for more is too optimistic. For others bookkeeping alone is not good enough for a full scientific treatment of a natural phenomenon. These authors insist that proper attention must be paid to the causal mechanisms responsible for the changes in gene frequencies (Wimsatt 1980, Sober and Lewontin 1982, Sober 1984a). Evolution is at least changes in gene frequencies but much more as well.

In part, the continuing disagreement between gene selectionists and organism selectionists results from an ambiguity built into the phrase "unit of selection." When gene selectionists such as Dawkins (1976) first claimed that single genes are the primary focus of selection, they were interpreted as referring to the entire selection process. As the controversy continued, they "clarified" their early views. They intended to be referring only to replication—the transmission largely intact of the information incorporated in the structure of the relevant entities. Certainly genes are the primary replicators in biological evolution, but the opponents of gene selectionism cried foul. Such gene selectionists as Dawkins were not merely explaining their views; they were changing them. They had been proven wrong and refused to acknowledge defeat.

Such organism selectionists as Ayala (1978: 64) in turn argue that "it must be remembered that each locus is not subject to selection separate from the others, so that thousands of selective processes would be summed as if they were individual events. The entire individual organism, not the chromosomal locus, is the unit of selection, and the alleles at different loci interact in complex ways to yield the final product." Ayala is not denying the role of genes in passing on information in the evolutionary process. However, he does want to point out that entities other than single genes interact with their environments so that the resulting replication is differential. Without replication there would be no evolution at all, but without differential replication evolution would not amount to much. As Mayr (1978a: 52) has emphasized tirelessly, "Evolution through natural selection is (I repeat!) a two-step process."

Both sides of this dispute have now come to distinguish clearly between the two processes that combine to produce selection. In his original exposition, Dawkins (1976) termed the entities which function in these two processses "replicators" and "vehicles." According to the terminology that Dawkins (1982a, 1982b) now prefers, evolution is an interplay between replicator survival and vehicle selection. Williams (1985) also recognizes this distinction but prefers Hull's (1980) terminology of "replicator" and "interactor." Although terminological distinctions do not solve empirical problems, they frequently facilitate their clear statement so that they can be solved (Brandon and Burian 1984, Brandon 1985, Mitchell 1987).

The disagreements among gene selectionists, organism selectionists, and species selectionists have given rise to a call for a "hierarchical" view of evolution, a view of evolution that takes ample account not only of the hierarchical organization of genomes, organisms, and species but also of the hierarchical organization of species in monophyletic clades (Rachootin and Tomson 1981, Arnold and Fristrup 1982, Gould 1982b, Ho and Saunders 1982, Sober and Lewontin 1982, Eldredge 1983, 1985, 1986, Sober 1984a, Vrba and Eldredge 1984, Salthe 1985, and Brooks and Wiley 1986). Quite obviously, these authors did not discover that plants and animals are hierarchically organized. Biologists have acknowledged this fact from before Aristotle, but this hierarchical organization has not played a very important role in the formulation of evolutionary theory itself, especially in the versions produced by population geneticists. In the vast literature of population genetics, the focus of attention has remained steadfastly on genes.

When molecular biologists began to discover how the genome of an organism functions in the production of proteins (heterocatalysis), to no one's surprise it turned out to be a very complicated affair. Genomes themselves turn out to be hierarchically organized systems (Doolittle and Sapienza 1980). If ever anyone thought that genes are like beads on a string, recent advances in molecular biology have laid that metaphor to rest. Several authors have suggested that the way in which genes function might influence the general character of evolution itself, possibly providing part of its driving force. Prior to the rise of molecular biology, no one would have predicted the existence of multigene families—genes repeated thousands of times in the genome. One explanation is that these genes are "junk," genes that perform no function but have proliferated the way that the selfish-gene hypothesis claims all genes should (Doolittle and Sapienza 1980, Orgel and Crick 1980). Shortly thereafter, Dover (1982) suggested several molecular mechanisms that might produce multigene families—unequal crossover, transposition, and gene conversion—mechanisms which he termed collectively "molecular drive." Molecular drive has two peculiar features—it operates even if it confers no reproductive or other advantage to the organisms involved and it can result in a new variant being proliferated throughout a population in a unitary, cohesive manner. Although molecular drive, to the extent that it exists, is subject to selection and drift, in its own right it supplies a nonselective mechanism to explain species cohesion (see Lewin 1982).

One major characteristic of the rise of genetics in the English-speaking world was its divorce from embryology, a divorce that did not occur, for instance, in Germany (Harwood 1985). Thus, when genetics was wed to traditional evolutionary biology in the Modern Synthesis, embryology was once again left out. No one doubted that genes in some sense code for the phenotypic characters of organisms, but no one knew very much about this process. We still do not, but on the basis of what little we do know, several authors are urging the reintroduction of embryology into evolutionary biology. Properties of these processes may explain some of the features of the evolutionary process that population genetics does not. For example, one explanation for stasis might be the existence of "developmental constraints." In any genome, alternative pathways exist, depending on the environment that happens to exist at crucial times. The reemergence and fixation of a developmental pathway that had been suppressed for thousands of generations might explain some of the staccato pattern of evolution.

Similarly, mutations in regulatory genes might be another mechanism that would lead speciation to be punctuational (see Lande 1985 for populational models for rapid speciation).

Demes, populations, and species have always been the subject matter of population genetics. The chief difference that the hierarchical view of evolution introduces into this area of biology is the suggestion that these groupings of organisms might be viewed as entities in their own right, entities with their own adaptations. Species selection is merely the most ambitious form of this program in evolutionary biology. Previously, I sketched the relevant issues with respect to species selection, but two notes of caution are called for. In order to have species-level adaptations, not only must species possess species-level properties, but also these properties must be heritable. Species might have species-level properties without these properties being species-level adaptations. They might simply be effects of selection occurring at lower levels of organization.

One of the ironies of the recent dispute over the metaphysical status of species is that the very people who have argued most forcefully for the real existence of species as individuals are among the most skeptical about species possessing the characteristics necessary for them to function as units of selection. As Ghiselin (1987: 141) presents this point:

> It would seem that species do very few things, and most of these are not particularly relevant to ecology. They speciate, they evolve, they provide their component organisms with genetical resources, and they become extinct. They compete, but probably competition between organisms of the same and different species is more important than competition between one species and another species. Otherwise, they do very little. Above the level of the species, genera and higher taxa never do anything. Clusters of related clones in this respect are the same as genera. They don't do anything either.

Eldredge (1985: 160) agrees that, once selection is properly understood, species result from the evolutionary process but are unlikely to function in it:

> Species do exist. They are real. They have beginnings, histories, and endings. They are not merely morphological abstractions, classes, or at best classlike entities. Species are profoundly real in a genealogical sense, arising as they do as a straightforward effect of sexual reproduction. Yet they play no direct, special role in the economy of nature.

Instead, Eldredge (1985: 7) construes evolution as involving two largely independent hierarchies—genealogical and ecological:

> Genes, organisms, demes, species, and monophyletic taxa form one nested hierarchical system of individuals that is concerned with the development, retention, and modification of *information* ensconced, at base, in the genome. But there is at the same time a parallel hierarchy of nested *ecological* individuals—proteins, organisms, populations, communities, and regional biotal systems, that reflects the *economic* organization and integration of living systems. The processes within each of these two process hierarchies, plus the interactions between the two hierarchies, seems to me to produce the events and patterns that we call evolution.

Eldredge explains the dissonance that seems to characterize our view of evolution by reference to the fact that these two hierarchies intersect but do not coincide. To the question, "What is the connection between the taxonomic hierarchy and ecological classification?" Ghiselin (1987: 140) replies, "Basically, there is no particular connection, though there are diverse correlations." Wiley (1978, 1981a) largely agrees but emphasizes an important distinction in Eldredge's genealogical, informational hierarchy between species and monophyletic taxa. According to Wiley (1981a: 75), species exhibit "both historical and ongoing continuity whereas supraspecific taxa have only historical continuity. Species are units of evolution, while higher taxa are units of history" (see also Hennig 1966: 147). Even so, various workers maintain that evolution can be understood in terms of patterns exhibited by higher taxa.

When paleontologists began to count the relative abundance of higher taxa in the fossil records, they began to discover unexpected patterns. First off, the earth's biota seems to have undergone occasional massive depletions—up to 96 percent of the taxa extant during one of these catastrophes became extinct. As a result, succeeding radiations are strongly influenced by the luck of the draw. Possibly the patterns that systematists recognize are more a function of these boom-bust cycles than of selection. Raup and Sepkoski (1984) have even suggested that these boom-bust cycles are not haphazard but regular, appearing about every 26 million years. Because nothing in our understanding of the evolutionary process as a biological process would lead us to expect any regular cycles, explanations must be sought elsewhere. Possible celestial causes for these apparent cycles have caught the imagination of the general public as well as segments of the scientific community, in particular the Death Star named Nemesis. More recently, Jablonski (1986) has suggested that these mass extinctions are not indiscriminate. Some general features of taxa increase the likelihood that they will make it through the bust period, characteristics that are of no significance in the boom periods. For example, clades with broad geographic distribution fare better in major extinctions than more provincial clades, no matter the number of species in the clades. Even more recently, Jablonski (1987) has presented data that imply geographic range is actually heritable (for a more general discussion, see Raup 1986).

Systematists and evolutionary biologists have long acknowledged that their concerns coincide at the species level *if* species are to be both the basic units of classification and the entities that evolve. However, no such unanimity exists with respect to higher taxa. If some systematists recognize grades in their higher taxa and others limit themselves just to clades, then anyone who simply counts higher taxa is likely to miss any patterns that actually exist. In order for the patterns discerned to mean anything, the entities being counted must be of the same sort, e.g., all grades or all clades. At the very least, consistent classifications are required. For example, one way to test Eldredge and Gould's (1972) model for speciation is to compare two clades that originated at roughly the same time, one species-rich, one species-poor. The former should exhibit much greater character spread than the latter. For this test, only clades will do. Van Valen (1984: 52), to the contrary, states that a cladistic classification would preclude any analyis like those he presents to back up his own Red Queen hypothesis, a hypothesis to the effect that within a relatively homogeneous higher taxon, subtaxa tend to become extinct at a stochastically constant rate.

Yet another controversy in evolutionary biology that has been carried on in parallel with the preceding disputes turns on the ease with which adaptive scenarios can be conjectured and how difficult these conjectures are to test, a tendency which has been dubbed "Panglossian" (Tattersal and Eldredge 1977, Gould and Lewontin 1979, Vrba 1984, and Rosen and Buth 1980). Perhaps evolutionary theory does not guarantee the best of all possible worlds, but the tendency among Darwinians has been to explain apparent adaptations in terms of natural selection, or sexual selection, or group selection, or *some* sort of selection. Only when all else fails, so the critics claim, do Darwinians concede that just possibly a structure might have no selective explanation at all. Perhaps a particular pattern is due to the constraints imposed upon it by the physical characteristics of the matter out of which it is made, to developmental constraints, or to the accidents of history. For example, Raup et al. (1973) have shown that the same sorts of phylogenetic patterns produced by traditional paleontologists such as Romer (1956), showing the relative abundance of organisms in particular taxa through time, can be generated by strictly random computer programs, while Cracraft (1982, 1985) has suggested that patterns of macroevolutionary development may not be the result of the selection of heritable variation but purely a function of different climatic and geological histories. In general, defenders of the adaptationist research program complain that their views have been parodied. No one thinks that virtually every aspect of an organism is a specific adaptation for some function, and as difficult as adaptationist scenarios are to test, they can and have been tested—extensively (Ridley 1983).

No survey of recent challenges and developments in evolutionary biology would be complete without acknowledging the most daring and ambitious program to date—a unified theory founded on the basic principles of nonequilibrium thermodynamics formulated by Dan Brooks and Ed Wiley (1986), two second-generation cladists who have already appeared in the course of this narrative. In Darwin's day, Lord Kelvin and his collaborators tried to show that, on the basis of the newly formulated science of thermodynamics, the earth could not be old enough to allow for the slow, stately evolution of species that Darwin proposed. Later developments in physics proved them wrong. More recently, physicists and molecular biologists have collaborated to devise possible ways in which the first rudimentary living molecules could have developed from nonliving matter purely by physical means. If the universe is, by and large, proceeding toward increased entropy (decreased order), how can pockets of decreased entropy crop up? Brooks and Wiley take this program a large step forward. The question is not just how rudimentary forms of life could have evolved but how the ongoing process of biological evolution proceeds the way it does. According to Brooks and Wiley, biological evolution does not occur in spite of the general universal tendency toward increased entropy but because of it. Their surprising conclusion is that entropy itself is the driving force in evolution. As Brooks and Wiley see it, their theory is an alternative to neo-Darwinism. Natural selection is not excluded from their theory, but it plays only a minor role, while the principles of population genetics are merely limiting cases of their own more general views.

By now, it should be obvious that evolutionary theory is in the process of undergoing fundamental reexamination. One main source of discontent is population genetics. As impressive as the accomplishments of mathematical population genetics have been, each additional increment of improvement seems to be purchased only at the cost of a geometrical increase in complexity. In spite of the vehement protests of such organismically oriented biologists as Mayr, population geneticists have been successful in getting other biologists to agree with them that population genetics forms the "core" of evolutionary biology (Ruse 1973, Michod 1981). Hence, any departure from the somewhat narrow restrictions of population genetics is viewed as a challenge to Darwinism or neo-Darwinism, even if this "departure" has a long history in traditional Darwinian studies.

The current disputes in evolutionary biology differ in no important respects from other scientific controversies. Accusations of rediscovering the wheel, beating dead horses, attacking straw men, and parodying the views of one's opponents have been ubiquitous. As Roger Lewin (1980) discovered in his description of a conference held on macroevolution at the Field Museum of Natural History in 1980, no disinterested, noncommittal, theory-free characterization of such events is possible. A host of biologists wrote to *Science* to complain of his "simplistic caricature of the modern synthesis" (Futuyma et al. 1981, Templeton and Giddings 1981, Olson 1981, Armstrong and Drummond 1981, and Carson 1981). A half dozen years later, Lewin (1986) returned to Eldredge and Gould's punctuational model of evolution. Levinton (1986: 1490) promptly complained that there "would have been no problem in the first place if the straw man of phyletic gradualism had not been invented" and concluded that the "theory of punctuated equilibrium as first stated by Eldredge and Gould appears now to be as dead as a doornail . . . . Like sociobiology and (the oxymoron) scientific creationism, punctuated equilibrium has become so diffuse that it is impossible to refute or even discuss it without in effect perpetuating the slogan. Any advertising executive would be envious!" Needless to say, Gould and Eldredge (1986) disagree.

Scientists in their more pious moments and philosophers of science nearly all the time insist that science is primarily a matter of reason, argument, and evidence. Exchanges such as the preceding are dismissed as unfortunate lapses from proper scientific etiquette. Occasionally scientists do descend to such unseemly name-calling, but happily it is so rare that it does not interfere with the ongoing process of science. Scientific rationality is preserved in spite of such high emotions. I think that there are several things wrong with this comfortable response. Such heated responses are common in science, not rare. The polemics that make it into print are but the residue of the actual exchanges that go on at meetings, in private correspondence, and in manuscripts before they are sanitized by editors and referees. More than this, scientists cannot view things in any other way. Scientists involved in scientific disputes cannot help but see the disputes from their own perspectives. They would not hold the views they do if they did not think that they were right. Hence, any credence shown to alternative views must necessarily be wrong-headed.

As morally admirable as pluralism and open-mindedness may be, equal time for nonsense makes no more sense in the disputes that constantly characterize science than it does in the controversy between defenders of science and advocates of so-called creation science. In the second half of this book, I argue that science proceeds as well as it does not simply in spite of the strong emotions generated by the ego involvement of individual scientists with the correctness of their views but in large measure because of it. As Lewin discovered, no descriptions by third parties of scientific disputes can possibly satisfy all sides. However, one thing that can be said for Lewin's (1980) discussion of the macroevolutionary conference: it transformed a meeting that both those who organized it and those who attended were forced to conclude had been a nonevent into a major watershed in evolutionary biology.

Sociobiology: Another New Synthesis

During the winter of 1934–35, a group of young French mathematicians decided that the available textbooks on methods of analysis were not good enough. A new, more encyclopedic work was needed. By the time that these young mathematicians had finished their first volume in 1939, the number of authors had grown to ten. They were put off by the prospect of having to refer over and over again to a list of ten names every time that they wanted to cite the work. Then it came to them. Why not make up a name for the collective group of authors? Besides, seeking individual credit for one's work was supposed to be somewhat tawdry. As an instance of the sort of wit to which very bright people are sometimes prone, they picked the surname of a French general who had been a consummate failure, Charles Denis Sauter Bourbaki (1816–97), and invented a nephew, Nicholas. To their surprise, the first volume sold very well, and everyone wanted to consult with the great Nicholas Bourbaki. Unfortunately, he never could be located. Quite rapidly the mathematics community discovered the ruse, but even so the group continued to function and publish under the fictitious name. By 1968 the series had run to thirty-four volumes (Fang 1970).

In the early 1960s several young evolutionary biologists found themselves in a similar position. They were extremely discontent with the state of evolutionary biology at the time. They became convinced that a more analytic and unified approach was needed. Among the members of this informal group were Robert MacArthur, Richard Levins, Richard Lewontin, L. B. Slobodkin, and E. O. Wilson. As Wilson (in Segerstrale 1986: 62) recalls:

> In the early 60's we gathered at MacArthur's place in Vermont. We were about a half dozen people, all the same age. We formed a little group, a self-conscious little group in the early 60's. . . . We talked deliberately about how one would create a new population biology based on modeling and how one would go into these areas that were unformed and make order for the first time.

Some of these young biologists found science, as they found society at large, too competitive and individualistic. They were convinced that science would function much better if only scientists would submerge themselves in more genuinely cooperative activities. As the young mathematicians had done thirty years earlier, these evolutionary biologists adopted a fictitious name, "Isidore Nabi," after the physicist Isidor Rabi.

In 1967 Leigh Van Valen joined Levins and Lewontin at the University of Chicago. During this period Lewontin joined Levins in his active opposition to the Viet Nam War and his open advocacy of Marxist politics. Thus, when Mayr and Wilson wanted to bring Lewontin to Harvard, Wilson had to convince the administration and several of his colleagues that Lewontin was able to keep his politics and science separate. Lewontin did join the staff at the Museum of Comparative Zoology at Harvard, and Levins soon thereafter took a position at the Harvard School of Public Health. In the

1960s, opposition to the Viet Nam War was quite strong, especially on college campuses. Numerous antiwar groups sprang up. One of these groups, Science for the People, was organized to expose the collaboration of science and capitalism. At the 1969 meeting of the American Association for the Advancement of Science held in Boston, members of Science for the People confronted speakers with objections more pointed and impolite than usual, demonstrated against the involvement of scientists in the war effort, and in general politicized the meeting. The scientists present were bemused, bewildered, angry. What did the war in Viet Nam and the military-industrial complex have to do with scientific research?

The following year, the Boston chapter of Science for the People took over publishing the newsletter of the organization and transformed it into a journal, entitled *Science for the People*. The pluralist, libertarian bent of the early newsletter was replaced by a more militant Marxist tone. The "system" in general and the "scientific establishment" in particular were identified as the enemy. As might be expected, given their political convictions, Lewontin and Levins joined the Boston chapter of Science for the People. Other faculty members from Harvard also belonged, including Steven Jay Gould.

Although the major thrust of Science for the People was against imperialist exploitation of workers, subgroups were formed to study more specialized issues. One of these subgroups took up the issue of genetic screening for XYY abnormalities. During meiosis, homologous chromosomes sometimes fail to separate after pairing so that one gamete gets both chromosomes of a single pair. In most cases the resulting fetus aborts naturally. On occasion, however, a viable individual results with an abnormal chromosome number. Because the Y chromosome is relatively inconsequential save in determining sex, individuals born with an extra Y chromosome (misleadingly termed supermales) are usually viable. The existence of XYY males was brought to the attention of the public when a man who had raped and murdered several nurses in Chicago was reported to exhibit such a chromosomal aberration. (As it turned out, this report was erroneous. Richard Speck was found not to be XYY.) A subsequent study showed that certain populations, such as those in prison, exhibit a higher frequency of XYY males than does the general population. The implication was that XYY males are more likely to commit crimes, especially crimes of violence, than are other males.

Members of the genetic engineering group of Science for the People discovered that a study was being conducted at the Boston Hospital screening all newborn males to detect those who were XYY. The goal was to follow these males into later life to see if they exhibited a higher frequency of antisocial behavior. The genetic engineering group waged a campaign to get the screening stopped. The ensuing dispute centered on the conflict between the rights of patients (including newborns) and the rights of scientists to pursue investigations that they found promising. The problem was brought before the appropriate committee of the Harvard Medical School which, after six months of deliberation, concluded that the genetic screening program could continue. The decision served only to inflame the controversy. After considerable publicity, the issue was brought before the entire faculty on March 14, 1974. The motion to curtail genetic screening for XYY karyotypes failed 199 to 35. But this dispute was only the beginning of troubles at Harvard (for a discussion of these events from the perspective of one of the protagonists, see Davis 1986).

As part of their opposition to what they took to be an overly close relationship between the National Academy of Sciences and government policies related to the Viet Nam War, two members of the section on ecology and evolutionary biology of the National Academy of Science resigned in protest, Lewontin and Bruce Wallace. Because this section is relatively small, these resignations decreased the influence of "whole organism" biologists even more in the actions of the academy. The National Academy of Sciences in the United States does not have nearly the prestige or power that similar organizations have in other countries, but even so it does afford scientists an opportunity to have some influence on government policy. Thus, when E. O. Wilson submitted Levins as a candidate for the academy, Mayr wanted some sort of assurance from Levins that he would accept the position and not use it as an opportunity to declare his opposition to the war. Wilson assured Mayr that Levins would accept if elected. However, during the workshop on the origins of the synthetic theory of evolution in 1974, Mayr was informed that Levins had turned down his election to the National Academy of Sciences. Mayr was furious. The influence of "whole organism" biologists had been even further reduced by Levins's political gesture.

In 1961 Robert Trivers enrolled at Harvard on a scholarship to study mathematics and to prepare himself to become a civil rights lawyer, but his lack of interest in his mathematics courses resulted in a loss of his scholarship, and he switched to United States history. When the law schools to which he applied rejected him, Trivers turned to writing children's books. After becoming fascinated by animal behavior and the prospect of explaining it biologically, he enrolled at Harvard again, this time as a biology major. His progress was so rapid and his abilities so apparent that he was appointed an assistant professor in 1972. Trivers was one of the junior faculty who were allowed to attend the workshops on the origins of the synthetic theory of evolution in 1974 on the understanding that they say nothing.

Even before earning his degree, Trivers (1971, 1974) published two papers on what he termed "reciprocal altruism" and "parental investment." As discussed earlier, Hamilton (1964) had explained the genetic benefit of cooperation of close kin, but in his papers Trivers tried to show how organisms at large, whether of the same or different species, could evolve various sorts of cooperative behavior. Trivers interpreted such cooperation as a matter of each organism trying to cooperate only with other organisms that would reciprocate. When Trivers (1971, 1974) turned his attention to the relationship between parents and their offspring, the results were anything but heartwarming. According to Trivers the genetic goals of parents and offspring do not always coincide. Parent-offspring conflicts are common and not infrequently resolved in less than appetizing ways. Even a mother's love was no longer sacred.

At this time, E. O. Wilson (1971) was just finishing his magisterial book on insect societies. In this work he argued that not only organisms but entire colonies can function as units of selection. Although Wilson was aware of Hamilton's paper, he was not fully convinced by it. Trivers's extensions of the sort of reasoning employed by Hamilton fired Wilson's imagination. Perhaps the sort of work he had done on insect societies could be extended to other organisms, including human beings. Wilson launched into a project even more ambitious than the one he had just completed—the development of an entirely new branch of science.

While Wilson was spending ten-hour days working on his massive project, Michael Ghiselin submitted a manuscript to Harvard University Press. In this manuscript. Ghiselin argued that species, though not superorganisms, are not classes either. He had a second goal in this manuscript as well—to extend an individualistic view of the evolutionary process to human behavior and social organization. According to Ghiselin, all apparently altruistic behavior must be explained in terms of individual selfishness. In cases of close relatives, kin selection can promote cooperation, but human societies consist primarily of individuals who are not very closely related to each other. The implications of this perspective for human societies are fairly obvious, if unpalatable:

> The evolution of society fits the Darwinian paradigm in its most individualistic form. Nothing in it cries out to be otherwise explained. The economy of nature is competitive from beginning to end. Understand that economy, and how it works, and the underlying reasons for social phenomena are manifest. They are the means by which one organism gains some advantage to the detriment of another. No hint of genuine charity ameliorates our vision of society, once sentimentalism has been laid aside. What passes for cooperation turns out to be a mixture of opportunism and exploitation. The impulses that lead one animal to sacrifice himself for another turn out to have their ultimate rationale in gaining advantage over a third; and acts "for the good" of one society turn out to be performed to the detriment of the rest. Where it is in his interest, every organism may reasonably be expected to aid his fellows. Where he has no alternative, he submits to the yoke of communal servitude. Yet given a full chance to act in his own interest, nothing but expediency will restrain him from brutalizing, from maiming, from murdering—his brother, his mate, his parent, or his child. Scratch an "altruist," and watch a "hypocrite" bleed (Ghiselin 1974a: 247).

Six months after submitting his manuscript to Harvard University Press, Ghiselin received two reviews from the editor. Both recommended acceptance but only after extensive revision. Ghiselin was very unhappy with the criticisms contained in these reports. Although the refereeing process is officially confidential, Ghiselin thought that he knew the identities of the referees—Gould and Wilson or possibly Trivers. Ghiselin wrote to Gould and Wilson. Instead of reworking his manuscript, he sent it to the University of California Press. While it was out for review for the second time, Ghiselin received responses from Wilson and Gould. Wilson had been asked to review Ghiselin's manuscript by Harvard University Press, when the first two reviewers did not give it a totally clean bill of health. However, Wilson had been unable to get his review to the press in time for it to influence the evaluation procedure. Because Wilson thought that it might help Ghiselin in revising his manuscript, Wilson included a copy of the review he had written for Harvard.

Gould responded that he had been one of the original reviewers. When he was approached by the University of California Press to read the manuscript a second time, Gould sent them a copy of the review he had written for Harvard. Tom Schopf of the University of Chicago was also approached. He, too, found the manuscript original and extremely important but objected, as had the earlier readers, to Ghiselin's polemical style of writing. After the inevitable delays imposed by modern technology, Ghiselin's *The Economy of Nature and the Evolution of Sex* was finally published by the University of California Press in 1974, three years after he first submitted it to Harvard University Press. It received little notice in spite of Hamilton's (1975) highly laudatory review.

A year later, when Harvard brought out Wilson's *Sociobiology: The New Synthesis* (1975a), the reception could not have been more different. In fact Wilson's book began to cause comment even before it was officially released. In May a front-page article appeared in the *New York Times* proclaiming the impending birth of a new science, followed by largely positive reviews in all the leading papers and journals from the *New York Times* and the *New York Review of Books* to *Science* and *Nature*. The chief sour note was sounded in *Newsweek* by the economist Paul Samuelson (1975), who likened sociobiology to Social Darwinism. Wilson (1975b) himself wrote a popular piece for the *New York Times Magazine*.

Members of the sociobiology study group of Science for the People reacted to all this publicity with alarm. Turgid commentary in learned journals is one thing, but only those who have not lived in the Washington-New York-Boston corridor can wonder at the importance placed by the people who do on the *Times* and the *NYRB*. Perhaps not everyone in the world reads these publications, but everyone who counts surely does. While members of the study group drafted a collective response, Lewontin issued a 5,000-word position paper which he released to the press. In this release, Lewontin was quoted as saying that "it is not surprising that the model of society that turns out to be natural, just and unchangeable," according to Wilson, "bears a remarkable resemblance to the institutions of modern industrial society, since the ideologues who produced these models are themselves privileged members of just such societies." (Rensberger 1975: 16). The collective response of the study group (Allen et al. 1975) was issued as a response to Waddington's (1975) lukewarm review in the *New York Review of Books*. According to the members of this group, Wilson's chief weakness was his assumption of genetic determinism:

> The reason for the survival of these recurrent determinist theories is that they consistently tend to provide a genetic justification of the *status quo* and of existing privileges for certain groups according to class, race or sex. Historically, powerful countries or ruling groups within them have drawn support for the maintenance or extension of their power from those products of the scientific community (Allen, et al. 1975: 43).

Because Waddington had died soon after composing his review of Wilson's book, he was unable to reply to the comments that it elicited, but Wilson (1975c) immediately protested this "openly partisan attack." He especially resented his views being lumped with those that led to the establishment of gas chambers in Nazi Germany. He was also wounded because he had been on friendly terms for years with several of the people who had signed the letter, in particular Lewontin and Gould. If only they had come to him with their doubts and reservations before publishing their response, perhaps some of the confusion and misinterpretation might have been avoided. All he could do at this juncture was to protest their "self-righteous vigilantism."

The editors of *Dialogue* solicited a paper from the sociobiology study group and a response from Wilson. In their longer paper, Wilson's critics spelled out their objections to sociobiology in greater detail. They complained of the postulation of genes for particular behaviors, the prevailing adaptationist "bias" exhibited in Wilson's book, and his sloppy use of such terms as "slavery" to refer to such widely different phenomena as ant and human slaves. Slave ants belong to species different from their masters; human slaves and their masters belong to the same species. Slave ants are no more "slaves" than are such domesticated animals as cows. As far as they were concerned, "we know of no relevant constraints placed on social processes by human biology" (Allen et al. 1976: 186). They emphasized, however, that the issue:

> is not the motivation of individual creators of determinist theories, but the way these theories operate as powerful forms of legitimation of past and present social institutions such as aggression, competition, domination of women by men, defense of national territory, and the appearance of a status and wealth hierarchy (Allen et al. 1976: 182).

Wilson was not especially placated by being demoted from a sexist and racist to merely a dupe for these positions. He also disagreed about what the real issue was. As Wilson (1976: 183) saw it, the real issue was academic vigilantism, which was "the judgment of a work of science according to whether it conforms to the political convictions of the judges, who are self-appointed. The sentence for scientists found guilty is to be given a label and to be associated with past deeds that all decent persons will find repellent." Wilson objected especially to the label "genetic determinist." Regardless of what the sociobiology study group might claim, he believed that the causes of human behavior are "closer to the environmentalist than the genetic pole." But the issue for the study group was not environmental versus genetic determinism but determinism. Any sort of determinism was anathema to the members of the group (see Lewontin 1977 and the Epilogue).

In the process of composing their initial objections to Wilson, the members of the sociobiology study group discussed the likelihood that their attack might serve only to bring more attention to a set of views that they thought had already received too much publicity, but they decided that things had already gone too far. Someone had to point out the political causes of Wilson's research program. These doubts proved to be well-taken. The scientists whose disciplines Wilson proposed to cannibalize tended to object hotly. A few, however, became enthusiastic converts, sending their graduate students out into the jungles to find correlations between investment and coefficients of relationship. At long last the social sciences were going to become scientific. However, the explicit introduction of politics into the discussion served primarily to work in Wilson's favor. Wade's (1976: 1153) response was typical:

> In short, the Sociobiology Study Group has systematically distorted Wilson's statements to fit the position it wishes to attack, namely that human social behavior is wholly or almost wholly determined by the genes. Such a degree of distortion though routine enough in political life, is perhaps surprising from a group composed largely of professional scholars. Nevertheless, the group probably deserves some credit for pointing out that the territory Wilson is broaching is fertile ground in which to sow all sorts of social and political dragon's teeth.

The conflict over sociobiology continued on many fronts. The academic parties did battle in conventional ways—symposia were held, entire issues of learned journals were devoted to arguing the pros and cons of sociobiology, tenure cases became elevated to referenda on sociobiology, and so on. The animosity at the Museum of Comparative Zoology became palpable. Not only was the Wilson side arrayed against Lewontin, Gould, and their supporters, but everyone else was angry at both sides for letting this dispute interfere with the running of the museum and hence with their own research. In a detailed study of the conflict between Wilson and Lewontin, Segerstrale (1986) finds the causes of this conflict among colleagues stemming from differences in their scientific-cum-moral agendas. According to Segerstrale, Wilson considers science as a creative, risk-taking adventure, while Lewontin tends to be a more critical, purist sort. Wilson finds Lewontin "too safe," while Lewontin considers Wilson's efforts as "not serious." In fact, Lewontin's chief feeling for Wilson's work in sociobiology is "one of disdain. I don't know what to say, it's cheap" (Segerstrale (1986: 75). After comments such as these, the continued ill-feelings that prevail at the Agassiz Museum are not surprising.

In June 1977 the National Endowment for the Humanities sponsored a conference at San Francisco State University on the implications for human studies of sociobiology. Because the organizers feared the worst, the speakers were given special means of identifying themselves and were spirited into the auditorium through a side door so that they could avoid the numerous groups picketing the front entrance. When the first speaker rose to present his paper, shouting erupted in the back of the auditorium. Immediately, lights for the television cameras flashed on, but before anyone in the audience could figure out what was going on, the lights went out, the shouting stopped, and the dozen or so demonstrators followed the cameras out of the hall. They would have their three minutes on the evening news. The speakers then proceeded to present their papers without incident (Gregory et al. 1978).

Not all manifestations of the dispute over sociobiology were limited to verbal abuse. For example, in February 1978 yet another symposium on sociobiology was held, this time at the annual meeting of the American Association for the Advancement of Science in Washington, D.C. The speakers for the first session were seated at a long table on the stage of a large auditorium. Only as the session was about to begin did Wilson limp up onto the stage, one foot in a cast. Nothing much happened as the speakers rose to present their papers, until it was Wilson's turn. Then about twenty members of the International Committee Against Racism marched from the back of the hall onto the stage chanting slogans and carrying placards. Because of past disturbances of this sort, the AAAS had established a policy according to which a spokesperson for any protesting group would be allowed to state the case for the group if the protesters would then allow the meeting to proceed. However, this time one of the demonstrators stepped up behind Wilson and poured a cup of water on his head. A howl of protest arose from the audience. Instead of turning around and slugging the attacker, Wilson simply took out his handkerchief and began mopping his hair. When the moderator tried to invoke the AAAS rule allowing protesters to have their say, the audience would have none of it and shouted down the spokesman for the protesting group. Eventually the protesters marched out of the hall the way that they had marched in. When Wilson finished delivering his paper, he received a standing ovation. Gould, who had been sitting so close to Wilson on the stage that he had been splashed with some of the water, voiced his disapproval of the demonstrator's behavior, while defending the position that they represented, but to no avail. The mood of the audience was firmly set (Barlow and Silverberg 1980).

The conflict between biology and the political left was not confined to the United States. A parallel controversy was proceeding apace in Great Britain. In several publications, leaders of the conservative National Front cited the writings of John Maynard Smith and Richard Dawkins to justify racial nationalism and anticommunism. Steven Rose (1981: 335) wrote to *Nature* requesting these two men to "dissociate themselves from the use of their names in support of this neo-Nazi balderdash"; both men promptly complied, but in his reply Dawkins (1981: 528) complained that members of the far Right were not the only ones dragging science into "parochial British politics." Rose and the far Left were just as guilty. This correspondence elicited a letter complaining of the apparently contradictory statements being made by Dawkins and Wilson. Sometimes these men wrote as if we are all helpless robots controlled by our genes; sometimes as if we can have the sort of society we would prefer in defiance of our genes. Which is it? The letter was signed "Isadore Nabi" and the return address was the Museum of Comparative Zoology at Harvard.[3]

Wilson (1981) promptly wrote to inform the readers of *Nature* that the signatory of the letter was fictitious and to challenge its author to make future statements in his own name. The editors appended a short note to the effect that Nabi was believed to be a pseudonym of Richard Lewontin. Lewontin (1981: 608) in turn wrote to state "categorically that any assertion that Isidore Nabi is none other than R. C. Lewontin is incorrect." Although Lewontin's denial was categorical, it was also ambiguous, implying either that he had not written the letter or that he alone had not written it. Van Halen (1981) wrote to *Nature* under the name Isidore Nabi denying that he had written the letter, and later using his own name, he wrote a brief history of the Nabi group. The editors of *Nature* were not amused.

Conclusion

This chapter differs from the preceding chapters on systematics in that I make only passing reference to the professional relations among the scientists involved. Instead I concentrate on the views that successive generations of evolutionary biologists have had about the evolutionary process. The discussions in this chapter serve two functions: they show exactly how various are the combinations of doctrines that have gone under the name of "Darwinism" and they illustrate present-day views about the nature of biological evolution. Both topics are relevant to the second half of this book. If conceptual systems are to evolve in anything like the way that biological species do, then they must be as variable as are biological species, both at any one time and through

3. Nabi's first name is sometimes spelled "Isidore" in this literature, sometimes "Isadore." It began as "Isidore" when the fictitious scientist was recommended for membership in the New York Academy of Science, but it appeared as "Isadore" in 1980 as the author of a "Satyrical Comment on the Reductionist Theory of Edward O. Wilson & Co." that appeared in *Science and Nature,* pp. 71–73. This spelling was continued in the exchange that appeared in *Nature* in 1981 until Isidore Nabi at the University of Chicago wrote to complain that he was not "Isadore Nabi." Thereafter, the name of the fictitious character returned to "Isidore." (I owe several of the preceding references to Leigh Van Valen.) And as Elliott Sober informs me, "Navi" in Hebrew means "prophet."

time. In addition, my view of the evolutionary process will have to exhibit at least one possible combination of these ideas. It cannot exhibit all of them. I must choose. For example, I argue that in the early stages of conceptual change, science exhibits the demic structure described by Wright. Later, as the views involved become more widely assimilated, Fisherian mass selection takes over.

The final section on sociobiology serves quite a different function. Sociobiologists propose to extend the general principles which characterize biological evolution to social evolution, in particular to the social evolution of the human species. Because scientists are human beings, any general principles that apply to all human beings should apply to them as well. Thus, it might seem as if this book is a contribution to the sociobiological research program when I depart from sociobiologists in one important respect. I do not think that changes in gene frequencies have anything to do with conceptual change in science. Rather I present a general analysis of selection processes which, I argue, applies equally to biological, social, and conceptual change. My primary example of this thesis is science.

Another topic of this book is the role that social influences play in science. By now it should be clear that I think that the professional relations among scientists are important in the ongoing process of science. Unless one pays attention to them, one cannot begin to understand science. However, externalist students of science are more ambitious. They insist that larger social forces such as a scientist's socal class influence his or her scientific views. Certainly the controversy over sociobiology should exemplify this more ambitious claim. Are advocates of evolutionary taxonomy inherently elitists, while cladists are egalitarians? I see no evidence to this effect, but then there seems no plausible connection between a preference for a particular form of monophyly and any significant social issues. Sociobiology is quite another matter. Here possible connections are quite apparent.

7

Down with Cladism—Long Live Cladism

From the first, Smith seems to have treated the [Linnean] Society as very much his private fief, and candidates for Fellowship of whom he happened to disapprove were liable to be blackballed (one such victim for a long period of years, being poor J. E. Gray, an eminent all-rounder on the staff of the British Museum, who was alleged to have sinned unforgivably on one occasion by missing out Smith's name in referring to "Sowerby's *English Botany*"). Those honoured by being admitted were expected to observe the full Linnaean creed, down to the last capital and hyphenated epithet; to revere Smith as the chosen apostolic successor, as the defender of the one true tradition no less than as the guardian of the sacred relics. Linnaeus had left behind a legend: Smith and his admirers turned it into a contemporary intellectual bludgeon.

D. E. Allen, *The Naturalist in Britain*

SEVERAL PRESENT-DAY philosophers of science insist that the ongoing process of science can be understood only if it is analyzed in terms of units more inclusive than single ideas or even theories. Lakatos (1971) terms these more inclusive units "research programmes." Research programs are important in science because they are the entities that scientists evaluate when deciding which problems to pursue, which techniques to use, and so on. Scientists prefer to commit their careers to a progressing research program rather than to one that is stagnating or degenerating. The problem is, of course, to assess accurately far enough in advance whether a particular research program is progressing, stagnating, or degenerating. These are also the categories which Lakatos recommends that students of science use in evaluating scientific research programs (see also Laudan 1977). Because research programs are so important both in science and in our understanding of science, the way in which they are individuated is absolutely crucial.

Scientists themselves expend considerable effort in delineating the boundaries of their research programs. Such effort is not idle. For example, Sokal and Michener gathered around them at Kansas a group of workers who wanted to make systematics more quantitative and methodologically rigorous. To accomplish these goals, Sokal found it necessary to prohibit theoretical input, at least during the initial stages of classification. Sokal was joined by Sneath in this conviction. However, after some initial wavering, Michener retained serious doubts on this score. How should someone studying the research program begun at the University of Kansas individuate it? In terms of its commitment to quantitative methods, its antipathy to the intrusion of theoretical considerations into the early stages of classification, or both? The answer one gives to this question is important because it strongly biases the decision as to whether this research program has been progressing, stagnating, or degenerating.

Sokal and Sneath preferred to call their research program "numerical taxonomy" in order to emphasize its quantitative aspect, and they used this appellation in naming their yearly meeting as well as their two major books. By making the use of quantitative methods the essence of their research program, Sokal and Sneath put themselves in a position to claim any advances in the quantitative aspects of systematics as contributing to their program, regardless of the affiliation of the authors. Thus, Sokal considers the burgeoning literature on numerical cladistics a sub-branch of numerical taxonomy, even though one of its chief architects is his arch-foe, James S. Farris. Of the twenty significant landmarks in numerical taxonomy between 1961 and 1971 listed in Sneath and Sokal (1973: 15), four concern cladistics.

Other systematists such as Mayr prefer to term the research program that Sokal and Sneath are promoting "phenetics" or "numerical phenetics," to emphasize its stringently empirical antitheoretical philosophy. Opponents of this program note that long before Sokal and Sneath came on the scene, systematists had made contributions to numerical taxonomy. For example, chapter 7 of Mayr, Linsley, and Usinger's *Methods and Principles of Systematic Zoology* (1953) concerns quantitative methods of analysis and contains twenty-nine references, including the classic 1939 book by Simpson and Roe—later Simpson, Roe, and Lewontin (1960). One consequence of picking opposition to the intrusion of theoretical considerations in systematics as the essence of Sokal and Sneath's research program is that any advances made by anyone who thinks that scientific theories necessarily enter into the construction of classification must be excluded from the program no matter how quantitative the techniques used might be. If both evolutionary systematists and cladists can claim a quantitative side to their research programs, then they have gone a long way toward counteracting the claims of success made by numerical taxonomists. Conversely, if the numerical taxonomists succeed in garnering all achievements in quantifying systematics for themselves, then they can make their research program appear spectacularly successful. Of course, those involved in these disputes rarely discussed the effects that the way in which they individuated their research programs would have on estimates of their success. They probably did not even think of such issues all that often.

During the past two decades, numerical techniques have become increasingly prominent in systematics, both in the theoretical literature and in actual taxonomic monographs. In this sense "numerical taxonomy" has been extremely successful. It has been and continues to be a Lakatosian progressive research program. However, when one looks at the role of the early opposition of Sokal and Sneath to theoretical speculation in the construction of general-purpose classifications, the fate of their research program looks quite different. It seems to have degenerated precipitously. For example, in the proceedings of the 1982 NATO Advanced Study Institute on Numerical Taxonomy held in Bad Windsheim, West Germany, nearly the entire volume concerns quantitative techniques, including an entire chapter on reconstructing phylogenies. The introductory discussions of phenetic philosophy are at best perfunctory. Although Sokal and Sneath retain as strongly as ever their early distrust of the role of theoretical considerations in the construction of classifications, this particular tenet has gradually decreased in importance in the literature which their work generated. Thus, my emphasis in this book on taxonomic philosophy has made the research program generated by Sokal and Sneath look much less successful than it actually has been.[1]

To complicate matters further, the antipathy expressed early in the history of phenetics to the intrusion of theoretical considerations in the construction of classifications

1. One measure of the success of Sokal and Sneath's school of taxonomy is the number of pages published each year in *Systematic Zoology* arguing for their phenetic philosophy. The best years for the pheneticists in this respect were under the editorships of Johnston (55 pages in 1967 and 52 pages in 1969, respectively), Smith (52 pages in 1981), and Schnell (105 pages in 1983, 56 pages in 1984, and 70 pages in 1985). A second measure of the success of phenetic taxonomy is the number of pages published which simply apply phenetic methods without arguing for them. The best years for papers with titles such as the "Numerical Taxonomy of the House Mouse" were under Johnston (138 pages in 1968) and Rowell (102 pages in 1972). Under cladist editors applications of phenetic methods all but disappeared from the pages of *Systematic Zoology*. After Schnell took office, applied papers filled the pages of the journal once again, but by this time both pheneticists and cladists were engaged in devising quantitative techniques for producing trees and cladograms. As a result, the high figures for applied papers under Schnell indicate very little about the relative success of pheneticists versus cladists. Similarly, in the early years, there is some justice for counting all quantitative papers as contributing to Sokal and Sneath's school, regardless of their intent. However, as numerical cladistics came to the fore, this justification rapidly disappeared, as those allied with Sokal and Sneath did battle with Farris and his colleagues.

Number of pages devoted to phenetic philosophy, applications of the methods of numerical taxonomy, and quantitative methods as such in *Systematic Zoology*, from 1964 to 1985.

Editor	Year	*Phenetic Philosophy*	*Applied Numerical Taxonomy*	*Quantitative Techniques*	Total
Byers	1964	28	52	0	80
	1965	28	0	9	37
	1966	0	78	53	131
Johnston	1967	55	60	27	142
	1968	22	138	41	201
	1969	52	58	61	171
	1970	28	62	44	134
Rowell	1971	33	35	72	140
	1972	9	102	110	221
	1973	14	58	34	106
Nelson	1974	9	70	13	92
	1975	10	18	5	31
	1976	0	19	47	66
Schuh	1977	4	46	26	76
	1978	0	37	51	88
	1979	32	36	26	120
Smith	1980	25	0	17	42
	1981	52	30	71	153
	1982	4	32	25	61
Schnell	1983	105	28	37	170
	1984	56	47	21	124
	1985	70	0	46	116

has reappeared in an unlikely place—in the writings of some of Hennig's intellectual descendants. In Hennig's own formulations of phylogenetic systematics, both the reconstruction of phylogeny and beliefs about the evolutionary process play central roles right from the beginning. According to Hennig, scientists do not and cannot start from scratch in their investigations. Nor can they proceed in any prescribed order. Instead, systematics in particular and science in general is a matter of "reciprocal illumination." Advances in one area of investigation allow improvements in another area, and these improvements in turn allow further refinements in the original subject matter. For example, Darwin used work in the first half of the nineteenth century on the filiation of languages to develop his theory of biological evolution. After Darwin published his *Origin of Species,* philologists such as August Schleicher (1863) refined their views about the evolution of languages by reference to Darwin's theory. In connection with classification itself, the message is that pattern recognition and our understanding of the processes that produce these patterns cannot be strictly separated. According to Hennig (1966: 8), "systematics fundamentally means any investigation of relations between natural things and natural processes insofar as they have the character of conformity to law."

Hennig's opposition in these matters were ideal morphologists. According to such ideal morphologists as Naef (1919), morphological classifications are both historically and logically prior to phylogenetic classifications. "Idealistic morphology has been the prerequisite for introducing phylogenetics not only in the history of science, but even today must precede it on logical grounds" (quoted in Hennig 1966: 10). Hennig (1966: 11) rejected in no uncertain terms the claims of priority made by ideal morphologists for their classification:

> The idea expressed in all these utterances, that phylogenetic systematics is based logically and/or historically on purely morphological or at least nonphylogenetic systems, and the view often derived from this idea that a pure (idealistic) morphological or at least nonphylogenetic system therefore merits precedence over the phylogenetic one because it stands closer to the natural facts and contains fewer hypothetical elements, is absolutely wrong.

Hennig (1966: 11–12) found two main faults with the arguments presented by his idealist opponents. The first was their belief that science can be carried on without assumptions, as if scientists can approach nature with totally empty minds and classify accordingly, sticking with observed facts and observed facts only. The second error that ideal morphologists make, according to Hennig, is to view the primary relationships between living entities as being "similarity relationships" when actually they are temporal relations—both embryological and genealogical. The similarities between the views expressed by Hennig's ideal morphologist opponents and those of such pheneticists as Sokal and Sneath are obvious, but it is surprising to discover some of Hennig's intellectual descendants expressing almost identical views. For example, Platnick finds perfectly cogent the arguments which Hennig (1950: 15) judged "durchaus irrig." According to Platnick (1982: 283), anyone who "argues for the primacy of evolutionary theorizing over systematics has the cart before the horse both historically and logically."

More generally, a group of systematists, who have come to be known as "pattern cladists" or "transformed cladists," insist that cladistic classifications should be as theory-free as possible, as theory-free as pheneticists claim that phenetic classifications should be. Of course, such pheneticists as Sokal and Crovello (1970) along with such pattern cladists as Nelson and Platnick (1981) admit that all the stages in the life cycle of a single organism must be included in the same species as well as males, females, and neuters, but neither side views the intrusion of such "theoretical" considerations as being incompatible with their general views about the role of theories in classification. They see a difference between letting beliefs about the importance of mating and ontogeny influence their classifications right from the start and letting considerations of gene flow or phylogeny do the same. The former are a matter of "look-see," while the latter are not. As is the case with many scientists before them, pheneticists and pattern cladists insist that the role of observation in science is not merely to check hypotheses, regardless of how these hypotheses are generated, but to serve as a sufficient "basis" for all of science.

Not that pheneticists and pattern cladists are in total agreement. As Platnick (1979: 544) sees it, the chief difference between pheneticists and cladists is that pheneticists are willing to reflect in their classifications any sort of pattern that they might happen upon, including overlapping clusters, and to count both the presence of a character as well as its absence as "characters," while cladists insist on perfectly nested, non-overlapping hierarchical classifications based solely on the presence or loss of true characters (i.e., synapomorphies). If putative characters do not fall into perfectly nested transformation series, then they are not actually "characters." Pheneticists themselves have noticed the similarities between their views and those of certain cladists, but see them somewhat differently. According to McNeill (1982: 338), both phenograms and cladograms are "produced not to reconstruct phylogeny but to describe character state distributions." Cladistic analysis and phenetic analysis are "alike in that both are made possible by evolution (or by some analogous 'external influence') but that neither provide evolutionary trees *per se* and neither, by themselves, permit phylogenetic reconstruction." The chief difference between phenetics and cladistics, as McNeill sees it, is that phenetic methods start from overall similarity, "whereas cladistic techniques use only character states considered to show derived patristic similarity (synapomorphies)."

Simpson (1978b) and Hull (1979a) were among the first to note in print the striking similarities between some of the views of pheneticists and some of Hennig's descendants. Platnick (1979: 541) responded that some improvements had been made by later cladists on the views expressed by Hennig. For example, he found Hennig's prescription that species be delimited only at speciation events to be "irrelevant to cladistic practice." However, in spite of such modifications, Platnick (1979: 538) insisted that cladistics remained "essentially unchanged" because the essence of Hennig's system is his methodology, and no changes have been made in it. Later, two other authors claimed to note similar transformations in the views of certain cladists (Ball 1981 and Beatty 1982). Platnick (1982) and Patterson (1982) promptly objected to such allegations. In a somewhat different context, Halstead (1981) reiterated the claim that cladistics had undergone a fundamental transformation, while Charig (1982: 378) complained that Platnick treated evolution and phylogeny as "optional extras" when they are central to Hennig's approach to systematics. Once again, Platnick (1985: 87) insisted that cladistics has remained essentially unchanged because "Hennig, Patterson, and myself would all arrive at the same cladogram for any data set we examined." McNeill (1982: 339), an advocate of phenetic methods, draws a similar conclusion with respect to phenetics and cladistics because much of the time these approaches "give the same answer." Hence cladistics and phenetics are essentially the same.

Farris (1985a) agrees that pattern cladism is a myth but for reasons opposite to those of Platnick. Platnick uses certain beliefs which he takes to be common to all cladists to argue that cladistics has not been transformed since Hennig's earliest writings. Farris argues that cladists have not speciated into two separate groups because no significant correlation in beliefs can be discerned within and between these putative groups. According to Farris, those cladists who are supposed to belong to this new school of systematics disagree with each other on numerous issues while nonpattern cladists can be found agreeing with every tenet of the so-called pattern cladists. Although both Patterson and Platnick are supposed to be pattern cladists, Patterson rejects a falsificationist view of science while Platnick is partial to it. Conversely, Brooks is a falsificationist but not a pattern cladist. Nelson supposedly maintains that the ontogenetic criterion for inferring plesiomorphy cannot be misleading, while Patterson agrees with Brooks and Wiley that outgroup analysis can refute inferences from ontogeny. Yet Nelson and Patterson are supposed to be pattern cladists; Brooks and Wiley are not. Parsimony is used by practically all cladists, including such antipattern cladists as Brooks and Wiley, while Patterson doubts its value. Similar discordances exist, according to Farris (1985a), with respect to the dispute over species being classes or individuals, not to mention the role of phylogenetic analysis in the study of evolutionary processes. Because Farris cannot find a sharp break with respect to substantive issues between pattern cladists and nonpattern cladists, he concludes that none exists (for further discussion, see Ridley 1986).

The continuing controversy over the essence of cladism, or the essence of Darwinism, should have an extremely familiar ring to systematists. Similar controversies have been commonplace in systematics from its inception. An exchange between Linnaeus and one of his disciples over the essence of the order Umbellatae is typical:

Linnaeus: Can you give me the character of any single Order?

Giseke: Surely, the character of the *Umbellatae* is, that they have an umbel.

Linnaeus: Good; but there are plants which have an umbel, and are not of the *Umbellatae.*

Giseke: I remember. We must therefore add, that they have two naked seeds.

Linnaeus: Then, *Echinophora,* which has only one seed, and *Eryngium,* which has not an umbel, will not be *Umbellatae;* and yet they are of the Order.

Giseke: I would place *Eryngium* among the *Aggregatae.*

Linnaeus: No; both are beyond dispute *Umbellatae. Eryngium* has an involuerum, five stamina, two pistils, &c. Try again for your Character.

Giseke: I would transfer such plants to the end of the Order, and make them form the transition to the next Order. *Eryngium* would connect the *Umbellatae* with the *Aggregatae.*

Linnaeus: Ah! my good friend, the *Transition* from Order to Order is one thing; the *Character* of an order is another. The Transitions I would indicate; but a Character of a Natural Order is impossible. (From Whewell 1847, 2d ed., vol. 3: 357).

The reaction of those who have been labeled "pattern cladists" has been surprisingly impassioned. Knuckles whiten; lips curl back from incisors. One explanation for such strong emotions is that pattern cladism *is* a myth, and cladists are simply expressing their exasperation at being told that something which they know to be patently false is nevertheless true. Another alternative is that acknowledging any dissension in the ranks of the cladists plays into the hands of their enemies. To the question of why the myth of pattern cladism has become so popular, Farris (1985a: 199) answers, "Anti-cladists find it useful." Certainly some of those who have pointed out the emergence of what they take to be a new form of cladism have done so because they would like to see the whole enterprise fail. However, others are themselves cladists. They object because they disagree with the views of pattern cladists and fear that the negative reactions on the part of many systematists to pattern cladism threaten the entire enterprise. In response, pattern cladists claim that their views are being misunderstood.

Some positions are easier to misunderstand than others. Perhaps those systematists who have reacted so negatively to pattern cladism have misconstrued the basic positions that they reject, but misunderstandings are just as effective as more accurate interpretations in subverting scientific research programs. A constant response of scientists to their critics is that their views are being caricatured. Apparently, natural languages are not sufficiently powerful to clear up the confusions. Whether in Hennig's native German, in English, or in French, no one who is not himself a cladist (and not all of them) has been able to express the views of Hennig and his descendants with sufficient clarity to avoid egregious misinterpretation. Although I have done my best, I suspect that my own discussions will turn out to be no exception to this general rule (see for example Nelson's 1987 review of Ridley 1986).

When Nelson (1971a) began to publish, he objected to cladism being referred to as a "school." Cladism did not exist except in the minds of certain anti-cladists. Later he and Platnick were equally insistent that pattern cladism did not exist. One thing is certain: for not existing, it has garnered its share of names—canonical cladism (Simpson 1978b), Cladism with a capital "C" (Hull 1979a), transformed cladism (Platnick 1979), methodological cladism (Hill 1981), pattern cladism (Beatty 1982), natural order systematics (Charig 1982), and modern cladism (Patterson 1982).

Curiously, the literature on the essence of phenetics is quite sparse. When Sokal began to present papers on taxonomic philosophy, he was greeted with the objection that his principles were "typological" (Inger 1958, Simpson 1961). Because "typology" was a dirty word among English-speaking systematists, Sokal initially felt obligated to reject the allegation, but when he actually began to read the works of such ideal morphologists as Naef (1919) and Zangerl (1948), he discovered that his views were not all that different from those of typologists, as long as typology was shorn of its metaphysical baggage. His was an empirical, statistical typology. Unlike some typologists, Sokal (1962: 250–51) did not deny that inferences can be made about phylogenetic descent. However, "all consistent empiricists must emphasize that these deductions should be made *after* the classification has been established, ***not during*** the process of classification."

Later Pratt (1972) complained about "phenetic" being used as a weasel word by the pheneticists. Sometimes it seemed to imply that pheneticists are willing to allow only direct, theory-free observations in the early stages of classification. At other times, they claimed that phenetic classifications can be based on our knowledge of the "fine structure of DNA" (Sokal 1962: 244). Pratt (1972) found it difficult to reconcile the two uses of the term "phenetic."

This topic arose again in a review by Farris (1977a) of the proceedings of the Eighth International Conference of Numerical Taxonomy held in Oeiras, Portugal, the conference that he refused to attend when Estabrook declined to include Mickevich in the program. The proceedings included two papers in traditional population genetics. Although Sokal (1975: 229) admitted that these two papers are "only peripherally of moment to this conference," they counted as contributions to "population phenetics." Farris (1977a: 229) objected:

> It seems that the definition of "phenetic" has been broadened to the extent that the word means little more than simply "empirical." I see no merit in such ambiguity, and I suspect, in fact, that a deliberate equivocation is intended: if "phenetics" is coextensive with empiricism in biological systematics, then surely "phenetic" taxonomy is the only defensible approach to classification, and such non-"phenetic" *scholia* as phylogenetic systematics can be dismissed as unscientific!

The topic arose again in an exchange between Farris (1980, 1982) and McNeill (1982). Farris (1980) argued that cladistic methods are superior to those of the pheneticists by the criteria which the pheneticists themselves propose, in this case Gilmour naturalness. McNeill (1982) responded that Farris had misconstrued Gilmour naturalness. On McNeill's own preferred measure, phenetic methods prevailed. In response, Farris (1982: 427) objected to McNeill subverting criticism of phenetics by redefining the evaluative criteria of phenetic taxonomy in midstream. "Phenetics cannot be criticized," Farris (1982: 429) complained, "if no one can discover exactly what 'phenetics' is." Farris (1982: 415) also noted that "pheneticists seldom deny that their methods classify the products of evolution. Their position is instead that classifying products ought to be free of suppositions about the process." To this contention, Farris (1982: 413) replied that scientific vocabularies are "not only a means of description, but also of discussion and explanation in terms of current understanding of the natural processes that gave rise to observations."

If "phenetic" means any observation whatsoever, then all empirical scientists are "pheneticists." However, if phenetic characters are "pure observations," then no scientist can possibly be a pheneticist because of Popper's well-known claim about how theory-laden even the most direct observations actually are. What Farris does not mention is that theory-ladenness poses just as serious problems for pattern cladists. Both pheneticists and pattern cladists claim that "observations" form the foundation of biological classification. "Inferences" and "speculations" of all sorts must be postponed until later. Farris also does not mention Nelson and Platnick's (1981) terminological convention according to which any activity that produces branching diagrams counts as "cladistics."

As the preceding discussion indicates, scientists themselves are in strong disagreement about the existence and delineation of their research programs. What is someone studying these controversies to do? Is there no fact of the matter? One of the main tenents of this book is that scientific research groups must be defined in terms of professional relationships, in particular the alliances and allegiances that scientists form. Certainly Sokal, Sneath, and Ehrlich disagreed with Nelson, Rosen, and Farris on a variety of issues. These disagreements in turn certainly had an effect on the alliances that were formed, but alliances can survive disagreements about substantive issues, and essential agreement does not entail the formation of professional alliances. Farris's aligning himself with Nelson and Rosen at the American Museum rather than with Sokal at his own institution was caused more by problems in professional relations than by any disagreements about substantive issues (see also Mishler 1987).

In this book, I have taken the professional relations between scientists as primary in delineating research groups and I have defined research programs accordingly. In this connection, I am at odds with the scientists whom I am studying. Scientists occasionally acknowledge the existence of professional relations among scientists, the formation of schools, and the influence that the politics of science can have on science, but they are strongly disposed to dismiss such matters as peripheral if not totally extraneous to science. Science would proceed better if none of this existed. For example, in response to Nelson's (1971a) defense of Brundin against Darlington's (1970) attack, Howden (1972: 129) contrasted science with politics. According to Howden, science, in developing theories and in testing them, is "based on observations, the accumulation of facts, and the attempts to utilize these facts." Politics, to the contrary, is the use of "words, promises, and even perhaps a few facts, to convince people that your point of view is correct." Nelson (1972b) responded that Howden's comments exemplified his own definition of "politics."

Years later, Farris (1985a: 191) returned to the topic of politics in science, decrying the "political tactics" used by evolutionary taxonomists and pheneticists alike and insisting that the only legitimate way to settle scientific issues is by "reason and analysis." He then proceeded to chide certain of his fellow cladists, in particular Brooks and Wiley, for resorting to tactics that were no less political than those employed by their enemies. Apparently Brooks had contacted both Farris and Mickevich to enlist them "in a crusade to stamp out pattern cladism" (Farris 1985a: 199). "Cladistics should have been purer than this. I hope it may still live up to its promise" (Farris 1985a: 200). However, my utilization of socially defined groups to delineate research programs in no way commits me to the view that the decisions which scientists make are inherently or at bottom "political." Nor does it impugn the crucial role played by reason, argument, and evidence in science. At bottom all that I am doing is distinguishing between agreement and alliances. I must admit, however, that grouping scientists initially according to the alliances they form and only then seeing the extent to which they agree or disagree does imply that I think cooperation is more important in the scientific process than agreement. This position is not quite equivalent to might makes right.

The best evidence in favor of my position is the effects of treating ideas in the same acausal, ahistorical way that ideal morphologists would have us treat morphological characters. If research programs are defined solely in terms of similarity of ideas, then phenetics and pattern cladistics are in danger of being reduced to the same program. Any set of principles which leads to such a classification has to be mistaken. Regardless of how similar the ideas of pheneticists and pattern cladists may (or may not) be, these two research programs can be kept distinct by tracing conceptual descent. Conceptual descent in turn must be traced by its vehicles, and chief among these vehicles are individual scientists. As different as the views of pattern cladists may (or may not) be from those enunciated by Hennig, they are connected by descent. Although intellectual descent does not follow perfectly the development of socially defined groups of scientists, there is reasonably good concordance.

As the preceding discussion indicates, I have a second reason for treating socially defined research groups as primary and conceptually defined research programs as secondary—my general views about the nature of science as a selection process. Because selection can take place only when the entities being selected are related by descent, I have treated descent as primary in my analysis of developments in systematics and evolutionary biology. In my own work I have not been able to disengage issues of pattern and process. In the ongoing conduct of scientific investigation, patterns are in no sense prior to processes. In this connection I find myself in agreement with Hennig: science is a matter of reciprocal illumination. Improvements in classification permit improvements in our understanding of natural processes, and increased knowledge of these processes allows us to refine our classifications. If anything, processes are more fundamental to science than patterns. The importance that I place on process is reflected in the title of this book. It is *Science as a Process*, not *Science as a Pattern*.

On this score I am at odds with some of the scientists whom I am studying. Advocates of both phenetics and cladistics can be found arguing that classifications can be and should be constructed in such a way as to exclude as many theoretical considerations as possible, in particular beliefs about phylogenetic descent and the evolutionary process. Some cladists continue to agree with Hennig (1966: 8) that "nothing at all is achieved by completely nontheoretical ordering of organisms." Others are willing to *justify* their preference for hierarchical patterns by reference to descent with modification, but that is all. Pheneticists do not allow even this much. Contrary to the importunings of pheneticists, I did not begin my own research with a phenetic, general-purpose classification of systematists and only then proceed to discover their professional and historical relationships. Nor did I begin by classifying them cladistically by looking for perfectly nested sets of issues upon which there was total agreement. In my investigations I wandered back and forth between pattern and process—the way everyone else does.

The dangers of this relatively informal manner of conducting research are obvious. Scientists are certainly right to make their methods as orderly and coherent as possible, but I do not think that the actual conduct of science can possibly begin with the facts and nothing but the facts in the way urged by pheneticists and pattern cladists alike. Certainly no pheneticist has ever conducted a purely phenetic study, nor a cladist a purely (pattern) cladistic study. To put the same point differently, I am committed to showing that any study reputed to be purely phenetic or purely (pattern) cladistic actually departs from these methods and that these departures are not merely unfortunate lapses but inherent in science. Like it or not, the characters which systematists use are so thoroughly laden with long-forgotten theoretical considerations that the task of uncovering and eliminating all of them prior to beginning a taxonomic study is prohibitive. Systematists have better things to do with their time than this. The other alternative is for systematists to begin their studies from scratch, ignoring all previous work. A worse alternative is difficult to imagine. One of the major features of science is that it is cumulative. Any method which precludes scientists from building on the accomplishments of their predecessors is surely mistaken. Theory-neutral classifications are impossible, but even if they were possible, they would not be desirable.

To the extent that the systematists whom I have studied insist that scientific classifications should be theory-neutral or that scientists should conduct their empirical investigations in a prescribed temporal order, I disagree with them about their own investigations as well as my own. This disagreement to one side, I still must classify the scientists I am studying somehow. Should I classify them phenetically, cladistically, or what? In a phenetic classification, systematists would be grouped into clusters according to how similar their views happen to be regardless of causal influence, professional or otherwise. Phenetically Nelson and Platnick might turn out to be more similar in their views to Sokal and Sneath than to Brooks and Wiley. In this book I have quite obviously not adopted a phenetic classification of systematists. Regardless of its adequacy for classifications in other areas, phenetics will not do for the study of science itself. Both history and the professional relations among scientists are too important to ignore. Even if a phenetic general-purpose classification of scientists existed, a special-purpose classification would be required as well. Pheneticists are likely to agree. In point of fact, in commenting on an early draft of this book, Sokal complained that I had not been sufficiently "historical" in my treatment. The father of phenetics raising such an objection is as ironic as Farris complaining about political tactics in science, ironies which are liable to be lost on no one, least of all Sokal and Farris.

Even though I have not classified the scientists whom I am studying phenetically, other alternatives remain. Should I classify them cladistically or according to the more elastic principles of evolutionary systematics? According to my reading of the history of systematics, several of Hennig's followers have diverged significantly from the principles enunciated by their intellectual ancestor. Many of the principles which Hennig thought were central to his phylogenetic system have been rejected or demoted in importance by his descendants. In fact, the principles espoused by such pattern cladists as Nelson and Platnick (1981) are more similar to those argued by Hennig's chief opponents than to Hennig's original views. If I am right, phylogenetic systematics has evolved into its opposite.

According to the principles of systematics set out by such evolutionary systematists as Simpson and Mayr, Hennig and these intellectual descendants should be classified into two distinct "schools" because their views have diverged. According to Hennig's own principles, his followers can diverge as much as they please from his original formulations and still remain part of his research program, just so long as no speciation has occurred. As I see it, cladists have speciated both conceptually and socially. Brooks,

Wiley, and numerous other cladists have remained relatively primitive, insisting that phylogeny must play a central role in phylogenetics, while Nelson, Platnick, and several other of the most innovative systematists working today form the derived group. If this observation is correct, then on Hennig's own principles Hennigian phylogenetic systematics must be considered extinct. Mishler (1987) agrees about the conceptual heterogeneity that characterizes present-day cladists, but thinks that, socially, cladists still form a single group. As evidence for this conclusion, Mishler (1987: 59) notes that "we all attend meetings together." If this chapter has a message, it is that Mishler is mistaken in his perception.

On these matters I can hardly claim an Archimedean independent perspective from which I can view these disputes. The scientists involved have their interests, but so do those of us who study science. In most cases, however, these interests do not conflict. Few commentators on science have vested interests in the proper definition of "monophyly" or the cost of meiosis. In my case, however, there is a significant overlap between my concerns and those of my subjects because I am attempting to present an evolutionary account of science. Although opting one way in science and another way in meta-science is not necessarily inconsistent, there is a strong pull to make parallel decisions. If too many discrepancies crop up between biological evolution (the subject matter of many of my subjects) and conceptual change in science (my subject matter), then my goal of presenting a single analysis of selection processes which is adequate for both sorts of evolution is jeopardized. Any claim that I might make for an independent, third-party perspective on these issues is further brought into doubt by the fact that, on occasion, I have taken on the role of a protagonist in the events I am chronicling. On the first count, I claim no lack of commitment. I have consciously construed biological and scientific change in ways calculated to reduce the apparent differences between them. However, on the second count, I have done my best to step back from those events in which I have participated to view them from the same sort of third-party perspective that I have attempted to bring to the other episodes that I have studied. I have no way of knowing how successful I have been.

In summary, in this book I am urging two positions on the nature of science that differ from those of some of the scientists whom I discuss. I have individuated research groups in terms of their professional relations. That way the members of two research groups can be in general agreement without their research programs collapsing into one. Conversely, members of the same research group can disagree with each other over important issues and still contribute to the same research program. The analogs of these phenomena in biological evolution are sibling species and polytypic species. Scientists themselves are so engrossed in their research that they rarely take much notice of the underlying professional relations that influence the course of their conceptual development. To make matters worse, when such issues arise, scientists are strongly disposed to dismiss them as being irrelevant to science. Hence, they are forced to engage in the politics of science while pretending to themselves that they are doing something else (see, for example, Wade 1981, Bliss 1982, Hallam 1983, Rudwick 1985).

Historians commonly structure their narratives so that they form lineages just because they are committed to lineages. They neither have a justification for such a preference nor feel the need for one (for an exception, see Richards, 1981, 1987). The emphasis I place on research groups in the individuation of research programs follows from my analysis of the scientific process as an instance of the operation of selection processes. In order for selection processes to operate, the entities must be organized into populations integrated through time by descent. These considerations did not function in my research as an after-the-fact justification or as a derivative speculation but informed it right from the start. As a result, those systematists who prefer a more theory-neutral stance in science will find my own research seriously flawed. I should have constructed either a general-purpose phenetic classification or else a general cladogram for all the episodes that I have studied and only afterwards speculated about causal and historical relations.

The Transformation of Cladistics

In his own writings, Hennig opposed the empiricist pretensions of his idealist opponents. In no significant sense are the patterns reflected in phylogenetic classifications prior to the processes that produce them. According to Hennig (1966: 93), " 'Transformation' naturally refers to real historical processes of evolution, and not to the possibility of formally deriving characters from one another in the sense of idealistic morphology." Brundin was just as explicit on the central role of our beliefs about the evolutionary process in phylogenetic systematics. According to Brundin (1966: 23), "It is apparent that the rule of deviation and concepts like plesiomorphy and apomorphy, indeed all the concepts and principles of phylogenetic systematics, are and have to be based on the speciation process," its premises and phylogenetic meaning. In his earliest writings, Nelson (1970: 375) seemed to agree. For example, in his outline of a general theory of comparative biology, he objected to "those who view science simply as 'ordered knowledge.' These persons, prone to embrace a philosophy of radical empiricism and to find the accumulation and ordering of comparative data a satisfying life's work, obstruct the progress of science, insofar as they may discourage and condemn as 'speculative' efforts to generate hypotheses and theories of scientific legitimacy." To the contrary, comparative biology is "concerned with the theory of evolution in all its detail" (Nelson 1970: 374).

However, in these early writings, Nelson also expressed reservations not to be found in the writings of Hennig and Brundin. For example, Nelson (1969) insisted that ancestor-descendant relationships cannot be demonstrated and cast doubt on the role of fossil distributions in biogeography; more than this, he defended the use of "morphotypes" in comparative biology and objected to such authorities as Simpson (1961) turning "typological" into a pejorative equivalent of "unscientific." Early in his career, Nelson developed a deep antipathy to the synthetic theory of evolution, especially the versions produced by population geneticists. Eventually he came to believe that any intrusion of evolutionary theory into classification is a mistake. He laid the groundwork for this conclusion in an unpublished paper in which he distinguished between cladograms, phylogenetic trees, and evolutionary scenarios.

Earlier Mayr had made similar but significantly different distinctions. According to Mayr (1969: 254–55), branching diagrams in general are to be termed "dendrograms."

He then subdivided dendrograms into three kinds—phenograms, cladograms, and phylograms (see fig. 7.1). All that phenograms portray is degree of similarity with no commitment to any special sort of similarity. Cladograms express degrees of relationship by means of the position of the branching points on the ordinate axis as well as degrees of difference on the abscissa. Although the angles in Mayrian cladograms are all the same, the branching points can be spaced closer together or further apart. Phylograms attempt to convey three sorts of information—degree of difference on the abscissa, geological time on the ordinate, and degree of divergence by the angle of divergence. In cladograms all angles are the same; in phylograms they vary.

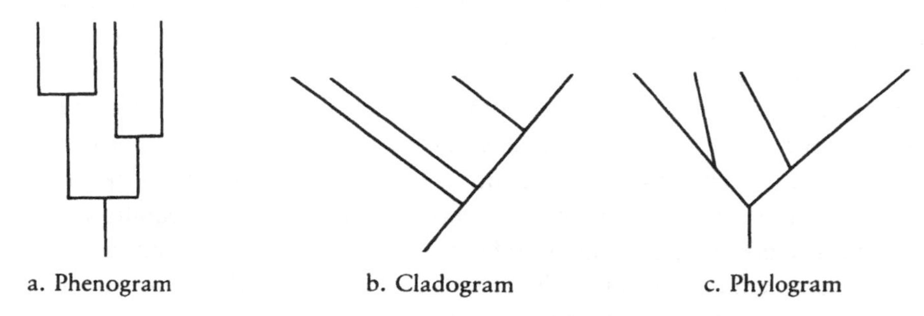

Figure 7.1. Mayr's division of dendograms.

Nelson's classification of branching diagrams was quite different from the classification set out by Mayr. At this stage in his intellectual development, Nelson considered cladograms to be synapomorphy schemes. That is, they represent sister-group relations. However, the abscissa in cladograms does not represent degrees of difference. It represents nothing at all. The only thing that cladograms indicate is order of presumed splitting. Branching points are spaced as evenly in the diagram as the topology allows. For Nelson, trees represent a variety of other relationships as well, including ancestor-descendant relationships and degrees of divergence. Scenarios are the adaptationist and selectionist stories that evolutionary biologists construct to explain the changes depicted in their phylogenetic trees. According to Nelson, cladograms are prior to trees, and trees are prior to scenarios. By this Nelson meant that every tree presupposes a cladogram, and every scenario presupposes a tree, but not vice versa. One can construct a cladogram with no particular tree in mind and a tree with no idea about the causal scenario that actually produced it. The progression of cladogram-tree-scenario is one of increasing information content. But Nelson also noted that this increase in putative information content is largely illusory because it is accompanied by a rapid decrease in the warrant for these claims. Cladistic relations are difficult enough to establish. The construction of trees is even more doubtful, while scenarios are largely unfalsifiable speculations. Although Nelson did not publish the preceding distinctions at the time, they quickly found their way into the literature (e.g., Tattersall and Eldredge 1977, Hecht, Goody, and Hecht 1977).

Early in the history of phenetics, Sokal and Sneath published a general work that became the "bible" of phenetic/numerical taxonomy. At the time, Sokal and Sneath could cite only sixty or so papers as supporting their emerging research program. Of these, they themselves had authored or coauthored twenty-four. Right from the start, anyone who wanted to know what phenetic/numerical taxonomy was could read their *Principles of Numerical Taxonomy* (1963). Hennig's *Phylogenetic Systematics* (1966) served a similar function in the emergence of cladistics, but as a general work it had several flaws. In the first place, it was extremely difficult to follow, possibly because it was a revision of an earlier work, possibly because Davis and Zangerl had condensed it considerably as they translated the German original. Four years later, R. A. Crowson (1970) published a much simpler and more comprehensible text. Although both Rosen and Nelson were enthusiastic about it and recommended it to their students, this work left little in the way of a record of any influence it might have had. Although several early cladists acknowledge reading it when it first appeared, references to it have rapidly dropped out of the literature and none of the particular tenets that Crowson suggested have been adopted. Nelson, Eldredge, and Cracraft decided that an updated general text was needed, a text which they titled tentatively *Principles of Comparative Biology.* The plan was for each man to write about a third of the book. As work proceeded, problems began to arise. As one of the authors of the punctuated equilibrium model of speciation, Eldredge quite understandably wanted to include a section on the speciation process. Perhaps cladistics is independent of any particular theory of speciation, but Eldredge saw no reason to exclude this traditional topic from their book. Omitting any reference to the species category from a general work on systematics was as unusual as writing a text on ancient Greek philosophy without mentioning fire, air, earth, and water.

Nelson to the contrary insisted that neither species nor speciation be treated at any length. As far as Nelson was concerned, cladistics was pure methodology, epitomized by his own system of "component analysis," the formal science of branching diagrams. Nelson was no happier with Cracraft's contribution. Cracraft's views can be gathered from an influential paper he published at the time setting out and comparing the principles of both phylogenetic and evolutionary systematics (see Appendix D). According to Cracraft (1974), all models of phylogeny must include certain basic components: statements about the kinship relations that integrate taxonomic units, the origination and diversification of taxonomic units into lineages, and the methods used to order such lineages. As far as the nature of these units themselves, Cracraft saw no difference in practice whether one looked upon them as species or as OTUs. In agreement with Nelson, Cracraft listed as two of the principles peculiar to phylogenetics that ancestors cannot be recognized or identified but can only be hypothesized, and that species are to be ordered into lineages solely on the basis of shared derived characters. As a result all higher taxa must be monophyletic in Hennig's sense. But he also included the principle that species originate only by allopatric speciation, assumed for methodological reasons to be dichotomous, and not by phyletic gradualism.

As far as Nelson was concerned, Cracraft's views were too "pluralistic," including too many bits and pieces from too great a diversity of authors and positions. Cracraft was no happier with Nelson's advocacy of Croizat's principles of biogeography; it was liable to reflect badly on cladistics. Nelson and Cracraft came to words more than once over these issues, both in person and in correspondence. The net effect was that

Eldredge and Cracraft split off from Nelson to write their own general text, *Phylogenetic Patterns and the Evolutionary Process* (1980). As the name implies, it concerned the evolutionary process, the resulting phylogenetic patterns, and principles by which phylogenetic patterns can be reflected in phylogenetic classifications. Even though Eldredge and Cracraft included an extensive discussion of species and speciation in their book, they agreed with Nelson that "phylogenetic patterns must be analyzed (as wholly independently from notions of process as humanly possible) and theory invented to explain pattern" (Eldredge and Cracraft 1980: 14). It may well be true that all cladists right from the beginning were in total agreement at least over fundamentals, but the less than fundamental disagreements that separated Cracraft and Eldredge from Nelson were sufficient to keep them from being able to cooperate in the production of a general text on the basic principles of phylogenetics.

In his early years at the American Museum of Natural History, whenever Nelson had a bright idea or a new enthusiasm, he rushed to share it with Donn Rosen. However, after Platnick arrived at the museum in 1973, Nelson increasingly sought him out for discussions. Platnick shared Nelson's enthusiasm for developing the formal structure of branching diagrams and his antipathy to letting process theories, especially the synthetic theory of evolution, enter into the process of pattern recognition. Platnick agreed with Nelson that pattern is necessarily prior to process. Platnick also had a strong predilection for philosophy, especially the views of Karl Popper. In collaboration with Eugene Gaffney, Platnick reviewed for *Systematic Zoology* every one of Popper's major works, from a 1968 English edition of his *The Logic of Scientific Discovery* (1959) to his autobiography, *Unended Quest* (1976). Popper was so impressed with these reviews that he wrote Platnick to compliment him on the understanding of his philosophical views that these reviews revealed.

When Nelson and Platnick (1981) published their synoptic book, *Systematics and Biogeography: Cladistics and Vicariance,* they acknowledged the influence of three earlier workers—Hennig, Croizat, and Popper. Although they admitted that Croizat (1976, 1978) found Hennig's principles of phylogenetic systematics antithetical to his own principles of biogeography and that neither Hennig nor Croizat had ever mentioned Popper, Nelson and Platnick still considered their own work as synthesizing and extending the views of these three men. The influence of Croizat can be seen in the organization of Nelson and Platnick's massive book. Croizat had termed his own 1964 book *Space, Time, Form: The Biological Synthesis.* Nelson and Platnick also subdivided their book into three sections—Form, Time, and Space. The difference in perspective between Croizat on the one hand and Nelson and Platnick on the other can be seen both in the change of order and in the amount of space devoted to each topic. For Nelson and Platnick, form is prior to time and space. They devote roughly the first half of their book—256 pages—to detailing the science of branching diagrams. According to Nelson and Platnick, a cladistic analysis of biological organisms is a necessary prerequisite to any work in biogeography. The second half of their book—186 pages—concerns biogeography. Sandwiched in between are a mere 22 pages devoted to a discussion of time in the context of Haeckel's biogenetic law. For Nelson and Platnick, ontogeny—not phylogeny—is the only source for a temporal dimension to patterns.

Nelson and Platnick's exposition differs in important respects from that of Hennig as well. According to Hennig, species as evolutionary units are central to the science of phylogenetics. Nelson and Platnick mention species and the evolutionary process in their book but primarily to dismiss such topics as irrelevant to cladistics. In a text of 543 pages, they devote less than two pages to the discussion of the species concept. According to Nelson and Platnick (1981: 12), "species are simply the smallest detected samples of self-perpetuating organisms that have unique sets of characters." As construed by Nelson and Platnick, species include the " 'subspecies' of those biologists who use that term." Although Platnick (1979: 541) denies that there has been any transformation of cladistics from Hennig to the present, he lists as one of the "changes in perspective" the abandonment of Hennig's requirement that species be delimited always and only by speciation events. Cladistics is not concerned with speciation events but with "detectable changes." Nelson and Platnick (1981: 35) are no less dismissive with respect to evolutionary theory as a process theory:

> Patterns such as this one are interesting for other reasons. One is the fashion these days to discuss processes of evolution. Some biologists expound on processes as if all worthwhile general knowledge is contained therein. Now what, one might ask, are processes of evolution? Do they not all presuppose the existence of a nonrandom pattern such as the one we have considered? No patterns—in general, no processes. No patterns, nothing to explain by invoking one or another concept of process. In short, a process is that which is the cause of a pattern. No more, no less. Pattern analysis is, in its own right, both primary and independent of theories of process, and is a necessary prerequisite to any analysis of process.

In the years between Nelson's first distinguishing between cladograms, trees, and scenarios and the appearance of his synoptic work with Platnick, his concept of a cladogram had become even more general. According to Nelson and Platnick (1981: 14), all that cladograms in the most general sense depict is "structural elements of knowledge." Cladograms in this sense are roughly equivalent to Mayr's dendrograms. Thus, the phenograms produced by pheneticists, gradograms produced by evolutionary systematists, and the traditional cladograms produced by cladists all become special sorts of cladograms in this more general sense. As Nelson and Platnick (1981: 171) summarize these relations:

> A cladogram, therefore, may be defined as a branching, or dendritic, structure, or dendrogram, illustrating an unspecified relation (general synapomorphy) between certain specified terms that in the context of systematics represent taxa. If the relation is considered common ancestry, there is a reason to call the structure a phylogram; or, alternatively, if the relation is considered phenetic or gradistic similarity (however measured), there is reason to call the structure a phenogram or gradogram, respectively. In either case, the structure itself need not change—only the interpretation of its significance.

Nelson and Platnick (1981: 172) acknowledge that "it may seem odd to think of a phenogram, for example, as a type of cladogram," but they note that a "phenogram does, after all, have a cladistic (branching) aspect that as *branching* is no different than the branching of a phylogram." However, it seems hardly less odd to view cladograms, as they had previously been characterized, as a type of cladogram in this more general sense. In a fully resolved cladogram in the traditional sense, all the end-points represent species and the characters used to construct these cladograms are part of transformation series in a historical, temporal sense (synapomorphies). If phenograms are to be a special type of cladogram and the basic entities of phenetics are Operation Taxonomic Units with no evolutionary content, then it follows that the terminal taxa in generic cladograms also can have no evolutionary content. They are no more than "diagnosable units" of the sort advocated by pheneticists. It also follows that the characters used to construct these cladograms need not be evolutionary homologies. In order for Hennig's system to become a general theory of taxonomy, Nelson and Platnick (1981: 142) note that "synapomorphy" must be redefined as an "element of pattern—a unit of resolution so to speak."

Once one realizes exactly how general Nelson and Platnick mean for generic cladograms to be, several apparently controversial claims which they make lose much of their force. Cladograms as dendrograms are more general than any specific sort of branching diagram, including traditional cladograms. In this sense generic cladograms are prior to all other sorts of branching diagrams. The sorts of cladograms that Hennig presented in his publications are not. However, cladograms in both senses are still prior to trees. Earlier, Sokal and Sneath traded off ambiguous interpretations of the term "phenetic" to claim that phenetics is necessarily prior to all systematic investigations. Phenetic characters supposedly can be observed directly in a theory-free sort of way. They exist out there in nature just waiting for an intelligent ignoramus or a scanning machine to record them. It is in this sense that phenetic characters and the classifications constructed from them are prior to all other classifications. But in the same breath, pheneticists claim that codons might prove to be the ultimate phenetic characters. Codons are anything but directly observable, and the concept of a codon is anything but theory-free.

Whether intentionally or not, certain cladists have traded off a similar ambiguity in the use of the term "cladistic." Once "cladistics" has been defined with sufficient generality as the science of branching diagrams, then it follows that cladistics is prior to all systematic investigations but only in the sense that formal science is prior to empirical science. Nelson and Platnick (1981) are clear in their intent to develop as general a science of branching diagrams as they can. They are much less clear about how divorced from empirical considerations this science is to be. Their basic units are diagnosable units, but these units must also have an "independent existence in nature" (Nelson and Platnick 1981: 11), although they mean by this only that ontogeny and reproduction matter. Systematists must continue to search until they find a diagnosis that groups "males, females, eggs, larvae, pupae, and adults together." According to Nelson and Platnick (1981: 12), each unit "must have a unique set of characters. It need not have even a single character that is unique to it, but the total set of its known characters must be different from that of all other known samples, or we will not be able to distinguish it." With respect to these smallest detected samples of self-perpetuating organisms, cladists are willing to adopt polythetic definitions of the sort that pheneticists utilize for all taxonomic levels. However, above the lowest level, cladists insist on perfectly nested characters.

But what of these characters themselves? For a science of branching diagrams to be totally general, the characters cannot be restricted to evolutionary homologies or even "synapomorphies," if this term is being used in a temporal, historical sense. Not all hierarchical patterns result from descent with modification. Some, such as those presented in the periodic table of elements, are a function of the structural relations among elements, regardless of their order of genesis. Nelson and Platnick's (1981: 140–2) discussion in this connection is worth quoting at length:

> The majority of the monophyletic groups specified in Hennig's type of trees (cladograms) might be considered evolutionary in nature, reflecting actual speciation events—branchings of the historical process. And so might the groups specified by phenetic trees. But *all* of Hennig's groups correspond by *definition* to patterns of synapomorphy. Indeed, Hennig's trees are frequently called synapomorphy schemes. The concept of "patterns within patterns" seems, therefore, an empirical generalization largely independent of evolutionary theory, but, of course, compatible with, and interpretable with reference to, evolutionary theory. The concept rests on the same empirical basis as all other taxonomic systems (the observed similarities and differences of organisms). But the concept is not wholly independent of evolutionary theory, for one of its basic elements (nature of evidence) is synapomorphy, or shared advanced character. The other basic elements, namely relationship (what is evidenced) and monophyly (what is resolved), are definable only with reference to the branching diagram, and carry no necessary evolutionary connotations. Indeed, the concept of synapomorphy may be definable purely as an element of pattern—a unit of resolution, so to speak. If so, Hennig's system would be understandable not merely as the theory of "phyletic" taxonomy but as the general theory of taxonomy of whatever sort. The general properties of Hennig's system—the basic elements and their logical interrelations as exhibited by the branching diagram—are, perhaps, the more interesting properties of this system.

Thus, in cladistics in the most general sense, "synapomorphy" can be interpreted in a purely empirical sense as an element of pattern without reference to evolution. All that is needed is the hypothesis of congruence. Whatever characters give rise to transformation series of patterns within patterns, they count as synapomorphies. No reference to evolutionary theory or phylogeny is needed. In recognition of the similarity of their views in this respect to early critics of both evolutionary systematics and cladistics, Nelson and Platnick (1981: 165, 151) remark that the cladogram, "as a summary of the pattern of synapomorphy, also satisfies, or nearly so, the need for empiricism stressed by Blackwelder, Sokal and Sneath." The "futile theorizing" to which these three systematists objected so strenuously belongs to "the three-step process of deriving a tree from a cladogram." In support of this contention, Nelson and Platnick (1981: 158) once again refer to the recognition of nested sets of characters by pre-evolutionary biologists, the same argument repeated so often by Hennig's idealist opponents. As much as Nelson and Platnick denigrate phylogenetic trees and even cladograms in the phylogenetic sense, in comparison to cladograms in the most general sense, they cannot very well dismiss them totally as being unscientific because the second half of their book devoted to biogeography depends on them (Nelson and Platnick 1981: 43).

One major area of confusion in the controversy over cladistics has been the character of cladograms. Hennig himself never used the term, but the concept appears in his works. Unfortunately, given his diagrams and some of his discussions, many early readers interpreted Hennig to be referring to highly stylized trees. Later workers cleared up these confusions. As it entered the literature, the term "cladogram" referred to synapomorphy schemes—branching diagrams that express sister-group relations between species and/or strictly monophyletic higher taxa. Eventually, Nelson and Platnick introduced a generic sense of "cladogram." All cladograms in the specific sense are cladograms in the generic sense, but not vice versa. The confusions engendered by this terminological convention have proven to be no less frustrating than the confusions introduced by Kuhn (1970) when he distinguished between two senses of "paradigm"—one referring to disciplinary matrixes and the other to refer to one element of these matrixes—exemplars. Thus, each time Kuhn refers to paradigms, one must ask which sense he intends, and it is not always possible to tell. The same can be said for Nelson and Platnick's two senses of "cladogram" and related terms. Ghiselin (1974b) and Hull's (1976) use of the term "individual" suffers from this same fault. In its most general sense, "individual" refers to any spatiotemporally localized entity, whereas in a more restricted sense it refers simply to organisms.

So far, I have detailed the respects in which the mature views of Nelson and Platnick differ from those of Hennig. Croizat was a second source of inspiration for their book. In chapter 5, I have already discussed Croizat's (1976, 1978, 1982) papers in which he complained of how thoroughly cladists had misrepresented his views on biography and biogeographic method. Platnick (1982) and Patterson (1982) countered Beatty's (1982) earlier claim that pattern cladists had significantly transformed Hennig's principles of systematics, and here was Croizat himself making comparable claims about his principles of panbiogeography. After Croizat's death, his cause was taken up by three young biogeographers from New Zealand—Robin Craw, John Grehan, and Michael Heads.

In the early years of phenetics, Sokal and Ehrlich struck more traditional taxonomists as too rude and abrasive. Later, both evolutionary systematists and pheneticists found Nelson, Platnick, and Farris to be abrasive and contentious, among other epithets. As Farris (1985a) remarked, memory is short. In the early years, pheneticists complained of their treatment at the hands of the editor of *Systematic Zoology,* Libbie Hyman, complaints that were to be repeated a decade later by Nelson and Brundin with respect to Bert Rowell. When the cladists took over the journal, charges of bias only increased. The story repeated itself once again with respect to the young Croizatians. They too were accused of being difficult, and they in turn charged Schnell with unfair treatment. This time, however, the cladists are cast along with the evolutionary systematists and pheneticists in the role of the conservative authorities who use their power to keep their own out-of-date ideas from being openly challenged (Craw 1982, 1983, Craw and Weston 1984, Grehan 1984, Grehan and Ainsworth 1985, Heads 1984, 1985).

Finally, Popper forms the third pillar of Nelson and Platnick's (1981) great synthesis, and once again the charge has been made that these two authors depart in important ways from the views expressed by Popper himself (Ruse 1979c, Panchen 1982). Nelson and Platnick have not simply adopted Popper's views; they have transformed them. In his early writings, Popper (1957, 1959, 1962) passed some extremely harsh judgments on evolutionary theory, chiefly that it is unfalsifiable and, hence, unscientific. Later he reevaluated evolutionary theory as a biological theory when he introduced an evolutionary basis for his own philosophical system (Popper 1972, 1974). According to Popper, Darwinism is best construed as a metaphysical research program. As such it is not falsifiable, but research programs are not supposed to be falsifiable. Popper introduced his principle of falsifiability to distinguish between those process theories which are genuinely scientific and those that pretend to be but are not. Metaphysical research programs are to be evaluated on other grounds. But what of evolutionary theory as a scientific theory? In response to his name being used to reject the synthetic theory of evolution, Popper (1978, 1980) explicitly retracted his early mistaken views (see Ruse 1977).

In spite of the title of his original *Logik der Forschung* (1934) being translated into English as *The Logic of Scientific Discovery* (1959), Popper did not think that there was such a thing as a logic of discovery. Psychologists might have something to say about the psychology of discovery, but philosophers have nothing to contribute to this subject. As far as Popper was concerned, science consists of conjectures, regardless of their source, and serious attempts to refute them. Popper's primary concern in this connection were theories and laws expressed in universal form about the fundamental processes in nature. They had to be universal in form because only with respect to universal statements is there an asymmetry between falsification and verification. Universal laws can be falsified in principle by a single counterinstance, but no finite number of confirming instances can verify them because they apply to indefinitely many particulars in the past, present, and future. Thus, laws and theories about natural processes are central to Popper's philosophy. By excluding process theories from cladistics in the generic sense, Nelson and Platnick have defined cladistics so that it, like Darwinism, is outside the realm of Popperian science.

If Nelson and Platnick are to make use of Popperian philosophy, it must be in a highly modified form, one that refers to theories about pattern, not process. Popper was not especially interested in scientific classification. What little he said on the subject was rather dismissive (Popper 1957: 106; 1958: 65). He really did not think that definitions are all that important. However, he did insist on the theory-ladenness of scientific concepts, and the theories that he had in mind, the ones that he explicitly mentions, are such process theories as Newtonian theory and relativity theory. Thus, anyone who insists that biological classifications to be truly scientific must begin with theory-free observation statements is departing significantly from Popper's philosophy as he himself set it out. Next to falsifiability, the theory-ladenness of observation terms is Popper's most important contribution to the philosophy of science. To the extent that cladistics is concerned with the "justification of particular scientific theories, such as the taxonomic hypothesis that spiders are more closely related to whipspiders than they are to scorpions" (Platnick 1979: 539), then a new philosophy of falsifiability must be formulated, a philosophy that will be at most analogous to Popper's philosophy of process (Hull 1983d).

Platnick (1986) sees this state of affairs quite differently. In reaction to a paper by Hill and Camus (1986) in which they point out discrepancies between pattern cladism and Popper's views, Platnick (1986: 291) responds that Hill and Camus have been misled by Hull (1983d). "Even the most cursory reading of *The Logic of Scientific Discovery* would have shown Hill and Camus that Popper regards all scientific hypotheses, from the broadest generalizations to the most trivial observation reports, as being necessarily subject to his falsifiability criterion; indeed, he argues that the point at which we decide to stop testing hypotheses, and merely accept them as unproblematic reports of observation, is entirely a matter of convenience and convention that may be altered at any time." In this matter, I find myself in agreement with Hill and Camus (1986). Observation reports, trivial or otherwise, are rarely universal in form. Hence, the asymmetry between verifiability and falsifiability does not apply to them. Observation reports are certainly testable but not "falsifiable" in Popper's technical vocabulary. Platnick has transformed Popper. The cladists' philosophy of falsifiability applies to the relation between phylogenetic hypotheses and putative synapomorphies.

As far as I can see, Platnick's claims about the essential identity between his views and those of Popper have the same source as his claims of essential agreement with Hennig—the need for patron saints. No transformations have taken place, just differences in perspective. For his part, Platnick (1986) is not an operationalist as Hill and Camus (1986) claim but only a "humble empiricist." The modifications necessary to make Popper an advocate of empiricism, humble or otherwise, are as great as those necessary to transmute Hennig into an ideal morphologist.

The preceding comments should not be interpreted as my objecting to scientists changing their minds or improving upon the work of past authorities. To the contrary, science would be impossible otherwise. When Nelson and Platnick (1981) synthesized the views of Hennig, Croizat, and Popper, they transformed them. They had to. The likelihood that these three men set out three extremely elaborate systems which are totally consistent is quite low. Nor am I complaining of Nelson and Platnick's attempting to make their views seem more similar to those of their patrons saints than they may actually be. As tactics in science, such maneuvers are frequently quite effective. However, in discussing the ongoing process of science, I am forced to distinguish between descent and similarity, and on my view descent is prior to similarity. Nelson and Platnick are conceptually descended from Hennig, Croizat, and Popper, but in this process considerable transformation has occurred. That cladists such as Platnick make the transformation of characters central to their systematic methodology in biology but refuse to apply it to their own conceptual development, I find a bit puzzling but far from inconsistent. If Collins can be a realist with respect to sociology and a relativist with respect to physics, then certainly Platnick can be a transformist with respect to the characteristics of plants and animals and an essentialist with respect to scientific change. Even the arch-foe of essentialism, Mayr (1983), cannot bring himself to reject essentialism totally when it comes to scientific theories and research programs.

Nelson and Platnick's (1981) treatise was only the second book on cladistics to appear at the time. Wiley (1981a) published a general work as well. One aspect of traditional systematics that first caught Nelson's eye was the huge gap between the information that supposedly goes into a classification and the minimal information that can be gotten out. If classifications really are information storage and retrieval systems, as everyone tirelessly claimed, then the information stored in a classification had better be retrievable. In the paper that he presented at the Boulder Congress in 1973, Nelson (1973e) showed exactly how little could be represented about phylogeny simply by ordering taxa in a particular sequence and by subordinating one taxon to another. If the only modes of representation used by systematists are sequencing and subordination, then very little about phylogeny can be unequivocally represented in hierarchical classifications. If more is to be represented, additional representational devices must be introduced. Hennig (1969) had done just this by devising a numerical code to indicate the position of a taxon in a classification. Patterson and Rosen (1977) followed suit, suggesting that extinct taxa be indicated in the usual way by means of a "dagger" symbol before their names, but it was Wiley (1979) who carried this alternative to its logical conclusion in his Annotated Hierarchical System. In his system Wiley presented a means by which anything that a systematist might think he knew about phylogeny can be represented explicitly and unequivocally in a classification. If a systematist thinks that he can recognize ancestral species as being ancestral, he can represent this relation in Wiley's system.

In the early years, Nelson devoted his time to converting his colleagues, usually senior colleagues. Wiley was the first convert that could actually be considered a student, in fact one of Rosen's students. On first meeting, Wiley was anything but imposing, but one soon realized that behind the Texas twang, the chipmunk grin, and good-old-boy manner was a very sharp mind. Although Wiley began as a student, he was the first cladist to resist publicly the increasing departures from Hennig's original views. He was also the first cladist who found it difficult to get his manuscripts refereed positively by other cladists. In 1981 he published his own synoptic work, entitled *Phylogenetics: The Theory and Practice of Phylogenetic Systematics*. As the title implies, Wiley's version of cladistics retained a close connection to phylogeny. According to Wiley (1981a: 97), a cladogram is a "branching diagram of entities where the branching is based on the inferred historical connections between the entities as evidenced by synapomorphies. That is, a cladogram is a phylogenetic or historical dendrogram." Wiley's book was also a genuine text, including chapters on species and speciation, higher taxa, tree construction, classification, biogeography, curating, quantitative techniques, and rules of nomenclature.

At the time Kluge and Farris were also working on a general text. Thus, in 1982 cladists found themselves in the uncomfortable position of having three "bibles" and the threat of a fourth on the way. To the extent that these works differed from each other, the cladist research program ran the risk of becoming pluralistic. Pluralism is one of those virtues that is easy to proclaim but difficult to practice. Nelson instituted a system of associate editors for *Systematic Zoology* to promote a more pluralistic editorial policy in the journal, and in the early years of this system the associate editors did represent the various factions in the systematic community. However, by the end of Smith's tenure in office, all of the editorial staff were cladists. This is pluralism of sorts. Calls for pluralism tend to be raised in science in the same two circumstances as in politics—by members of newly emerging groups when they are struggling to gain a toehold and by those in power when it looks as if they are about to be deposed. Those firmly lodged in power rarely can see the virtues of pluralism. In science both pluralism and the pressure to eliminate it are equally crucial. Without alternatives to be selected, scientific change cannot occur; but without constant pruning, science becomes a formless clutter. Both extremes have existed at different times in various areas of science.

When pluralism raises its head in different research groups, conflict ensues as each of the competing groups tries to become dominant. Pluralism within groups introduces a different sort of conflict. It is much easier for scientists to cooperate with each other when they are in fundamental agreement. Hence, when certain authors noticed what they took to be an increase in conceptual pluralism among the cladists, the first response on the part of cladists was to deny it. However, as the schism became wider and more evident, some cladists themselves have come to acknowledge its existence. Thus far, I have been concerned with the conceptual side of this issue—agreements and disagreements with respect to the cognitive content of cladistics. Must the terminal taxa in fully resolved cladograms be species in the evolutionary sense or not? Can transformation series be established solely on the basis of character covariation, or must temporal considerations enter in? If the latter, must recourse be made to the evolutionary process or will ontogenetic sequences do? Because those involved in this dispute tend not to present their positions in categorical forms but include occasional demurers, it is difficult to decide exactly how wide the conceptual gulf is between phylogenetic cladists and pattern cladists.

Matters are much clearer with respect to the professional relations among those concerned. The outlines of the research groups that have materialized are much clearer than the boundaries of their associated conceptual systems. It is the continued failure to distinguish consistently between scientific groups individuated in terms of cooperation and conceptual systems defined in terms of agreement about substantive issues that has introduced such deep disagreements over the existence of pattern cladists. Even if we were to accept Platnick's (1985) contention that all cladists produce the same classifications, given the same data, phylogenetic and pattern cladists might still exist as separate, even warring social groups. Conversely, even if Farris (1985a) is right about the intersecting agreements and disagreements both within and between these two groups, they still could form separate groups. It is to this topic that I now turn. (For the most recent discussion of the existence or nonexistence of the distinction between pattern and phylogenetic cladism, see Brooks and Wiley 1985, Platnick 1985, 1986. Ridley 1986, and Mishler 1987.)

Dissension in the Ranks

As cladism became more successful, cladists found it increasingly difficult to maintain their original cohesiveness, both because the number of people who considered themselves "cladists" had expanded considerably and because dissension among the founders of cladism in the United States was increasingly difficult to contain. An instance of the first sort of difficulty is provided by a workshop on the theory and application of cladistic methodology organized by Thomas Duncan and Tod Stuessy at the University of California, Berkeley, March 22–28, 1981. The plan was to have participants send data sets to the organizers in time for them to prepare them to be used on a variety of computer programs in an effort to see which programs are best able to discern any patterns inherent in the data. The mornings would be devoted to speakers expanding on the various general issues in taxonomic theory, while the afternoons would be spent working with the data sets.

Only four of the speakers were "cladists" in the narrow sense—Gareth Nelson, Arnold Kluge, Vicki Funk, and Dan Brooks. Brooks replaced Farris, who was unable to attend at the last minute because of a death in his family. Warren Wagner also presented a paper. Although Wagner was one of the sources for numerical cladistics and Farris named his "Wagner trees" after him, Wagner had never allied himself with the group centered at the American Museum. He did not oppose the cladists, but he also did not join with them in their controversies. This policy of nonalignment had already begun to unravel because of a paper he had just published with Duncan and Raymond Phillips. In this paper, Duncan, Phillips, and Wagner (1980) compared various methods of deriving branching diagrams from a hypothetical data set and concluded that no method was clearly superior to all others for all purposes—including Farris's method. Farris was not amused. None of the other speakers at the conference on cladistic methods could be considered "cladists." In general, the cladists were less than happy with the National Science Foundation for picking two noncladists to organize a major conference on cladistics.

Several speakers at the conference presented extremely sophisticated papers showing the wide variety of formal techniques that existed for clustering organisms into taxa. Each technique had its strengths and weaknesses. None was clearly superior. In response to these papers, Funk remarked, "Doesn't it make you want to run right out and become a systematist?" The cladists to the contrary presented one method and one method only, a method which they claimed was preferable on all counts. To those students present who were struggling to produce a dissertation in a proscribed period of time, the choice appeared obvious.

The afternoon sessions were somewhat less than successful, in part because the data sets that had been contributed by the participants tended to be extremely noisy and in part because not all the instructors were familiar with the programs that were being used. The competition between cladists and more eclectic types was reflected in the message T-shirts worn by the two sides. One read, "I'm a cladomasochistic character. Are you?" To which the eclecticists responded, "Hi, I'm a compatible character. Are you?" When Funk and Brooks (1981a) reviewed the conference for *Systematic Zoology*, they were less than enthusiastic. Too many of the participants lacked "even the basic understanding of the principles of cladistic analysis" and over half the speakers were "not cladists." In the future the National Science Foundation might well choose cladists to organize a conference on cladistics if they wanted to support a conference on cladistics rather than on eclectic taxonomy. (For the proceedings of this conference, see Duncan and Stuessy 1984.)

Those systematists who controlled the Numerical Taxonomy Group had never been very anxious to meet with the Society of Systematic Zoology. In 1980 there was some talk of the numerical taxonomists and cladists meeting together in 1981, but those engaged in these discussions were too suspicious of each other's motives for such a joint meeting to be organized. Nevertheless, Numerical Taxonomy 15 and Hennig 2 were held at the same time at the University of Michigan in Ann Arbor. Even though most of the papers at the numerical taxonomy conference concerned cladistic techniques, very few participants at either meeting crossed over to attend sessions at the other conference. In a joint paper at NT-15, Rohlf, Sokal and Colless continued to detail the problems they were having with Mickevich's data and results, but the highlight of the conference was Sokal's unveiling of Camin's "true cladogram" for his Caminalcules (McNeill 1982).

One of the symposia at Hennig II concerned problems that species of hybrid origin pose for the principles of cladistic analysis. This topic naturally led to questions about possible differences of opinion among cladists over the role of phylogeny and the evolutionary process in cladistics. Should reticulations in phylogenetic trees be represented in cladograms by reticulations or trichotomies? Those cladists who opted for the first alternative were including elements of trees in their cladograms, while those who limited themselves to trichotomies retained the purely branching character of cladograms.

In his presentation, Colin Patterson described the effect that his first exposure to phylogenetics had had on him. It had come as a revelation, like discovering logic for the first time. He was forced to admit to himself that nearly all his earlier work in systematics had been a total waste of time. According to Patterson, much of the confusion that had followed upon the rise of cladistics was caused by the mistaken belief that the branching diagrams used by cladists (cladograms) have something to do with evolution, when they do not. As far as he was concerned, cladistics in its purest form is neutral with respect to evolution. He emphasized that in his own summary statement of the principles of cladistics, he had not needed to use any of the terms that had caused so much consternation and argumentation in the preceding decade, e.g., monophyly, paraphyly, primitive, derived, phylogeny, speciation, dichotomy, ancestry, or adaptation. According to Patterson, all that cladists need to do is to analyze characters in such a way that they are perfectly congruent. The only temporal dimension that needs to be introduced into systematics is ontogenetic. Ontogenetic transformations can be used to determine character polarities because they, unlike phylogenetic transformations, can actually be observed.

Sara Fink (1982: 196) in her report noted that the "tug of war between the pattern viewpoint and the phylogenetic viewpoint in cladistics is leading on the one hand to an attempt to divorce hierarchical pattern from evolution and create a nearly assumption-free methodology, and on the other to new approaches to the study of the transformation of form." She herself concluded that, "when a methodology is divorced from any explanatory theory, the effort becomes purely operational and seems to have no scientific merit" (Fink 1982: 188). When the proceedings of Hennig II appeared (Platnick and Funk 1983), the book did not include Patterson's paper, but in his preface, Kluge (1983: vii) noted the dissension that had appeared in the ranks of the cladists:

> Another area of special interest involves the relationship of cladistics to the general theory of evolution. I see an ever-widening schism developing among cladists in this respect. Some treat their research as a direct extension of evolutionary theory—a test of the hypothesis of descent with modification, the search for natural groups, a reliance on synapomorphies. Others take the position that the relationship is unnecessary and is even to be avoided because it was our preoccupation with evolutionary mechanisms and processes that for so long retarded the de-

velopment of systematics. I wholeheartedly agree that the neo-Darwinian emphasis on genetic mechanisms and adaptation, and especially the dogmatic view that genetic similarity is *the* index to monophyletic groups, contributed little, if anything, to the discovery of patterns in nature. I would even concede that a lot of time was, and still is being, wasted through such misdirected research. However, to deny any connection to evolutionary theory seems to lead to some unacceptable propositions, not the least of which is that contemporary systematics is then not a science but an inductive endeavor in search of a general theory. Also, how would the methods and conclusions of the anti-evolutionists differ from those of the pheneticists? I contend that to entertain the concept of character state polarity requires the assumption of evolution and that the real problem with phenetics is that synapomorphies and symplesiomorphies are not distinguished. I don't pretend to understand all of the ramifications of the schism but I can see that they are far-reaching and deserve further debate.

The schism which Kluge acknowledged was fostered ever so subtly in the banquet address presented by Dan Brooks. Brooks had never planned a career in biology. He had gone to college on an athletic scholarship as a high-jumper. However, when an injury cut his career in athletics short, he began to take his studies more seriously. At the first meeting of the Hennig Society at Lawrence, Kansas, he had just been offered a job in the biology department at the University of British Columbia, in Vancouver, if he could obtain permission to emigrate to Canada. Now, as a rising young academic, he looked very much the well-groomed sort of man that pharmaceutical companies employ to urge doctors to use their products. During his first year as an assistant professor, Brooks had collaborated with Ed Wiley to expand work he had begun on information theory (Brooks 1981) to produce a radically new view of the evolutionary process. He was so enthusiastic about his brainchild that he was sure his fellow cladists would share his enthusiasm. Thus, instead of giving a humorous talk about cladists and their enemies in his banquet address, Brooks sketched this new theory. The flavor of his talk can be gathered from the following summary published the next year:

Evolution may be described as a nonequilibrium process involving the conversion of information from one form to another and the maintenance of old or forging of new reproductive networks. Species participate in nonequilibrium processes because they have properties of closure and because evolution is a historically irreversible phenomenon. Speciation is a process which consolidates available potential information into two or more stored information systems. Character change and the history of a clade are highly correlated because potential information is constrained by its own evolutionary history. Thus new and potential information may be converted into stored information only to the extent that this new and potential information is compatible with the ancestral information system or can find expression in an alternative ontogenetic pathway. All modes of speciation as well as anagenesis follow the second law of thermodynamics and may be described by a summary equation which charts the changes of entropy states of information and cohesion over time. We suggest that some empirical evidence corroborates our theory and that the research programs involved in further studying nonequilibrium evolution are largely in place now. (Wiley and Brooks 1982: 1)

The celebrants were anything but amused. After a cocktail party, too much wine, and a heavy meal, they did not feel up to following differential equations. To make matters worse, some of those in attendance saw Brooks and Wiley's theory as an attempt to tie cladistic analysis to a bizarre theory of evolution. No sooner had cladists freed themselves from the dead weight of the synthetic theory of evolution than Brooks and Wiley proposed to burden them with another speculative process theory. Nelson and

Rosen's early enthusiasm for Croizat's nonstandard theory of biogeography had been viewed by some cladists as being dangerous enough for their research program without Brooks and Wiley tying it to such bizarre ideas about the evolutionary process. All Brooks and Wiley actually claimed was that cladistic analysis is a prerequisite for viewing evolution as an entropic process, not vice versa (the same relation that Nelson, Rosen, and Platnick claimed for vicariance biogeography), but the enemies of cladistic analysis were not likely to pay sufficient attention to such niceties. Criticisms of Brooks and Wiley's theory were likely to reflect back on cladistics as surely as comparable objections to vicariance biogeography had before them. Of course, to the extent that either theory succeeded, this success would lend support to the principles of cladistics.

When Wiley and Brooks (1982) published a sketch of their theory, they entitled it "Victims of History—A Nonequilibrium Approach to Evolution." They chose their title to emphasize the constraints that historical development places on the evolutionary process. Løvtrup (1983) entitled his scathing response "Victims of Ambition." Donn Rosen, Wiley's former professor, characterized the efforts of Brooks and Wiley as the "synthetic theory written in Chinese" (Farris 1985a: 195). In spite of intense discouragement from their fellow cladists, Brooks and Wiley persisted and eventually published a full-blow version of their theory of evolution as entropy (Brooks and Wiley 1986).

Throughout its existence, the Numerical Taxonomy Group was an informal organization, the only "official unofficial officer" of this group being the convener who organized the meetings each year. Although Farris was the convener for Numerical Taxonomy 12, Sokal remained the de facto power behind the annual conferences. Because the organization had no formal procedures or constitution, it could not be taken over by parliamentary subversion. Even though Rosen and Nelson continued to appeal to their fellow cladists not to form an organization of their own, the consensus of those cladists who met at Ann Arbor was that they were tired of arguing the same old points with their opponents. As long as they presented papers at either the meetings of the Numerical Taxonomy Group or the Society of Systematic Zoology, that is what would continue to happen. At Ann Arbor an informal "council" met to discuss the structure of the proposed society. Farris strongly favored a society run by a small cadre of founders. Others preferred a more democratic society. At this meeting, Joel Cracraft was especially vocal in his opposition to Farris. Chris Humphries suggested, half in jest, that they adopt the constitution of the Linnean Society of London, with its system of councillors, fellows, and associates.

Cracraft left this meeting under the impression that he had been charged to organize the next meeting of the Hennig Society in Chicago, while Chris Thompson was supposed to start the paperwork necessary to incorporate the Willi Hennig Society as a nonprofit organization. Finally, in August 1982, Farris, Mickevich, Charles Mitter, and Chris Thompson met as the board of directors of the new society and installed themselves as president, vice president, secretary, and treasurer respectively. At this meeting, they also drew up a list of twenty-three founding fellows. From this list of founding fellows, they selected twelve to run for the council. Given the peculiar structure of the society, future officers had to be drawn from the council and councillors from the fellows. Although others could come to the meetings of the Willi Hennig Society and present papers, they were not voting members of the society. (For a list of the twenty-three founding fellows of the Willi Hennig Society, see Appendix E.)

In January of 1982, James Slater, the president of the Society of Systematic Zoology, began the process of electing a new editor of the society's journal. He asked Virginia Ferris and Charles Michener to join him on a search committee. By March they had come up with two names—Arnold Kluge and Charles Harper, Jr. Kluge was a cladist with respect both to his views and to his alliances. Harper was not a cladist in either sense, but he also was not an enemy, having engaged in an amicable exchange with Platnick over cladistics in 1978 (Harper and Platnick 1978). The council agreed almost unanimously with the choice of nominees. Then things began to unravel. Because the recession had hit Michigan especially hard, funds at the University of Michigan were cut back sharply. If Kluge were to become editor of *Systematic Zoology,* he would have to have financial support from the society. When Kluge was informed that the society was not in a position to supply any financial assistance, he withdrew his name from consideration. Soon thereafter Harper started having second thoughts about the amount of work facing him and withdrew his name from nomination as well. Suddenly Slater and the nominating committee found themselves right back where they had started six months earlier. At that time, one of Slater's former students was encountering problems in gaining tenure. In a telephone conversation with this former student over her difficulties, Slater asked her for suggestions of possible editors for the journal. The former student volunteered that she herself might be a good candidate. After more telephone calls to Olson as past president of the society and to the other two members of the nominating committee, Slater forwarded five names to the councillors as possible nominees for editor. The only name that failed to be approved by the council was that of Slater's former student—by one vote.

In the meantime, Slater heard rumors that some council members thought that he had acted improperly in suggesting the name of his former student as a candidate for editor. None too surprisingly, Slater was upset. By this time the nominating committee thought that it had done as much as it could, and the process was turned over to the council. The initial list of candidates produced by the council included five names—two cladists, two pheneticists, and Slater's former student. The two cladists were Farris and Platnick. They agreed that if both made it through the initial balloting, one would withdraw so as not to split the cladist vote. As it turned out, only Platnick received enough votes to make it to the final ballot. Both pheneticists, Gary Schnell and Wayne Moss, also made it onto the final ballot, and they had no prior agreement about one of them dropping out so as not to split the pheneticist vote. Slater's former student again failed to receive enough votes to become an official nominee for editor, but after a very bitter fight she did receive tenure in her department.

Because the council was so strongly divided on the candidates for editor, the highly unusual decision was made to poll the entire membership. It was their journal; let them elect the next editor. As the votes began to trickle into the office of the secretary of the society, Roger Cressey, rumors circulated about the results. At first, all three candidates were doing equally well, then Moss began to fall behind. Finally, when the official closing date arrived, Cressey was dismayed to discover that Schnell and Platnick were tied. The closing date was extended two days, until December 3, 1982. When the election was finally brought to a close, Schnell had won by seven votes, and *Systematic Zoology* was moving from Los Angeles back to the heartland, ruby slippers or no. Of greater significance, the new editor was once again a pheneticist.[2]

When Cracraft began to make plans for the next meeting of the Hennig Society, he discovered that he had been mistaken in thinking that he was to organize it. Instead Mitter was organizing Hennig III, which was to be held at the University of Maryland in November 1982. By the time that the third annual meeting of the Willi Hennig Society was convened, Beatty (1982) had published his paper in which he argued that pattern cladists were mistaken in thinking that the principles of cladistic analysis are entirely independent of evolutionary theory. Beatty claimed to the contrary that they are incompatible with it. The strain between pattern and phylogenetic cladists flared up in the hallway over coffee when Nelson found himself on one side of a circle of participants facing Arnold Kluge and Bill Fink on the other side. In response to Cracraft's espousal of pattern cladism, Kluge asked "How do you distinguish cladism from pheneticism?" Morris Goodman joined in by asking, "Why call us cladists if we give up evolution?" Nelson could not see what all the fuss was about.

A second confrontation occurred during one of the sessions when Wiley took to the stage to challenge Platnick's claim that no important differences of opinion divided the cladists. Several years earlier, Wiley (1977) and Platnick (1977a, 1977b) had disagreed over the appropriate way to apply the notion of monophyly, a concept central to phylogenetic systematics. Wiley now attempted to show that he differed with Nelson and Platnick over the number of trees compatible with a particular cladogram, again an issue that was hardly peripheral to cladistics. First Nelson and then Platnick rose from the audience to object to Wiley's contention that cladists might well produce different cladograms depending on the views they had on the individuation of species. As Platnick and Wiley argued, Nelson fiddled with his glasses until they snapped in his hands. By then the moderator, Vicki Funk, decided that the argument had gone on long enough and allowed Wiley one more comment. When he started on a second, she pulled the plug on the microphone. Wiley relinquished the stage but gave Funk a good-natured, though no doubt sexist, slap on the bottom as he did.

At the meeting of the Founding Fellows, the four officers and sixteen of the twenty-three fellows appointed by the officers attended. Funk showed up with a list of additional fellows only to discover that the constitution required that such lists be accompanied by seconds arranged in advance. The only list of nominees that fulfilled all the requirements was the list presented by Farris. It contained what Funk took to be several noteworthy omissions. Wiley moved that the rules be suspended so that both lists of nominees could be admitted. After a few heated exchanges, the motion was defeated. Additional arguments ensued. When Farris told Funk that, under the constitution, there was nothing that she could do, she replied, "You're wrong, Steve. I can leave." And she did. Eventually, those present voted to postpone the selection of additional members

2. Although Slater remembers that all three candidates agreed to have the closing day for the election extended, Platnick does not remember being asked for his permission. He discovered that the closing date had been extended when he called Cressey to find out about the results of the election.

until the next meeting, when everyone would have an equal chance to join in the nomination process.[3]

The Founding Fellows also elected eight of their own to serve on the council—Dan Brooks, Virginia Ferris, Bill Fink, Vicki Funk, Chris Humphries, Arnold Kluge, Norm Platnick, and Ed Wiley. After the meeting of the fellows, the council convened. The chief topic was future meetings. They decided to have a meeting to nominate additional fellows in August 1983 in conjunction with the meetings of the American Institute of Biological Sciences in Grand Forks, North Dakota. Hennig IV, however, would not take place until July 1984 in London, England. The fellows elected at Grand Forks would be formally inducted into the society in London.

In his banquet address at Maryland, Chris Humphries restored the good humor of the meeting by presenting the sort of personal, in-group talk that promotes intragroup cohesiveness. At the first meeting of the Hennig Society, Funk had retrieved the broken pointer which Farris had discarded in a wastebasket and in the interim had it mounted on a wooden plaque. Humphries presented the broken pointer award to Farris in recognition of the dubious achievement of his having broken it in the first place. After making a gracious acceptance speech, Farris immediately passed on the award to Brooks for his banquet address of the preceding year.

Soon after the third annual meeting of the Willi Hennig Society, the issue of *Systematic Zoology* appeared with responses to Beatty (1982) by Platnick (1982), Patterson (1982), and Brady (1982). These authors insisted that Beatty was wrong about the principles of so-called pattern cladism being at odds with evolutionary thinking, because the principles of cladistic analysis are totally independent of any process theories, including evolutionary theory. Platnick in particular rejected the priority which both Ball (1981) and Beatty (1982) claimed for causal theories in the construction of classifications. Perhaps causal theories need classifications, but classifications do not need causal theories. In the very next number of the journal, Farris addressed this same issue. In his paper, Farris (1982: 413) attributed to advocates of evolutionary systematics the view that the "descriptive and explanatory roles of the biological reference system should be kept separate, and that description and explanation impose conflicting goals on classification." Farris (1982: 415) also objected to the pheneticist view that classifications empty of theoretical content are not only possible but also desirable:

> Of course pheneticists seldom deny that their methods classify the products of evolution. Their position is instead that classifying products ought to be free of suppositions about the process. But process and product can scarcely be entirely independent. Sneath and Sokal did not—and could not—succeed in devising classificatory methods free of the effects of evolutionary phenomena. In attempting to ignore phylogenetic considerations, they merely left the relationship of process to product unrecognized, and so un-understood.

3. The relevant section of the minutes of the founding fellows of the Willi Hennig Society, 20 November 1982, read as follows:

> There was much discussion, and no little disagreement, on the question of the election of additional fellows. All agreed that there were many persons who should be added to the list of Fellows, but only one nomination had been received by the deadline and in the format prescribed by the Society's by-laws. Various emergency procedures for circumventing the latter and permitting the immediate election of new Fellows were raised and then dismissed. At length a motion was unanimously approved delaying the election of any new Fellows until the next meeting.

One would think that the relation between evolutionary processes and the character patterns that result from them is sufficiently straightforward that continued confusion over the relation would be easy to avoid. Such did not prove to be the case. In any event, the social realignments among the cladists did not accord perfectly with the conceptual distinction between pattern and phylogenetic cladists.

In his report of Hennig III, P. F. Stevens (1983: 287) took note of the increasing heterogeneity in cladistics.

> A number of papers were presented, bearing on more philosophical issues, dealing especially with what so-called "transformed" or "pattern" cladistics might or might not be. Here the smokescreens produced by the protagonists proved almost impenetrable, and the situation was complicated because nobody in the audience seemed to belong to what is already apparently an endangered species—or perhaps it has yet to evolve.

Did pattern cladists exist, or were they figments of overheated imaginations? Stevens also noted that Brady's (1982) argument to the effect that pattern recognition should be carried out independently of covering process theories, because pattern recognition is both historically and epistemologically prior to process theories, was reminiscent of the arguments presented thirty years earlier by Hennig's ideal morphologist opponents. The "wheel has turned full circle." Stevens (1983: 291) concluded his report by reiterating a point made earlier by Hardy Eshbaugh in connection with the deliberations of the systematics panel of the National Science Foundation: "The most damaging reviews are made by cladists of other cladistically oriented work, rather than by cladists of non-cladistic work; factions are attempting to kill off opposing factions in the process of peer review."

Normally the Society of Systematic Zoology met with the American Society of Zoologists in late December, but in 1983 it met with the American Institute of Biological Sciences in August in order for zoologists to have some interaction with systematic botanists. Unlike the ASZ, the AIBS includes botanical societies as members. Not only were flights to Grand Forks International Airport expensive and difficult to book, but also the weather was intolerably hot. In spite of the isolation, heat, and ubiquitous flies, the local hosts did everything that they could to make the participants comfortable. At least the lecture rooms at the University where the sessions were held, were cool. The motel at which the meeting of the Hennig Society was held was constructed to look like a Western ranch house, replete with wagon-wheel fences and a swimming pool built in the shape of a cowboy boot. This time those cladists in attendance were presented with a superabundance of candidates for fellows, and they elected an even fifty. There were two noteworthy omissions from the list—Donn Rosen and Joel Cracraft. Rosen had explicitly asked that his name not be proposed. He was still strongly opposed to the formation of the Hennig Society in competition to the Society of Systematic Zoology. Cracraft had not been on the list of Founding Fellows appointed by the officers of the Society. Nor was his name on this second list.

The council meeting of the Society of Systematic Zoology at Grand Forks was largely taken up with an examination of the draft of a new constitution prepared by a committee chaired by Walter Bock. After making a few changes, the council voted unanimously to adopt the new constitution and to send it out to the membership for ratification once it had been checked for any possible inconsistencies. Just as the council meeting was about to adjourn, Farris moved that Nelson be elected an honorary member of the society. In 1977, P. J. Darlington had been added to the list of original honorary members. In that same year, Hennig's name reappeared on the list, this time with a "dagger" in front of it indicating that he was deceased—the same symbol that Patterson and Rosen (1977) had suggested for extinct taxa. In 1979 Lars Brundin was also elected to this distinguished group. When Farris stood to nominate Nelson to be an honorary member, his fellow cladists were taken aback. Farris had given them no warning. Although Nelson was not present at the meeting, no one was anxious to speak against his nomination. Yet several had reservations. In the midst of the confusion, Wiley suggested that the vote be by secret ballot. Farris's motion failed by a vote of four in favor, eight opposed. When Farris polled the council members after the meeting to see how each had voted, five claimed to have voted in favor of Farris's motion.

By this time the first issues of *Systematic Zoology* edited by Gary Schnell had begun to appear. They included a response to Mickevich by Rohlf, Colless, and Hart (1983), a paper on butterfly nomenclature by Ehrlich and Murphy (1983) which Smith had previously rejected, and most noticeably four papers by Sokal on Caminalcules (Sokal 1983a, 1983b, 1983c, 1983d). The cladists were irate. The referees' reports which Smith had obtained for a single paper by Sokal on Caminalcules had not been very enthusiastic, and here was Schnell publishing a four-part paper on the topic running to a total of seventy-one pages. Although papers by cladists continued to appear in the journal, its complexion clearly changed after Schnell took over as editor. The ratio of pro-phenetic to anti-cladistic papers changed dramatically (for comparative figures, see chapter 8).

Part of the reason for this change was that a backlog of phenetic papers had piled up while cladists edited the journal, in part because of rejection, in part because of the perception that numerical/phenetic manuscripts had little chance of appearing at the time in *Systematic Zoology.* A second reason for the shift of emphasis in the pages of the journal was that for two years the cladists published a proceedings of the Hennig Society, draining off quite a few papers. The first two proceedings of the Willi Hennig Society (Funk and Brooks 1981b, Platnick and Funk 1983) contained twenty-seven papers that might have appeared in *Systematic Zoology.* Because the sales of these proceedings were not quite as good as the editors of the Columbia University Press had anticipated, they were not enthusiastic about a third volume. The officers of the Willi Hennig Society decided that it was time for the society to publish its own journal instead of issuing yearly proceedings. When the fellows of the society were polled by mail and telephone, most were favorably inclined.

In preparation for Hennig IV to be held in London, Farris called an extraordinary meeting of the officers and councillors of the Hennig Society in Washington, D.C. The letter went out June 1, 1984, and the meeting was called for June 15, 1984. In order to conduct such a meeting, only a quorum of five voting members was required, and precisely five officers and councillors were able to make the meeting—Farris, Platnick, Funk, Thompson, and Mitter. The vote to establish a new journal to be entitled *Cladistics: The International Journal of the Willi Hennig Society* passed unanimously. Platnick and Humphries were appointed editors. Although Funk's slate of officers, headed by Wiley as president, was late, she was allowed to present it. The others at the meeting promptly voted it down and opted to retain the original officers with only one change, the substitution of Arnold Kluge for Charles Mitter as secretary of the society. It was also time for four of the council members to be replaced. Virginia Ferris had already sent in her resignation, and Platnick as one of the editors of the new journal had to be replaced as well. Over Funk's protests, William Fink and Ed Wiley were added to the list of council members who were to be rotated off the council. Funk also complained once again of the exclusion of Cracraft from the society. He had been one of the earliest advocates of Hennig's system and had been extremely active in the founding of the society.

Funk left this meeting determined that no longer was the Hennig Society going to be the "Steve and Mary Show." When Funk informed several of the councillors of the society who had not been able to attend the meeting of the actions taken there, they were disconcerted. Brooks, Fink, Funk, and Wiley sent an open letter to the other fellows objecting to the short notice that had been given for the June meeting and the actions that had taken place there. Attached were proposed amendments to the bylaws of the society, especially the provision that new officers must come from the ranks of the councillors. Shortly thereafter, Mitter as secretary of the society sent a letter to the fellows detailing the actions taken at the meeting held in Washington and listing topics for discussion and possible action at the London meeting. Enclosed was a ballot for the retirement of the four councillors and their replacements, the nominations for officers, along with a list of twenty-nine candidates for associate membership in the society. The stage was set for the fourth meeting of the Willi Hennig Society, to be held in London at the British Museum (Natural History) and sponsored jointly by the Systematics Association and the Linnean Society. Funk's protestations had at least one effect. Cracraft was retroactively elected a founding fellow of the society.

A Recipe for Revolution

Nelson first introduced phylogenetics to English-speaking systematists while he was visiting the British Museum (Natural History). The early history of phylogenetic systematics at the British Museum was quite different from events at the American Museum of Natural History in New York, largely due to differences in the personalities and status of its advocates. Although Patterson was also in the early years of his career, he too took up the Hennigian cause. However, he did not confront his colleagues but gradually insinuated the new system into the intellectual community of his museum. Nelson's style was open confrontation. Even so, the relative peace at the British Museum was not to last for long. It was broken by the unlikely event of the introduction of new exhibits in the public rooms of the museum and resulted in charges of a Marxist takeover of this venerable institution.

Shortly before his death, Hans Sloane (1660–1753) willed his huge collection of plants, animals, rocks, minerals, fossils, scientific instruments, and library to king and country. It was eventually installed at Montague House in London, where it continued to increase in size and variety. Eventually, in 1845, this greatly expanded collection was transferred to the British Museum building in Bloomsbury, where it joined the collection of art and antiquities. No sooner had the natural history collection been installed in the British Museum than Richard Owen began to press his friend Gladstone for a separate building. In 1856 Owen became superintendent of the natural history collections, and in 1862 Gladstone attempted, without success, to raise money from Parliament for a new building. Not until 1881 was the huge edifice designed by Alfred Waterhouse (1830–1905) completed in South Kensington and Owen installed as its first director. It had been financed largely by means of a public lottery. The original building in Bloomsbury with its collection of art and antiquities continued to be known as the British Museum, while the new building in South Kensington was given the derivative and inelegant name, the British Museum (Natural History).

For a century after the natural history museum opened, generations of parents took their children to see the dinosaurs that populated the great vaulted halls and the schools of ichthyosaur fossils that lined the grey stone walls. It was not until 1963 that the two museums became administratively independent. Then in 1972 a report was presented to the trustees of the natural history museum claiming that the museum had ceased to fulfill its educational goals. The Victorian cabinets full of shells and insects were too static and old-fashioned. More modern exhibits were needed if new generations of students were to be attracted to the museum. In 1973 the Director of the museum put an expert on fossil fishes, Roger Miles, in charge of renovating the museum's exhibits. The result was the replacement of the Victorian legacies with bright plastic exhibits that could be made to light up by pushing buttons. Immobile dinosaurs were replaced with dynamic dioramas. Perhaps the modern exhibits would appear a bit out of place in Waterhouse's great pile of terra-cotta, but that could not be helped.

Just as universities attempt to fulfill the twin functions of teaching and research, museums also serve two functions—as repositories of great research collections to be used by professional scientists and as places for the edification and entertainment of the general public. Not infrequently the two functions conflict. Several members of the professional staff objected to the plans to modernize the public rooms and said so. Colin Patterson was one of the few members of the professional staff who were willing to cooperate in the production of the new exhibits. He saw these new exhibits as an opportunity to impress upon the general public that the results of scientific investigation are provisional; they are not infallible truth. As might be expected, he found cladograms to be the ideal vehicle for illustrating the relationships between species of organisms.

A couple of short pieces on the "sinister" goings on at the natural history museum appeared in *Nature,* but it was a paleontologist from Reading University, Beverly Halstead, who brought the issue out into the open. In a letter to *Nature,* Halstead (1978) reported the results of a conference held at Reading University on cladism and paleontology. The confrontation between cladists and traditional paleontologists produced the usual results. When the paleontologists present were told that their cherished ancestor-descendant relationships were unknowable and that sister-group relationships were prior anyway, they became irate. In his report, Halstead (1978: 760) opted for the principles of Simpson and Mayr and complained of the "religious fervour" of the cladists. What kind of changes would the likes of the cladists make in the venerable museum?

After a brief exchange between Halstead and several defenders of cladistics, including Patterson and Miles, the controversy subsided. Two years later, when Halstead saw the two new exhibits that had been completed, the dinosaur and fossil-man exhibits, the controversy was reopened. As Halstead (1980: 208) perceived these exhibits, they were sheer propaganda for cladism and punctuational theories of evolution. But of greater importance, the British Museum (Natural History) was playing into the hands of creationists and Marxists. "According to the stated assumptions of cladistics none of the fossil species can be ancestral by definition." As Halstead feared, the booklet that accompanied the fossil-man exhibit declared that *Homo erectus* was not our direct ancestor. "This presents the public for the first time with the notion that there are no actual fossils directly antecedent to man. What the creationists have insisted on for years is now being openly advertised by the Natural History Museum."

Nor were the creationists the only group subversive to science who were being aided by the new exhibits. According to Halstead (1980: 208), the key tenet of dialectical materialism is that "qualitative changes occur not gradually but rapidly and abruptly, taking the form of a leap from one state to another." Halstead declared, "This is the recipe for revolution." If the punctuational model favored by such cladists as Eldredge were accepted over more gradualistic models, then a "fundamentally Marxist view of the history of life will have been incorporated into a key element of the educational system of this country. Marxism will be able to call upon the scientific laws of history in its support, with a confidence that it has previously enjoyed." (As a young man, Halstead had himself been an enthusiastic Marxist, chairing the Student Communist Society while an undergraduate at Sheffield University.)

Patterson responded that Halstead's perceived connection between cladism and saltatory evolution was an illusion for the simple reason that there is no connection between cladism and any particular hypotheses about the evolutionary process, punctuational or otherwise, no more than there is a connection between Venn diagrams and the evolutionary process. Some cladists advocated saltatory forms of evolution; some did not. Patterson (1980: 430) concluded. "Whether they are Marxists is another matter, in my view an irrelevant one." The other respondents also found Halstead's connection between Marxism, punctuational speciation, and cladism "silly," smacking of "academic vigilantism." Others complained that he had Engels all wrong. Dialectical materialism does not propose mutational leaps but the conversion of quantitative change into qualitative change.

Halstead (1981: 106) responded that in his complaints about cladism he was not referring to Patterson's "transformed cladism" but to the sort of phylogenetic cladism portrayed in the public galleries of the natural history museum. As far as Marxism was concerned, Eldredge and Gould were the ones who had pointed out the connection between Marxism and punctuational modes of speciation (Eldredge and Gould 1972, Gould and Eldredge 1977). Gould (1981: 742) was finally roused to object to the tight connection that Halstead saw between Marxism, cladism, and punctuational models of evolution. Gould was not a cladist, and Eldredge was not a Marxist.

As the controversy continued, the emphasis shifted from politics to epistemology. In response to Halstead's complaints about the refusal of cladists to recognize ancestors, Malcolm McKenna (1981: 626) of the American Museum wrote in to congratulate his colleagues at the British Museum for "bringing epistemology into its exhibits and teaching the visitors that science is a method, not a body of revealed knowledge." But the editors of *Nature* (1981a: 735) saw things somewhat differently as they quoted from the brochure put out by the British Museum (Natural History):

> Biologists try to reconstruct the course of evolution from the characteristics of living animals and plants from fossils, which give a time scale to the story. If the theory of evolution is true, the features used to classify species in groups . . . were acquired by the common ancestor of the group inherited by the living descendants.

The phrase to which the editors of *Nature* (1981a: 735) took particular exception was "If the theory of evolution is true." They interpreted it as implying that there was some real doubt about the truth of the theory of evolution:

> If the words are to be taken seriously, the rot at the museum has gone further than Halstead ever thought. Can it be that the managers of the museum which is the nearest thing to a citadel of Darwinism have lost their nerve, not to mention their good sense? . . . Nobody disputes that, in the public presentation of science, it is proper whenever appropriate to say that disputed matters are in doubt. But is the theory of evolution still an open question among serious biologists? And, if not, what purpose except general confusion can be served by these weasel words?

As the author of the objectionable passage in the 1978 museum guide to the exhibits, Patterson (1981: 82) replied that Darwin himself used similar locutions in his own work. Twenty-two other members of the staff of the museum replied somewhat more hotly in a joint communiqué. Would *Nature* have them present a theory as fact? "This is the stuff of prejudice, not science, and as scientists our basic concern is to keep an open mind on the unknowable. . . . Are we to take it that evolution is a fact, proven to the limits of scientific rigour? If that is the inference then we must disagree most strongly. We have no absolute proof of the theory of evolution. What we do have is overwhelming circumstantial evidence in favour of it and as yet no better alternative. But the theory of evolution would be abandoned tomorrow if a better theory appeared" (Ball et al. 1981: 82).

Several readers wrote in to chide these worthy scientists for exaggerating the speed at which science changes. Abandoning the theory of evolution "tomorrow" was a bit unrealistic. Wiley (1981b: 720) insisted that one must distinguish between evolution having occurred somehow or other and the precise mechanism by which it occurs. Wiley found much less legitimate doubt about the former than the latter. Evolution surely has occurred; precisely how is another question. Nelson (1981) joined in to defend the way in which cladists treat fossils. In the midst of this discussion about what scientists can and cannot "know," the authority of Sir Karl Popper once again was raised. "Popper dismisses evolutionary theory as being unfalsifiable, and he should know." "Yes he should, and no he didn't, not really."

When Simpson (1981: 286) wrote to warn the readers of *Nature* that McKenna and Nelson "belong to a small group" at the American Museum of Natural History in New York "which is representative neither of the staff of the museum as a whole nor of the majority of American paleontologists," twenty-two members of the staff of the American Museum signed a letter claiming otherwise. Of the thirty-two professional zoological systematists at the museum, twenty-six used cladistic methods in their systematic work.

When Patterson explained himself more extensively, his stance with respect to phylogeny and the evolutionary process did not appear quite so radical. As a heuristic device, Patterson thought that it might prove to be productive for biologists to view the world a fresh, as if the patterns apparent in nature were not due to evolution. Eventually, they might return to their former views, but the experience would be rewarding. At least, Patterson had found the exercise worthwhile in his own case. At least one junior member at Patterson's institution, Christopher Hill (1981: 540), found statements such as these coming from the most prestigious paleontologist at the British Museum more than he could take and wrote a letter to *Nature* objecting to Patterson's position. Perhaps cladism as Patterson construed it was not about evolution, but the cladism which he practiced very definitely was:

> Basically I consider evolution and the fossil record as very much the proper context for cladistics and not the converse. It is surely commonsense to use historical evidence to examine history and not to view it askance through a haze of global aristotelian philosophy of character analysis, as transformed cladists appear now to be doing. Transformed cladists may justly have rediscovered Aristotle, but if they decide to wipe out Galileo they reject not only Darwin but all of modern science.

Several defenders of traditional religion also had their say. Three creationists from the University of Glasgow complained of their being "banished to the lunatic fringe" and of the close-mindedness of atheists for refusing even to countenance divine activity. McBride (1981: 90) saw no major conflict between science and religion. They merely provide "different languages to describe the same phenomenon." Baker (1981: 623) pleaded for pluralism. "Is there no room for Darwin, Hennig, God—and even Marx?" The answer was no. MacKie (1981: 403)) objected to "excessive tolerance" in the "failure to distinguish between scientific thinking and irrational myth-making," and Jukes (1981: 186) refused to accede to the precedent that threatened to become established in the United States of "equal time for nonsense."

After allowing Halstead one final chance to reply to his critics, the editors of *Nature* declared the controversy closed. No more letters would be published on plastic push-button exhibits, connections between cladism and Marxism, or the scientific status of the theory of evolution, lest the readers of the journal "become so bored that they will take to buying some other journal instead" (Anonymous 1981b: 1). During a two and one-half year period, *Nature* had published sixty of the hundreds of letters that it had received as well as six articles on the death of Darwinism in South Kensington. What had begun as an intramural squabble at the British Museum (Natural History) over some new exhibits had blown up into an unseemly public display. In an ironic conclusion to this dispute, Sir Andrew Huxley (1981) in his anniversary address as president of the Royal Society of London compared the controversy to the famous debate between his great-grandfather and Bishop Wilberforce (for a good summary of this sequence of events from a cladist's point of view, see Schafersman 1985).

By the time that cladists began assembling at the natural history museum in August 1984 for the meeting of Hennig IV, the new exhibits had already began to become appropriately dusty. Those attending the meeting paid no attention to the coveys of schoolchildren clustered around the dioramas, mimeographed sheets in hand, trying to find the right cladograms for the organisms depicted. Although the schoolchildren seemed to have no idea what they were doing, they soon discovered that if they punched enough buttons, they would eventually come up with the right answer. Two visiting academics did stop to try their hand at finding the right cladograms. However, the instructions proved to be so confusing that they were no more successful than the children in selecting the correct cladogram, and they returned to the more rarefied atmosphere of the lecture hall where such details could be safely ignored.

Attendance at Hennig IV was greatly expanded by numerous representatives from Britain, Sweden, and other European countries. The Americans could be distinguished from the other participants by their informal attire and even more informal behavior. They were used to the Hennig Society meetings being local, small-group affairs. The blunt discussions following some of the papers and the casual disregard for formalities were disconcerting to some of the Europeans who were unfamiliar with past meetings of the society.

On the evening that the council was to meet, Patterson hosted a party at his home for the participants. The councillors would join the party after they had reaffirmed the actions taken at the meetings in Grand Forks and Washington, but such proved not to be the case. The wrangling was intensified by two fifths of Jack Daniels that were passed around as the meeting dragged on. Those at the party chatted nervously, wondering what in the world was taking the councillors so long. Finally, late in the evening, they appeared, exhausted. They had voted six to three to nullify the actions taken at the Washington meeting and to start from scratch, but by the time all was said and done, the results were not materially different from those that had been rejected earlier. However, Farris agreed to set up a committee to study changes in the bylaws of the society. Several councillors had objected to the brochure announcing the appearance of *Cladistics*. The list which it included of the three axioms of cladistics sounded very much like a statement of the principles of pattern cladism:

1. Features shared by organisms (homologies) manifest a hierarchical pattern in nature.
2. This hierarchical pattern is economically expressed in branching diagrams, or cladograms.
3. The nodes in cladograms symbolize the homologies shared by the organisms grouped by the node, so that the cladogram is synonymous with a classification.

So many people signed up for the banquet that it had to be moved from one large room to two rooms in the basement of a restaurant. Thus, only half the people in attendance were able to hear Farris's banquet address, and even they had a difficult time because the public address system was inadequate. The fact that the room was decorated as an Italian garden replete with grape arbors separating the booths did not help matters. Farris used the occasion to chronicle the injuries that cladists had received through the years at the hands of their enemies. Although Farris modified his address before publishing it in *Cladistics,* he did not totally expunge its personal flavor. According to Farris (1985a: 191–92), evolutionary taxonomists adopted political tactics in their disputes with the pheneticists. As an example he cites Mayr's circulating the "rumor that Sokal's degree was in mathematics, not biology. Mayr and Brown, among others, became fond of taking note in private of Sokal's abrasive and contentious personality. Pheneticists capitalized on all of this by whining about it, playing the role of an oppressed and righteous minority. They appear to have regarded the attempts at suppression as proof of the validity of their position."

Later, when the pheneticists began to cooperate with evolutionary systematists to oppose cladistics, both Gould and Mayr commented on the "disagreeable qualities of cladists." Farris (1985a: 192) continued:

> Soon it was Nelson who was reputed to be abrasive and contentious. When he suggested in publication that Darwin's ideas might lead to phylogenetic systematics rather than evolutionary taxonomy, Mayr awarded him the additional property of "impudence." Sokal also proved eager to denigrate cladists (as well as cladistics). Interestingly, at about this time Sokal lost his earlier traits and became "dignified" and "gentlemanly," according to evolutionary systematists. When I questioned the descriptive superiority of phenetic and evolutionary systematics (Farris, 1977), I earned from Sokal (and other noncladists as well) the titles vicious, argumentative, underhanded, and insane (!). Mickevich irritated Sokal and Rohlf particularly, not just because of her papers, but also because she finished a degree at Stony Brook despite their attempts to prevent this by fabricating new (and quite original) procedural rules and requirements. She was provided with the epithets unprofessional and immoral.

Farris continued by supplying details on the accusation that Nelson had refused to publish papers on phenetics when he was editor of *Systematic Zoology,* on the controversy over the Janowitz paper, on Michener's complaints to the council of the Society of Systematic Zoology, and so on. Included in this list was a condemnation of Brooks and Wiley's entropic version of evolutionary theory.

Past meetings were concluded by the presentation of the Broken Pointer Award for dubious achievement. There was some talk about awarding it to Chris Humphries for suggesting that the Hennig Society adopt the constitution of the Linnean Society of London. Instead the pointer was retired, and Brooks returned the plaque to Farris, noting that he thought it inappropriate for scientists to honor the institutionalization of violence. When Humphries had originally made his suggestion that the Willi Hennig Society adopt the constitution of the Linnean Society, no one thought to read up on the effects that this constitution had had on the Linnean Society itself. If they had, they might have reconsidered. As it was, the ghost of the Linnean Society hovered over the London meetings of the Hennig Society. Not only had the complexities of the constitution promoted discord among the founding fellows as they attempted to plan the future course of their society, but also some of the new fellows took offense at having to swear allegiance to the principles of cladistics when they were inducted into the society, an oath mandated by the constitution.

James Edward Smith (1759–1828) had included in the constitution of the Linnean Society the requirement that each new fellow swear allegiance to the principles of the society. As one of Smith's contemporaries remembers a meeting of the Linnean Society which he attended in 1829:

> The President wore a three-cornered hat of ample dimension, and sat in a crimson arm-chair in great state. I saw a number of new Fellows admitted. They were marched one by one to the President, who rose, and taking them by the hand, admitted them. The process costs £25. (Allen 1976: 173)

Although Farris might have overlooked this vestigial practice, he did not. The swearing-in ceremony for the new fellows of the Willi Hennig Society was not nearly so ostentatious. It took place in a large, bare room. Farris wore no hat but sported his usual suspenders. He administered the oath to each of the new fellows and admitted each with a firm handshake. The cost was only five dollars. All too appropriately, one of the social functions at the fourth meeting of the Willi Hennig Society was even hosted by the venerable Linnean Society in its wood-paneled rooms—wine, canapés, and conversation under the stern portraits of long-dead systematists.

Chris Humphries and Josephine Camus (1986: 99) concluded their summary of the papers presented at Hennig IV with the observation that, despite the "rather circus-like atmosphere, cladistics was alive and well; methodological progress in data coding, transformation analysis, computer algorithms, and consensus techniques was very apparent. On the black side, certain clashes showed no sign of resolution. The attacks against, and the arguments for, 'transformed' cladistics, the stubbornness of the compatibility versus parsimony argument, and the modelling versus empirical methods debate all showed signs of wearisome persistence." On the bright side again, "With the publication of its journal *Cladistics,* and the prospect of exciting meetings in Miami (1985) and Stockholm (1987), the Society is almost beginning to look like the Establishment." (The Stockholm meeting was delayed to 1988.)

After the London meeting, William Fink wrote to Sadie Coats, whom Farris had chosen to chair the committee which was to look into revising the constitution of the society, with numerous suggestions, including abandoning the oath and changing the name of the journal to "reflect the interests (phylogenetic systematics) of the bulk of the membership of the Society." For their part, both Brooks and Wiley resigned their positions on the advisory board of the journal because of what they took to be Farris's ad hominem attacks in the published version of his banquet address. At the next meeting of the council in October 1985, three of the original officers were replaced. Once again, only one valid list of nominations had been received. Kary Bremer became president, Hans-Erik Wanntorp became vice-president, and Diana Lipscomb secretary. Thompson stayed on as treasurer. Platnick moved that the minutes of the council meeting in London not be accepted because that meeting had not been properly constituted and had acted improperly in suspending the bylaws. The motion passed unanimously. As a result, the decisions made at the Washington meeting were reinstated. Subsequently Funk resigned from the council. Thus, of the original eight councillors, four departed the council under less than pleasant circumstances.

As far as the Society of Systematic Zoology is concerned, Nelson took office as president-elect in 1984. In 1985 his candidate for editor of *Systematic Zoology,* Robert Schipp, was elected to start his duties at the end of 1986, when Schnell's tenure in office would be complete. In 1986 Ed Wiley became president-elect of the society, and Nelson moved up to president. Nelson's first act as president was to appoint Wiley to rework the constitution of the society.

Conclusion

In this chapter I have consciously distinguished between the cooperation that occurs within socially defined research groups and the sort of agreement that characterizes the members of such groups. In the early years both pheneticists and cladists formed tightly-knit groups, but they differed in one important respect: cladists were much more intent than pheneticists on maintaining that they were all in fundamental agreement over basics. Sokal and Sneath had two agendas, one epistemological and the other methodological. They wanted to make the science of classification as free as possible from theoretical input during the initial stages in the construction of classifications, and they wanted to make it as quantitative as possible. However, they welcomed anyone who was mathematically inclined, even if they did not adhere to their preferred epistemological strictures. Sokal and Sneath were also quite liberal with respect to which quantitative methods might be used and to the areas investigated. Their movement began in biological classification, but it soon branched out to include all sorts of classifications, from diseases to classifications in the social sciences.

Initially, cladists were much more intent on developing a single methodology for a narrowly defined area of science. Early converts thought they were joining a program to develop methods for inferring phylogenetic relationships in biological evolution and their representation in classifications. They were developing a single method for a single purpose. In addition, Nelson and Farris expected allegiance. Nelson was willing to spend long hours explaining his views, but eventually, if the newcomer was unable to join in the cause, Nelson lost interest. Nelson was vehement in his polemics against those who openly opposed cladistic analysis. He was prone simply to dismiss those who were unable to go all the way with him. Indifference was not part of Farris's psychological makeup. In short, cladists insisted on a greater degree of orthodoxy than did the pheneticists.

Early in the development of cladistics, most cladists thought they were engaged in Hennigian phylogenetics. However, Nelson had a more ambitious, hidden agenda. The methods that he was advocating could be extended to any pattern that results from descent with modification. As Nelson gradually revealed his more ambitious goals for cladistics, he also began to be more open about his anti-process-theory stance. The only theories involved in cladistic analysis were theories about patterns of character distributions, not genealogy. Like the pheneticists, Nelson promoted a strongly empiricist epistemology. In the order of knowledge, there was nothing more to genealogy than patterns of character distribution. However, as the pheneticists became less insistent on their epistemological program, Nelson became more insistent on his.

Both pheneticists and cladists tried to assert dominion over all of systematics, if not all of science, by means of the same sort of terminological maneuver. Pheneticists argued that all systematists are at bottom engaged in phenetics because they all begin their studies with phenetic characters. They might adulterate their classifications later by letting theoretical speculations intrude, but no one could deny the central role of characters in classification. Hence, all systematists, if not all scientists, are actually pheneticists in their practice. The equivocation that allowed the pheneticists to make their argument seem plausible is between "character" and "phenetic character." If characters are nothing but theory-free descriptions of the entities being studied, then the two are equivalent, but if theory-free descriptions are impossible, then the two are far from equivalent. It was not for nothing that pheneticists tried to show that intelligent ignoramuses recognize the same patterns as trained systematists.

Cladists argued, to the contrary, that all systematists are at bottom engaged in cladistic analysis because the patterns that they perceive in their data sets can be represented by cladograms. Perhaps not all systematists are as consistent as cladists in constructing their branching diagrams, but if cladistics is the science of branching diagrams, then all systematists are really engaged in cladistic analysis whether they realize it or not. Here the equivocation is over the term "cladogram." In the narrow sense, cladograms represent order of branching in phylogeny; "cladograms" in the broader sense refers to any branching diagram whatsoever.

The pheneticists and cladists were not alone in attempting to claim the entire field of systematics for themselves. Mayr (1981) proposed another "synthesis," this time among all the disparate views being urged at the time. Initially, systematics might well engage in activities not all that different from those proposed by pheneticists, but these activities are only alpha taxonomy, crude first approximations. Next, systematists must find the hierarchical structure buried in phenetic clusters. Here the methods of cladists come into play. The mistake that cladists make is to think that systematists should stop here. Instead, the final step is the application of the principles of evolutionary systematics to produce mature, robust classifications that reflect all aspects of the evolutionary process, not to mention phylogeny. Anything short of this ultimate goal is incomplete.

Civil war never broke out among the pheneticists, not because pheneticists found themselves in such massive agreement but because their social organization became increasingly diffuse. They did not remain cohesive enough, either socially or conceptually, to speciate. Cladists, to the contrary, remained socially very tightly-knit. As their views became more widely accepted, they fought off attempts to take over their research program from the outside. Perhaps the organizers of the Berkeley conference considered themselves "cladists," but the original cladists did not. Thus, as cladists began to distinguish between cladistics in Nelson's most general sense and traditional phylogenetics, differences of opinion about the relative value of these two enterprises produced increasing tension in the group. Because there was a group, the group was able to split in two.

The split that occurred in the ranks of the cladists was only partially a function of disagreements over the goals of systematics. Interpersonal relations also influenced the alliances that formed. In this connection, Farris was much more active than Nelson. Nelson was much more concerned with issues than with professional politicking. He needed only one or, at most, two collaborators. When he entered into the workings of the various societies, he did so directly and personally. If he did not like certain provisions of a new constitution, he contacted the relevant official and "urged" that these provisions be changed. He rarely attempted to anticipate events before they occurred so that he could marshall support for his own preferences. Farris enjoyed professional politicking. Even if he could be assured of having his way openly and directly, he still preferred behind-the-scene maneuvering. When he was frustrated in his attempts to take over the Numerical Taxonomy Conference and the Society of Systematic Zoology, he formed his own society and established himself as president. At first, others were more than happy to let him do all the work necessary for such a venture. If he was willing to do most of the work, others were willing to let him make most of the decisions.

As the conceptual split between pattern and phylogenetic cladism began to emerge, Farris was put in the position of having to choose. Although he was conceptually closer to the phylogenetic cladists than to the pattern cladists, he eventually joined forces with Nelson and Platnick. Among his reasons for siding with one side over the other were long-standing animosities toward the second-generation cladists who were emerging as the leaders of the conservative faction. If the choice was between Nelson and Platnick on the one hand, and Brooks and Wiley on the other, there was no contest. Joel Cracraft, for one, was caught in the middle. He found himself conceptually on the side of Nelson and Platnick, but he was consistently rebuffed as he tried to participate in the activities of the Hennig Society.

The pheneticists branched out rapidly and extensively. The cladists were more cautious and selective. When Nelson proposed to combine Hennigian principles of classification with Croizatian principles of biogeography, several of his allies were less than enthusiastic. The increasing importance of Popperian philosophy of science received a more enthusiastic response, Colin Patterson excepted. When Brooks and Wiley proposed to do the same for their entropic version of evolutionary theory, they caused considerable consternation among their fellow cladists. J. D. Smith was one of their few allies. Cladists were wise to scrutinize potential conceptual linkages. Success would reflect on cladistics favorably; failure would count against it. Conceptual connections with both Croizatian biogeography and thermodynamics were high-risk ventures. Both Croizat and the association of biological evolution with thermodynamics had very bad reputations.

If one gauges success in terms of influence, regardless of explicit recognition, both the Kansas group of pheneticists and the American Museum cladists have been very successful. Kansas pheneticists were extremely influential in making biological systematics more quantitative. They have also had some influence outside biological systematics. Social scientists assume that numerical taxonomy is the received view in the biological community. Biologists themselves think differently. The value of cladistic techniques is widely acknowledged, although many, possibly most, systematists maintain reservations about particular tenets of cladistic analysis. Cladistics also has its numerical side, and here members of both schools have contributed. However, as far as explicit recognition is concerned, the credit typically goes to the cladists. We hear of the contributions of individual numerical taxonomists but not of the contributions of *the* numerical taxonomists. In this respect, the cladists are the clear winners. When credit is assigned, it is to cladists as cladists.

Although widely based inductive inferences are dangerous enough, inductions on the basis of two or three examples are even more questionable. Even so, one of the goals of this book is to see if any lessons can be learned from the study of two or three small groups of scientists. Looking back, it would seem that Sokal made two mistakes. He did not insist strongly enough on conceptual orthodoxy and branched out into too many other areas before his power base in biology was secure. Cladists made neither of these mistakes. Given the power of hindsight, the major mistake that Farris made was to hold on too tightly to the running of the Hennig Society. Conceptual orthodoxy and social cohesion are beneficial, just so long as they are not overdone. The trouble is no one knows how much conceptual orthodoxy and social cohesion are too much, or how little is too little. To the extent that scientists are disinclined to acknowledge the relevance of either consideration to science, we are unlikely ever to have even rough estimates of the effects that these factors have on science as a process. It is to topics such as these that I now turn. In the preceding chapters, I have chronicled particular sequences of episodes in science, highlighting certain aspects of these events. How are they to be explained? Are explanations of the scientific process even possible, or is science inexplicable?

8

The Need for a Mechanism

As many more individuals of each species are born than can possibly survive; and as, consequently, there is a frequently recurring struggle for existence, it follows that any being, if it vary however slightly in any manner profitable to itself, under the complex and sometimes varying conditions of life, will have a better chance of surviving, and thus be *naturally selected.* From the strong principle of inheritance, any selected variety will tend to propagate its new and modified form.

Charles Darwin, *On the Origin of Species*

Scientific ideas compete in an open marketplace. Each offers the possibility of a plausible solution to what might be a potentially significant problem. In its promise, an idea will attract other scientists—fellow explorers who will articulate, criticize, and ultimately determine the idea's actuality. While these explorers can breathe life into an idea, their absence or defection leads to its death. Ideas without recruits become like Bishop Berkeley's proverbial unheard falling tree.

David L. Krantz and Lynda Wiggins, "Personal and Impersonal Channels of Recruitment in the Growth of Theory," *Human Development*

IN MY earlier discussions of evolutionary biology and systematics, I have not attempted to hide my partiality for those scientists who provide mechanisms for the processes under investigation, no matter how mistaken the mechanisms turn out to be. As far as I can tell, this partiality has been widespread, though far from universal, among scientists throughout the history of science. The chief exceptions can be found among certain French and German idealists in the eighteenth and nineteenth centuries, critics of genes as material bodies at the turn of the century, as well as pheneticists and pattern cladists today. The importance that most scientists place on mechanisms can be seen in their reactions to various phenomena through the years prior to any suggestions about possible mechanisms. The interlocking rings of five circles which the Quinarians found in the living world were intriguing, but in the absence of even the hint of a mechanism capable of producing such pentamorphous patterns, most scientists remained skeptical.

The rough correspondence between the east coast of the New World and the west coast of the Old might have resulted from the two land masses having once been joined, but how in the world could they have drifted apart? Later, two events made the idea acceptable: new data about the earth's magnetic field and the theory of plate tectonics. Although the mechanism suggested to explain the drifting of the plates in which the continents are embedded is not much of a mechanism, it at least holds out the promise that some purely mechanical explanation exists. No sooner did Raup and Sepkoski (1984) report the existence of 26-million-year cycles of mass extinctions in the fossil record than physicists began to suggest possible mechanisms to produce these cycles, from the existence of an unknown planet outside the orbit of Pluto to Nemesis, the Death Star (Raup 1986).

Theories that explain phenomena are necessary to raise them above the level of curiosities, but theories also need data if they are to be taken seriously. Not only do scientists have to present mechanisms, but these mechanisms have to be up to the task and presented in scientifically reputable ways. The transmutation of species is a good case in point. Scientists periodically toyed with the idea of species evolving. On occasion some even suggested mechanisms for evolutionary change, e.g., Lamarck, Chambers, Darwin, and Wallace. Without a mechanism, few scientists were willing to take the idea of evolution seriously. But more than simply stating a mechanism, the author was expected to follow out the implications of his proposed mechanism and present evidence to show that the mechanism was actually operative. None of Darwin and Wallace's "precursors" fulfilled these requirements.

Owen and Spencer are sometimes listed as early evolutionists. Both did advocate "evolution" of sorts, but their respective slogans did not count as mechanisms. In response to Owen's (1860: 500) claims for priority because of his reference to the "ordained continuous becoming of organic forms," Huxley (1863: 106) remarked sarcastically that "it is obvious that it is the first duty of a hypothesis to be intelligible," and Owen's formula "may be read backwards, or forwards, or sideways, with exactly the same amount of signification." Although Huxley was much gentler with his friend Spencer, Spencer's (1857) general law of the "transformation of the homogeneous into the heterogeneous" was hardly better. Perhaps Victorian scientists were too "positivistic," but few were impressed by such vague allusions. Even though both Owen and Spencer claimed to have anticipated Darwin and Wallace with respect to the transmutation of species, neither man generated any interest in his hypothesis.

Chambers (1844) provided somewhat greater detail in his discussion of his embryological mechanism for evolution. According to Chambers, God built into the first living creatures a master program for all subsequent forms of life—what Chambers termed a "higher generative law." Normal reproduction is periodically interrupted, and a more advanced species is produced. That species, however, was in some sense already contained in its ancestor. When his early work was hit by a barrage of objections, Chambers (1845) did try to meet them, but he lacked the professional knowledge to do so with much success. With a couple of noteworthy exceptions, professional scientists were not very impressed with the mechanism that Chambers suggested for evolution (Egerton 1970, Yeo 1984).

Lamarck (1809) discussed his mechanisms for the transmutation of species somewhat more fully. According to Lamarck, two forces are involved: an internal, hydraulic process that leads to an upward surge to complexity and the inheritance of acquired characteristics. The former gave evolution a direction, while the latter explained the particular paths that various sorts of organisms actually took. Few of Lamarck's contemporaries doubted that acquired characteristics are inherited, but they were far from convinced that this mechanism was capable of causing ancestor-descendant sequences of organisms to progress from one species to another. Although a belief in progress was almost pandemic at the time, few of Lamarck's contemporaries put much stock in the mechanism that Lamarck suggested to give direction to evolution, and Lamarck said almost nothing about the precise way in which these fluids were supposed to operate. Instead, he (1809: 212) contented himself with repeated references to subtle fluids that are in a "constant state of agitation and expansion, from which they derive the faculty of distending the parts in which they are insinuated, of rarefying the special fluids of the living bodies that they penetrate, and of communicating to the soft parts of these same bodies, an erethism or special tension which they retain so long as their condition is favourable to it." According to our standards today, such vague references hardly count as mechanisms at all. They did not sound so foreign to Lamarck's contemporaries, but they also led nowhere. Which phenomena should we expect to find, according to Lamarck's theory, which not? How did Lamarck plan to take care of problem cases? To these questions, Lamarck gave only the sketchiest of answers.

In retrospect we are surprised that the Darwin-Wallace papers read at the Linnean Society and published in its journal did not cause much of a stir. But Thomas Bell (1792–1880) was right when he observed in his presidential address that nothing much had happened in the Linnean Society that year. Certainly the year had not "been marked by any of those striking discoveries which at once revolutionized, so to speak, the department of science on which they bear" (Gage 1938: 56). How could Bell have been so blind? Had not Darwin and Wallace announced their revolutionary theory of organic evolution? Our amazement is a function of hindsight. If all Darwin and Wallace had done was to publish their Linnean papers, it is very unlikely that biology would have been revolutionized. These papers were mere sketches, very good sketches, but hardly more complete than the suggestions made by Lamarck and Chambers.

The *Origin of Species* deserves all the emphasis it receives because in it Darwin converted a promising sketch into a scientific theory. Darwin treated natural selection seriously. He scanned the wide range of phenomena that his theory had to explain and showed which cases it could handle without any difficulty, which were doubtful cases, and which anomalies. For example, if species evolve from common ancestors, one should expect to discover the characteristics of present-day organisms falling quite naturally into groups within groups. If species evolve primarily by natural selection, then one should expect to find organisms largely but not perfectly adapted to their stations in life. One of the most convincing events in the development of a scientific theory is the conversion of an apparent anomaly into a confirming instance. For example, Darwin (1859: 236) discussed at great length "one special difficulty, which at first appeared to me insuperable, and actually fatal to my whole theory. I allude to neuters or sterile females in insect-communities." If these neuters did not breed, how did they develop such complex adaptations? In order for other scientists to take Darwin's theory seriously, he had to show how his theory could account for this class of phenomena. His solution was to explain the adaptations of neuter insects in terms of selection at the level of the family.

Popperians claim that in order for something to count as a genuine scientific theory, the advocates of this theory must specify conditions under which they would abandon it. Although Darwin did not go this far, he did list four sorts of phenomena that would pose difficulties for his theory if they were discovered (Darwin 1859: 189, 199, 201, 56):

> If it could be demonstrated that any complex organ existed, which could not possibly have been formed by numerous, successive, slight modifications, *my theory would absolutely break down*. . . .
>
> [Opponents of the utilitarian doctrine] believe that very many structures have been created for beauty in the eyes of man, or for mere variety. This doctrine, if true, would be *absolutely fatal to my theory.*
>
> If it could be proved that any part of the structure of any one species had been formed for the exclusive good of another species, it would *annihilate my theory,* for such could not have been produced through natural selection.
>
> It is not that all large genera are now varying much, and are thus increasing in the number of their species, or that no small genera are now varying and increasing; for if this had been so, it would have been *fatal to my theory;* inasmuch as geology plainly tells us that small genera have in the lapse of time often increased greatly in size; and that large genera have often come to their maxima, declined, and disappeared. (Italics added.)

Each of the cases that Darwin mentions have bears on a different part of his theory—adaptations in neuter insects on his strong principle of inheritance and its implications for individual selection, the impossibility of forming a complex organ by small steps on his preference for gradual evolution, structures created to appear beautiful to human beings on the naturalistic character of his theory, an adaptation in one species for the exclusive good of another species on natural selection, and the fate of small versus large genera on his principle of divergence.

Of course, Darwin is exaggerating. Scientists do on occasion specify conditions under which they would abandon even their most cherished beliefs, but such claims must be taken with a grain of salt. Even if the phenomena that Darwin lists above had been discovered, he would hardly have given up his theory. He would have finagled a bit, the way he did when Galton showed that transfusing blood from one rabbit to another did not influence heredity. The way that early Mendelians did when certain transmission ratios did not fit the simple Mendelian hypothesis. The way that advocates of the classic theory in population genetics did when gel electrophoresis revealed much more genetic heterogeneity in natural populations than their theory implied should be present. The way that gene selectionists and organism selectionists did when the results of certain experiments could not be accounted for by their parsimonious theories. The way that pheneticists did when it appeared that on the basis of actual data sets cladistic methods were superior to phenetic methods on their own criteria.

As Darwin himself remarked, he was a master "wriggler." Any scientist who is incapable of wriggling a bit will never succeed in science, but there are also limits to the wriggling. If it becomes too pervasive, the scientist ceases to be a "scientist." There are more ways than one for science to degenerate into nonscience. Nearly all the views that we tend to dismiss as not being "scientific" because they are not "falsifiable" are actually quite falsifiable. The problem is that their proponents are not interested. Scientists need not abandon their most fundamental views in the face of a single apparent counterinstance, but they cannot totally ignore data either. Although the Popperian specification of conditions under which the advocates of a theory would be willing to abandon it is to a large degree a public relations ploy, there is a point for anyone developing a scientific theory to attempt to anticipate which phenomena might cause them the greatest difficulties. If a theory handles a wide range of phenomena and only a few anomalies crop up, a scientist can afford to set these counterinstances to the side for the moment. Perhaps the data are in error or possibly a slight modification of the theory can account for them. However, if through time enough phenomena turn out to be sufficiently recalcitrant, he or she might well be led to abandon the theory altogether.

Anyone proposing a theory about the scientific process itself must treat it as seriously as Darwin and other scientists treat their theories. Vague gestures were not good enough in the nineteenth century; they are certainly not good enough today. In the second half of this book, I set out a mechanism that accounts for the general structure of the scientific process. I also specify which sorts of phenomena are expected, given the theory, which count against it, and which remain as mere curiosities, counting neither for or against it (Laudan 1977). The mechanism that I propose rests fundamentally on the relations which exist in science between credit, use, support, and mutual testing. Science functions the way that it does because of its social organization. It is not enough to specify the social norms that characterize it. Why do scientists adhere to these norms? This mechanism is an instance of a selection process, but it is social, not biological.

The desire for recognition by one's fellows is very close to a cultural universal among human beings. Harré (1979: 3) notes that, "though the means by which reputation is to be achieved are extraordinarily various . . . , the pursuit of reputation in the eyes of others is the overriding preoccupation of human life." Once basic subsistence needs are met, people strive for recognition as for nothing else save sex. One would think that something as pervasive as social status would be positively correlated with genetic inclusive fitness. The offspring of tribal chiefs should have more offspring than others in the group. In many societies, this expectation is met. However, in more affluent, developed societies, the correlation is frequently inverted. No matter how one determines rank, the higher the rank the fewer the offspring. The desire for status can become maladaptive. In any case, in this book I take it for granted. In particular, I do not attempt to explain why scientists want credit for their contributions. Instead, I discuss which sorts of credit are considered most important as well as the effects that this striving for credit has on science.

From before Darwin, the possibility of extending a strictly biological theory of evolution to include social and cultural change has seemed promising, but until recently this possibility has remained just that, largely promissory. With the rise of sociobiology, a series of workers have tried to do more than make vague gestures; they have presented detailed theories (e.g., Alexander 1979, Cavalli-Sforza and Feldman 1981, Lumsden and Wilson 1982, Richerson and Boyd 1985). Some of the models presented by these authors are strictly biological in the sense that the relevant transmission is genetic. Organisms are not just genetically predisposed to learn something or other but to hold certain beliefs or to behave in certain ways. Conceptual predispositions are as "programmed" into the genes as are more pedestrian phenotypic characteristics. These theories are "evolutionary" in the strongest sense possible. Thus, the decrease of fertility that seems to be contingent upon the rise of status in present-day societies poses a problem for them—exactly how serious a problem is debatable (Vining 1986). That sociobiologists do not immediately abandon their research program in the face of apparent anomalies is but one more instance of a general principle about scientists. However, other models presented by the preceding authors involve "dual inheritance," i.e., both genetic and cultural transmission.

In some of the works that come under the general rubric "sociobiology," transmission is not biological at all but cultural. The relevant entities are not genes but "memes" (Dawkins 1976). Hence, for these theories, the inverse relation between status and reproductive fecundity is irrelevant. The parallel problem for cultural evolution would be the discovery that striving after social status is counterproductive; for instance, the more that a scientist attempted to gain credit, the less he or she got. According to some theories of cultural transmission, biological evolution is taken to be fundamental, with cultural evolution being only analogous to it. In this book I do not treat meme-based conceptual change as being *analogous* to gene-based biological change. No system has any priority over any other. Biological, social, and conceptual changes are all equally instances of the same sort of process (Toulmin 1972).

In this book I am concerned with only one special sort of status: conceptual inclusive fitness in science—the recognition that scientists receive from other scientists. Being a successful scientist is as likely to enhance survival and reproduction as being successful in any other human endeavor, but, as with other sorts of status, the two can conflict as scientists become so wrapped up in their work that either they do not have children or else they shirk their responsibilities as parents. Although I know of no data on this score, I would not be at all surprised to discover that scientists in a society have fewer offspring than the population at large, or even that the more productive a scientist is, the fewer children he or she might have.[1]

1. If the biographies and autobiographies of great scientists are an accurate indication, scientists exhibit the same range of family life as other people, from Darwin's idyllic Victorian household to the chaos and disharmony of Fisher's brood. Near the end of the interviews that I conducted, I asked "How important is your work to your life as a whole—on a scale of one to ten, one being just a job and ten being a compulsion?" With one exception, all answered nine or ten. Roe (1952: 58) came up with the same results. A frequent comment of the scientists whom she interviewed was, "I have no recreation. My work is my life." Many of my subjects volunteered that their families were just as important to them as their research and that they had been very fortunate in both respects. Simpson remarked that, after the misery of his first marriage, his years with Anne Roe had been very happy even though in recent years he and his wife had been forced to take turns occupying a bed at the local hospital.

The sort of status with which I am concerned is a special sort of "credit" that scientists confer on each other in their roles as scientists. Scientists can be "successful" in many different ways. They can become media figures, they can have their views officially endorsed by political leaders, they can be put in charge of large research institutes. Although these sorts of status are not irrelevant to science, they are not the sort that makes it function as it does. The form of "status" that matters in science is the credit that accrues when one scientist makes use of another scientist's work, in particular, its incorporation in his or her own research, preferably with an explicit citation. Increasing one's conceptual inclusive fitness in science means increasing the number of replicates of one's contributions in the work of successive generations of other scientists.

Even if a significant correlation did exist between genetic and conceptual inclusive fitness among scientists, I do not think that very many features of science are going to be explicable on strictly biological grounds. Certainly both the content of particular scientific theories and our understanding of the proper way to go about doing science have changed too rapidly for changes in gene frequencies to have played much of a role in either. According to evolutionary epistemologists, the only legitimate justification for our knowledge of the empirical world can be found in our evolutionary history. Certain authors are pleased when scientists and ordinary people recognize the same groupings. For example, Mayr has noted that the local inhabitants of a region frequently recognized the same nondimenional species of birds as he, a trained ornithologist, did. Apparently, this concordance implies something about the "reality" of species. I fail to see the connection. Sometimes common sense and scientific conceptions coincide. Sometimes they do not. It does not matter.

On closer inspection, the concordance between species of plants recognized by professional systematists and by the local inhabitants of a region varies from 95.7 percent to 52.1 percent (Brown 1985). Species are hardly less real because of such cultural variation. Some of our conceptions may have a genetic basis. To the extent that they coincide with the needs of the best scientific theories we have at the moment, we should be pleased. Our students are liable to have easy access to these theories. They should "make sense." However, to the extent that our genetically influenced modes of conceptualization are inadequate or mistaken, we will be forced to struggle generation after generation to overcome these misconceptions. Anyone who has attempted to explain biological evolution knows the scope of this difficulty. Species do not *seem* to be genetically heterogeneous. All character distributions within a species

---

In general, the older scientists seemed to have very good marriages of the sort considered traditional at the time; i.e., while the husband devoted himself to his career, the wife performed all the duties of a wife and mother as well as most of the duties of the man of the family. The wives seemed to take considerable pride in their husband's careers and the indirect contributions that they were making to them. The husbands in turn expressed considerable admiration for their wives' accomplishments. The marriages of the next generation of scientists were markedly different. Few wives were willing to take on the double duty of their predecessors. To the contrary, husbands were expected to share in the duties at home. Divorce was quite common, followed by marriage to a fellow scientist. Roe (1952) found differences in divorce rates of physicists, biologists, and social scientists. They were respectively 5 percent, 15 percent, and 41 percent. (For additional information, see Mitroff, Jacob, and Moore 1977).

must be unimodal. Grouping plants into trees, bushes, and plants (or herbs) may seem natural, but it is not. Time and again, scientists have shown that certain ways of viewing nature, no matter how fundamental and widespread they may be, are inherently mistaken. It just might be that we are forever condemned to begin the knowledge game with fundamentally the same set of skewed perceptions and partially inappropriate conceptions willed to us by our evolutionary past, but this supposition should be cause for dismay, not rejoicing. It means that generation after generation, we will have to overcome our biological predispositions if we are to approach an adequate understanding of the world in which we live (for a more sympathetic treatment of common conceptions, see Atran 1985).

Our evolutionary past may bias not only what but also how we think. In certain limited situations human beings are predisposed to reason in appropriate ways. Given our limited knowledge, the shortness of time that we have to make our decisions, and our limited ability to process data, some of the rules of thumb which we use are good enough often enough. Some of these techniques can even be applied with some profit in science, but they are far from good enough. The sorts of reasoning that go on in science are not in the abstract extremely sophisticated, but they are far removed from anything that might be considered our genetic predispositions on how to reason. If anything, scientists must guard against our apparently innate tendencies to reason in certain ways, e.g., to be influenced most strongly by our most recent experience, to discount exceptions, and so on (see Nisbett and Ross 1980).

Science *is* a cognitive process. Scientists must have the cognitive abilities necessary for them to do what they do, but strictly biological considerations do not take us far enough in understanding conceptual change of the sort that goes on in science. Of all the cognitive abilities latent in our species, each person can realize only a small fraction. The notion of "cognitive resource" is much more relevant to our understanding than "innate cognitive abilities." If they are to succeed, scientists must use the cognitive resources which they have available to them, either as individuals or as groups of individuals (for further discussion, see Giere 1987). For example, members of Morgan's Fly Room were especially fortunate in their mix of talents. To some extent, the competition that arose in those tight quarters led to a differentiation of function (Cock 1983). Dobzhansky first pooled his conceptual resources with Sturtevant, then Wright. Sokal collaborated first with Michener and then with Sneath to initiate numerical taxonomy. The combining of individual knowledge and abilities was repeated in the case of Farris and Kluge. But none of this pooling involved genes. In short, evolutionary theory as a gene-based biological theory is not good enough to explain very much about either the content or the conduct of science (Hahlweg 1986).

In chapter 11, I present a general analysis of selection processes which I take to be applicable equally to biological, social, and conceptual change. Such an endeavor is certainly ambitious. Providing an analysis of selection processes in any one of these areas is difficult enough without attempting to provide one that applies equally to all three. Sociologists of science are first and foremost sociologists, and sociology has not proven to be among the most successful sciences, in the sense that sociologists have not been able to discover much in the way of regularities that characterize social systems as such. Attempts to do so have tended to elide either into characterizations that are so abstract that they imply nothing in particular about any social system or into descriptions of how a particular system happens to be functioning at the moment. Advocates of general systems theory tend to err on the side of excessive generality as, for instance, when they inform us that systems are composed of interrelated parts. Economists not infrequently find themselves at the other extreme, explaining how the economic system in the United States functioned between 1932 and 1942 instead of explaining how economic systems in general function. The economic systems in the United States and Great Britain used to be Keynesian; now they are not. Comparable objections might well be lodged against Freudian theory.

I intend my account of how science functions to be general: whenever the conditions I specify obtain, the result should be the growth of scientific knowledge. However, I do not think that it is so general that data cannot be brought to bear on it. In fact, some data remain recalcitrant. Hence, according to the standards of the naive falsificationist, my views are not only falsifiable but also falsified. Nor am I committed to the view that the elements I specify are unique to science. Quite obviously they are not. The sort of mutual citation that goes on in science also characterizes other academic disciplines, including biblical scholarship. The sort of curiosity and innovation so important in science can also be found in the creative arts. Composers, painters, sculptors, and other kinds of artists have more in common with scientists than with any other group. Appropriately, the term "scientist" was coined by analogy with "artist" (Whewell 1834). However, I do claim that the combination of the elements which I specify is unique to science. To some extent this claim is definitional: to the extent that these elements are present, an activity counts as "scientific"; otherwise not.

Producing general analyses of concepts is one of the traditional activities of philosophers of science. Providing a general analysis of selection processes is not all that different from providing a general analysis of functional systems or scientific explanations, two subjects on which philosophers of science have written at some length. In this chapter, however, I also do something that is usually not considered appropriate for a philosopher. I present a mechanism which, I argue, is adequate to explain a great deal about the way in which science works, a mechanism which flows naturally from viewing science as a selection process. Whenever the conditions for the operation of this mechanism are met, the result will be science as we know it. The mechanism that I describe does not explain everything about science. Certain preconditions must be met, and several features of science remain as curiosities, due possibly to other factors. If presenting mechanisms to explain why science operates the way that it does is not a proper activity for a philosopher, then so be it. I stand convicted. But if science is to be viewed as an evolutionary process, the specification of a mechanism up to the task is necessary, and this mechanism must be spelled out in sufficient detail.

What Needs Explaining

The episodes in science that I have traced in the first half of this book exemplify a variety of aspects of science that cry out for explanation. Most strikingly, the interrelations between scientists are not as harmonious and balanced as one might expect given the public face of science. I have been concerned primarily with systematists, evolutionary biologists, and geneticists. No one is surprised to discover members of opposing schools have a go at each other, but the periodic disharmony within individual research groups may not be all that expected. The members of none of the successive "triumvirates" that produced the Synthetic Theory got along all that well, the relations between Morgan's young workers in his fly room were not always that harmonious, nor did the cladists always treat each other with kindliness and consideration. No matter the area of science that anyone studies, the results are the same. For instance, Mitroff (1974: 585) in his study of the Apollo moon rocket project discovered the "often fierce, sometimes bitter, competitive races for discovery and the intense emotions which permeate the doing of science."

Science is both a highly competitive *and* a highly cooperative affair. The need to cooperate with one's competitors adds a pervasive tension in the professional relations among scientists. The role of informal alliances among scientists is also obvious. Although scientists frequently generate new ideas in relative isolation, they make greater use of their fellow scientists in developing and testing these ideas, while dissemination is inherently a social process. Although science in the beginning functioned in the absence of informal groups of scientists working together to develop a particular cluster of views, such groups soon emerged and have characterized science ever since. More than this, those scientists who participate in such groups tend to be the most productive (Blau 1978, Andrews 1979). Scientific research groups in the narrowest sense are most productive when composed of three to five participants. The informal groups that scientists form must be doing some good or else there would not be so many of them. Scientists sometimes agree with each other and sometimes disagree. Both need explaining (Laudan 1984). They also cooperate and compete. This too needs explaining.

In biological evolution, kinship relations are very important. An organism can pass on its genes directly in reproduction or indirectly through its siblings and other close kin. In some cases, more than kin selection is involved. Structured kin-groups form. Species of sexual organisms also tend to break down into semi-isolated demes, resulting in increased gene exchange among members of the same deme and decreased transmission outside the deme. Science began primarily in an "asexual mode." Early scientists worked largely on their own, building almost exclusively on the work of past scientists. There were too few scientists working on the same problems in sufficiently close proximity for them to cooperate to any great extent; but, as the numbers of scientists increased, they began to "clump," both geographically and conceptually, forming highly structured research groups and more diffuse but still largely closed demes. One effect of the formation of research groups is that scientists behave differently toward other scientists depending on whether they are members of their own research group or a competing research group. Mutual positive citations within a research group serve to promote social cohesion within that group. Negative citations of one's opponents also promote cohesion within the group as well as demarcate it from other groups. In biological evolution, competition tends to promote diversification. Groups of organisms that are attempting to exploit the same niche at the same time can enhance their chances of survival by utilizing different elements in their environment. As Edge and Mulkay (1976: 239) discovered in their study of the emergence of radio astronomy in Britain, scientists behave in much the same way. Two research groups that found themselves in too close competition reduced this competition by differentiating their research interests.

The preceding may strike some as little more than a facile analogy, but my intention is not to reason analogously from biological to social and conceptual evolution, but to identify general features of selection processes as such. The effects of small-group size on rates of change should be the same regardless of whether these small groups consist of organisms exchanging their genetic material or scientists exchanging the results of their research. Although the explanation which I offer for the functioning of science is not especially sociobiological, it does take as one of its cues the type of explanation that evolutionary biologists present for the evolution of apparently altruistic behavior. If the disinterested search for truth is anything, it is altruistic. It costs the agent massive amounts of time, labor, and—prior to the rise of public financial support for science—money, and it affords benefits to anyone who can use the knowledge produced. Giving credit where credit is due appears to be no less altruistic. When evolutionary biologists are confronted by behavior that is phenotypically altruistic (i.e., it requires the expenditure of energy or puts the organism in danger and the apparent benefactor is another organism), the first thing that they do is to look for the genetically selfish ends that this behavior serves. In biological evolution, energy flow should by and large follow gene flow. I propose a parallel explanation for the apparently altruistic behavior of scientists, but in the case of science energy is expended to promote the transmission of a scientist's contributions to his or her discipline. Scientists should behave in ways to increase their conceptual inclusive fitness. In conceptual evolution, energy flow should follow the flow of ideas.

The polemics so characteristic of science arise both because scientists care about the knowledge they generate and because they want credit for having generated it. Scientists are prepared for their ideas, especially those that are significant departures from received wisdom, to meet considerable resistance. The chief weapon of the scientific community to new ideas is a conspiracy of silence. For example, when Colless (1967b) began to attack the principles of evolutionary systematics, Mayr consciously decided that the best way to deal with the man was to ignore him. Certainly Simpson and Mayr chose not to dignify Croizat's outlandish views by referring to them in print. Advocates of vicariance biogeography are not rushing to publicize the complaints of their Croizatian critics.

Needless to say, highly innovative scientists become extremely frustrated when confronted with such indifference, whether real or feigned. They cannot see why other scientists refuse to drop what they are doing and pay attention to *them*, to *their* ideas, to the truth. Darwin discovered upon his return from his voyage that his fellow scientists were very busy. They were not strongly predisposed to interrupt their own research to curate his specimens. Mendel discovered that von Naegeli was interested in his breeding experiments on peas only to the extent that they bore on his own research into *Hieracium*. To some extent, Nelson and Platnick were interested in the views of Hennig, Croizat, and Popper for the sake of those views themselves, but more centrally they were interested in how they could use the ideas to support one another in the construction of their own research program.

Scientists know that the struggle for acceptance is likely to be arduous, but they are not prepared for how long the process actually takes, nor for the amount of misunderstanding and misconstrual that inevitably ensues. Tenacity is as essential as originality in science. The same objections are raised over and over again. It does no good to say that one has already answered that objection elsewhere. Anyone promoting a nonstandard view must be prepared to write the same paper, fight the good fight, answer the same criticisms over and over again. Scientists cannot simply publish their views and let it go at that, the way that Crowson (1970) did, or to take off for Cuba in the midst of the battle as did Macleay. Both Lundberg (1973) and Beatty (1982) published what should have been seminal papers and then left it at that. Failure in science is more often a function of the lack of resolve than anything else.

One of the most frustrating aspects of science is the alacrity with which scientists seem to be able to misunderstand ideas that seem patent to their authors. For example, Max Planck (1910: 1188; 1970: 48) eventually gave up any hope of convincing his adversaries: "on the contrary, I must be prepared to be reproached for having again misunderstood everything. So I will calmly let the approaching flood flow over me, and wait until something factually new appears." Several referees of Croizat's critique of Darwin complained that he attacked straw men and beat dead horses. As Turner (1983: 159) remarked in connection with Eldredge and Gould's (1972) criticisms of gradualist versions of evolutionary theory, "Ism-Schisms may be unprofitable, but they are seldom stale or flat. Scientists, let us be frank, find straw men useful for structuring their own arguments and developing their careers." Levinton (1986) repeated this objection.

But the situation is not this cut-and-dried. Reason, argument, and evidence are supposed to decide controversies in science, but when scientists have to make choices, evidence is never totally determinate, nor arguments overwhelmingly convincing. More than one alternative is not just possible but also plausible. The appropriate conclusion to such realizations is not that anything goes. There has to be some middle ground between one and only one possible answer and total arbitrariness. To make matters more complicated, when new ideas are first introduced they are rarely expressed with sufficient clarity. Scientists do not know what it is that they intended to say until they find out what other scientists think that they have said. Misconstrual verging on parody is part of the process of clarification. If the scientists under scrutiny did not mean what their critics think they meant, then what *did* they mean?

Darwin and Huxley caricatured creationists as maintaining that organisms arise fully formed from the dust of the earth, as if a rhinoceros could arise from the sudden concurrence of a half-ton of inorganic molecules. When the issue was put this bluntly by Huxley, few creationists at the time admitted to postulating such miraculous occurrences, but if this is not what they had in mind, how did they think new species arose? Darwin's critics in turn caricatured his theory as a combination of life arising by the chance colligation of atoms and natural selection—nature red in tooth and claw. Mendelians were portrayed as postulating absolutely discrete, monolithic genes, each controlling an equally discrete character. More recently, population geneticists have been accused of treating genes as beans in a bag, while they in turn dismiss more organismic versions of evolutionary theory as mystical obscurantism. The synthetic theory has been characterized by its opponents as assuming strict gradualism and Panglossian adaptationism, while punctuationalists have been interpreted as asserting that more gradual forms of change never occur and that natural selection plays no role in evolution.

In response to the position paper published by the Sociobiology Study Group criticizing E. O. Wilson, Wade (1976: 1153) remarked that such a "degree of distortion, though routine enough in political life, is perhaps surprising from a group composed largely of professional scholars." But distortion is more than routine in science. It is a traditional mode of argumentation, and a mode that is not entirely counterproductive. It forces scientists to commit themselves. Initially Nelson and Platnick were a bit ambiguous about the role of scientific theories in general, evolutionary theory in particular, and phylogeny in one's proposed principles of classification. As Platnick (1979: 545) asked in response to the suggestion that he and other cladists had modified Hennig's principles substantially, "does all this represent a transformation of Hennig's position, or have these ideas been part and parcel of cladistics all along, even if they were perhaps not very carefully or clearly enunciated (or indeed, even if they were perhaps very carefully *not* clearly enunciated)?" Early on, Nelson was rather cryptic in setting out his more idealistic views. Perhaps he was not especially sure how completely he was committed to these views; perhaps he felt that his fellow systematists were not quite ready for them.

But, eventually, repeated accusations that one's principles of classification presuppose that classifications can be theory-neutral must be answered. Some pheneticists responded by insisting that scientific classifications can be and should be totally theory-neutral, at least initially (Sneath 1983, Sokal 1985), while others responded by restricting their prohibitions just to the effects of previous classifications (Colless 1985). Those cladists who both acknowledged only the most minimal role for process theories in their preferred methodology and strongly advocated the philosophy of Karl Popper have been forced into a particularly tight corner because one of Popper's major theses is that scientific terminology, no matter how observational, cannot be totally free of theoretical assumptions and connotations. In reaction, some cladists have undertaken to interpret Popper's claims about theory-neutrality in an extremely attenuated form (Platnick 1985, 1986), while others openly accept the central role of scientific theories right from the start in the process of classification (Wiley 1981a, Hill and Camus 1986).

The efficacy of parody in conceptual change is not limited to politics and science. Philosophers practice this art in its most refined form. Kuhn (1962), for example, portrayed his predecessors as insisting that to be "rational" theory-choice had to be deductive. Since deduction plays such a minimal role in the replacement of one scientific theory by another, Kuhn appeared to conclude that scientific revolutions are at bottom arational. Did he mean it or didn't he? As D. T. Campbell (1979a: 182) notes, Kuhn's students, at least, consistently read him as arguing that scientific communities are "self-perpetuating mutual admiration societies whose social systems prevent reality testing, stifle innovation as heresy, and suppress disconfirming evidence." Although the "real" Kuhn may have actually held a position only slightly at variance with the one he was criticizing, the Kuhn who has influenced our understanding of science is a cultural relativist when it comes to matters of truth (Reingold 1980). No doubt some of the authors whom I criticize in this book will complain that I have caricatured their views. This complaint is liable to be voiced most frequently by those whom I term "philosophers," but that is only to be expected because I am a philosopher myself. If I am lucky, I fully expect my views to be caricatured with equal enthusiasm. One plausible parody is that I am arguing that science is nothing more than scientists engaged in a struggle for their own career advancement. Straw men and dead horses are as useful in the philosophy of science as in science (for parallel observations with respect to the work of Merton, see Gieryn 1982).

Science is also characterized by a mixture of secrecy and openness. While scientists work on their ideas, they tend to keep them to themselves, in part because they are not sure that they will pan out, in part because they hope that they will. They do not want to declare themselves publicly and then be shown to be mistaken. As the episode involving Farris and Janowitz indicates, this preference is not reciprocal. A scientist does not mind his or her enemies making mistakes in public. Watson and Crick were delighted to discover that Linus Pauling's proposed model for DNA (deoxyribonucleic *acid*) was not an acid. Instead of writing to Pauling to warn him of his error, they used the time it took for him to uncover his mistake to continue work on their own model. However, those on the outside of a particular dispute object, as in the exchange between Janowitz and Farris. If they unwittingly use erroneous research, they pay in lost time and effort, regardless of points scored by one warring faction over another. Scientists should, they are told, submerge their petty bickering to the greater good.

Scientists do not want to make mistakes in public, but they also do not want to be scooped. As counterproductive as the practice may seem, nearly all the credit for a contribution in science is supposed to go to the first author who makes it public, even though others may have invested years in working on the same problem and were themselves on the verge of publishing. When two or more investigators present a series of very similar findings at roughly the same time, priority disputes are liable to erupt. Issues of priority can arise in two quite different settings—when two scientists are working in social isolation from each other and when they are working in the same research group. In the first instance, these scientists may have depended on the same published sources, in some cases on one another's publications, but they usually have only limited access to unpublished work.

Priority disputes can be resolved by the larger scientific community when the issue is merely priority of publication. Priority disputes become stickier when scientists have access to each other's work prior to publication. Scientists working together on the same research team quite obviously have access to the products of each other's labor prior to publication. The messiest sort of priority dispute occurs among scientists working together, as for example in the discovery of insulin (Bliss 1982). But even scientists not working in consort frequently have access to each other's findings prior to publication. For example, one of the loudest priority disputes in recent memory resulted from Luc Montagnier of the Pasteur Institute in Paris sending a sample of the AIDS virus he had isolated to Robert Gallo at the National Cancer Institute in the

United States. Another way that scientists who belong to different research groups can gain access to unpublished research is through the refereeing system. Research proposals are supposed to include information about scientists' most recent discoveries as well as their current best estimates of the direction of their future research, just the sort of information that would be most useful to their competitors. And it is to these competitors that research proposals are sent for evaluation. The opportunities for illicit use are ideal. Scientists are most vulnerable immediately prior to publication, and if Garvey (1979: 20) is right, a flurry of information exchange occurs at this juncture. Although professional allies are in a better position to appropriate one's contributions than are strangers or enemies, allegations of both sorts of theft are common. Scientists do cooperate with each other, but such cooperation is not easy.

One surprising feature of the stories that I have related thus far in this book is how rarely issues of priority became public. Most of the grumbling is done in private. For example, Crovello complained to Sokal and Sneath that they did not give him sufficient credit for his contributions to numerical taxonomy in the second edition of their book. Nelson objected that much of the substance of the talks that he gave at the American Museum and in manuscripts he circulated, especially his distinction between cladograms, trees, and scenarios, was published by others before he himself got around to publishing. At the first meeting of the Hennig Society, Farris complained of the excessive credit that Eldredge and Cracraft were getting for the notion of a cladogram, just because they were first to get a book out on the topic. Craw complained to Croizat about his lack of recognition. But none of these irritations erupted into public disputes over priority.

Some ill feelings, however, did see print. For example, Croizat repeatedly objected to how little credit he received from other biogeographers. He even went so far as to claim that Hennig had stolen at least one of his keys ideas, the principle of dichotomy, from Rosa and threatened to accuse Nelson and Rosen of plagiarizing his own work. Considerable confusion also exists over the proper parentage of the punctuational view of speciation. Stanley (1979) traces the idea to a paper published by Mayr (1954). Eldredge and Gould (1972) are not so sure that Mayr's views are actually ancestral to their own. Mayr has his doubts as well.

The amazing thing about the two most famous examples of the simultaneous announcement of an important scientific innovation—Wallace's familiar bolt from the blue and the so-called rediscovery of Mendelian genetics—was that neither of them gave rise to priority disputes. Darwin certainly did not complain about lack of recognition for his views on evolution, but he did object to Lyell's constantly referring to Lamarck when he discussed the topic. But Wallace also never complained about not getting the credit that was his due. Such complaints arose only later as others jockeyed for position during the continuing controversies over the nature of the evolutionary process. As Merton (1973: 291) notes, this feature of priority disputes is quite general. Nor did the two, possibly three, rediscoverers of Mendel's laws fall at each other's throats. Bateson short-circuited the impending priority dispute by conferring credit retroactively on an obscure Moravian monk. Correns and Tschermak were happy to concur. They were not about to get credit anyway. De Vries was much less cooperative.

Even though priority disputes do not always break out when circumstances are ripe for them, they are common in science. Richard Owen made an occupation of engaging in them (Desmond 1982), and Louis Agassiz was officially charged with failure to give sufficient credit to his co-workers. As the episode of Lyell's derivative discussion of the antiquity of man indicates, such a failure can raise greater animosity within groups than between them (Bynum 1984). Complaints about misrepresentation, pigheadedness, and political maneuvering have been frequent in the episodes I have related; accusations of theft have been extremely rare. Even rarer were claims of outright fabrication. Why are infractions of the mores of science so rare, and why are certain infractions rarer than other?

Closely connected to issues of priority with respect to contributions are issues of secrecy and openness with respect to the identities of the scientists conducting the investigations. Anyone reading the preceding pages of this book is likely not to be surprised at the frequency with which scientists identify themselves as authors of their own work. Unless scientists append their names to their work, they cannot receive the credit due them. Periodically in the history of science, book reviews have been published anonymously. One of Darwin's favorite pastimes after the appearance of the *Origin of Species* was guessing who the authors of the various reviews might be. Today book reviews are almost always signed. Evaluations of manuscripts prior to publication are also traditionally confidential. They remain so today. With the rise of government support of science, the refereeing process has expanded to include research proposals. Book reviews have gone from being commonly anonymous to usually acknowledged, without destroying the fabric of science. Why not make science more open by making the names of referees part of the public record? Conversely, might not the petty bickering that goes on among scientists be reduced if research were published anonymously or under pseudonyms? If the identities of referee reports are kept secret, why must the authorship of original research be so prominently displayed?

No less an authority than Darwin (1899, vol. 1: 452) remarked in a letter to his friend Fox on February 22, 1857, that he wished that he "could set less value on the bauble fame, either present or posthumous, than I do, but not I think, to any extreme degree: yet, if I know myself, I would work just as hard, though with less gusto, if I knew that my book would be published for ever anonymously." Three months later Darwin wrote to Wallace acknowledging that he had read his 1855 paper in the *Annals of Natural History* and that he too was working on the species problem. For all the defamation of the "bauble fame," very few scientists in the history of science have published anonymously or under pseudonyms, and when they have, it was usually to avoid censure for their heretical views; e.g. Whewell's (1853) *Of the Plurality of Worlds* and Chambers' (1844) *Vestiges of the Natural History of Creation.*

The two main exceptions are Nabi and Bourbaki. The varying group of authors who attributed short notes to the fictitious Nabi did so to be cute. However, given the critical nature of some of these notes, their targets were not amused. They wanted their accusers to present their accusations out in the open and not to hide behind a pseudonym. Bourbaki poses a more significant problem. Numerous, first-rate mathematicians published under this pseudonym. If gaining credit from one's peers is as central to the effective functioning of science as I claim it is, Bourbaki is more than a curiosity. It is an anomaly that must eventually be converted into a confirming instance. Unfortunately, those students of science who possess the linguistic and mathematical abilities necessary for such an undertaking have yet to investigate this episode in science as anything other than a curiosity. Of course, according to current theories about the nature of science, it *is* only a curiosity. Of course, on current theories about the nature of science, nearly everything about science is classed as a curiosity.

The same interplay between observations and theories that characterizes the rest of science has also characterized evolutionary biology and systematics. Sometimes a field is invigorated by a theoretical innovation, a theory in search of data. Hamilton's (1964) effective formulation of kin selection and inclusive fitness as well as Eldredge and Gould's (1972) theory of punctuated equilibria are two good cases in point. It took a while, but eventually scientists began to put these theories to the test. Conversely, sometimes empirical investigations serve a complementary function, providing data that cry out for a theoretical explanation. The discovery by Hubby and Lewontin (1966) and Harris (1966) of extensive genetic heterogeneity in natural populations is a good example. For years Muller and Dobzhansky had quarreled about how heterogenous natural populations should be. Quite suddenly both sides were presented with extensive data. Not that this new class of data instantly resolved the dispute. Instead the lines were redrawn, certain alternatives were precluded, and the theoretical debate became significantly more precise and sophisticated.

No one can deny that the methodological innovations introduced by the pheneticists and cladists radically altered systematics, but the connection between these innovations and data is more tenuous than usual. After the pheneticists formulated general techniques for classifying organisms through the use of explicitly stated algorithms, they published paper after paper showing the sorts of classifications that actually resulted from the consistent application of their principles. The same can be said for the cladists. If nothing else, systematists could look at the results of competing methodologies and compare them. Mickevich's (1978) work was especially important in this connection because she claimed to show that on six data sets cladistic methods fulfill the pheneticists' own goals more adequately than any of the phenetic methods. For anyone who doubts the role of data in science, the collective intake of breath was almost audible when Mickevich published her work. Of course the pheneticists did not instantly abandon their views. Instead they attacked the data and Mickevich's handling of it (Sokal 1985).

Darwin set out on his voyage knowing very little. He was not quite the unlettered underling whom Bacon foresaw as the ideal fact-gatherer or the intelligent ignoramus whom Sokal and Sneath envisaged as the ideal systematist, but he was close, and he paid for his ignorance. Not having any general views in mind that need testing certainly decreases the likelihood that these general views will bias one's selection or interpretation of data, but it also makes data-collecting hopelessly haphazard. Wallace made much better use of his travels. One lesson that Dobzhansky learned from Wright was that he better have his theories straight before he set up experiments in the field to test them. Our beliefs do bias how we conceptualize and evaluate our observations but not so completely that observations have no reciprocal influence on our beliefs. Time and again scientists end up refuting the very views that they set out to support. Both Darwin and Morgan are good cases in point. The difficulties that Darwin had in classifying his specimens upon his return to England shook his confidence in the immutability of species, and Morgan's attempt to confirm de Vries's mutation theory led him to reject it.

Another feature of the scientific episodes that I have related is the extent to which the conceptual systems being developed were heterogeneous at any one time and how much they became transformed through time. Science is not a matter of monolithic conceptual structures persisting through time. The most that one can say about the numerous versions of particular hypotheses, theories, and even more inclusive conceptual systems is that they share a "family resemblance" to each other. Multiplicity, not homogeneous simplicity, characterizes the conceptual development of science. Rarely are any of the elements of a system totally new, but the array of combinations of familiar ideas is overwhelming and the sort of heterogeneity they exhibit is difficult to grasp. In the most typical case, families of conceptual systems form multimodal clusters.

In the past, certain philosophers have emphasized that theory choice is always a matter of comparison. Scientists construct their theories with one eye toward their most likely competitors because that is how other scientists will evaluate them. Sometimes they react to their immediate competitors, e.g., Mayr's (1942) reaction to Goldschmidt's (1940) attack on "Darwinism" or Hennig's (1950) response to the resurgence of idealism in his native Germany. Sometimes they reach further back; e.g., Darwin was very strongly influenced by Lyell's critique of Lamarck in the second volume of Lyell's *Principles of Geology* (1832) and Chambers' (1844) defense of evolution in his *Vestiges of the Natural History of Creation*. When Darwin finally published, he constantly found himself being compared to Lamarck and the Vestigiarian. Cladistic analysis was not evaluated solely on its own terms but also in contrast to the recent controversy over phenetics.

The view that theory choice is always a matter of comparison with current alternatives depends, of course, on how one individuates theories. Even when only one global theory is available at a particular time in the history of a discipline, choices must be made among numerous versions of that theory, versions that are so different that they might easily be judged to be instances of different conceptual systems. There is frequently more heterogeneity within a particular research program than between two different programs. Hence, comparisons are constantly being made between competing global theories, when more than one is extant, as well as between different versions of the same theory.

Conceptual heterogeneity is both exaggerated and obscured by semantic plasticity. Scientists show extreme partiality to their own preferred terminology. To some extent, the appearance of conceptual heterogeneity in science is spurious: sometimes scientists make very much the same claims in quite different terms. For example, little difference exists between saying that taxa should be defined in terms of shared derived characters or in terms of synapomorphies. Sometimes the appearance of conceptual homogeneity is equally spurious: two scientists might seem to be making precisely the same assertion, when the content of their claims is quite different. For example, both evolutionary systematists and cladists might be found insisting that all higher taxa must be monophyletic but mean very different things by the term "monophyletic." But even cladists, when they urge the construction of monophyletic taxa, might have different things in mind. Platnick and Wiley, for instance, disagree about the paraphyly of ancestral species. Conversely, the taxa that cladists term monophyletic, evolutionary systematists term holophyletic. The two systems of terms map neatly onto each other, e.g.:

(a) Ashlock (1971)	(b) Nelson (1971b)
Monophyly	Monophyly
Holophyly	Non-monophyly
Paraphyly	Paraphyly
Polyphyly	Polyphyly

Usually, however, the systems of terminology preferred by scientists in conflict are not perfectly isomorphic. It is easier to make the claims that one wishes to make in one's own terminology than in the terminology devised by one's opponents. As a result, the adoption of one terminology over another is important. Scientists tend to be extremely rigid when it comes to terminology, and this rigidity is not misplaced. But even when different terminological systems do include perfectly isomorphic sets of terms, there is still good reason for different scientists to prefer their own sets. At least superficially, success in science is gauged as much by the spread of a particular terminology as by the ideas signified by this terminology. For example, in the midst of the current AIDS epidemic, scientists have seemed as intent on naming the virus as on finding some means to halt its spread (Connor 1987). However, the issue is not rally the "proper" name of the virus but priority for discovering it. The research group that first discovers a new kind of organism usually has the right to name it. Hence, the name one uses implies priority. To sidestep this problem, a panel of viral systematists suggested replacing Montagnier's LAV (lymphadenopathy associated virus) and Gallo's HTLV-III (human T-cell leukemia virus three) with HIV (Human immunodeficiency virus). A similar controversy arose over the name of the substance that could cure diabetes. The adoption of "insulin" was both a cause and an effect of the credit awarded Banting (Bliss 1982). A major goal of scientists is to convert their idiolect into common usage.

The rigidity with which scientists insist on their preferred terminology is matched only by the semantic plasticity of that terminology. Two sorts of "wriggling" go on in science. Scientific theories are not easily confronted with data. All sorts of messy decisions must be made, operational definitions adopted, and so on. Initially, no observations or experiments are absolutely crucial. The process of theory testing gives scientists ample opportunities to try out different combinations of limiting assumptions and the like. But scientists also "wriggle" when it comes to what it is precisely that they mean. Throughout the history of science, scientists have repeatedly complained about how difficult it is to pin down one's opponents so that they can be refuted. A caterpillar cannot escape the collector's needle by metamorphosing into a butterfly, but no sooner is a scientist skewered than he slips off the point by claiming that he never held the position being attributed to him or that it was only of peripheral importance.

For example, opponents objected to the ease with which gene selectionists changed from "genes are the primary units of selection" to "genes are the primary units of replication." The pheneticists' use of the term "phenetic" and the cladists' use of the term "cladogram" are two more important examples of semantic wriggling.

Much of the conceptual change in science does not occur out in the open but is hidden by verbal smoke screens. Sometimes scientists are aware of what they are doing when they throw up such smoke screens; usually they are not. In the ongoing process of science, the meanings of scientific terms are rarely perfectly clear. It does not take much effort to convince oneself that one's current views are totally consistent with one's past utterances. Terms are out in the open for anyone to inspect. Meanings are more elusive.

The reader might also have been surprised by the recurrent role of more "philosophical" issues in science. Chief among these is the nature of science itself. Sometimes the scientists who are engaged in a scientific dispute agree basically over the rules of the game and disagree merely over who has fulfilled the relevant requirements, but when major changes are introduced into science, the nature of science is itself a point at issue. In fact, which changes in science count as being "major" or "fundamental" are in part a function of the changes that they necessitate in our understanding of science. As Cohen (1981) argues, Newton not only transformed the science of mechanics but also changed the way that subsequent scientists went about doing science. For example, Buffon, Geoffroy St.-Hilaire, and Darwin among many others thought of themselves as doing science in the Newtonian mold (Hall 1968). The dispute between Cuvier and Geoffroy in the early nineteenth century was as much about the nature of science as the number of archetypes needed to characterize the animal kingdom. Cuvier insisted that science must begin by the accumulation of positive facts out of which patterns would necessarily emerge. Geoffroy saw nothing wrong with the introduction of highly speculative views right from the start.

Time and again scientists proclaim their views to be "inductive," "empirical," or "falsifiable," depending on the terminological fashion at the time, while their opponents' hypotheses are not. Because scientists as different as Cuvier and Croizat repeat such incantations, the temptation is to dismiss them all as so much propaganda, but some scientists *are* more empirical than others, and sometimes a "return to the data" has proven to be helpful. For example, the attempt by pheneticists to estimate overall phenetic relations among organisms taught us quite a bit about these relations. If nothing else, it taught us that numerous such relations can be discerned. The continued efforts of cladists to strip methods for discerning branching patterns to their basics have forced us to see which assumptions are primary in this activity and which secondary. Now that evolution is accepted by all serious biologists, attempting to look at the relationships among organisms without presupposing that they evolve can be highly instructive. Croizat's dogged plotting of tracks revealed patterns that more traditional biogeographers missed. I have presented the preceding examples as if they were merely strategies in scientific investigation, but their advocates presented them as if they were permanent features of science.

As Rudwick (1972) has documented with respect to geology, Lyell was as concerned to redefine the science of geology as he was to support his own view of particular geological phenomena. In order for the identification and ordering of geological strata to count as "science," historical reconstructions had to count as being scientific (R. Laudan 1987). The success of Darwin's theory of evolution extended the realm of totally naturalistic science by defeating creationism. To the extent that creationism entailed non-naturalistic origins of species, it was not science. It also succeeded in halting the spread of idealism in Great Britain. Flesh-and-blood ancestors annihilated ethereal archetypes.

At the turn of the century, the debate over Mendelism and Darwinism also turned in significant measure on differing views of the nature of science. Is the postulation of genes as real material bodies scientific, or must true scientists limit themselves to correlations of observed phenomena? Can observations in the field count as part of science, or is science necessarily experimental? By the time that the founders of the Modern Synthesis began the task of pulling together the separate strands of development that had taken place in genetics, paleontology, evolutionary biology, and so on, idealism had once again become prevalent among comparative anatomists and paleontologists on the Continent. Once again the Darwinians had to defeat this view of science. The process continues to the present as pheneticists and cladists alike argue that their methods of classification are genuinely scientific while those of their opponents are not. Pheneticists and pattern cladists urge a highly empirical view of science not unlike that urged by certain idealists, while evolutionary systematists and phylogenetic cladists insist on the necessary role of process theories in even the earliest stages of the classificatory process.

Although science and philosophy of science as ideal types can be distinguished, the activities cannot be. Scientists frequently engage in philosophical discussions, while somewhat less frequently people who are officially philosophers try their hand at doing a little science. If this were all there was to the intersection of science and philosophy, it would not cause any special problems. However, a more fundamental interplay exists between the two activities. In the cases I listed above, the success of the particular views championed by these scientists was an important factor in getting their more general views about the nature of science accepted. Reciprocally, acceptance of their preferred conception of science lent support to their particular scientific theories. Laudan (1981b: 9) sees the relation between science and philosophy of science as going only one way: "it is shifting *scientific* beliefs which have been chiefly responsible for the major doctrinal shifts within the philosophy of science." As I see it, the relation is reciprocal. Sometimes views about the nature of science have an influence on the acceptance of scientific theories and sometimes the success of a particular scientific theory lends support to a particular view of science. The nature of science is constantly under negotiation, and the currency of these negotiations is success.

Precisely parallel observations hold for any theory of scientific development. Philosophers have tended to content themselves with analyzing such meta-scientific concepts as explanation, theory, rationality, and justification. For example, to count as being "scientific" an activity has to exhibit a certain set of characteristics. To bolster one preferred list of characteristics over another, philosophers appeal both to our intuitions about science and to certain episodes in the history of science. Recourse to reason, argument, and evidence is rational, while reliance on revelation, infallible proclamations in holy books, and beliefs about the morally preferable state of the universe are not. This is what "we" mean by the term "rational." For example, central to Laudan's (1977) theory of scientific growth is the demand that methodologies of science at the very least capture as rational certain choices that scientists made in the course of science which anyone who studied them would see intuitively as instances of rational theory choice. But more recently, Laudan (1986: 123) notes that grounding the philosophy of science in our intuitive judgments makes "epistemology of science as nothing other than the ordinary-language philosophy of a special linguistic community, viz., the users of scientific language."

There is some point to philosophers analyzing the scientific terms that scientists use; e.g., gene, mass, species, and Id. What scientists have to say about these terms has a certain priority over the opinions of anyone else, including philosophers. Philosophers can contribute to the evaluation of scientific terms, but the ultimate arbiter in such matters is *use*—not the undifferentiated use that pheneticists propose for their general-purpose classifications, but use in particular theories. Those theoretical terms that appear most centrally in our most powerful theories take precedence over all others. Scientists use other terms as well, terms like "theory," "truth," "explain," and "falsifiable." Here scientists and philosophers are more on a par. For their part, scientists have the advantage over philosophers in knowing one present-day area of science firsthand, but philosophers have the edge in that they tend to have knowledge of various branches of science and pay sustained attention to questions about the nature of science.

As in the case of science itself, the ultimate arbiter for such metascientific terms should be the roles that they play in particular theories about science. If these theories conflict with our intuitions, that cannot be helped any more in the study of science than in science itself. We are not born with our intuitions about science intact. We acquire them as we learn about science, but each student of science is liable to have his or her own peculiar point of entry into this field of study, and this idiosyncratic history might influence his or her intuitions thereafter. In the past, those philosophers who knew anything about science tended to be most familiar with physics, primarily celestial mechanics, Newtonian theories of light and gravity, and quantum mechanics. They viewed science accordingly. Any area of science that departed from these paradigms was not really science. My background is in evolutionary biology and systematics. Hence, my intuitions are quite different from those of earlier generations of philosophers of science. Hence, I cannot put much stock in intuitions. We all *start* with them, but this is not where we should *end*.

I have no objections to all the effort that philosophers put into analyzing metascientific concepts, but the only way to evaluate these analyses is in the context of a general theory about the nature of science. The next step is to confront these general theories with empirical data. In science no data are ever totally "raw." The data that bear on metalevel hypotheses about science are no less theory-laden. Even though all our observation statements are to some extent theory-laden, in many cases they are not so theory-laden that it is impossible to use them to choose between competing theories. If so, then the same state of affairs should obtain with respect to metascientific theories. For example, Lakatos (1971: 100) recommends what he terms a methodology of scientific research programs in which these programs are evaluated according to whether they are progressing, stagnating, or degenerating:

A research programme is said to be *progressing* as long as its theoretical growth anticipates its empirical growth, that is, as long as it keeps predicting novel facts with some success (*"progressive problemshift"*); it is *stagnating* if its theoretical growth lags behind its empirical growth, that is, as long as it gives only *post-hoc* explanations either of chance discoveries or of facts anticipated by, and discovered in, a rival programme (*"degenerating problemshift"*). If a research programme progressively explains more than a rival, it "supersedes" it, and the rival can be eliminated (or, if you wish, "shelved").

Lakatos (1971: 104) goes on to note that a

favourite hunting ground of externalists has been the related problem of why so much importance is attached to—and energy spent on—*priority disputes*. This can be explained only *externally* by the inductivist, naive falsificationist, or the conventionalist; but in the light of the methodology of research programmes some priority disputes are vital *internal* problems, since in this methodology *it becomes all-important for rational appraisal which programme was first in anticipating a novel fact and which fitted in the by now old fact only later.* Some priority disputes can be explained by rational interest and not simply by vanity and greed for fame. (Italics in the original.)

According to Lakatos, priority disputes between advocates of *different* research programs can be explained internally in terms of rational interest because they contribute to the status of the respective research programs as progressing, stagnating, or degenerating, while priority disputes between advocates of the *same* research program can be explained only externally in terms of vanity and greed for fame. Lakatos's distinction between rational and irrational interests with respect to priority disputes turns on an overly monolithic conception of research programs. According to Lakatos, each research program can be analyzed into its hard core and protective belt. Extensive modification of the protective belt can be made while the research program remains the *same* program, while any modification in its hard core transmutes it into a *different* program. The distinction is strongly reminiscent of Aristotle's distinction between essence and accident.

Aristotle has been termed the philosopher of common sense. From the human perspective, the distinctions which Aristotle makes seem so right (Atran 1985). Aristotle's distinction between essence and accident is among his most intuitively correct. Certain properties an entity possesses make it what it really is; others are only incidental. The latter but not the former can vary while the entity remains the same sort of entity. Some variation in the world of our experience is undeniable, but if everything is variable, then knowledge would seem to be impossible. The power of this way of viewing the world can be seen time and again in evolutionary biology. The compulsion to treat something about particular species as invariant is all but irresistible. For example, Carson (1975) suggests that species of diploid organisms have two different systems of genetic variability: an open system that varies within a species and is subject to phyletic evolution, and a closed system of co-adapted, internally balanced blocks of genes that vary between but not within species and, hence, cannot evolve phyletically.

Just because an idea originated with Aristotle does not mean that it is mistaken—though this is not a bad rule of thumb. Sometimes the world might very well be what it seems to be, but caution is advisable. Perhaps two different systems of variability do exist in diploid species. Perhaps elements that make up scientific research programs do come in two sorts—hard-core elements that never vary and all the rest that do. However, the most common objection to Lakatos's important conception of research programs is that, in fact, no such hard cores can be found. In most cases research programs can be made to possess eternal, immutable "hard cores" only retrospectively, by fiat. Like it or not, research programs are extremely heterogeneous. They include within them more limited programs. If research groups are defined in terms of the relevant social relations, it is perfectly possible for members of the same research group to disagree with each other even over fundamentals, and they do. Even the members of the smallest research groups have interests which are not perfectly coincident.

But this much being said, Lakatos does not go on to examine the relative frequencies of within-group versus between-group priority disputes. On his views which should be more prevalent? Lakatos evaluates analyses of science in terms of how much of science can be accounted for by a particular analysis (and is therefore "internal" with respect to that analysis) and how much must be counted "external." According to Lakatos (1971: 126) priority disputes between Newtonians and Cartesians were a matter of rational interest, while the priority dispute between Adams and Leverrier over who first discovered Neptune was merely an issue of vanity and greed for fame because no matter "who discovered it, the discovery strengthened the same (Newtonian) programme."

On Lakatos's criterion, a methodology that makes priority disputes between individuals who are working in the same as well as different research programs a matter of "rational interest" would be superior to one that did not. The account that I present in this book does precisely this. Again, according to Lakatos, research programs that predict unanticipated facts are preferable to those that do not. In my view of the science of science, the behavior of scientists serves as evidence. If a certain sort of behavior is prevalent among scientists, one possible explanation is that it is serving some function, shades of Dr. Pangloss notwithstanding. Just because *sometimes* adaptive assumptions are mistaken, it does not follow that structures *never* serve functions. If those scientists who exhibit a behavior most frequently are among the most successful scientists, then the possibility that this behavior is scientifically adaptive is only increased. Although philosophers are right to bring their critical faculties to bear on scientific practice, no matter how pervasive and entrenched that practice might be, the success of science as it has been conducted in the past should not be taken lightly.

I also am not using "success" here in the sense of crass career advancement. Scientists themselves distinguish between scientists who are "successful" in making important contributions to science and those who are merely on the make, although the two kinds are not incompatible. Sometimes scientists on the make can and do make important contributions to science. Watson and Crick are only one example of this coincidence.

By and large those scientists who make important contributions to science are rewarded in exactly the way they most value—not by being given an impressive office and high administrative post but by *use.* Thus, on the view of science that I am advocating, success is to be gauged in terms of use, not the undifferentiated use to which pheneticists appeal but a very specific sort of use. As Raup (1986: 211) notes, "one's success as a scientist can be measured more by the number of people he or she puts to work on new problems than by the correctness of specific research results." Science as a selection process, like all selection processes, is only locally optimizing. Such localization increases the role of contingency in the course that science takes. Scientists reason from the contingencies of their local conceptual environments to the world as such, a mode of reasoning that is as subject to error as to innovation. However, the best way discovered to date to progress toward global optimization in knowledge acquisition is this system of iterated local optimizations. Striving after truth for its own sake in the absence of the social structure of science that has grown up to foster this search is about as effective as Don Quixote's efforts to help humanity.

I am not suggesting that those of us who study science should reason inductively from scientific practice to scientific principles. Quite obviously the same sort of interplay between theory and observation that characterizes other empirical investigations must also characterize ours, but the practice of taking certain principles as a priori "scientific" and criticizing scientific practice accordingly can lead to extremely foolish observations. For example, Mahoney (1979: 350) compiled a list of the characteristics of the "ideal" scientist from introductory texts and scientific biographies: e.g., objectivity and emotional neutrality, rationality as evidenced by superior reasoning skills, open-mindedness, superior intelligence, integrity in data collecting and reporting, and the open and cooperative sharing of knowledge. He then proceeded to show that scientists are victims of perceptual errors, prone to confirmatory bias, and are frequently quite emotional where their research is concerned, for example engaging in "unbridled jubilation" when they make important discoveries and "depression and disappointment" when they fail.

Some of the observations which Mahoney presents about science have some point, e.g., Wolin's (1962) difficulty in obtaining the raw data on which other scientists had based their publications. But most are incredibly naive. For example, in deductive logic reasoning from "if p then q" and "q" to "p" is fallacious, the fallacy of affirming the consequent. When scientists and nonscientists were tested, scientists tended to commit this error more frequently than ordinary people. They also tended to reason quite rapidly from minimal data to possible explanations. The only subjects who were errorless in conforming to proper scientific method were two ministers! That the principles of proper scientific reasoning on which these experiments were based might themselves be deficient never occurred to Mahoney. After all, scientists are involved primarily in nondemonstrative forms of inference, and by definition nondemonstrative inferences fail the canons of deductive logic. As fallacious as affirming the consequent may be in deductive logic, it is central to science. Perhaps ministers can afford to wait around until "all the evidence" is in—whatever that might be—but scientists cannot.

Mahoney reasoned from extremely simplistic notions of proper scientific method to the misbehavior of professional scientists. My mode of inference in this book is just the opposite of that utilized by Mahoney. If generations of scientists behave in certain ways, then possibly there is something to be said for this behavior. The inference is even stronger if there is a correlation between the prevalence of a behavior and the success of the scientists who exhibit it. Scientists are far from infallible in the methods they employ in their investigations, but they must be doing *something* right.

Most of those workers who have studied science are surprised at how effective the various mores and norms of science are when compared to those of other social institutions. Zuckerman (1977: 98) concurs in this general impression but emphasizes that it is only an impression. Thus far, both those who argue for the unusual efficacy of the institutional norms in science as well as those who have their doubts have been forced to rely primarily on anecdotes—examples of instances in which scientists conformed to a particular norm, examples of when they did not. In order for these lists to be of any use, the nature of the norms being evaluated must be specified clearly and the data collected must be both relevant and collected systematically. The correct way in which the norms are to be specified depends, of course, on the general theory that one holds about the nature of science. In the next two chapters, I discuss various sorts of behavior that are considered improper in science, paying close attention to the punishment that results when infractions are discovered. If no punishment results, then possibly this "norm" is no norm at all.

One of the strengths of the view that I present in this book is that it explains why lying (publishing fabricated research) is so much rarer than stealing (failing to give credit where credit is due). It also explains why misconduct in general is so rare. All professions have their codes of conduct. Why do scientists seem to adhere so much more closely to theirs than do members of other professions? Politicians, judges, and others in comparable positions of trust frequently violate that trust. Why don't scientists take the money which they receive for research and salt it away for an early retirement? Long after they have received tenure, long after they have gained a position in the scientific hierarchy sufficiently secure so that they could coast into retirement, they keep working. Why do scientists work so hard? Self-policing professions are infamous for not policing themselves—all except science. Anyone who peruses scientific journals is sure to be struck by the bizarre things that capture the imaginations of scientists. How can anyone become so passionately interested in such obscure subjects? In general, why does science work so well?

These are the sorts of questions that someone who happened upon science for the first time might well ask. These are also precisely the sorts of questions that philosophers of science dismiss as not being relevant to philosophical analyses of science. If so, then so much the worse for philosophical analyses of science. Whenever possible, I present systematic data that bear on the answers I provide to these questions. Like Zuckerman, I am suspicious of general impressions. For example, a widely held belief in Darwin's day that has come down to the present is that within ten years after the publication of the *Origin of Species,* nearly everyone in Great Britain who could be considered a genuine scientist rapidly came to adopt at least the central tenet of Darwin's theory that species evolve. Precise figures are never given, but the implication is that only a handful of aged scientists and religious bigots held out for more than a few years. By

1869 Darwin had triumphed (Kingsley 1863, Bennett 1870). After gathering fairly extensive data for the period, Hull, Tessner, and Diamond (1978) discovered that as late as 1869 only 75 percent of Darwin's fellow scientists had accepted even this minimal proposition. Although 75 percent is a healthy plurality, 25 percent is more than a handful.

The preceding view about the prevalence of a belief in evolution in 1869 is often coupled with the claim that numerous scientists believed in evolution prior to Darwin. Darwin exaggerated the prevalence of a belief in the immutability of species to bolster his own claims to originality. As my discussion in earlier chapters indicates, such issues are not easy to decide. If one defines such terms as "mutability" and "evolution" broadly enough, nearly everyone from Aristotle to Agassiz becomes an evolutionist. A few biologists prior to Darwin did advocate evolution of a more significant sort. Some of these biologists were quite famous. However, Hull, Tessner, and Diamond (1978) were able to find only a half dozen of Darwin's contemporaries in Great Britain who believed in anything like the evolution of one species into another.

A Mechanism for Scientific Change

Conflicts arise in societies because the interests of individuals often come into conflict with each other, not to mention with the manifest goals of the society. Human beings seem to have a strong desire to believe that genuinely cooperative, altruistic behavior in human societies can be both long-term and widespread, and several leading ideologies set this sort of behavior as their ultimate goal, Christianity and Marxism being two of the most prevalent. People *are* capable of the sorts of cooperative, altruistic behavior urged by so many, but the problem is that they *do* need to be urged. Massive effort is expended in societies to encourage genuinely altruistic behavior towards one's "neighbors" or more generally the "people." We are constantly being entreated, if not enjoined, to submerge our individual interests to the good of the group. Such efforts do some good, but the fact remains that effort must be expended to encourage cooperation and altruism. Self-interest of the crudest sort seems not to need such massive encouragement. It emerges all on its own. In certain societies, nepotism has been considered perfectly acceptable, even admirable. In such situations, very little effort needs to be expended to promote such behavior, while constant effort is required to discourage nepotism in those societies which officially disparage it, usually with only marginal success. The moment that vigilance is relaxed, the payrolls are padded with relatives once again.

Although everyone acknowledges that people do behave in ways that they themselves perceive to be in the larger public interest, nearly all theories of rational choice assume that human behavior can be explained in terms of narrow self-interest, at least as a first approximation. Margolis (1982) is unique in introducing a second utility function for those instances in which people make contributions to what they perceive as the public good where the return to the individual appears inconsequential, e.g., in voting. As Margolis notes, people do seem to act as if they have two different utility functions, one individual and one social. Sometimes these two perspective coincide but not always. The question is the relative importance that these two utility functions have in various societies and under different situations.

Sociobiologists insist that the asymmetry in the ease with which human beings cooperate and compete in various situations has a biological basis. Social good and biological good do not always coincide, and when they do not, sociobiologists argue, biological good inevitably prevails. Group selection for group characteristics is possible, but the conditions for it to have a sustained effect are liable to be quite rare. Most social scientists, to the contrary, present social explanations for the asymmetry between the amount of conditioning needed to elicit altruistic, cooperative behavior and the amount needed for selfish, competitive behavior. I do not know who is right on this score. I suspect that human beings are genetically predisposed both to cooperate and to compete. Either sort of behavior can be elicited. However, the sort of cooperation that our genetic makeup is likely to encourage is limited to relatively small groups. Like it or not, human beings now live in very large groups. The problem is to extend forms of small-group cooperation to huge masses of people. Similar observations hold with respect to altruism and selfishness.

In this book I do not address the ultimate causes of human cooperation and altruism at large. I do not know how prevalent such sorts of behavior are or the degree to which they are influenced by our genetic makeup. For my purposes, it does not matter. All I am assuming is that individual and group goals can conflict and that, when they do, agents must choose. But whatever views one has on these larger issues, one thing is sure. Whatever is true of people in general had better apply to scientists as well. Scientists are people. One cannot claim simultaneously that people in general cannot sustain long-term, widespread altruistic behavior and that scientists can, without offering some explanation for this peculiar state of affairs. Why are scientists so special? Because no one yet has suggested that any unique features of science as a social institution are due to the peculiar genetic makeup of scientists, the only plausible answer must lie in the peculiar structure of science as a social institution. If scientists succeed better than the population at large in cooperating with large numbers of other people for long periods of time, then the explanation for this difference cannot be genetic. It must be social.

All social institutions have norms that characterize them. The peculiar feature of science is the remarkable degree to which scientists seem to adhere to these norms and the spectacular success which science has had in attaining its manifest goals. Conflicts arise in other social institutions because the manifest goals of the institution and the goals of the individuals functioning in this institution frequently come into conflict. The most important feature of science from its inception is that it has not been utopian. Scientists need not sacrifice their individual interests for the larger good. What is good for General Motors is not always good for the nation, but once science is properly understood, it turns out that what is good for the individual scientist is by and large good for science. In other institutions such coincidences have been rare and haphazard. For example, Duffy (1976: 289) noted that in the early years of the professionalization of medicine in the United States, the quality of medical care markedly improved. In this instance, "Virtue and self-interest went hand-in-hand." But in general, virtue and self-interest do not go hand in hand. The chief exception is science. The institutional norms of science are so structured that they tend to facilitate rather than frustrate the goals of individual scientists. As I hope that the first half of this book has amply demonstrated, a scientist need not be a good person to do good science. Some of the most impossible people have been among the most productive scientists. In fact, one is tempted to claim that the most impossible scientists have been the most productive.

The coincidence of individual interests and the overall goals of science is reflected in biological contexts in those situations in which individual genetic advantage is reconciled with the good of the species. Sometimes an allele spreads through a population by means of meiotic drive with a resulting increase in the likelihood that the population or entire species will go extinct. Asexual organisms can develop from a sexual species, increasing their own individual genetic contributions to future generations but only at the cost of future adaptability. Leigh (1977) argues that selection can favor the survival and multiplication of species whose genetic systems or social organization favors the evolution of mechanisms reconciling individual with group advantage, but in doing so he is forced to rely on species selection. Individual and group good need not always conflict, but usually they do. The rarity of altruistic behavior, whether metaphorically at the level of individual alleles or more literally at the level of organisms, indicates how rarely the two coincide. The greatest strength of science is that it is so organized that individual and group interests tend to support each other.

As multifarious as science has been and continues to be, a great deal about it can be explained by reference to just three elements: a desire to understand the world in which we live, the allocation of responsibility for one's contributions (both credit and blame), and the mutual checking of these contributions; in short, *curiosity, credit,* and *checking*. Scientists are curious about the world in which they live. They want to understand what makes it tick. For example, in his autobiography, Simpson (1978a: 274) notes that there is "currently much discussion about the motivation of scientists and other professionals. The motivations are numerous and usually complex even for any one individual. Those of us so oriented ponder our motivations. In the main, mine is surely that I was born with or somehow very early acquired an uncontrollable drive to know and to understand the world in which we live."

Most scientists enjoy research. It is their life. They feel especially fortunate to be paid to do what they want to do more than anything else in the world. The wow-feeling of discovery, whether it turns out to be veridical or not, is exhilarating. Like orgasm, it is something anyone who has experienced it wants to experience again—as often as possible. When electrodes are implanted in the orgasm center of male rats and these rats are allowed to produce orgasm by pressing a bar, they keep pressing it until they experience a seizure. When they come to, they drag themselves over to the bar and start pressing it again. Scientists are not quite this predisposed to the exhilaration associated with discovery, but they are close. In this book I pay very little attention to this aspect of science and instead concentrate on intermediate causes.

Any question in biology can be answered in terms of successively more distal but more ultimate causes (Mayr 1961). Once again, orgasm is a good case in point. Why do human beings engage in sexual intercourse? One answer is to have children, but this answer is far from complete. Throughout most of the history of the human race, our ancestors were very unlikely to have noticed any causal connection between sexual intercourse and reproduction. Until recently, so anthropologists tell us, the members of several extant societies still fail to recognize such a connection. If the conscious desire to have children had been the only mechanism to promote intercourse, we would not be here. But even among those people who are fully aware of the connection between the two, the actual motivation for intercourse is not always procreation. The most proximate cause for people engaging in sexual intercourse is that it feels good. Anthropologists add that it also enhances pair-bonding in a species in which considerable parental investment is needed to raise offspring. Population geneticists add that it also produces genetic diversity which facilitates evolution. Although these various causes can run at cross-purposes, they usually conspire to produce the same net effect. The same can be said for such factors as the joy of discovery, the desire for credit, and the search for truth in science.

Beetles, butterflies, and the magic of numbers have all served to fire curiosity in budding young scientists. Which natural phenomena happen to interest a particular scientist is largely a contingent affair. That scientists are curious is not. It is play behavior carried into adulthood. Once a young scientist is seduced into the profession, this curiosity is sometimes squelched by the tedium of academia, but for those who survive, a second motivation is added to the first—the desire for credit.

Although the desire for credit may not be as admirable as seeking knowledge for its own sake, it is nevertheless a powerful spur to action. Lightfield (1971) found a strong correlation between how early a scientist is cited and how productive he or she turns out to be. Of the eighty-three sociologists he studied, 73 percent who were cited during the first five years of their careers continued to publish, while only 6 percent of this group dropped out of sight. Only 2 percent of those who were not cited during the first five years after receiving their degree had productive careers. None too surprisingly, Zuckerman (1977: 145) discovered that the most successful scientists not only publish a lot but also begin to publish early. It would seem that reinforcement is as effective with scientists as with everyone else.

As Alexander (1979: 104) sees it, the aim of sociobiology is to find "reasonable ways to interpret all human behavior as either actually maximizing inclusive fitness or else representing the surrogates of such behavior under dramatically or rapidly changed environmental conditions." In science, Alexander's surrogate is credit. Scientists receive various sorts of credit both from without and from within the discipline. Scientists tend to disparage any credit they receive from outside the profession. Scientists must seek first and foremost to have their work accepted by their peers, not by government officials, science reporters, or the general public. Any other behavior is demeaning, embarrassing, inappropriate.

Lamarck detracted from his own reputation by issuing weather reports from his apartment in Paris. Geoffroy St.-Hilaire damaged his reputation by appealing to the general public for approbation. Some of the negative reaction to Wilson's sociobiology resulted from the publicity it received in the popular press. Dawkins has found his own career affected negatively by the popular success of *The Selfish Gene*. Gould's reputation also suffered when he caught the attention of the mass media. His column in *Natural History* was acceptable, with some reservation, but a cover story in *Newsweek* was not. Raup (1986: 164) terms the loss of professional reputation that occurs when a scientist's name becomes displayed too prominently in the popular press "saganization," after an especially extreme instance of this process, Carl Sagan. Both Ridley (1986) and Dawkins (1986) in turn complain of the attention paid to cladistics in the popular press.

The disapprobation attached to outside attention applies even when scientists, as they periodically do, take up various noble causes, causes sometimes closely connected to a scientist's own area of research. Ehrlich's involvement in ecology and population control generated all sorts of grousing from some of his fellow scientists. "He couldn't make a name for himself in real science, so he goes on the Johnny Carson Show." When the distance between a scientist's own area of expertise and the social cause increases, the damage to the scientist also increases. For example, Haldane's advocacy of numerous left-wing causes, especially his refusal to dissociate himself from the Communist party during the Lysenko affair, impeded his career in science. Similarly, all the fuss over sociobiology has damaged the reputations of those scientists who became too intimately caught up in its interconnections with social policy, whether left- or right-wing.

The sort of validation and approbation that is supposed to count in science must come from within the scientific community. Medals and awards are all right just so long as scientists are the ones who decide who receive these honors. Even so, scientists feel called upon to disparage these more superficial forms of recognition. Immediately upon commenting on his own uncontrollable drive to know, Simpson (1978a: 274) observes:

> I have not aspired to promotion and office; on the contrary I have refused opportunities for them. I have received many tokens of recognition such as honorary degrees, medals, and other awards, honorary memberships in learned societies, and other such things. I could not quarrel with anyone who felt that I have received more recognition than I have deserved, and I cannot deny that I enjoyed it. But I have never done anything motivated by desire for recognition and I have never sought it.

The tokens of recognition that scientists receive serve several functions. Sometimes they are awarded to young scientists to encourage them to continue careers in science, especially in the absence of monetary rewards. For example, when Huxley returned from his voyage on the *Rattlesnake* in 1850, he was penniless. During the next four years, while he searched frantically for some way to support himself in science, he was made a fellow of the Royal Society in 1851 at the age of twenty-five and a year later was awarded a medal of the Royal Society. Such honors served not only to keep his spirits up but also to enhance his reputation and hence his chances of obtaining one of the few positions in science available in Great Britain at the time. Over a century later, the American Society of Parasitologists awarded Dan Brooks its Henry Baldwin Ward Medal toward that same end.

The really prestigious awards, however, tend to be reserved for aging scientists as a reward for a lifetime of contributions. Perhaps aging scientists are no longer capable of producing first-rate work, but at least they can busy themselves with writing introductions to anthologies and can enjoy an occasional honorary degree or gold medal.

These awards also serve to legitimize the positions which these scientists have championed. For example, in 1864 the president of the Royal Society, Sir Edward Sabine (1788–1883), attempted to except Darwin's theory of evolution from the grounds for the Copley Medal the society was about to award him, but Darwin's supporters were having none of it. When Sabine expressed his intention in his awarding speech, Huxley rose and in a surprisingly temperate voice asked for the minutes of the meeting to be read to see if they included such reservations. They did not, but Sabine (1864) published his address as delivered (Darwin 1899, vol. 2: 213, Huxley 1901, vol. 1: 275).

A century later, this same story was replayed at the American Museum of Natural History. Nelson and Rosen considered Hennig, Croizat, and Popper to form the three pillars of their view of, respectively, classification, biogeography, and scientific method. In the spring of 1973, McKenna and Rosen proposed that Croizat be made a corresponding member of the American Museum. Several of the staff were taken aback. One wrote to the chairman of the council to object to Croizat's candidacy on the grounds that he "is one of the wackiest parodies of a scientist of this or any time. He is not quite in the class of Velikovsky . . . but almost." He also sent copies of his note to Rosen and McKenna, asking that, as a personal favor, they toss out the memo after reading it because Croizat was "precisely the kind of fanatic who would enjoy suing someone." In spite of such protests, the management board of the American Museum was unanimous in electing Croizat a corresponding member.

Croizat's supporters at the Museum immediately set about trying to obtain the Museum's prestigious Gold Medal for him, but to no avail. As Rosen (1981: 3) recalled this sequence of events:

> That year [1975], however, battle lines were instantly reformed when Croizat's panbiogeography (and its consequences for Darwinian dispersalism) was brought into the debate. Following so closely on the heels of the stern challenge by cladistics, the advocacy of panbiogeography and its explicit criticism of traditional dispersalism proved to be intolerable to the museum's ardent Darwinians. Accusations were plentiful, but mostly they were about the morality and intellectual competence of the advocates of vicariance theory. It was to be many months before there were useful discussions about the conflicting ideas. In fact, one letter from an irate curator to the chairman of the museum's scientific council of curators expressed incredulity at the nomination of Croizat as a Corresponding Member of the museum and asked that the letter be disposed of lest it be used in a libel suit. Following a later recommendation by eleven staff members to award Croizat the museum's Gold Medal, another curator threatened resignation. A third curator stopped me one day in the museum's halls to ask why I (and others) continued to raise issues that were so clearly disrupting staff harmony. In turn, I asked if he would keep to himself, or share, new ideas that might be viewed as fundamental to our science. There was no reply then, or since, and that was years ago.

Nelson and Rosen were more successful in their attempts to honor Hennig and Popper. Gold medals are made prestigious in part by the people who receive them. Hennig and Popper were prestigious; Croizat for all his 10,000 pages of published work was not. Scientists are not supposed to aspire to such superficial symbols of accomplishment, but Croizat repeatedly railed in print and in his correspondence against Hennig receiving a gold medal while he himself was denied this honor.

Although being elected an honorary member of the Society of Systematic Zoology is not as prestigious as being awarded a gold medal by a major institution, it is nevertheless an honor. Nelson, Rosen, and Platnick were more successful in obtaining this honor for their candidates—Croizat and Hennig in 1975 and Brundin in 1979. The indirect validation that accrues with such awards was neutralized somewhat in this instance by the presence of the names of their chief opponents on this list as well—Mayr, Simpson, Sneath, Sokal, and P. J. Darlington. Farris did not succeed in getting Nelson elected an honorary member at the 1983 meeting of the society in Grand Forks, but Nelson successfully pushed the candidacy of Blackwelder in 1985 at the Baltimore meeting. At first, Nelson's nominating Blackwelder to be an honorary member of the society might seem to conflict with the validation function of such official honors. After all, Blackwelder (1977) opposed Hennig's principles of phylogenetic systematics as vigorously as he opposed those of the pheneticists and evolutionary systematists. According to Blackwelder, systematists should not clutter their classifications with idle speculations about phylogeny and the evolutionary process. Instead, the grand object of classification is to group objects according to their "essential natures." In the interim, Nelson had come to see the virtues of Blackwelder's position.

The overall effects on science of such honors are difficult to gauge. On the main, they seem harmless enough. Although Simpson admitted that he rather enjoyed the tokens of recognition he had received through the years—the shelves and walls of his office were filled with plaques and gold medals—he also claimed never to have done anything motivated by the desire for this sort of recognition. But some scientists talk about "their" future Nobel Prize with all the possessiveness that young women once reserved for the anticipation of "their" ring. Unless Wade's (1981) study of the race between Roger Guillemin and Andrew Schally to work out the molecular structure of the thyrotropin-releasing factor in the brain and Bliss's (1982) history of the discovery of insulin are extremely atypical, competition for Nobel prizes can become so intense that it can frustrate the scientific process as the contestants abandon avenues of research that would have presented greater long-term payoff and go for quick and dirty results (see also Taubes 1987). The recent history of AIDS research at the National Institute of Health in Washington, D.C., shows how much damage the prospect of a Nobel Prize can do to research (Connor 1987).

However, I have introduced these external and superficial sorts of credit primarily to dismiss them as not being central to the machinery that makes science go. Rather the sort of credit that really matters is *use*. Individual scientists want credit for their contributions, as much credit as possible. Scientists exhibit considerable resistance to many of the observations students of science make of the scientific enterprise, but they readily acknowledge that they crave for recognition, not the recognition of the public at large or even of that amorphous hodgepodge termed the scientific community, but of the few scientists working in their area whom they genuinely, if sometimes grudgingly, respect. Neither feature of science is of recent development. For example, in his autobiography, Darwin (1899, vol. 1: 55) remarked that, "though I cared in the highest degree for the approbation of such men as Lyell and Hooker, who were my friends, I did not care much about the general public."

Although Darwin does not explicitly say so here, even the recognition given by other scientists is not of equal value. The most important recognition that scientists can receive for their work is its use by those scientists working in the same area, including of course their closest competitors. No scientist can get very far without using the work of his or her fellow scientists and in using it tacitly admits its value. Furthermore, if a scientist wants to *support* his or her conclusions by reference to the work of other scientists, that work must be cited. Science is to structured that scientists must trade credit for support. In any one instance, they can get credit or support, but not both. But scientists need both. They must find the right balance between claims of originality and citations for support. Croizat tended to sacrifice the support he might have received in order to claim as much originality as possible.

Recently sociologists of science have used explicit citations as operational definitions of "influence." Explicit citations are far from infallible guides to influence, but they are good first approximations because they are part of the mechanism that makes science operate the way that it does. At the very least, positive citations indicate which work a scientist thinks lends greatest support to his or her own research. Scientists give credit not so much where credit is due but where it can be useful. In a word, science is a function of *conceptual inclusive fitness*. Scientists behave in ways calculated to encourage other scientists to use their work, preferably with ample acknowledgment of that use. Conversely, the best thing that one scientist can do for another is to use his or her work, preferably with explicit acknowledgment. Winning in science is getting your name on the cup, but the cup itself is never retired. Medawar (1972: 104) finds the idea that scientists ought to be indifferent to matters of priority to be "simply humbug. Scientists are entitled to be proud of their accomplishments, and what accomplishments can they call 'theirs' except the things they have done or thought first?"

Numerous commentators on science have noticed that it is a matter of competitive cooperation. Merton (1973: 273) argues that the competitive cooperation that is so characteristic of science is a product of the institutional accent on originality, while Popper (1975: 78) sees control in science stemming from "friendly hostile cooperation of scientists which is partly based on competition and partly on the common aim to get nearer the truth." In biological evolution, cooperation and competition are hardly simple and straightforward. For example, the need to reproduce forces various sorts of cooperation upon organisms—cooperation between mates in some sexual organisms, cooperation between neuters and their sexual congeners in eusocial organisms, and sometimes cooperation between parents and offspring. Because the interests of the organisms involved in such relations are not always coincident, cooperation can blend imperceptibly into conflict, whether between males and females, siblings, or parents and offspring. The most problematic sort of cooperation among organisms is, however, between those that are not closely related genetically. Social organisms must cooperate with their sexual competitors if they are to survive and reproduce. No problems arise for neuter organisms because the only way that they can reproduce is by means of their sexual congeners. However, sexual organisms are presented with a problem. The only organisms with which they can cooperate are also their sexual competitors. To the extent that tendencies for cooperation have a genetic basis, too much cooperation can result in the elimination of these genes. However, the sorts of biological relations that evolve are as accurately characterized as mutual exploitation as cooperation. Each organism gives as little as possible while trying to get as much as possible in return.

Hypocrisy and romanticism to one side, the same can be said of scientists. Scientists cooperate to the extent that they do because it is in their own self-interest to do so. The degree of cooperation among scientists varies directly with the degree of benefit that the scientist receives. Similarly, the severity with which an infraction in science is punished varies directly with the number of scientists it harms. For example, scientists punish fabricated research so severely because it harms everyone who uses this research. Appropriating someone else's contributions is punished much less severely because it harms only the scientist whose work has been appropriated. The asymmetry results from the fact that numerous workers can use the results published in a particular paper. If these results are mistaken, every one who uses them has their research set back. Hence, all of those scientists who were burned by a particular worker are much less likely to use his or her work in the future. Some scientists also get reputations for producing results that are not trustworthy. Other scientists need not wait until they themselves have personally suffered before they learn to avoid using the results of an unreliable worker. Social learning is possible.

The situation is markedly different with respect to priority. If a particular data set is extremely good, it makes no difference to the scientists who use it whether the author listed first did the research or merely appropriated the labor of one of his or her assistants, whose name may be listed last or not at all. As far as *use* is concerned, it makes no difference who the real author of this data set actually is. As far as *support* is concerned, the more prestigious the author the better. Hence, the misassignment of credit can be quite useful to other scientists. In 1859 citing Darwin for natural selection was likely to generate more support than citing Wallace, questions of justice aside. However, if the disparity between contributions and credit become too great, the system will break down. By and large, it is to everyone's general benefit if published results are dependable and if credit is assigned accurately, but as is the case with other general benefits, the effects are so diffuse that actual punishments and rewards are correspondingly diluted. It was certainly not nice of Agassiz to appropriate the work of his junior colleagues, but no one stopped citing Agassiz's work for that reason.

The hubbub a few years back over Sir Cyril Burt is a good case in point. Because certain workers did not like the hereditarian bent to Burt's research, they began to look more carefully than usual at the papers he published alone as well as with collaborators. The succession of consistently high correlations looked somewhat suspicious. Burt was dead, but the inexplicable disappearance of his collaborators was even more suspicious. Finally, it came out that he had fabricated not only his collaborators but also much of his later data. No one was very excited that Burt took the credit later in life for the work done by more junior workers. A contribution is no less valuable because its authorship is misassigned. However, when Burt's fabrications were uncovered, the validity of the work of everyone who had cited Burt's suspect papers in support of their own conclusions was automatically brought into doubt—and this was a large number of papers indeed.

If this hypothesis is correct, then publications in active areas of research should be cited more frequently than those in moribund areas—a claim that would be a tautology if "active area" were defined operationally in terms of numbers of citations. None too surprisingly, the cases of fabrication that take the longest to surface are those that occur in scientific backwaters (Broad and Wade 1982). One message, then, is that, if you want to fabricate scientific contributions and not get caught, you should pick an area that no one is interested in. Of course, one consequence of such a choice is that no one will use and/or cite your work. Scientists who want to succeed in science must take the chance of being proven wrong.

Large areas of science are interconnected. A mistake or outright fabrication in one area is liable to have effects in other areas. One reason scientists cheat so rarely is that they suspect that they are very likely to get caught. This conviction is illustrated by an anecdote told by Lewis Branscomb (1985: 421) about a graduate student who needed some way to determine the temperature of the atmosphere at 1,000 kilometers above the earth's surface. As luck would have it, he happened across just such a technique:

> Delighted to find from the literature that his thesis problem could be successfully attacked, the student set about reproducing the experiment described in *Nature*. After months of fruitless effort, he became suspicious that the results reported were in error and even that the photograph published with the text was not a picture of the Schumann-Runge spectrum at all. Indeed, it appeared that the results might have been fabricated from the proverbial whole cloth. In any case, six months of a predoctoral fellowship were lost, and another way to tackle the thesis problem had to be found.

That graduate student was Branscomb himself. Anyone raised in science is aware that armies of graduate students, not to mention fully accredited professionals, are out there searching the literature for something that they can use in their own research. If they light on your work and it turns out to cost them time and effort, you are in trouble. Dismay is sometimes registered about all the attention that is paid to *who* made a particular discovery. Scientists are not all that interested in giving other scientists credit when things work out right, but they surely want to know the identities of scientists whose work has led them astray, so that they will not make the same mistake twice. One reason why the increase in multi-authored papers disturbs scientists is that responsibility is diluted. When the research turns out to be first-rate, everyone listed as an author is happy to accept credit. However, when things start going wrong, author after author signs off. He or she really did not have much of a hand in doing the research, or not that part of the research, etc. Perhaps attaching one's name to publications encourages vanity and greed for fame, but it also allows the focusing of responsibility, for good or ill. As the system is now structured, one cannot pursue credit without risking blame; nor can one cite the work of another author in support of one's own research without conferring credit (Grinnell 1987).

Periodically, surprising anomalies do crop up. For example, one would think that a significant error in anything as fundamental to physics as the speed of sound would not go undetected for long. However, recently, a senior research officer of the National Research Council of Canada, George S. K. Wong, discovered that the figure of 741.5 miles per hour that had appeared in all standard handbooks and textbooks since 1942 stemmed from a calculation error. The speed of sound was actually a half mile per hour slower. More recently, Hsui (1987) has argued that the geophysically determined Newtonian gravitational constant is consistently larger than the laboratory value by 1 to 2 percent.

Biologists are hardly immune to such errors. In 1923 T. S. Painter counted forty-eight chromosomes in human beings, a mistake not discovered until 1953. Early papers on the discovery of what came to be known as insulin by F. G. Banting and C. H. Best provide another example of the role of testing in science. They announced their "discovery" in 1921 and published the results of their experiments in 1922. By the end of that year, a British researcher, F. Roberts, published a note in the *British Medical Journal*, pointing out serious problems with these experiments. According to Roberts (1922: 1193), "The production of insulin originated in a wrongly conceived, wrongly conducted, and wrongly interpreted series of experiments." J. J. R. Macleod, Banting's superior who would eventually share the Nobel Prize with Banting, responded to Roberts. He did not deny any of Roberts' specific charges. Instead he responded that, despite all the errors and misunderstandings revealed in these papers, the net effect was the discovery of insulin. In the next issue of the journal, a letter by H. H. Dale appeared seconding Roberts' position. Banting and Best deserved credit for committing a "productive blunder" (Bliss 1982: 203–11).

A few years back, two journalists, following in the tradition of Charles Babbage's (1830) *Reflections on the Decline of Science in England,* documented as many cases as possible of "fraud and deceit in the halls of science."[2] Although Broad and Wade (1982) discuss only thirty-four cases, they take them to be the tip of the proverbial iceberg. They argue that fraud and deceit have been characteristic of science from its inception. More seriously, in recent years both seem to be increasing. Of the thirty-four cases which Broad and Wade discuss, seven involve very early scientists, from Hipparchus and Ptolemy to Newton and Dalton. The rest are scientists during the past two centuries. However, in their discussion, Broad and Wade run together several importantly different sorts of "fraud and deceit" in science.

Five of the examples of fraud that Broad and Wade (1982) discuss can be ignored because they involve people who can be counted only marginally as scientists—an explorer, a parapsychologist, and several unknown hoaxsters. Four additional examples can be discounted because they involve plagiarism, not the falsification of research findings. When commentators on science, such as Merton (1973: 276), remark on the "virtual absence of fraud in the annals of science" and the "unparalleled" degree to which scientists police themselves, they are referring to a particular sort of "fraud." It is certainly true and remarkable that scientists have been so rarely accused, let alone convicted, of financial embezzlement. With rare exception, the money that scientists get for research goes for research (Shapley 1978). In one instance, when a scientist lost outside financial support for his research, he supported it out of his own pocket until his money ran out and then proceeded to make and sell drugs (Smith 1980). Sometimes scientists use money obtained for one project on another. Sometimes they put in for

2. In his book, Babbage (1830) was mainly concerned to indict the pitiful amount of public money that was being spent in England at the time, in comparison with such countries as France, as well as the sorry state of the Royal Society. His chief target was Captain Sabine, the man who in his presidential address to the Royal Society tried to exclude Darwin's theory of evolution as one of the grounds for Darwin's receiving the Copley Medal. Babbage found Sabine's scientific achievements paltry and the rewards he received excessive. In fact, Babbage devoted very little space to discussing fraud in science. For a critical reading of Babbage's book, see Hall (1984).

money for a fundable project when they are reasonably sure that they are going to spend it on another project, a project that they take to be more important though possibly less fundable. But the money that they obtain for research does go for research. The fact that so little is made of this fact indicates something about the general attitude we have about scientists. We expect a fairly high rate of financial fraud in other professions, but not in science. One major exception to the claim that money obtained for research actually goes for research is the increasing percentage of research money that goes for maintaining the infrastructure of science, especially in areas of biomedical research. Although this infrastructure may well be necessary for big science, a very large chunk of money spent for research contributes to it only indirectly.

Embezzlement, however, is not the sense of "fraud" that Merton has in mind when he notes the remarkable lack of fraud in science. Nor does he intend to be referring to the failure of giving credit where credit is due—the theft of someone else's contribution. Priority disputes and claims of plagiarism are fairly common in science. In fact, graduate students grudgingly expect not to get as much credit for their work as they are likely to get once they are out on their own and are no longer dependent on the offices of their research adivsor. Although not giving credit where credit is due is less than honorable, it is not an instance of the sort of "fraud" that sociologists claim is rare in science.

Although percentages vary from one area of science to another, the vast majority of citations in scientific journals are positive, while only a very small fraction are negative. That is, most citations are used to support one's own findings, while only a few works are cited as being mistaken. For example, of 575 papers published in the *Physical Review* between 1968 and 1973, 87 percent of the citations were positive (Moravscik and Murugesan 1975). In the works studied by Chubin and Moitra (1975), 95 percent of the citations were positive, and of the remaining 5 percent, none were totally negative. The primary purpose of positive citations is not so much to give credit where credit is due but to gain support for one's own work (Goudsmit 1974). The only way to gain support is by citation. Appropriating someone else's work without citing it may inflate one's own claims to originality, but needless to say, such acts generate no support.

Hence, positive citations should be primarily to the work of those scientists whose support is worth having, and this number does not include graduate students. In addition, the easiest work to appropriate is that of one's graduate students. However, such practices have their costs, since one's graduate students are also one's scientific heirs. As in biological evolution, parent-offspring conflicts can and do occur. The Merton school of sociology of science serves as a nice exemplification of its own principles. In a citation analysis of Merton's work, Cole (1975) found that 42 percent of citations to Merton were largely "ceremonial," for the legitimation of the authors' own work. Few Mertonians bothered to test Merton's ideas; they just used them. Of 123 articles dealing with Merton only 7 were critical. In general, when the Mertonians have investigated the social organization of the Mertonians, they discovered that everything was just rosy (Cole and Zuckerman 1975). If they had found otherwise, they would have falsified their own view of the nature of science.

Negative citations also serve important functions in science, chiefly to establish boundaries between us and them, boundaries that are reinforced by different terminologies. For example, within a year of each other, several cladists published three general works on cladistic analysis—Eldredge and Cracraft (1980), Wiley (1981), and Nelson and Platnick (1981). If one looks at the patterns of citations in these three works, two trends are obvious: the positive citation of one's own work and the work of one's allies coupled with the negative citation of the work of one's opponents (see Appendix F). Of the eleven most cited works in Eldredge and Cracraft (1980), six are by Simpson, Mayr, and Bock—opponents of cladistics. Hennig comes in fifth. Of the fourteen most frequently cited works in Wiley (1981), five are by Simpson, Mayr, and Ashlock, but in Wiley's book Hennig leads the list of most frequently cited works. Nelson and Platnick's (1980) book differs from the other two in that it emphasizes the history of systematics and biogeography much more strongly, and these two authors consciously keep references to contemporary disputants to a minimum. Even so, of the ten most cited works, two are by opponents—Darlington as well as Sneath and Sokal. Two of the most frequently cited works are also by Hennig. If one excludes self-citation, five of the seven most frequently cited works in these three books are by opponents of their authors. Obviously the views that one opposes play an important role in any general exposition. For all the criticism that citation analyses have received of late, the summary list of the most frequently cited authors in the three works under discussion accords well with my own intuitions of the scientists whose work has been most influential in disputes over cladistics—Mayr, Simpson, Hennig, Nelson, Farris, Patterson, Bock, and Rosen (see also Appendix F).

A citation analysis of the literature of cladistic analysis reveals an additional pattern—a division between those who are interested in constructing computer programs for generating cladograms and nonquantitative cladists who are intent on working out the basic logic of cladistic analysis. Farris and Nelson epitomize these two groups. Although these two men joined forces in the middle 1970s to promote cladistic analysis, they rarely cited each other's work. On the basis of a citation analysis of the writings of these two men, one would conclude that they did not belong to the same conceptual deme, let alone research group. However, this diremption in citations has not been accompanied by a social division. Just as Sokal and Sneath were the chief spokesmen for phenetics, Farris and Nelson remain the titular heads of the cladists. Sokal and Sneath, however, are separated by the Atlantic Ocean, while Farris and Nelson reside only sixty miles apart.

The sort of fraud that is most relevant to science is the publication of results that one knows to be false or else has no right to expect to be accurate. After nonscientists and cases of failure to give adequate credit are eliminated from Broad and Wade's list, twenty-five cases remain. Of these, five, possibly six, are not examples of outright fabrication but of the sort of "fudging" and "finagling" that are so central to science. If observations and the results of experiments came in discrete atoms that are either clearly true or clearly false and nothing in between, scientists could report the facts and nothing but the facts, but this is not what the world is like at the cutting edge of science. Although scientists cannot hope to succeed if they finagle at all costs, they also cannot dumbly report their research findings in as raw a form as possible and hope to contribute much of value. For example, Westfall (1978) has shown that even the great Newton fudged his data on occasion, and Holton (1978) has shown in great detail how Robert Millikan succeeded in estimating the magnitude of the charge of the electron when Felix Ehrenhaft failed because Millikan carefully selected the data that he used while Ehrenhaft doggedly reported all of his.

The borderline between the sort of finagling that is absolutely essential to progress in science and the sort that frustrates this goal is not sharp. Those concerned with socializing future generations tend to take an extremely conservative position. For example, Ian Jackson (1984: 14), the executive director of Sigma Xi, in a special publication of the society, concludes his discussion of episodes such as the preceding by declaring that "Whether or not you agree that trimming and cooking are likely to lead on to downright forgery, there is little to support the argument that trimming and cooking are less reprehensible and more forgivable. Whatever the rationalization is, in the last analysis one can no more be a little bit dishonest than one can be a little bit pregnant. Commit any of these three sins and your scientific research career is in jeopardy and deserves to be." If one dishonest act makes one dishonest, then everyone, including scientists, is dishonest. In this night all cows are black. If the notion of dishonesty is to be of any use, degrees of dishonesty must be recognized. One dishonest act differs from several thousand, and certain sorts of dishonesty are more serious than others.

Zuckerman (1977: 116) also takes an extremely conservative view of such matters, commenting that if Newton actually did fudge his calculations the way that Westfall claims, then this would be a historic case of deviance. But if the most successful scientists in the history of science stand repeatedly convicted of "deviance," especially in those instances in which they are most successful, then possibly the relevant norms need reworking. At the very least, anyone studying science must distinguish between professed norms and those that are actually operative. As it turns out, only about half of the cases Broad and Wade (1982) discuss actually concern legitimate scientists publishing research that they knew was unwarranted if not downright false.

In ordinary moral judgments, intentions usually are considered by professional philosophers and the general public alike to be extremely important. Getting in a car and intentionally killing someone is considered a serious crime, while intentionally getting drunk, intentionally getting in a car, and unintentionally killing someone is considered much less serious. On this view, the best way to stay morally unaccountable is to make sure that one's intentions rarely influence one's behavior. In her discussion of deviant behavior and social control in science, Zuckerman (1977: 115) plays down the role of intentions—"fraud is fraud whether intentional or not." Schmaus (1981, 1983), a philosopher, responded that terms such as "fraud" carry moral connotations and, hence, for moral claims intentions do matter. The publication of data one knows to be mistaken is morally different from the publication of sloppy work. One is fraud, the other negligence.

Zuckerman (1984) responded that, regardless of appearances, she was not using such terms as "fraud" and "negligence" in a moral sense. But on one point Zuckerman and Schmaus were in agreement: intentions are irrelevant to the *effects* that the publication of erroneous results have on science. Scientific investigations are well calculated to uncover factual errors. The intentions of the scientists themselves are quite another matter. Although I know of no data bearing on this issue, I suspect that scientists do not much care about the intentions of the scientists whose work they use. All they want is for it to hold up under scrutiny. The damage done to their own research is the same whether the errors were introduced intentionally or unintentionally (for one example, see Holden 1987a).

Of the eighteen examples of genuine fraud that Broad and Wade (1982) discuss, it is striking how many of these fabrications were perpetrated by students, especially students trained originally in countries without much in the way of a scientific tradition. A likely hypothesis is that these students had not internalized the system of institutional norms that sociologists such as Merton claim gives science its special character. An unusually high percentage of these cases of fraud also occurred in medical research. Apparently, when we ourselves are the research organism, pressure increases and with increased pressure comes an increase in fabrication.

In 1985 at a meeting of the American Association for the Advancement of Science, William F. Raub, Deputy Director for Extramural Research and Teaching of the National Institute of Health, presented a paper in which he reported that only 50 cases of serious fraud were detected out of 20,000 projects supported by the institute during a five-year period. Most of these cases involved such things as sloppy record keeping, the rest outright falsification of data. In response to one especially egregious case, that of John R. Darsee, the National Institute of Health set up a panel to see what could be done to stem the rising tide of fraud in the science. Chief among the recommendations that ensued were increased supervision and formal acknowledgment of responsibility. Once again, individual responsibility seems to be the key to appropriate behavior. If one is willing to take the credit for producing results that other scientists find useful, one must risk discredit if these results turn out to be faulty and mislead the scientists who used them (see Culliton 1983, 1986, 1987, Norman 1984, 1987, Marshall 1986, Crawford 1987, Koshland 1987, Laderman 1987, and White 1987).

Even though several commentators on science regard a rise in fraud and plagiarism as a serious problem, apparently scientists themselves do not share this concern. Of the 4,113 scientists and engineers who responded to a poll conducted by Sigma Xi, a scientific research society, 43 percent thought that the two most important issues facing the scientific community are the interruption of funding and the lack of public understanding of science. Even when fraud and plagiarism were lumped together, only 6 percent of the scientists who responded were especially worried about it (Jackson 1986).

Among the episodes in science that I have studied, allegations of outright fabrication were extremely rare. One is famous—the case of Gregor Mendel. Thus far I have discussed Mendel only in connection with the influence of his papers on the development of genetics at the turn of the century. The current consensus among Mendel scholars is that the founders of Mendelian genetics not only stumbled upon Mendel's paper in time for it to influence their own productions but also that it did. Granted, early Mendelians read Mendel from their own perspective, ignoring those parts of his theory that were not central to their own concerns, and reading into Mendel positions that Mendel did not actually hold. For example, Mendel thought that when the "elements" which he postulated are different, they remain materially distinct. When they are the same, they fuse. Hence, he expressed his ratios as "1A : 2Aa : 1a" instead of the familiar "1AA : 2Aa : 1aa." In this respect and several others as well, Mendel is no Mendelian (Olby 1979, Brannigan 1979).

During the early years of Mendelian genetics, the advocates of this new research program accepted Mendel's work at face value—a face seen in the mirror of their own interests. It was left up to one of the opponents of Mendelian genetics, Weldon, to expose Mendel's experiments to careful inspection. Even though he would like to have found fault with them, he could not (Kevles 1980). However, much later, motivated by repeated claims that Darwin's influence was responsible for the neglect of Mendel's work and that Mendel himself was hostile to Darwin's theory of evolution, Fisher (1936) reexamined Mendel's original research and concluded that his results were too good to be true. Is Mendel's 1866 paper "one of the great classics of the scientific literature, a model scientific report, clearly setting out the objectives, concisely presenting the relevant data, and cautiously formulating truly novel conclusions" (Mayr 1982: 710) or were the "data of most, if not all, of the experiments" that Mendel reported "falsified so as to agree closely with Mendel's expectations" (Fisher 1936)? Fisher himself discounts conscious falsification by Mendel, though possibly not by an assistant, shades of Kammerer. Instead, he concludes that Mendel unconsciously biased his later experiments to agree more closely with his expectations.

Several defenses have been mounted to charges that Mendel "sanitized" his data. One defense is that Mendel's results were similar to those reported by other pea hybridizers at the time, scientists who can hardly be accused of biasing their data to accord more closely with Mendel's expectations. If any "sanitizing" was going on, it was independent of Mendel's theory (Weiling 1966, Orel 1971). A second defense is that garden peas, because of peculiarities in the way that they produce pollen, tend to give overly good results (Thoday 1966). Thus, Mayr (1982: 720) concludes that "there was really nothing drastically wrong with Mendel's figures; indeed, Mendel was an almost pedantically precise recorder of data" (see also Campbell 1976).

More recently, Root-Bernstein (1983) has emphasized two related explanations for the overly good concordance between Mendel's expected ratios and the actual numbers he recorded. The first is that the statistics which Fisher used to evaluate Mendel's work are inappropriate in several respects (see also Weiling 1986). One source of this inappropriateness is that the character states which Mendel had to score are not always sharply distinguishable. In 1911 Raymond Pearl crossed two varieties of maize which differed on two characters—yellow and white kernels, starchy and sweet endosperm. He gave the same 532 kernels to fifteen trained scientists and received different results from all fifteen in how they classified them. Root-Bernstein repeated this experiment, using fifty undergraduates at Princeton University enrolled in an introductory biology course, and had similar results. Too many kernels were intermediate. However, when these students were forced to shift these indeterminate kernels into one of the preestablished character states, the results were similar to Mendel's—too good.

A fairly common assumption of Mendel's defenders is that he needs defending. A spectrum exists between the sort of blatant fabrication of data to support one's views apparently practiced by Burt in his later years and the inevitable biasing of data that results from the sort of simplification, structuring, and "common sense" necessary in science. Science cannot be carried on effectively by intelligent ignoramuses. Purely objective, unbiased collection of raw data is a myth that has distorted our understanding of science for long enough. Scientists cannot wipe their minds clean. Even if they could, they would be unwise to do so. If Mendel let his accumulated understanding of inheritance in garden peas influence both the construction and interpretation of his experiments, he is in good company. If this is fraud, then fraud in science is widespread, possibly universal. It is especially prominent in the work of the scientists who have made the greatest contributions to science. It is just as prominent in my own work. Classifying systematists into patterns and phylogenetic cladists is no easier than classifying kernels of corn into starchy and sweet.

The other example of problems with data that I relate in the first half of this book concerns the data sets that Mickevich and Johnson (1976) and Mickevich (1978) presented to support the superiority of cladistic methods over those promoted by pheneticists. No one has claimed outright fabrication, but doubts surfaced with respect to the reliability of these data sets. However, the seriousness with which the scientists treat data can be seen in the scrutiny to which Mickevich's studies were exposed—by her opponents. For example, Sokal (1985: 16) warns that "many of Mickevich's results are not reproducible and that, in fact, when the computations are repeated in what is believed to be a correct mode, there is little difference in congruence between phenetic and cladistic classifications. Thus, the citation of her work by a number of cladist authors (Farris, 1979, Wiley, 1981, Schuh and Farris, 1981) in support of cladistic classifications may be unfounded."

Conclusion

In summary, science is a matter of competitive cooperation, and both characteristics are important. The most important sort of cooperation that occurs in science is the use of the results of other scientists' research. This use is the most important sort of credit that one scientist can give another. Scientists want their work to be acknowledged as original, but for that it must be acknowledged. Their views must be accepted. For such acceptance, they need the support of other scientists. One way to gain this support is to show that one's own work rests solidly on preceding research. The desire for credit and the need for support frequently come into conflict. One cannot gain support from a particular work unless one cites it, and this citation automatically confers worth on the work cited and detracts from one's own originality.

Just as organisms in general behave in ways likely to increase their own genetic inclusive fitness, scientists tend to behave in ways calculated to increase their own conceptual inclusive fitness. In neither case are the entities involved necessarily aware of what they are doing. Flour beetles are totally unaware that such a thing as genetic inclusive fitness even exists let alone capable of performing the required calculations. Scientists are not appreciably different in their quest for conceptual inclusive fitness.

The functioning of science had better not depend crucially on widespread self-awareness among scientists about their own motivations and the effects of their actions, because most scientists are no more self-reflective than the public at large. They are intensely aware of their subject matter. As far as they are concerned, they themselves do not count. Scientific publications are always written in the passive voice. Are all human actions genetically selfish? "Yes." Including your current disputes with the Marxist opponents of sociobiology? "Oh, I never thought of that."

Once one identifies the operative norms of research science, the explanation for the high frequency with which individuals adhere to these norms becomes obvious. It is in their own self-interest to do so. In saying this, I do not mean to deny that scientists really do value good research, want to behave honorably and feel good about themselves, but comparable feelings and motivations are commonplace throughout human societies without much effect. College professors in their roles as professors feel no less benign than scientists in their roles as researchers. Even so, such benign feelings lead us to teach as little, rather than as much, as possible. We want fewer classes, smaller enrollments, teaching assistants to grade our exams, and so on. No one had a higher opinion of their own beneficence than the upper classes in Victorian England, a high opinion that we today do not share. People are very good about not seeing the effects of their actions. I do not know how operative the lofty attitudes professed by scientists actually are in making science as an institution as productive as it is. If scientists in general did not hold these ideals, perhaps the institution would not function so well. However, I do know that scientists do not have to rely solely on calls to do their duty. The institution is so structured that by and large they are rewarded for doing what they are supposed to do and punished when they do not.

In evaluating the preceding claim, two distinctions must be made: between failure to give credit where credit is due and publishing faulty work; and, with respect to the latter, between the sort of judicious finagling that is so central to science and outright fabrication. Priority disputes are common in science. Sometimes outright theft is involved, but usually it is not. Each scientist tends to see his or her contributions in the most positive light possible. Because their contributions are so important to them, they naturally read this inflated importance of their own work into the work of others. Various shades of finagling are also quite common in science. Science is not an activity that lends itself to rote procedures. Time and again in the course of investigation, scientists are forced to exercise their judgment with respect to which data are good and which can be rejected, which factors must be taken into account and which ignored, and so on. They cannot tell the whole truth and nothing but the truth because at the cutting edge of science no one can be sure precisely what this truth is. Scientists are liable to make enough honest errors without adding conscious fabrications to their vitae.

Excessive finagling as well as outright fabrication are rare. They are rare because in areas of active scientific research, erroneous claims are likely to be discovered, and whether they are intentional or not, punished severely. Scientists are so conscientious about producing dependable work because their allies tend to incorporate that work into their own, usually without testing it, and their opponents are just as likely to expose it to careful scrutiny. Erroneous views are liable to hurt one's opponents, but they are even more likely to damage one's allies. Prior to publication, scientists expose their manuscripts to the evaluation of their allies. Afterwards, critical scrutiny comes from their opponents. The self-correction so important in science does not depend on scientists presenting totally unbiased results but on other scientists, with different biases, checking them.

9

Secrecy and Bias in Science

If you have an apple and I have an apple and we exchange these apples then you and I will still each have one apple. But if you have an idea and I have an idea and we exchange these ideas, then each of us will have two ideas.

George Bernard Shaw

I am concerned about the effect of prizes on the practice of science. My main objection to prizes is that they promote an inappropriate metaphor. They bring the mentality of the sports world into the domain of the scholar. They suggest that there are winners in science, and that therefore, there must also be losers, that many are called but few are chosen. Prizes suggest that competition is the driving force of scientific achievement. I believe that the prize-fighter imagery is a perversion and wish to recall an alternative attitude: *In science, we are all winners.*

David L. Nanney

ONE POSSIBLE explanation for scientists, insisting that they receive credit for their contributions is that it is an accident of history. Because science arose in individualistic societies in which personal property rights were of central importance, scientists merely carried over into science the acquisitiveness programmed into them by early socialization. If science had arisen in more genuinely cooperative societies in which there was no strong sense of personal property, then perhaps science would have lacked this feature. Or, quite possibly, the desire for individual credit serves a positive function in science. If scientists did not desire and obtain individual credit, science would not work as well as it does. A strain does exist in science between various sorts of "ownership": individual ownership in the sense that only the scientist who comes up with a new idea or empirical discovery can use it, individual ownership in terms of the credit that scientists get for their contributions, and the communal ownership that results from scientists making their views known so that any scientist so inclined can use them.

Secrecy is the antithesis of communal access to knowledge. Early in the history of science, scientists were extremely secretive, in part because science grew out of traditions in which secrecy was embedded—wizardry and technology. Alchemists and the like kept their knowledge as secret as possible because only in this way could they benefit from it. Many commentators on science decry the emphasis that scientists place on getting credit for their contributions to science and the ensuing priority disputes, but some way had to be found to get early scientists to share their findings with their fellow scientists. Without such sharing, science would lack one of its most important characteristics. It would not be cumulative.

Early in the history of science, Bacon (1620) urged that knowledge be genuinely communal, scientists working for the general good and not for individual credit. Several early groups of scientists tried to incorporate the structure of Bacon's House of Salomon (Hall 1981) into the institutional structure of science, including the French Academy of Science. However, within twenty years of the founding of the French Academy in 1666, the academicians had concluded that undifferentiated credit was simply not working. As a result, in 1699, the new constitution of the Academy included the following article:

> It having been found by experience that there are disadvantages in the tasks to which the academicians apply themselves in common, each one shall choose a particular object for his studies, and by the account he shall give of it in the meeting, he shall endeavor to enrich the Academy by his discoveries and improve himself at the same time.

Enriching science while improving oneself at the same time was the format that the Royal Society of London adopted right from the start. One of the first devices invented to insure priority was the anagram. Scientists such as Galileo, Newton, Hooke, and Huygens translated statements of a new law or principle into an anagram and then made it public. While others tried to unscramble the anagram, the author could continue working undisturbed. Later, when he was ready or when his priority was threatened, he could reveal the solution to the anagram. Because the anagram method did not lend itself to the protection of mechanical contrivances, the first secretary of the Royal Society, Henry Oldenburg (1615–77), offered to seal any new invention in a box to be opened at the author's discretion. Such methods helped to insure credit, but they had one drawback—others could not use the content of anagrams and sealed boxes. The problem was how to make a contribution public so that other scientists could use it while insuring credit to its author. Oldenburg's solution was rapid publication in the *Philosophical Transactions* of the Royal Society. With all its shortcomings, this is the method that has come down to us for promoting individual ownership while allowing communal use.

The Refereeing Process in Science

As long as Oldenburg published everything he received, priority was easy enough to achieve, but under the influence of Thomas Sprat, an Anglican bishop and the author of the first history of the Royal Society (Sprat 1667), the *Philosophical Transactions* took on a second function as well—authentication. The authors who submitted their work to be published had no doubt about its value, but so many questionable papers had been published by then that those in charge gradually decided that they needed some assurance about the quality of the work that was receiving their imprimatur. Because Oldenburg did not know enough in all the various areas then being investigated to pass knowledgeable judgment on all of them, he began seeking the advice of outside experts, some of them close competitors of the authors whose works were at issue. Thus, publication took on three functions: making discoveries public so that anyone who chose to could use them, awarding credit for the contribution to the author of the paper, and conferring some authenticity to the publication. The members of the French Academy took authentication as one of their chief functions. Authors submitted their work, and members of the academy actually repeated the relevant observations and experiments. The academy was much like a tribunal before which scientists came to receive an official imprimatur. Refereeing, as it emerged, was not quite so rigorous. Referees were willing to read a manuscript and evaluate it in the context of their own knowledge and expertise, but they were not about to stop their own research and check all the raw data and rerun the experiments. Referees can detect implausible findings but not plausible fraud. They can also weed out errant nonsense, not to mention its very close relative—radical innovations.

However, scientists were reluctant to be held publicly responsible for the rejection of a manuscript by one of their colleagues. Professional relationships are sensitive enough without adding that extra weight. Right from the beginning, Oldenburg kept the identities of his "referees" confidential in order to insure candor. As is usually the case when the same structure attempts to fulfill more than one function, conflicts arose. The possible abuses of the refereeing system are well known. If referees are working on the same problem as the author of the manuscript under review, they can postpone returning their reports while they continue to work on their own research with the aid of the information obtained from the manuscript under review. And once they have made as much use of it as possible, they can then even recommend rejection because the work is out of date. In addition to fears of theft, recurrent charges of cronyism have characterized the refereeing process. There is always the suspicion that an editor who is a professional friend of the author will send the manuscript to referees who are also close professional friends; conversely, unfriendly editors can pick referees whom they know are hostile to the manuscript under review.

Two solutions have been offered to this problem—total confidentiality and total openness. According to the first solution, if the referees' identities are to be kept secret, then so should the identities of authors. That way the referee cannot know if the author of the manuscript under review is friend or foe, a major figure in the field or a graduate student. According to the second solution, the identities of both the author and the referees are made public. Like it or not, referees must stand behind their positions. Neither of these suggestions has met with much enthusiasm. The first has been dismissed as being largely ineffectual and somewhat detrimental. It is ineffectual because in the vast majority of cases, anyone qualified to referee a manuscript is also likely to be able to identify its author even if the name is omitted. One hint as to the identity of the author of a manuscript is the author most frequently cited in the bibliography, especially if the work is always cited positively. At the very least, the research group can be identified. In the early years of numerical taxonomy as well as cladistic analysis, most manuscripts submitted to *Systematic Zoology* could be placed with little difficulty in one camp or the other even without the aid of the identity of the author. Double-blind refereeing is ineffectual at best. It also has one serious defect. Although scientific research is published in droplets, each paper is part of an ongoing research program. A manuscript cannot be judged adequately outside the past work of the author. Keeping the identity of the author confidential interferes with the knowledgeable evaluation of a manuscript. Who the author of a manuscript is sometimes does make a cognitive difference.

The other alternative is to make the identities of the referees public. If someone wants to sabotage the work of one of his enemies, let him do it out in the open for everyone to see. The two most common objections to open refereeing are that keeping the identities of referees confidential is no more effectual than trying to keep the indentities of authors confidential and that it would discourage candor. It is certainly true that authors can frequently guess the identities of some of the referees of their manuscripts. Some authors have very distinctive writing styles; others distinctive points of view. For example, Colless had no trouble identifying Farris as the referee of his manuscript criticizing Mickevich. One dead giveaway is the list provided by the referee of the work that the author has not properly cited. It usually contains several papers by the referee. However, in general, referees' reports contain fewer clues to the identities of the authors than do manuscripts submitted for publication. If nothing else, withholding the identities of referees allows scientists to maintain at least the pretense that they do not know who gave a negative evaluation of one of their manuscripts or research grants.

Of greater importance than such guessing games is the relation between confidentiality and candor. Are hostile critics likely to tone down their appraisals once their identities are known? If so, this might not be a bad thing. In reading six years of referees' reports for *Systematic Zoology,* I was struck by the gratuitous nastiness that characterized so many of the reviews. Authors are at their most vulnerable during the refereeing process. They must turn over their brainchild to one or more anonymous referees to be judged. Even careful, judicious rejection is painful enough without enduring snide remarks made under the cover of confidentiality. For example, one referee of a paper by a young cladist agreed with the message of the manuscript but concluded that the author's writing style had all the flair of picking one's nose and eating the snot. In his history of the Nemesis affair, Raup (1986: 69) includes his own evaluation of the manuscript submitted by Luis and Walter Alvarez, Frank Asaro, and Helen Michel to *Science* claiming that a collision with a meteor was responsible for the mass extinction at the end of the Cretaceous period. He concluded his review of the efforts of this Nobel laureate physicist and his co-authors with the following observation: "If a graduate student gave me this manuscript to read, I would see it as a brilliant piece of work (indicating that the student has enormous potential) but I would give it back to be done right."

One might suspect that such pointed comments would be less frequent if the identities of referees were known. In this case, Raup did acknowledge authorship as well as note the likelihood that his comments might have been interpreted by the authors as somewhat sharp. One fact that counts against the hypothesis that making refereeing open would reduce overly negative comments is that the very same scientists who are so polemical in their confidential evaluations can be found making no less acerbic comments in print. Huxley was famous for his skillful invective. For example, he began his review of the tenth edition of Chambers' *Vestiges of the Natural History of Creation* (1853) with Macbeth's lament about the pertinacity of the murdered Banquo: "Time was, that when the brains were out, the man would die." In an exchange over the existence of atoms, Max Planck and Ernst Mach were no less vituperative. For example, in response to a critique by Planck, Mach (1970: 37–38) notes:

> The only real point of difference which has so far come to light concerns the belief in the reality of atoms. Here again, Planck can hardly find words degrading enough for such wrongheadedness. Anyone who enjoys psychological conjectures must read his lecture for himself, and is welcome to do so. After exhorting the reader, with Christian charity, to respect his opponent, Planck brands me, in the well-known biblical words, as a "false prophet." It appears that physicists are on the way to founding a church; they are already using a church's traditional weapons. To this I answer simply: "If belief in the reality of atoms is so important to you, I cut myself off from the physicist's mode of thinking, . . . I do not wish to be a true physicist . . . , I renounce all scientific respect—in short: I decline with thanks to the communion of the faithful. I prefer freedom of thought."

Farris is also well-known for his knack of turning a truculent phrase. For example, in his review of the proceedings of the NATO numerical taxonomy conference held in Bad Windheim, Germany, Farris (1985b: 97) remarked that "Felsenstein's discussion of computers in systematics is introductory, at a level that one might expect in *Family Computing*." As the preceding chapters indicate, editors, referees, and colleagues constantly enjoin authors to tone down their polemics. The committee that the council of the Society of Systematic Zoology formed to look into the editing of its journal strongly urged the censoring of all "abusive language."

Scientific opponents have shown themselves willing to pass negative judgments on each other as readily in print as in private in spite of repeated entreaties to hold down the polemics. The candor that would be sacrificed if the identities of referees were made public would be that of friends, not foes. Giving a manuscript an easy ride through the refereeing process does the author no good. Because everyone is working on his or her own research, serious criticism prior to publication is difficult to obtain. The refereeing process forces scientists periodically to provide careful evaluations of the work of their fellow scientists prior to publication. What is more, the criticisms that are most valuable to an author are those from someone who shares the same general outlook. Some professional friends can take considerable disagreement and remain friends; others can take anything but criticism. The chief function of confidentiality in refereeing is to allow scientists who are working in the same research program to offer criticisms while maintaining a good working relationship.

The hypothesis is plausible. Is it true? Because the refereeing process is confidential, evidence is difficult to obtain. Ideally one would like to compare evaluations made both publicly and confidentially by scientists working in the same and different research programs. Do the judgments passed by opponents differ significantly when made confidentially versus out in the open? How about professional supporters?

A second belief that is quite widespread among scientists is that editors can bias the outcome of the refereeing process by their choice of referees. For example, Garvey (1979: 84) in his book on communication in science remarks:

Like most scientists, most editors have attitudes about what constitutes quality in their journal. These attitudes are influenced by their theoretical bias, their methodological preferences, etc., and they tend to select consulting editors and referees who share them.

This hypothesis is also plausible, but is there any evidence to back it up? To answer both questions, once again I must return to the six years of referees' reports for *Systematic Zoology,* in particular the six manuscripts that Nelson sent out for multiple reviews. Did Nelson and Schuh send manuscripts they favored primarily to referees who were likely to respond positively and those they opposed to referees who were likely to respond negatively? How did Nelson and Schuh react to the referees' reports they received? Did they publish manuscripts by cladists that the referees voted against and reject manuscripts by noncladists even if the referees favored them? The preceding questions concern the possible influence of bias on the part of editors. Another set of questions concerns bias on the part of referees. Are there correlations between the allegiances of referees and the recommendations they make for the disposition of manuscripts sent to them for evaluation? Once again, because of the availability of evidence, I concentrate on the six years when Nelson and then Schuh edited *Systematic Zoology.*

Nelson saw 198 manuscripts, both articles and points of view, through the refereeing process during his tenure as editor. Of these, he published 133 and rejected 32. An additional 33 manuscripts also failed to appear, not because Nelson rejected them but because their authors never resubmitted after receiving the referees' comments. Nelson listed no manuscripts in this category for 1976 because he willed all these manuscripts to Schuh. Most of these did not appear. Hence, Nelson's acceptance rate of 67 percent is made to look a good deal higher than the eventual disposition of the manuscripts initially submitted to him might warrant. Of the 265 manuscripts processed by Schuh between 1977 and 1979, he published 156. In his editor's reports, Schuh does not distinguish between the manuscripts he rejected and those he returned for revision. Instead, he presents a single figure—109. On my count he rejected 75 manuscripts, and an additional 34 did not appear because the authors did not rewrite and resubmit.

Of the 32 manuscripts that Nelson rejected during his tenure as editor, most were rejected on the unanimous recommendations of the referees. There were seven exceptions to this claim. Nelson rejected two manuscripts for which the referees recommended only minor revision, because he judged them inappropriate for the journal. They were good papers but not the sort usually published in *Systematic Zoology.* He rejected two other manuscripts for the same reason on the advice of the referees. Even though the authors of these latter two manuscripts complained, Nelson held firmly to his original decision. The other three manuscripts that he rejected received mixed reviews, but in each case at least one referee recommended rejection. Only one of these authors rewrote his paper and attempted to get it published in the journal. He failed.

Conversely, Nelson overruled the negative recommendations of his referees in nine cases. In three instances he published manuscripts by cladists even though all or most of the referees found them unpublishable, but he did the same for manuscripts by two evolutionary systematists and even one numerical taxonomist.

No records exist for fourteen papers. In the final number that Nelson edited, he published ten papers presented at a symposium on marine mammals sponsored by the Society of Systematic Zoology. From all indications, he did not have these manuscripts refereed. Of the four remaining papers, three were responses to Mayr's (1975) critique of cladistic analysis by Rosen (1974), Nelson (1974a), and Hennig (1974). The fourth was a short paper by Nelson (1974b) on historical biogeography. Because Hennig's paper was merely a translation of Hennig's response to Mayr in German, there was no need to have it refereed. Records do exist for seven other papers that Nelson published without having them refereed. Three were short replies or notes, two concerned the profession, and two were substantial though largely descriptive papers by major figures in the field. Only Nelson's own two papers and the paper by Rosen raise any serious questions about the equitable treatment of manuscripts submitted to the journal.

Of the remaining manuscripts submitted to Nelson that did not appear in the journal during his editorship, the authors failed to return their revised manuscripts. Most manuscripts submitted to *Systematic Zoology* were judged by the referees to need either major or minor revision. Nearly all those authors who did revise and resubmit were successful. Strangely enough, the referees' reports for those manuscripts never revised and returned did not differ materially from the reports for those manuscripts that eventually did appear. Two papers which were returned to their authors for only the most minor revision were never resubmitted.

The 65 papers that were submitted to Nelson but did not appear during his tenure in office exhibited a second peculiarity as well—45 were by authors who were relative unknowns at the time and who never published any papers subsequently in *Systematic Zoology*. If they made a name for themselves, it was elsewhere. Of the remaining twenty papers, six were authored by evolutionary systematists, seven by pheneticists/numerical taxonomists, and seven by cladists. Of those that Nelson rejected, all but four were rejected on the advice of at least one referee belonging to the same school as the author. Of these four, one was submitted by an author at Kansas on numerical taxonomy. The single referee for this manuscript was also a numerical taxonomist and recommended outright rejection or major modification. The author rewrote and resubmitted. This time Nelson sent the manuscript to one pheneticist and one cladist. The pheneticist recommended major modification prior to acceptance, while the cladist recommended rejection. After numerous letters back and forth between Nelson and this author, Nelson rejected the paper.

A second pheneticist submitted a paper critical of cladists. Nelson sent it to two referees—an evolutionary systematist and a cladist. The evolutionary systematist found the paper ready for publication while the cladist recommended rejection. The author rewrote and resubmitted. Nelson sent this revised manuscript to a second evolutionary systematist who strongly urged rejection because the manuscript was merely a rehash of past work. Thus, Nelson rejected the paper without a negative evaluation by a referee belonging to the same school as the author.

Nelson also rejected two manuscripts by a leading cladist. The cladist referee for the first manuscript recommended only minor modification, while the numerically inclined referee presented several detailed objections that convinced the author to withdraw the manuscript. A second manuscript by this same author received high marks from its cladist referee, but the evolutionary systematist recommended major modification. The author never resubmitted the manuscript. Whether or not it appeared later in another guise is impossible to say.

If a manuscript bore on systematic philosophy, Nelson nearly always sent it to at least one cladist referee, but by and large the summary judgments of the referees did not break down along party lines. Cladists found papers by noncladists worth publishing and vice versa. For example, one second-generation cladist submitted three manuscripts to Nelson during the final year of his editorship. Nelson sent the first manuscript to two cladists, who judged it unpublishable. He sent the second to two noncladists who recommended rejection. He sent the third to seven referees of various persuasions who, in the main, were impressed by it and urged publication. The chief exception in this connection was a manuscript submitted to Nelson by a prominent cladist. The two cladist referees judged it ready for publication. The evolutionary systematist who refereed it found it not publishable. During this period, Nelson actually rejected a manuscript submitted by Hennig himself!

The boundary between Nelson's and Schuh's tenure in office is not sharp because Nelson willed Schuh numerous manuscripts in process, including a couple of serious problem cases. Of the 109 manuscripts that Schuh eventually rejected or that simply died while he was editor of *Systematic Zoology,* Schuh rejected 73 outright. Of these, he rejected 23 as inappropriate for the journal. The most common reason that Schuh gave for rejecting a paper as inappropriate was that it was too descriptive and had too little general import. He rejected five numerical studies of particular groups simply because they were applied numerical taxonomy. He rejected three others because the referees found them poor numerical taxonomic studies. He also rejected two papers because they dealt with problems in mathematics that seemed to have no clear bearing on systematics. He rejected most of these manuscripts without sending them out to be refereed—14 in all. The message was clear—descriptive manuscripts, especially applications of numerical techniques, no longer stood much of a chance of being accepted for publication in *Systematic Zoology.*

In general, Schuh was much more likely than was Nelson to reject a manuscript rather than to return it to be rewritten. Of the 36 papers that he returned for revision, 34 were never resubmitted. Schuh rejected eight papers by prominent evolutionary systematists, three papers by well-known pheneticists, and seven papers by leading cladists. The papers which he rejected were refereed by cladists and noncladists alike, and these referees found themselves in large agreement regardless of their philosophical differences. Schuh rejected two manuscripts by the Canadian paleontologist who had been the first to receive multiple reviews under Nelson's editorship. One was on the nature of cladograms, the other on ancestry. Schuh sent the cladogram manuscript to four referees—two cladists and two noncladists. Both cladist referees recommended rejection, one of the noncladists agreed that it should be rejected, while the other recommended acceptance as is. When Schuh rejected the paper, the author wrote back asking for the names of the referees. Schuh refused.

As in the case of Nelson's editorship, a high percentage of Schuh's referees were cladists, but by now there were many more cladists to choose among. Of those manuscripts dealing with the principles of classification, Schuh sent nearly all to at least one cladist referee, and, by now, the recommendations of the referees were beginning to split along party lines. For example, one manuscript co-authored by two leading cladists was judged ready for publication by the cladist referees, while the noncladist referee agreed that it was ready for publication, given the prevailing standards of *Systematic Zoology*. By any other standards, it should be rejected. A second paper by one of these authors received a clean bill of health from cladist referees, but the two noncladists disagreed, one finding it an "incoherent, illogical rave."

In a few cases, Schuh sent manuscripts by cladists solely to other cladists to referee and in just as few instances a manuscript by a noncladist just to cladists. But the taxonomic community had not become totally factionalized. Cladists still found fault with the work of their fellow cladists and voted in favor of noncladist manuscripts on occasion—and vice versa. For example, Schuh sent Colless's (1977) "A Cornucopia of Categories" to three cladist referees. One suggested publication after minor change, one publication after major change, and the third outright rejection. As a result of the negative tone of this third referee's comments, Colless wrote to Schuh requesting that no future manuscripts of his be sent to this referee. Schuh did not honor this request. He sent the next manuscript submitted by Colless to four cladists, including the referee whose comments Colless found so offensive. One of these referees gave no summary recommendation, two recommended publication after minor change, and Colless's nemesis urged rejection. After some correspondence, Colless withdrew his manuscript.

Cladists were hardly immune from criticism while Schuh was editor. A leading cladist submitted a major manuscript to Schuh which Schuh sent to two referees, one cladist and one noncladist. Both found the manuscript heavy going, especially because of its unconventional and sometimes downright misleading terminology. Schuh returned the manuscript and referees' reports along with extensive comments of his own. A year later, the author resubmitted the manuscript which Schuh sent to six referees—two evolutionary systematists, two pheneticists, and two cladists. One of the evolutionary systematists recommended major modification, while the other concluded that it was not for him to reject the manuscript. One of the pheneticists objected to the unconventional terminology and suggested major modification, while the other strongly urged rejection to protect the reputations of both the author and the journal. Finally, one of the cladists recommended major modification, while the other thought that only minor modifications were needed prior to publication. Eventually, the paper did appear but not until Smith had taken office.

Schuh overrode a majority of negative votes six times, twice in favor of cladist authors. In one case, the negative recommendations were by noncladists, in the other case by cladists. In three instances, Schuh overruled negative reviews by cladist referees to publish noncladist manuscripts, and once he published a critique of phenetics by a noncladist even though two of the three pheneticists who refereed it recommended rejection. In two of the preceding cases, Schuh went along with the recommendations of his associate editors who solicited the referees' comments in the first place.

Of the authors who submitted manuscripts to *Systematic Zoology* while Nelson and Schuh were editors, I was able to identify the taxonomic affiliation of approximately 200. Of the manuscripts submitted by these authors, Nelson and Schuh eventually rejected 38. Nelson rejected six manuscripts submitted by evolutionary systematists, seven by pheneticists, and seven by cladists. Schuh rejected eight manuscripts submitted by evolutionary systematists, three by pheneticists, and seven by cladists. Very roughly, the rejection rates for the three most prominent schools of taxonomy during this six-year period are 24 percent for evolutionary systematists, 19 percent for pheneticists, and 16 percent for cladists. (I must emphasize that my identification of taxonomic affiliation in some instances was, by necessity, very impressionistic. Most of these authors were clearly identified with a particular group at the time, but some were not. In fact, some of these authors contributed so rarely to the taxonomic literature that I had to decide their preferences on the basis of a single manuscript.)

By and large, those authors who resubmitted their manuscripts after revision succeeded in getting them published. A few failed. Nelson rejected four revised manuscripts, two of which were by cladists. Of the five revised manuscripts which Schuh rejected, two were by cladists. In general, those authors who fought to get their manuscripts published succeeded. One characteristic of the authors who did not take negative editorial judgments quietly is worth noting. All were either well known at the time or eventually became prominent.

There are two problems with the preceding observations. One is that the high acceptance rates obscure differences of opinion among referees. The second is that they concern a variety of papers each refereed by only a very few referees. In such cases, the eventual disposition of particular papers can depend very much on the luck of the draw. As inappropriate as Nelson's sending a half dozen manuscripts to multiple referees might have been, it provides a more balanced picture of the effects of professional allegiance on the refereeing process. In table 5.2, I presented the referees' summary judgments of these six manuscripts without any attention to the referees' taxonomic allegiance. Table 9.1 presents the same data distinguishing between cladist and noncladist referees.

Of the six manuscripts that Nelson sent out for multiple reviews, three concerned Croizatian biogeography: A by Croizat, B by a Canadian paleontologist, and C by a New Zealand biogeographer. Three concerned cladistic analysis: D by two young cladists, E by a critic of cladistic analysis from the American Museum, and F by Sokal. Manuscripts A and C were refereed primarily by noncladist referees, manuscripts B and E were refereed primarily by cladist referees, while the referees of manuscripts D and F were divided more evenly between cladists and noncladists. During this period, cladists were no longer the rare subspecies that they had been when Kiriakoff (1966) first realized that Hennig's views differed materially from those of Simpson and Mayr, but even so they certainly did not make up half of the systematics community.

Table 9.1. Summary recommendations for six manuscripts which received multiple reviews in *Systematic Zoology.*

	Biogeography			Cladistics		
Recommendations	*A*	B	C	D	E	F
Ready	1–0	0–4	2–3	0–3	1–0	8–7
Minor modification	3–3	2–2	4–1	3–1	2–1	5–1
Major modification	9–0	0–0	5–3	1–2	3–5	0–1
Reject	3–1	2–4	0–0	1–2*	1–7	0–2*
No rating	0–0	1–2	0–0	0–0	0–1	0–2
Totals	16–4	5–12	11–7	5–8	7–14	13–13

Note: The first figure in each column indicates the opinions of noncladist referees. The second figure indicates the opinions of the cladist referees. In two cases marked by asterisks, the referees recommended alternative actions—rejection or major modification.

In the case of the manuscript by Croizat (A), three of the six referees who thought that it should be published after only minor modification were cladists, while one cladist found Croizat's manuscript sorely out-of-date and complained that Croizat was still fighting battles that had been won long ago. When this cladist received a revised version of Croizat's manuscript to referee, he still urged rejection. Once cladist referees are removed from the list, the weight of opinion with respect to Croizat's manuscript was toward major revision. These revisions were carried out by Nelson and Rosen.

Manuscripts B (by the Canadian paleontologist) and C (by the biogeographer from New Zealand) both opposed the system of biogeography which Nelson was championing. Of the seventeen referees for manuscript B, twelve were cladists. All four of the referees who recommended the manuscript for publication were cladists and two found it needed only minor modification, while four cladists recommended outright rejection. Obviously, cladists at the time were strongly divided about the virtues of Croizatian biogeography. Noncladists were also evenly divided between minor revision and rejection. At first glance, cladist and noncladist referees seem to exhibit the same spectrum of opinions about manuscript C, but it should be recalled that three of the referees who recommended publication did so for unusual reasons. Two of the cladists who recommended publication of the manuscript voted the way they did so that they could attack it once it appeared, while one of the noncladist referees thought that as terrible as the manuscript was, it was no worse than other manuscripts Nelson was publishing. On first reading, one cladist recommended publication after minor modification. When he was sent the author's revised version, he recommended that it be published. In general, however, the estimations of the cladists and noncladists did not differ greatly from each other.

The remaining three manuscripts are better examples for our purposes because they concern the principles of cladistic analysis, not Croizatian biogeography. Nelson sent manuscript D by two young cladists to thirteen referees, eight of them cladists. One might expect cladists to be universally enthusiastic. They were not. The three who recommended publication without modification were all cladists, but one cladist recommended major revision, a second major revision or outright rejection, while a third voted to reject the manuscript. The revised manuscript was sent to a single referee, a cladist, who recommended publication after minor revision. Once again, the opinions of the cladist and noncladist referees did not differ greatly from each other.

Of the twenty-six referees to whom Nelson sent manuscript E, which was critical of cladistic analysis, fourteen were cladists. The cladist referees were extremely critical of the manuscript, both in its original and revised forms. The weight of opinion was toward outright rejection or major modification. The seven noncladist referees were somewhat more positively predisposed toward the manuscript but not much. The only referee who recommended publication thought it was a terrible paper but, once again, no worse than other papers that Nelson was publishing. Just the opposite state of affairs obtained with respect to Sokal's defense of Mayr against his cladist critics. Half the referees were cladists, but most of them found the paper either ready to publish or publishable after only minor revision. Only two opted for rejection.

The net effect of these summary data is that the referees did exhibit some of the expected bias, but it was far from total. Cladists were willing to recommend publication of papers with which they disagreed strongly and rejection of papers by their fellow cladists. But summary data are not everything. When one reads the more detailed comments made by the referees, strong preferences are obvious. If the referees attempted to hide their allegiances, they did not succeed. The referees presented their differences with the authors quite bluntly, but poor exposition elicited more negative comment than did disagreements about substantive issues. Cole, Rubin, and Cole (1978: 106) found this same discrepancy between qualitative, anecdotal data and quantitative summary data concerning the influence of the personal and social characteristics of the principal investigators submitting proposals to the National Science Foundation on the resulting peer review. On the basis of qualitative data, scientists who have a good record of publishing and obtaining research funds stand a better chance of being funded than those that do not, but this difference does not show up in the summary results.

The same pattern can be found for those manuscripts that received only two or three reviews. Of the 90 or so cladists who submitted manuscripts to *Systematic Zoology* between 1974 and 1979, I was able to identify the taxonomic affiliation of the referees for 73. Of the 160 referees for these manuscripts, 78 were themselves cladists. As table 9.2 indicates, cladist and noncladist referees presented fairly much the same range of summary evaluations of the manuscripts submitted by cladists. The chief

Table 9.2. Summary evaluations of cladists and noncladists of the 73 manuscripts submitted by cladists to *Systematic Zoology* between 1974 and 1979.

Evaluations	*Cladists*	*Noncladists*
Accept	17 (22%)	12 (15%)
Minor modification	17 (22%)	23 (28%)
Major modification	18 (23%)	25 (30%)
Reject	18 (23%)	14 (17%)
Abstain	8 (10%)	8 (10%)
Totals	78	82

difference is that cladists opted more frequently to both reject and accept manuscripts by their fellow cladists than did noncladists, while noncladists tended to want more extensive modification of the manuscript before publication.

To return to the questions with which I began this chapter, does confidentiality serve mainly to allow scientists working in the same research program to evaluate each other's work candidly? Are editors influenced in how they process papers by their theoretical biases? Are these same biases apparent in the judgments that referees make of the manuscripts sent them to referee? As the preceding data indicate, cladists were somewhat less critical of the manuscripts submitted by their fellow cladists than they were of those submitted by noncladists, but they were far from totally biased. They were willing to pass harsh judgments on their fellow cladists under the cloak of confidentiality. Would they have opted differently if they had been forced to make their decisions in the open?

As is usually the case in present-day journals, book reviews appearing in *Systematic Zoology* are signed. Between 1974 and 1979, 45 percent of the book reviews that appeared in *Systematic Zoology* were favorable. The rest were mixed or negative. During this same period six cladists reviewed books arguing for the principles of cladistic analysis in the journal, and all six were uniformly favorable. These data, of course, are far from conclusive. Referees' reports and book reviews are quite different things. A referee can hope to improve the manuscript under review, while it is too late to do much about books that have already appeared. However, the data just discussed lend some support to the hypothesis that scientists working in different research programs are willing to pass negative judgments on the work of their opponents in public as well as in private, while those working in the same research program are willing to present negative evaluations of the work of their allies only in confidence.

One important feature of such research groups as the pheneticists and cladists is social cohesion, and social cohesion is easier to maintain when the members of the group play down their areas of disagreement, especially in public. If this is true, then making the refereeing process open would pose problems for the continued functioning of the informal groups that are currently so characteristic of science. To the extent that these informal groups serve a positive function in science, the potential benefits of openness would be purchased at a price. The system as it now functions is far from entirely confidential. Word does get out. Scientists frequently have suspicions about the way that a particular scientist voted with respect to the publication of a manuscript or the funding of a research grant, but they cannot be sure. Even if confidentiality is sometimes a fiction, it is a useful fiction (for numerous comments on the refereeing process, see Harnard 1982).

But on my view, shouldn't scientists working to promote the same general set of views give uniformly high marks to every manuscript by an ally that he or she referees? If scientists working in the same research program were in total or even large agreement about fundamentals, perhaps so; but, as I have argued at some length, they are not. Each scientist has his or her own preferred set of views that he or she takes to be typical. Any departures from these fundamental tenets is sure to elicit a negative response, whether the offender is an ally or not. In fact, deviance is more serious when it comes from an ally. If, for instance, a pheneticist writes a stupid paper, it only helps the enemies of phenetic taxonomy. That is why Farris did not discourage the publication of Janowitz's (1979) paper. He took it to be an easy target. Several referees of the papers previously discussed voiced this same conviction. When someone in your own research program writes a stupid paper, it damages everyone else working in that program. That is why cladists were so willing to recommend rejection for manuscripts submitted by their fellow cladists.

But at this juncture, a few words need to be said about the reliability of the data that I have presented. I had expected to have the usual trouble with borderline cases. In connection with the overly good results that Mendel obtained for his garden peas, Root-Bernstein (1986) noted the role of borderline cases, i.e., those cases in which contrasting character states could not be clearly distinguished. If these borderline cases are omitted in studying character transmissions, one obtains the sort of figures one should expect from the laws of chance. If they are forced into one category or other, unusually good results can be obtained, depending on the decisions one makes. One of the main contentions of this book is that any observations that hold for my subjects should also hold for me. During the early years of phenetics (or numerical taxonomy) and cladistics (or phylogenetics), membership in these research groups was relatively easy to ascertain. While Nelson and Schuh were editing *Systematic Zoology*, it was very easy to distinguish between cladists and noncladists. Any systematist was either on the bus or off the bus. It was beginning to get more difficult to individuate pheneticists. As these research programs continued to develop, membership became increasingly diffuse. By the end of Schnell's tenure as editor of *Systematic Zoology*, the percentage of authors who count as borderline cases had become so large that any figures obtainable were all but useless. Mass selection was beginning to swamp the results of demic selection.

A second weakness of the preceding data is my lumping noncladists into a single group, regardless of their affiliation. Contrasting cladists with noncladists is like contrasting vertebrates with invertebrates. Cladists and vertebrates are genuine groups; noncladists and invertebrates are not. Perhaps if I had subdivided noncladists into evolutionary systematists, pheneticists, and those systematists who were unaffiliated, the results might have been importantly different. The problem is that in most instances the data were so sparse that further subdivision was likely to obliterate any patterns.

In spite of the preceding reservation, the alacrity with which cladists were willing to vote against their fellow cladists, albeit in secret, casts some doubt on the efficacy of professional alliances in science. If the differences between how cladists and noncladists evaluate the work of cladists are not all that great, then possibly socially defined groups are not all that important in science. Just as Bourbaki poses a serious problem for the claim that credit for contributions is one of the main elements in the mechanism that is responsible for science having the general characteristics it has, the fact of cladists voting to reject papers by other cladists poses a problem for the importance of socially defined groups in science. But the best thing about anomalies is that sometimes they can be converted into confirming instances. When I originally started collecting data on refereeing practices in *Systematic Zoology*, the division between pattern and phylogenetic cladists was only beginning to materialize. After its outlines had become clearer, I went back to see which cladists voted against which cladists. The division had already begun to materialize.

Finally, were Nelson and Schuh biased in how they edited the papers sent to them? The general impression that I received from studying the referees' reports for six years between 1974 and 1980 is that both Nelson and Schuh were strongly committed to cladism. Nelson especially seemed to feel as if the cladists were an embattled minority fighting powerful forces of darkness. He was so convinced he was right that he could not understand how others could be so blind. If Schuh was equally convinced, he was better able to disguise his feelings. Nelson seriously misjudged how authors would respond to the highly critical letters he sent them, even if he did acknowledge that he would publish their manuscripts no matter how wrongheaded they were. Certain cladists were given very easy access to the pages of *Systematic Zoology,* an ease of access that was hardly an unmixed blessing. They were denied the benefits of the refereeing process. Publishing a long, rambling diatribe is one thing; getting other people to read it is quite another.

The committee of the Society of Systematic Zoology that investigated the editorial policies under Schuh and Smith concluded that in a half dozen cases, authors were not fairly treated. The correspondence concerning these cases makes for very painful reading. My own impression is that convictions about bias throughout the tenure in office of any editor stem from a relatively few cases, so few that the effects of this apparent bias are lost in summary data. However, in suggesting that editorial bias may well have been less pervasive than it may have seemed at the time, I do not intend to diminish its importance. A victim of a miscarriage of justice is not in the least impressed by the observation that such miscarriages are relatively rare.

The Public Face of *Systematic Zoology*

Although data on refereeing in *Systematic Zoology* are available for only six years, the results of the refereeing process can give some idea of the effect that an editor's allegiances have on publication patterns in the journal which he or she is editing. During the early years of numerical taxonomy, Sokal was certain that Hyman was biased against papers with numbers in them and that his budding research program suffered under her editorship. Nelson was equally sure of Rowell's bias. However, if one actually charts the course of these research programs, the results are quite surprising. Table 9.3 indicates the percentage of pages concerning taxonomic philosophy published during the tenure of office of each of the editors of *Systematic Zoology* from its inception in 1952 to 1985.[1] Except for the inexplicable dip during the editorships of Rowell and Nelson, the fraction of papers arguing the pros and cons of various issues in taxonomic philosophy initially increased and then held steady at about a third:

1. In order to count as a contribution to taxonomic philosophy, a paper had to argue the pros and cons of various principles of classification. The most difficult cut to make in assigning papers in this respect was between papers arguing for a particular taxonomic philosophy and those that applied the principles of a particular school. Because officially *Systematic Zoology* is committed not to publish papers that are "merely" descriptive, authors of more descriptive papers tried to make them look as theoretically significant as possible, with varying degrees of success. In general, the data which I present should be treated with caution. There is no such thing as "raw" data.

Table 9.3. Percentage of pages during the tenure of each editor devoted to taxonomic philosophy.

Editor	*% of Pages*
Brooks	17
Hyman	22
Byers	25
Johnston	30
Rowell	17
Nelson	10
Schuh	32
Smith	33
Schnell	34

If the papers arguing the pros and cons of taxonomic philosophy are subdivided according to particular schools—the results reflect the careers of the three chief schools of taxonomic philosophy in the pages of *Systematic Zoology.* As table 9.4 indicates, papers arguing the virtues and vices of evolutionary systematics remained a constant feature of the journal until Nelson took over, when their number took a sharp drop, but this low figure is due primarily to the few papers which Nelson published on taxonomic philosophy in general. Evolutionary systematics staged a comeback under Schuh only to all but disappear under Smith and Schnell. Phenetics came on the scene quite abruptly during the last three years of Hyman's tenure as editor. The frequency of manuscripts discussing phenetics increased when the journal moved to Kansas but began to drop under Rowell while the journal still resided at the home of phenetic taxonomy. This decrease was accentuated under Nelson, again largely due to the small number of papers that he published on taxonomic philosophy. Under Schuh phenetics began to be discussed again in the pages of the journal, a trend that was continued under Smith and Schnell. Finally, cladistics was a slow starter. Not until Schuh became editor was cladistics debated extensively in *Systematic Zoology.* However, once it gained prominence, it remained the main point at issue in the journal even after Schnell became editor.

Table 9.4. Percentage of pages devoted to each taxonomic philosophy during the tenure of each editor.

Editor	*% Evolutionary Taxonomy*	*% Phenetic Taxonomy*	*% Cladistic Taxonomy*
Brooks	16	—	—
Hyman	16	11	—
Byers	20	16	7
Johnston	12	19	8
Rowell	11	9	9
Nelson	5	1	8
Schuh	14	9	30
Smith	6	15	30
Schnell	3	16	41

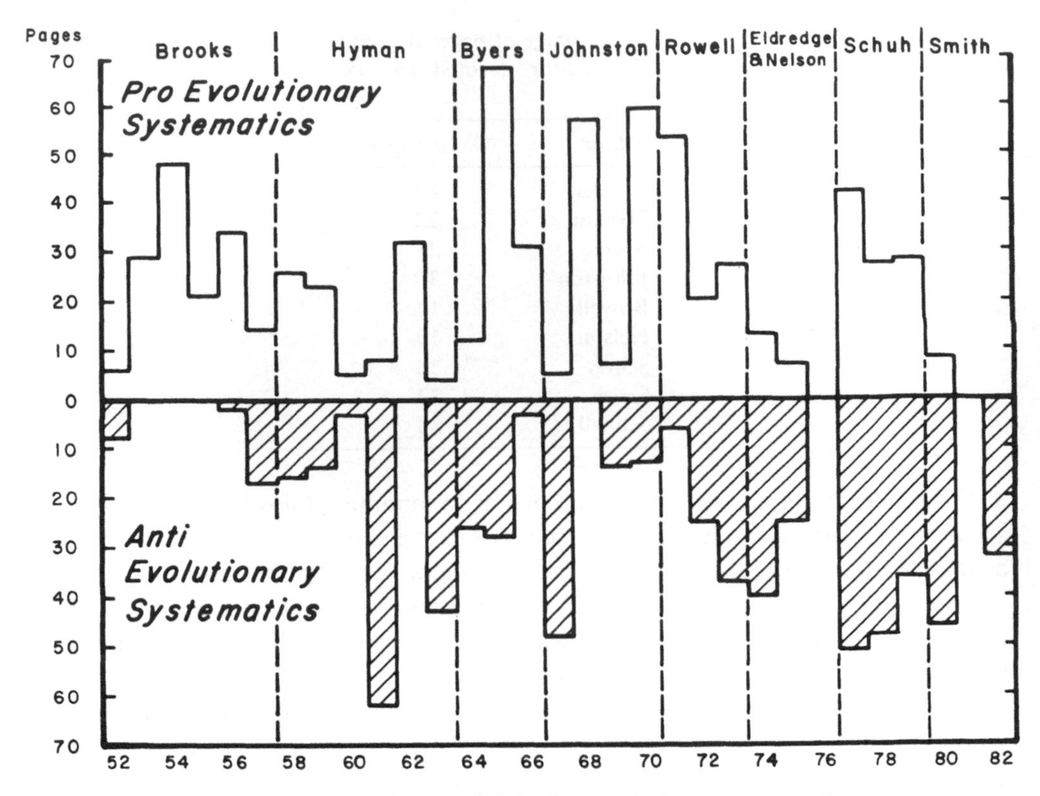

Figure 9.1. Number of pages attacking and defending evolutionary systematics.

More important than just the percentage of pages devoted to discussing a particular taxonomic philosophy is a comparison of those arguing for it versus those opposed. As figure 9.1 indicates, initially evolutionary systematics received little criticism, but in 1961 advocates of numerical taxonomy launched an attack. The next year evolutionary systematists defended their views; followed by attacks and counterattacks seesawing through the years. The same pattern of attacks and defenses can be seen in the figures for phenetics (see figure 9.2). However, because many of the criticisms of evolutionary systematics at the time were also defenses of phenetics and vice versa, the patterns are not independent of each other. When cladistics came along, the disputes became three-cornered, and the pattern of attacks and defenses is less clear (see figure 9.3).

By the time that Schnell became editor, very few papers were written by evolutionary systematists defending their preferred methodology or attacking others. Hence, the antagonists were primarily pheneticists and cladists, but both groups had changed significantly in the interim. Pheneticists no longer formed a tightly knit group, and the emphasis had shifted decidedly from phenetic philosophy to numerical techniques. Numerical pheneticists argued with numerical cladists over compatibility methods versus Wagner trees, and workers who considered themselves to be "numerical taxono-

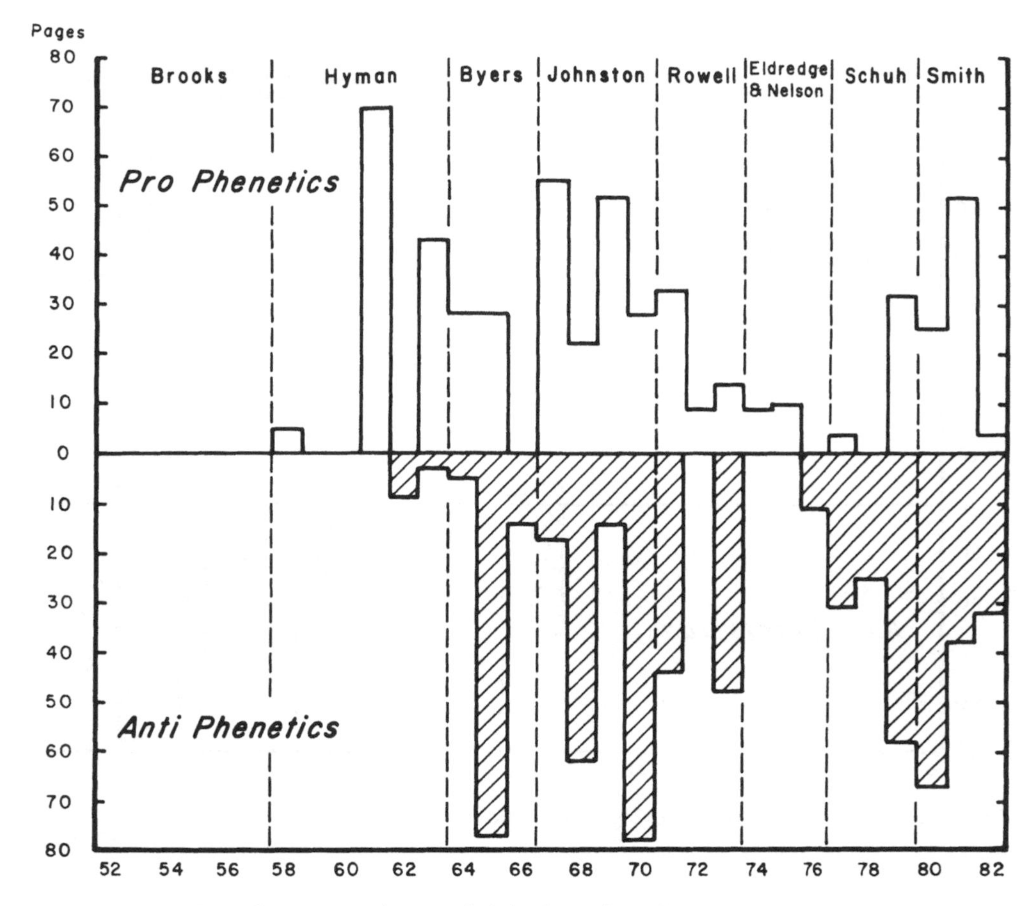

Figure 9.2. Number of pages attacking and defending phenetic taxonomy.

mists" in a loose sense debated quantitative techniques. The issue of phenetic philosophy rarely arose. Cladists had also become a good deal more diffuse as they were joined by numerous workers who were committed to using cladistic techniques but owed no special allegiance to the original cladists. To complicate matters further, the cladists speciated and began to argue with each other over proper taxonomic method. Six of the papers published by Schnell concerning cladistic methods were written by cladists against cladists.

When ratios of favorable to unfavorable papers are computed for each taxonomic philosophy, striking differences materialize (see table 9.5). Evolutionary systematics did especially well during the early years of the journal, even after the journal moved to Kansas. However, when the journal returned to the American Museum, evolutionary systematics suffered an eclipse, an eclipse from which it has yet to recover. Perhaps the principles of evolutionary systematics continue to predominate among the silent majority, but very few authors are debating these issues in the pages of *Systematic Zoology.* (The positive ratio under Schnell is due to small numbers—20 pages pro to 17 con).

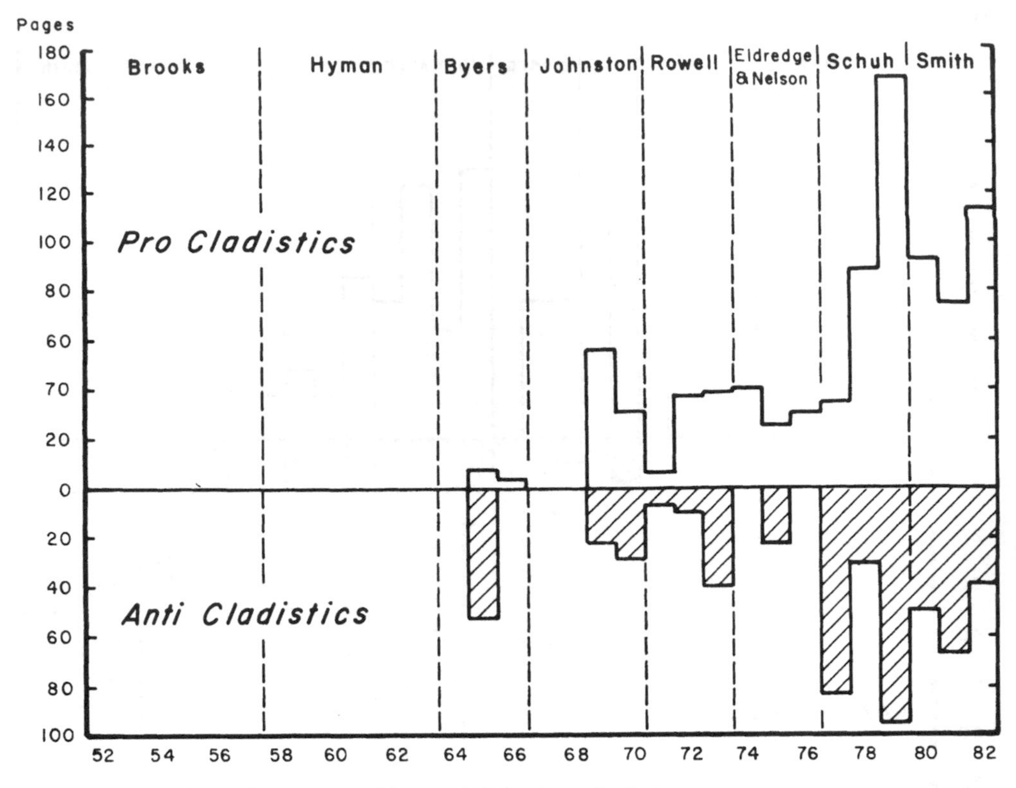

Figure 9.3. Number of pages attacking and defending cladistics.

During Hyman's tenure in office, phenetics did quite well—118 pages favoring it to 12 opposed for a ratio of almost 10 to 1. It did not do so well once the journal moved to Kansas. Although phenetics and numerical taxonomy were the chief topics addressed at the time, more pages were consistently published opposing than supporting this taxonomic philosophy. The favorable ratio under Nelson results, once again, from small numbers (19 pages pro to 11 con). The ratio dipped again under Schuh, rose a bit under Smith, and then exploded under Schnell. During Schnell's tenure in office, many authors who opposed numerical taxonomy as a taxonomic philosophy published in the journal, but no one treated this topic at any length. Perhaps Schnell rejected manuscripts that openly objected to his preferred taxonomic philosophy. There is no way to know. In general, at the time, the papers critical of numerical taxonomy appeared in the proceedings of the Hennig Society as well as in its journal, *Cladistics*.

The story for cladistics is somewhat different. During Johnston's final two years, he published 87 pages in favor of cladistic taxonomy and only 51 pages opposed, while Rowell published 81 pages in favor and 57 pages opposed. Contrary to impressions at the time, cladistics did quite well under Johnston and Rowell—better than phenetics. When Nelson took over as editor, cladistics did extremely well, but once again primarily because of small numbers. Under Schuh and Smith, the ratios returned to their previous

Table 9.5. Ratios of positive to negative papers for each of the taxonomic philosophies under each of the editors.

Editor	*Evolutionary Taxonomy*	*Phenetic Taxonomy*	*Cladistic Taxonomy*
Hyman	1.03	9.83	—
Byers	1.95	0.58	0.21
Johnston	1.70	0.92	1.71
Rowell	1.47	0.61	1.42
Nelson	0.31	1.73	4.13
Schuh	0.72	0.32	1.38
Smith	0.10	0.60	1.80
Schnell	1.18	38.50	0.78

levels. Under Schnell, cladistics came in for a good deal more criticism, but it should be remembered that six of the papers objecting to the work of certain cladists were written by other cladists. Of course, the objects of these criticisms were claiming at the time that their cladist opponents were no longer really cladists.

In general, the pattern that emerges from the preceding data is not especially correlated with changes in the allegiance of the editor of the journal. Instead what one discovers are cycles of attacks and defenses with time-lags of one to two years, cycles that are oblivious to the taxonomic preferences of the editors. The main exceptions to this claim are Nelson and Schnell. Nelson's deviations were due to his publishing so few papers on taxonomic philosophy of any sort, Schnell's to the paucity of papers critical of numerical/phenetic taxonomy as a general taxonomic philosophy as well as to cladists beginning to attack their fellow cladists. Perhaps various editors were biased in their treatment of the authors who submitted papers to their journal, perhaps outsiders had to fight harder to get their papers published than insiders, but the end result is that all sides succeeded in being heard. During the past two decades, *Systematic Zoology* acquired the reputation of being a highly contentious journal in which warring factions did battle with each other. This impression is certainly correct, but it is also true that several very important issues that other journals were disinclined to address were thrashed out in the pages of this journal. Systematics, evolutionary biology, and biogeography would have been much the poorer without the forum provided by *Systematic Zoology.*

Policing the Police

Thus far, I have discussed only two of the elements that make science work as well as it does in realizing its manifest goals—curiosity only briefly and credit quite extensively. My emphasis has been deliberate. Although curiosity is a necessary condition for science, I take it for granted as a prerequisite for the mechanism which drives science. The desire for credit is half the mechanism. The mutual checking of research that goes on in science is the other half. Sir Peter Medawar (1982: 46) characterizes scientific reasoning as an "interaction between two episodes of thought—a dialogue between two voices, the one imaginative and the other critical; a dialogue, as I have put it, between the possible and the actual, between proposal and disposal, conjecture and criticism, between what might be true and what is in fact the case."

One reason why cheating is likely to be uncovered in science is that it is in the self-interest of individual scientists to check work that threatens their own. A second is that scientific claims *can* be checked—tested. Scientific claims are not *easily* testable or *instantly* testable, but scientists are drawn to those areas in which their beliefs can be brought into contact with the empirical world. Other areas of investigation may be in principle amenable to scientific investigation, but they are not yet "ripe." The subject matter of science is surrounded by a penumbra of potentially but currently not actually testable beliefs about the empirical world. As science has progressed, scientists continue to encroach upon the inside boundary of this penumbra and expand its external margins as they become more ambitious.

An important asymmetry exists between what scientists do with statements they take to be true and those they take to be false. True statements have to be integrated into theories, while those that are supposedly false just get dumped in a heap. Thus, within those systems of statements that scientists take to be true, errors should ramify. A theory of any scope at all includes hypotheses from a wide variety of sources. No one scientist can have firsthand knowledge of every sort of natural phenomenon that bears on his or her own research. Scientists must rely on each other. For this reason, one operational definition of "truth" in science is consensus, but it is only *one* operational definition of this emotionally charged term and it is only an *operational* definition (Ziman 1968: 9). Regardless of the impression that one might get from reading the recent literature in the sociology of knowledge, scientists really do make extensive observations and run exhaustive and exhausting experiments. All this effort is not mythical behavior designed to camouflage the causal factors that are actually operative in their arriving at their conclusions. For example, Morgan was eventually forced to accept the very views that he set out to refute, and the results of his experimental studies seemed to play a crucial role in this transformation.

Except for operationism, no other philosophical doctrine has been so thoroughly misunderstood or caused more damage in science than falsifiability. Popper (1962: 34) introduced his principle of falsifiability to distinguish between genuine scientific theories such as Newtonian theory and relativity theory and such pseudo-scientific theories as Freudian theory and Marxist economics. (Later Popper added Darwinism to his list of pseudo-scientific theories only to remove it still later.) Popper's choice of falsifiability rather than verifiability turns on a peculiarity of statements of universal form, e.g., "All A's are B's." In principle, statements of this form can be falsified by a single counterinstance. Because such statements refer to indefinitely many instances, past, present, and future, they are not verifiable, even in principle. However, in practice, this asymmetry is somewhat neutralized by the fact that scientific laws do not exist in isolation but are organized into theories. In any instance of apparent falsification, too many alternative sources of error are not only possible but plausible. *If* one could be absolutely sure that a particular observation is veridical *and* that no modification elsewhere can save a particular hypothesis, then single-minded attention to falsification would be justified. However, in the real world, scientists must balance apparent confirming instances against apparent disconfirmations and make their decisions accordingly.

Falsifiability in the Popperian literature concerns theories, not theorists; science, not scientists. One is tempted to dismiss so-called creation science as unfalsifiable. Certainly some of the claims made by creationists are unfalsifiable in the propositional sense because they concern the supernatural, but many are straightforward assertions about the empirical world, and as such they are eminently falsifiable. To be precise, they have been falsified. Creation science, in this sense, is not pseudoscience but abysmally bad science. Ruse (1982) thinks falsifiability in this abstract sense is not enough. Perhaps a theory can be scientific or not, independently of the behavior of anyone who holds that theory, but science depends crucially on how scientists actually behave. Perhaps the claims made by creation scientists are testable, but creation scientists themselves actually do little or nothing by way of genuine test. More than this, when others expose their views to tests, creation scientists ignore the results, "refusing to allow their position to be falsified" (Ruse 1982: 75). For example, time and again, physicists have explained why thermodynamics is not incompatible with current versions of evolutionary theory. Creation scientists simply ignore these protests and doggedly repeat their contrary claim. As far as creationists are concerned, professional physicists know no more about physics than professional biologists know about biology.

The differences between creation scientists and genuine scientists should not be exaggerated. Before scientists publish their work, they frequently expose it to severe tests, but the frequency and severity of these tests tend to decline precipitously thereafter. Nor do they instantly acknowledge the validity of the falsifying instances produced by others. They finagle for a while first. Raup and Sepkoski's behavior with respect to the 26-million-year-cycles of mass extinctions is typical. At first they were highly skeptical. Somebody or other was always coming up with life cycles of species, not to mention societies and scientific theories. They did not want to be banished by their colleagues to the lunatic fringe. Before publishing, they did their best to kill their brainchild but to no avail. Eventually they published. Thereafter, their primary concern was to defend their hypothesis against apparent refutations (Raup 1986: 121). Supposedly, Luis Alvarez remarked to William Clemens, "Bill, if you find one dinosaur bone above the iridium level, you've got me." Even though a half dozen paleontologists have since claimed to discover the remnants of dinosaurs four feet above Alvarez's iridium layer, Alvarez has not capitulated (Patrusky 1986–87: 4). Falsifiability as an inferential relation is timeless. Testing in science is not. It has a crucial temporal dimension.

According to Ruse, there is more to science than inference. How putative scientists actually behave is also relevant. Critics of Ruse's stance on creation science acknowledge the preceding distinctions, but disagaree about their relevance and importance. For example, if the issue is the nature of science, Laudan (1982: 17) concludes, "What counts is the epistemic status of Creationism, not the cognitive idiosyncrasies of the creationists." Quinn (1984: 47) agrees. "The requirement is that a scientific theory be testable, not that its proponents actually test it." Such issues as cheating and the falsification of data are irrelevant to creationism being a genuine science and creationists' views being genuinely scientific. Although Quinn (1984: 48) does not want to defend, mitigate, or excuse the intellectual dishonesty one finds in the writings of creationists, he warns that "integrity, like tentativeness, is a characteristic of persons either individually or as groups." Ad hominem arguments in science as elsewhere are irrelevant.

The disagreement between Ruse and his critics does not concern creationism or the nature of science so much as the nature of philosophy of science. When philosophers of science analyze "science," what is their subject matter? As far as Quinn is concerned, it is propositions and inference. As much as Laudan has criticized logical empiricists' analyses of science, he apparently agrees with Quinn's estimation. There is much wrong with the logical empiricist analysis of science according to Laudan, but its single-minded concentration on inference is not on the list. As one might imagine from all the attention that I pay to scientists and their professional relations, I think that reference to inference is not enough. What scientists actually *do* is also relevant to an activity counting as being genuinely scientific. A theory or proposition may be testable in principle, but if, for too long, no one bothers test it, or if those who hold a particular view refuse to acknowledge the results of these tests, then in-principle testability is of little consequence to the ongoing process of science. In this connection I agree with Thagard (1978). Astrology did not begin its career as pseudoscience, but it ended that way, not because various versions of it were untestable but because its practitioners ceased to be interested in such matters. Similar observations hold for phrenology. Phrenology began its career as genuine if marginal science. It rapidly degenerated into pseudoscience but not before it left its mark on psychology (Hull 1979b). A similar story can be told for psychoanalysis.

Just as sociologists of science have found a marked asymmetry in the number of positive versus negative citations in the literature of science, they have also discovered an equally great asymmetry in the publication of positive versus negative results. In reviewing over 300 articles in four psychological journals, Smart (1964) found that less than 10 percent of these articles reported negative results. All the rest confirmed the hypothesis under investigation. Mahoney (1979: 355) is appalled:

> Review articles tend to be tallies of successful predictions and confirmatory experiments; a positive verdict is returned when the tally shows that a theory has been "successful" more often than it has "failed." This common practice is illogical, not only because of the selective publication of positive results, but also because of the epistemological discrepancy between the information content of "positive" and "negative" results.

Given Popper's technical definition of "information content" and scientific theories as ideal entities divorced from any pragmatic considerations, Mahoney's dismay has some justification, but in the real world both positive and negative results are valuable. Nothing limits the invention of scientific theories with incredibly high "information content" save the imagination of the scientist. Scientists cannot spend the few years each has available for an active career attempting to falsify one hair-brained theory after another. They have to use all sorts of heuristic ploys in order to make their job manageable. In Giere's (1987) terminology they must employ the cognitive resources actually available to them.

For example, when the presence of unusually high concentrations of iridium were discovered in a geological stratum immediately prior to one of the "mass extinctions" in the fossil record, physicists did not proceed to test the geologic column at random to see if a correlation materialized between iridium deposits and mass extinctions. Instead they searched for similar deposits at other points in the geologic column where mass extinctions were known to occur. As Raup (1986: 94) remarks, "This is not the best science, but scientists are human and compromises have to be made." Chemical analysis for iridium is both time-consuming and costly. Once the crucial areas of the column had been tested, then, depending on the results of those tests, there might be a point to conduct random tests. In order for a theory to be worth refuting, it has to have some support. Scientists spend a great deal of time trying to accumulate evidence in favor of their views and they publish the results of their research even if they are confirmatory. "Negative results" in the sense of experiments that did not turn out, rarely deserve publication. However, apparent refutations of widely held views certainly should be published, but such refutations are hard to come by. That may be one reason why so few are published.

Because commentators on science have noted the startling difference between the sorts of policing that go on in science and in other professions so frequently, we take it for granted, but this anomaly cries out for explanation. Other professionals are decent people. If asked, they defend the honor of their profession, apparently in all sincerity. Bad publicity is due to a few bad apples. But self-delusion is easy, in fact much easier than the delusion of others. Zimmerman (1978: 23) goes so far as to list the asymmetry between self-delusion and the delusion of others as yet another psychological trait which may serve to maximize individual fitness in multiparty altruistic systems:

> we are quicker to suspect and perceive the motives of others than our own, and it is much easier to convince ourself of something than it is to convince others. This curious combination of traits—and the hypocritical contradiction between personal and public morality it engenders—make it possible to both engage in deception and perceive the same in others.

The columnist Sydney J. Harris (1979: 15) provides an excellent example of Zimmerman's asymmetry when he notes that among "professions only the scientists do not try to protect one another: When a scientist is wrong, or malpractices, his colleagues call him on it, publicly and immediately—unlike law or medicine, which are organized largely to conceal defects and resist criticism." As an outsider, Harris can supply at least third-party objectivity in his comparison of science, law, and medicine, and he opts for scientists being more virtuous with respect to self-policing. However, years later, when Harris (1986: 13) turned his attention to his own profession, he found a "high degree of personal honesty among its practitioners. I don't believe that the most respected elements in American life—doctors or lawyers—can even approach the incorruptibility of newspapermen." During his more than forty years in the profession, Harris knew of only three colleagues who he was reasonably sure were "on the take."

Commentators on science, such as Merton (1973: 276), do not claim that scientists police themselves with complete efficiency and accuracy, only that they do so to a degree "unparalleled in any other field of activity." Given the near universal failure of other professions to police themselves according to the stated goals of the profession, these standards are not difficult to surpass. Scientists need not police themselves very vigorously to exceed the sort of self-policing that occurs in medicine, law, police enforcement, and university teaching. Every time anyone investigates these professions, they discover that the sort of behavior that is supposed to be prohibited is widespread and the number of cases in which members of these professions have been penalized minuscule.

Whether or not a profession counts as policing itself depends on the grounds being assumed. For example, all Harris (1986) demands is that his colleagues not take money to bias their reporting; but as Harris also notes newspaper men, as a class, do not seem to be highly motivated by money. Instead, what they enjoy most is "knowing things, and knowing things that the public at large does not know." If reporters were evaluated on publishing ill-informed, misleading, sensational stories in order to further their own careers, I suspect that they would appear no more virtuous than doctors or lawyers.

Broad and Wade (1982) complain that scientists are lax in checking the raw data collected by other members of their research teams. Hence, fabrications can slip through. Apparently they think that individual scientists should not only conduct their own research but also check the results of their co-authors. I cannot think of worse advice. The whole point of collaborative research is to pool cognitive resources. Testing raw data takes almost as much time as collecting this data in the first place. Scientists who do not trust each other are not likely to collaborate in the first place. And when one scientist begins to doubt the trustworthiness of the results of a colleague's work, the way that Sturtevant did with respect to Dobzhansky, the appropriate response is to sever professional ties. As in all areas of human endeavor, one antidote for hypocrisy is self-reference or, in this case, a healthy application of the goose-gander principle. Certainly, Broad and Wade were not prepared to take their own advice. Although they collaborated in writing a book on fraud and deceit in science, they did not bother to check each other's raw data. When the editors of *Nature* finally agreed to publish a paper by Stewart and Feder (1987) questioning the integrity of scientific literature at large, they prefaced it with an editorial in which they suggested that Stewart and Feder were guilty of some of the very same failings that they point out in the work of others (for further discussion of science reporting, see Friedman, Dunwoody, and Rogers 1986).

The sorts of honesty that matter in professions are those internal to the profession. Anyone can be "on the take." Doctors and lawyers are constantly being convicted of financial irregularities of one sort or another, but one can give first-rate service to one's clients while nevertheless embezzling money from some federal program. That scientists have been accused of financial dishonesty so rarely is certainly admirable (Holden 1987b), but this is not the sort of honesty that is relevant to science. University professors in their roles as teachers provide yet another example of the distinctions which must be made with respect to the sorts of "honesty" relevant to the self-policing in a profession.

Universities in the United States serve two functions—teaching and research. Once again, when a single structure is saddled with performing two functions, sacrifices must be made. By and large, university professors are rewarded by their institutions for research, not teaching. In fact, the chief reward for success in research is a reduced teaching load. Thus, one should not be surprised to discover that teaching is sacrificed more often than research when the two conflict. The only place where the two overlap is in the sort of instruction that goes on in graduate seminars and with respect to one's research assistants. Otherwise teaching is largely extraneous to research and can be shirked at the sole cost of vague feelings of guilt.

Officially, of course, universities are committed to good teaching, and the one study I discovered that addresses the issue comes to the comfortable conclusion that good researchers tend to be good teachers and vice versa (Bresler 1968). The fact remains that university professors are rarely penalized for not teaching very well, especially once they have attained tenure. In the entire history of the University of Wisconsin, not a single professor has lost tenure for failing to teach adequately, although three did have their tenure removed for disrupting the operation of the university while demonstrating against the invasion of Cambodia during the Vietnam War. Either the level of teaching at the University of Wisconsin is extraordinarily high or else good teaching is not one of the duties of a university professor.

The situation is no different in the medical profession. For instance the Federation of State Medical Boards reported in 1973 only ninety-six judgments against doctors in the United States, ranging from reprimands to loss of license. Of course, these figures might simply reflect the high level of performance by the American medical profession. However, in its first year and a half of operation, a board staffed by nonphysicians as well as physicians resulted in eighty judgments against doctors in Pennsylvania alone (Rottenberg 1980; for additional data, see Bosk 1979 and Wallis 1986). The basic principle is that if too many features of an institution seem inexplicable on the basis of a particular view of the goals of that institution, perhaps those are not the institution's operative goals. For example, if one thinks that hospitals are structured primarily for the good of the patient, then one will come across anomaly after anomaly. If one changes one's perspective and views hospitals as being organized primarily for the immediate convenience and benefit of doctors and the hospital itself, then many features which seemed inexplicable before make much more sense. The good of patients is a factor in the way in which hospitals are organized, but it is no more than third on the list. The ease with which physicians and hospital administrators can convince themselves that the patients come first indicates the ease with which we can delude ourselves when our own self-interest is at stake. The same story can be told for profession after profession—including science.

Several of the characteristics of the sort of policing that goes on in science call for some discussion. First, many experiments and observations are not replicated. To count as scientific, a finding must be replicable. It need not actually be replicated. Scientists cannot spend their time testing every research finding they use in their own research. If they did, they would still be deciding whether the earth or the sun is in the center of what we now term the solar system. In many empirical, largely experimental areas of science, such as certain more pedestrian areas of chemistry, replication is common.

Rows of chemists need something to do, and rerunning experiments already conducted by other chemists at least hones one's skills, and there is always the outside chance that an error can be uncovered or an avenue for new research stumbled upon.

However, pedestrian science is only one form of science. In the more exciting areas, scientists dread the possibility that their work will turn out to have duplicated someone else's findings. Exactly how deflating this experience is can be seen in the reaction of other scientists working on the structure of DNA to Watson and Crick's (1953) announcement of their success. Maurice Wilkins, who would eventually share in the Nobel Prize with Watson and Crick, responded to them upon receipt of their manuscript as follows:

> I hope you won't mind me being a trifle awkward on 2 points. First I think Frazer should have the opportunity of publishing his 3 chain model, we rather stopped him doing so a year ago because we thought we could do better and didn't. . . .
>
> Second [we] should like to publish a brief note with a picture showing the general helical case alongside your model publication. Frazer's thing can be tucked away anywhere just for record purposes (Olby 1974: 417–18)

Although indiscriminate testing does take place in science, in general scientists must have some reason to test the findings that someone else publishes. As Price (1965) has noted, scientists spend very little time in haphazard reading of the literature, even the literature in their own area of expertise. There is much too much published for that to be a productive activity. Instead they scan the literature looking for findings that bear on their own research. As Sandler (1979: 177) notes, "The working scientist has neither the time nor, evidently, the inclination to consider work that is not essential or not of immediate use to his particular problem." Scientists seem always to be asking of their fellow scientists, "Will your ideas, techniques, or data prove useful to me in obtaining solutions to my own problems?" Scientists incorporate into their own work those findings that support it, usually without testing. Testing is reserved for findings that put one's own research in jeopardy. Popper is right that falsification plays a more important role in science than verification but this is not because of any inferential asymmetry between the two. The asymmetry results from differences in commitment of different scientists. Scientists spend a large chunk of their time trying to support their own hypotheses. If anyone can show that a particular scientist is right, it is that scientist. Scientists also spend another large chunk of time trying to refute the hypotheses of those scientists with whom they disagree. If anyone can show that you are mistaken, it is your opponents.

Once again, the social structure of science plays an important role in testing. The immediate attention that pheneticists paid to Mickevich's findings is an excellent case in point. Cladists did not put the research that supported their own position to the test; their enemies did. The areas in science in which the workers are subdivided into warring factions are precisely the areas in which claims are put to the most severe tests, not infrequently tests that are unrealistically severe. Efforts to subject one's own pet hypotheses to severe tests, to attempt to falsify them, are always warranted. The likelihood that one will impose unrealistically high standards is quite low. The same cannot be said for the standards that scientists set for the work of their opponents.

For example, Simberloff (1983) has made a career of showing that the data which other scientists have put forth in support of their particular theories are as readily accounted for by a null hypothesis. As unseemly as factionalism in science may be, it does serve a positive function. It enlists baser human motives for higher causes. It would be nice if large numbers of human beings were able to contribute more frequently to the global good without the encouragement of local rewards, but as Friedland and Kappel (1979) note, humanity at large does not constitute a genuine clientele. Even the scientific community as such is too large and diffuse to function as a causal entity. Individual scientists and small research groups are quite another matter. They are human-sized.

Nelson viewed his role in the early years of cladistics as an "initial irritant" to prick the complacency of his senior colleagues. If he caused them some discomfort, that could not be helped. For Nelson the quest for knowledge was more important than decorum. He did not view politeness, respect for authority, and the like as virtues in science. As he saw it, mavericks play important roles in science and should be at least tolerated no matter how badly they might seem to misbehave (see also Raup 1986). Nelson could not understand why Simpson and Michener took offense at the "tone" of some of his letters and papers. Later, when cladistics became the "orthodoxy," the defenders of true Croizatian biogeography felt called upon to smite it. Nelson did his best to maintain his allegiance to his earlier attitudes about polemics and mavericks, but it was not easy.

Polemics can be an important factor in one individual or group prevailing over another as long as it is not completely indiscriminant. Scientists must have their allies. For example, Raspail was so abrasive to all of his contemporaries save Geoffroy Saint-Hilaire that his work was studiously ignored, and eventually he dropped out of science. In 1830 Lyell (1881, vol. 1: 260) warned John Fleeming (1785–1857) to "keep in well with some party" or the nickname the "Zoological Ishmael" might stick. As polemical as Farris periodically became, he was always careful to keep well with at least one party. During his early years at Stony Brook, he was somewhat circumspect in his criticisms of his more phenetically inclined colleagues. However, once he became part of the cladists, he turned on the pheneticists with a vengeance. When the cladists began to speciate, he had to choose sides. He opted for Nelson and Platnick against Brooks, Wiley, and Funk. In addition to pooling their cognitive resources, Farris and Nelson have been careful not to let their very different interests generate social dissension.

Slashing criticism is only one strategy out of many in science. Adopting a lofty tone can also be effective. For example, Nelson's usual line of action was attack, to go for the jugular, which is precisely what he did in his dispute with Michener. In response to Nelson's (1978a: 105) accusation that he advocated "taxonomy by decree," according to how well it complemented his "Royal Judgment," Michener (1978) carefully avoided any comments that could be interpreted as being ad hominem. In defending sociobiology, Wilson went on the offensive; Alexander took the higher road. One style is officially decried, while the other is highly recommended, but no matter. As long as both are efficacious, scientists will use both.

Thus far, I have not discussed one motivation which has been commonly attributed to scientists, that of helping humanity. From Bacon to the present, science has been justified in terms of both the intrinsic value of understanding and the power that such understanding gives us over the world in which we live. Young scientists are raised to believe, possibly half believe, that good scientists selflessly pursue science both for its own sake and for the good of humanity. As effective as these appeals to the higher good have been in encouraging idealistic young men and women to become scientists as well as in persuading society at large to support science, the desire to help humanity is not as prevalent among scientists as many would have us believe. During the hundreds of hours of interviews that I conducted with systematists and evolutionary biologists, numerous motivations were offered for why they became scientists and continued to pursue their labors. Not once did anyone mention the good of humanity. When I raised the issue, they tended to respond, "Oh, yes, yes, I certainly want to help humanity." Of course, anyone wanting to help humanity in any direct way is unlikely to pick systematics or evolutionary biology as their area of concentration, but in her study of the scientific elite Roe (1952: 232) reached the same conclusion:

> This leads me to comment upon the opinion, held by many people, that the scientist is a completely altruistic being, devoting himself selflessly to the pursuit of truth, solely in order to contribute to the welfare of humanity. I do not intend it as a derogation of men whom I cherish when I say that this is, in my experience, not really the basic motivation for any of them, and as an additional motivation it is more often absent than present.[2]

To be sure, scientists are willing to take credit for applications of basic research which people take to be good. They are less willing to accept responsibility for the undesirable effects of their work. In any case, good intentions are not adequate to explain the peculiar success of science. If the desire to help humanity were very efficacious, humanity would be well served. From all indications, it is not. Perhaps the desire for individual credit is not as admirable as innate curiosity or the desire to help humanity, but by all indications it is a good deal more efficacious. If the only way that scientists can get credit for their work is to make it public, then they will make it public.

2. In her study of sixty-four eminent men of science, Roe (1952) found very few regularities. The scientists whom she studied tended to come from upper middle-class homes, their fathers were professionals, they tended to be brighter than average, and with the exception of the social scientists they were sexually late bloomers. She also found Catholics statistically under-represented. Of the sixty-four scientists she interviewed, six were Jewish, none Catholics. One possible explanation for the absence of Catholics is that adherents of dogmatic ideologies are not likely to be able to succeed in science (Roe 1952: 61). Another explanation which I find somewhat more plausible turns on immigration patterns in the United States. First-generation immigrants are not likely to choose a life of science, and Catholics were at the time among the more recent immigrants. Roe also limited her study to American-born scientists, excluding any Catholic scientists who may have immigrated to the United States.

Although thirty-four of the fathers of the scientists whom she studied were professional men, only one father was himself a scientist. As Outram (1984) has documented, extensive nepotism characterized the Natural History Museum in Paris while Cuvier was in charge, and the Hookers were also successful in keeping Kew Gardens in the family; but in general, nepotism in a literal sense has not proven very prevalent in science.

Sometimes humility is also listed among the norms of science. For example, Merton (1973: 303) counts as a sign of humility the practice of scientists "acknowledging the heavy indebtedness to the legacy of knowledge bequeathed by predecessors" (see also Ziman 1968: 96). As an example, he quotes Newton's remark in a letter to Robert Hooke (1635–1703) that, "If I have seen further, it is by standing on the shoulders of giants." Merton also notes that the main purpose of this letter is for Newton to defend his own priority for his theory of colors. If "humility" is taken to be a psychological characteristic of scientists, I have not noticed that scientists are any more humble than anyone else. Some scientists are fairly humble; others not. I do not see that it makes much difference. No one has shown any correlation between humility and scientific productivity. The chief function of referring to one's predecessors is not to do them homage but to gain their support and the support of others in the attempt to take their place. Newton stood on the shoulders of giants, but he also humbly offered his own shoulders for the next generation to stand upon. If in the process, he was acknowledged as a giant, well, that could not be helped. "Humility" seems a peculiar choice of words to characterize this attitude.

That scientists themselves feel under obligation to defend themselves against charges of arrogance can be seen in David Raup's response to a complaint lodged by a traditional paleontologist, Robert T. Bakker, when physicists and chemists began explaining mass extinctions in terms of such phenomena as collisions with meteors. According to Bakker, "The arrogance of those people is simply unbelievable. They know next to nothing about how real animals evolve, live and become extinct. But despite their ignorance, the geochemists feel that all you have to do is crank up some fancy machine and you've revolutionized science." Raup (1986: 105) found Bakker's statement "more than a little appalling and I am glad to say that the charge of arrogance is untrue." He himself had found the geochemists with whom he had come in contact "uniformly humble about their ignorance of paleontology and genuinely eager to learn from any 'primitive rock-hound' willing to give them some time."

The attempt by an astrophysicist and two chemists, even if in collaboration with a geologist, to explain mass extinctions in an entirely novel way takes "arrogance" of sorts, the same sort of arrogance that Raup (1985) exhibited when he invaded physics to suggest his own purely physical explanation for the 26-million-year cycles of mass extinction. This is the sort of arrogance that all outsiders must have if they are to bring what they know to an area of science of which they know little. But scientific disciplines also exhibit an arrogance of a more literal sort. Those scientists working at more fundamental levels tend to feel that their conclusions should carry more weight than those of scientists working at higher levels when the two groups come into conflict. Kelvin's crew exhibited this arrogance in their dispute with geologists and evolutionists over the age of the earth. Generations of biologists and geologists have taken perverse delight in the physicists turning out to be wrong. Raup himself (1986: 132) mentioned another instance of geologists being right and physicists wrong, a period in the 1950s when geologists, using their techniques, claimed that the earth was older than physicists thought the universe could possibly be, using their techniques. As a result of a number of these examples, Raup declares a "moral authority" of geology over astrophysics.

This too is humility of sorts. In any case, the call to tone down polemics and curb arrogance are about as effective on scientists as on everyone else, which means not very effective at all.

Conclusion

The impetus of personal credit in promoting the sharing of knowledge may be only a contingent feature of science depending solely on an accident of history. If so, it is an accident that was repeated several times over (Hull 1985c). However, given science as it has been structured since its inception and the sorts of human beings who become scientists, individual credit has served as a powerful spur to effort. Scientists must have some motivation for putting in the long hours that they do. Curiosity is one powerful motivation in science. Without it, science would be impossible. For those who have never come under the sway of a compulsion, it is difficult to understand the intensity with which scientists pursue their investigations. Trains hardly move silently on their rails, and yet H. F. Strickland (1811–53) was so engrossed in collecting fossils on a track bed that he failed to hear an oncoming train and was killed. Because Darwin had used Strickland for his chief source for data on species, he had to lean thereafter more heavily on Hooker, which meant a shift in emphasis from birds to plants, from individuals as "varieties" to groups as varieties (Rachootin and Thomson 1981).

The practice of giving nearly all the credit to the first scientist who publishes a particular contribution arose in order to force scientists to make their findings public so that others could use them. How long would Darwin have waited to publish in the absence of this incentive? Darwin thought that he had a "grand enough soul" not to care who got credit for natural selection, but when Wallace's paper arrived he discovered that he was wrong. Waiting too long to publish is one danger in science; the other is rushing into print. One of the dysfunctions of paying so much attention to priority is that scientists feel pressured into publishing before they are actually ready. The only deterrent to excesses of this sort is the reputation one gets upon the repeated publication of ideas that do not pan out. In connection with a meeting he was about to attend, Raup (1986: 122) lists the considerations relevant to his decision to make his controversial views public—should he mention the 26-million-year cycles of mass extinctions or not? On the one hand:

> the group of people in attendance was ideal for my purposes. I wanted feedback and help. If Jack and I were off base, this group could set us straight. On the other hand, I was hesitant to talk about the research too much—it was so new and raw, and possibly embarrassingly wrong. I was not concerned about giving away secrets and getting scooped, because it was unlikely that anyone else was working along the same lines, and Jack's data base in its computerized form was not generally available at the time. Also, most of us have learned that the gains that come from sharing information and getting feedback far outweigh the possible loss of priority.

Note that Raup does not say that he was not concerned about getting scooped. To the contrary, the reason why he was not concerned was that he and Sepkoski were far ahead of everyone else and had preferred access to the necessary information. Hence, they could share their findings and get feedback with little danger of losing credit. As it turned out, they still decided against disclosure at this meeting.

Another motive which may seem even less admirable than wanting credit for one's contributions but which is equally efficacious is to desire to refute those scientists with whom one disagrees. For example, James Hutton (1726–97) stewed over his theory of the earth for twenty years before discussing it in public at a meeting of the Royal Society of Edinburgh in 1785 and publishing a short summary of it in 1788, and that is where matters may well have remained had it not been for a criticism published several years later by Richard Kirwan (1794). The very day after receiving a copy of Kirwan's critique, Hutton sat down to begin work on his massive *Theory of the Earth, with Proofs and Illustrations* (1975). Time and again, scientists whom I interviewed described the powerful spur that "showing that son of a bitch" supplied to their own research. When Sokal first toyed with the idea of making systematics more quantitative so that computers could be used, he was not all that committed to the project, but when he sketched his ideas to Mayr, Mayr dismissed Sokal curtly with the observation that it had been tried before with no success and supplied him with several references to back up his assertion. Sokal's response was not to defer to the judgment of the older and more powerful Mayr but to show him exactly how wrong he was. Sokal also vowed never to behave in such a fashion to younger scientists when he became powerful. Joel Cracraft's response to a review of some of his work by David Wake (1980) was literally "I'm going to get that son of a bitch." Resistance from C. D. Michener served much the same function for Nelson.

The division of scientists into small research groups as well as large "demes" facilitates testing. Scientists are not likely to expose their own views to the sort of rigorous testing that their opponents are likely to lavish on them. If anything, their opponents are apt to be too critical. One danger of the demic structure of science is that intergroup competition can become too intense. Several commentators on recent disputes in systematics have found them unusually nasty. If participants in these disputes are able to evaluate each other's work with even a vestige of understanding and objectivity, then science in these respects is even more resilient than even its staunchest defenders have claimed. The data presented in this chapter on refereeing patterns in *Systematic Zoology* indicate increased factionalism in the systematics community. However, even at its height, differences in the evaluations between cladist and noncladist referees were not all that great. No one knows what the most advantageous mix of factionalism and cooperation actually is in science, but the balance toward factionalism in the systematics community seems not to have been exceeded, because progress has been marked.

10

The Visible Hand

The analogy between Darwin's argument and Malthus' is superficially very close: the tendency to fecundity and the struggle to survive—competition for food—which results from it, and the eventual control of population growth by disease and starvation. Like the invisible hand of Adam Smith, the invisible paw of nature works efficiently and leads inexorably toward its end, as do the forces of economics and reproduction in Malthus.

Elizabeth Wolgart, The Invisible Paw

THE ORGANIZATION of science that I have set out in the preceding two chapters is, broadly speaking, functional in character and has very much the appearance of providing "hidden hand" explanations for the success of science. Nowadays both sorts of explanation are definitely out of fashion in the social sciences. The most common objection to functional explanations is sorely misplaced. According to critics, ordinary causal explanations supply sufficient conditions for the events that they explain, while functional explanations purport to supply necessary conditions for the events that they explain. Because of the prevalence of functional equivalents in nature, rarely are functional items necessary for the functions they perform. Hence, rarely do functional explanations supply necessary conditions for the events they purport to explain. If they do not, then how can they possibly explain these events? The sense in which necessary, as distinct from sufficient, conditions can be said to explain events is problematic enough without adding the extra twist that functional items are usually not even necessary for the phenomena they are said to explain. The trouble with this line of reasoning is that just as rarely do ordinary causal explanations supply sufficient conditions for the events they purport to explain. Chlorophyll does function as a catalyst in photosynthesis even though it is not necessary for this process (other catalysts will do, though not as well). Smoking does cause lung cancer even though it is not a sufficient condition for this affliction (see Sober 1984a).

Anyone who doubts how rarely causal explanations actually provide sufficient conditions for the events they are said to explain can test this claim by checking the number of times the word "cause" is used either in ordinary discourse or in scientific contexts, then seeing if the cause mentioned is in and of itself a sufficient condition. The conditions or sets of conditions listed in the explanans of causal explanations invariably require extensive supplementation, *ceteris paribus* clauses, and the like. Once one appreciates the complexities that result from overdetermination and the like, causal explanations do not look so different from functional explanations. Functional equivalents in functional explanations are mirrored by alternative causes that are equally capable of bringing about the effect in causal explanations. Functional explanations look illegitimate only when evaluated on standards so stringent that ordinary causal explanations cannot meet them either. More adequate explanations of both sorts become increasingly complex (for further defense of functional ascriptions, see Rosenberg 1985).

Although no one has complained that functional explanations in biology encourage the status quo, for some reason this is a common objection in the social sciences. Supposedly, viewing social institutions and societies as functional systems entails, or at least encourages, the belief that they cannot change. This criticism is especially puzzling to a biologist because the clearest examples of functional systems are living organisms, and if anything is capable of change, an organism is. Anyone who follows a parasite through a dozen or so metamorphoses soon realizes how much change functional systems can undergo. However, ontogenetic development is "programmed" in a way absent from other natural phenomena. At the very least, the genotype of an organism sets limits to its possible development. In general, characteristics are structured more by the information encoded in the genome of the organism than by any structure provided by the environment.

The number of social scientists who have succeeded in persuading themselves that social change is also in some sense "programmed" is staggering (Blute 1979). Organisms possess a genome to direct the changes that they undergo. Nothing like a genome exists for societies. In this respect societies are more like species than organisms. The array of genomes instantiated in a species at any one time places local constraints on the immediate development of that species. If gradualists are right, genomes place no global constraints on possible change. One species can gradually evolve into another, but according to gradualists species do not go through anything like life cycles or stages. However, one characteristic of evolutionary explanations since Hardy (1908) and Weinberg (1908) has been that equilibrium conditions are taken as the base line against which natural populations are compared. According to the Hardy-Weinberg law, gene frequencies remain the same in a population unless some factors such as immigration, emigration, selection, or drift cause them to change. Because such changes are difficult to handle both empirically and computationally, equilibrium conditions have loomed unrealistically large in the literature of evolutionary biology (Levins and Lewontin 1985). If, to the contrary, Eldredge and Gould (1972) are right and stasis typically characterizes species once they have succeeded in evolving, then global constraints also exist for species, although the precise specification of the mechanisms that preclude unlimited development remains to be provided. But Eldredge and Gould do not maintain that species go through stages. No change followed by no change hardly counts as a stage theory of species development.

This much being said, the functional perspective does lead one to be somewhat cautious in attempting to change a system, whether that system is as tightly organized as some organisms or as loosely organized as an ecosystem. To the extent that a system is functionally organized, changes are sure to ramify, and these ramifications may well be extensive, not to say unpredictable. Unless one is willing to risk the destruction of the system that one wants to change, caution is called for. Even if the functional perspective did encourage the status quo, this tendency is undesirable only if one thinks that the current state of the system needs fundamental change. If not, then the conservative character of the functional perspective is not in the least a drawback. (For the purported political implications of inclusive fitness theory, see Masters 1982.)

A third objection to functional explanations is that they are not testable. As in the preceding examples, this objection stems from an unrealistic view of how easily testable hypotheses must be to count as scientific. Scientific hypotheses must be testable, but such testing is frequently quite indirect and never easy. Because matters are not crystal clear on the cutting edge of science, scientists cannot simply present data—clearly and simply. For example, two competing explanations for the purported 26-million-year cycles of mass extinction are the oscillation of the earth through the galactic plane and the orbit of Nemesis, a companion star to the sun. However, the most recent mass extinction was 12 million years ago and yet we are currently quite near the galactic plane. Raup (1986: 138) remarks, "Something is wrong! Not surprisingly, the proponents of the galactic-plane explanation have wiggled out of the paradox, while the advocates of a companion-star explanation find the mismatch in timing to be a devastating problem."

Although the systems under investigation are in no way functional, testing of alternative explanations is far from easy. Everyone agrees that ad hoc explanations are undesirable, but during scientific disputes the arguments presented by the other side always look ad hoc while one's own justifications appear perfectly cogent. To make matters worse, time and again explanations which are ad hoc on even the narrowest definition of this loaded term turn out to be correct. When Mendelian geneticists introduced epistatic genes, their opponents complained that these genes were postulated just to save Mendelian ratios, and they were; but as it turns out, epistatic genes do exist, illegitimate as this mode of explanation may be. The postulation of a companion star to the sun and an unknown planet outside the orbit of Pluto to explain the 26-million-year cycles of mass extinction found by Raup and Sepkoski are equally ad hoc (Raup 1986: 139). One or the other may nevertheless turn out to exist.

Williams (1985: 17) takes the use of ad hoc assumptions to be central to science. If a prediction turns out to be mistaken, perhaps the culprit is hiding among the most fundamental axioms of one's science—or possibly in some minor simplifying assumption, peripheral auxiliary hypothesis, or technique of measurement. Modifying the parts of one's explanatory scheme that have the most restricted ramifications to accommodate apparent anomalies is a good sign of a scientist's sanity. To put the point somewhat less strikingly, any attempt to produce atemporal criteria for distinguishing between ad hoc hypotheses in the pejorative sense and legitimate modifications of one's explanatory scheme is guaranteed not to coincide with our retrospective evaluations. There is no such thing as totally raw data. If there were, they would be useless in science. All data are structured to some extent. To count as *evidence,* data must count *for* or *against* some hypothesis or other. One of the most important talents a scientist can have is a feel for the judicious structuring of data.

The question then becomes, on a more realistic view of testing, can functional claims be tested? The answer has to be yes because biologists have been testing functional claims for centuries. We now know that brains do not serve to cool the body, that hearts pump blood, and so on, even though in many instances functional equivalents exist. When the functional systems under investigation are not so tightly organized and have feedback loops that are long, relative to the duration of scientists, functional claims are much more difficult to test. However, they are not beyond testing. For example, I have presented a functional explanation for science having the general features it does. One axiom is that scientists want credit and that science is so structured that this desire leads to increased knowledge of the empirical world. On occasion a scientist not caught up in this system of credit for contributions might make a significant contribution to our understanding of the empirical world, but such exceptions must be rare and sporadic. Although mathematics is hardly an empirical science, Bourbaki still presents a significant anomaly, one that had better be convertible into a confirming instance.

Hidden-hand explanations are a special sort of functional explanation. Geertz (1973: 206) aptly characterizes hidden-hand explanations as follows:

> a pattern of behavior shaped by a certain set of forces turns out, by a plausible but nevertheless mysterious coincidence, to serve ends but tenuously related to those forces. A group of primitives sets out, in all honesty, to pray for rain and ends by strengthening its social solidarity; a ward politician sets out to get or remain near the trough and ends up mediating between unassimilated immigrant groups and an impersonal governmental bureaucracy; an ideologist sets out to air his grievances and finds himself contributing, through the divisionary power of his illusions, to the continued viability of the very system that grieves him.

My goal in this book is to show that the coincidence between the professional interests of individual scientists to gain credit and the institutional goals of science to increase our knowledge of the empirical world is not in the least "mysterious," nor the mechanism that produces this coincidence in the least "hidden." Science is so organized that once a person who is curious about nature gains entry into a particular scientific community and begins to receive credit for his or her contributions, the system of mutual use and checking motivated by self-interest comes into play. Science is so organized that self-interest promotes the greater good. As I have mentioned before, other institutions and human activities are also organized to some extent in this same way, but science differs from these other social systems in one important respect—the domain of science is limited to those areas of human inquiry that lend themselves to testing. The boundary between those claims that are testable and those that are not is neither stationary nor sharp, but it is nonetheless real.

Geertz (1973: 206) complains that hidden-hand explanations name phenomena whose reality is not in question rather than explain them. It is not enough to allege causal feedback loops in science. They must be demonstrated. The mechanism that I have suggested as the driving force in science may not seem like much of a mechanism, but when one thinks about it, natural selection is not much of a mechanism either, and yet it is responsible for the fantastic range of adaptations that we see in the living world. How could totally caused "random" variation and subsequent selection through successive generations produce anything as intricately adapted as the human eye? As implausible as it may seem, the interplay between the replication of the genetic material and the interaction of the organisms produced with their environments is the predominant cause of adaptation in biological evolution. Although the drifting and reshaping of the continents on the face of the earth pale in comparison to organic adaptations as phenomena in need of explanation, the mechanism provided in the theory of plate tectonics is not much of a mechanism either. One feature of science from the beginning has been the attempt to explain a lot by a little, to bring order out of chaos, to reduce multiplicity to simplicity, and in the process, using Nelson's phrase, "to beat back the fog." Seldom has so little done so much for so many as it has in scientific explanations (see also Elster 1979 and Douglas 1986 for replies to Geertz).

Merton's Norms of Science

The preceding discussion about the institutional structure of science overlaps extensively with the classical analysis provided by the founding father of sociology of science in the United States—Robert K. Merton. According to Merton (1942, reprinted in 1973), science is characterized by four norms that enjoin certain sorts of behavior—universalism, communism, disinterestedness, and organized skepticism. My analysis emphasizes at least the last three. By "communism" Merton (1973: 273) is referring to the communal availability of the contributions made by scientists even though those who first make these contributions public receive the lion's share of the individual credit. Merton emphasizes the asocial character of keeping new discoveries secret. Scientists have the right to work up their ideas for a while before making them public, both as a legitimately earned edge in the race for discovery and as a safeguard against foisting half-developed views on the scientific community. Only if scientists make their work known can other scientists use it and produce the cumulative character of science. The need by other scientists of the productions of their fellow scientists is so strong, that John Aubrey, a seventeenth-century biographer, suggested that it is "better to have scientific goods stolen and circulated than to have them lost entirely" (Merton 1973: 317). One price that we pay for using the leverage of credit to force scientists to make their views known is giving nearly all the credit to the scientist who first makes a contribution public, regardless of how much work others may have put in on the problem and how close they themselves were to announcing. A second price is unseemly priority disputes. A third is an increasing tendency toward premature publication of poorly developed studies, experiments, and hypotheses.

When Merton (1973: 276) claims disinterestedness as an institutional norm of science, he certainly does not mean to be asserting that scientists are not interested in their research or its fate in the scientific community at large. They are passionately interested in both. Nor is Merton committed to scientists being able to evaluate the work of their peers in a totally disinterested fashion. Scientists not infrequently have an inflated idea of the importance of their own contributions relative to those of their colleagues. Invariably they think that their contributions are correct and important.

Although Popper (1975: 93) is concerned to argue for the rationality of scientific revolutions, he suspects that "should individual scientists ever become 'objective and rational' in the sense of 'impartial and detached', then we should indeed find the revolutionary progress of science barred by an impenetrable obstacle." Unless scientists are reasonably good at research, we never hear of them. However, not infrequently, scientists who know how to go about putting nature to the test are extremely naive about what is necessary to get their views noticed, let alone accepted. They are reasonably good at picking a journal that reaches the relevant audience (Gordon 1984), but time and again they act as if they think that publishing an idea is equivalent to making it public. Communication is a relation. Not only must a message be sent, but also it must be received (Manten 1980).

In the early decades of this century, two different views on the contributions of scientists to the growth of scientific knowledge were suggested: one that a small percentage of scientists are responsible for most progress in science, the other that all scientists contribute their bit even though only a few scientists gain much in the way of recognition—Lotka's law and the Ortega hypothesis, respectively. When the mathematician Alfred J. Lotka (1926) counted the number of names in the index of two abstract journals, he discovered considerable differences in frequency of references among the scientists listed. Many years later, Price (1963) drew what he took to be the obvious conclusion from these and other data: a very few scientists produce the majority of contributions in science. J. Ortega y Gasset (1932) expressed just the opposite contention: a majority of scientists, even the most mediocre, contribute to the general advance of science. Science is likened to one great edifice to which each and every scientist contributes at least a brick or two. As reassuring as the Ortega hypothesis may be, all the evidence points in the other direction. For example, half of the physiologists that Meltzer (1956) studied published fewer than four papers during a three-year period. Of the 238 chemists receiving doctorates between 1955 and 1961 that Reskin (1977) sampled, 7.5 percent published nothing during the first decade after receiving their degree, and 11 percent published only one article. In any given year, 60 percent of chemists publish nothing whatsoever—and this in a field in which bits and pieces of research can be published with little difficulty. During the two-year period studied by Ladd and Lipset (1977), almost a third of the physicists teaching in American universities and colleges published nothing. The studies that social scientists have conducted on themselves show that they are even less productive than physiologists, chemists, and physicists (Babchuck and Bates 1962, Yoels 1973). Garvey (1979) discovered that only 10 percent of psychologists published at least one article a year.

Of course, there is more to making contributions to science than the number of papers published. A scientist might publish only a single paper and that paper nevertheless might have massive effects on the future course of science. However, when we turn from publication to the effects of publication, the disparity becomes even greater. For example, Menard (1971: 99) discovered that 2 percent of the federally employed scientists working in oceanography accounted for 65 percent of all citations, while Garvey (1979) found that only 10 percent of the psychologists who publish are cited once a year. Only 10 percent publish at least one paper a year, and only 10 percent of these are cited! Blau (1978) estimated that 43 percent of the high-energy physicists whom she studied contributed little if anything either to the literature of the discipline or to administration. In the face of figures such as these, it is difficult not to conclude that publishing a paper is roughly equivalent to throwing it away.

Most scientists publish very little during the course of their careers and of the few who do, most have no discernible influence on the course of science (Cole and Cole 1973: 43). Regardless of what one thinks of citations as a mark of influence or inherent worth, very few unpublished papers are likely to have much impact, no matter what their intrinsic value. Most scientists contribute very few bricks to the scientific edifice, and the occasional brick that most scientists ostensibly add is never noticed by later scientists who might have been able to use it had they been aware of it. But even when earlier papers are rediscovered, frequently their contents are not usable because the author neglected to note precisely what the later author needed to know. Unlike mass-produced row houses, scientific theories are works of art and their elements made to order.

MacRoberts and MacRoberts (1984), however, sound a note of caution. On the basis of their own experience, they were acquainted with very few scientists who were as unproductive as the above figures indicate. Of course, their experience might be atypical. Those of use who teach at research-oriented universities forget the huge numbers of scientists who work at institutions devoted almost exclusively to teaching or industry. A chemist who spends eight hours a day testing the oil content of peanuts for a candy company is still a chemist, but he or she is not likely to publish much in the way of research. In any case, MacRoberts and MacRoberts warn that the methodology used by abstracting services is likely to underestimate the absolute productivity of scientists, especially if they publish in more than one area as defined by the abstracting service. Even so, if the sociologists who have studied scientific productivity and influence are right, great advances in science are made by only a very few scientists. More disheartening still, even most normal science is the product of a very few scientists. As in the case of the performing and creative arts, science is extremely elitist (Cole and Cole 1972, Hallam 1983: 159; for a critical review, see Fox 1983).

Some commentators on science find its elitist character undesirable. Egalitarianism is intrinsically good; elitism intrinsically bad. Furthermore, because society in general should be as egalitarian as possible, all institutions in a society should be equally egalitarian. The fact that only a tiny percentage of those who study a musical instrument or take ballet lessons ever get to perform in public is as damning of a society as the fact that only a very few scientists contribute measurably to science. I happen to prefer egalitarian societies, but one need not be a cultural relativist to have doubts about reading one's preferences about desirable societies into all times, ages, and institutions. In Victorian England, scientists were enjoined to be humble, Christian, and above all else gentlemen. In the writings of several contemporary commentators, "gentleman" has taken on a pejorative tone (Morrell and Thackray 1985, Rudwick 1985).

In Victorian England a particular sort of behavior was acceptable. In the United States today, quite different sorts of behavior are acceptable. However, in the midst of such changes, the professional behavior of scientists has not changed that much. The dispute between Huxley and Owen is indistinguishable in its conduct from the dispute between Farris and Sokal, the softening effect of time notwithstanding. Perhaps all people are of equal worth, but not all ideas are equally good. Cognitive change in science is necessarily elitist. Any restructuring of science that neutralizes the highly selective character of this process is sure to be destructive. Perhaps all scientists might receive the same salaries without impeding conceptual progress in science, but an equal distribution of research funds is liable to have serious effects, and treating all scientists as if they are equally good at what they do is sure to be disastrous. Societies need not be perfectly homogeneous. Even egalitarian societies can contain elitist institutions. Jacques Monod was a political radical, but his politics did not extend to his own laboratory (Abir-am 1982). Just as affirmative action stops at the doors of the Congress of the United States, egalitarianism stops at the laboratory door even for the most radically egalitarian scientists.

Because selection in science is so intense, science is extremely elitist. However, nothing about an evolutionary account of conceptual change in science entails that selection must be intense or that the contributions made to science must be skewed toward a very few scientists. After all, the relative frequency of gradual, global change versus punctuational change involving a very few organisms in biological evolution is still a matter of some dispute. The relative frequency of these two sorts of change in science is likely to be no less problematic, and there is no reason to expect the relative frequencies to be the same in both. Whether Lotka or Ortega is right about science, an analysis in terms of selection is still applicable. The only clear implication is that if change is especially rapid, small groups are liable to be the site of such change. A new view or combination of views becomes established in a small group of scientists and from that power base becomes disseminated more widely, possibly to a single scientific deme, possibly to a much more inclusive group of scientists.

Although transmission is necessary for success in science ("success" in the technical sense that I am developing), success is a function of more than sheer numbers. After all, few scientific ideas get diffused throughout the entire scientific community. Most are limited to a single scientific domain, possibly even to a single deme in that domain. If the principles of cladistic analysis become widely accepted and practiced among biological systematists, the cladists will have been successful even though most scientists have never heard of them, even if the scientists who use the taxa produced for their own purposes are unaware of the considerations that led to the genesis of these taxa. In the operation of the system of credit for contributions in science, local, rapid recognition of the more explicit sort is important, global, long-term implicit recognition much less so.

On the view of science that I am developing, success and failure is a function of the transmission of one's views, preferably accompanied by an explicit acknowledgment. This transmission can be only local, or local and then global. In order for the system to work, the speed of transmission is important. If science were a simple matter of a particular scientist happening upon the truth, fully formed, then rapid dissemination might be desirable but hardly necessary. However, as I see it, scientists *never* present their views fully formed. Science is a social process in which scientists evaluate and criticize each other's work, leading to successive improvement. The appearance of some great scientist presenting a particular revolutionary view fully formed in a single seminal work is primarily a function of retrospective reconstruction. Thus, on my view, truly unappreciated precursors are the most extreme form of failure. In the annals of science, they are villains, possibly victims, but hardly heroes.

One way to fail in science is not to publish at all. Another is to publish in ways calculated to get other scientists to ignore or reject your work. A recurrent refrain in the episodes presented in the first half of this book is that mode of presentation in science matters. Perhaps it should not, but it does. Publishing highly speculative theories expressed in an idiosyncratic manner is one way to increase the likelihood that one's work will be ignored. Lamarck, Geoffroy Saint-Hilaire, Wallace, and Croizat come immediately to mind. All four men ruined their reputations by their failure to present the impression of being good solid workers, setting forth their theories only reluctantly in full knowledge of the dangers of speculation.

Presenting one's views in an irresponsible manner is one way to decrease their influence. Another is to write in a verbose, opaque style. Numerous scientists have suffered because either they were unable to write in a clear and effective manner or else they did not care to. Michel Adanson, the eighteenth-century systematist who was to become the patron saint of numerical taxonomy in the early days, apparently was among this number. Not until John Playfair (1802) rewrote James Hutton's prolix *Theory of the Earth* (1795) did it begin to have an impact. As Hallam (1983: 15) remarks, "Hutton's great work had little impact on the geological community, principally it appears because of his somewhat verbose and obscure prose style and lack of a coherently organized structure, together with his failure to publish the critical evidence for the intrusive character of granite. Such impact had to await a far more lucid account by Playfair at the beginning of the nineteenth century." Swainson (1835) in turn popularized Macleay's (1819) Quinarian system. Victorians, it seems, could say nothing in less than three volumes, and given the low cost of printing, they could afford to indulge themselves; more importantly, their contemporaries had the leisure to read such lengthy works. For example, Spencer published volume after volume of prose that even his contemporaries found soporific, but they did read his works. The tolerance of present-day scientists for multivolume treatises is considerably reduced, as Sewall Wright has discovered. Among even the most intensely interested population biologists, few can claim that they have read all four of Wright's four volumes cover to cover. Like the Bible, they are books that one reads parts of or looks things up in.[1]

The most common complaint voiced by referees of papers submitted to *Systematic Zoology* was that they were written in extremely poor English. The complaints did not merely concern style and excessive polemics but also matters of basic English. In this respect, systematists are not unusual. For example, Raup (1986: 67) felt guilty that he let the poor exposition of the manuscript by Alvarez and company on extra-

1. I realize that the preceding comments aptly apply to this book itself. I could plead that I have made this book so ponderous in an attempt to ensure that the dangerous ideas expressed in its pages are read only by those few professionals who are serious students of science, but I have not. Instead I have found myself unable to resist the same forces that I describe in the careers of the scientists whom I have studied.

terrestrial causes for mass extinctions influence his evaluation, but it did. Poor writing style does not guarantee that a work will be ineffective. Sometimes a scientist is willing to take up the cause and present another scientist's view in ways that have some effect, and sometimes readers are willing to wade through even the most turgid exposition. For example, the British geneticist E. B. Ford (1980: 336) recalled his own reaction as an undergraduate at Oxford to Morgan's (1919), *The Physical Basis of Heredity.* He found it to be

> one of the worst written books I had ever encountered and I think so still. Having, in my classical studies, read books by people who could write, I was horrified. How dreadful are the blots on the pages that are called figures. If anything could have stopped me from taking an interest in genetics, it would have been that book of monumental dullness and incompetent presentation.

Thirty years later, systematists whose first language was German nevertheless found Hennig's (1950) *Grundzüge* extremely difficult to follow. Nelson initiated the Hennigian revolution among English-speaking systematists never having seen Hennig's classic work, let alone having read it. Even the English translation of Hennig's revised manuscript did not have the impact that Brundin's more compact and simple exposition had. Early cladists obtained their Hennig from Brundin, not Hennig. Conversely, a flowing, graceful writing style can sometimes conceal serious conceptual rifts. Lyell, Simpson, and more recently Gould provide excellent cases in point. Readers are so charmed by the easy flow of their prose that they run the danger of not being sufficiently critical of the views expressed. Those scientists who do not write well are at a decided disadvantage compared to those who do. Sometimes their works have an impact all on their own, but more often as the result of the more lucid presentations of a disciple.

A more common and serious cause of ineffectiveness in science is the failure to push one's views. Sometimes this failure stems from the author failing to see his or her contributions in the same light as later workers see them; sometimes just from insufficient aggressiveness and combativeness. For example, anyone who has heard of the Glacier theory immediately thinks of Louis Agassiz's (1840) *Études sur les Glaciers,* not the work of Jean de Charpentier published five years earlier. As Hallam (1983: 69) notes, "Charpentier was a rather shy, mild-mannered, cautious man who lacked the temperament to engage with any enthusiasm in polemical battles with his scientific adversaries, preferring to pursue with great care and scruple his detailed researches. Agassiz was quite different, and quickly became an active proselytizer for the cause, developing a comprehensive glacial theory that extrapolated boldly beyond the available evidence in a way that the more conservative Charpentier found difficult to tolerate."

Mendel presents the most famous example of the effects of modesty and caution in science. Mendel had a deep understanding of the implications of his experiments for inheritance in garden peas. How far he thought that his results could be generalized and the possible implications he saw for evolution are difficult to ascertain. But one thing is certain: he made only the most ineffectual gestures at getting his views noticed by his contemporaries. Even though Mendel's mathematical form of expression was somewhat unusual for the literature of the time, it was not so nonstandard that his colleagues would reject it out of hand on that account. Mendel's main problem was that he was too humble: a couple of papers, a couple of letters to a famous colleague, and that was that.

Wallace provides another case of a scientist who did not receive the credit that he might have. To be sure, Darwin had been working on his version of natural selection for two decades when he received Wallace's paper. It is unlikely that Wallace could have managed to supersede Darwin as the primary author of evolutionary theory. However, he could easily have received more credit than he did. If he had returned to England immediately upon hearing of the joint publication of his paper along with those of Darwin and tacitly asserted himself by publishing on evolution, he might easily have avoided his subsidiary role as Darwin's moon, but he did not. In the decades immediately after the publication of the *Origin*, only a very few people were working on the evolutionary process, as distinct from phylogeny reconstruction. The depth of Wallace's understanding of evolution was surpassed at the time only by that of Darwin. If Wallace had continued to contribute to the development of evolutionary theory right from the start and had repeatedly asserted his co-authorship, today we might refer to the Darwin-Wallace theory of evolution as we talk about the Hardy-Weinberg law or the Watson-Crick model of DNA.

In 1970, when systematists in the English-speaking world were looking for an elder statesman to take up the cause of phylogenetic systematics, Hennig was the most likely candidate, but the language barrier was a formidable obstacle. In addition, Hennig was an extremely shy man. At this same time, R. A. Crowson published his *Classification and Biology* (1970). Nelson wrote him enthusiastic letters, attempting to enlist his aid, but Crowson was not interested. He had written his book, and that was good enough. Less senior converts were also greeted with enthusiasm. For example, early on John G. Lundberg, a zoologist at Duke University in Durham, North Carolina, published a paper using Farris's Wagner trees to reconstruct phylogeny. Nelson's (1973a: 87) initial response was positive. The work of Lundberg (1972) and Farris (1970) was the first step in showing that "numerical taxonomy is not replacing phylogenetic systematics, but rather is being absorbed by it."

In this same paper, Nelson also indicated that the embryological method was the only real candidate for establishing transformation series. Unlike phylogenetic development, embryological development can be observed. Lundberg (1973) disagreed. The embryological method is not up to the task because it is inherently ambiguous. None too surprisingly Nelson (1973d) responded somewhat testily, but Lundberg declined to pursue the matter. He found Nelson's response too blunt and unfriendly. As future events clearly indicate, the issues raised by Lundberg turned out to be central to the ongoing controversy over cladistics, but these issues were pursued by others. Lundberg's paper was not resurrected until much later in the dispute between Kluge (1985) and Nelson (1985) over the split between pattern and phylogenetic cladists (see also Gould 1973, 1977b, 1979, Løvtrup 1978, Nelson 1978b, Rosen 1982, Voorzanger and van der Steen 1982).

A similar story can be told with respect to Beatty's (1982) paper pointing out what he took to be differences in the views of cladists over the role of phylogeny and the evolutionary process in cladistic analysis. His contention was immediately attacked in no uncertain terms by several cladists. Platnick (1982: 283) ventured that "philosophers" should do their homework before they enter into scientific debates. Although Beatty had been warned that his paper might elicit some strong responses, he was not prepared for the hostility that characterized the letters he received from several of the aggrieved parties. Although Beatty's name continues to punctuate the debate over the distinction between phylogenetic and pattern cladism, usually in a derogatory aside, Beatty himself has declined to respond. His silence is interpreted by those concerned as an admission of defeat. If he could answer the rebuttals to his 1982 paper, he would.

The study I conducted of refereeing practices between 1974 and 1979 for the journal *Systematic Zoology* serves only to reinforce the role of aggressiveness and persistence in science. During this six-year period, roughly 65 percent of the manuscripts submitted to the journal eventually appeared. Of those that did not appear, about 60 percent were rejected, while the remaining 40 percent simply died when the authors did not bother to resubmit. As far as resubmission is concerned, it made no difference whether the manuscript was returned for major or minor revision. Equal percentages of both failed to appear in the journal because the author declined to resubmit. If the author reworked and resubmitted, the manuscript almost always was accepted. Under Nelson's editorship, only four papers were rejected after resubmission; under Schuh only five (see earlier discussion in chapter 9). Most authors took rejection in silence, but a few fought back. Of those who did, almost all succeeded in getting their manuscripts published. The authors who fought rejection either were already big names in the field or eventually became such. Of those who accepted their fate in silence, not one became well known in systematic philosophy. If they were successful in science, it was in some other area.

The message of the preceding figures is patent, though rarely acknowledged officially by scientists. If one is not willing to stand up and fight for one's views, don't bother. Other scientists complained of the aggressive behavior exhibited by Simpson, Mayr, Sokal, Ehrlich, Nelson, and Farris, but in the absence of such behavior these scientists might well have been relegated to the role of unappreciated precursors of other scientists later, scientists who were not so virtuous. One conclusion that Hallam (1983: 159–69) derived from his history of geological controversies is the "desirability of persistent and persuasive advocacy in order to change thought. It is simply not enough to have a novel idea, however good it may be. The eloquence of Werner and the enthusiasm he transmitted to his star pupils played a major role in establishing neptunism as standard doctrine across Europe and North America."

If we are genuinely interested in educating students who are most likely to contribute to the growth of science, we might well give applicants to graduate school aggressiveness tests as well as achievement tests. When Roe (1952) did administer the Thematic Apperception Test to the members of the National Academy of Science, she discovered that none of the three groups of scientists in her study scored very highly on "aggression." To the extent that this test has any validity, it indicates that scientists are basically not very aggressive. If to succeed in science, scientists must behave aggressively, it follows that scientists are being called upon to behave in ways that they do not find especially easy. People who are not especially aggressive are being forced to behave aggressively while they pretend to be doing something else.

Evolutionary biology during the past century in France is one final example of inefficaciousness in science. When Darwin's work began to appear in French, it was not totally ignored, but it was not greeted with the same attention that it received in other countries that also had strong scientific traditions. It certainly was not adopted. Not until 1878 was Darwin elected a correspondent of the Botany section of the French Academy of Science and then only in spite of his theory of evolution. Mendelian genetics received no warmer a welcome from the French scientific establishment. When T. H. Morgan was put up for the same honor in 1924, he received not a single vote. Finally, in 1931, just two years before he won the Nobel Prize, did he make it. Not until after the Second World War did a Darwinian succeed in getting an academic post in France. When I mentioned the curious resistance of French biologists to change in certain areas of science to members of the staff of Cuvier's Musée d'Histoire Naturelle in 1982, I was informed with some pride that history had proved French biologists right. Recent developments in evolutionary biology had shown that evolutionary theory as set out by Darwin, through Weismann, to Simpson and Mayr is deeply flawed. Perhaps so, but being right and being efficacious are two very different things. French biologists played no role in uncovering any of the putative flaws either in Darwin's original theory or in the later Modern Synthesis. They carped a bit from the sidelines and that was it. If one waits long enough, one can turn out to be right about any great achievement in science. They all turn out to be deficient in one way or other. The issue is whether you want history to prove you right or to participate in the ongoing process of science.

Unless a scientist is content to play the role of an unappreciated precursor, he or she had best do more than just publish. Where to publish, in what form, how often, who to cite, who not to cite, and so on all enter into the process of making one's work public. The scientific community is not a seamless whole. It is broken down into nested and overlapping informal groups, what I have termed the demic structure of science. The smallest unit in science is the individual scientist. The next largest is the research team whose members are in daily contact. Scientific "demes" are much larger, consisting of dozens to several hundred scientists, depending on the discipline. Invisible colleges are the largest groups of all. These groups of increasing scope serve several functions in science. One function of the smaller groups is to provide sympathetic criticism while a scientist develops his or her ideas. Darwin presented his theory at various stages in its development to several sympathetic readers. Their reactions helped prepare Darwin for the sort of reception that his theory would receive once he made it public. He was able to uncover possible misunderstandings and likely criticisms. Chambers was in no such position. He concluded his infamous book with the plaintive remark that it was "composed in solitude, and almost without the cognizance of a single human being, for the sole purpose (or as nearly so as may be) of improving the knowledge of mankind, and through that medium their happiness." His book suffered accordingly.

Some scientists see no value in setting out their ideas with potential readers in mind. They write for themselves. If others can understand their exposition, fine; if not, that is their loss. Suggestions for revision or rewording are met with extremely hostile reactions. Croizat published his most important works at his own expense. One effect of this practice was that he did not have to submit to the editorial process. He could publish what he pleased as he pleased. As a result, the very readers whom he needed to convince found his works idiosyncratic, prolix, difficult to follow. Although Croizat is an extreme case, the resistance of authors to modifying their prose is widespread. Goldschmidt consistently declined to have his manuscripts read by others prior to publication. Although the referees for manuscripts submitted to *Systematic Zoology* had some effect on the papers that appeared, their advice on presentation was ignored more often than not. Time and again, equivocations and confusions that emerged only much later in the literature, after considerable back and forth, were pointed out by referees and ignored by the authors. The courage to go one's own way enhances originality almost as much as it is destructive of successful communication.

One also increases the chances of having one's work noticed by becoming part of one or more informal groups of scientists. When Darwin (1899, vol. 1: 244) returned from his voyage, he was dismayed that professional zoologists initially showed so little interest in curating his collections. In a letter to Henslow, he complained, "I have not made much progress with the great men. I find, as you told me, that they are overwhelmed with their own business." When Mendel wrote to von Naegeli, the great man quite naturally suggested that Mendel take up von Naegeli's own group, *Hieracium,* the hawkweed. Although Mendel's paper on *Hieracium* did not lend much support to Mendel's own budding research program, at least it was noticed (MacRoberts 1985). It was noticed because others were able to use it. Unless there are other scientists who can use your work, it will not be used. A scientist working in social, though not conceptual, isolation on occasion is noticed, but visibility is greatly enhanced by being part of a group of scientists working on the same cluster of problems.

Members of these groups also come to each other's aid after publication. Not only can they keep the ideas of their allies from being ignored, but also they can guarantee that at least a few of the published responses will be positive. The Darwin-Wallace papers were presented at a meeting of the Linnean Society by two of the most influential scientists of the day, and among the first reviews of the *Origin of Species* were several written by Darwin's allies. When hostile criticism began, Darwin did not have to meet it alone. He had several powerful allies to help him in his confrontations with his opponents. Lamarck, Chambers, Goldschmidt, and Croizat stood alone against their critics. Croizat, Brundin, and Løvtrup made tentative gestures at forming an alliance, but found it impossible not to attack each other's work. Darwin's idealist contemporaries and the later neo-Lamarckians found themselves in the same position. The immediate effect of such isolation is rejection. However, everything has a price, and the price we pay for the demic structure of science is intergroup warfare of the sort that everyone officially decries.

If science is to be cumulative and self-correcting, then it is inherently social. However, it does not follow that it must at all times exhibit a demic structure. After all, throughout the first half of life on Earth, evolution proceeded in the absence of species. Thus, science might proceed even though scientists did not form research groups to formulate and disseminate conflicting scientific views. However, whether in biological evolution or in science, change should be much slower. In fact, science has been and continues to be characterized by research groups—phrenologists, Quinarians, Darwinians, Kelvin's group, neo-Darwinians after Weismann, Mendelians, the MacArthur group, numerical pheneticists, cladists, vicariance Croizatians, true Croizatians, and so on. In other cases, views were championed only by isolated individuals, e.g., the British idealist opponents to Darwinism as well as later neo-Lamarckians. As these examples indicate, the formation of a group may guarantee that a particular view will not be ignored, but it is not sufficient for widespread acceptance. Most research groups fail, so much so that they leave little in the way of records. A list of failed research programs would be a list of names few readers would recognize. However, successful research programs will by necessity generate research groups. Success requires adherents, and it is very difficult, possibly impossible, for numerous scientists to adopt a cluster of related views without cooperation emerging.

For a host of reasons, scientists are wise to join a scientific research group or to form one themselves. How does one go about doing this? Krantz and Wiggins (1973) studied the four founders of behavioral psychology—Clark Hull, B. F. Skinner, Kenneth Spence, and E. C. Tolman—to ascertain the influence of personal and impersonal channels on recruitment of students and followers. What impact did the personal characteristics of these early psychologists have on their followers in contrast to their published works? To answer this question Krantz and Wiggins (1973) sent a questionnaire to former students and colleagues of these four men. They received 127 usable responses—22 for Hull, 24 for Skinner, 53 for Spence, and 28 for Tolman (see Appendix G). The main conclusion reached by Krantz and Wiggins (1973: 133) was that, "while impersonal recruitment produces the greatest number of followers, personal recruitment increases and maintains affiliation."

Their data can be used for other purposes as well. We all pick up by social osmosis certain beliefs about the personal and intellectual traits of "good" scientists. Scientists should be creative, inspiring, open-minded, accept criticism well, be willing to change their minds in response to criticism, and treat their graduate students as equals, allowing them to have ideas of their own, being supportive in the way they criticize, helping them find positions, and giving them credit for their work. Scientists should not be authoritarian, aggressive, cold, and distant. Although there is nothing wrong with scientists being convinced of the value of their own positions, they should not insist on a strong commitment from their students to their own particular approach, nor should they teach their own position to the exclusion of others. They should teach a fair representation of views and let their students pick.

One might expect that if scientists behave the way that they should, they would be more efficacious. The quality of the views they are pushing surely has an effect, but proper conduct as a research scientist and as a teacher should also make a positive contribution. The results of Krantz and Wiggins's questionnaire do not support this supposition very strongly (see table 10.1). Skinner was much more successful in getting students to follow in his own footsteps than any of the other founders of behavioral psychology. Spence was the next most successful, while Tolman and Hull tied for last place (Krantz and Wiggins 1973: 135). All four men were successful. Their views became widely disseminated. They are figures in the history of psychology. However, they were not equally successful, and I doubt that professional psychologists today would evaluate the quality of the work of these four men in this same order as the degree to which they were successful. More than this, the personal attributes which

Table 10.1. Summary data for the perceptions of former students of the four founders of behavioral psychology.

Factors		*Hull*	*Spence*	*Skinner*	*Tolman*
1. Aggressive, authoritarian ideological commitment	M SD	6.93 0.95	8.09 0.62	6.17 1.05	3.89 0.75
2. Emotional and instrumental professional aid and support	M SD	5.39 2.42	5.29 1.86	4.49 1.85	6.09 2.06
3. Willingness to encourage intellectual autonomy and two-way interaction	M SD	6.33 1.91	5.33 1.56	6.34 1.79	8.07 0.87
4. Intellectual lability	M SD	4.89 1.50	5.14 1.58	3.35 1.48	5.61 1.50
5. Broad, abstract vs. narrow, concrete approach to knowledge	M SD	5.86 2.03	6.97 1.33	8.14 0.92	8.30 0.88
6. Interpersonal openness and expression	M SD	6.05 2.01	5.50 1.44	5.80 1.78	7.83 0.91

Note: The range of ratings varies from 1 (inaccurate) to 9 (accurate). The 32 questions were grouped into six categories by factor analysis and Verimax rotation. The number of responses for Hull was 22 except factor 2 for which n = 21, for Spence 53 except factor 5 for which n = 52, for Skinner 24, and for Tolman 28 (from Krantz and Wiggins 1973).

apparently enhanced their personal following are not those that we are likely to find most admirable. For example, D. T. Campbell (1979a: 187–88) notes that, although Tolman was the "most personally beloved of all the four theorists" and his theory "can be seen to have been clearly the best," Tolman's students were least loyal.

According to his students and colleagues, Tolman adhered most rigorously to proper scientific conduct. He was the most humble and unpretentious of the four men, the most questioning and flexible in regard to his work. He was also much more supportive of his students, interpersonally open, and allowed them considerable intellectual autonomy. He was also the most willing to change his mind, but only slightly more so than Spence and Hull. Tolman did not want or need disciples. Skinner to the contrary was aggressive, authoritarian, and ideologically committed. He was not very supportive of his students either personally or professionally, and he was significantly less willing to change his mind than the other three. One bright spot: he did encourage a certain degree of intellectual autonomy among his students. Spence in his turn suffered from being too cold and distant, too aggressive and authoritarian. Although Hull was aggressively committed to his views and intellectually not very labile, he was least broad and creative in his approach to knowledge. On the basis of biographical details, Krantz and Wiggins (1973) suggest that he was not more successful in influencing his students and associates because he lacked Skinner's sureness and self-confidence. Except for a few disciples, few of his associates followed in his footsteps but instead adopted Spence's version of the Hullian theory.

Krantz and Wiggins (1973: 153, 139) conclude that Skinner was much more successful than the other founders of behavioral psychology both because his theory held out more "promise"—there was so much work to be done, so many experiments to be run, numerous variables to be explored and procedures to try—and because he was an "enthusiast." An enthusiast is

> one who believes that the "Cause" is first and foremost, demanding the dedication and blind faith of the followers. The enthusiast is inner-directed, having full confidence in the truth of his convictions. Consequently, he does not show great concern for the opinions of others, and he operates under the principle that if some should refuse to share his vision, so much the worse for them. To press his views, to place himself on the dictator's podium, would be a compromise—and when it comes to the "Cause", there can be no compromises.

Perhaps one would not want one's child to marry an enthusiast, but if this is what it takes to bring about significant change in science, perhaps such behavior should not be condemned too strongly. There is no reason that every subgroup in a society has to be characterized by the same social and moral norms that are widely valued in society at large. But perhaps the preceding conclusions follow only with respect to the founders of behavioral psychology. After all, behaviorism was only one research program out of many. To see if similar trends might be discernible among leading systematists, I sent a similar questionnaire to students and colleagues of Robert R. Sokal, Gareth Nelson, and George Ball. I chose Sokal because he was one of the founders of numerical/phenetic taxonomy, Nelson because he was one of the founders of cladistics, and Ball because he had been teaching the principles of cladistic analysis at least as long as Nelson, had influenced numerous students, but had not taken on the role of a "leader." Krantz and Wiggins designed their questionnaire for the founders of behavioral psychologists long after they had become grand old men of the movement. When I sent out my questionnaire in 1981, Sokal and Ball were well into their careers, though not quite "grand old men," while Nelson was only just becoming established. The number of questionnaires sent were 11 for Ball, 13 for Sokal, and 13 for Nelson. There were eight responses for each man. Of these, five counted themselves students of Ball, whereas only two respondents for Sokal and three for Nelson thought of themselves as students. The rest were colleagues.

The results of my questionnaire accord closely with those of Krantz and Wiggins (see Appendix G). As the preceding chapters have indicated, both Sokal and Nelson have been successful in bringing certain issues to the forefront in the systematics community, and both are identified as leaders of their respective research programs. Although Ball had probably trained more systematists than the other two men put together and has just as strong views about taxonomic method, he has not developed nearly the following that the other two men have. Yet he like Tolman is perceived by those who know him professionally as embodying the best characteristics of a genuine scientist. Ball, Sokal, and Nelson are all convinced of the value of their own systematic position and set high standards for the work of others (questions 1 and 12). None of them has been overly influenced by others' views of his approach (question 21). However, Ball differs from Sokal and Nelson on a host of counts. He does not take himself as seriously as the other two men and is more openminded, humorous, and outgoing

(3, 16, 19, 25). All three men were viewed as being broadly versed and tended to present all positions in their teaching, but on both counts Ball surpassed the other two men (7 and 15). The respondents also viewed Ball as being more supportive than Sokal and Nelson in his criticisms, more receptive to the ideas of others about his work, more concerned with the positions of others in their departments, more understanding of the needs of others, as well as more helpful in obtaining financial support for their research and in publishing the results (13, 17, 18, 22, 26, 32). Ball also allowed greater autonomy in his students' choice of research problems, treated them as intellectual equals, and was more ready to let others know him as a person (2, 14, 28).

Not all the correlations were quite so one-sided. Sokal stood out on four questions. He was judged somewhat less inspiring than either Ball or Nelson, less likely to reformulate his position, less variable in his moods, and somewhat less likely to confer appropriate credit to the contributions of others to his systematic approach (questions 6, 11, 29, 30). Nelson tended to stand alone on several questions. He expected a stronger commitment to his approach from others, accepted criticism less well, and was more critical of others' systematic position (4, 5, 9). All three men were judged aggressive, creative, and strong-willed, but Nelson more so than the other two men (23, 24, 31). On four questions the three men were ordered sequentially. Ball was the most concerned with the professional future of his students, followed by Sokal and then Nelson. Of course, when the poll was taken, Nelson was not in a position himself to be able to help others in this respect (8). Sokal was judged most authoritarian, followed by Nelson, and then Ball a distant third (10). Nelson was judged most philosophically oriented, followed by Ball and then Sokal (20). Although those polled responded that Ball allowed them to know him as a person much more readily than either Sokal or Nelson, they distinguished between Sokal and Nelson when asked if these men were distant interpersonally. As might be expected, Ball was least distant, Sokal next, and Nelson most distant (27).

Mitroff (1974: 586) came up with much the same results in his study of forty-two scientists involved in the Apollo Moon Project:

> The three scientists most often perceived by their peers as most committed to their hypotheses and the object of such strong reactions were also judged to be among *the* most outstanding scientists in the program. They were simultaneously judged to be the most creative and the most resistant to change. The aggregate judgment was that they were "the most creative" for their continual creation of "bold, provocative, stimulating, suggestive, speculative hypotheses," and "the most resistant to change" for "their pronounced ability to hang onto their ideas and defend them with all their might to theirs and everyone else's death."

The message of the preceding discussion is that, as far as professional relationships are concerned, certain behavior which most of us are likely to find admirable is not very effective in getting one's views adopted by others, while other sorts of behavior that we tend to decry promote recognition. If recognition for one's contributions is as important in science as Merton and others claim and scientists ever come to recognize this feature of science, then scientists have some hard decisions to make. They are going to labor assiduously, regardless of the outcome. Do they want their contributions to be disseminated widely or not? If they do, are they willing to behave in ways calculated to increase their visibility? If views could be noticed without any attention being paid to their authors, this question would not be so poignant, but apparently this is not possible. The price that reclusive scientists must pay if their views are to have an impact is personal recognition. This is also the reward that their more extroverted colleagues receive under the same conditions.

After emphasizing how important recognition is in science and how differentially it is bestowed, Merton (1973) immediately tries to ascertain whether the credit that scientists receive is "deserved." Do eminent men of science simply publish more papers than their run-of-the-mill colleagues, or are their papers also better? Does credit accrue primarily because of the inherent worth of the contribution or do all sorts of other factors such as institutional affiliation, prestige, professional connections, and the like play important roles? The conclusion that Merton and his students have reached is that by and large the reward system in science recognizes quality research (Cole and Cole, 1973, Zuckerman and Merton 1972 in Merton 1973, Gaston 1978). In my analysis, I spend little time trying to distinguish between those papers that should have had more attention or less attention paid to them from some third-party perspective. Instead, I content myself with the use that scientists actually make of a particular publication. I fail to see what point there is for a present-day student of science second-guessing the scientists of a particular period with the power of hindsight.

Most research programs progress for a while and then disappear, leaving a literature as calcified and causally inert as the fossils of extinct species. The scientific process comes close to being as wasteful as its biological counterpart. All one has to do to check this claim is to pick at random any field of inquiry and read the literature of a generation or two back. If one reads retrospectively, concentrating primarily on the literature that led to currently accepted views, science seems moderately efficient. Many issues keep being reexamined, each time with some improvement. However, if one reads the literature at large, the picture is very different. A few of the problems addressed led somewhere, but most were just dropped. Present-day scientists have never heard of them, and show no interest in taking them up again.

Discovery versus Dissemination

As romantic as unappreciated precursors may seem to us, they simply do not count. The most that can be said for them is that they warrant footnotes in scholarly, if slightly Whiggish, histories of science. However, on occasion one of these precursors is resurrected to serve either as a patron saint or as an evil demon. Anyone who studies science is liable to be surprised by the ferocity of opinion that the views of long-dead scientists can engender in scientific disputes. For example, in the nineteenth century, Lyell as well as the Darwinians made Cuvier their "whipping boy" (Desmond 1982: 173), while Geikie "canonized" Hutton (Davies 1969: 179). A century later, Raup (1985) reversed the roles of Lyell and Cuvier, making Cuvier into a maligned precursor and Lyell into the Prince of Darkness for his single-minded rejection of anything that might smack of catastrophies.

During the formation of the Modern Synthesis, Simpson and Mayr settled on Goldschmidt as the paradigm of an evolutionist gone wrong. More recently, Gould (1977c,

1982a) has attempted to refurbish Goldschmidt's reputation once it was pointed out to him that his own views might well be interpreted as neo-Goldschmidtian (see also Piternick 1980). Løvtrup (1973) reached back to Chambers as a precursor for his theory of epigenetics. Dawkins (1978) unhesitatingly named W. D. Hamilton as the founding genius of the selfish-gene approach to ethology, Ghiselin (1974a) opted just as unhesitatingly for Darwin as the author of his own more organism-oriented theory, while Stanley (1979: 22) traced the punctuational view of speciation all the way back to Mayr (1954).

In systematics, pheneticists chose two patron saints—Buffon's contemporary, Michel Adanson, as well as J. L. S. Gilmour. Both had advocated theory-free classifications based on degrees of overall similarity. Cladists none too surprisingly chose Hennig as their patron saint but also added Darwin. Although Nelson tends to view Darwin as the author of infinite error in evolutionary biology and biogeography, he found Darwin's views on classification prescient. In response to Mayr's (1969) claim that Darwin advocated the same principles of classification Mayr himself preferred, Nelson (1971a: 375) insisted that Darwin was really a cladist. "If indeed there is a 'cladistic' school, Darwin is its founder and its chief exponent." Darlington (1972) and Mayr (1974) thought otherwise. The resulting dispute was remarkably protracted and acrimonious (e.g., Nelson 1971a, 1973e, 1974a, Ghiselin and Jaffe 1973, Ghiselin 1985, Mayr 1985).

The penchant of scientists for insisting that earlier scientists actually anticipated their own preferred views is puzzling on several counts. Religious and political propagandists are infamous for claiming that all truth lies buried in the writings of such great figures as Jesus or Marx, but in science experiment, not exegesis, is supposed to be the source of truth. In connection with the main thesis of this book, scientists are supposed to want credit for their contributions, as much credit as possible. Why then should they attribute their views to someone else, implying that this other scientist, not they themselves, deserves the credit? Among contemporary scientists working on a particular cluster of problems, assigning credit is both relevant and important, but what difference can it possibly make if some earlier authority did or did not hold a particular view, especially if this authority is long dead and no one engaged in the present-day dispute was aware of the apparent similarity between their views and those of this authority?

One possible explanation for the erratic and occasional resurrection of a long-neglected scientist is that scientists feel a genuine obligation to recognize early, unappreciated work. It is only fitting and proper. However, when one looks at the uses to which such references are actually put, this explanation begins to pale a bit. More often than not, these references are used to promote the author's own contributions. If a scientist thinks that his epoch-making views are being ignored or treated unfairly, it is difficult for him to say so without sounding self-serving. Unappreciated precursors are especially useful under such circumstances. One can in all propriety rail against the persecution of earlier geniuses. For example, when Geoffroy Saint-Hilaire leapt to Lamarck's defense, he was actually defending himself. Thereafter, any biologist who thinks that his or her ideas are not being treated with the consideration they deserved has happily repeated the myth of Lamarck's persecution (Elliot 1914, Steele 1981; for a review, see Hull 1984a). That historians of science find every element of this indictment inaccurate is of little consequence (Burkhardt 1977, Outram 1984, Appel 1987). A similar fate has befallen Michel Adanson, the patron saint of phenetic taxonomy. When Adanson's (1763) works were actually read with any care, it was discovered that his views really were not all that similar to those being urged by the pheneticists (Lawrence 1963, 1964, Burtt 1966, Nelson 1979, Mayr 1982).

Reference to earlier scientists can serve a second function as well. If for some reason a scientist does not want to attack his contemporaries directly, he or she can attack them indirectly by attacking one of their precursors. For example, Huxley (1854) used his infamous review of the tenth edition of Chambers' *Vestiges of the Natural History of Creation* (1853) to attack Owen. When the objects of these attacks have been adopted as patron saints by one's enemies, such polemics are even more effective. For example, Lord Kelvin declined to attack Lyell's uniformitarianism directly but instead vented his hostility upon Playfair (Hallam 1983). One thing can be said for scientists who are dead: they cannot respond to calumnies.

But there is more than propaganda involved in the frequent references made by scientists to their ancestors. The actual course of science is extremely complicated and contingent. Just as evolutionary biologists prune evolutionary bushes to form phylogenetic trees, scientists rewrite history in ways to make their own tradition stand out in bolder relief. If research programs formed kinds characterizable in terms of a list of essential tenets, scientists could locate themselves by means of definitions. As the various debates over the essence of Darwinism, neo-Darwinism, cladism, and so on, indicate, this is precisely what scientists attempt to do, but they also half realize that such efforts are guaranteed to fail. If science is provisional, as all scientists admit that it is, then it is bound to change. The only way that advocates of a particular research program can maintain that their views are both immutable and correct over any period of time is to change them surreptitiously. The operative entities in conceptual change are traditions integrated by historical connections. In this respect science is like politics and religion. The relevant historical entities can retain their individuality and identity while undergoing change only if they do so gradually without too much internal disruption. Patron saints and evil demons serve as exemplars or signposts in this activity. If borderlines are fuzzy and variable, at least one can point to clear cases.

In the individuation of scientific approach programs, actual historical connections matter, but they can also get lost in propaganda wars and idiosyncratic details. If one's opponents pick a particular figure as a patron saint, one can always attempt to co-opt their choice. Darwin and Popper are two good cases in point. When Simpson (1961) and Mayr (1969) claimed that Darwin was an evolutionary systematist and Bock (1973) enlisted Popper as showing that evolutionary systematics, unlike its competitors, was genuine science, the next move was obvious. Both men had to be appropriated by the cladists (Nelson 1971a, 1973e, 1974a, Ghiselin and Jaffe 1973, Wiley 1975, Kitts 1977, 1980, Platnick and Gaffney 1977, 1978, Settle 1979, Nelson and Platnick 1981, Hull 1983d).

Another example of the role of patron saints in science is presented in the dispute between Nelson and Gould over Haeckel's biogenetic law. Nelson (1973a) proposed to circumvent what Colless (1967b) termed the "phylogenetic fallacy" by means of the study of ontogeny. Although ontogeny does not recapitulate phylogeny in precisely the way proposed by Haeckel, Nelson argued that it can be used to confer polarity on transformation series. In response to Nelson's appeal to Haeckel, Gould (1973: 322) responded:

> The past can stage no active protest; it is available for our use in any way we choose. At worst, we rape it by extracting distorted sentences to suit our present purposes. At best, we understand it aright and enlighten our current proceedings. Somewhere in between, we may simply misrepresent it as Nelson has done.

As Gould himself acknowledges, he was not raising the issue simply to set the historical record straight but to argue that Haeckel actually exemplifies his own pluralistic commitment. In the development of cladistics, cladists assigned Gould the role of an evil demon because he both opposed cladistics and stood astride recapitulation, the primary area of empirical inquiry that Nelson and Platnick (1981) were willing to let intrude into cladistics. Other cladists allow a wider spectrum of empirical considerations to influence their decisions (see Gould 1977b, Løvtrup 1978, Nelson 1978b, 1985, Fink 1982, Alberch 1985, Kluge 1985).

When the Modern Synthesis was in the making in the English-speaking world, ideal morphologists under the lead of Schindewolf picked Goethe as their patron saint. Although ideal morphology in the guise of the comparative method in biology has undergone a resurgence in Great Britain and the United States, Goethe has not been resurrected as a patron saint of this movement. However, one of the earliest participants in what came to be known as cladistics, Ron Brady, is partial to Goethe. The dispute between Goethe and the Newtonians over the science of optics was a crucial juncture in science, and according to Brady (1987) scientists took the wrong turn in opting for the complex machinery of theoretical science over careful attention to the phenomena. As might be expected, Brady (1979, 1982a, 1982b, 1985) is extremely critical of Darwinian versions of evolutionary theory and urges scientists to make their classifications as independent of process theories as possible. In response to Beatty's (1982) paper, Brady does not deny the existence of the view Beatty terms "pattern cladism." To the contrary, he favors at least the "pattern" part of pattern cladism. Cladism itself is quite another matter (for additional discussion of precursors in science, see Sandler 1979).

This story is currently being replayed by those who are arguing that species should be construed as spatiotemporal individuals. Once Ghiselin forced this view to the conscious attention of his fellow biologists, numerous precursors popped up, including Willi Hennig (1950, 1966). With some justification Mayr (1987) notes that he had long objected to species being treated as classes or natural kinds. They are "populations," which are clearly not classes and not quite individuals. He also insists that nearly all naturalists from before the Modern Synthesis to the present viewed species as populations while only outsiders such as philosophers conceived of them as anything like classes. One of the main messages of Mayr's (1982) massive history of biology is that naturalists have been basically right on nearly all fundamental questions in biology, while he reserves error for physicists, mathematicians, and the like. Strangely enough, no one has been very intent on arguing that Darwin considered species to be individuals, although he toyed with the idea in 1835 (Hodge 1985).

Officially, in science, the person who first makes a discovery, solves a problem, or presents a new theory is supposed to be given credit for his or her contributions. Discovery is more important than dissemination. According to the model that I am proposing, both discovery and dissemination are necessary, and if they occur in close proximity, discovery is locally more important than dissemination. However, the more distant in space and time an undisseminated discovery is, the less important it is. As Lamarck (1809: 404) ruefully concluded his *Philosophie zoologique*, "Men who strive in their works to push back the limits of human knowledge know well that it is not enough to discover and prove a useful truth previously unknown, but that it is necessary also to be able to propagate it and get it recognized." In his autobiography, Darwin (1899, vol. 1: 72) was just as resigned in recognizing this law of science:

> Hardly any point gave me so much satisfaction when I was at work on the "Origin," as the explanation of the wide difference in many classes between the embryo and the adult animal, and of the close resemblance of the embryos within the same class. No notice of this point was taken, as far as I remember in the early reviews of the "Origin," and I recollect in expressing my surprise on this head in a letter to Asa Gray. Within late years several reviewers have given the whole credit to Fritz Müller and Häckel, who undoubtedly have worked it out much more fully, and in some respects more correctly than I did. I had materials for a whole chapter on the subject, and I ought to have made the discussion longer; for it is clear that I failed to impress my readers; and he who succeeds in doing so deserves, in my opinion, all the credit.

On Darwin's principle, he deserves all the credit for his theory of evolution because it was he, and not the army of his precursors that later authors discovered, who convinced scientists at large. On this same principle, he is forced to admit that he does not deserve the credit for what came to be known as the biogenetic law, no matter how much his own work anticipated it, because he failed to impress his readers. Historians of science are to some extent interested in intellectual justice; scientists much less so. According to the model I am proposing, local credit is very important. It is the driving force of science. Credit of a more global sort is not, and retroactive credit allotted to unappreciated precursors, when it does not serve some other agenda, is a misfiring of the mechanism. There is an important sense in which Columbus actually did discover the New World even if it was already inhabited by numerous tribes of human beings, and even if Eric the Red did land on the shores of Newfoundland four centuries earlier.

There can be discovery without dissemination but no dissemination without discovery. When the discovers are also operative in dissemination, credit is not very problematic. However, sometimes scientists other than discoverers are most operative in dissemination, Playfair, Bateson, and Nelson being only three of many examples. Why are they content to take second place to some unappreciated precursor instead of asserting their own originality and priority? One answer is that they are not. They are not promulgating the views of the scientist whom they have designated their unappreciated precursor but their own. Perhaps they began their research because of the impetus of stumbling upon the work of an earlier scientist, but that work is soon left behind as these disseminators pursue their own avenues of research. Bateson and Nelson are two excellent cases in point. By exaggerating the importance of discontinuity in Mendel and playing off an ambiguity inherent in the contrast between continuous and discontinuous variation, Bateson used Mendel to foster his own research program. Nelson did the same with Hennig and Croizat. Although Hennig never registered any complaints about how much Nelson was transforming his principles of phylogenetics, several of Hennig's disciples did. Croizat lived long enough to register his objections himself.

If science is a selection process, transmission is necessary. Disseminators are operative in this process. Perhaps they do not get the ceremonial citations that patron saints do, but they are liable to get much more in the way of substantive citations. For example, cladists frequently refer to Hennig's *Grundzüge,* but usually only in the first paragraph of a paper and then only in general, not with respect to particular tenets. Substantive citations to the English translation of Hennig's seminal work have through the years decreased as they are replaced by those of his "disseminators." To the extent that disseminators substitute their own views for the patron saints whom they cite ceremoniously, they are functioning as germ-line parasites—the cowbirds of science (see Appendices D and F).

In the preceding pages I have been concerned to discuss two of Merton's four norms—communism and disinterestedness—in an attempt to show in which respects my views are similar to his and in which respects they differ. I discussed a third of Merton's institutional norms of science—organized skepticism—in the preceding chapters. One difference between my views on organized skepticism and those of Merton is that I emphasize somewhat more strongly than he does the selective nature of scientific skepticism. Scientists, after some initial skepticism, become extremely attached to their own views and argue for them with a certainty that dismays the unconverted. Not infrequently they continue to hold onto them long after others think that they have been definitively refuted. Although this pigheadedness often damages the careers of individual scientists, it is beneficial for the manifest goal of science.

When new views are initially introduced, the majority of the considerations that are supposed to influence the decisions scientists make frequently count against them. These views are not as well articulated as those of their opponents, and the overwhelming mass of evidence collected from the perspective of the entrenched views is incompatible with them. About all that novel ideas have going for them is that they hold out "promise." Unless the advocates of new views are willing to push them for all they are worth, they will not get any evaluation at all, let alone a fair one. New sets of views need time to show their worth. Because of the contingencies associated with their original development, all sorts of views become "linked." Some of them turn out to be more fruitful than others. For example, Huxley became intrigued by Macleay's Quinarian system because the groups he was studying fit so neatly into it. This linkage had to be severed before Huxley could hook up with Darwin. If anything is true of the conceptual lineages I have traced through the years, it is that they consist of variable elements. Certain tenets are dropped, others added, still others transformed. Such a winnowing process takes time.

Conversely, adherents to received views are just as obligated to support their positions and attack those of the Young Turks. In the absence of such defense, fads could sweep science as readily as they do in several other areas of intellectual endeavor. Scientists share a common interest in the elimination of error by testing scientific hypotheses, but they differ with respect to the effects on their own careers of uncovering particular errors. They are not disinterested with respect to the outcomes of the testing procedure. One of the chief functions of the subdivision of science into intellectual demes is to allow, even encourage, severe testing.

Merton's final institutional norm for science, and the one I find most problematic as well as least central to science, is universalism. According to Merton (1973: 270),

> Universalism finds immediate expression in the canon that truth-claims, whatever their source, are to be subjected to *preestablished impersonal criteria:* consonant with observation and with previously confirmed knowledge. The acceptance or rejection of claims entering the lists of science is not to depend on the personal or social attributes of their protagonist; his race, nationality, religion, class, and personal qualities are as such irrelevant. Objectivity precludes particularism. The circumstances that scientifically verified formulations refer in that specific sense to objective sequences and correlations militates against all efforts to impose particularistic criteria of validity.

Merton's norm of universalism is stated prescriptively, as norms *should* be stated, but is it descriptively accurate? Do scientists behave universally enough, often enough, for universalism to count as an operative norm in science? Every scientist lives in some society or other. Are the beliefs that scientists have about societies at large influenced by their personal experiences with their own society? Similarly, every scientist works in some scientific community or other. Are the beliefs that scientists have about science at large influenced by their personal experiences with their own scientific community? The answer in both instances is quite clearly yes. But reasoning from one's own experience is hardly a mistake. The mistake is reasoning too facilely. Just because your society is male-dominated, it does not follow that societies as such are essentially male-dominated. Just because Western science is inherently competitive and individualistic, it does not follow that science as such is inherently competitive and individualistic.

The bias that our own personal experience can introduce into our general conceptions can be shown in example after example. For instance, what percentage of the world's population do you think has blue eyes? One answer made common by a basic misunderstanding of Mendelian genetics is 25 percent. But once this error is pointed out, the people with whom I have much contact are liable to guess 10 percent to 15 percent. The actual figure is less than 1 percent. Even those people who pride themselves on being sensitive to such issues, occasionally stumble in the most embarrassing ways. For example, during the flap over IQ tests, Noam Chomsky (1976: 296) chastised Richard Herrnstein for publicizing his views on race and intelligence:

> As to social importance, a correlation between race and mean IQ (where this is shown to exist) entails no social consequences except in a racist society in which each individual is assigned to a racial category and dealt with, not as an individual in his own right, but as a representative of this category. Herrnstein mentions a possible correlation between height and IQ. Of what

social importance is that? None, of course, since our society does not suffer under discrimination by height.

Anthropologists and sociologists have amply documented that "heightism" is pervasive in societies from the most "primitive" to our own (Gregor 1979). On the average, taller men occupy the more prestigious positions in societies. When a firm that recruits lawyers for openings in business surveyed the chief executives of the nation's largest industrial firms, more than half were over six feet tall; the average height of American men at the time was five feet nine inches. If anyone doubts that scientists can be as affected by the social biases of their societies as are other people, all they have to do is to read a little history of science. For example, Edward S. Morse (1838–1925), in his vice-presidential address to the American Association for the Advancement of Science condemned not only the Turks and lower races but also the Mormons and other advocates of free love—and this from the morally neutral stance of a scientist (Morse 1876: 176).

Such selective blindness is just as characteristic of our beliefs about science. For example, students of science have argued that a significant correlation exists between scientists' age and the alacrity with which they adopt new ideas. If anything is irrelevant to the inherent worth of a scientific view it is the age of scientists. Hence, these students conclude, Merton's norm of universalism is mistaken. Scientists themselves tend to share in the perceptions of these students of science about the role of age in science. The older a scientist is, the less likely he or she is to accept new ideas. In fact, new ideas triumph only through attrition, by the opponents dying off and their places being taken by younger scientists brought up in the new orthodoxy. The conceptual rigidity of aging scientists is often held to be as psychologically inevitable as the stiffening of their joints. Although statements to this effect can be found in the writings of such diverse scientists as Lavoisier (1862, vol. 2: 655), K. M. Lyell (1881, vol. 2: 253), and Darwin (1859: 481), the most frequently quoted passages are by Max Planck (1950: 33, 97):

> a new scientific truth does not triumph by convincing its opponents and making them see the light, but rather because its opponents eventually die, and a new generation grows up that is familiar with it.
>
> An important scientific innovation rarely makes its way by gradually winning over and converting its opponents: it rarely happens that Saul becomes Paul. What does happen is that its opponents gradually die out, and that the growing generation is familiarized with the ideas from the beginning: another instance of the fact that the future lies with youth.

Such prejudice against the aged is all the more amazing because it is based on only the most casual observations. For example, in a paper concerning the reactions of Louis Agassiz, Asa Gray, and J. D. Dana to Darwin's theory, Loewenberg (1932: 687) remarks:

> Darwin's characteristic perspicacity is nowhere better illustrated than in his prophecy of the reaction of the world of science. He admitted at once that it would be impossible to convince those older men ". . . whose minds are stocked with a multitude of facts, all viewed . . . from a point of view directly opposite of mine. . . . A few naturalists endowed with much flexibility of mind and who have already begun to doubt the immutability of species, may be influenced by this volume; but I look with confidence to the young and rising naturalists, who will be able to view both sides with impartiality."

Loewenberg's own examples do not fit his "ageist" prejudices all that well. In 1859 Agassiz was fifty-two, Gray was forty-nine, and Dana was forty-six. None of these men can be considered especially old. Gray was one of Darwin's earliest converts to a belief in the evolution of species, Dana held out until 1871, while Agassiz remained opposed until his death in 1873. If these three men show a significant correlation between age and the alacrity with which they were able to accept new ideas, I fail to see it. More than anecdotes are necessary to test such claims.

A few years back, Hull, Tessner, and Diamond (1978) investigated the reception of the evolution of species by scientists in Great Britain between 1859 and 1869 (see Appendix H). In order to be included in the study, the scientists had to be at least twenty in 1859 and to have lived at least until 1869. They had to declare in print or in their correspondence that they accepted the evolution of species, though not necessarily Darwin's mechanisms. If the ages of those scientists who accepted the mutability of species between 1859 and 1869 are compared to those who still rejected this minimal tenet of Darwin's theory after 1869, the difference is significant. The average age of those scientists who came to accept the mutability of species prior to 1869 is forty, while the average age of those who did not is forty-eight. However, if these scientists are subdivided into ten-year cohorts according to their age in 1859, an interesting pattern emerges (see table 10.2). Those scientists in their fifties were most reluctant to change their minds, followed by scientists sixty and above. The youngest scientists, surprisingly, came in third. Hence, both older scientists and the very young were less likely to change their minds on the species question.

If one looks just at those scientists who did accept the evolution of species prior to 1869, no correlation can be found between age in 1859 and how quickly they came to accept the mutability of species (see table 10.3 and Appendix H). Those scientists who continued to opt for the immutability of species after 1869 were on the average older than those who changed their minds sometime between 1859 and 1869, but age explains less than 10 percent of the variation in this latter group. As far as such studies go in the social sciences, 10 percent is impressive but hardly sufficient to explain the widespread perception of a marked correlation between age and conceptual rigidity. These figures are also interesting in that almost half of those who accepted the evolution of species did so within the first year after publication of the *Origin of Species*. Change of mind came more slowly thereafter.

Table 10.2 Numbers of British Scientists Accepting or Rejecting the Mutability of Species prior to 1869, by Age in 1859 (suggested by Ron Westrum)

Age in 1859	*No. in Group*	*No. of Holdouts*	*% of Holdouts*
20s	17	4	24
30s	16	1	6
40s	16	2	13
50s	12	6	50
60s or more	7	3	43
Totals	68	16	24

Table 10.3. Average Ages in 1859 of British Scientists Accepting Evolution of Species in Successive Years from 1859 to 1869.

Year of Acceptance	*No. of Scientists*	*Average age in 1859*	*Average age in four-year aggregates*
1858	4	35	
1859	2	31	
1860	15	43	
1861	3	47	41
1862	1	43	
1863	2	50	
1864	8	35	
1865	2	30	35
1866	3	37	
1867	1	62	
1868	4	42	
1869	7	37	40
Totals	52	40	

Several additional studies have also been conducted comparing age and alacrity with which scientists accept new ideas, and the results have been mixed. Diamond (1980) investigated the acceptance of "cliometrics" (the use of quantitative methods in the study of history) among historians of economics and came up with the same result as Hull et al. (1978). Less than 10 percent of the variation is explained by reference to age. When McCann (1978) studied the adoption of the oxygen theory of combustion, he discovered a negative correlation between age and use of the oxygen paradigm in papers published between 1760 and 1785. However, Nitecki et al. (1978) came up with findings that seem to support Planck's principle. They sent a questionnaire to 293 fellows of the Geological Society of America and active members of the American Association of Petroleum Geologists. They were asked, among other things, if they accepted the theory of continental drift and plate tectonics. If so, when did they first come to accept it? Nitecki et al. (1978: 663) discovered that in general, "the earlier the degree, the longer the delay between encounter and acceptance." More recently, Stewart (1986) found that age alone was not a significant predictor with respect to the acceptance of continental drift. Allen (1979) in turn studied age differences in the scientists engaged in the controversy between the evolutionists (biometricians) and the Mendelians soon after the turn of the century in England and the United States. Of the thirty-five scientists whom he studied, he discovered that the average age in 1900 of the evolutionists was fifty-five and that of the Mendelians was forty. Because the biometricians were established at the time and the Mendelians were the new boys on the block, age does seem to have made a difference.[2]

2. The number of authors who have accepted some version of Planck's principle is amazing, e.g., Whewell (1851: 139), Huxley (1901, vol. 2: 117), Loewenberg (1932: 687), Barber (1961: 596), Kuhn (1962: 150), Samuelson (1966, vol. 2: 1517), Greenberg (1967), Feyerabend (1970: 203), Dolby (1971: 19), Richter (1972: 80), Paul (1974: 412), Wisdom (1974), Bondi (1975: 7), Gunther (1975: 458), Knight (1976: 14), Barash (1979: 240), and Broad and Wade (1982: 135). Only a few authors have advised caution in accepting this cynical view, e.g., Hagstrom (1965: 283), Scheffler (1967: 28), Merton and Zuckerman (1973: 514),

All the authors who have bothered to check Planck's principle realize that alternative explanations can be given for any differences discernible in the ages of scientists who jump on a bandwagon and those who do not. Age may simply be correlated with something else that is actually operative. Instead of conceptual inertia being a psychological characteristic of aging scientists, it may be the result of career interests. For example, Allen (1979: 195–6) suggests, "Thus, it could be that evolutionists, all well on in their careers by the time Mendel's principles were rediscovered, were simply too old or too channeled into fields of special interest, to look with favor on any new theory. The younger school, however, was less committed through extent of training to any given set of ideas or traditions." Knight (1976: 14) concludes similarly that "scientists do not seem to be any more keen than anybody else to drop a whole world-view upon which their life's work has been based because somebody has proposed a new and wide-ranging theory" (see also Diamond 1980, Hallam 1983, and Stewart 1986).

Hull, Tessner, and Diamond (1978) came to much the same conclusion. When scientists such as Lavoisier, Darwin, and Planck complained about older scientists being so closed-minded, they were not referring to old scientists in general but to a few experts in the field, the experts whose views were under attack. These experts tended to be both old and difficult to convert, but their age was not the operative factor in their resistance to new ideas. Instead, they dug in their heels because it was *their* view that they had to abandon. Darwin grew increasingly exasperated with Lyell for his reluctance to come out publicly in favor of evolution. Part of his reluctance may well have been the implications of Darwin's theory for human beings, but, as Lyell himself admitted, another reason was that he would have to rewrite the entire second volume of his *Principles of Geology.* Anyone who thinks that this is a reason of little moment has never written a book.

Aging scientists do change their minds, but such changes are easier in areas somewhat distant from their own life's work. In the interviews I conducted, I asked, "Do you consider science provisional, that is, any belief which scientists have about the world can be challenged and possibly abandoned?" Everyone answered in the affirmative. Later, I asked if they could recall any belief that they held in their early years which they had been forced to abandon since. A momentary look of panic usually resulted. They were committed to science being provisional, but off the top of their heads they could not recall ever having to revise any of their own views. Then, a smile of relief as they recalled an example. The peculiar feature of these examples was that they tended to concern issues peripheral to the scientist's own research.[3]

Cole (1975: 181), Cantor (1975a: 196), Brush (1974), Blackmore (1978: 347), Hufbauer (1979), and Hallam (1983: 157). Several authors acknowledge that perhaps the correlation between age and ability to change one's mind is not as great as it once appeared to be, but even so the studies conducted to date indicate that there is some connection, e.g., Nitecki, Lemke, Pullman, and Johnson (1978), Mahoney (1979), and Cock 1983).

3. Ernst Mayr suggested that a division of the scientists who accepted or rejected the evolution of species prior to 1869 into those who had some claim to being naturalists, or at least biologists, and those who did not might prove instructive. The results are as follows: the average age in 1859 of the 35 biologists who

As several authors have also pointed out, conservatism in science is not all bad (Barber 1961, Lakatos, 1970, Knight 1976, Laudan 1977, Lugg 1980, Diamond 1980). Hindsight is a powerful tool. We investigate those research programs that turn out to be "right", or at least successful, to see why those who opposed them were so blind. But in the vast majority of cases, those scientists who rejected a bright new idea turn out to be wise after the fact. No one complains that scientists, young or old, resisted novel theories that we now take to be mistaken. The first major work in phrenology appeared in 1810 (Gall and Spurzheim 1810). In the ensuing years this doctrine accumulated a list of illustrious adherents, including four early evolutionists—Chambers, Geoffroy Saint-Hilaire, Spencer, and Wallace. A comparison of age and alacrity with which scientists accepted phrenology would no doubt produce the same sorts of results as the preceding studies, depending on the methodology used.

If nothing else, the studies that have been conducted on putative correlations between age and the alacrity with which scientists adopt new ideas show exactly how difficult even the easiest hypotheses about the scientific process are to test. One would think that if *any* correlation could be tested, it would be one in which the variables are easily discernible, and what could be easier to discover than the age of scientists? Their views on particular scientific issues are more difficult to discover, but scientists do commit themselves in print and in their correspondence. The trouble is that evidence about failures is all but impossible to discover. Young scientists who commit their careers to research programs that never get off the ground leave no trace.

For example, soon after the appearance of the *Origin,* seven liberal Anglicans published a collection of essays in which they looked at the Bible in the critical fashion introduced by German scholars. In reaction, six chemists began circulating a religiously conservative declaration for their fellow scientists to sign. It declared that science and religion, properly understood, could not conflict. Five of these chemists were under twenty-five in 1859, while one was fifty-five. Did they eventually change their minds on evolution? Available documents are not adequate to answer this question. None of them turned out to be important enough to leave much in the way of records. Incidentally, Brock and MacLeod (1976), in their study of this declaration, came to the conclusion that age did not seem material in determining who did and did not sign the declaration.

If age did play a significant role in the reaction of scientists to new ideas, then Merton's norm of universalism would have at least be undermined somewhat. Although age is not the sort of "particularistic" factor that Merton's critics are most concerned to show influences the decisions that scientists make, it is nevertheless "particularistic." Merton himself is well aware that in point of fact a plethora of "particularistic" factors

accepted the evolution of species prior to 1869 is forty, while the average age of the 16 nonbiologists who accepted the evolution of species prior to 1869 is thirty-nine; among those who rejected evolution in 1869, the average age in 1859 of the 12 biologists is forty-eight and a half and nonbiologists is forty-four. The only interesting feature of these figures is the high figure for naturalists who rejected evolution. Instead of biologists being more receptive to the evolution of species, as Mayr anticipated, they were less so, possibly because of a greater career investment in the old view.

have influenced the spread of scientific knowledge, including language barriers, a variety of social prejudices, not to mention numerous historical contingencies. As the preceding episodes in the history of science indicate, the decisions that scientists make can be and frequently are influenced by factors other than reason, argument, and evidence, but Merton views these particularistic influences as being perturbations of the basic process—as noise and not the message. Such particularistic static may obscure the message but not totally drown it out. In the long run, if a scientific community is universalistic enough, such particularistic influences are neutralized. "Sooner or later, competing claims to validity are settled by universalistic criteria" (Merton 1973: 271). Thus, universalism must be interpreted to mean that the more universalistic science happens to be the more successfully it will attain its manifest goals. A little bit of universalism iterated over the years can have a pronounced long-term effect. Externalists interpret the causal situation in just the opposite way. Particularistic factors are not in the least "particularistic." They are the chief driving force of science.

The dispute between Merton and his "externalist" critics concerns causation. Which factors predominate in science? In the main, do scientists opt one way or the other on substantive issues because of the internalist trinity of reason, argument, and evidence or some complex of other causes? Numerous examples of both sorts of factors influencing the corpus of science can be given. Students of the human species have flipped back and forth with respect to the relative importance of our genetic makeup to our psychological traits in almost total independence of any significant changes in our knowledge of the human genotype or the specific actions of many genes (Provine 1973). If such changes have been caused primarily by internalist considerations, they must have had their sources in changes in our general understanding of genetics and embryology, not in any specific knowledge of human reaction norms. But if internal considerations are inadequate to account for these changes, which of the many possible external factors are responsible? Pastore (1949) suggests political persuasion. He discovered a strong correlation between the political leanings of British and American scientists between 1900 and 1940 and their views on the nature-nurture issue. Out of twelve environmentalists, Pastore judged all but one, J. B. Watson, to be liberal to radical in their politics. Out of twelve hereditarians, he found all but L. M. Terman to be conservative. According to Pastore, Terman was a political liberal. Strangely enough, Terman was the man who introduced the Stanford-Binet IQ test in 1916 for use in the United States Army and championed eugenics to the end of his life.

Externalists agree that the predominant influences on the course of science are external. However, with respect to particular episodes, the disagreements about which external factors were actually operative and how important each was is extensive, but that is only to be expected. The same observation holds for internalists. Internalists agree on the general efficacy of reason, argument, and evidence in science but disagree extensively about particular internalist scenarios. For example, no sequence of events has been studied in greater detail than Darwin's development of his theory of evolution. Yet extensive disagreement continues about the particular internal factors that actually influenced Darwin. The reconstruction of particular causal scenarios of whatever sort is extremely problematic and endlessly contentious. The same observation holds for evolutionary scenarios—phylogenetic trees. History is extremely difficult to reconstruct; if pattern cladists are right, impossible. (For disagreement over the appropriate external factors that influenced the reception of phrenology in the early decades of the nineteenth century in Edinburgh, see Parssinen 1974, Cantor 1975a, 1975b, DeGiustino 1975, Shapin 1975, 1979a, 1979b, Cooter 1976, Young 1968, 1970; for Darwin's intellectual development, see the papers in Part One of Kohn 1986).

Questions about particular causal chains to one side for the moment, a second question concerns the *regularity* of both sorts of causal factors. Can the beliefs of scientists about the world which they study and social factors be analyzed into kinds such that some sort of regularity exists between these kinds? The answer that seems to be emerging in the externalist literature is a resounding No. The rise of phrenology, or evolutionism, or hereditarianism during the early decades of the nineteenth century in Edinburgh, London, or Paris has a set of external causes in each case, but these causes vary in a haphazard way (Shapin 1979a). The rise and fall of a particular scientific view such as Darwinism in any one area through time also have their external causes, but they keep changing in erratic ways.

The appeal of externalism rests on the "logic" exhibited by the causal connections alleged. Political conservatives should oppose the evolution of species, while revolutionaries should favor it, but as Outram (1984) argues, the paradigm example of this purported correlation, Cuvier, does not even fit. With respect to the putative connections between the nature-nurture issue and the justification of political systems, Harwood (1979: 234) concludes, "Thus, 'nurture,' no less than 'nature.' can be of legitimatory use to the power behind the welfare state." For those people who feel the need for intellectual justifications, perhaps Darwin's theory of evolution might appear to "justify capitalist exploitation" (Bernal 1971, vol. 2: 644), but thus far the available data do not strongly confirm this logical connection (Glick 1974, Bannister 1979), notwithstanding endless repetition of this facile assertion.

As Halstead (1980) discovered when he claimed a connection between Marxism, cladism, and punctuational theories of speciation, human beings are not this "logical." Gould favors punctuationism and Marxism of sorts but opposes cladism, Eldredge is predisposed to punctuationism and cladism but not Marxism, and so on. More recently, Tassy (1981) claims that cladism is egalitarian and evolutionary taxonomy elitist. The use of game theory is supposed to encourage the status quo (Martin 1978), but the biologist who introduced game theory into evolutionary biology, Richard Lewontin (1961), and the man who has developed it most extensively since, John Maynard Smith (1974), are anything but political conservatives. The source of Richard Dawkins's (1976) gene selectionist version of evolutionary theory is G. C. Williams (1966), but Williams is as politically liberal-to-radical as his opponents. If connections of the sort alleged by Halstead exist, they are extremely sporadic, tenuous, and contradictory. About the only correlation discernible in the sociobiology controversy is that several of the leading opponents of this research program when it was first emerging were not only Marxists but Jewish.

Another apparently "logical" connection that has been alleged is between the disproportionate attention which has been paid by ecologists to competition in nature and the capitalist worldview of the societies in which they reside; but when May and Seger (1986) checked the two leading ecological journals for a ten-year period, they found no such correlation. Of the 474 papers that appeared between 1965 and 1975 in the *Journal of Animal Ecology,* only 20 concerned interspecific competition. Of the 1,665 papers in *Ecology* during this period, only 98 concerned interspecific competition. May and Seger (1986: 260) concluded, "Not only is the argument inconsistent with the actual record of published work, but it also seems strangely illogical; we find it easy to imagine that dedicated capitalists would rather hear about prey-predator relations than about mere competition." But adding the number of papers on prey-predator relations to the figures on interspecific competition does not help much. The two together make up 15 percent of the papers in the *Journal of Animal Ecology* and less than 10 percent in *Ecology.*

To make matters worse, there are many areas of scientific investigation in which no "logical" connections between social causes and scientific views suggest themselves. How should political conservatives choose with respect to the cost of meiosis or the principle of dichotomy? If one looks at the backgrounds of the wide variety of people who contributed to the early years of pheneticism at the University of Kansas, no regularities seem apparent. Given the racial and social makeup of the United States at the time, nearly all early pheneticists were white males, but this is equally true of their opponents. Both sides included immigrants who had fled Europe, including a few Jews. At the time, the pheneticists, especially Sokal and Ehrlich, were considered rude and contentious. Later, when cladists came on the scene, they were considered even ruder and more contentious, especially Nelson and Farris. But that is about all (however, see Tassy 1981).

Those of us who study science would very much like to find regularities of some sort reflected in its development, but to the extent that the sorts of external factors discussed by externalists influence the decisions which scientists make, the connections are highly "particularistic." They are particularistic in two senses: they are not a matter of reason, argument, and evidence, and, more than this, they are highly idiosyncratic as well. If this is all there is, then the only sort of science of science that is going to be possible is the serial listing of which external factors happened to influence which scientists in each case—one damned thing after another—and this seems to be the current state of the Strong Programme in the sociology of knowledge. As Barnes and Bloor (1982: 23) put their view, "the incidence of all beliefs without exception calls for empirical investigation and must be accounted for by finding specific, local causes of this credibility" (see also Barnes 1977: 58 and Bloor 1982: 164). Whenever anyone, scientist or otherwise, comes to adopt a belief, there will be some local conditions or other. Externalists will insist that these conditions, whatever they happen to be, caused the change of belief.

But to fill out the story, less than perfect correlations also exist between sorts of internal factors and sorts of scientific beliefs, and both data and arguments are to some extent malleable. If one compares the reception of a particular view such as species evolving in various countries around the world, external factors vary wildly, but the reasons, arguments, and evidence adduced for and against this view also vary, although not quite so extensively (Glick 1974). Various areas of science develop differently in different language communities. In the nineteenth century Germans worked extensively on embryology, while the French took the early lead in what came to be known as paleontology. The implications of each for organic evolution were quite different. At the time that Darwinism was introduced into France, Cuvier had succeeded in convincing French scientists that science must be based on positive facts, not flights of fancy, while German scientists were still intrigued by the highly speculative views of the *Naturphilosophen*. Most paleontologists found the fossil record to be inherently progressive; Lyell and Huxley saw no clear progress. The implications for evolution are obvious. Physicists thought that thermodynamic considerations decisively refuted Darwin's theory. Evolutionists and geologists were not so sure. They thought that considerations of elevation and denudation, as crude as they might seem to physicists, indicated that something was wrong with the physicists' reasoning. Although evidence may have borne uniformly on the decisions which scientists made with respect to Darwin's theory, the evidence that was taken to be relevant varied extensively.

Because internal and external factors are so entangled in any one causal situation, they are all but impossible to separate. Ideally, what one needs is something like a twin study at the level of scientific communities. In the absence of such an unlikely eventuality, it would be nice to find a case in which the predominance of external factors led in one direction while a predominance of the internal factors led in the opposite direction. Which way did scientists opt? Such ideal situations are liable to be extremely rare. One feature of science does lend some support to the internalist position—scientists do tend to reach a consensus through the years that is correlated with a convergence of internal but not external factors. Right now professional scientists with any competence in the relevant literature acknowledge that species evolve. There is also a strong consensus on which internal factors support evolutionary theory and which count against it, but one word of caution needs to be raised. The internal factors that we now take to be relevant to the belief that species evolve are not perfectly coincident with those that initially led to its acceptance.

For example, I began chapter 8 with a series of quotations from Darwin about which phenomena, if observed, would be fatal to his theory. The problems that would be posed for evolutionary theory by the existence of a complex organ that could not have been formed by numerous, slight variations or which functioned for the exclusive good of an organism belonging to a different species will strike us today as relevant, but the relations between large and small genera have dropped out of the literature. Darwin put considerable store in his "botanical arithmetic" (Browne 1980), but later evolutionists have not. Internal histories of science are somewhat less particularistic than are external histories, but these are differences in degree.

The message of the preceding discussion should not be taken to be that no regularities exist in scientific development. Rather it is that decisions made by individual scientists or groups of scientists with respect to particular research programs is the wrong level at which to look for regularities. Why did William Bateson hold out against the chromosome theory long after others working in the area had come to accept it? Coleman (1970) cites his political conservatism, antimaterialism, anti-Benthamism, intuitionism, aestheticism, and organic conception of society. Cock (1983) disagrees. He cites defects in the evidence, embryological difficulties, the fact that Bateson had a theory of his own, his distaste for microscopy, his iconoclasm, a particular historical accident, and his personal dislike for Morgan. Both lists are miscellaneous collections, but this is the most that one can hope for in explaining anything as particularistic as an individual scientist holding a particular view. If regularities exist in scientific change, they must exist at a much higher level of generality.

But before this conclusion is taken as condemning the science of science, one should note that exactly parallel observations hold for biological evolution. There can be no law of particular species. Whatever regularities exist in the evolutionary process, they occur at a much higher level of generality. The evolution of sexual reproduction is the sort of thing that can receive a theoretical explanation, but not the evolution of the horse *qua* horse or any other taxon as that particular taxon. Taxa are particulars, not kinds. Of course, horses reproduce sexually and thus must exhibit whatever regularities obtain with respect to sexual reproduction. Horses are genetically heterogeneous not because they are horses but because they are a species that reproduce via meiosis. If follow-the-genes is a good rule in biological evolution, then follow-the-ideas may be equally applicable in the evolution of ideas. If the demic structure of biological species increases speciation rates in biological evolution, then possibly the demic structure of scientific communities serves a parallel function. If "universalistic" explanations are to be found for scientific development, they are *not* going to be at the level of particular research groups or research programs. Rather they must refer to kinds of groups and programs. The evolution of horses can be explained only by invoking a whole series of regularities and contingencies that apply to other species as well. *Any regularities about evolutionary processes, whether biological, social, or conceptual, must concern convergences, not particular evolutionary homologies.*

When externalists urge the importance of external factors on the course of science, they nearly always have "larger" social factors in mind, such as the rise of the mercantile middle class, but there are certain "external" factors that are actually internal to science, in particular, professional allegiances and alliances. If universalism is taken to preclude the systematic effects of such allegiances and alliances, then universalism is widely flouted in science. More than this, these allegiances and alliances have a systematic effect on the content of science. One of the main purposes of the first half of this book has been to show how important the groups that scientists form are to the ongoing process of science. Views are evaluated in science not only on the basis of their assertive content but also in terms of *who* enunciates them. In the United States, not everyone has equal access to a career in science. Many of us feel guilty that all sorts of subtle social factors discourage women, blacks, and certain religious groups from entering science. We have no reservations about deselecting dull, indolent students, or those who, at the moment, are too busy having a good time. But the doors remain cracked for future reentry. Several of the scientists discussed in this book were late bloomers, from Charles Darwin to Ed Wiley. Others, such as Lewontin, Ehrlich, and Trivers were clearly bright enough but early on lacked the necessary commitment. All of them became professional scientists anyway.

To the extent that entry into science is not based strictly on merit, the development of science is likely to be retarded. Station in society has influenced science to some extent. The fact that Darwin inherited enough money so that he could devote himself almost exclusively to science certainly gave him the edge over Wallace, but then Darwin suffered throughout his adult life from illnesses that were as incapacitating as they were mysterious. Wallace was blessed with good health. The independently wealthy geologist, Roderick Impey Murchison (1792–1871), had a decided advantage over his chief rival, Henry De la Beche (1795–1855), who did not have the financial resources even to attend professional meetings in London, let alone traipse all over Europe and Russia gathering evidence to support his theory (Rudwick 1985). On the basis of his own experience, Huxley proclaimed the "democracy of science" (Paradis 1978). In what other area of Victorian society could a poor boy with no connections rise to such eminence entirely on the basis of his own efforts? But in Huxley's democracy, suffrage was hardly universal. Not everyone had equal access to the profession, and once one became part of it, "one man, one vote" clearly did not obtain. The votes of certain scientists count much more than the votes of others. Science was no less elitist in Huxley's day than it is today. Thus far, even though the same sorts of prejudices that permeate the rest of society have served to discourage certain groups from contributing to science as fully as they might, enough white, middle-class, males have possessed sufficient talent and drive to fulfill the goals of science.

Of greater significance than ease of entry for the structure of science is the role of social status in the *use* of scientific contributions. Assuming that there is some way for other scientists to know the sex, race, religious affiliation, and the like of the authors of the papers and books they scan, do they evaluate these contributions accordingly? How often has a scientist refused to use a bit of data, a new technique, or a theoretical innovation that would allow him or her to make a major contribution to science, possibly even win a Nobel Prize, just because the author was of the wrong sex, race, or religion? Systematic data are difficult to obtain on such questions. Certainly J. D. Watson was happy to use Rosalind Franklin's X-ray-diffraction pictures. It never occurred to him not to, even though she was a woman. Of the twenty-nine most frequently cited papers published in *Systematic Zoology* between 1953 and 1976, the major author of one was a woman. That the Maxson of Maxson and Wilson (1975) was female did not dissuade those scientists who could use her study of albumin in the evolution of tree frogs from using it (see Appendix D). Of the first ten recipients of the Ernst Mayr Student Award for best paper presented at the meetings of the Society of Systematic Zoology, four were women (see Appendix I), while one of the nine editors of *Systematic Zoology* was a woman (see Appendix C).[4]

4. Of the groups currently classed as "minorities' whose members have suffered discrimination in the United States, only two are represented in any numbers in the scientists whom I studied—Jews and women. Prior to World War II, quite a few of the scientists working in the United States were Jewish. This number was augmented markedly by Jewish scientists fleeing persecution immediately prior to and during World War II. With one major exception, I noticed very little anti-semitism in the professional dealings among the scientists whom I studied. The only exception was Mayr being referred to once as that "dirty little Jew."

Although women are less likely than men to go into most branches of science, Cole and Cole (1972, 1973) claim that, once women begin to publish, their contributions are treated no differently from the

Scientists tend to hold much the same social prejudices as other citizens in their societies especially when these prejudices do not impinge on their own areas of research. Scientists who think that women are intellectually inferior to men are likely to place less emphasis on the research produced by women, assuming that they are aware of the sex of a paper's author, but if the mechanism that I am suggesting is correct, such considerations should be of decidedly secondary importance in their choice of views to use in their own research. Scientists have to use each other's contributions and in doing so lend at least tacit recognition of the worth of the contributions they use. Scientists on occasion would prefer not to give credit to another scientist, but on the view I am urging it will not be because this scientist is male, Muslim, or Marxist but because he or she belongs to a competing research group. Perhaps scientists largely ignore the race, sex, religion, and so on of other scientists when they scan the literature for views which either support or threaten their own, but they are not impervious to the professional allegiances and alliances of other scientists.

Scientists treat exactly the same thesis differently depending on who presents it, whether friend or foe. A view presented by a friend is given the benefit of the doubt and, if rejected, rejected gently and usually in private. The same view presented by someone outside one's circle of allies is greeted quite differently. For example, Darwin was muted in his disagreements with Lyell, not simply because Lyell was a powerful scientist whose opposition could be a serious drawback but because Lyell had been his mentor. Several of the issues over the role of geographic isolation in speciation that Darwin and Wallace debated in their private correspondence were raised again by Moritz Wagner (1868). Darwin replied to Wallace in sadness that they would never agree. His response to Wagner was one of hostility. Darwin and Huxley also disagreed sharply over the nature of the fossil record, but they avoided bringing these conflicts out into the open. At the turn of the century, Mendelian geneticists and evolutionary

---

publications of their male colleagues. In this book, I have ignored the claims made by some feminists that science is itself sexist. By this they do not mean simply that scientists have been and continue to be sexists, i.e., that their views about sexual dimorphism in the human species are frequently biased, but that scientific methods are themselves in some significant sense male-biased. On one interpretation, this claim is that such things as clear expression and careful observation are somehow a male plot against females. I find this view not only silly but, worse still, insulting to women. However, if I am right about the central role of competition and aggression in science and if these characteristics are more common among males than females (regardless of why), then there may be a sense in which the social organization of science is male-biased. For two recent discussions, see Longino and Doell (1983) and Harding (1986).

Although the results are sometimes awkward, I have followed the practice of attempting to reduce the sexism inherent in English by avoiding the neuter use of the male pronoun. A good example of how this ambiguous usage can lead to ridiculous results can be found in the opening paragraph of Desmond Morris's *The Naked Ape* (1967: 9): "There are one hundred and ninety-three living species of monkeys and apes. One hundred and ninety-two of them are covered with hair. The exception is a naked ape self-named *Homo sapiens*. This unusual and highly successful species spends a great deal of time examining his higher motives and an equal amount of time studiously ignoring his fundamental ones. He is proud that he has the biggest brain of all the primates, but attempts to conceal the fact that he also has the biggest penis."

One group whose members have consistently suffered discrimination in nearly all societies but is not currently classified as a "persecuted minority" in need of protection in the United States is homosexuals. Little in the way of systematic data exist for the extent to which homosexuals have been discriminated against in science and what effects such putative discrimination has had on the content of science.

biologists refused to use each other's work in part because of the superficial understanding that the members of each group had of the work of the other, but also in large measure because of intergroup hostility. A generation had to pass before the two disciplines could merge in the English-speaking world. This diremption never arose in Russia.

Similar matters of intergroup hostility and intragroup loyalty arose time and again more recently in systematics. When cladistics began to have an impact in the English-speaking world (under the name phylogenetics), the enemies of the cladists were the evolutionary systematists, but later, under the influence of Colless's attacks and Farris's problems with Sokal, the lines were drawn even more sharply between cladists and pheneticists. When Nelson, Patterson, and Platnick began to argue that classification should be as independent as possible of both phylogenetic reconstructions and ideas about the evolutionary process, the conceptual distance between their branch of cladism and phenetics began to look very small indeed. Professionally, however, the gap was unbridgeable. Perhaps members of these two warring camps did learn from reading each other's work, but if so, no one explicitly mentioned any positive influence and, if asked, refused to acknowledge any. The most that Nelson and Platnick (1981: 151) said on behalf of Sokal and Sneath is that they were partially justified in their concern about the sort of "futile theorizing" that characterized the work of evolutionary systematists. Farris has not been able to grant them even this much.

Medawar (1972) finds one of the attractions of the scientific life the element of "camaraderie or mateyness." When things are going well in a research group, the personal relations among the members of the group can be very pleasant. One reason why Funk reacted so negatively to the handling of the Hennig Society by its officers was that she found her early associations with the founders of cladistic analysis so joyous. This is what science is supposed to be! Scientists need each other. They need to share each other's conceptual resources—what they know, the skills they have developed, their access to data, etc. They also need each other's professional resources—their reputations, positions in a particular scientific community, control over the editorial policies of journals, influence on funding agencies, etc. They can meet these needs most easily when they happen to agree about substantive issues, but as I have tried to show, cooperation can occur even in the presence of significant disagreement.

There are also limits to mateyness and camaraderie. Scientists have rarely seconded a view they take to be mistaken simply because they are close professional associates of its author. Nelson, Eldredge, and Cracraft were unable to coauthor a book because they disagreed about the worth of some of the views that each of them wanted to include. At the time, Cracraft and Eldredge thought very little of Croizat, while Nelson had a very low opinion of evolutionary theory, not to mention Eldredge and Gould's (1972) theory of punctuated equilibria. When Brooks and Wiley (1986) proposed to tie cladism to their theory of entropic evolution, other cladists tried to discourage them, first in private and, when that failed, in public. A century earlier, Romanes feared the negative effects that the neo-Darwinian extremists would have on his own, pluralist, true Darwinism. In general, the good of individual scientists coincides with the good of science, but conflicts arise much more frequently between individual interests and the interests of particular research groups, sometimes within a group, sometimes between groups. In such circumstances, professional allegiances and alliances interfere with use and hence the ongoing process of science as an institution.

Finally, even though "preestablished impersonal criteria" may in the long run be the final arbiters in science often enough for science to fulfill its traditional goals, some note must be taken of the fact that these criteria have changed through time. At times explanations in terms of fairly simple geometric devices are required, at other times compatibility with Scripture, while later only experimentally tested hypotheses are acceptable. Some scientists accept introspective evidence as "evidence"; others not. Correlations satisfy some scientists; others insist on causal connections and the specification of mechanisms. Some scientists are willing to accept as genuine testing only deductions from universal laws; others are content with probabilistic inferences. The list is not infinite, but it is extensive. Many philosophers of science see as their task the construction of a list of impersonal criteria to which scientists should have appealed from the beginning of time, but thus far no list has received wide, let alone universal, assent among professional philosophers. The other alternative is to admit that both particular scientific research programs and science as such have evolved. Both are historical entities. In any case, the message of this chapter is that lists of defining properties of science are of secondary importance to the continued action of the mechanism that I have set out. If there is an essence to science, this mechanism is at its core.

In an article investigating the effects of the widespread acceptance of classified research after World War II on academic values, Rosenzweig (1985: 42) concludes that "inhibitions on free communication in teaching and research, whether imposed on an unwilling academy, as was McCarthyism, or accepted voluntarily, as was classified research, are undesirable and to be avoided. We can also say with some confidence that the latter may be harder to deal with because they come clothed in self-interest." If my underlying perspective is correct, then Rosenzweig's confidence is justified. If the self-interest of scientists is redirected, then the basic structure of the scientific enterprise is likely to be modified and the results altered accordingly. We cannot take the structure of science for granted. For a long time it was largely absent in human affairs. It came into being; it could just as readily cease to exist.

In the preceding discussions, I have moved easily from the motivations of individual scientists to the institutional structure of science and back again. Such transitions are made easier by the fact that in ordinary English the same adjectives can frequently be applied to individual people and to social institutions with equal facility. For example, no linguistic habits are compromised by saying that Darwin was objective and that science is objective. The two, however, are quite different sorts of claims. Although Merton frequently mentions the motivations that scientists have in his discussions of science, for him institutional norms are what really count. For example, Merton (1973: 275–76) insists

> Disinterestedness is not to be equated with altruism nor interested action with egoism. Such equivalences confuse institutional and motivational levels of analysis. A passion for knowledge, idle curiosity, altruistic concern with the benefit to humanity, and a host of other special motives have been attributed to the scientist. The quest for distinctive motives appears to have been

misdirected. It is rather a distinctive pattern of institutional control of a wide range of motives which characterizes the behavior of scientists. For once the institution enjoins disinterested activity, it is to the interest of scientists to conform on pain of sanctions and, insofar as the norm has been internalized, on pain of psychological conflict.

The motivations of individual scientists are important, especially in histories that follow the careers of individual scientists. They are also important because of the incredibly complex relation between individual motivations and institutional norms (Holton 1973: 383). Certainly social groups are more than sets of people. They are people organized in particular ways. The issue is how this organization can influence the behavior of the agents so organized. Can norms function as causal agents? Can entities at all levels of organization be understood solely in terms of their constituent elements and the relations between them? Are claims about individual organisms performing acts such as mating or fleeing a predator merely shorthand expressions for the interrelations of thousands of cells? Are claims about species splitting, merging, and going extinct merely shorthand expressions for the interrelations of thousands of organisms?

In all the preceding cases, the issue is reduction. Reductionism is one of those big questions that perennially divides philosophers. As with so many issues in philosophy that begin as genuine problems, with the two sides opting for significantly different positions, once the opponents become sufficiently sophisticated the various sides become all but indistinguishable. They present the same facts about the world. Then one side concludes that reductionism has been decisively refuted; the other that it has been strongly supported. Although the interconnected set of questions that constitute the problem of reduction have yet to be resolved to anyone's satisfaction, one thing is reasonably certain: characteristics at higher levels of organization are not *easily* reducible to entities and processes at lower levels of analysis.

For example, according to all appearances, Mendelian genetics has been reduced to molecular biology. However, when one attempts to set out this reduction explicitly by specifying the derivations of modern versions of Mendel's laws from the basic principles of molecular biology, no kind-kind correlations can be found (Hull 1982a, Rosenberg 1985; but see Balzer and Dawe 1986). Similar conclusions follow for the relation between fitness in evolutionary theory and the genetic basis of evolution (Sober 1984a, Rosenberg 1985). Just as we need a hierarchical theory of biological evolution (Gould 1982b, 1982c, Brandon and Burian 1984, Eldredge 1985), we need a hierarchical theory of science (Levins and Lewontin 1985). Neither theory is going to be easy to formulate, and expecting a multilevel theory may well be asking for too much (Williams 1985). But this is the topic of the next two chapters.

Conclusion

Every explanation takes certain things for granted and explains other things in terms of them. I have taken for granted that scientists are by and large curious about the world in which they live. I provide no explanation for this curiosity. At the very least, the process by which young people are introduced into science avoids destroying this curiosity in enough students to allow the continuance of science. In some cases, an education in science even enhances it. Nor do I explain why scientists want credit for the contributions they make to science. They simply do, and the structure of science enhances this desire. Even if a budding young scientist enters science not caring about something as paltry as individual credit, he or she will find it very difficult not to get caught up in the general enthusiasm. Nor do I say very much about the possibility of intersubjective checking. The world might have been sufficiently haphazard so that each time an experiment was run, no matter how carefully, a different outcome would result. In such a world, science would be impossible.

But given curiosity, a desire for credit, and the possibility of checking, the structure that I claim characterizes science can explain quite a bit about the way in which scientists behave. Scientists want other scientists to recognize their contributions, and the most fundamental form of recognition is use, preferably with a formal acknowledgment. Scientists in turn need to use the work of other scientists and in using it give them at least tacit credit. Scientists would very much like to be able to use the work of other scientists while keeping their own advances to themselves as long as possible in order to gain as much advantage as possible; but to get credit scientists must make their findings public. The reason that so much emphasis is placed in science on the person who first announces a contribution is the need to force scientists to make their results available for others to use. With the possibility of credit comes the possibility of blame. Scientists cannot spend very much time checking the work of other scientists if they themselves are to make contributions. They reserve checking for those findings that bear most closely on their own research, chiefly those that threaten it. Because different scientists are committed to different views, the checking that goes on in science rarely degenerates into empty show. When scientists refute their own favorite hypotheses or their opponents confirm them, one can place considerable confidence in the results.

Scientists need very little encouragement to adhere to the institutional norms of science. Scientists do not band together in weekly support groups to urge each other to mend their sinful ways. They need not because by and large their individual goals coincide with the institutional goals of science. Scientists adhere to the norms of science because it is in their own self-interest to do so. Outright fabrication of data is very rare in science because it is in the self-interest of other scientists to uncover fraud. Because scientific propositions are organized into inferential clusters, errors ramify. Even though scientists may not set out to test a particular result, in their efforts to find out what is going wrong in their own research they are likely to zero in on the source of error. Fraud is penalized so severely because it harms everyone who uses a fraudulent result. Infractions of this sort gradate imperceptibly from outright fraud to incompetence in one direction and to finagling in another.

Scientists are not nearly so good about giving credit where credit is due. Such issues are not always clear either, especially when scientists are working in close collaboration. Who thought of what first? It is not always easy to say. Prior to publication, scientists communicate with each other informally. Theft is always a possibility, whether conscious or unconscious. One price that we pay for awarding most of the credit to the first person to make his or her findings public is priority disputes. Another is the rush to publish. A second source for theft also exists. Scientific journals cannot very well publish every manuscript they receive. Some sort of selection process must be devised. Referees are used in the hope that some correlation will exist between the judgments of the referees and the eventual worth of the paper. The purpose of the peer review system is to increase the quality of the work published. Peer review also has its costs. It provides one more opportunity for theft, and on occasion it weeds out the work we find in retrospect to be the most valuable. It is very difficult to distinguish between crackpots and geniuses. Failure to give appropriate credit is punished much less severely than the publication of fraudulent research because it harms only the person who has not been given sufficient credit. However, unless awarding credit is not sufficiently equitable, the self-reinforcing structure of science would collapse.

Science on a small scale would be possible if scientists worked in professional isolation, that is, if they conducted their research all on their own and interacted only in print. This system would have its virtues—no cliques, no intergroup warfare, no scientific nepotism. But science has become too massive an undertaking for pure individualism to survive for long. It really helps to develop one's ideas with the aid of friendly critics. Several scientists working together can pool their cognitive resources in developing a particular research program. They can also increase the visibility of each other's work. The greatest danger for any bright idea is for it to be ignored. Other scientists are too busy with their own research to take much notice of a contribution, no matter how innovative, if it does not clearly bear on their own work. For example, after coming up with 26-million-year cycles of mass extinctions, Raup (1986) wonders why his predecessors were ignored. The answer is simple: nearly all work in science is ignored.

When several scientists collaborate, they automatically constitute an audience for their own publications. The demic structure of science provides conceptual niches for the development of new ideas. They need not confront the entire scientific community totally naked. The price of small research groups is intergroup polemics. Views are not evaluated solely on their own merits but also because of the allegiances of their authors. Individuals can compete in isolation, but once cooperation gives rise to informal research groups, scientists compete as members of these groups. The groups themselves compete. When scientists working in a research group make contributions, they enhance not only their own conceptual inclusive fitness but also that of their research group and its associated program. Conversely, any disastrous error also reflects both on the individual and on the group. The chief reason for the confidentiality in the evaluations that scientists make of each other's work with respect to both publishing and research funds is to allow allies to evaluate each other's work candidly while still being able to cooperate. As in any functionally organized system, this practice is both a cause and an effect of the social organization of science.

Many commentators on science find one or more features of the institution unpalatable. Scientists are too polemical, aggressive, and arrogant. They are too anxious to publish so that they can scoop their competitors. They seem more intent on enhancing their own reputations than helping humanity. They should turn their attention from the problems they find interesting to those that currently confront the human species. The mechanism I have described in this chapter should explain why scientists do not behave the way that these commentators think they should behave. From an operational point of view, behavioral psychologists are right on at least one point: organisms tend to do what they are rewarded for doing, pious hortatory harangues notwithstanding. If a massive educational program were instituted, perhaps scientists could be made to be less polemical in their exchanges and more humble in their opinions of the value of their own work. I am not sure the effort would be worthwhile. The continuing campaigns for scientists to study those problems whose solutions are likely to help humanity in a reasonably direct way have had mixed results. By and large, political leaders are the ones who get to decide which problems need solving, and they are no better than the rest of us in this endeavor. The temptation for those who are empowered to decide which problems need solving is also to decide what the solutions are, scientific opinion to one side.

11

A General Analysis of Selection Processes

He opened the geography to study the lesson; but he could not learn the names of places in America. Still they were all different places that had different names. They were all in different countries and the countries were in continents and the continents were in the world and the world was in the universe.
He turned to the flyleaf of the geography and read what he had written there: himself, his name and where he was.

Stephen Dedalus
Class of Elements
Clongowes Wood College
Sallins
County Kildare
Ireland
Europe
The World
The Universe

James Joyce, *A Portrait of the Artist as a Young Man*

IF ANYTHING about the living world has seemed obvious to biologists, it has been that it is organized hierarchically. Hierarchies are intriguing structures. Generations of school children have been struck at one time or another by their position in an increasingly more inclusive geographic hierarchy. On the inside cover of his geography book, James Joyce's alter ego, Stephen Dedalus, listed just such an egocentric hierarchy leading from himself in the Class of Elements at Clongowes Wood College to the Universe. In his biology book, he might well have listed a second hierarchy radiating out from himself just as egocentrically—the taxonomic hierarchy of *Homo sapiens*, Homo, Hominidae, Primata, Mammalia, Vertebrata, Chordata, and Animalia.

Hierarchies can be divided into two kinds—part/whole and class inclusion. Part/whole hierarchies consist of individuals of increasing size. County Kildare is part of Ireland, Ireland part of Europe, and so on. Class-inclusion hierarchies consist of classes of increasing inclusiveness. The more inclusive classes are, the more members they have. However, this inclusion relation need not be spatial. Traditionally the taxonomic hierarchy was interpreted as being one of class inclusion (to use contemporary terminology). Taxa are collections of organisms of increasing scope. Every organism that belongs to Mammalia also belongs to Vertebrata, but not vice versa. Most vertebrates do not belong to the class Mammalia. In part/whole hierarchies, all the entities are of the same sort. They belong to the same metaphysical category, that of individuals. Class-inclusion hierarchies require two sorts of entities—elements and classes. At bottom the elements of class-inclusion hierarchies are individuals such as organisms. As the taxonomic hierarchy was construed for centuries, organisms are members of their species. They are also members of increasingly more inclusive higher taxa—genera, families, tribes, etc. Species in turn are included in genera, genera in families, and so on. All taxa on this account are equally classes but classes of increasing scope.

Although part/whole hierarchies differ from class-inclusion hierarchies with respect to the metaphysical character of the entities that exemplify them, many of the same sorts of inferences can be made with respect to both. For example, if County Kildare is part of Ireland, and Ireland is part of Europe, then County Kildare is part of Europe. Similarly, if Primates are included in Mammalia, and Mammalia is included in Vertebrata, then Primates are included in Vertebrata. This sort of transitive relation also applies when one moves from the level of element to that of class. For example, if Gargantua is a gorilla and gorillas are primates, then Gargantua is a primate. However, when we add a third metaphysical level, the relations cease to be transitive. For example, *Gorilla gorilla* is a species, i.e., it is a member of the class of all species. Thus, one might be tempted to reason as follows: Gargantua is a member of *Gorilla gorilla,* and *Gorilla gorilla* is a species; hence, Gargantua is a species. But something has gone wrong. Gargantua is an organism, not a species. An organism can belong to a species taxon but not the species category. The membership relation is intransitive.

Neither example of the hierarchies that I have discussed has a temporal dimension. Although the political subdivisions in Europe certainly arose through a very complicated series of historical events, the geographic part/whole relation at any one time is atemporal. Prior to Darwin, similar observations applied to the taxonomic hierarchy. For some biologists, such biological taxa as Mammalia and Vertebrata were held to be spatiotemporally unrestricted. For example, Aristotle viewed classes of plants and animals as secondary substances. As such they are eternal and immutable. They always were and always will be. Later, Christian systematists introduced a temporal dimension. God created the first members of every species in the Garden of Eden. Thereafter the organisms belonging to these species propagated themselves true to type, an occasional monster notwithstanding. However, within the time-span of creation and the second coming, species are eternal and immutable. As the remains of extinct species began to be discovered, systematists allowed that certain species no longer had any extant members. For some that meant that species are not eternal; for others it meant merely that, for the moment, a particular species was not exemplified. Even though Lamarck argued for the transmutation of species, he continued to view them as being in a sense eternal and immutable. Members of the most rudimentary species are constantly created from inorganic substances and through subsequent generations proceed up one of his trees of life. Species can go extinct, but they can also re-evolve. Organisms proceed through successive generations from one species to another, but the species themselves do not evolve.

With the introduction of Darwin's theory of evolution, however, a tension was introduced into biology, a tension that lay all but unrecognized until recently. If species are viewed as segments of the phylogenetic tree and higher taxa as chunks of this tree, then the genealogical hierarchy is historical. It is not only historical but also part/whole. To count as a dodo, an organism must be part of the genealogical nexus denoted by *Didus ineptus.*" Because this twig of the phylogenetic tree has terminated, no future organism can belong to this species no matter what might happen elsewhere in the phylogenetic tree. Parallel observations hold for higher taxa. To count as a primate or a mammal, an organism must be part of the appropriate chunk of the phylogenetic tree. In genealogical classifications, extinction is necessarily forever. Taxa are individuated on the basis of their insertions in history, not because of any general properties that they might have. On this view, organisms, species, and higher taxa are all the same sort of thing—historical entities.

If one follows this line of reasoning to its logical conclusion, certain problems arise for the traditional taxonomic hierarchy. Prior to Darwin, organisms were viewed as being spatiotemporal entities, primary substances, while species and higher taxa were viewed as classes, secondary substances, or some such. On this view, organisms are members of their species. The relation between species, genera, families, and so on from this perspective is one of class inclusion (to use modern terminology). However, from an evolutionary perspective, organisms, species, and higher taxa are all the same sort of thing—historical entities—and the paradigm example of a class-inclusion hierarchy is converted to a part/whole hierarchy. Just as organisms are part of their species, species are part of higher taxa (see figure 11.1). Although the change is metaphysically quite drastic, it does not alter any traditional inferences. It still follows that

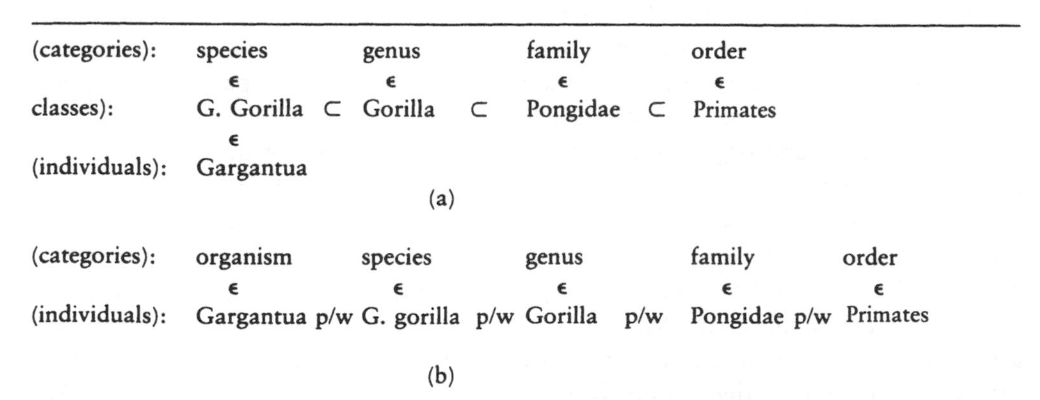

Figure 11.1. (a)The traditional taxonomic hierarchy in which organisms are members of taxa, taxa are related by class inclusion, and these classes in turn are members of their categories. (b) The same hierarchy modified to treat both organisms and taxa as historical entities. Organisms are part of their species, species part of their genera, and so on. The symbol ∈ stands for class membership, ⊂ for class inclusion, and p/w for the part-whole relation.

if all men are mortal and Socrates is a man, then Socrates is mortal. The only difference is that everything has been moved down a membership level. On the old view, categories are classes of classes, while on the new view, they are merely classes.

The taxonomic hierarchy is not the only hierarchy in biology. Just as apparent is the organizational hierarchy of genes, cells, organs, and organisms. This organizational hierarchy is obviously part/whole. Genes are as much parts of their cells as County Kildare is part of Ireland. Although the entities that make up the organizational hierarchy are historical entities, the levels of this hierarchy are not determined on historical considerations the way that they are on the evolutionary interpretation of the taxonomic hierarchy. A particular segment of DNA in a particular cell is an individual. DNA itself is a class. Any molecule anywhere in the universe made of the appropriate atoms and exhibiting the proper structure counts as DNA. The same can be said for cells, organs, and organisms. Gargantua is a historical entity. The class of all organisms is not. Terrestrial life apparently is monophyletic. Life, in the sense of all living creatures, probably is not.

On the traditional interpretation of the taxonomic hierarchy, the organizational hierarchy stops at the level of organisms. Genes are part of cells, cells part of organs, and organs part of organisms. One might even add colonies to the sequence, noting that some organisms are part of a colony or some other kind of kinship group, but that is it. On the traditional view, organisms are members of their species.

A recent controversy in evolutionary biology concerns the levels at which selection can occur. Gene selectionists argue that ultimately all selection can be treated as if the only units of selection are genes. Organism selectionists insist that the organism is the primary focus of selection. Not to be outdone, a few daring biologists have argued that species themselves can be selected. On the traditional interpretation, species selection would not be false but sheer nonsense. Regardless of the size of the individuals which function in selection processes, they are all individuals. If species are classes, selection could not possibly act on them. However, on the evolutionary interpretation, species selection is at least a possibility because the organizational hierarchy is expanded to include entities more inclusive than organisms and kinship groups. The part/whole progression of genes, cells, organs, organisms, and colonies is expanded to include demes, populations, species, and higher taxa.

The taxonomic and organizational hierarchies are only two of the hierarchies apparent in the living world. Ecologists classify organisms in a variety of ways according to their rolls in the ecological web. One of these groupings, the trophic pyramid, is hierarchical with a few carnivores at the top and numerous primary producers at the bottom. This ecological hierarchy is not genealogical and does not coincide with the genealogical taxonomic hierarchy. Not all organisms that belong to the class Carnivora are carnivores (e.g., pandas), and there are thousands upon thousands of species of carnivorous organisms that do not belong to this class—including a few plants! According to Eldredge (1985), the evolutionary process can be understood only if genealogical and ecological hierarchies are distinguished. The genealogical hierarchy consists of genes, chromosomes, organisms, demes, species, and monophyletic higher taxa. Eldredge's ecological hierarchy consists of proteins, organisms, populations, communities, and ecosystems. The reason that the evolutionary process appears so complex is that these two hierarchies do not map neatly onto each other. The most that can be said is that they intersect at the organismic level. Organisms play roles both in the genealogical and the ecological hierarchies. The other elements do not.

Levels of Selection

I began this chapter by observing how apparent hierarchies are in the living world. As apparent as they are, as much importance as biologists place on them, they have not played a very important role in modern versions of evolutionary theory until quite recently. Plenty of lip service is paid to the "hierarchical" character of biological evolution, but that is about all. As the evolutionary process is usually characterized, it consists of two processes (mutation and selection) that give rise to a third (evolution). On this view, genes mutate, organisms are selected, and species evolve. Other levels of the organizational hierarchy (e.g., cells, organs, colonies, etc.) and the genealogical hierarchy (genera, families, etc.) are rarely mentioned let alone incorporated into the basic structure of the theory. Some authors have made evolutionary theory even more parsimonious by attempting to account for all evolutionary phenomena entirely in terms of gene frequencies. Quite obviously, organisms and species exist, but they need not be mentioned in the axioms of evolutionary theory. Gene selectionists are adamant in their insistence that genes will have to do if we are ever to have a coherent, reasonably simple formulation of evolutionary theory. Introducing entities at other levels of organization inevitably produces conceptual chaos. Others are just as insistent that any theory of evolution couched entirely in terms of changes of gene frequencies is inadequate. It leaves too much out. Just as God must love poor people—that is why he made so many of them—entities such as organisms and species must play some sort of a role in the evolutionary process or else there would not be so many of them.

The controversy in evolutionary biology over the levels at which selection occurs is to some extent empirical, but it also has a large conceptual element. One of these conceptual elements is the hold that the traditional organizational hierarchy has on the minds of most biologists. Whether one limits the hierarchy just to genes, organisms, and species or expands it to include other levels such as cells and demes, it is the background against which subsequent discussion is conducted. My main contention in this chapter is that as long as the traditional organizational hierarchy is taken as fundamental, then selection will be found to wander erratically from level to level and, consequently, explanations in terms of selection will be highly variable and contingent. Selection will be discovered to occur at different levels in different taxa not to mention at different stages of the life cycles of individual organisms. Sober (1984a) is willing to accept this consequence; I would like to find a way to avoid it.

For example, Bonner (1967) points out that the sort of organization present in such "social" organisms as the cellular slime molds changes as they undergo their ontogenetic development. At one stage, they exist as numerous, free-living cells, and selection is liable to occur primarily at the cellular level. Later, these cells come together to form a single multicellular organism, and selection is liable to occur primarily at this higher level of organization. Wilson (1971) makes comparable observations about colonies.

In certain species of bee, each hive initially includes several queens. As the hive develops, all but one queen is eliminated. Initially selection cannot operate at the level of the entire hive; later it can.

The general message of the preceding observations is that levels of the traditional organizational hierarchy are not, with respect to the evolutionary process, natural kinds. What is it in the evolutionary process that all and only genes do? What is it in the evolutionary process that all and only organisms do? The answer to these and other questions couched in terms of the levels of the traditional organizational hierarchy is—nothing. As the organization of the entities functioning in the evolutionary process changes, so does the level at which selection can occur. Thus, any generalizations couched in terms of these entities must mirror these vagaries. In short, as long as the traditional organizational hierarchy is taken as basic and selection is defined in terms of it, selection will wander from level to level.

In this chapter I adopt a different strategy. I define the entities that function in the evolutionary process in terms of the process itself, without referring to any particular level of organization. Any entities that turn out to have the relevant characteristics belong to the same evolutionary kind. Entities that perform the same function in the evolutionary process are treated as being the same, regardless of the level of organization they happen to exhibit. Generalizations about the evolutionary process are then couched in terms of these kinds. The result is increased simplicity and coherence.

This simplicity and coherence, however, exact a cost. Considerable damage is done to commonsense conceptions. Instead of organisms always being compared to organisms and species to species, sometimes organisms are compared to species, genes to organisms, and so on. However, the benefits are worth the price. One benefit is that, once properly reformulated, evolutionary theory applies equally to all sorts of organisms: prokaryotes and eukaryotes, sexual and asexual organisms, plants and animals alike. "But do not present-day versions of evolutionary theory already do that?" The answer is, surprisingly, not really. For example, all too often organisms that reproduce asexually are either dismissed as anomalous exceptions or counted as actually being sexual. Even if the traditional organizational hierarchy, biased as it is toward large, sexual animals, were adequate for the purposes of evolutionary theory, the inclusion of clonal organisms would require fundamental changes. As Jackson (1986: 298) remarks, "Cloning inevitably necessitates a rather nightmarish reexamination of familiar concepts and terms."

A second benefit is that the analysis of selection processes I present is sufficiently general so as to apply to sociocultural change as well. In this chapter I spend a great deal of time discussing various unsolved problems about the "levels" at which selection can occur in biology. One reason why I devote so much time to problems that are currently so controversial is to give the reader a feel for the range of issues that arises for selection in purely biological contexts in preparation for the next chapter, when these questions are addressed again but in the more problematic context of conceptual change in science. People tend to reject selection models in conceptual change out of hand because they have a simplistic understanding of biological evolution. Most objections to selection operating in conceptual change, if cogent, would count just as decisively against selection models in biological evolution. As it turns out, the amount of increased generality needed to accommodate the full range of biological phenomena turns out to be extensive enough to include social and conceptual evolution as well.

In his classic paper on units of selection, Lewontin (1970: 1) characterizes the selection process in terms of three basic principles:

1. *Phenotypic variation*—different individuals in a population have different morphologies, physiologies, and behaviors.
2. *Differential fitness*—different phenotypes have different rates of survival and reproduction in different environments.
3. *Fitness is heritable*—there is a correlation between parents and offspring in the contribution of each to future generation.

If one thinks that evolution is Weismannian as distinct from "Lamarckian," one might add a fourth principle as well:

4. *Weismannian Inheritance*—there is no correlation between the needs of an individual and the variations that occur.

Although Lewontin's three principles are read most naturally as referring to organisms and their phenotypic traits, Lewontin (1970: 1–2) acknowledges that they can apply to other levels as well. "For example, if we replace the term *individual* with the term *population* and interpret *phenotype* to mean the distribution of phenotypes in a population, then Principles 1, 2, and 3 describe a process by which one population may increase its proportional representation in the species relative to other populations. Similar reinterpretations of these principles could be made for species rather than populations and even communities rather than species." Hence, Lewontin's general characterization of selection processes does not provide an answer to the question of the levels at which selection can occur. All it does is set the framework within which this question can be addressed coherently. Mary Williams (1970) makes comparable observations about her axiomatization of evolutionary theory (see also Rosenberg 1985).

In its most literal reading, Lewontin's first principle states that organisms vary with respect to their phenotypic traits. In a particular litter of puppies, some will be bigger than others, some more vigorous, and so on. But parallel observations can be made for entities at lower and higher levels of the organizational hierarchy as well. For example, in a particular ejaculation some sperm might be able to swim faster, some better able to penetrate the envelope of the ovum, and so on. At a higher level, a particular hive of bees might split successively in time until a family of related hives exists in which some have a greater percentage of workers than the others, some with a different spatial distribution of workers in the hive, and so on. All these properties are characteristics of the hive, not individual bees. Even species have "morphologies." Some might be subdivided into demes; others not. Some might have extensive, highly convoluted peripheries; others not.

Lewontin's second principle can also be interpreted to apply to several levels of the organizational hierarchy once one expands a bit on the notion of survival and reproduction. If one takes a typical vertebrate as paradigm, it must survive long enough to reproduce, and more often than not it itself survives reproduction. For example, a dog that dies before it becomes sexually mature cannot reproduce. Hence, its direct fitness is zero. Its inclusive fitness is another matter. If it does survive to reproduce, it can have successive litters of various sizes. The organisms in these litters in turn can have variable rates of survival. Other organisms can reproduce only at the expense of their own survival. For example, salmon spawn only once per generation and then die. A few organisms survive for thousands of years (e.g., organisms belonging to certain species of plants), but most survive for only a few minutes or days (e.g., most unicellular organisms such as bacteria and blue-green algae). However, in many sorts of organisms, the distinction among individual survival, reproduction, and growth are not all that clear. For example, in asexual reproduction, it is difficult to distinguish between growth and reproduction. The processes are the same. One cell divides into two. If these cells stay in contact, we tend to consider that growth: these cells belong to the same organism. When the cells drift apart, that is reproduction: each cell is a separate organism. On the face of it, this way of distinguishing between growth and reproduction contains at least an element of arbitrariness (see Jackson et al. 1986). In any case, since entities other than organisms can have "phenotypes," they can have differential fitness. For example, hives with a very few drones might proliferate more quickly than those that have more, and sexual species might speciate more rapidly than species that reproduce asexually. The percentage of drones in a hive is a phenotypic characteristic of *hives,* not *organisms,* and sexual reproduction is a relation *between* organisms, not a simple property *of* organisms.

Lewontin's third principle is so obvious that its importance can easily be overlooked. It is not enough for entities to exhibit phenotypic traits and for these traits to have different rates of survival and reproduction. Generation must occur. Parents must produce offspring in successive generations. Genes do produce genes, and organisms produce organisms. In order for selection to occur at higher levels of organization, not only must these entities exhibit traits of their own, they must produce "offspring." Sometimes these parent entities survive reproduction. Just as a pair of dogs can produce successive litters, a species can produce successive peripheral isolates. In some cases, however, the only way that an entity can reproduce itself is by losing its own identity in the process. When an amoeba reproduces by fission, it loses its own individuality. The same observation applies when a single species splits in two.

Although Lewontin does not emphasize it in this paper, "correlation" is not strong enough for heritability. The correlations must be causal. The reason that certain phenotypes proliferate and others are eliminated is that they facilitate survival and reproduction (Sober and Lewontin 1982). In short, these phenotypes must be adaptations. Not all phenotypic traits are adaptations—some are only effects of selection operating on other characters—but some phenotypic traits must be adaptations if the changes are to count as being the product of selection. To use the usual example, chins increased in size through the course of human evolution but not because they were being selected. Rather this increase was merely an effect of selection occurring elsewhere (Gould and Vrba 1982, Vrba and Eldredge 1985). Presented grammatically, the distinction is between "selection of" and "selection for" (Sober 1984a). It follows, therefore, that if species are to be selected, not only must species exhibit phenotypic traits of their own but also these phenotypic traits must count as adaptations (Vrba 1984). Of course, those who acknowledge no difference between correlation and causation are unlikely to acknowledge any difference between effects and adaptations.

Although Lewontin's characterization of selection leaves open the question of the levels at which selection occurs, it does take for granted that this question is to be answered in terms of traditional hierarchical levels. Lewontin's answer is that the level at which selection occurs varies from situation to situation but that selection tends to be concentrated at the lower levels, primarily at the levels of genes, cells, and organisms; rarely at the levels of populations or species. Although later authors have questioned this conclusion, arguing that selection can and does occur quite frequently at higher levels, they too address this problem in terms of traditional hierarchical levels (Wade 1978, D. Wilson 1980, 1983). In this chapter I argue that a truly general analysis of selection processes is possible only if traditional hierarchical levels are ignored and the relevant levels are defined in terms of the selection process itself. When this is done, certain long-standing disagreements among biologists can be seen as only apparent, a function of conceptual ambiguity. Ample disagreements remain, but at least they can be expressed clearly. However, I do not propose to resolve the levels of selection controversy by definitional fiat. The most that appropriate definitions can do is to increase conceptual clarity.

The term "gene" has received successive definitions as it has developed in the context of Mendelian genetics and then transferred to molecular biology. Using the techniques of transmission genetics, geneticists distinguished between such units as mutons, recons, codons, cistrons, and operons. With increased knowledge of the molecular mechanisms actually at work, molecular biologists transferred these genetic units to the molecular level and introduced additional units such as introns, exons, and junk DNA. However, as central as these various units are for transmission genetics, developmental genetics, and molecular biology, they are not appropriate for evolutionary theory. A more general unit is needed.

In his seminal book, G. C. Williams (1966: 25) redefined "gene" in evolutionary contexts as "any hereditary information for which there is a favorable or unfavorable selection bias equal to several or many times its rate of endogeneous change." Thus, in species of organisms that are genetically quite heterogeneous and in which crossover is common, only very small chunks of the genetic material count as "genes." In genetically homogeneous species, species in which crossover is rare or does not occur at all, and in asexual organisms, much larger structures count as evolutionary genes, e.g., entire chromosomes and even entire genomes. One additional consequence of Williams's evolutionary definition of "gene" is that neutral genes cannot exist. Of course, nothing precludes Mendelian or molecular genes from being neutral with respect to selection processes, but Williams defines "evolutionary gene" in terms of selection processes. Any entity for which there is no favorable or unfavorable selection bias with respect to a particular selective episode cannot count as an evolutionary gene with respect to that episode. Perhaps it functioned as an evolutionary gene in past episodes and may do so again in the future, but with respect to that particular sequence of events, it is not functioning as an evolutionary gene. Although some commentators have found this characteristic of Williams's definition a serious shortcoming, it is a consequence of *any* functional definition. For example, if hearts are defined in terms of pumping blood, then when they are not performing this function, they are not hearts.

For all the hostile criticisms that Dawkins's *The Selfish Gene* (1976) has received, both for its purple prose and for its apparently unrelenting gene-selectionist perspective, it also contains an excellent discussion of the nature of selection. Dawkins accepts Williams's characterization of evolutionary genes and extrapolates from it to a more general notion—replicators. According to Dawkins, genes are at least the primary replicators in biological evolution, possibly the only replicators, but his general definition of "replicator" leaves this question open. His definition also allows for replicators to function in other sorts of evolution. For example, Dawkins (1976: 206) terms the replicators that function in cultural evolution "memes."

As Dawkins characterizes replicators, they have three important properties—longevity, fecundity, and fidelity. In each case, these properties refer to *copies,* not just to similar tokens. Copies require descent of some sort. For example, particular genes are not very long-lived. As material bodies they are destroyed at each replication. All that actually survives intact is their *structure* (for a general analysis of structure, see Harré 1979). They "survive" only in the form of copies. The fidelity to which Dawkins refers is copying fidelity. In replication the structure of genes survives *largely* intact. Occasionally it is altered slightly. Fecundity concerns the number of copies produced. Thus, when segments of DNA are functioning in metabolic processes, they are not functioning as replicators. Only in mitotic division does the sort of replication relevant to evolution take place, and in most cases these replication sequences are dead-end. The only mitotic divisions that enter directly into the evolutionary processes are those that are open-ended, e.g., those that occur in germ cells. In order for evolution to occur indefinitely, replication sequences must be open-ended (Dawkins 1978). However, not all selection processes must be able to eventuate in indefinite change. For example, selection also occurs in the functioning of the immune system, but Steele's (1981) claims notwithstanding, the effects of these selection processes cannot be passed on from parent to offspring (for a more general discussion, see Darden and Cain 1988).

A recurrent problem in evolutionary biology is the relative priority of descent to similarity. If the two always went together, there would be no problem, but they do not. Sometimes descent produces dissimilar replicators; sometimes structurally very similar replicators arise independently. The distinction is between homology and homoplasy. This distinction is crucial for inferring phylogeny, but it is also relevant to selection processes. In selection processes, both descent and structural similarity are required because the only way that changes in gene frequencies can build up in successive generations of the same population is via these generations being related by descent. In fact, the notion of generation itself requires descent. Organisms living sequentially in time do not count as "generations" unless they are related by descent. More subtly, as Mayr (1982) has emphasized repeatedly, the notion of a population as it functions in biology itself presupposes descent. To belabor the obvious, Lewontin's heritability principle requires descent.

The sort of identity that is central to selection processes is *identity by descent*. For example, a species that is homozygous for allele M speciates to produce two new species each homozygous for allele M. Now, quite independently, M undergoes the same structural change in both daughter species. In both cases M mutates into M′. Are these two new alleles the same or different alleles? Should both of them be designated M′, or one M′ and the other M″? For the purpose of reconstructing phylogeny, the answer to this question is important. M, M′, and M″ must be kept distinct. From the perspective of intraspecific selection, it makes no difference whether or not these alleles are distinguished because their fates in their respective selection processes are independent of each other. However, the same sort of mutation can occur within a single species. Here it can make a difference because of the strong local effects that kin selection can have. If M mutates to M′ within the confines of a small isolated population, it can become rapidly fixed within this population. In general, the identity that is crucial to Wright's shifting-balance theory is identity by descent. However, in species that lack the appropriate population structure, descent is still required but identity by descent is not very important. If the same structural change in M were to occur twice, the resulting mutant alleles would have the same effects even though they are not identical by descent. For mass selection, descent is required but not identity by descent narrowly construed (Sober 1984a, Grafen 1985).

Replication, Interaction, and Selection

Confusion over the nature of selection is in part terminological. Time and again in the history of science, scientists have transferred a term developed in one context to another context: the transmutation of "evolution" from its embryological to its evolutionary usage, the use of de Vries's "mutation" to refer to numerous slight changes in the genetic material, the adoption by evolutionary biologists of Owen's term "homology," in spite of its origin in idealist metaphysics, and the recent attempt by pattern cladists to retrieve it. The inevitable result of such terminological transformations is a period of confusion as certain scientists object to the new usage and others adjust to it. Similar observations hold for the term "population" in statistics and population biology. As Jerzy Neyman (1967: 1456–57) remarked in his appreciation of Fisher, originally the term "population" included no historical component:

> The era in the history of statistics that preceded Fisher's may be labeled by the excellent German term *Kollektivmasslehre* invented, I think, by Bruns. Its development followed the realization, at the end of the 19th century, that the treatment of the then novel subjects of scientific investigations, namely studies of what we now call "populations," be it populations of stars or of molecules, of plants or of humans, require a new "collective" mathematical discipline. A population is characterized by a distribution of one or of several "individual" characteristics of this population's members.

Statisticians can deal as readily with "populations" of stars as with "populations" of organisms, but in doing so they are using "population" in two radically different senses. On occasion a star can give rise to two or more stars by some sort of massive upheaval, but stars do not reproduce themselves indefinitely through time the way that organisms do. Stars have histories, but they do not form populations of the sort that organisms form. Organisms can be grouped together in a variety of ways—parasites, carnivores, males, and so on. None of these ways require descent. The populations of population genetics do. When Mayr (1987) belabors the difference between biological populations and other sorts, he is not simply insisting on the autonomy of biology. He is emphasizing an important characteristic of populations as they result from and function in selection processes.

Which is the "correct" usage of the term "population"—the generic usage that includes groups of entities related by descent as well as groups defined in terms of the common possession of some property, or the narrower usage preferred by Mayr that requires descent? I do not think that etymological investigations are especially relevant. Terms, like species, evolve. However, if serious confusion is to be avoided, the distinction must be consistently made between these two senses of "population." To this end, two different terms would certainly help.

Similar observations hold with respect to Williams's (1966) use of the term "gene." Just as the genes delimited by the methods of transmission genetics (Mendelian genes) do not map neatly onto the genes delimited by the methods of molecular biology (molecular genes), neither of these sorts of gene map neatly onto evolutionary genes. As might be expected, Williams's redefinition of the term "gene" has met with some heated opposition from those working in the areas of biology from which he borrowed the term. Stent (1977: 34) for instance is beside himself. According to Stent, all working geneticists employ the term "gene" in its molecular sense. "This perverse definition" in terms of selection "denatures the meaningful and well-established concept of genetics into a fuzzy and heuristically useless notion." Memory is short. Mendelian geneticists were no less irate when molecular biologists introduced their perverse definitions. To paraphrase Noah Webster (1831: 110), in science as well as in farming it all depends on whose ox is gored.

One way to avoid ambiguity is to introduce neologisms. For a while at least, a term introduced to serve a particular function is liable to remain univocal. Sooner or later, however, if this new term gains any coinage at all, ambiguity will inevitably creep in. After all, "gene" was at one time a neologism. As much as we may complain, this is the way that natural languages evolve. Thus, Dawkins's (1976) replacement of Williams's (1966) evolutionary gene with "replicator" is an improvement, both because it avoids the confusion of Mendelian and molecular genes with evolutionary genes and because it does not assume the resolution of certain empirical questions about the levels at which selection can occur in biological evolution; but it too carries with it opportunities for misunderstanding. Among these is the confusion of replication with what I term "interaction" and in turn the confusion of these two processes with selection. Thus, in an effort to reduce conceptual confusion, I suggest the following definitions:

replicator—an entity that passes on its structure largely intact in successive replications.
interactor—an entity that interacts as a cohesive whole with its environment in such a way that this interaction *causes* replication to be differential.

With the aid of these two technical terms, selection can be characterized succinctly as follows:

selection—a process in which the differential extinction and proliferation of interactors *cause* the differential perpetuation of the relevant replicators.

Replicators and interactors are the entities that function *in* selection processes. Some general term is also needed for the entities that result from successive replications:

lineage—an entity that persists indefinitely through time either in the same or an altered state as a result of replication.

In order to function as a replicator, an entity must have structure and be able to pass on this structure in a sequence of replications. If all a gene did was to serve as a template for producing copy after copy of itself without these copies in turn producing additional copies, it could not function as a replicator. Although genes are well adapted to function as replicators, it does not follow from the preceding definition that genes are the only replicators. For example, organisms also exhibit structure. One problem is the sense in which they can be said to pass on this structure largely intact. From the human perspective, populations do not seem to exhibit much in the way of structure. Even so, population biologists recognize something they term "population structure"—any nonrandom associations of organisms that belong to a population during any part of their life cycles, but in particular any deviation from panmixia that results in nonrandom associations between genotypes during mating (Michod 1982). Thus, if populations can pass on this structure during successive replications of these populations, then they too might function as replicators.

Many cohesive wholes exist in nature, but only a very few of them function in selection processes. Hence, only a very few count as interactors. In order to function as an interactor, an entity must interact with its environment in such a way that some replication sequence or other is differential—the "relevant" replicators. Organisms are paradigm interactors. They are cohesive wholes, they interact with their environments as cohesive wholes, and the results of these interactions influence replication sequences in such a way that certain structures become more common; others rarer. However, many other entities also function as interactors—even genes. Genes have "phenotypes." DNA is a double helix that can unwind and replicate itself. In doing so it interacts with its cellular environment. In the beginning, the same entities had to perform both functions necessary for selection. They had to replicate, and they had to interact with their environments in such a way that replication was differential. However, because replication and interaction are fundamentally different processes, the properties which facilitate these processes tend also to be different. None too surprisingly, these distinct functions eventually were apportioned to different entities. As I have mentioned several times previously, when different functions must be performed by a single structure or entity, none are performed very well. Interaction occurs at all levels of the organizational hierarchy, from genes and cells, through organs and organisms, up to and possibly including populations and species.

As I have characterized it, selection is an interplay between two processes—replication and interaction. Usually the entities functioning in these two processes are different, although in a pinch a single entity can perform both functions. The interaction that affects replication must occur at the same organizational level at which replication is occurring or higher, usually higher. Selection is composed of several causal processes: replicators producing other replicators, as well as, sometimes, interactors, and interactors interacting with their environments. The net effect of all these causal processes is the differential perpetuation of the replicators that produced them. Vrba (1984: 319) presents very much the same definition of "selection" that I do, albeit without using the technical terms I am proposing. According to Vrba:

> Selection is that interaction between heritable, emergent character variation and the environment which causes differences in birth and/or death rates among variant individuals within a higher individual.

In earlier versions of the preceding definitions, I did not sufficiently emphasize the causal character of all of the processes involved in selection (Hull 1980). As a result, Brandon and Burian (1984) as well as G. C. Williams (personal communication) have complained that my definition of "selection processes" mistakenly includes drift as a form of selection, when traditionally drift has been clearly distinguished from selection. However, when the causal character of my definition of "selection" is noted, drift is excluded. An entity counts as an interactor only if it is functioning as one in the process in question. Thus, if changes in replicator frequencies are not being caused by the interactions between the relevant interactors and their environments, then these changes are not the result of selection. Drift is differential replication in the absence of interaction. There are replicators but no interactors. Hence, there is no selection (see also Brandon 1985, Mitchell 1987, Lloyd 1988).

Many entities persist indefinitely through time. Of these, some change; some do not. However, the only entities that can count as lineages are those that are formed by sequences of replicators. "Lineage" is inherently a genealogical concept. For example, the solar system has changed through time. However, it does not count as a lineage because no replication was involved in such changes. The general notion is that of an historical entity—a space-time worm. Lineages, as I am using the term, are a special sort of historical entity—historical entities formed by replication series. And to the extent that selection is operative, these lineages will be formed as a result of selection as well.

Genes form lineages; so do organisms. In most cases gene lineages are wholly contained within organism lineages (see Goodman et al. 1979). According to more gradualistic versions of evolutionary theory, species do not *form* lineages; they *are* lineages, paradigm lineages. They are the things that change gradually through time without losing their individuality. But nothing about selection requires that evolution be gradual. It might be the case that species are incapable of indefinite change, that speciation is always saltative. If so, then species are not themselves lineages but form lineages through ancestor-descendant sequences of species. It might even be the case that for particular organisms and at particular periods, species as higher-level entities integrated by gene exchange do not exist. Evolution via selection had better be possible in the absence of species, because for most of the history of life on Earth, species did not exist—that is, if significant amounts of gene exchange are necessary for the existence of species. Perhaps quite early in the history of terrestrial life, cells began on occasion to fuse (Bernstein et al. 1985), but even if one counts such a phenomenon as "gene exchange" in the relevant sense, all indications are that it was extremely rare until the beginning of the Cambrian, about 650 million years ago (Carr and Whitton 1973, 1982, Fogg et al. 1973, Schopf 1978, and Jackson et al. 1986). Even though species are not a necessary consequence of replication, lineages are.[1]

By now, one thing should be clear: everything involved in selection processes and everything that results from selection are spatiotemporal particulars—individuals. Both replicators and interactors are unproblematic individuals. To perform the functions that they do, they must have finite durations. They must come into existence and pass away. Replicators must exhibit structure, and interactors must interact with their environments as cohesive wholes. These are all traditional characteristics of individuals. That individuality is at the heart of the selection process can be seen by the frequency with which biologists who want to argue that species can be selected begin by arguing that they have the properties usually attributed to individuals, ordinary perceptions and conceptions notwithstanding (Eldredge and Gould 1972). Conversely, those who argue against species selection frequently begin by arguing that species lack these very characteristics. In fact, those who argue that not even organisms can function as units of selection begin by casting doubt on their status as individuals, superficial appearances to one side (Dawkins 1976, 1978).

Lineages are also individuals but of a special sort. In order to function as a replicator, an entity can undergo only minimal change before ceasing to exist. In order to function as an interactor, an entity can undergo considerable but not indefinite change. Lineages are peculiar in that they can change indefinitely through time. Thus, any entity that can function either as a replicator or as an interactor cannot function as a lineage, and vice versa. Lineages are continuous through time. They also exhibit a certain amount of cohesiveness but not so much that they cannot change. To the question, "Can X evolve?" one must respond, "How well integrated is it? If it is too tightly knit, it cannot evolve, but if it is too diffuse, it will also be unable to evolve." Jumbled collections can change through time but not "evolve" in the technical sense that I propose. Lineages must exhibit at least a minimum degree of "cohesiveness," and the cohesiveness that is relevant is genealogical (Mayr 1975, Wiley 1981a). All sorts of factors can enhance or destroy the cohesiveness of a lineage, but in the absence of genealogy, an entity cannot count as a lineage.

In any system that is evolving through selection, sooner or later, a point occurs at which the genealogical fabric is rent and networks become trees. At whatever level in the traditional hierarchy that this occurs, the resulting entities are lineages. Hence, among strictly asexual organisms, no lineages exist more inclusive than organism lin-

1. In an earlier discussion of lineages, I treated them as if they had to result from selection processes. Dan Brooks pointed out to me that this assumption is too restrictive. Replication will do. Hence, "historical entity" is the most general notion, "lineage" less general, and "lineages resulting partly through selection" less general still.

eages. Among sexual organisms, some may form lineages no more inclusive than sequences of populations. In some, gene exchange may be sufficiently extensive and sustained to integrate entire species into lineages. As Arnold and Fristrup (1982: 116) observe, "The central theme of our hierarchical approach to evolutionary theory is that branching and persistence are the essential components of fitness and evolutionary success at all levels."

In sum, being a spatiotemporal individual is necessary for any entity functioning as either a replicator or an interactor. Collections won't do. Extensive organization is required. However, being a spatiotemporal individual is far from sufficient for an entity to function as either a replicator or as an interactor. Many spatiotemporal individuals do not and cannot function as either. Because of the spatiotemporal character of replication, lineages are also spatiotemporal particulars. A fortiori those lineages that result from the appropriate interplay between replication and interaction (i.e., selection) are also spatiotemporal particulars. Hence, all the entities that function in selection processes as well as those that result from them are spatiotemporal individuals—historical entities. Some are so well-integrated that they are incapable of indefinite change (replicators and interactors). Others are to some extent "cohesive" but not so well-integrated that they cannot change indefinitely through time (lineages). Of equal importance, some of the structure passed on in replication must count as "information" (see chapter 12). Even if terming lineages "individuals" jars one's linguistic sensibilities, at the very least one must admit that they are more than mere collections of organisms. They have a certain degree of particularity.

Before turning to specific applications, I must make one final observation about the technical terms that I have introduced. They are defined in terms of actualities rather than potentialities. In order to count as a replicator, an entity must not only have the potential to replicate, it must do so. To count as a replicator, an entity must be either the first, intermediate, or final member of a replication sequence. Any entity that has the characteristics necessary to enter into such sequences but does not is only a potential replicator. Parallel remarks hold for interactors and lineages. Although I use my terminology entirely in terms of actualities rather than potentialities, I place no great weight on this convention. I could say everything I need to say if I were to adopt the opposite convention, albeit with greater difficulty. All that is necessary is that one does not surreptitiously equivocate between these two terminological conventions, unconsciously wandering between talk of actualities and potentialities (see the earlier objection to Williams's functional definition of "evolutionary gene").

The Organism under Siege

As simple as the distinction between replicators and interactors is, it is sufficient to eliminate a recurrent disagreement about selection processes by showing that it is only apparent, not real. At least sometimes gene selectionists and organism selectionists are not disagreeing with each other but referring to two different aspects of the same process. For example, such organism selectionists as Ayala (1978: 64) emphasize that "it must be remembered that each locus is not subject to selection separate from the others, so that thousands of selective processes would be summed as if they were individual events. The entire individual organism, not the chromosomal locus, is the unit of selection, and the alleles at different loci interact in complex ways to yield the final product." Dawkins (1978: 69) disagrees:

> Of course it is true that the phenotypic effect of a gene is a meaningless concept outside the context of many, or even all, of the other genes in the genome. Yet, however complex and intricate the organism may be, however much we may agree that the organism is a unit of *function,* I still think it misleading to call it a unit of *selection.* Genes may interact, even "blend," in their effects on embryonic development, as much as you please. But they do not blend when it comes to being passed on to future generations. I am not trying to belittle the importance of the individual phenotype in evolution. I am merely trying to sort out exactly what its role is. It is the all important instrument of replicator preservation: it is *not* that which is preserved.

To put this disagreement succinctly, we can say that, for Ayala, organisms are interactors. Maybe so, Dawkins responds, but they are not replicators. And I must add that neither genes nor organisms are literally *preserved.* Tokens of both organisms and genes are extremely ephemeral entities, but some beget similar tokens. All that is needed in either case is the preservation of structure.

Since the preceding dispute over the units of selection broke out, all sides have come to accept the distinction between replicators and interactors, albeit not necessarily in the terms I am urging. For example, Sober (1984a) prefers distinguishing between "selection of" versus "selection for." Sober's favorite example of this distinction is a child's toy that sorts marbles according to size. If smaller marbles are red and larger marbles are blue, one might think that these marbles were being selected for color when actually they are being selected for size. Replicators get passed on, but they are passed on differentially because of the relative success of the interactors with which they are associated. Dawkins (1982a: 60) now tacitly acknowledges that "selection" is used more appropriately in connection with what he terms "vehicles" and not replicators:

> My main concern has been to emphasize that, whatever the outcome of the debate about organism versus group as *vehicle,* neither the organism nor the group is a *replicator.* Controversy may exist about rival candidates for vehicles, but there should be no controversy over replicators *versus* vehicles. Replicator survival and vehicle selection are two aspects of the same process.

Dawkins introduced the terms "replicator" and "vehicle" because of their generality and because of the common connotations of such terms as "gene" and "organism." Genes may not be the only replicators, and organisms may not be the only vehicles. But the terms he devised also have connotations. As far as I can see, the connotations of the term "replicator" are entirely appropriate, while those of "vehicle" are not. Vehicles are the sort of thing that agents ride around in. More than this, the agents are in control. They steer, and the vehicles follow dumbly. Although Dawkins explicitly assigns distinct evolutionary roles to both replicators and vehicles, the picture that Dawkins's terminology elicits is that of genes controlling helpless and hapless organisms. As Sober (1984a: 255) has emphasized, the "units of selection controversy began as a question about causation," and it remains a question of causation. Hence, a terminology that tends to exaggerate the causal role of replicators at the expense of the different though equally important causal role of what I term interactors is less than perspicuous.

Needless to say, I prefer my own terminology, in part because it is mine but also because it is less misleading (see also Williams 1985). A second reason for preferring "interactor" to "vehicle" is that this term is already used in the literature of evolutionary epistemology in connection with replication in both biological and conceptual evolution. According to D. T. Campbell (1979a), genes are the physical vehicles that transmit the information in biological evolution, while everything from stone tablets to electronic chips serve as the physical vehicles in conceptual evolution. Using "vehicle" to refer to both replication and interaction is sure to generate confusion, especially when on occasion one and the same entity is functioning as both.

Once terminological issues have been clarified, the units-of-selection controversy resolves into two different questions: what are the units of replication, and what are the units of interaction? Once the phrase "units of selection" has been disambiguated in this way, differences of opinion still remain. According to Dawkins, he never intended to deny that organisms are vehicles (my interactors). To the contrary, he acknowledged this role right from the beginning. Needless to say, Dawkins's opponents interpret the situation differently. As they see things, Dawkins was confused on all issues from start to finish. His switching from "replicator selection" to "vehicle selection" amounts to total capitulation. Given my general goals, my best strategy is to select from both sides those positions that I can utilize in my own analysis, emphasizing the continuity of my analysis with elements on both sides of this especially emotional dispute. My emphasis on Dawkins is also in part idiosyncratic because it was through my efforts to find out what was missing in Dawkins's analysis that I came up with my own.

Dawkins continues to insist that in biological evolution genes of various sizes are the only entities that function as units of replication. Replicators must be able to pass on their structure largely intact, and genes can certainly do that. As Williams (1966) pointed out and Dawkins (1976, 1978) has repeatedly emphasized, crossover during meiosis constantly tears the genome apart and reassembles it. Such recombination matters only in genetically heterogeneous populations because what is at issue is the integrity of the information encoded in the structure of the genetic material, not the integrity of the genetic material itself (see chapter 12). In genetically homogeneous populations, recombination can occur but has no effect. Thus, in genetically homogeneous populations and in instances of asexual reproduction, the entire genome functions as a single replicator. It is interesting to note in this connection that, if Eldredge and Gould (1972) are right, speciation occurs primarily by the establishment of small, peripheral isolates. Small populations tend toward increased homogeneity, both because very small populations cannot incorporate the genetic heterogeneity of larger populations and because of the effects of increased inbreeding. As a result, at speciation, when it really counts, replicators tend to be quite large chunks of the genetic material, possibly entire genomes.

Single cells, whether they exist in multicellular organisms or as free-living, can also function as replicators. Cells exhibit structure of their own and can pass on this structure quite directly and largely intact. For example, when a paramecium splits longitudinally to produce two new paramecia, both the genetic material and the organism itself are replicated. In fact, modifications of the organism's phenotype can themselves be transmitted in this way, not only transmitted but also transmitted in the absence of any alterations of the genetic material. For example, if a portion of the cortex of a paramecium is surgically removed and reinserted with the cilia facing in the opposite direction, this modification is transmitted to later generations even though the genetic material remains unchanged. In this case, one would think that the organism is functioning as a replicator. Dawkins (1982b: 177) rejects the idea. Instead, he is willing to consider the basal bodies of the cilia to be part of the germ-line of the organism even though they are proteins, not nucleic acids. I think that Dawkins is willing to do too much damage to the notion of a replicator just to preclude single-celled organisms from functioning as replicators.

Even multicellular organisms can reproduce by fission. The most serious problem cases for multicellular organisms functioning as replicators is provided by sexually reproducing organisms. In sexual reproduction, organisms do not produce other organisms directly but only indirectly via gametes. In nearly all forms of sexual reproduction, the structure of the parent organisms is lost and reappears only upon maturation. Because the overall structure of the parent organisms detours through the genetic material and the genomes of the offspring are likely to differ from those of the parents, the transmission of organismal structure is neither very direct nor very faithful.

Another reason for arguing that organisms cannot function as replicators turns on two different senses in which genes and organisms can be said to "contain information" in their structure. Each genome, each "genotoken," to use Wimsatt's (1980) felicitous term, has the structure that it has and no other, just as each organism, each "phenotoken," has the structure that it has. Given any one genotoken, numerous alternative phenotokens may be produced, depending on differences in the environment. All possible translations of the structure in a genotoken into various phenotokens are known collectively as a reaction norm. Thus, the "information" encoded in the genome is largely "potential." In any one instance of translation, the reaction norm is narrowed to one eventuality, one phenotoken. None of the other potentialities equally "programmed" into the genotoken are realized. The net effect is the loss of nearly all the potential information in the genotoken. Replication at both the genetic and organismal levels requires environmental input. However, in the two cases, the effects of the environmental contributions are quite different. In cases of fission, organisms pass on their structure just as surely as genes pass on theirs. However, the only "information" that an organism can pass on in replication is the information realized in its structure.

In asexual reproduction, organisms can pass on their structure both directly and largely intact; in sexual reproduction, the transmission of the organism's phenotype is never direct and frequently not very faithful. In both cases, however, the information that is transmitted is minimal when compared to the information passed on when the genetic material replicates. In sum, organisms do not function as replicators very often or very well, but they cannot be ruled out quite as automatically as Dawkins claims (for a discussion of genetic information, see chapter 12).

Dawkins (1978, 1982b) is not content to argue against organisms functioning as replicators. He goes on to attack the idea of organisms as unified wholes functioning as interactors—his vehicles. He thinks that "the organism" has exercised too strong a hold over the way in which evolutionary biologists conceptualize evolution. He proposes to extend the limits of the phenotype beyond the traditional limits of organisms. According to Dawkins, nests and mating dances are as much part of an organism's phenotype as are its hooked talons or webbed feet. The song a bird sings is as much an adaptation as the shape of its beak. Once this perspective is adopted, then it is much easier to accept the organism as a mosaic of separate characters. Thus, in response to Gould's (1977d) claim that selection cannot see genes and pick among them directly but must use bodies as intermediaries, Dawkins (1982a: 58) retorts, "Well, it must use *phenotypic effects* as intermediaries, but do they have to be bodies? Do they have to be discrete vehicles at all?" Dawkins answers no to both questions. In doing so, he does not deny the importance of interaction in selection, only the necessity of having discrete bodies perform this function, and here he is once again in direct opposition to such organism selectionists as Ayala and Gould. Dawkins proposes not only to extend the phenotype but also to disassemble it.

I agree with Dawkins that the conception of organisms as adult vertebrates has played too central a role in the way in which evolutionary biologists conceptualize evolution. Many organisms are not very well-organized, at least not throughout their entire life cycle. For example, organisms that undergo considerable metamorphosis become dedifferentiated between stages, losing nearly all their internal organization. During such periods, the parts of an organism can be rearranged quite extensively without doing much damage. Nor are the spatiotemporal boundaries of all organisms especially sharp. Some organisms go through stages during which they dissolve into separate cells. When they do, it is impossible to decide whether one organism is present or hundreds. Zoocentrism notwithstanding, not all organisms are animals. A patch of crabgrass may look to us like a series of separate plants. However, if we could see below the surface of the earth, these separate plants would merge into a single network connected by runners. A crabgrass patch is as much an organism as is a tree, just without the trunk. Most species of dandelions are apomicts. They reproduce asexually, with no sexual union whatsoever. Common aphids undergo sexual reproduction periodically, but in between they undergo massive increases in numbers by parthenogenesis. To the extent that all these separate organisms are genetically identical, Janzen (1977) argues, they should be considered parts of single evolutionary individuals.

Dawkins (1982a) agrees. "Generations" should be counted only between instances of sexual reproduction. The production of numerous commonsense generations, after fertilization, in organisms such as aphids should be considered "growth" and not "reproduction." In fact, asexual reproduction should not be considered "reproduction" at all. Just as a multicellular organism is composed of cells which in most cases are genetically identical to each other and to the cells in the germ line, aphid clones are composed of organisms which in most cases are genetically identical to each other and to the female that produced them. According to Dawkins (1982b: 255), the relevant distinction is not between sexual and asexual reproduction, nor even between reproduction and growth, but between the divisions that can in principle go on indefinitely and those that are dead ends.

As foreign as these conceptions may be to zoologists, especially vertebrate zoologists, botanists distinguish between tillers and tussocks, ramets and genets. For example, many sorts of grass grow in tufts (tussocks) composed of numerous sprouts (tillers) growing from the same root system. Which is the "organism," each tiller or the entire tussock? More generally, botanists term each physiological unit a ramet, all the ramets that result from a single zygote, a genet. Sometimes all the ramets that compose a single genet stay attached to each other; sometimes not. And, according to botanists such as Harper (1977), natural selection acts on genets, not ephemeral ramets. As Cook (1980: 91) remarks:

> Through the eyes of a higher vertebrate unaccustomed to asexual reproduction, the plant of significance is the single stem that lives and dies, the discrete, physiologically integrated organism that we harvest for food and fiber. From an evolutionary perspective, however, the entire clone is a single individual that, like you and me, had a unique time of conception and will have a final day of death when its last remaining stem succumbs to age or accident.

If selection is a process of differential survival and reproduction of certain "units," then we had better get clear about which entities to count—each tiller, each tussock, or perhaps an entire field of several acres. Tillers and tussocks are usually quite small and short-lived, while genets can be quite large and survive for a long time. For example, one botanist estimated that a particular genet of the common red fescue stretched for more than 240 yards and was more than a thousand years old. Another estimated that the clones of a bracken fern extended over 474 square meters and were at least 1,400 years old. Evolution will be perceived as proceeding very quickly or very slowly, depending on which entities one counts. As Cook (1980: 93) concludes, "the laws of natural selection are not constrained by human notions of individuality." In a collection of papers investigating clonal organization, the authors consistently alternate substituting genets for ramets in their equations to see what the results turn out to be (Jackson et al. 1986).

Although both Dawkins and I do our best to cast doubt on the absolutely sharp boundaries in the organizational hierarchy that make the organism stand out so starkly, we do so for quite different reasons. He wants to extend the phenotype beyond commonsense boundaries of organisms so that he can play down the importance of organisms as discrete bodies in selection processes. My reason for arguing that organisms are not the unproblematic individuals that our relative size, duration, and perceptual acuity would lead us to suppose is to persuade the reader that a more general notion is needed—interactor. Strongly entrenched verbal habits are likely to preclude biologists from treating clones as organisms. No such verbal habits stand in the way of conceptualizing such things as clones as interactors.

Just as genes are not the only replicators, organisms are not the only interactors. Just as variable chunks of the genetic material function as replicators, entities at different levels of the organizational hierarchy can function as interactors. It does not take much to show that entities from genes to organisms can function as interactors. More inclusive entities are more problematic. As long as one retains the vertebrate bias that informs common sense, the task is hopeless. For example, both Kitcher (1984: 617) and Williams (1985: 8) note a difference between organisms and species which they take to be significant. If one were to remove an organ from a highly organized organism or reorganize its parts, it would be destroyed. However, similar alterations of a species are liable to affect it very little. Most species could lose 10,000 or even 100,000 of their constituent organisms without serious disruption of their population structure. It is certainly true that most organisms are more highly organized than most species, but these degrees of organization gradate into each other. As Mayr (1987: 159) notes, in "lower invertebrates and in many kinds of plants . . . , a seriously mutilated individual can be restored as quickly as a decimated species."

I also have a second reason for wanting to "liberate" our understanding of the organizational hierarchy from our vertebrate biases. In the next chapter, I propose to argue that conceptual change in science can profitably be viewed as a selection process. One common objection to this undertaking is that in biological evolution the relevant units are discrete and unproblematic—genes are genes, organisms are organisms, species are species, and that is that—while in conceptual evolution, the units are extremely amorphous and variable. One purpose of the preceding discussion is to show that certain commonsense assumptions about genes and organisms are simply mistaken. There certainly are differences between the two, but they are of degree, not kind. In the next section, I propose to extend this line of reasoning to such "groups" as colonies, schools of fish, populations, and species. The hierarchical boundary between organisms and groups of organisms is no sharper than that between genes and organisms, in fact much less so.

Liberating the Species

From the earliest days of evolutionary biology, evolutionists referred quite casually to various adaptations being for the good of the species, as if species themselves can be selected. In 1964 Hamilton took an important step in resolving the problems surrounding group selection by distinguishing between kin selection and kin-group selection. Kin selection is just a matter of increasing one's inclusive fitness and, hence, counts as individual selection. Only if organisms form groups which are themselves selected is one justified in postulating group selection. But it was Williams (1966) who exposed the notion of group selection to its closest examination. The chief problem with group selection is that if it occurs at all, it proceeds so slowly when compared to selection at lower levels of organization that it is unlikely to have much effect. Thus, Lewontin (1970) concluded that selection at the level of populations or even entire species is possible but probably not very effective.

In the interim, Wade (1978) has countered that one reason why group selection seems such an inefficient process is that several of the simplifying assumptions made in most mathematical models do more than just make calculations more tractable; they also make selection at higher levels look less likely. More than this, Wade was presented experimental evidence that selection at the level of groups of organisms does in point of fact occur, while D. Wilson (1980, 1983) has enumerated situations in which selection in what he terms "trait groups" is likely to be efficacious. In particular, Wilson (1983: 173) concludes that "abundant evidence for the existence of female-biased sex ratios has accumulated, especially among small arthropods in subdivided habitats . . . a category that includes more species than all the vertebrates combined." Just because the conditions necessary for interdemic selection may not be very common among vertebrates, it cannot be dismissed out of hand.

Stanley (1975, 1979) has argued that selection can occur at even higher levels of organization, including entire species. According to Stanley, the prevalence of sexual reproduction can be explained in terms of the rapidity with which sexual species speciate. In such a process, species themselves are being selected. More recently, Vrba and Eldredge (1984) have emphasized that species are individuals. As such, they can have properties of their own and might be selected in virtue of possession of these "emergent" properties. For example, if speciation is primarily a matter of the formation of peripheral isolates, then those species with longer, more convoluted peripheries are liable to speciate much more frequently than those species that lack these characteristics. However, both Eldredge and Vrba are forced to conclude that even though species have such emergent properties, they are rarely selected for them. In order for group selection of this sort to occur, such emergent properties must be heritable and this seems implausible (but see Jablonski 1986).

However, if the disagreements about the prevalence of group selection are to be untangled, we have to distinguish once again between replication and interaction. If such things as colonies, demes, populations, and species are to function either as replicators or as interactors, then they must be construed as individuals. Being an individual is necessary for any entity to function in either process. In neither case is it sufficient. Many individuals play no roles whatsoever in selection processes (Hull 1980, Sober 1984a, Vrba 1984). If by "group selection" one means that such things as demes and possibly entire species can function as *replicators,* then these groups must be divisible into "generations" and exhibit structure that is passed on largely intact from generation to generation. More strongly, this structure must count as "information." If groups are to function as *interactors,* then they must form cohesive wholes that interact as such with their environments so that replication is differential. What is more, these group-level characteristics must count as "adaptations," not just effects. Differences between such groups must be heritable.

In the context of the usual organizational hierarchy, replication is concentrated at the lowest levels, primarily at the level of the genetic material, while interaction occurs at a much greater variety of levels, from single genes to gametes and organisms up to colonies and other kin-groups. Whether or not populations and entire species can function as interactors is a good deal more problematic. Thus, one would expect replication at these higher levels to be even more problematic. However, Williams (1985: 7), for one, argues that entire gene pools have what it takes to function as replicators at least in the bookkeeping sense. "The bookkeeping consists of the proliferation of some sets of gene-frequency values and the disappearance of others. A gene pool, no less than a gene, can potentially persist forever and is therefore a suitable medium for writing a record." Thus, for Williams, replication occurs at the level of single evolutionary genes, as well as entire gene pools, but at no intermediary level.

For such species selectionists as Vrba (1984), species purportedly function as interactors. Replicators *code* for adaptations, while interactors are the entities which actually *exhibit* them. For example, DNA is peculiarly well-adapted for replication. The way that it can "unzip" and fill in the missing nucleotides is a marvel of engineering. In this respect, DNA molecules are functioning as interactors. They are adapted for replication. DNA molecules also code for the production of an incredible array of other molecules which, in consort, produce and constitute more inclusive entities, in particular organisms. These more inclusive entities exhibit the sorts of adaptations which have so fascinated generations of biologists—the operculum of snails, the nictitating membrane in sharks, and the stomata in plants. These characteristics have arisen during the course of evolution to enable organisms to cope with their environments. DNA *codes* for these adaptations, but it does not *possess* them or *exhibit* them. If more inclusive entities are to perform these same functions, then these same distinctions must apply to them as well. They must possess group-level properties that function as adaptations.

According to Vrba (1984: 324), species can exhibit a variety of species-level properties, including population size, spatial and genetic separation between populations, and the nature of the species' periphery. However, she goes even further. Species selection in its strongest sense occurs when these species-level properties are "emergent":

> The mere presence of emergent species characters is not sufficient to make species units of selection. The characters also need to be heritable, to be variable within a monophyletic group, and to interact with the environment to cause differential speciation and extinction.

According to Vrba (1984: 319), a property is emergent if it is more than the "summed properties of entities at lower levels." The distinction is the same as that drawn two decades ago by Williams (1966: 108) between a "population of adapted insects and an adapted population of insects."

More recently, Damuth (1985) has presented yet a third alternative—higher-level entities that function as both replicators and interactors, entities he terms "avatars." According to Damuth, avatars are made up of the organisms belonging to a particular species in a particular community. A community in turn is composed of all the various avatars that make it up. As a result, each avatar is a geographically localized entity, and the only relations that are relevant to its success or failure are those it has with the other avatars in its community as well as with local environmental conditions. The only organisms that can belong to a particular avatar belong to the same species, but once that requirement is met, "avatar" is a purely ecological concept. The entities that evolve as a result of selection among avatars are ecological communities.

The important feature of avatars and communities is that they are ecologically and geographically localized. Only in such situations can natural selection take place. Thus, in a particular community, a particular avatar might go extinct and the community respond accordingly. Later, via immigration, another avatar of the same species might become established. Because of the genetic heterogeneity of most species, this new avatar is likely to be genetically different from the avatar it replaced. It in turn could send out emigrants to other communities. According to Damuth (1985: 1138), the causal processes of change involve "only localized ecological entities" and changes in genealogical entities such as clades are only an "indirect reflection" of the changes occurring in these localized entities.

One strength of Damuth's analysis is that avatars are spatiotemporally localized entities in a much narrower sense than are species. Species have ranges, but these ranges need not be continuous. For protracted periods of time, populations belonging to the same species frequently remain geographically isolated from each other. Species are also said to occupy their own peculiar niches, but to the extent that this claim is not definitional, it is false. Many species range over extremely varied environments, often occupying significantly different niches in each. Any broadening of the definition of "niche" so that all populations of a cosmopolitan species occupy the "same" niche results in an all but vacuous sense of "niche."

Vrba (1984) treats species selection in terms of the variation of emergent species-level properties within a monophyletic group. Some species in a clade exhibit this property; some do not. Damuth departs from Vrba in two important respects. According to Damuth, the higher-level variation that matters in biological evolution is the variation of avatar-level properties within a community, not a clade, and these higher level properties need not be emergent in Vrba's sense. In my terminology, Damuth is arguing that avatars, like genes, can function as both replicators and interactors. They function as replicators because they exhibit structure and can pass on this structure through successive "generations" of avatars. They can also function as interactors because sometimes at least the avatars that proliferate do so because of the relations they have to other components of their communities.

Of the numerous issues that divide the preceding scientists, some are sufficiently fundamental to count as philosophical. Chief among these is "reduction." Reductionists and antireductionists alike acknowledge the importance of what Wimsatt (1980: 230) has called the "bookkeeping" aspect of selection, i.e., changes of gene frequencies. They differ with respect to the adequacy of this bookkeeping aspect of selection for explaining evolution and the feasibility of including additional aspects of selection in the formulation of any general theory of biological evolution. It is possible to interpret some of the things which Williams (1966) and Dawkins (1976) say in their early publications as claiming that replication alone is sufficient for selection. Because they also argue that genes are the primary replicators, both men have gotten the reputation of being "gene selectionists," as if they thought that reference to changes in gene frequencies is sufficient to characterize the evolutionary process. However, as the relevant distinctions have emerged more forcibly, both men have made it clear that they acknowledge the existence of interaction as well as replication.

This much being said, they go on to argue that evolutionary theory can be and in the last analysis must be couched entirely in terms of replication. Any causal processes that do not eventuate in changes in replicator frequencies (usually gene frequencies) simply do not matter. Changes in replicator frequencies *record* the effects of everything that is relevant to the evolutionary process without making direct reference to all these various sorts of events. Introducing any reference to what I term "interaction" in the basic axioms of evolutionary theory would complicate it prohibitively. Opponents of gene selectionism, such as Sober and Lewontin (1982), admit that reference to changes of gene frequencies is adequate for this "bookkeeping" aspect of selection but insist that any theory of evolution that is limited to this single aspect of the evolutionary process is inadequate (but see Lloyd 1988).

Gene selectionism is an extreme form of reductionism. A traditional reductionist insists that a species is nothing more than all the organisms that make it up and the relations among them. An organism, in turn, is nothing more than all the cells, tissues, and so on that make it up and the relations among them. According to gene selectionists, species must be treated as if they were nothing more than gene pools and organisms as if they were nothing more than genomes. Reference to all the other parts is omitted. Williams and Dawkins are well aware that organisms exist and function in the evolutionary process. Their preference for gene selectionism turns on their guess as to which research strategy is likely to prove most successful. Just as Mendelian geneticists formulated theories of hereditary transmission leaving out all reference to embryological development, gene selectionists propose to formulate their theories of evolutionary change without any reference to the causal relations I refer to by the rubric "interaction." This strategy has proven very successful in transmission genetics. It has proven just as successful in evolutionary biology. Pious disclaimers notwithstanding, most work in population genetics since its inception has been implicitly gene selectionist.

Critics of both fields think that now is the time for biologists to become more ambitious. A more general theory of evolution is needed that incorporates the aspects of the evolutionary process that earlier workers have omitted. As successful as traditional transmission genetics and population genetics have been in the past, their rates of improvement have slowed in recent years. Now is the time for major revisions, not continued tinkering. Williams (1985: 2) suspects that the biologists attempting to present more inclusive versions of evolutionary theory are going to be disappointed. Their criticisms of his own less ambitious version of evolutionary theory are based on "unrealistic expectations."

The issue of reduction arises again in connection with Vrba's notion of emergent properties. Although the dispute between reductionists and emergentists is perennial, the distinction between those characters that are emergent and those that are not can be made in a fairly straightforward way. Some higher-level properties are simple functions of their constituent parts. For example, the mass of an organism is literally nothing but a sum of the masses of its constituent parts. The relations of these parts to each other are irrelevant to its mass. In other cases, the organization of the parts plays an important role. The higher-level property may be nothing but a function of the parts and their interrelations, but the function is not very simple. For example, water is an extremely simple molecule. Yet physicists are still struggling to derive the gross properties of water from quantum mechanics. Such a derivation must be possible in principle. After all, none of the gross properties of water are in any sense immaterial or supernatural. They involve nothing so fancy as new ontological levels of the sort that are frequently invoked for the emergence of life and mind. Yet, in practice, no one has been able to accomplish this derivation.

Vrba (1984) argues for species selection in the strongest sense possible. Not only must a species possess emergent character, but also these characters must be heritable and cause differential speciation and extinction. Eldredge (1985: 160) notes the irony of the fact that those biologists who have been most concerned to argue that species are individuals also go on to deny them any causal role in selection. Because species are individuals, they could function in the evolutionary process, but under most circumstances they do not. According to Eldredge (1985: 160), species result from selection processes but do not function in them:

> Species, then, do exist. They are real. They have beginnings, histories, and endings. They are not merely morphological abstractions, classes, or at best classlike entities. Species are profoundly real in a genealogical sense, arising as they do as a straightforward effect of sexual reproduction. Yet they play no direct, special role in the economy of nature.

Ghiselin (1987: 141) agrees:

> It would seem that species do very few things, and most of these are not particularly relevant to ecology. They speciate, they evolve, they provide component organisms with genetical resources, and they become extinct. They compete, but probably competition between organisms of the same and different species is more important than competition between one species and another species. Otherwise, they do very little. Above the level of the species, genera and higher taxa never do anything. Clusters of related clones in this respect are the same as genera. They don't do anything either.

Thus, Eldredge and Ghiselin conclude that, on Vrba's criteria, species rarely if ever function as interactors. The problem with emergent properties is not their existence, nor their in-principle reducibility, but as Cracraft (1985) has pointed out, their heritability. In the absence of heritability, any property that a species might have cannot count as an adaptation. In the meantime, Jablonski (1986) has presented summary data from the fossil record indicating that geographic range is a species-level trait that affects extinction rates and that it is heritable. If differences in the emergent properties of groups turn out to be heritable, as Jablonski claims, then possibly group selection in the strongest sense does occur.

A second irony in the recent literature in evolutionary biology is that the expectations of gene selectionists and their more holistic critics are reversed when it comes to adaptive scenarios. The most vocal defenders of the essential role of organisms in the evolutionary process are also among those who are most skeptical about evolutionary "Just-So Stories" that purport to provide evolutionary explanations of organismic adaptations (e.g., Gould and Lewontin 1979, Gould 1980), while several of the strongest advocates of gene selectionism see nothing wrong with evolutionary scenarios (Dawkins 1982b, Ridley 1983, Williams 1985). Critics of adaptive scenarios warn that many of the characteristics that evolutionary biologists claim are adaptations might well be nothing but effects. In addition, the ease with which adaptationist scenarios can be constructed to explain particular adaptations casts considerable doubt on the entire program. Hence, in general, organisms and their adaptations are so central to the evolutionary process that reference to them cannot be omitted from any adequate theory of evolution, but any reference to particular organisms and their adaptations is scientifically unjustified.

Williams (1985: 15) disagrees, arguing that "attempts to understand organisms as problem-solving complexes with the aid of a reductionist view of the origin and maintenance of the problem-solving machinery have been richly rewarding, despite the pessimism expressed by Gould and a few others." Although Williams recommends omitting any explicit reference to organisms and their adaptations in the axioms of evolutionary theory, he sees nothing wrong with explaining organismic adaptations in terms of adaptive scenarios. Certainly any sort of scientific mode of explanation can be abused. Some may be more liable to abuse than others. Even so, Williams argues, there is nothing intrinsically "unscientific" about so-called Just-So Stories (see also Ridley 1983). At the risk of putting too fine a point on the dispute, Williams thinks that his critics are too optimistic about the potentialities of a general theory of evolution and too pessimistic about its applications to particular cases.

I have sketched the preceding views about the levels at which replication, interaction, and selection occur in biological evolution primarily to show that biological evolution is not nearly as neat and clean as critics of all attempts to view conceptual evolution as a selection process suppose. Yes, conceptual evolution can occur at a variety of levels, and, no, the levels are not sharply distinguishable. But by now it should be clear that exactly the same state of affairs exists in biological evolution. If atomistic genes, monolithic organisms, and discrete species are necessary for selection to function in biological evolution, then we are in real trouble. However, I cannot leave the preceding topics without indicating my own preferences on these issues.

I agree with Vrba that species selection of the most significant sort occurs when species function as interactors, i.e., when species proliferate or go extinct because of species-level adaptations. However, like Eldredge, Ghiselin, and Cracraft, I have my doubts as to how often species function in this way, though Jablonski's data make this alternative look viable once again. I also think that other sorts of "group selection" probably have a greater influence on the course of evolution; for a good taxonomy of various sorts of group selection, see Sober (1984a), Mayo and Gilinsky (1987), and Damuth and Heisler (1988).

As congenial as I find Damuth's analysis of selection in terms of avatars, I am forced to depart from it in one respect. I agree with Damuth that from the point of view of ongoing processes, the past is past. All that matters in particular selection processes is the conditions current at the time. The past "constrains" the future only to the extent that it has structured the present. "Genealogy" in this sense is not relevant to selection processes. Selection in the limiting case can take place within a single generation, but for evolution ancestor-descendant sequences of entities are necessary. Without such sequences, there can be no replication and no differential perpetuation. Hence, in this sense, "genealogy" is central to selection processes. As I interpret avatars, they have both a genealogical and an ecological component. They are parts of species (genealogical entities) and communities (ecological entities). As such, they embody the sort of intersection between the genealogical and ecological hierarchies that Eldredge (1985) recommends.

With respect to Williams's doubts about how ambitious evolutionary biologists can become without introducing intractable complexities into evolutionary theory, I find myself in partial agreement with both sides of the dispute. The inclusion of details about particular interactions between interactors and their environments in the axioms of evolutionary theory would be lethal. Although the basic constraints imposed on organisms by the physical makeup of bone, muscle, chitin, and the like are as constant as those that govern steel and concrete, the possible combinations of these materials within these constraints is incredibly diverse and highly variable. The various functions that organisms must perform in order to survive to reproduce are not all that numerous, but the structures that they have evolved to perform these functions are staggering in their variety.[2] Actual examples of adaptations in nature are infinitely more bizarre than any that one is likely to meet in science-fiction. Darwin (1877), for example, devoted an entire volume to the various contrivances by which orchids are fertilized by insects. If such vagaries must be included in the basic axioms of evolutionary theory, the results are sure to be hopelessly complex. I think that any adequate theory of evolution must include reference to the interactor/environment interface (Plotkin and Odling-Smee 1984), but the inclusion of such reference need not complicate evolutionary theory any more than including reference to replication does.

It is certainly true that particular interactions are extremely varied and contingent, as varied and contingent as the myriad causal situations that give rise to the incredible array of adaptations that makes the study of biology so endlessly fascinating, but the information contained in the genetic makeup of organisms for these adaptations is just as multifarious. The introduction of either sort of complexity into the general characterization of the evolutionary process would be lethal, but no such introduction is necessary. In this connection, all that must actually be included in formal statements of evolutionary theory are the general characteristics of replicators and interactors, not the specification of the information contained in a particular gene or the causal scenario that gave rise to a particular adaptation. Only when this general theory is applied to particular cases must such details be introduced. Scientific theories are general. Their applications are contingent and highly particularized. Of course, the axioms of evolutionary theory must include reference to more than the general characterizations of replicators, interactors, and lineages, but it need not and cannot include reference to the characteristics of particular replicators, interactors, and lineages.

With respect to applications, I see no reason to shy away from claiming that a particular characteristic arose as an adaptation to a putative environment even though frequently such claims may actually be false. Adaptive scenarios are subject to error, but as Boyd and Richerson (1985: 282) point out, so are their alternatives. Past history is difficult to reconstruct, but I see nothing intrinsically "unscientific" about historical reconstructions. Physicists have reconstructed the first three minutes of the universe nanosecond by nanosecond. After the first one-hundredth of a second, the temperature of the universe was 100,000 million degrees Kelvin and contained, among other particles, protons and neutrons in a 50–50 ratio. After 0.11 seconds, the universe cooled down to a mere 30,000 million degrees Kelvin and the ratio of protons to neutrons

2. One reason that functions appear to be fewer in number than structures is that we tend to individuate functions in a much cruder fashion than structures. If we were to pay comparable attention to both, they might well turn out to be equally numerous.

shifted to 62–38, and so on (Weinberg 1977). If physicists can infer such minute occurrences during the first three minutes of the formation of the universe, perhaps biologists can be forgiven for attempting to reconstruct the course of evolution during the past few billion years. Nelson and the cladists are certainly right about the decreasing certainty of cladograms, trees, and scenarios. Gould is willing to draw the line at trees. Adaptive scenarios are too likely to be erroneous, but the ancestor-descendant relations represented in phylogenetic trees are acceptable. Cladists such as Nelson are even more skeptical than Gould. Where one draws the line depends on how certain one thinks that scientific conclusions must be.

Adaptive scenarios are one sort of historical reconstruction, perhaps an especially difficult sort, but in most cases nothing much rides on the correctness of a particular adaptative scenario. These scenarios cannot be incompatible with our general understanding of evolution. For example, recently the claim has been made that the mass extinction which finally finished off the dinosaurs at the boundary between the cretaceous and tertiary periods was caused by a large meteor hitting the earth. Paleontologists and evolutionary biologists have been slow to accept such a hypothesis because it throws off their reconstructions based on less dramatic terrestrial causes, but it is in no way incompatible with evolutionary theory. Raup and Sepkoski's (1986) claim that mass extinctions occur in 26-million-year cycles is another matter. If this periodicity is to flow from evolutionary theory itself, there are problems. Nothing about any present-day theory of evolution implies periodicity of any sort. If it occurs, then the theory needs improving. As it turns out, all the explanations that have been suggested to explain these 26-million-year cycles are physical—the earth oscillating through the galactic plane, a companion star to the sun with an extremely elliptical orbit, and an unknown planet lying beyond the orbit of Pluto (Raup 1986). None of these suggested explanations is incompatible with any present-day version of evolutionary theory. We would like to have extensive, accurate phylogenetic reconstructions, but they are not absolutely necessary for evolutionary theory.

Showing that processes other than replication and interaction are actually responsible for the biological part of biological evolution would be of primary importance. Specifying the various levels at which selection can occur and under what general conditions is also important. But detailing the difficulties in applying evolutionary theory in particular cases is much less important. To many scientists reference to unseen companion stars, solar planets, and Oort clouds seems a bit "unscientific," no better than the Just-So Stories of the adaptationists, but both are in the best traditions of science. All scientific theories are difficult to apply. Inference to particular cases must be possible if evolutionary theory in general and adaptationist scenarios in particular are to be testable, but such testing need not be easy or automatic. Critics are not content with a theory being falsifiable. They want it to be easily falsifiable, when no scientific theory is easily falsifiable. Critics of adaptationist scenarios are often right with respect to particular instances but wrong in their blanket condemnation of scenarios, adaptive or otherwise (see chapter 13).

One of the main theses of this book is that both process and pattern are fundamental to science. Constructing phylogenies, inferring adaptive scenarios, and working out the basic structure of the evolutionary process are all necessary components of evolutionary biology. I must admit, however, that phylogeny reconstruction and adaptationist scenarios are so fascinating that they often seduce biologists into ignoring other questions, in particular questions about the evolutionary process. Darwin himself avoided, whenever possible, committing himself to particular phylogenies, let alone evolutionary scenarios. He was interested primarily in the evolutionary process. But most of his contemporaries were more interested in history—which species gave rise to which—to the detriment of continued improvement of Darwin's theory.

There is also the temptation to think that adaptationist scenarios have greater warrant than they actually have. For most species, such misplaced confidence is likely to do little harm. The recent upsurge of criticisms of evolutionary scenarios has had two, largely independent sources. One is a general distaste for process theories. Real science is nothing but pattern recognition, whether phenetic clusters or nested transformation series. The other is a fear of the damage that particular scenarios concerning the human species might have on organisms belonging to that species. What if human males really are inherently more aggressive and dominant than human females? What if groups of human beings really are territorial?

The axioms of any scientific theory are extremely general—Newton's law of universal gravitation refers to highly abstract constructs—mass points. Such constructs as replicators and interactors are just as abstract. Evolutionary biologists must also produce second-level generalizations that are at least compatible with the basic axioms of the theory, preferably derivable from it. For example, Lewontin (1961: 397) hypothesized that in diploid sexually-reproducing organisms, homozygotes are more specialized in their adaptive properties than heterozygotes; Dobzhansky (1970: 167) argued that genetic variability is maintained in populations as an adaptation to environmental heterogeneity; Shields (1982) suggested that philopatry should occur in, and only in, low-fecundity organisms, while outbreeding through vagrant dispersal should characterize high-fecundity organisms; and Schopf (1973) argued that specialists are more likely to evolve in stable, predictable environments, while generalists—and the monomorphic colonies they form—do best in rapidly changing environments.

The point of the preceding examples is not that these regularities actually exist, or that they will all eventually find their way into a more inclusive theory of biological evolution, but that they refer to *kinds* of entities, not particular *instances*. The natural kinds referred to in second-level evolutionary laws are such things as diploid, sexually reproducing organisms, low- versus high-fecundity organisms, specialists versus generalists, and stable, predictable environments versus those that are changing rapidly. The waggle dance in bees may serve the function of communication in bees, but do not expect "waggle dancing" to turn up in any formulation of evolutionary theory—communication perhaps, but not waggle dancing (Cavalli-Sforza and Feldman 1981: 5). Comparable observations hold for theories of conceptual change.

Summary

In this chapter, I have argued that concepts as general as *replicator, interactor,* and *lineage* are necessary if evolutionary theory is to be sufficiently general to account for evolutionary change of the most narrow, biological sort. Replication as a causal process is confined primarily to the genetic material, albeit to chunks of this material that vary extensively in their size. Organisms can function as replicators in only very special circumstances. If gene pools are distinguished from populations and species, then there is a sense in which gene pools can fulfill the bookkeeping aspect of biological evolution, but in this role they really are nothing but the summations of the information contained in their constituent genomes. Populations and species are quite another matter. They do not have the characteristics necessary to function as replicators sufficiently well developed so that they can fulfill this function very often or very well. Avatars are more likely candidates for higher-level replicators. Interaction occurs at a much wider range of levels of organization, from genes, cells, organs, organisms, and colonies to avatars and possibly even entire species. Because selection as a causal process results from the interplay of replication and interaction, any adequate characterization of selection must include reference to both processes. Because both replicators and interactors exist at a variety of levels of organization, these interactions are likely to be extremely complex.

Replicators form lineages, and when the conditions for selection are present, lineages evolve as a result of the interplay between replication and interaction. Just as replication and interaction occur at a variety of levels of the traditional organizational hierarchy, the level at which evolution is occurring also wanders from level to level. In strictly asexual organisms, the most inclusive level at which splitting followed by little or no merger occurs is the single organism. The only lineages that they form are organism lineages. With the advent of gene exchange, higher-level lineages came into existence. Replicators both split and merge. In some situations, the largest genealogical unit that exists even among sexual organisms may well be local populations. In such cases, lineages are composed of sequences of such populations. In other situations, gene exchange may be more extensive, encompassing traditional biological species. Gradually evolving species are the traditional lineages of the synthetic theory. To the extent that speciation is punctuational, particular species themselves do not count as lineages. Only sequences of species form species lineages. Finally, in many plants, gene exchange continues at even higher genealogical levels, resulting in lineages of even higher levels (Mishler and Donoghue 1982). Although avatars are primarily ecological units, successions of avatars, if they are related by descent, may also form lineages.

The fact that all three processes—replication, interaction, and evolution—occur at a variety of levels in the traditional organizational hierarchy is one very good reason to abandon this hierarchy for the purposes of capturing evolutionary regularities in nature. In the evolutionary process, what role do all and only Mendelian genes play? How about operons, organisms, colonies, populations, and species? The answer to each of these questions is—none. In a sufficiently general theory of biological evolution, like must be compared to like—replicators to replicators, interactors to interactors, and lineages to lineages, regardless of the traditional level. Thus, in some circumstances, a single cistron must be compared to an entire genome—both are the largest unit of replication. In certain circumstances, a single cell must be compared to an entire multicellular organism—both are the most inclusive entities functioning as interactors.

Similarly, genealogical lineages of organisms, populations, species, and possibly even more inclusive entities may turn out to be comparable.

What difference does this change in perspective make? Let me give two examples. One of the most recalcitrant problems in evolutionary biology today is the cost of meiosis. The two fundamental premises that combine to produce this problem is that meiosis exacts a 50-percent cost and that meiosis is extremely common, nearly universal. Numerous solutions have been suggested to this problem, but no one solution or combination of solutions has gained wide acceptance. The premise that slips in with little comment is that meiosis is quite common. However, estimates of how common meiosis is depend on what gets counted. The *only* way that meiosis turns out to be common is if sexual species are compared to asexual "species," when asexual organisms *do not form species*. Certainly taxonomists need to classify both sexual and asexual organisms. Hence, taxospecies must be provided for both, but nothing is accomplished by terming both "species" with respect to the evolutionary process except a misleading appearance of univocality.

Sexual reproduction produces higher-level lineages but not many of them, when compared to the organism lineages of asexual organisms. Sexual reproduction in the simplest case of the merging of two formerly distinct cells may have evolved quite early (Bernstein et al. 1985), but from all appearances, it took a much longer time for meiosis to evolve, much longer than for life to evolve in the first place. When lineages are compared to lineages, meiosis is still quite rare, as rare as it should be if it exacts such a high cost. In fact, it turns out to be rare on every measure suggested by evolutionary biologists—number of organisms, biomass, amount of energy transduced, and so on. The only way that it turns out to be common is if sexual and asexual "species" are compared.[3] In very special cases, the cost of meiosis can be outweighed by the various benefits suggested by biologists. In others, such as in low-fecundity vertebrates, it may be a maladaptation for which these organisms lack the preadaptations necessary to

3. From past experience I have learned that vertebrate biases run deeply. As a result, a few words need to be said about the comparative rarity of meiosis. Cosmologists tell us that the earth as a separate planet originated about 4.6 billion years ago. It took only slightly more than a billion years for life to evolve. As far as we can tell, the earliest forms of life were most similar to present-day prokaryotes—bacteria and blue-green algae. According to best estimates, it took much longer for sexuality and multicellularity to evolve—in fact, another three billion years (Schopf et al. 1973). If the time it takes to evolve is any measure of how difficult it is for a particular innovation to develop, sexuality and multicellularity posed greater difficulties than the evolution of life in the first place. In any case, for the first three-quarters of life on Earth, the primary, possibly sole form of reproduction was asexual. Even today many organisms still reproduce asexually.

But, one might object, some signs of parasexual reproduction have been discovered in certain groups of extant blue-green algae. According to one estimate, one cell in 240,000 replications gives any indication of any sort of gene exchange, while according to another, the frequency is only one in 20 million (Carr and Whitton, 1973; Fogg et al. 1973). In the first place, such forms of parasexual reproduction do not involve meiosis, and the issue is the cost of meiosis. In the second place, if gene exchange that is this rare is sufficient to claim that a group reproduces sexually, then the even more frequent occurrence of monozygotic twins, triplets, etc. would make Vertebrata asexual. No less an evolutionary biologist than John Maynard Smith (1983) warns of the dangers of formulating evolutionary theory primarily with higher eukaryotes in mind. Interestingly enough, in his own application of game theory to biological evolution, he takes asexual reproduction as the standard case and claims that it can be generalized without any difficulty to sexual reproduction (Maynard Smith 1982). Austad (1984) is suspicious.

eliminate it (Williams 1975: 103). Such piecemeal explanations look suspicious for widely distributed phenomena but not for occasional rare cases (see Shields 1982 for a more traditional explanation).

Evolutionary biologists can make numerous sorts of predictions, but one sort that they cannot make is the course of the development of particular species through time (Williams 1985). Numerous authorities have taken this to be a serious shortcoming of evolutionary theory (e.g. Smart 1963, 1968, Popper 1974), but if species are lineages, and lineages are historical entities, no one should expect to make predictions about a particular species qua that particular species. Species are particulars and as particulars they must be assimilated to an appropriate reference class before predictions can be made about them. If generalists evolve in a rapidly changing environment and *Homo sapiens* is a generalist, then the environment in which we developed had better be characterizable as having been rapidly changing. To be sure, evolutionary theory is "about" lungs as evolutionary homologies and Primata as a monophyletic taxon but only in the sense that Newton's laws of celestial mechanics are "about" Mars. All three are instances mentioned in applications of these theories, not natural kinds that function in them.

One important characteristic of individuals is that they can change the kinds to which they belong. For example, in a species of polychaete of the genus *Ophryotrocha,* each individual starts out male and becomes permanently female at maturation unless it meets another female. Then one becomes male, and the two mate. After mating, they reverse their sexes and mate again. Or an organism might be free-living when it is young, parasitic when sexually mature. A species might begin monotypic and genetically homogeneous and later become polytypic and genetically quite heterogeneous. Because the ontogenetic development of organisms is to some extent determined by their genetic makeup, it is to some extent cyclical and regular—not as cyclical and regular as most people think, but nonetheless cyclical and regular to an extent not exhibited by species and lineages in general. The most that can be said about lineages at all levels is that their development is locally constrained. Thus, on the theory that I am proposing, our inability to predict the future development of a particular lineage through time is not a curiosity but a necessary consequence of this theory.

Williams (1985: 2) remarks that one day historians may well marvel that it took so long for us to realize that "life history attributes are subject to natural selection, and evolve no less than teeth and chromosomes, that the prevalence of meiosis presents a major theoretical challenge, or that many of the formulations of natural selection are applicable to cultural evolution." One reason for our tardiness in treating sociocultural evolution as a selection process is that most of us know a great deal about the vagaries of sociocultural transmission and have an overly simple view of biological transmission. If biological evolution were the neat process of genes mutating, organisms being selected, and species evolving, then sociocultural change is nothing so simple. One purpose of this chapter has been to show that biological evolution is not so simple either. In this chapter I have shown how general the characterization of selection processes must be if they are to apply to biological evolution. In the next chapter, I extend this analysis to include one special sort of sociocultural change—conceptual change in science. Just as lineages are transformed by means of the interplay of replication and interaction in biological evolution, scientific theories and more inclusive research programs change in this same way. The specific mechanisms involved in biological and conceptual evolution are quite different. Conceptual change does not depend on DNA, competition for mates, and what have you. Instead, it depends on the interplay between curiosity, giving and receiving credit for contributions, and the mutual checking of results.

12

Science as a Selection Process

To express the matter more vividly, in the world of our model organisms, whose behaviour is determined strictly by genotype, we expect to find that no one is prepared to sacrifice his life for any single person but that everyone will sacrifice it when he can thereby save more than two brothers, or four half-brothers, or eight cousins . . . .

W. D. Hamilton, "The Genetical Evolution of Social Behavior"

The explanatory power of evolutionary theory rests largely on three assumptions: that mutation is nonadaptive, that acquired characters are not inherited, and that inheritance is Mendelian—that is, it is atomic, and we inherit atoms, or genes, equally from our two parents, and from no one else. In the cultural analogy, none of these things is true. This must severely limit the ability of a theory of cultural inheritance to say what can happen and, most importantly, what cannot happen.

John Maynard Smith, "Natural Selection of Culture?"

According to Hamilton an organism should behave in ways to increase its inclusive fitness. Because closely related organisms contain duplicates of each other's genes, varying degrees of cooperation should be expected among genetic relatives. Some organisms produce numerous offspring, as many as possible. Their investment in the future is solely in terms of numbers. Others have many fewer offspring than they might and make up the difference by caring for their young. This latter sort of behavior is phenotypically "altruistic" but genetically "selfish." Similar observations hold for siblings, nieces, nephews, cousins, and so on. Numerous problems confront this elegantly simple perspective, not the least of which is sex. In most cases, organisms mate with individuals that are not very closely related to them genetically. From the selfish-gene point of view, sexual reproduction involving meiosis must have been very difficult to evolve, and if current estimates are correct, it was. It took three times as long for the mechanisms necessary for sexual reproduction to evolve as it took for the first living creatures to evolve here on Earth in the first place. The step from nonlife to life was apparently much easier than the step from asexual to sexual organisms.

Although inclusive fitness can explain only a small percentage of cooperative behavior among organisms, it has played such a large role in the recent excitement in evolutionary biology that Dawkins (1978: 61) declares Hamilton to be the "founding genius" of sociobiology, especially in connection with the centrality of the "gene's eye" view of evolution. But how about all the apparently cooperative behavior among organisms that are not especially closely related, both within and between species? Trivers (1971) has suggested a second mechanism to account for cooperation among organisms other than close kin—what he terms "reciprocal altruism." The idea that one human being may aid another in the expectation that the person receiving the aid might well reciprocate in the future is commonplace. Trivers's treatment of reciprocal altruism is unusual in two respects: it applies to even the most lowly creatures, entities that can in no sense anticipate the future, and it assumes a genetic basis for the behavior. In fact, the only reason organisms that are incapable of conscious deliberation can engage in acts of reciprocal altruism is that the mechanisms involved *are* genetic. In some organisms learning plays a role, but in all organisms genetic predispositions are central. Trivers's (1971) main message is that the net effect of reciprocal altruism has to be genetically "selfish."

Phenotypic cooperation among relatives can be enhanced by two mechanisms—kin selection and reciprocal altruism. All other instances of cooperation depend solely on reciprocal altruism. Cheating in the context of reciprocal altruism is quite easy, much easier than in cases of kin selection. In short, the versions of evolutionary theory promoted by Hamilton, Trivers, Dawkins, and others treat all phenotypic characters, including behavior, as eventuating in an increased likelihood that an organism will pass on replicates of its own genes directly via offspring or indirectly via their duplicates in close kin. Each organism exists at the center of its own little eddy of inclusive fitness in a very shallow sea of reciprocity.[1]

In most species an important sort of organization exists between individual organisms and entire species—demes or local populations. Ideally a deme consists of organisms in sufficient proximity to each other that they all have equal probability of mating with each other and producing offspring, provided they are sexually mature, of the opposite sex, and equivalent with respect to sexual selection. To the extent that these conditions are met, the organisms belonging to a deme share in the same gene pool. Of course, in natural populations, some mating occurs between adjacent demes, and not all organisms within a single deme have precisely equal probability of mating, but the isolation between demes is met often enough and well enough for demes to play an important role in biological evolution. In fact, they form the foundation for Wright's shifting-balance theory.

If science is anything, it is cooperative. It is also highly competitive. One would think that something like inclusive fitness and demic structure might apply to the

1. As Robert Brandon has pointed out to me, Hamilton (1964) does not limit his theory of kin selection only to kin. The important relation is the regression of the genotype of the organism that receives the altruistic behavior on the genotype of the altruist. Kinship is just one way to bring about such a regression. However, in the vast majority of cases, identity by descent is the mechanism that actually produces such an end. Either identity by descent or some equivalent mechanism is necessary for inclusive fitness.

professional relations among scientists, and the weight of this chapter is to show that it can. Scientists are organisms. They are caught up in the same struggle to pass on their genes as are other organisms. Scientists are also human beings. As such, they exist in cultures. Being a successful scientist is likely to have the same range of effects on scientists as being successful in other human activities has on other human beings. For example, in certain sorts of societies, high-status individuals delay parenthood and have fewer offspring than do low-status individuals. In such societies, if science is a high-status profession, scientists should behave accordingly.

However, the replicators in conceptual change in general and scientific change in particular are not genes. What functions as the replicators in science? The answer is not very surprising: elements of the substantive content of science—beliefs about the goals of science, proper ways to go about realizing these goals, problems and their possible solutions, modes of representation, accumulated data, and so on. Scientists in conversations, publications, and classroom lectures broach all these topics. These are the entities that get passed on in replication sequences. Included among the "vehicles" of transmission in conceptual replication are books, journals, computers, and of course human brains. As in biological evolution, each replication counts as a generation with respect to selection.

In biological evolution, variable chunks of the genetic material are the primary replicators. As such, they pass on the information they contain in their structure. They also function as interactors in the production of more inclusive interactors. Conceptual replicators cannot interact directly with that portion of the natural world to which they ostensibly refer. Instead they interact only indirectly through scientists. The ideas that these scientists hold do not produce these scientists in the way that genes produce organisms, but they do influence how they behave. Scientists are the ones who notice problems, think up possible solutions, and attempt to test them. They are the primary interactors in the conceptual development of science. Hence, in conceptual change, agents such as scientists function as "vehicles" in both Campbell's sense and Dawkins' sense. Scientists serve as both vehicles of transmission and as the chief vehicles for conceptual interaction (for differences in the term "vehicle," see Kawata 1987).

Scientists learn from their experience, but they do not confront the world of their experience as isolated individuals. If they did, science could never have developed. Scientists must use each other's ideas, pass them on, and improve upon them. In the space of a single lifetime, a scientist holds a whole series of different ideas in different combinations. The struggle is to find just the right combination to produce a comprehensive and coherent explanation of natural phenomena. Replication sequences also pass through series of other scientists who repeat the process. Although genuinely novel ideas crop up once in a while—quite rarely—most progress in science occurs by means of recombination. All scientists have intellectual interests, but these interests rarely coincide. Different scientists push different ideas as well as different combinations of ideas, keeping in mind that ideas in selection processes are individuated in terms of descent. Two tokens of the same type count as different in the competition for credit if they occur in different conceptual lineages.

If the only interests that scientists have were intellectual interests of the purist sort, the interrelations between scientists would be complicated enough, but scientists have other interests as well, and the actions that would maximize these various interests do not always coincide. In addition to purely intellectual interests, scientists also have career interests. They belong to scientific groups, societies, and academic departments. They also are members of families, political groups, social classes, races, and so on. Externalists have argued that in fact these extra-scientific interests are not all that "extra-scientific." They are part and parcel of the scientific process, influencing its outcome significantly. In its most extreme form, this position is that such things as social class determine the ideas that particular scientists hold, as if scientists were helpless victims in the maw of society.

Advocates of a significant role in science for reason, argument, and evidence in fixing the belief systems of scientists emphasize the connections between words and the world. Scientific theories mirror the world, and the scientist as an agent in this process becomes transparent. Although internalists are likely to be dismayed at the comparison, in this context they are analogous to genetic reductionists. Just as gene selectionists think that the evolutionary process can be analyzed entirely in terms of gene frequencies, internalists with respect to science act as if science can be understood entirely in terms of the substantive content of science. Both positions are reductionist in character. Gene selectionists are forced to their position by the fear that any explicit reference in evolutionary theory to interaction will introduce hopeless complexities. They justify their parsimonious view by observing that any interactions that are not reflected in replicator frequencies cannot influence future generations. Internalists with respect to science have much the same reaction to claims about all the social factors that supposedly influence the substantive content of science, but they also add that social factors of the more interesting sort are neutralized by the methods of investigation adopted by scientists. In the short run, such things as racial prejudice may influence our ideas about race, but through successive sequences of hypothesis formation, expansion, integration, and testing, such influences are weeded out.

The view I am pushing treats science as a social process, but the social processes that I have found to be most relevant are those that are more narrowly connected to science itself. For example, a scientist can in principle subject his or her own views to testing, but if science had to depend solely on individual scientists proving their own ideas wrong, it would be in real trouble. The most important testing that occurs in science is one scientist testing the views of another. Scientists have conceptual kin—those scientists who share ideas that are identical by descent. Part of the cooperation that occurs in science is similar to kin selection. For example, in the early years of what came to be known as numerical (or phenetic) taxonomy, Sokal coauthored papers with Michener, Camin, Rohlf, and Sneath. During this first dozen years or so, the professional relationships among these scientists were so close that the transmission of their contributions can profitably be viewed as exhibiting the same sort of structure as kin selection.

Rather rapidly, however, pheneticists formed a much larger group—a conceptual deme. Scientists in this "deme" used each other's ideas with a much higher frequency than they used ideas originating outside this deme; the frequency of intrademic positive citation was also much higher than interdemic citation, but the social organization was not as tight as within the original research group. Similar observations hold for terminology. For example, the spread of "OTU" can be traced from its origins in the work of Sokal and Sneath to the work of other members of their school. Although the pheneticists had their origin with Sokal and Michener at Kansas and Sneath in England, all the scientists involved were concerned with their own conceptual fitness. Each wanted credit for his or her contributions but in the context of their conceptual kin-group and larger deme.

Given the finite and relatively short duration of the most important vehicles in science—individual scientists—conceptual replication requires sequences of scientists. No matter how good a scientist's views may be in the abstract, if no one notices, they might just as well have never been produced. The feeling of discovery with respect to unappreciated contributions is no doubt just as exhilarating and, if true, just as true as those that are appreciated, but if science is viewed as a selection process, a "new truth" must find its way into a replication sequence if it is to count as a genuine contribution. As Toulmin (1972: 206) puts this point:

> For us to speak of a genuine "conceptual variant", it is not enough to find some obstinately honest individual entertaining a conceptual innovation; it takes something more than the personal reflections of open-minded individuals to create an effective pool of conceptual variants in science.

Conceptual replication is a matter of information being transmitted largely intact from physical vehicle to physical vehicle. Certain conceptual vehicles are much more ephemeral than others. The spoken word is extremely transitory, human beliefs incorporated in individual brains can last for longer periods of time, but people die. If a belief is to survive, it must be replicated. In books and journals, ideas find a much more durable medium. They can enter into a replication series and then lie fallow for generations, until someone else happens to stumble upon them to initiate a new series. Or they can be passed on unnoticed, the way that certain atavistic genes are, until they begin to function again. Hennig's notion of species as systems is a good example of an idea that was transmitted but unappreciated for years. The term "vehicle" appropriately captures this feature of replication sequences. But if science is to be a selection process, interaction must take place as well. Scientists' brains can serve as vehicles for replication sequences, but scientists themselves are anything but passive vehicles for such sequences. Without scientists, no conceptual replicator could ever be tested, and testing is essential to science. As a shorthand expression, we frequently talk about the meanings of words and sentences, but neither words nor sentences mean anything. People do. We mean something by what we say. Individual scientists are the agents in scientific change. Hence, I find the term "interactor" much more appropriate for scientists in this role than "vehicle."

In sum, conceptual replication is a matter of ideas giving rise to ideas via physical vehicles, some of which also function as interactors. Replicators are generated, recombined, and tested by scientists interacting with the relevant portion of the natural world. Because I see a ball accelerate initially as it rolls down an inclined plane, I come to hold beliefs about the motion of balls as they roll down inclined planes. Something in the nonconceptual world has initiated a replication sequence in the conceptual world. These sequences of events in the nonconceptual world are the sorts of causal connections that natural science is designed to discover. Social scientists study the perceptual connections between individual organisms and the rest of the world, including other organisms.

Causal connections exist among scientists. Causal connections also exist between scientists and the rest of society. Both sorts of causal connections influence what scientists believe. In order for science to have the characteristics it is traditionally held to have, it is not necessary that this second sort of causal influence be totally eliminated. What makes science work the way it does is that the effects of the career interests of scientists coincide often enough with the manifest goals of science. It is this coincidence that allows scientists to notice and overcome the effects that other sorts of social influence have on their belief systems. These other sorts of social influence are not always of the sorts that need to be "overcome," but when they are, the social organization of science permits it. In this chapter I discuss conceptual replication as completely as possible without reference to conceptual interaction, the topic of chapter 13. Precisely how closely connected conceptual replication and interaction are is reflected in the difficulty of this task. Because selection is an interplay between replication and interaction, attempts to treat each separately are sure to be strained.

Conceptual Replication

The similarity between genetic and memetic replication is enhanced by the apparent appropriateness of talking about the transmission of "information" in both. As people traditionally conceptualize these phenomena, conceptual transmission of information forms the literal usage, while the usage in the genetic context is analogical; but it is an analogy that is extremely apt. Information requires structure, but not all structure in the empirical world has what it takes to count as "information" without this term being deprived of all significance. Planets travel in ellipses, gases expand when heated, molecules are transported differentially through semipermeable membranes, and crystals form very regular shapes. None of these regularities count as "information" in the appropriate sense. The order of bases in molecules of DNA does count as information because of the character of this order, its origins, and its effects. Much of the structure of DNA is strictly lawful. Given certain general constraints, it has to be the way it is. For example, if the external "backbones" of a DNA molecule are to be kept equidistant down its length, the "rungs" between these backbones must be of equal length. These rungs happen to be made of molecules of two different sizes—purines and pyrimidines. Thus, each rung must be made up of one purine and one pyrimidine. As its name implies, deoxyribonucleic acid is an acid. Hence, it must have the general characteristics of acids.

These are the considerations that helped such molecular biologists as Linus Pauling, Francis Crick, and J. D. Watson work out the "phenotypic" structure of DNA molecules. The order of bases in such molecules, to the contrary, cannot be unraveled in this way. The purines and pyrimidines that make up the rungs of the DNA ladder occur in two forms each: the purines adenine and guanine and the pyrimidines thymine and cytosine. Adenine can bind only with thymine, and guanine can bind only with cytosine, but, with minor exception, any of the four bases can precede or succeed any of the other bases as well as itself. All orders are equally likely from the perspective of physical law. That is why the order that happens to exist can function as a code. With equally minor exceptions, any letter in a natural language such as English can precede or succeed any other letter. The genetic code forms a systematic though degenerate relation between the messages encoded in the DNA and particular amino acids. For example, UUA along with eight other codons code for leucine. In translation, the linear array of bases supplies the initial instructions for the construction of highly complicated proteins. Energetically minimum causes have energetically major effects (Engelberg and Boyarsky 1979).

The distinction between which features of a molecule follow lawfully from the fundamental character of the physical world and which are contingent is important. The distinction between a code (or language) and particular messages expressed in that code is also quite important. Both languages and messages are built up historically and can be used to infer history. Numerous genetic codes were possible when life evolved on Earth. Which code evolved is primarily a function of the particular circumstances that obtained at its origin. Because all terrestrial organisms, with minor, slight variations, use the same code, the assumption is that all life here on Earth had a single origin. (The logic of this sort of inference is set out in Sober 1988). Although the genetic code is relatively simple when compared to a language such as English, it is still sufficiently complicated that the likelihood that exactly the same code could have evolved twice is quite small. But this is about all that can be inferred from the nature of the genetic code itself. The various messages in the genetic makeup of extant organisms allow for much more detailed reconstruction, again because their structure is a function of the sequences of events that led to them. By means of an interplay between replication and interaction through a succession of generations, increasingly lengthy messages have been built up. Perhaps we will never know which contingencies gave rise to the various messages exhibited by extant organisms, but the records of these events allow us to infer increasingly detailed lineages.

In a sense, any "record" of the past can serve as "information" about the past. For example, unusually high concentrations of iridium in thin layers of the geological strata can be used to infer past impacts with meteors, the footprints of a dinosaur in the hardened mud of a river bed imply the existence of these organisms at the time when the layer was formed, and masses of charred wood indicate an earlier forest fire. This sort of "information" can, as is the case with all empirical phenomena, *initiate* replicator sequences, but the entities involved do not themselves count as replicators. Organisms can learn about the world in which they live either directly by interacting with it or indirectly by observing some other organism do the interacting. Learning from experience in the first sense may *initiate* a replication sequence but does not itself count as replication.

In their investigations, scientists learn about the structure of the empirical world. They record this knowledge in a language of some sort. This characterization of the natural regularity counts as information but the natural regularity itself cannot without making the notion of information vacuous, as Lewis Carroll (1898) pointed out so nicely in his fable, *Sylvie and Bruno, Concluded*. According to Carroll, the inhabitants of a particular land prided themselves on their map-making ability. They began by making maps with a proportion of one inch to the mile, then one foot to the mile, culminating in a map whose proportion was one mile to the mile. The only trouble was that they could not unroll this map without blotting out the sun and killing their crops. Then it occurred to them. Why not use the country as its own map?

The point of this fable is that the ultimate map is no map at all. The point I wish to make with respect to information and replication sequences is that the natural regularities which sometimes initiate replication series can no more count as replicators than a country can serve as its own map. Just as a series of analogies presupposes literal usage, information must be information *about* something. However, the genetic code and the messages that such replicators incorporate add an interesting twist to the story. The messages incorporated in the genetic material are *about* something—the phenotypes of the various structures that they help produce. They are more than just part of the causal chains that produce these structures. In large measure, they are responsible for the structure of these structures. Scientists in turn can describe the base-pair sequences incorporated in the genetic material and produce a linguistic replicator that is isomorphic to the genetic replicator—a message about a message. Most records of the past are simply that—records of the past—and no more. A footprint in the sand is a record, but it does not have what it takes to be a replicator. Some records are also replicators.

Organisms can learn about some feature of their environment by interacting with it, for example, a cat might learn by touching a hot coal that fire burns. Organisms can also learn about their environment by watching some other organisms learn about its environment. A cat might learn that fire burns by watching another cat exhibit pain behavior when it touches a hot coal. Human beings have carried both sorts of learning to their extremes. These are the abilities that scientists have exploited to such striking effect. Students can run experiments themselves or watch their instructors do so. But most learning in science comes from reading or hearing about the activities of others. Only a small portion of what a scientist knows about the world arises by means of this scientist interacting with the relevant natural phenomenon. Each scientist has only a few decades to contribute to science. Time cannot be wasted checking every single knowledge-claim before it is accepted. Accepting without testing makes scientific progress possible, but it also increases the likelihood that some of the knowledge-claims accepted by a scientist will be mistaken. However, one should not forget that knowledge by acquaintance is also far from infallible. Some of the erroneous views that scientists come to hold are of their own making. Whether the knowledge acquired is first, second, or third hand, it can always be mistaken. If science required infallibility or absolute certainty, it would be in trouble. We can be happy that it does not.

In the preceding discussion, I have used the terminology of analogy because this is the traditional usage. In discussing information, I have taken human languages as the paradigm and the genetic code as the analogue. The genetic code is like human languages. In discussing selection processes, the priority is reversed: changes in gene frequencies are the paradigm and conceptual change the analogue. Conceptual change in science is like gene-based biological evolution. However, the main goal of the preceding chapter was to present an analysis of selection processes sufficiently general to accommodate all sorts of selection processes. In the development of this analysis, I gave no special priority to any particular sort of selection. The character of gene-based selection has been worked out in great detail, but so has the nature of selection in immune responses even though the effects of these latter selection processes cannot be passed on to successive generations by means of the genetic material. The "generations" that matter in immune responses are cellular. Genetic inheritance and immune responses are both biological, but only the former is accomplished via genetic replication (Darden and Cain 1988). If talk of selection is appropriate only when selection is accomplished in the context of genetic replication, then quite obviously conceptual change cannot be analyzed in terms of selection, but then neither can immune responses. Anyone who insists that conceptual change is a unique phenomenon in nature, akin to no other including biological evolution, has to find some other justification for this stand.

Critics of attempts to assimilate biological and sociocultural change to a common mode of explanation have produced a long list of differences between the two, a list that is designed to show that these phenomena are so different that no analysis could possibly apply to both. Among the differences most commonly cited between biological and sociocultural change are: (1) that sociocultural evolution occurs much more quickly than biological evolution, (2) that the entities selected in biological evolution are atomistic while those selected in sociocultural evolution are not, (3) that biological evolution is always biparental while sociocultural evolution rarely is, (4) that cross-lineage borrowing is more frequent in sociocultural evolution than it is in biological evolution, (5) that sociocultural evolution is Lamarckian while biological evolution is Darwinian, i.e., Weismannian, (6) that conceptual evolution at least in certain areas of science is progressive while biological evolution is not, and (7) that sociocultural evolution is to some extent intentional while biological evolution is not. I take all these claims save the last to be at best confused, at worst mistaken. In the remainder of this chapter, I show that these mistakes stem from a misconstrual of biological evolution, sociocultural evolution, or both. In the process of explaining why these differences are misplaced, the sense in which sociocultural change is a selection process should become increasingly clear.

Rates of Evolutionary Change

If the only metric appropriate for biological and conceptual evolution were physical time, then quite obviously sociocultural evolution, especially conceptual development in science, occurs much more quickly than biological evolution in multicellular organisms. On one rough estimate, a new allele that arose in the time of Julius Caesar would only now be becoming widespread in the human species even under the most rigorous selection pressures possible. During this same period, several cultures have undergone massive changes. Fortunately, physical time is not the only appropriate metric for either biological or sociocultural change. Generation time is also relevant. As biologists from T. H. Huxley (1852) to J. T. Bonner (1974) have shown, the notion of a single generation in plants and animals is not as straightforward as one might think from the perspective of our vertebrate biases, but the living world includes, in addition to vertebrates, dandelions, aphids, slime molds, liver flukes, and corals. These difficulties to one side, one relevant parameter in population biology models is number of generations. Each instance of selection acts in the generation time of the organisms involved, whether it is the few seconds needed for certain viruses to replicate or the scores of years required for century plants.

Cavalli-Sforza and Feldman (1981: 14) remark that in cultural evolution there is no natural unit analogous to biological generations. I fail to see the problem. Each time a practice or idea is replicated, that is a conceptual generation. Each time an old man shows a young boy how to make a slingshot, that is a conceptual generation. Each semester in colleges and universities around the world, the Pythagorean theorem is taught. Each replication forms a conceptual generation. In physical time, conceptual generations are variable (Boyd and Richerson 1985), but so are biological generations. Viruses replicate much more quickly than any conceptual change has ever been disseminated even in the most active areas of science. Elephants reproduce more slowly. With respect to physical time, rates of replication in biological and sociocultural evolution overlap extensively. Generations exist in both, sometimes overlapping, sometimes not.

Although the appropriate metric for replication is generation time, selection processes also involve the interaction in physical time of entities with different speeds of replication. Viruses that infect vertebrates have a decided advantage because of the rapidity with which they can replicate and, hence, evolve. Several new strains of a virus can develop in the space of a single vertebrate generation. As a result, their hosts cannot keep up solely through the mechanism of selection acting at the organismic level. That is why vertebrates evolved their immune systems, systems that can react as rapidly as viruses can replicate. If we vertebrates had not been able to evolve this system or some functional equivalent, we would not be here. Because such interactions *are* interactions, a fuller discussion must be postponed until the next chapter. However, if any sense is to be made of conceptual selection processes, one must continually strive to overcome one's vertebrate biases. A human being can continue to exist during only two or three generations of its biological offspring. Scientists are human beings. As such they can continue to exist during hundreds of generations of their conceptual offspring. Although differences in degree do exist, it is simply not true that all genes in an organism have the same generation time, while concepts have very different generation times relative to their vehicles.

The Need for Particles in Selection Processes

If conceptual evolution is to be understood as a selection process, then it must be analyzable into replicators that increase and decrease in frequency relative to other replicators. A common objection to construing conceptual change as a form of selection is that in biological evolution there are discrete units that are all of the same sort and size, units that can be counted, while this is not the case in sociocultural evolution. With respect to Lumsden and Wilson's (1981) "culturgens," Caplan (1982: 9) remarks that the "greatest difficulty confronting the concept of a culturgen is that it is not clear that culturgens must be particulate units. For Darwinian evolution to occur, the units of heredity must remain constant and relatively immune to environmental alteration if selection is to pick and choose among genotypes." Hallpike (1982: 13) agrees and adds that sociocultural phenomena are "not just aggregates or heaps of elements but structures, with rules, relationships, and transformations of these, at differing hierarchical levels."

Those who object to applying selection models to conceptual change and to considering the genetic code a genuine language usually have a reasonably good understanding of conceptual change and the nature of human languages but an overly simple view of biological change and the genetic code. They are well aware that natural languages do not (and formal languages cannot be made to) map onto the world in the simple one-to-one fashion proposed a couple of generations ago by logical atomists. There are no atomic sentences, no atomic facts, and no one-to-one correspondence between the two. Our understanding of the world cannot be subdivided into units of equal size and treated in isolation from all other conceptual units. Our understanding of the world and our theories about it are much more holistic than this. Therefore, these critics reason that since "particulate" inheritance is necessary for selection to have any effect in biological evolution and sociocultural units are not "particulate," a selection model cannot be applied to sociocultural change.

As Boyd and Richerson (1985) point out, there has to be something wrong with this line of reasoning. In the first place, prior to the turn of the century, most evolutionary biologists assumed theories of inheritance that were "particulate" in the sense that they viewed the hereditary material as consisting of particles, but according to these theories inheritance was "blending" because so many particles influenced any one character and these particles were not thought to be transmitted in pairs. More importantly, present-day versions of evolutionary theory are not "particulate" in the way being assumed. If one conceives of genes as discrete beads on a string, as they were pictured for one brief moment in the history of genetics, then of course nothing in sociocultural evolution answers to them, but then nothing answers to them in biological evolution either. Quite rapidly the atomistic one gene/one character hypothesis was soon shown to be overly simple (Carlson 1966). If one turns from the units distinguishable in traditional breeding experiments (Mendelian genes) to the molecular level, the situation gets even more problematic. Genes do not function in isolation but as parts of integrated gene complexes, resulting in what Mayr (1975) terms the "unity of the genotype." Because of such complexities at the level of both transmission and developmental genetics, Williams (1966) introduced the notion of evolutionary genes as carriers of fitness values. Such genes are defined so that they function as units in the evolutionary process regardless of any one-to-one correspondence between them and either Mendelian or molecular genes. These are the genes that function as replicators in biological evolution.

Just as selection models can be applied in biological contexts only if the complexities of developmental genetics are bypassed, they can be applied in the context of sociocultural evolution only if comparable complexities are ignored. Thus, any conceptual or cultural entity for which there is a favorable or unfavorable selection bias equal to several or many times its rate of endogenous change counts as a single unit in sociocultural change. The "size" of a conceptual replicator depends on the severity of the selection pressures, the fidelity of transmission, and the homogeneity of the conceptual population of which it is part. "The idea" is as much of a myth as "the gene." From other perspectives, cultural units may appear to vary all over the place, but from the perspective of sociocultural selection, all these units are of the same size. Just as biological evolution does not require atomistic genes mapping isomorphically onto atomistic phenotypic characters, conceptual evolution does not require atomic sentences that map neatly onto atomic states of affairs.

As Rosenberg (1982: 22–23) notes, anyone who proposes to provide a selection model of sociocultural change will not be able simply to adopt the "units" identified and individuated by social scientists, conceptual historians, and philosophers and treat them as units of sociocultural selection. Instead, they will have to individuate the relevant units through attempting to apply their models to real cases. Just as Mendelian genes and molecular genes are not good enough for the units of replication in biological evolution, the phonemes and morphemes of linguistics, the stimuli and responses of behavioral psychology, and so on are not adequate for the purposes of an evolutionary account of sociocultural change. They may form points of departure for particular investigations, but that is all.

As in the case of Williams's (1966) definition of an evolutionary gene and Dawkins's (1976) parallel definition of a replicator, the "size" of a conceptual replicator is determined by the selection processes in which it is functioning. From the point of view of replication alone, units are not needed. Entities can pass on their structure largely intact even if this structure is not subdivided into smaller units. But selection also requires interaction, and this process is far from undifferentiated and homogeneous. Once the distinction between replication and interaction is made, one can distinguish four possibilities: changes in replication frequencies due to interaction (directional selection), no change in replication frequencies because the effects of the relevant variations happen, by chance, to balance each other out (balancing selection), changes in relative frequencies that are not due to any environmental interactions (e.g., drift), and replication sequences in which there are neither changes in replicator frequencies nor significant environmental interactions (stasis).

By definition there can be no neutral units of selection because selection is a combination of replication and interaction. There also can be no neutral interactors because interactors are those entities that interact as cohesive wholes with their environments in such a way that replication is differential. However, there can be neutral replicators—replicators that either engage in no interactions or else engage only in interactions that have no effects on differential replication. Furthermore, some changes in replicator frequencies can occur independently of interactions. Hence, the definitions that I set out in chapter 11 do not exclude neutral evolution. Definitions cannot exclude empirical possibilities. However, the entities involved in selection processes can be distinguished and their limits determined only when they are engaged in selection processes. Because selection is variable, the size of these units is also variable.

Biparental Inheritance and Lineages

The discreteness of the units of sociocultural replication to one side, another common objection to the application of selection models to sociocultural change is that in biological evolution variations of the same gene—alleles—obey Mendel's laws, while nothing like this exists in sociocultural evolution. Hence, construing these phenomena as instances of selection is likely to obscure rather than elucidate their most interesting features (Maynard Smith 1986). For example, Quinton (1982: 220) remarks that it is "perfectly all right to talk of the evolution and of the natural selection of these social characteristics. But the limits of their analogy with genetically inherited characteristics should not be forgotten. In the first place our social or cultural inheritance is not biparental: in this respect we are the direct heirs of all ages."

Two considerations count against this particular objection to an analysis of conceptual change in terms of selection. First, modern versions of Mendelian genetics have been part of Darwinian versions of evolutionary theory starting in the early decades of this century, but selection can hardly require this particular family of genetic theories since Darwin and his successors got along without them for half a century. A second consideration, however, is more telling. Although sociocultural evolution has probably been largely "vertical" throughout most of its existence (i.e., learning proceeding primarily from parents to offspring), it is clearly not "biparental." It is not biparental in two senses. Ideas do not always have two sources. Sometimes they have one, sometimes two, sometimes more. Secondly, ideas do not always exist in just two forms, just two conceptual "alleles." Numerous alternatives exist.

But then, similar observations hold for biological evolution as well. If selection requires that genes always be transmitted from two parents, one male and one female, to one or more offspring, then selection could not possibly apply to the more interesting sorts of sociocultural transmission, but as anyone who has read the preceding chapters already knows, no such requirement exists. Throughout most of the history of life on Earth, the only organisms available for selection to act upon were asexual. Mendel's binomial distributions do not apply to strictly asexual forms of reproduction. If biparental inheritance is required for selection, selection is inapplicable to biological evolution throughout most of its history.

In its narrowest sense, selection occurs only among alternative alleles residing at a single locus. Alleles are different versions of the same gene, "sameness" being defined here in terms of occupying the same locus on a chromosome. Although the term "allele" was coined in the context of transmission genetics, it was generalized for the purposes of population genetics. In selection models in population genetics, alleles are the things that compete with each other. In populations of organisms, many alleles can exist at a single locus. This does not mean that any one organism can have all these alleles. In haploid organisms only one allele can exist at a particular locus in a particular organism, in diploid organisms at most two, in triploids three, tetraploids four, and so on. But in the context of a particular population, all the alleles that reside at the same locus can be said to compete with each other regardless of the chromosome complement that this particular species of organism happens to exhibit. It is in the population genetics sense that different conceptual "alleles" can be said to coexist in the same conceptual pool. Quite obviously, more than two conceptual variants can exist in the same conceptual pool, but nothing about biological evolution implies otherwise.

The "size" of conceptual replicators is highly variable. For instance, scientists make statements of varying degrees of generality. Some are quite narrow and descriptive, e.g., this culture of mouth protozoa is unaffected by concentrations of one milligram of stannous flouride per cubic centimeter of water. A variant of this same idea is that this same culture is unaffected by concentrations of two milligrams of stannous fluoride per cubic centimeter of water, and so on. At a higher level of generality, a biologist might believe that DNA produces RNA, and RNA produces proteins, but never the opposite. A variant of this view is that sometimes RNA codes for DNA, but changes in proteins are never read back into molecules of RNA. Similar observations hold for biological replicators.

In addition to conceptual variants being of greater or lesser generality, they are also bound together to varying degrees into conceptual systems. Given a particular conceptual system, alternative conceptions can perform the same function. For example, in the early years of Mendelian genetics, several alternative mechanisms were suggested for explaining dominance and recessiveness. They all were part of the same family of conceptual systems. They were all part of Mendelian genetics. Just as genes are organized into more inclusive functional systems (i.e., genomes), ideas are organized into more inclusive inferential systems, commonly termed "theories." Different versions of the same theory are "different" because at one or more conceptual loci, different conceptual variants exist.

In conceptual change, we frequently think of conceptual variants of limited scope replacing each other within a more global conceptual entity, but we also conceive of conceptual change in terms of these more global structures competing with each other. Although Copernican astronomy grew out of Ptolemaic astronomy, the two eventually came to exist alongside each other as competing theories. They were no longer versions of the same theory but different theories. Eventually Copernican astronomy prevailed over Ptolemaic. It is difficult not to conceive of competition in conceptual change occurring at a variety of levels simultaneously. But comparable observations hold for biological evolution, current limitations of population genetics notwithstanding. Many biologists are unhappy with traditional selection models in biological contexts. They find "fitness" as applied to alleles as unsatisfactory as the assigning of intelligence quotients to individual people. Both are in important ways "fictions." They are urging the introduction of a hierarchical view of biological evolution in which selection operates at several levels of organization simultaneously. Thus far, considerable headway has been made in certain areas of this research program, while other areas remain as promissory as they are in conceptual evolution. All I can say in defense of the sketch of multilevel selection that I provide for conceptual evolution is that more stringent standards should not be introduced when selection models are applied to sociocultural change than can be met even by theories of biological evolution. Treating biological and conceptual replicators as units with respect to fitness may be inherently inadequate, but no more so in one context than in the other.

One important feature of both biological and conceptual replication is that the relevant variants must be organized into "families" and "versions" by means of propinquity of descent. Thus, alleles count as "alleles" only if they are modifications of genes residing at the same locus. In evolutionary biology, the coefficient of relationship is defined in two different ways—both as the fraction of genes that two individuals share that are identical by descent and the likelihood that any two individuals will share genes that are identical by descent at a particular locus—but in both definitions identity by descent is required. Thus, in any of the equations of population genetics in which the coefficient of relationship occurs, identity by descent is required.

Numerous senses of the "same" and "different" elide imperceptibly into each other, but two are central to selection: same structure versus same origin. The two can go together, but they need not. Alleles with the same structure can result over and over again by means of distinct mutations. They have the same structure but different origins. Conversely, alleles with different structures can result from a single mutation. Though they arise from the same ancestral gene, they comprise different alleles because they have different structures. Parallel observations hold for traits as evolutionary homologies and species as lineages. Homoplasies are traits with very similar structures but distinct origins. Evolutionary homologies, to the contrary, can exhibit very different structures even though they have the same origins. Finally, phenetic species can arise over and over again. Once a set of characters and a coefficient of similarity has been settled upon, any time an organism arises that fulfills these conditions, it belongs to this species. As Sneath and Sokal (1973: 54) note, "An allopolyploid might originate repeatedly, giving rise to phenetically identical new species each time." All these hybrid species might be phenetically identical, but they have separate origins and form separate lineages.

Similar observations hold for versions of the "same idea" or "same theory" in connection with the notion of conceptual inclusive fitness. To the extent that individual conceptual entities can be isolated and identified, rarely does one count as being totally new. Invariably dozens of unappreciated precursors can be found. However, from the point of view of conceptual inclusive fitness, the origination of a new idea counts as "new" no matter how many times it may have cropped up before. Nelson was hardly the first biologist to have doubts about the identification of extinct species as common ancestors, but this particular token of this conviction occurred for Nelson in the context of his attempt to integrate Hennig's system into a general theory of comparative biology. To understand the development of cladistics, one must treat this conviction as distinct from other similar convictions that dot the history of evolutionary biology. Anyone who came to accept Nelson's view on the unknowability of ancestors increased Nelson's conceptual inclusive fitness. Similar observations hold for the overall phenetic similarity of Sokal and Sneath, Farris and his notion of parsimony, the entropy of Brooks and Wiley, and so on. For both genealogical and ecological reasons, propinquity of descent is required for selection to produce evolution. If variants are to compete and replace each other, they must come to coexist in the same location (Damuth 1985). One may choose differently on these matters than I have, but anyone who is cursed with one of those small minds that insists on consistency cannot opt for the usual practice of slipping back and forth between these two different senses of "sameness" as the occasion demands.[2]

2. More generally, according to the semantic theory of theories, a theory is a family of models (Giere 1984; Lloyd 1988). This view of scientific theories is compatible with a selection model only if the members of these families are related by descent. Families of models that have nothing in common save similar structure are not good enough.

Because individual scientists are essential links in conceptual replication sequences, ideas come to "belong" to particular scientists. From the perspective of replication in its most general sense, the "ownership" of ideas can be ignored. Identity by descent remains identity by descent. However, with respect to the social mechanism for scientific change that I set out in the preceding chapters, the identification of scientists with their ideas is crucial. It is this self-interest that is harnessed in the system of credit for contributions that has characterized science from its inception. Exactly how quickly the effects of "kinship" are dissipated can be seen in the relationship between Lotka and Volterra. Lotka published his predator-prey equations before Volterra, and when Volterra continued work in this area, he not only built on Lotka's work but also gave him credit. Thereafter, Volterra's students built on their teacher's work and tended to cite Lotka's earlier contributions much less frequently. Hence, those aspects of Lotka's work that were not reflected in Volterra's work were lost. In the sense of impersonal replication sequences, later developments in mathematical ecology were predicated on Lotka's original work. They were connected by descent. However, Lotka grew increasingly frantic because increasingly Volterra was getting most of the credit; he had students and Lotka did not (Kingsland 1985). Identity by descent is necessary for an analysis of science in terms of selection. It is not sufficient. Both explicit and implicit credit are necessary.

Scientific theories as they are commonly construed are treated ideally as being integrated by deductive inference. Axioms entail theorems, and that is that. Inferences either follow or they do not. If the axioms of a theory contain a contradiction, then any proposition whatsoever follows from these axioms regardless of whether anyone ever exploited this contradiction. Similarly, the axioms of a theory entail all possible theorems derivable from these axioms regardless of whether anyone ever makes these derivations. One theory can entail another, contradict it, or be independent of it. But on the view of conceptual change that I am urging, only those inferences that are actually made enter into the process. In retrospect, later workers have on occasion discovered that particular versions of earlier theories have included in their axioms implicit contradictions. A contradiction might entail all possible propositions, but if a particular contradiction was never exploited, it does not matter.

The requirement of deductive inference in science is not stringent enough because it does not preclude those inferences that were never made. As philosophers such as Toulmin (1972) have argued, the requirement of deductive inference in science is also too stringent. If "systematicity" is confined to the simpler forms of inference, then the most that one can expect for particular versions of scientific theories are "pockets of systematicity." Weaker inferential relations also play a role in science. Even though one theory might not entail another, the two might in some sense mutually support one another. Similarly, even though two theories might not strictly contradict each other, the acceptance of one might make it more difficult to accept the other. But the substantive content of science includes more than just theories. It includes methods and aims as well, and they too are related. As Laudan (1984: 63) characterizes these relations, theories "constrain" methods while methods "justify" theories. Theories and aims must only "harmonize" with each other, while aims in turn "justify" methods, and methods must "imply" the realizability of aims. As the quotation marks indicate, the preceding notions remain to be explicated.

As problematic as such loosely organized conceptual systems may be, the relations that integrate them are all matters of inference broadly construed; but if conceptual systems are to function in selection processes, inference alone will not do. The inferential relations both within and between theories must be supplemented by causal relations. Just as there are increasingly more inclusive lineages in biological evolution, there are increasingly more inclusive lineages in conceptual evolution, termed variously disciplinary matrices (Kuhn 1970a), research programmes (Lakatos 1970), disciplines (Toulmin 1972), theories (McMullin 1976), and research traditions (Laudan 1977). Although these more inclusive conceptual entities as they are explicated by these philosophers are different in many important respects, one thing that they have in common is that they are *historical*. As Laudan (1977: 95) characterizes his research traditions:

> Research traditions, as we have seen, are *historical* creatures. They are created and articulated within a particular intellectual milieu, they aid in the generation of specific theories—and like all other historical institutions—they wax and wane. Just as surely as research traditions are born and thrive, so they die, and cease to be seriously regarded as instruments of furthering the progress of science.

Some of the more inclusive conceptual entities postulated by the preceding philosophers are historical merely in the sense that later components come after earlier components, but others are historical in a more significant sense—later components are causally related to earlier components. Descent is, after all, a causal relation. Whether at the level of fairly limited concepts (such as acid turning litmus paper blue) to more inclusive conceptual systems (such as Mendelian genetics) the units of selection must be ordered by descent. All sorts of combinations of ideas are possible, but at any one time only a limited number actually exist. Competition occurs only among those that are actually entertained, and competition is itself a causal relation. Predators and prey may influence each other's evolution, but they do not "compete" with each other in the sense that organisms within the same species and species with sufficiently similar niches do.

Similarly, two scientific theories can compete with each other only to the extent that they are trying to explain roughly the same class of phenomena. Darwin's theory of evolution replaced creationism because in large measure they both purported to explain the origin and termination of species in time. As inherently incommensurable as they might have been on a host of counts, they had enough in common to come decisively into conflict. Darwinism and idealism were much more difficult to bring into conflict because the senses in which they were attempting to "explain" the "same phenomena" are much more elusive. They explained phenomena in very different ways, and the phenomena that they explained were themselves quite different. Archetypes are not ancestors. What is Darwinism, Mendelism, phenetics, cladistics, and sociobiology? Because I have already discussed the answers to these questions extensively in the first half of this book, I will not repeat myself here, except to note two characteristics of the answers to these questions—they involve much more than inferential relations and the elements of research programs can and do change piecemeal (Laudan 1984: 73).

According to Toulmin (1972), the conceptual analogues to species are intellectual disciplines, composed of families of theories, methods, and techniques. Richards (1981: 57), while acknowledging the promise of evolutionary models of scientific development, thinks that disciplines are too heterogeneous to count as species. They seem "more like evolving ecological niches, consisting of symbiotic, parasitic, and competing species. The proper analogue of species is, I believe, the conceptual system." Even though species are themselves genetically quite heterogeneous, the organisms that belong to a species are all fundamentally the same sort of thing. The chief exceptions are sexes, stages in life cycles, and eusocial castes. Ecosystems are composed of representatives of different species (avatars).

The primary difference between species and ecosystems is genetic continuity. Members of a single avatar pass on genes from generation to generation, while different avatars that are part of the same ecosystem remain genetically distinct. They influence each other's genetic development but not genetically. The parallel distinction in conceptual development is between conceptual entities that form genuine replication sequences and those that merely influence each other. For example, Darwin's consistent adherence to gradual evolution forms a straightforward replication series in his writings. So does his commitment to naturalism. As conceptually distinct as these two replication sequences are, they did influence each other without merging into a single sequence. Separating out mergers from influences without mergers is not easy in conceptual evolution; it is not easy in biological evolution either.

One of the most important features of selection processes, especially when the entities being selected form more inclusive organized systems, is the degree to which earlier systems place developmental constraints on their descendants. Elements can change piecemeal but only to the extent that they do not interfere too markedly with the functioning of the larger system. The existence of such constraints is obvious in both biological and conceptual contexts. The clearest examples are presented by maladaptive vestigial organs, e.g., the human epiglottis and the typewriter keyboard. In both cases, the structures willed to us do not work very well. Little thought is needed to think of significant improvements, but bringing about these changes is quite another matter. The extent to which an adequate theory of biological evolution can and should incorporate such considerations is currently under debate in evolutionary biology. Because those workers concerned with temporal changes in conceptual systems have, until recently, not taken their own research program literally enough, the question of conceptual constraints has hardly been broached (for earlier discussions, see Shapere 1974, and Nickles 1974).

In summary, in both biological and conceptual evolution, replicators exist in nested systems of increasingly more inclusive units. There are no unit genes or unit ideas. Selection models in the narrowest sense apply to the smallest units, the alleles that replace each other at a chromosomal locus as well as the different versions of the same idea in a scientific theory. But selection also acts in an extended sense at the level of these more inclusive entities. Whether or not selection at these higher levels can be treated entirely in terms of the fitness of the minimal units depends on how content one is with the bookkeeping aspects of evolution. Anyone who thinks that treating biological evolution solely in terms of changes in gene frequencies is good enough should be no less content with treating conceptual change in the same way. Conversely, if assigning fitness values to isolated ideas seems too impoverished a view in conceptual change, then possibly the parallel maneuver in biological evolution is no less impoverished.

Cross-Lineage Borrowing

One of the chief advocates of evolutionary epistemology, Donald Campbell (1979b: 41), finds the amount of cross-lineage borrowing that occurs in sociocultural evolution to be the major disanalogy between it and biological evolution. By "cross-lineage borrowing," Campbell means elements from one lineage being integrated into another lineage. In biology that would mean genes from one lineage finding their way into another lineage, an event that Campbell assumes is rare. In conceptual evolution that would mean concepts from one research program being adopted by advocates of another program. Campbell assumes that such borrowings are common in sociocultural evolution. In the first place, cross-lineage borrowing of the most extreme sort does occur in biology. Apparently, viruses can pick up genes from their hosts and transport them from host to host not only within a single biological lineage but also between distantly related lineages. In addition, considerable gene exchange takes place between separate biological lineages, especially in plants; and I repeat, if a theory of biological evolution is to be adequate, it must apply to both plants and animals. Of course, if too much borrowing occurs between two lineages, they cease being distinct and merge into one, an occurrence that again is not all that uncommon among plants. In all organisms, sooner or later, a level is reached at which cross-lineage borrowing ceases. As important as this level is, it is also highly variable when viewed from the perspective of the traditional biological hierarchy of genes, organisms, and species.

In principle, we are direct heirs to the cultures of all ages, and this transmission need not be strictly "vertical" from the perspective of gene transmission. Most cultural diffusion has usually been accompanied by the diffusion of genes as well. Immigrants from one society to another bring not only their cultural traits but also their genes. But culture can diffuse much more extensively and rapidly than any accompanying genetic diffusion. For example, the transmission of science from its original home in Europe to the rest of the world far outstripped gene flow. Because the flow of cultural traits has become increasingly "horizontal" through the years, Cavalli-Sforza and Feldman (1981: 33) suggest that a better model for the diffusion of innovation is the epidemic model (see also Shrader 1980). In epidemics, diseases can be transmitted independently of the gene flow of the host organisms. However, these infections are not independent of the gene flow of the disease agents, and we are right back at selection in a literal sense.

If biological lineages are individuated on the basis of gene flow, then it follows by definition that relatively little cross-lineage gene flow can occur. If such gene flow becomes too extensive, what were formerly two lineages become one. The situation with respect to scientific change is complicated somewhat by the existence of two lineages that are semi-independent—social and conceptual lineages. Social groups can split and merge independently of conceptual systems. When Campbell (1979b) talks of "cross-lineage borrowing" in science, he is referring to two scientists belonging to different social lineages using each other's ideas. The lineages are social, while the borrowings are conceptual. For example, the Kansas pheneticists and New York cladists formed two independent social networks. The relevant professional relations were much more frequently realized within each of these lineages than between them. The question then becomes how much "borrowing" actually occurred between these two groups.

If one views concepts as classes of tokens belonging to types because of their substantive similarity, then the two groups held many similar ideas, but similarity is not enough for "borrowing." For cross-lineage borrowing to occur, a term-token in one lineage must give rise to a descendant term-token in another. For example, both pheneticists and pattern cladists insist that in the initial stages of character recognition and classification, evolutionary theory as a process theory must be rigorously excluded. As similar as these conceptual tokens may be, pattern cladists insist that they are not even tokens of the same type. Pheneticists and pattern cladists might utter the same words under the same circumstances, utterances which have precisely the same behavioral manifestations, but nevertheless insist vehemently that these are *different* claims. From a historical perspective, as far as I have been able to discover, they are right. These two tokens of this epistemological view are as independent as causal chains can get, allusions to precursors notwithstanding. In evolutionary terms, they are parallelisms.

How much cross-lineage borrowing occurs in science? After all, one of the goals in science is unification, synthesis. As its name might imply, the "Modern Synthesis" in evolutionary biology was meant to be the merger of several areas of biology into one grand synthetic theory. Croizat characterized his *Space, Time, Form* (1964) as a "biological synthesis." Wilson (1975a) was just as quick to label sociobiology as another "new synthesis." Similarly, Nelson and Platnick's (1981) main goal was to combine the comparative method of ideal morphology with Hennig's principles of classification, Croizat's theory of biogeography, and Popper's principle of falsifiability.

My general impression is that the merger of lineages is more common in biological evolution and much less common in sociocultural evolution than superficial appearances would lead one to expect. Was Mendelian genetics actually merged with Darwinian theories of evolution? Yes and no. Certainly, neither Mendel's theory nor Bateson's theory was merged with the theories of the biometricians. Both theories were extensively modified. In fact, even this claim is too strong. Those who produced the merger were not very familiar with some of the theories that they were merging. They did not take a reasonably accurate understanding of Mendelian genetics as conceived of by Mendel or Bateson and modify it. The founders of the new synthesis did not know much about Mendel or Bateson—or care to. At most they took a very few tokens of these theories, construed them as typical, generalized them, and then merged these generalizations.

Similarly, Nelson had not even read Hennig's *Grundzüge* prior to forming his own ideas. His borrowings came from Brundin, and Brundin's views differed in many respects from those of Henning. Nor was Nelson all that conversant with Popper's views as Popper and his disciples construe them. Platnick was, but as later authors have noted, Platnick at least transformed Popper's views, Popper's complimentary comments notwithstanding. If we are to believe Croizat and his disciples, Nelson and Rosen totally misconstrued panbiogeography. It seems that cross-lineage borrowing is more a matter of reinvention and transformation than borrowing.

In any case, if we restrict ourselves to conceptual lineages, when too much borrowing occurs, two lineages merge into one. From this same perspective, all sociocultural inheritance is vertical. Claims that one sort of replication is "horizontal" result from viewing one replication sequence from the perspective of another. A viral infection is horizontal with respect to the host's genes; vertical with respect to its own. If the social lineages formed by socially defined research groups are taken as primary, then from the perspective of those lineages, some cross-lineage borrowing does occur in science. Conceptual lineages do not coincide perfectly with social lineages. The boundaries of neither are perfectly sharp, let alone perfectly coincident. However, given the close identification between scientists and their ideas, the two sorts of lineages tend to parallel each other closely. Neither pheneticists nor cladists are rushing to join in Mayr's call to synthesize their disparate methodologies under the ecumenical banner of evolutionary systematics.

Lamarckian Inheritance

As much disagreement as there is about sociocultural evolution, everyone who writes on the topic agrees on at least one point—biological evolution is not "Lamarckian" while sociocultural evolution is. Certain authors think that this difference invalidates any attempt to provide a general analysis of selection processes that applies equally to both sorts of change; others think that it does not (Maynard Smith 1961, Waddington 1961, Popper 1972, Toulmin 1972, Cohen 1973, Losee 1977, Rescher 1977, Campbell 1979, Thagard 1980; Richards 1981). The Lamarckian character of sociocultural evolution is so commonly referred to that no one bothers to say very much about what they mean by it. For myself I have been unable to devise an interpretation that is both coherent and makes sociocultural evolution in any important sense "Lamarckian." I am not complaining about historical accuracy. Expecting these authors to be referring to what Lamarck actually said would be expecting too much. "Lamarckian" has taken on a life of its own. I would be content if sociocultural evolution were Lamarckian even in an extended sense. However, as far as I can tell, sociocultural evolution is "Lamarckian" in only two senses, one trivial, the other a caricature. It is certainly true that some organisms learn from their experience, but I fail to see in what sense learning from one's experience is "Lamarckian." It is also true that some forms of learning, especially as exemplified in science, are intentional. Scientists do strive to solve problems, and they would not solve their problems as frequently as they do if they did not. To the extent that intentional behavior is "Lamarckian," it is Lamarckian in only the most caricatured sense of this term. Just as giraffes have long necks because they strive to reach leaves at the tops of tall trees, scientists solve problems because that is what they try to do (Hull 1982b, 1984a).

In biological contexts, the inheritance of acquired characters involves both a particular mechanism and a particular correlation between environmental causes and phenotypic effects. Changes in an organism's environment produce changes in its phenotypic makeup. This alteration in the phenotypic makeup of the organism produces a change in its genetic makeup. These altered genes in turn are transmitted to the organism's progeny, resulting in this acquired characteristic appearing once again in the next generation, this time without the environmental stimulus. In order for a form of inheritance to count as Lamarckian in the present-day sense, the acquired character must be inherited. Nongenetic transmission is not good enough. Of equal importance, a correlation must exist between the nature of the environmental factor that altered the organism's body in the first place and the nature of the ensuing change. For example, growing a shade plant in the sun might cause it to have narrower leaves. This alteration is then transmitted genetically to its offspring so that they are born with a predisposition to produce narrower leaves. If this process is kept up in successive generations, a species of broad-leafed plants could be converted quite rapidly into a species of narrow-leafed plants.

As long as biologists thought that modifications in the phenotype of an organism could be transmitted to the hereditary material, there was some reason to expect that organisms, through successive generations, might get and be able to transmit to their progeny the variations that they need. Not only would organisms be able to adjust to their environments, but also they would be able to pass on these adjustments to their offspring. Heritable changes would be preserved by the very conditions that produced them in the first place. The converse inference is not nearly so warranted. If one discovers a correlation between environment and character change that appears to be Lamarckian, it does not follow that it is. Other mechanisms can produce the same result, albeit by more circuitous routes.

In the past, the term "Lamarckian" has been so abused that it has frequently degenerated to functioning solely as a propaganda device. Some defenders of Lamarck have argued that inheritance really is Lamarckian because the causal chains that produce mutations in the germ cells pass through the organism's body on the way to the gonads. I assume that this parody calls for no discussion. Another version of Lamarckian inheritance is that changes in the phenotype of an organism get passed on to its progeny without touching the genetic material. For example, Dawkins (1982b) terms the transmission of alterations in the pellicle of a paramecium through fission to its progeny a form of "Lamarckian" inheritance. The propaganda advantage of such terminology is obvious. Anyone who thinks that organisms can function as replicators is saddled with a notion of very poor repute. But such transmission no more counts as Lamarckian inheritance than does a mother congenitally transmitting syphilis to her baby or a female dog giving her puppies fleas. In none of these cases is the genetic material altered. Hence, in none of these cases is the phenomenon "Lamarckian."

In order for sociocultural evolution to be Lamarckian in a literal sense, the ideas that we acquire by interacting with our environment must somehow become programmed into our genes and then transmitted to subsequent generations. Some sociobiologists do maintain that certain very fundamental behavioral predispositions are literally innate (as innate as any other characters), but they do not claim that these predispositions have been programmed into the genetic material by Lamarckian means. To the contrary, sociobiologists are strict Darwinians (i.e., Weismannians). So far I have been unable to find anyone who thinks that sociocultural evolution is *literally* Lamarckian. Instead, those who argue that sociocultural evolution is Lamarckian maintain that the relevant transmission is cultural, not genetic. In this *metaphorical* sense, the analogue to genes is ideas, concepts, practices, and so on—in short, memes. Thus, the transmission of ideas or practices from person to person is an instance of the inheritance of acquired *memes,* not acquired *characteristics.* From the perspective of genetic transmission, memes are characteristics; from the memetic perspective, they are the analogues to genes. Social learning can be made to look Lamarckian only by mixing the literal and metaphorical interpretations. This casual equivocation can be seen in the following quotation from Medawar (1977: 14).

> Human beings owe their biological supremacy to the possession of a form of inheritance quite unlike that of other animals: exogenetic or exosomatic heredity. In this form of heredity information is transmitted from one generation to the next through nongenetic channels—by word of mouth, by example, and by other forms of indoctrination; in general, by the entire apparatus of culture. . . .
>
> Apart from being mediated through nongenetic channels, cultural inheritance is categorically distinguished from biological inheritance by being Lamarckian in character; that is to say, by the fact that what is learned in one generation may become part of the inheritance of the next. This differentiates our characteristically human heredity absolutely from ordinary biological heredity.

I agree with everything that Medawar says about biological and cultural heredity except that the latter is in any sense "Lamarckian." From the literal perspective, ideas might well count as acquired characteristics. If they were transmitted genetically, then acquired characteristics could be inherited, but Medawar is clear that knowledge is not transmitted genetically in cultural evolution. Hence, it cannot literally be a form of Lamarckian inheritance. From the metaphorical perspective, ideas are interpreted as being analogous to genes. To the extent that they are transmitted in cultural evolution, they count as an instance of the inheritance of acquired "genes," not characteristics. Only if such things as ideas and practices are interpreted simultaneously as literal characters *and* metaphorical genes can cultural transmission be viewed as an example of Lamarckian inheritance. Perhaps biological and cultural evolution differ in several important respects. However, terming any of these differences "Lamarckian" serves only to mislead.

A related problem that has been noted with respect to cultural evolution is the distinction between cultural genotypes and phenotypes, between germplasm and soma. Once again, this criticism stems from not taking the idea of sociocultural evolution seriously enough. One might be tempted to suggest that the axioms of a theory are analogous to the germplasm and the theorems derivable from these axioms to the soma. But axioms and theorems are equally conceptual, equally replicators. As long as information is passed on from physical vehicle to physical vehicle largely intact, then these transmissions all count equally as replication. Theorems are more restricted in their assertive content than are axioms, but that is all. Only when this information is translated and confronts the nonconceptual environment in the form of interactors is the "phenotypic" level reached.

According to Wimsatt (1981), the conceptual phenotype is the *applications* of theorems in testing. If anything is true of testing, it is that at the interface between concepts and the world, considerable slippage occurs. Rarely in science can axioms be tested very directly. Instead, much more limited statements derivable from them are brought to bear on natural phenomena. Theorems may contain less information than the axioms from which they were derived, but they need be no less explicit and determinant. When theorems are "operationalized" prior to testing, additional information is lost and considerable indeterminacy is introduced. Statements about litmus paper turning blue, precipitates forming in test tubes, and diplococci staining violet are the conceptual entities that stand at the interface between genomes and phenomes in conceptual evolution. A description of an experiment or an observation statement is part of the conceptual genome; the experiment itself or the observation itself is the phenome. In testing, a sentence token is correlated with an instance of a natural phenomenon. An experiment is run. An observation is made. Thereafter, replicates of this sentence token can be traced back to this particular event in link-on-link chains. As much as we might wish otherwise, these link-on-link chains are not always reference preserving, contra Kripke (1972). The correlation is as direct as possible, but the implications of the results of this correlation are far from straightforward (see chapter 13).

Thus, if conceptual replicators and interactors are distinguished in the preceding way, the claim that conceptual evolution is Lamarckian amounts to nothing more than that some organisms learn from experience and then pass on this learning in conceptual replication sequences. In this metaphor, the interaction between an agent's beliefs about some limited area of the empirical world and the relevant phenomena counts as the "phenotype" and these beliefs as the "genotype." Descriptions of a particular experiment as it was run at a particular time can be passed on in replication sequences, but not the experiment itself. In these sense, conceptual phenotypes do not get passed on. Hence, learning from experience and then the transmission of this learning is not Lamarckian because there is no inheritance of acquired characters.

The upshot of the preceding discussion is that, in the only sense that conceptual replicators are characteristics, they are not inherited; in the only sense in which they are inherited, they are not characteristics. What then can those authors who claim that sociocultural evolution is "Lamarckian" mean by this appellation? Toulmin (1972: 337) explicates what he means in terms of "coupling." According to Toulmin, mutation and selection are "uncoupled" in biological evolution, "coupled" in sociocultural evolution. For example, in scientific development, "novel variants entering the relevant pool are already pre-selected for characteristics bearing directly on the requirements for selective perpetuation," while in biological evolution, the "factors responsible for the selective perpetuation of variants are entirely unrelated to those responsible for the original generation of those same variants." In short, conceptual variations can be preserved by the very considerations that produced them in the first place. For example, a protozoologist knows that amoebas usually undergo proliferation by means of a cyst stage. No cyst stage has been observed in the parasitic amoeba that lives in the human mouth. Hence, while studying slides taken from human subjects infected with this amoeba, the investigator keeps an eye out for anything that might be a cyst and eventually finds one. If this investigator had not been, among other things, looking for something that resembled an amoebic cyst, she is unlikely to have noticed the cyst when it appeared on one of her slides.

Novel variants in conceptual evolution are "pre-selected" in the sense that they have to fit into what a particular investigator already believes. For all the beliefs that are tested in a literal sense, there are hundreds that have been tested only via "thought experiments." A scientist notices a white-eyed male fly produced from a cross between two red-eyed parents. "But that is impossible." Something has to give: either some mistake was made in experimental procedure or else one or more of this scientist's beliefs about genetic transmission are mistaken. Just as genes must be able to function adequately in genomes, ideas must form coherent systems of beliefs. But consistency by itself is not enough. Every once in a while, interaction must take place.

I see nothing wrong with characterizing scientific investigations in terms of "coupling" and "uncoupling," but Toulmin (1972: 339) then goes on to remark that if mutation and selection were in fact coupled in biological evolution gametes would have "some clairvoyant capacity to mutate, preferentially, in directions pre-adapted to the novel ecological demands which the resulting adult organism are going to encounter at some later time." Toulmin's critics agree. Gametes have no "clairvoyant capacity" (Cohen 1973: 47, Losee 1977: 350). But something has gone desperately wrong. The coupling of variation and selection in sociocultural evolution has no more to do with clairvoyance than it does in biological evolution. On the basis of their current understanding of the empirical world, scientists make predictions about the future; some of these predictions turn out to be true, others false. Scientists are better predictors of the future than are gametes, but they are no more clairvoyant than are gametes. To make matters worse, even if biological evolution were Lamarckian, it would be neither programmed nor prescient. It might be less wasteful and take place more rapidly, but it would be no less reactive to changes in the environment than is Darwinian evolution. Neither Lamarckian nor Darwinian evolution involves anything that can be termed clairvoyance. Neither does science.

The main fault with terming sociocultural evolution "Lamarckian" is that it obscures the really important difference between biological and sociocultural evolution—the role of intentionality. Campbell (1979b: 41) acknowledges that social evolution "involves the social inheritance of learned behavior, or taught behavior or socially acquired adaptations," but he objects to terming social evolution "Lamarckian" on this account. "This is a distracting usage for those of us for whom Lamarckianism also connotes self-conscious purposiveness and insightful knowledge on the part of the organism as to what it needs, which we deny as essential." In sociocultural evolution, correlations do exist sometimes between environmental causes and their conceptual effects. Because snow is white, people come to believe that it is. One goal of science is to make these correlations as frequent as possible.

Scientific change is surely in some significant sense "directed" (Boyd and Richerson 1985), but the mechanism responsible for these correlations is "Lamarckian" in only the most caricatured sense of this term. Many organisms besides human beings "strive" in the literal sense of this term. They strive to escape predators. In some species females choose among their suitors, resulting on occasion in highly exaggerated secondary sex characteristics in the males. But none strives to change the physical makeup of its future generations so that they will be better adapted to future environments. Only human beings, and very few of them, entertain such notions. Of these only a very few have put them into practice, with mixed results. Depending on how one distinguishes between biological and conceptual evolution in the first place, they either do or do not involve "self-conscious purposiveness" and "insightful knowledge." Although scientists are far from clairvoyant, at the very least they are self-consciously purposeful and frequently quite insightful. But it remains to be seen whether, as a result, science is progressive.

Conceptual Replication and Progress

In a nice turn of phrase, Rescher (1977: 133) contrasts *natural* selection with *rational* selection:

> Rational selection is a process of fundamentally the same *sort* as natural selection—both are simply devices for elimination from transmission. But their actual workings differ, since elimination by rational selection is not telically blind and bio-physical, but rather preferential/teleological and overtly rational.

Both advocates and critics of an evolutionary account of science agree that conceptual evolution is purposive while biological evolution is not. They disagree about the effects that this difference has on any attempt to view both sorts of change as resulting from selection processes. Advocates of an evolutionary account of science, such as Campbell (1965), Toulmin (1972), and Rescher (1977), insist that natural selection and rational selection are processes of fundamentally the same sort, intentionality notwithstanding. Critics such as Elster (1979) and Thagard (1980) think otherwise. For example, Elster (1979: 1) sees an unbridgeable gulf between the two:

> in spite of certain superficial analogies between the social and the biological sciences, there are fundamental differences that make it unlikely that either can have much to learn from the other. The difference, essentially, lies in the distinction between the *intentional* explanations used in the social sciences and the *functionalist* explanations that are specific to biology.

Thagard (1980: 190) agrees:

> theoretical variation is substantially different from biological variation. The main differences have concerned blindness, direction and rate of variation, and coupledness of variation and selection. It is ironic that the great merit of Darwin's theory—removing intentional design from the account of natural development—is precisely the great flaw in evolutionary epistemology. The relevant difference between genes and theories is that theories have people trying to make them better.

I would be the first to agree that science is as purposive and intentional an activity as can be imagined. If anything counts as intentional, science does. But I fail to see why the prevalence of intentional behavior in science rules out an account of scientific modes of reasoning in terms of selection processes. The argument seems to turn on how we distinguish between biological and sociocultural change. Brandon and Hornstein (1986: 176) distinguish between the two as follows: "Thus biological evolution results from natural selection acting on biological entities and cultural evolution results from cultural selection acting on cultural entities." By "natural selection acting on biological entities," I take Brandon and Hornstein to mean the survival of organisms and their reproduction via their genetic material. By "cultural selection acting on cultural entities," I take them to mean the proliferation of ideas, practices, and the like in part by means of the conscious deliberation of agents. But what happens when the two do not go together, e.g., how should we classify the selection by plant and animal breeders of new biological variants? The activity is "cultural," as cultural as scientists' selecting conceptual variants, but the entities involved are biological. But before this question can be addressed, a series of related issues must be clarified, namely, the connections between intentional behavior, functional systems, directional change, and progress. The issue of progress is relevant because the most prevalent view about science is that, at least in limited areas, scientific change is clearly progressive, while the progressive character of biological evolution is not nearly so clear.

From the point of view of the evolutionary process, systems can be subdivided into those that are homeostatic, those that are homeorhetic, and those that are locally constrained but open to indefinite change. Homeostatic and homeorhetic systems together count as functional systems; systems that are only locally constrained do not. A functional system consists of a large causally inclusive system and a much smaller subsystem which is causally a part of it. (Although the smaller subsystems are usually physically part of the larger systems, they need not be. Causal inclusion is sufficient.) A change in the environment of the larger system alters one or more variables in that system. One minor effect of these changes is a change in the smaller, included system. Alterations of the smaller system are slight, rapid, and involve very little energy relative to the entire system. Changes in the larger system are correspondingly more massive, ponderous, and involve large amounts of energy. One effect of the changes in the smaller system is a change in the larger system, which compensates for the original modification induced by the environment. In a homeostatic functional system, the larger system can be kept at a relatively steady state in the face of considerable though not unlimited environmental variation. Functional systems can also be homeorhetic. Instead of the feedback loops returning the larger system to a particular state, they result in the larger system changing to a different state, sometimes quite extended sequences of states. For example, organisms go through life cycles as a result of the three-cornered interplay between their genetic makeup, their phenotype at the moment, and their environment.

In both homeostatic and homeorhetic systems, the smaller system functions as a "program" relative to the larger system (Mayr 1976a). Sometimes this program is as rudimentary as a bimetal bar in a thermostat or as complicated as the genome of an organism. In homeostatic systems, the smaller system varies in its actions, but the structure of the system does not change relative to its predispositions for action. The bar bends and straightens as the temperature varies, but changes in the larger system and the environment consistently have the same effects. In homeorhetic systems, the behavior of both systems is sequential. The program is partially "open," represented most appropriately as a decision tree. As the entity undergoes its development, it has access to a reduced area of the tree. Each fork taken cuts off possibilities. In its interplay with the larger system and the environment, the action of the program is sequential, but the structure of the program itself does not change. In this connection, a homeostatic system is the limiting case of a homeorhetic system.

Organisms are both homeostatic and homeorhetic systems. For periods of their existence, their internal mechanisms keep the system as such within limits, but at the same time it is undergoing ontogenetic development. Functional organization in turn is the limiting case of a selection process. Selection results from the interplay of two subprocesses—replication and interaction. The chief difference between selection processes and feedback mechanisms in functional systems is that this interplay is not programmed but at most locally constrained. Through each cycle of replication and interaction, the structure of the replicators themselves changes so that, at each turn of the wheel, a new decision tree is produced. In homeostatic systems, each round of action and reaction feeds back into structurally the same systems. In homeorhetic systems, both the smaller, included system and larger system change through time in their actions, but this sequence of events is more than constrained. Many alternative paths are possible, but there is a sense in which the reaction norm of the system is set and cannot be changed. The systems involved in selection processes have no reaction norms. Their programs are so open that they can no longer count as "programs."

Both homeostatic and homeorhetic processes involve feedback loops. Quite obviously, feedback loops do not go back in time. They are characterized in these terms because the two systems persist while causal chains proceed from one to the other. In selection processes, feedback loops never become established because with each replication-interaction cycle, genuinely new systems are produced. For example, certain organisms in a population are better able to cope with their environments than are other conspecifics. They are able to reproduce themselves. Because survival to reproduce is selective, each gene pool contains different information than its predecessor. With each generation, only a limited array of possibilities exist for the next generation. Hence, development in selection processes is locally constrained, but each new generation is a new game. One likely scenario for the generation of functional systems is by entities undergoing selection evolving into them. All that it takes for entities engaged in selection to become functionally organized is for the systems to remain intact long enough for feedback loops to become established. On this scenario, selection is temporally prior to functional organization (Hull 1982a).

Periodically, evolutionary biologists have claimed that evolutionary development is homeorhetic—that it is directional, possibly even progressive. That evolutionary development produces lineages which have directions of sorts is unproblematic. For a while, organisms in a particular lineage get larger and more complex, while in another lineage they are getting smaller and less complex. After a while, the directions of these two lineages might well become reversed. The interesting questions are: (a) is there some overall direction to evolution, (b) is there anything about currently accepted mechanisms of evolutionary change that would lead us to expect it to be directional, (c) are any of these putative directions to evolution in any sense "programmed" into the structure of the lineages that are evolving, and (d) can any of these directions count as being "progressive" in any significant sense?

If one takes a quick glance at the fossil record, it looks clearly directional and, making allowances for anthropocentrism, probably progressive. However, two considerations should suggest caution, one circumstantial, one more directly relevant. The first is that, historically, opinion on both counts has fluctuated widely. Strangely enough, those among Darwin's contemporaries who were most strongly disposed to see the fossil record as progressive were opposed to evolution, e.g., Sedgwick, Buckland, and Agassiz, while Lyell and Huxley argued that it lacked even a direction of any significance. Darwin was equivocal. In his more technical passages, he (1859: 314) could see no reason for evolution through time being especially progressive, although in his more poetic passages, he sometimes sounded as if he thought that it was (Darwin 1859: 489).

However, through the years, various "stage laws" of evolutionary development have been popular. Evolution supposedly leads to larger organisms, or larger organisms in colder climates, or some such. Rensch (1971: 131–39) lists an even hundred stage laws that have been proposed for biological evolution at one time or another. Simpson (1961) evaluates the six most plausible candidates and finds them all wanting. Does the fossil record reflect any of these stage laws? Before this question can be answered, two sorts of stage laws must be distinguished—those that concern connected lineages and those that concern unconnected events that simply follow each other in time. Although subsequently Marshal Sahlins (1976) has opposed the use that sociobiologists have made of biology in reaching conclusions about social organisms such as human beings, in his younger days he himself was guilty of this grievous sin. According to Sahlins and Service (1960: 22–23), "specific evolution" must be distinguished from "general evolution":

> specific evolution is "descent with modification," the adaptive variation of life "along its many lines"; general evolution is the progressive emergence of higher life "stage by stage." The advance or improvement we see in specific evolution is relative to the adaptive problem; it is progress in the sense of progression along a line from one point to another, from less to more adjusted to a given habitat. The progress of general evolution is, in contrast, absolute; it is passage from less to greater energy exploitation, lower to higher levels of integration, and less to greater all-around adaptability.

Evolutionary biologists are in fairly wide agreement that progress along particular lineages does exist. However, any trend can be and often is reversed. Claims about progress in general evolution of the absolute sort claimed by Sahlins can be interpreted so that they are dismayingly vacuous. Pick a character, any character. There will be an organism early in the history of life that exhibits it to a particular degree, e.g., efficiency in utilizing light in photosynthesis. This lineage might well die out. Nevertheless, the ability to utilize light in photosynthesis may evolve again later. Many of these new instances may well be less efficient than the original instance, but sooner or later a more efficient method of photosynthesis may evolve. If these instances need not be connected, and if no time limits are set to how long one must wait for the next "improvement," any character must sooner or later "progress" in this vacuous sense. Needless to say, "character" is being used here in an ahistorical sense—traits that "twinkle" rather than evolutionary homologies.

A partial list of the characteristics that supposedly increase in general evolution includes number of organisms, the overall biomass, the number of species, the complexity and adaptability of organisms, the variety of structural plans exhibited, the amount of energy transduced, the efficiency of energy conversion, and the number of adaptive zones and/or niches occupied. Initially, all of these characteristics, it would seem, had to increase. Initially, evolution had to be progressive in almost every sense imaginable, but it took a long time for much in the way of progress to take place. Not a whole lot happened between the first origins of life and the Cambrian three billion years later. Only then did evolution take on anything that might be termed a direction.

Biologists currently disagree about the implications of the fossil record since the Cambrian for the directional character of evolution. Even though Sahlins and Service (1960: 13) think that life inevitably diversifies, Gould (1977a: 19) thinks otherwise. According to current estimates, "there are no intrinsic trends towards increasing (or decreasing) diversity. Ecological roles are, in a sense, 'preset' by the nature of environments and the topological limits to species packing; they are filled soon after the Cambrian explosion. Thereafter, inhabitants change continually, but the roles remain." On this view, life may inevitably diversify but not since the Cambrian.

Initially in the history of life, when so much of the environment was empty, selection was not very rigorous and everything increased but very slowly. As the earth filled up, selective forces became more rigorous and species packing increased until there was little room for improvement without a compensating loss occurring elsewhere. Hence, since the Cambrian, the history of life is better characterized as steady-state, punctuated by mass extinctions which set the system going again until another steady-state is reached. On this view, there was a little progress during the first three billion years of life on Earth, then in the Cambrian massive progress, which has been followed ever since by peaks and valleys. Are there trends such that each peak is a little higher than earlier peaks? Not noticeably. About the only instance in which evolution has had a direction during the past 500 million years is the surprisingly late exploitation of air as an adaptive zone. It was used periodically but not extensively until quite recently.

About the only way that evolution can be made to look progressive in a specific sense is to pick some characteristic that is more highly developed in an extant species than in any known extinct form and show how it increased. Although this procedure can be used for any highly developed characteristic of any extant species, the only one for which we have an inclination to run through this exercise is *Homo sapiens*. Because the human species has the most highly developed central nervous system in the history of life on Earth, all evolution has been zeroing in on us (Goudge 1961, 133–37, Simpson 1961: 175, Dobzhansky, et al., 1977: 428). If just once the result of such an exercise was that some species other than the species conducting the exercise was left standing at the pinnacle of evolution, I would be somewhat less suspicious. It seems that the main reason for so much concern over evolution having a direction is so that we can be its crowning achievement.

Evolutionary biologists currently debate these issues (e.g., Ayala 1974, Flessa and Levinton 1975, Waddington 1975, Hahlweg 1981). Recently, Signor (1985) has argued that species richness has increased throughout the Phanerozoic, regardless of what one might gather from looking at trends among higher taxa. However, the point I wish to make is that neither directionality nor progress is a *clear* feature of the fossil record. Of equal importance, according to traditional views of the evolutionary process, there is no reason to expect continued directionality in evolution, let alone progress. According to Lewontin (1982: 165), biological evolution is opportunistic. Species are constantly adapting to their local environments, only to have these environments move out from under them. "Evolutionary theory in general no longer incorporates notions of progress or of unidirectional change. Evolution, at least in the modern view, is going nowhere in particular. Older ideas that evolution has led and is leading to greater complexity, greater homeostasis, greater stability, are now seen as vestiges of a 19th century progressivism that is without empirical foundation in the history of life." The main exception to Lewontin's claim about the mechanisms of evolutionary change is the entropic theory of Brooks and Wiley (1986). According to these second-generation cladists, entropic forces produce increasing biological information through time.

Similar arguments have taken place over the existence of sociocultural progress, in particular with respect to science. Evolutionary biologists may have been of two minds about the existence of progress in biological evolution from Darwin and Huxley to the present, but for most of this period social scientists were in broad agreement that sociocultural evolution is clearly and ineluctably progressive. Every day in every way we are getting better and better. In his *Cours de philosophie positive,* August Comte (1830) provided a classic statement of progress in human societies. At first Comte's work elicited very little attention, but by the time that Spencer (1852) began to publish his own brand of evolutionary progressivism, Comte had become extremely popular. According to Spencer, all natural processes can be explained by his fundamental law that "matter passes from relatively indefinite incoherent homogeneity to a relatively definite, coherent heterogeneity" (Richards 1987). After Darwin, stage laws became even more popular in sociology and anthropology than in biology. Blute (1979: 47) presents a partial list:

> Societies or cultures were held to develop from despotism through monarchism to republicanism (Montesquieu), from the theological through the metaphysical to the scientific (Comte), from status to contract (Maine), from the primitive through the feudal and capitalist to the socialist (Marx), from savagery through barbarism to civilization (Morgan), from gemeinschaft to gesellschaft (Toennies), from the ideational through the idealistic to the sensate (Sorokin), from folk through feudal to industrial (Redfield), from mechanical to organic solidarity (Durkheim), etc.

Blute shows that when social scientists today object to "sociocultural evolutionism," they are not referring to attempts to formulate a selection model which is applicable to sociocultural phenomena but to stage theories. Like Christianity, sociocultural evolutionism in the first sense has not been tried and found wanting; it has hardly been tried at all. Because the notion of "progress" in biological evolution is so equivocal when not downright anthropocentric and because stage laws in sociology and anthropology have been so abused, a strong tendency exists to reject progress in all areas—including science.

That science is progressive has long been axiomatic in philosophical analyses of science. Is there some warrant to this view or is it just a carryover from the prejudices of earlier ages? Popper for one thinks that considerable warrant exists for considering science progressive. According to Popper (1972: 287, 1975: 83), progress in science results from the formulation of tentative theories to solve current problems, the testing of these theories to eliminate errors, and the formulation of the next generation of problems and their solutions—conjectures and refutations. Popper's goal is not truth itself but increasing verisimilitude: given any pair of theories, scientists should adopt the one that is nearer the truth. For most of his career, Popper claimed to shun epistemological questions, complaining that epistemologists keep cleaning their glasses and cleaning their glasses but never put them on to see anything. For most of his career, Popper (1957: 51, 106; 1962: 340) also disparaged evolutionary theory as a scientific theory. Both in its early and contemporary formulations, it was unfalsifiable, historical, etc. Thus, it came as quite a surprise when Popper (1972) adopted a form of evolutionary epistemology. In the meantime, however, he has acknowledged that some of his early comments on evolutionary theory were misinformed (Popper 1978; see also Ruse 1977, Hull 1983d).

The source of Popper's confusion is instructive. Popper reasoned that if biological lineages (species) are natural kinds, then descriptions of them should count as natural laws, but they do not. At most these generalizations are only historical laws, and one of Popper's main contentions at the time was that historical laws are not laws at all. For Popper, genuine laws of nature do not and cannot change. Our understanding of them might change, but they themselves do not change. On the basis of similar considerations, Sahlins and Service (1960) reached the opposite conclusion. Because species evolve and descriptions of species through time count as scientific laws, such laws themselves evolve. For Sahlins and Service, at the time, evolving laws of nature were perfectly acceptable. Thorson (1970) reached the same conclusion with respect to political systems. Because political systems evolve, the political laws governing them also evolve.

But neither argument is cogent. Neither particular biological species nor particular political systems can be construed as natural kinds. They are the particulars over which the relevant laws range. As particulars, they can change while the relevant laws remain unchanged. Mars, the human species, and American society undergo change; the classes consisting of planets, species, and societies do not.[3] Thus, the historical sequence of political stages through which societies are supposed to evolve, commonly attributed to Marx, is Asiatic, ancient, feudal, capitalist, and communist. As such, these stages are kinds through which particular societies may or may not progress. These kinds themselves do not change.

Kuhn has consistently been interpreted as being an epistemological relativist, but he insists (1977: 508) that he is not and in defense cites one of the concluding passages

3. A third alternative is that species are numerical universals, i.e., sets with membership limited by some sort of spatiotemporal circumscription; e.g., all the coins in my pocket, all the fruit in Smith's garden, or all solar planets. The only spatiotemporal circumscription not precluded is "in the universe throughout all space and time." The only trouble with treating species as numerical universals is that it detracts considerably from their importance. Numerical universals do not have very important roles to play in science. Scientists do perform assays; e.g., the number of cars per household in New York City, the number in Baltimore, etc. If species are as important to our understanding of evolutionary processes as many biologists insist they are, then they cannot be numerical universals.

in his *The Structure of Scientific Revolutions* (1962) where he compares progress in science to progress in biological evolution. According to Kuhn (1970b: 264):

> An answer to the charge of relativism must be more complex than those which precede, for the charge arises from more than misunderstanding. In one sense of the term I may be a relativist; in a more essential one I am not. What I can hope to do here is to separate the two. It must already be clear that my view of scientific development is fundamentally evolutionary. Imagine, therefore, an evolutionary tree representing the development of the scientific specialities from their origin in, say, primitive natural philosophy. Imagine, in addition, a line drawn up that tree from the base of the trunk to the tip of some limb without doubling back on itself. Any two theories found along this line are related to each other by descent. Now consider two such theories, each chosen from a point not too near its origin. I believe it would be easy to design a set of criteria—including maximum accuracy of predictions, degree of specialization, number (but not scope) of concrete problem solutions—which would enable any observer involved with either theory to tell which was the older, which the descendant. For me, therefore, scientific development is, like biological evolution, unidirectional and irreversible. One scientific theory is not as good as another for doing what scientists normally do. In that sense I am not a relativist.

As Toulmin (1972: 323) has argued before me, Kuhn may be right about scientific development, but he is wrong about biological evolution. If one follows evolutionary lineages through time, all sorts of things happen. For a while they become more "specialized" or more "articulated" (Kuhn 1962: 172), but then they can just as well reverse these trends. Evolutionary biologists currently disagree about overall trends in the fossil record, e.g., whether through time species richness has, on the main, increased in spite of occasional setbacks. However, no present-day evolutionary biologists believe that particular lineages necessarily continue along certain lines of development. For a while some do; then they do not.

For those authors who construe biological evolution as being progressive, the progressive character of scientific change poses no problem. Selection processes produce global progress in both. But as I understand the current state of evolutionary biology, a respectable number of evolutionary biologists think that biological evolution is no more than locally progressive. Thus, if scientific change is both locally and globally progressive, an important difference exists between the two. One alternative is to argue that science is no more globally progressive than is biological evolution. As far as I can see, the history of science precludes such a conclusion. Although progress in science is not a simple matter of accretion or accumulation, scientific theories can be arrayed in sequences that show improvement on a variety of counts; of greater importance, these series correspond very closely to the temporal order in which the theories were promulgated. No hopscotching around the history of science is needed. Of greater importance still, earlier members of these sequences actually influenced later members. Later theories are not just better than earlier theories, they are built on them.

It is certainly true that scientific progress is marked by losses as well as gains. A confirming instance of an early version of a theory can become neutral or disconfirming with respect to later versions of that theory. Subtle alterations in meaning can change a paradigm application of a theory to a borderline case. It is also true that a scientific research program that initially made considerable progress and showed even more promise can become stultified and degenerate. Later, when similar problems are taken up again, these later theories can be even less successful than the earlier theories long since forgotten. Even the paths that conceptual lineages take can vary. For example, because ambient temperatures here on Earth are not all that extreme, Boyle and Gay-Lussac were able to set forth their simple ideal gas law. If the ambient temperatures that had confronted them had been much hotter or much colder, physicists would have been forced to reject this law out of hand. They would have had to find some other path to more complex formulations that apply to all temperatures. As a result of these and other considerations, progress in science is far from simple, but it is no less progressive on that account.

Claims about the progressive nature of science are usually made with respect to the more "mature" areas of science; to be frank, physics. Genetics, evolutionary biology, and systematics are rarely presented as paradigm examples of progress in science, but I think that progress in these areas is just as apparent as in physics. The progress in transmission genetics during the early decades of this century was phenomenal. Every Ph.D. dissertation turned into a major contribution. After this initial burst of development, transmission genetics settled down to slow, unexciting growth. In the second half of this century, another burst of activity occurred as molecular biology finally came into its own. Nobel prizes came hot and heavy. Although the relationship between transmission genetics and molecular biology is far from simple (Hull 1982a, Rosenberg 1985), progress is nonetheless clear. Transmission studies are still good enough for the initial crude analysis of genetic structure, but if further refinement is needed, recourse to the techniques of molecular biology are necessary. Increasingly, biologists go directly to the molecular level without any initial Mendelian investigations.

A similar story can be told for evolutionary biology. As great as Darwin's original achievement was, present-day versions of evolutionary theory are superior to Darwin's on a variety of counts. We now know that most species are genetically quite heterogeneous, and the mechanisms that can produce and maintain this heterogeneity have been set out in some detail. We now see more clearly than early biologists why selection at higher levels of organization is so difficult and have distinguished between kin selection and kin-group selection. I could expand on this list almost indefinitely (see Michod 1981). As in the case of genetics, two relations are involved: the transformation of earlier theories into later theories and the reduction of theories at higher levels of organization to those at lower levels. The former is a temporal, causal relation; the latter inferential. In the case of evolutionary biology, the organismic level is still usually the point of entry for research.

Although progress in a science such as systematics is more difficult to document, I think that it can also be discerned in the controversies detailed in the first half of this book. If nothing else, the issues are clearer. No longer do systematists confuse taxa and categories. They see clearly the difference between grades and clades as well as between patristic and cladistic relationships. The computer programs that have been devised allow systematists not only to conduct research on large numbers of specimens and characters but also to play with the data to discern alternative patterns. For example, Colless (1985) subjected the characters and classification produced by D. C. F. Rentz by traditional, more or less intuitive methods to analysis by numerical procedures.

He was surprised to discover that, even though the author "sanitized" his characters considerably, omitting "bad" characters, extensive homoplasy remained. If one accepts any one character as accurately reflecting phylogenetic relationships, then numerous *others* are incompatible with this implication. Even so, nearly all of Rentz's genera turned out to be extremely robust. They emerged as clusters no matter what method Colless used.

Both biological and conceptual evolution (especially in science) are locally progressive. However, differences of opinion exist over global progressiveness in both areas. One resolution of this apparent difference between biological and conceptual evolution is that thus far science can count only as a local phenomenon. It arose only about two hundred years ago as an isolated phenomenon and has diffused into other cultures. Progress has been made in increasing numbers of areas of empirical inquiry, but eventually, as all the fundamental problems in that area are solved, rates of progress will decrease until a steady state is reached. After all, twenty years ago, Stent (1969) predicted an end to progress in biology once molecular biology had fulfilled its promise. Advances at the molecular level have resulted in improved understanding of biological phenomena, but other areas of biology contributed as well, and the end is not even in sight yet. At the very least, science is more clearly progressive both locally and globally than biological evolution is.

If this difference were due to differences in the mechanisms involved in biological versus conceptual change, then it would count against any attempt to produce an analysis of selection processes that is equally applicable to both sorts of change, but the genuine differences between rational selection and natural selection that I have discussed thus far do not. Instead, the source of this difference lies in differences in the relevant "environments" of these two processes. All organisms at all times must survive, compete, reproduce, and so on within the confines set by the basic regularities that characterize the empirical world. Perhaps an organism is so small that surface tension is more important to it than gravity, but no organism can "contravene" the laws of nature. The differences we see in organisms result primarily from changes in the variable parts of their environment, and these environmental variables continue to change, sometimes slowly, sometimes quickly, sometimes regularly, sometimes haphazardly. For this reason, evolutionary biologists are led to argue that biological evolution can be only locally progressive. It may also be the case that biological evolution has thus far been globally progressive as well, but if so, this is only because the upper limit of species-packing here on Earth has yet to be reached. Even so, an upper limit must exist. When it is reached, global progress must cease until some sort of mass extinction occurs.

Scientists are interested in both the variable and the constant aspects of the universe. They describe stars exploding, continents drifting, species evolving, and political leaders being assassinated. The descriptions that scientists provide of these variable aspects of nature can be improved in several ways. They can be made more extensive, more detailed, more accurate. The mark of a good description, however, is relevance to a powerful scientific theory. The more intimately connected a description is to the most powerful process theory available at the time, the better it is. Descriptions have spatial and temporal indices. About three and a half million years ago, the Isthmus of Panama was formed. In 1914 Archduke Ferdinand was shot at Sarajevo. But scientists also strive to discover regularities in nature that are in the relevant sense spatiotemporally unrestricted. If true, they are true anywhere and at any time that the relevant conditions obtain (Rynasiewicz 1986). As Rescher (1977: 130) notes, "If the world is not a stable cosmos ordered by perduring laws, the entire enterprise of evolutionary explanation is vain. We cannot have it both ways, that is, one cannot combine an evolutionary account of knowledge with a rejection of anything approaching a principle of regularity in nature." Even if the regularities that scientists currently think are eternal and immutable turn out themselves to change through time, if they change regularly, then these regularities form the foundations for new laws of nature. Only if all change in nature is inherently haphazard and time-dependent must this view of nature be abandoned.

All species at all times are sufficiently well adapted to their environments or they would not be able to survive. Some parts of their environments are highly variable. The relations between organisms and these variable parts of their environments are primarily responsible for the myriad adaptations which continue to fascinate so many biologists. Because so many aspects of their environments change so rapidly and haphazardly, species seem to be forever chasing their changing environments. The beliefs of scientists also change, beliefs about both limited, contingent states of affairs and eternal, immutable regularities. It is the successive approximations of scientific theories to the latter which are responsible for global progress in science. If no such regularities exist, scientists could not approximate them.

Gellner (1985: 51) sees another difference between natural and rational selection. According to Gellner, "Natural selection tests organisms in quite a different sense from the sense in which science tests a hypothesis. A hypothesis can contain or constitute a generalization which is simply false. By contrast, an organism, a 'carrier', cannot literally embody a falsehood; no living creature itself literally *defies* a law of nature. A *theory* can contradict a valid law of nature, an organism cannot."

Organisms cannot embody falsehood or defy laws of nature, but they can die and/or fail to reproduce because they are not sufficiently well adapted to one or more aspects of their environment, whether at the variable, contingent end of the scale or toward the eternal, nomic end. Scientists cannot defy laws of nature either, but they can be mistaken about them. They can hold beliefs that are false. Sometimes false beliefs lead to death, but usually in trial-and-error learning, our beliefs live and die in our stead. However, in both cases, the issue is "fit," i.e., how well an organism (or more generally an interactor) fits the relevant aspects of its environment and how well a belief fits its subject matter.

In *Progress and Its Problems,* Laudan (1977) presents another alternative. Instead of taking truth and rationality as basic, Laudan analyzes science in terms of comparative problem-solving effectiveness. If scientific theories are ordered according to their effectiveness in solving problems, later theories are by and large better than earlier theories. Thus, Laudan acknowledges the progressive character of science and makes it axiomatic to his entire analysis, but he does not explain it. It is a brute fact about science. In this respect, I find realist philosophies more satisfying than the nonrealist views now being explicated by such philosophers as van Fraassen (1980) and Cartwright (1983). In general, I find such philosophical disputes, once they become sufficiently sophisticated, irrelevant to anything a scientist might or might not do. For that reason they are not of special relevance to my concerns.

From a heuristic point of view, more simplistic versions of these two extreme metaphysical positions have proved fruitful at different times in the history of science. Sometimes scientists who are good, solid, table-thumping realists succeed in cutting through the fog; sometimes a more instrumental view of current theories and laws helps to liberate scientists who are trapped in old ways of thinking. For example, in the early years, many Mendelians did not bother themselves with questions about the existence of genes. Instead, they worked out Mendelian ratios. But further progress was made by those who insisted on a material mechanism to account for these regularities in transmission. Both pheneticists and pattern cladists have urged systematists to liberate themselves from speculations about mechanisms, causes, and processes in order to concentrate on patterns. Such reappraisals did produce new insights. Even so a majority of scientists retain their dogged interest in the material basis of natural phenomena.

If nothing else, a realist perspective provides an important motivational foundation for science (Fine 1986a). If scientists did not think that regularities exist in nature, they would have no reason to search for them, and if they did not search for them, science would have had a very different history from the one it has had (Leplin 1986). However, it is now time to investigate the primary difference between rational selection and natural selection which supposedly makes any general analysis of the two processes impossible—intentionality. It is this difference that Gellner seems to have in mind when he sees such a gulf between natural and rational selection.

The Role of Intentionality in Science

The world can be classified in many different ways, depending on one's interests and principles of classification. These classifications in turn determine which comparisons seem natural and which unnatural, which literal and which merely analogical. For example, a common classification of living creatures throughout the history of science has been into three cognate groups—plants, animals, and man. According to this classification, human beings are not a special kind of animal, nor animals a special kind of plant. Thus, any comparisons between the three cognate groups are strictly analogical. Reasoning from inheritance in garden peas to inheritance in fruit flies and from these two species to inheritance in human beings are all equally analogical, sheer poetic metaphors. Another mode of classifying living creatures is commonly attributed to Aristotle. Instead of plants, animals, and man being cognate groups, they are nested. All living creatures possess a vegetative soul that enables them to grow and metabolize. Of these, some also have a sensate soul that enables them to sense their environments and move. Of these, one species also has a rational soul and is capable of true cognition. Thus, human beings are a special sort of animal, and animals are a special sort of plant. Given this classification, reasoning from human beings to all other species with respect to the attributes of the vegetative soul is legitimate, reasoning from human beings to other animals with respect to the attributes of the sensate soul is also legitimate, but reasoning from the rational characteristics of the human species to any other species is merely analogical. According to both classifications, the human species is unique. In the first, it has a kingdom all to itself; in the second, it stands at the pinnacle of the taxonomic hierarchy.

A major change in the history of classification of plants and animals, for which Linnaeus is usually given credit, was classifying the human species along with other species. Plants and animals were treated as cognate groups, and *Homo sapiens* became just one animal species out of many. It had taxonomic neighbors. Thus, if reasoning from zebras to horses is legitimate because of their taxonomic propinquity, reasoning from various species of ape to humankind is also legitimate. The introduction of evolutionary theory further encouraged the legitimacy of considering human beings to be part of nature. Taxonomic propinquity reflected propinquity of descent. Even so, the tendency has remained to view human beings as somehow special, somehow "outside" the natural order. The dams built by beavers are natural and enhance the environment, while the very same dam built by a human being is unnatural and inevitably degrades the environment. This same romantic tendency continues to the present in the insistence that certain characteristics possessed by human beings are not only unique to them but also insulate the human species from comparisons to other species. Perhaps it is all right to use rabbits to decide whether a woman is pregnant but any interspecific comparisons with respect to the "higher" faculties of human beings are necessarily illegitimate. Among these unique, higher faculties are consciousness, self-consciousness, use of tools, possession of a "true" language, and expression of intentional behavior.

Homo sapiens is unique. All species are. But this sort of uniqueness is not enough for many, probably most people, philosophers included. For some reason, it is very important that the species to which we belong be uniquely unique. It is of utmost importance that the human species be insulated from all other species with respect to certain modes of explanation. Human beings clearly are capable of developing and learning languages. For some reason it is very important that the waggle dance performed by bees not count as a genuine language. I have never been able to understand why. I happen to think that the waggle dance of bees differs from human languages to such a degree that little is gained by terming them both "languages," but even if "language" is so defined that the waggle dance slips in, bees still remain bees. We still form unproblematically distinct species. None of us is about to mate with a bee. It is equally important that no other species use tools. No matter how ingenious other species get in the manipulation of objects in their environment, it is absolutely essential that nothing they do count as "tool use." The various contrivances that naturalists have discovered in various species are fascinating, but I fail to see what difference it makes whether any of these probes, balloons, etc. are "really" tools. All the species involved remain distinct biological species no matter what decisions are made. Similar observations hold for rationality and anything a computer might do.

One possible explanation for the importance that people place on the human species possessing a character that sets it off sharply from all other species is essentialism. For a species to be a genuine species, it is not enough that it be reproductively isolated from all other species (a relational property). It also must possess a character that all and only human beings possess (a one-place predicate). But something more is involved. Numerous definite descriptions pick out the human species, including "featherless biped." No, the character must be not only unique but also "higher." Hence, the usual predilection for essentialism is not sufficient to explain the fervor with which we are inclined to defend the status of the species to which we belong.

Another source of the compulsion to insulate the human species from the rest of nature can be found in the traditional hierarchical organization of science. Because physics and chemistry deal with properties that all material bodies exhibit, they are the most fundamental sciences at the base of the pyramid of scientific specialties. Here and there, in various nooks and crannies of the universe, certain entities exist that are alive and can reproduce themselves. They are the subject matter of biology. Of these entities, a few can also sense, behave, have their behaviors reinforced, and possibly even think. Psychology deals with this peculiar complex of attributes. Of these creatures, only one sort forms "true" societies, the province of sociology. Once again, human beings find themselves as the pinnacle of a hierarchy of their own construction.

However, problems arise if we allow that organisms other than human beings form "true" societies, or that entities that are in no sense alive can "think," and so on. Social insects have nervous systems that are so rudimentary it is difficult to see how they can be capable of much in the way of thought, but they exhibit very complicated social behavior. Computers are not alive, but they can perform many of the functions that have commonly been associated with thinking. However, if "sociality" is attributed to social insects and "thinking" to computers, levels in the hierarchy are being skipped. An entity that is not in the province of psychology falls within the purview of sociology, and an entity that is not part of the subject matter of biology is nevertheless a proper subject of study for psychologists. If too many phenomena are discovered that skip or invert levels in the traditional hierarchy, the appropriateness of this hierarchy is called into question. It is one thing to extend "thought" or "intentionality" to other living creatures, especially if they are closely related to us. It is quite another to extend such properties to inanimate objects. The former threatens our uniqueness; the latter undermines our hierarchical organization of the world and our knowledge of it. One possible reason for the importance which we attach to the human species possessing not only an essential trait that sets us off from everything else in the universe but also a "higher" essential trait is to guarantee our position at the pinnacle of the traditional pyramid of scientific investigation.

Elster (1979) takes the traditional organization of the sciences for granted and argues that a diremption exists between the social and biological sciences. By definition, according to Elster, theories in the social sciences and only in the social sciences concern intentional behavior. It may be that only human beings are capable of intentional behavior or, as Dennett (1983) argues, that the "intentional stance" may be extended to other species, not to mention computers, but wherever this cut comes, it cleaves the social sciences from the rest of natural science. This particular method of classifying the sciences has some peculiar consequences for evolutionary biology. For example, earlier I quoted Thagard's (1980: 190) remark that theories, unlike genes, have people trying to make them better, but some genes also have people trying to make them better. The sort of selection practiced by horticulturists and animal breeders is intentional even though the traits being selected are biological and the mode of transmission genetic. Thus, if we accept Elster's classification, the main argument in Darwin's *Origin of Species* is fallacious because Darwin reasoned from artificial selection to natural selection. He reasoned that, if plant and animal breeders could do so much with so few organisms over so little time, nature could do so much more with all the organisms at its disposal during such great expanses of time.

Darwin did view his argument from artificial to natural selection in the *Origin* to be analogical; it can hardly be an essential part of the argument because Wallace never utilized it, but it can hardly be dismissed as one gigantic blunder (Waters 1986). What is more, contemporary biologists no longer view the relation between artificial and natural selection as an analogy. Instead, in present-day text books, artificial selection is treated as a special case of natural selection, intentionality notwithstanding (Dobzhansky et al. 1977; Endler 1986). Even though plant and animal breeders are engaged in intentional behavior, their science belongs among the natural, not the social, sciences. But, one might object, people can intentionally produce conceptual variations at will, whereas plant and animal breeders must wait around for an appropriate mutation. I doubt that this is a serious objection because, in the near future, scientists will also be able to produce desired mutations. When they do, I doubt that all objections to comparing artificial and natural selection will cease.

Another way of classifying selection processes is first according to the character of the replicators and only then according to the presence or absence of intentional behavior. In biological evolution, the replicators are primarily segments of the genetic material. In conceptual evolution, the relevant vehicles in Campbell's sense are books, brains, and magnetic tapes. According to this classification, artificial and natural selection are the same sort of phenomena because the relevant replicators are genes, even though the selective mechanism in the case of artificial selection includes selection by intentional agents. This is not to say that only human beings are capable of intentional behavior. Many other organisms "strive" in a variety of senses of this term, from the most literal to the most metaphorical. For example, a rabbit strives to elude a fox chasing it just as surely as the fox strives to catch it. However, with respect to selection, neither organism is striving to change its future course of evolutionary development, although their behavior may well have this effect.

The preceding considerations are sufficient, I think, to show that Elster's preferred classification of the sciences is not quite as "natural" as he makes it out to be. On the first classification, the development of a black tulip is the same sort of phenomenon as a compulsive eater staying thin by an exercise of willpower. Both involve intentionality. On the second classification, Mendel's breeding experiments are the same sort of phenomenon as the spread of the latest flu epidemic. Both involve patterns of gene transmission. If both modes of classification resulted in natural phenomena being divided along similar lines, it would not matter which one chose, but they do not. Hence, one must choose, but on what grounds?

Arguments can be presented for both modes of classification, but as far as I can see, one consideration overrides all others—effect on the resulting explanatory theories. The classification that facilitates the most powerful theories about the world in which we live is preferable. Some philosophers and social scientists have attempted to construct theories of intentional behavior. Several authors have also attempted to present an analysis of biological and sociocultural change that is equally applicable to both sorts of phenomena. The relative success of these two endeavors will determine which classification is preferable and, hence, which inferences legitimate. Thus far, neither research program has proven to be all that successful, but at least the rate of progress in "evolutionary epistemology" has recently been on the increase, while theories of intentional behavior seem to have stagnated. Needless to say, this book is designed to increase this difference between the two research programs, and to the extent that I succeed I will justify my preferred mode of classification over its intentional competitors.

In opting to classify phenomena initially in terms of the character of the entities selected and only then according to the character of the selection process itself, I do not mean to deny the importance of the mentalistic character of science. Scientists are conscious, rational, intentional agents. The point at issue is the effects that these characteristics have on science. Does the increased role of such factors in rational selection when compared to natural selection entail that a single analysis of the two processes is impossible or at least not profitable? In this connection, Rescher's (1977) distinction between Popper's (1972) thesis Darwinism and his own method Darwinism is helpful. As the names might imply, Popper purposes a variation/selection model for choosing between alternative hypotheses about particular phenomena, while Rescher proposes such a model to choose between different methods of making such choices. Rescher (1977: 153) objects to Popper's thesis Darwinism because, for any phenomenon, there are "*just too many* imaginable hypotheses to be gone through in any inductively blind trial-and-error search." However, with respect to methods, Rescher (1977: 159) argues that the number is sufficiently limited that trial-and-error just might be good enough:

> The human imagination is fertile enough that [at] any given stage the range of theoretically envisageable hypotheses is "more plentiful than blackberries." But the range of human experience is such that where the solution of our cognitive problems is concerned, the range of available investigation and explanatory *methods* is emphatically limited. At this level, the prospects of a trial-and-error evolutionism are vastly improved.

The distinction that Rescher draws between thesis and method Darwinism is helpful, but the asymmetry he perceives between the two with respect to number of envisageable alternatives stems from his applying different standards to the two cases. In the absence of any plausibility requirements, the range of alternative hypotheses to explain any class of natural phenomena is indefinitely large. As a result, Feyerabend (1975) has urged as a methodological principle in science "let a thousand theories bloom." What precisely scientists would do with so many theories, he does not say. They certainly could not test them. In fact, at any one time in science, the number of alternative explanations for a particular class of phenomena actually entertained by scientists is remarkably small, certainly no larger than the methods available for testing them. Of course, nothing stands in the way of scientists letting a thousand methods bloom—save good sense. Rescher (1977: 157) recognizes that selection in science occurs "not as a blind groping amongst all *conceivable* alternatives, but as a carefully guided search among the *really promising* alternatives." This observation holds equally for theses and methods.

In selection processes of all sorts, selection takes place among actual, not possible, alternatives. Given the number of loci and currently extant alleles in the human species, more possible combinations of these alleles exist than there are atoms in the universe. Although the number of combinations that actually exist at any one time is quite large, it is still a tiny fraction of the possible combinations, and selection occurs among actual, not possible, combinations. In science, alternative hypotheses and methods that are in some sense "envisageable" are limited only by the imagination and good sense of scientists. That Venus came close to colliding with the earth is conceivable, that Mars came in near collision with the earth is also conceivable, and so with Mercury, and Jupiter, and Alpha Centauri. Reasoning by simple induction with one finger raised is a conceivable mode of inference, with two fingers raised is also conceivable, and so on. Happily evolution has given most of us only ten fingers, but if the existence of brains in a vat and a twin earth containing a clear tasteless fluid with chemical constitution XYZ are genuine hypotheses that we must seriously entertain, then the alternative methods I have suggested are no less serious. Human beings, scientists included, are capable of seriously entertaining what to others seem like very silly hypotheses, whether about natural phenomena or about ways of investigating these phenomena. However, on the model of science that I am urging, those hypotheses that are never actually envisaged play no role. Hypotheses that are envisaged but never pursued also play no role.

Previously I have argued that both biological and conceptual evolution are only locally progressive, while some disagreement exists about global progress. At the very least, conceptual evolution in certain areas of science is more clearly progressive in a global sense than is biological evolution. Elster (1979: 9, 87) argues that the reason for this difference is a difference in the two mechanisms involved. Natural selection is a process capable of only local maximization, whereas people (and only people) are capable of global maximization—they can sacrifice immediate small gains for increased likelihood of long-term gain. Compromises are constantly made in biological evolution. Chief among these is balancing the needs of personal survival with the needs of reproduction. If either side is shorted too drastically, the effect is the same—extinction. Even the most gradualistically inclined evolutionary biologists acknowledge that, on occasion, speciation can occur in fairly large, discrete steps, but no matter how large or discrete these steps may be, each step must be viable. Similar observations hold in conceptual change. Sometimes conceptual changes are fairly abrupt and large. In science, such periods have been dubbed "revolutions." But scientific change can also occur quite slowly and gradually in very small steps. Regardless of the speed of conceptual change, each step must also be "viable," that is, it must actually be a link in a conceptual replication sequence. Hypotheses that are envisageable but never envisaged are irrelevant to conceptual change.

People can sacrifice immediate small gains, within limits, for increased likelihood of long-term gain. The gratification of eating an extra dessert can be sacrificed for the more extended gratification of good health and admiration from one's sex objects. But what does this distinction mean when applied to science and scientists? Scientists want to understand the natural world. They want to know why fire goes up and rain comes down, why organisms seem so marvelously adapted to their stations in life, why Peeping Toms rarely molest. Most scientists play it safe, making small, uncontroversial contributions. A few attempt to revolutionize some area of science. They risk failing big. Most times they do fail. A scientific career might survive one fiasco of this sort, as did Goldschmidt's, but not many. His early errors in counting chromosome numbers were excusable, but his postulating but not pursuing drastic modifications of each new consensus resulted in his later work being ignored.

But this does not seem to be what Elster has in mind. Rather he thinks that science is globally maximizing because scientists can anticipate the future. Although science is as intentional as an activity can get, the effects of this intentionality are minimal (Giere 1987). Although scientists strive to solve unsolved problems, all that this striving influences is the relative speed of change (Zimmerman 1978), not its global direction. As Popper (1959) has repeatedly argued, scientists cannot know the results of their research in advance. All scientists are trying to understand the world in which they live, but only a small percentage succeed. Science as a process looks so directional because of the retrospective bias that we bring to it. It can be made to *look* globally maximizing by careful editing. Most research fails or leads nowhere. Of the research that finds its way into print, most has no appreciable effect. Once we weed out all the failures and false steps, science looks very efficient, goal-directed, and globally maximizing. However, if we expand our perspective to include all of science and not just the rare successes, it takes on quite a different appearance. Scientists share a general goal, but most of their specific goals turn out to be illusory.

People and possibly only people are capable of global maximization. But rarely is this ability exercised, and when it is, to no great effect. At the very least, it has had little effect on science. The increased role of intentionality in conceptual replication makes it somewhat more efficient than genetic replication—but just barely. If scientists did not recognize certain phenomena as being problematic and strive to solve these problems, science would not have the character that it does. However, the scope of the effects of this striving should not be exaggerated. Even if one interprets the presence of intentionality in rational selection as introducing a difference in kind between it and natural selection, it is a difference in kind that does not produce much of a difference in degree. It influences the speed and efficiency of the process somewhat, but it is not responsible for science being globally progressive. Our moral behavior is just as intentional as our attempts to understand the empirical world. People are just as capable of global maximization in both. But the effects of the exercise of these abilities in the two areas have been significantly different. Hence, the causes of these differences must lie elsewhere.

Conclusion

In this chapter I have addressed the various claims about essential differences between the ways in which selection operates in biological and conceptual evolution, paying special attention to science. As far as generation time is concerned, conceptual evolution occurs at the same rate as biological evolution—by definition. As far as physical time is concerned, conceptual evolution lies somewhere in the middle. Lineages of viruses evolve much faster than conceptual lineages and lineages composed of sequoia trees much more slowly. Neither biological nor conceptual replicators are all that "particulate." In both cases, the relative "size" of the entities that function either as replicators or as interactors is highly variable. For a large number of organisms, inheritance is biparental; for most it is not. In conceptual evolution, rational agents sometimes combine ideas from precisely two sources; however, polyploidy seems much more common in conceptual evolution than in biological evolution. However, when one switches from the level of individual entities to populations, no significant differences can be found between the two. If lineages are defined in terms of replication sequences, extensive cross-lineage borrowing is ruled out, once again by definition. When conceptual and social lineages are distinguished in science, extensive cross-lineage borrowing becomes possible. However, it does not seem to be as common as all the talk about "syntheses" would lead one to expect.

No one claims that conceptual evolution in science is literally Lamarckian, as if the basic axioms of quantum theory are somehow going to find their way into our genetic makeup. At most conceptual change in science is only metaphorically Lamarckian, but the only metamorphical sense in which conceptual change in science is Lamarckian is that it is to some extent intentional. If natural phenomena are divided initially into those that are intentional and those that are not, then artificial selection and rational selection are the same sort of activity because they both are intentional activities even though the things that are altered in artificial selection are gene frequencies and those in rational selection are meme frequencies. If, to the contrary, natural phenomena are divided first according to the sort of replicators involved, then artificial and natural selection are the same sort of activity because they both effect changes in gene frequencies. Sociocultural practices that eventuate in changes in frequencies of cultural entities belong in a second category, regardless of whether these changes result from intentional or unintentional behavior. In this book, I have adopted the second classification.

A common response to Laudan's (1977) decision to treat truth as of secondary importance to progress in his analysis of science has been that he has totally rejected truth and thus has adopted some form of instrumentalism (Jardine 1978). I expect a comparable response to my arguing for the secondary importance of intentionality in understanding selection processes. The critics are likely to reason that, if intentionality is of secondary importance, then it is of no importance at all, and I am arguing for radical behaviorism. I am afraid that nothing that I can say can preclude such a parody. Even though I must admit that in the preceding discussions of the role of intentionality in science I have carefully begged all the philosophically interesting questions, one of the basic premises of this book is that any issue that has little in the way of effect on the actual conduct of science can safely be ignored, and I do not see the effects of the intentional character of rational selection as being massive.

That science is globally progressive in a way that is problematic in biological evolution is more important. In order for development to be progressive, it must be at least directional. Conceptual development in certain areas of science (as well as elsewhere) has been directional, even progressive, but not because it is intentional. Intentionality can speed up conceptual development, but it cannot assure directionality, let alone progress. If everything about the natural world were in a state of haphazard flux, scientific theories would also continue to change indefinitely, not just because scientists continue to change their minds about nature, but because nature itself is changing. Goal-oriented behavior can have a direction in a global sense only when the goals stay put.

Whenever the conditions are right, evolution by means of natural selection will occur. The global goal of natural selection may well be increased adaptation, but for particular lineages the particular goals keep changing, not because genetic variation is "blind," not because natural selection is nonintentional, but because so many of the aspects of the environment to which organisms must adapt keep changing. Conceptual evolution, especially in science, is both locally and globally progressive, not because scientists are conscious agents, not because they are striving to reach both local and global goals, but because these goals exist. Eternal and immutable regularities exist out there in nature. If scientists did not strive to formulate laws of nature, they would discover them only by happy accident, but if these eternal, immutable regularities did not exist, any belief that a scientist might have that he or she had discovered one would be illusory.

13

Conceptual Interaction

The reciprocal relationship of epistemology and science is of noteworthy kind. They are dependent upon each other. Epistemology without contact with science becomes an empty scheme. Science without epistemology is—insofar as it is thinkable at all—primitive and muddled. However, no sooner has the epistemologist, who is seeking a clear system, fought his way through to such a system, than he is inclined to interpret the thought-content of science in the sense of his system and to reject whatever does not fit into his system. The scientist, however, cannot afford to carry his striving for epistemological systematic that far. He accepts gratefully the epistemological conceptual analysis; but the external conditions, which are set for him by the facts of experience, do not permit him to let himself be too much restricted in the construction of his conceptual world by the adherence to an epistemological system. He therefore must appear to the systematic epistemologist as a type of unscrupulous opportunist.

Albert Einstein, *Albert Einstein: Philosopher-Scientist*

SELECTION PROCESSES are so complicated because they involve the interplay between two processes acting on entities at a variety of levels of organization. Replication is concentrated at the lower levels while interaction occurs at a much wider variety of levels from the lowest to the highest. Interaction ceases when the entities lack the requisite organization to interact as cohesive wholes with their environments. Almost the only general observation that can be made about the levels at which replication and interaction are occurring is that the interaction that results in the differential perpetuation of the relevant replicators occurs at the same level or higher levels than that at which replication is occurring. Some biologists are content to treat biological evolution solely in terms of replication; others do so but only grudgingly. They would like to include interaction in their causal accounts, but the task appears to be too difficult. However, even those biologists who are perfectly content with genic selection in biological evolution are unlikely to find its correlate in conceptual evolution sufficient because it leaves scientists out of the causal picture, and scientists are likely to find inadequate any account of science that omits them.

The primary focus of interaction in science is at the level of individual scientists, albeit increasingly modified by the organization of scientists into research groups.

Selection must be possible in the absence of such higher-level entities because it proceeded in their absence for the first half of life on Earth, not to mention the early years of science. In the absence of some form of sexual reproduction, biological evolution was a simple matter of replicators and interactors splitting successively from generation to generation. When mechanisms for genic recombination arose, the character of biological evolution changed drastically. As species came into being, the speed of biological evolution increased. Not until scientists began to build on the work of their predecessors did "science" come into existence. As they began to become organized into social groups, science changed its character. As cooperation among contemporaneous scientists emerged, competition between individual scientists became overlaid with additional levels of cooperation and competition.

Scientists are the entities that run the experiments, publish the results, read the literature, and pass on conceptual replicates. Textbooks, journal articles, and the like tend to be relatively passive vehicles in conceptual replication sequences. Scientists are anything but passive. In scientific evolution, scientists function both as elements in replication sequences and as the primary interactors. They are "vehicles" in both senses of this term. One of the great strengths of scientists is the interest they take in their work, an interest that is reflected in long hours, emotional involvement, and periodic unseemly dissension. Those commentators on science who claim that all behavior can be explained entirely in terms of interests term everything from conscious states to social forces "interests." In this book I have all but ignored the latter sorts of "interests."

Science is not impervious to the course of human history at large. The two world wars influenced the development of science in Europe, Great Britain, and the United States in several straightforward ways. Because the First World War did not have a very profound impact on the United States, scientists were able to resume their labors quite rapidly after the armistice. None too surprisingly, the locus of research in several areas, including Mendelian genetics and population genetics, shifted to the United States. The terror in Russia as well as the rise of fascism in Europe led many scientists to flee their homelands, including Dobzhansky, Mayr, Croizat, and Sokal. Later, Willi Hennig fled East Germany to West Germany in order to pursue his research in greater freedom. The formation of such organizations as the Rockefeller Foundation and the National Science Foundation also influenced the development of science. But even episodes in science that bear more narrowly on the cognitive content of science might also be explicable in terms of social forces, e.g., why have French scientists remained so conservative in so many areas of science? Why were the versions of evolutionary theory that became widely accepted in the second half of the nineteenth century so different from the version that Darwin set out? If Darwin's detailing a mechanism for the transmutation of species legitimized a belief in biological evolution, why did so few of his contemporaries accept the important role of natural selection in biological evolution? Part of the answers to these questions may be in terms of reason, argument, and evidence, but part may also be more external.

The trouble with such social forces is that they appear to influence the conceptual development of science in such an idiosyncratic and haphazard fashion that the only way to deal with them is in the context of highly particularized historical narratives.

However, they do belong in such narratives. If evidence can be obtained to suggest that a particular social force did influence a particular belief of one or more scientists, then it should be mentioned.

For example, if Darwin's position in the emerging middle class of Victorian England was operative in his coming to hold the "gradualist" views he did about the evolutionary process, then this connection needs detailing. But the justification of such a causal connection will be difficult because it involves such a particularized situation. It is not an instance of the general claim that scientists brought up comfortably in a society will oppose (or even tend to oppose) revolutions, whether physical, biological, or social. Sometimes they do; sometimes they do not. Wallace and Huxley came from poor backgrounds. Wallace favored gradual evolution just as much as Darwin, while Huxley opted for more saltative views. Even if one agrees that social forces influence the substantive views that scientists come to hold about natural phenomena, it turns out that certain scientists are influenced by one subset of forces, certain scientists by others, and so on. Victoria, after all, was queen to Darwin, Wallace, and Huxley alike (Gingerich 1984: 338). But such communal experiences do not eventuate in substantive agreement.

In this connection, one must remain skeptical of popular myths that fit one's predilections. For example, from the secondary literature on the subject, one would think that social Darwinism was a major influence among American intellectuals. After all, did not John D. Rockefeller, Sr., justify the rapacious character of American capitalists by reference to survival of the fittest? However, when historians have actually looked into the historical record, they discover that social Darwinists were noteworthy for their rarity. As Wilson (1967: 93) states, "No more than a handful of American business leaders or intellectuals were 'social Darwinists' in any sense precise enough to have a useful meaning." In addition, it was not John D. Rockefeller Sr. who attempted to justify capitalism by reference to biological evolution. It was his son. Like the story of Marx asking permission to dedicate *Das Kapital* to Darwin, such myths die slowly because they fulfill social needs, factual support to one side. It is certainly true that, in the second half of the nineteenth century, great numbers of Americans were ideologically committed to the desirability of competition in the business world and biologists considered biological evolution to be no less competitive; but much more than correlation is required to show causal connection (see also Bannister 1979).

Defenders of the effects of the more interesting sorts of social causes on the substantive content of science are not likely to look at the sequence of feuds that characterized systematics in the past two decades for ammunition in their externalist program. No one found Halstead's alleged connection between Marxism and cladism in the least convincing. Marxists, non-Marxists, and anti-Marxists could be found in equal numbers on all sides of this dispute. Perhaps Marx thought that fundamental social change was likely to come about only by means of political revolutions, and perhaps advocates of punctuational theories of speciation hold comparable views about the evolution of new species, but superficial similarities to one side, any causal connections seem unlikely. However, sociobiology would appear to be an ideal place to look for the effects of social influences. Advocates of sociobiology in the broad sense should be politically conservative, and opponents should be somewhat Left. After all, several of the most prominent opponents of sociobiology were Marxists (e.g., Levins, Lewontin, and Gould), but several of the chief architects of this research program were no less Left in their political leanings (e.g., G. C. Williams, Trivers, and Maynard Smith). Although applications of game theory have been viewed as conservative plots against the people (Martin 1978), Lewontin was one of the first to apply the techniques of game theory to the evolutionary process, while Maynard Smith developed the techniques that Dawkins subsequently popularized. Historically, the origins of the explanatory tools to which critics of sociobiology object most strenuously had their source in the writings of scientists whose politics were to the left of center. If politics played a role in the selective use of game theory, it was in the putative misuse of these techniques.

Scientists from different social backgrounds sometimes hold similar views, while scientists from the same social background sometimes hold different views. Until significant correlations between *kinds* of social forces and *kinds* of substantive scientific beliefs can be established, externalism of the more interesting sort remains only a hypothesis about science, a promising hypothesis perhaps, but that is all. And one must be prepared to discover causal connections that are not all that "logical." Political categories in the broad sense, such as liberal, conservative, middle class, and the like, may not be the appropriate natural social kinds to explain scientific change. Why has evolutionary theory tended to be so gradualistic from Darwin to the present? In part because of the historical legacy of Darwin. Darwin was raised a Lyellian uniformitarian and willed this predilection to succeeding generations of Darwinians. This tendency was only accentuated by the fact that the opponents of Darwinism tended to be saltationists. As a result, evolutionists who viewed themselves as Darwinians felt obligated to play down saltative mechanisms. The political factors that influence the substantive content of science most directly concern the politics of science.

The preceding observation about the highly particularized character of causal explanations for scientists' holding the views that they do should come as no surprise. The same observation holds for biological evolution. Organisms by and large are adapted to the environments in which they live, but the particular forms that the interactions between living creatures and their environments take are extremely variable. This variability is what causes certain evolutionary biologists to exclude reference to interaction from their formulations of evolutionary theory. In fact, some biologists are so dismayed by the array of adaptations that actually exist in nature and the lack of any significant regularities in this process that they are led to deny adaptationist scenarios the status of "scientific" hypotheses. Curiously, some scientists who dismiss adaptationist scenarios about dinosaurs and trilobites as being unscientific offer exactly the same sorts of scenarios for scientific development. According to these scientists, hypotheses about connections between the drying up of pools in river beds and the evolution of lungs are not properly scientific, while comparable hypotheses about connections between social class in Victorian England and the beliefs of Victorian scientists are perfectly scientific. In general, I see little difference in the warrant of such externalist Just-So stories and their adaptive counterparts.

To those who dismiss historical narratives as being unscientific or at best only explanation sketches, my referring larger social forces to elements in historical narratives about science will be read as a dismissal. On this view, if the laws governing conceptual change in science do not refer to such things as the rise of the mercantile middle class or if no such laws exist, then such events are irrelevant to our understanding of science. I disagree. Even though historical explanations do not involve derivations from laws of nature, as the covering-law model of scientific explanation requires, they are still explanatory. The explanatory force in historical explanations comes from the coherence and continuity of the historical entities whose development they chronicle. In historical explanations either an entity is shown to be part of a more inclusive historical entity or else the course of this entity is described. In the first instance, an entity is placed in its historical context—a star in its galaxy, a fossil in its lineage, a scientist in his or her research group, or a term-token in its research program. In the second instance, an historical entity is traced through time. Historical narratives chronicle the development of historical entities through time. Although which entities count as entities is theory-dependent, one need not always have a highly articulated theory to distinguish historical entities and follow them through time. Hence, even though historical explanations do not involve derivations from laws of nature, they are not theory-independent either (for further discussion, see Hull 1975, 1981, 1984b).

One major stumbling block in explaining scientific change is that it is carried on at the wrong level of analysis, as if one could present a lawful explanation of Mendel devising his view of inheritance in garden peas, his work being ignored, later formulations being widely accepted, and so on. But such things as Mendel's laws are particulars, analogous to tail-wagging in bees, and not kinds. One might reasonably expect to find laws governing communication. If so, then tail-wagging in bees might well be an instance of this regularity, but that is all. Similarly, any laws about conceptual development in science might well make reference to such things as ideas which threaten the common consensus but not something so particularized as Mendel's laws. There can be no law of *Homo sapiens qua Homo sapiens* because as a biological species *Homo sapiens* is a particular. Similarly, there can be no law of Mendel's laws as Mendel set them out (i.e., his particular pronouncements) because these pronouncements are conceptual particulars, tokens in conceptual lineages.

Regularities do exist between conceptual change in science and two sorts of factors—one traditional in philosophy of science, the other traditional in sociology of science. By and large, philosophers of science from Herschel, Whewell, and Mill to the present have agreed with scientists that their ideas about the natural world are strongly influenced by the results of their investigations. Geneticists came to hold the views they did about the relation between dominance and recessiveness in heredity in large part because of the results obtained in their breeding experiments. Just because all scientific theories are underdetermined by anything that might be called "the facts," it does not follow that evidence is irrelevant. As in the case of more external factors, considerable contingency also characterizes the connections between the reasons, arguments, and evidence that influence the beliefs scientists come to hold and their coming to hold these beliefs—once the effects of retrospective biases are eliminated. Scientists do reach periods of consensus, evidence does influence scientists in their reaching these periods of consensus, but which sorts of evidence influenced which scientists turns out to be more variable than one might expect. However, the sorts of factors referred to by internalists in their explanations of scientific change have a good deal more coherence and continuity than those postulated by externalists.

Certain sociologists of science have noted that a second regularity exists with respect to the views that scientists come to hold and a second sort of factor—obtaining credit from other scientists. Words are cheap; credit is not. In order for science to be a seletion process, scarcity is required. Credit is so important because it is so rare (Gellner 1985: 160). If science is a selection process, it should be no less influenced by the complex interplay between replication and interaction than is biological evolution, but conceptual change in science is complicated by the fact that ownership is not incorporated in the structure of the entities being replicated. All that matters in genetic replication is the sequence of bases incorporated in the structure of the genes. Hence, one organism gains nothing by stealing another's genes, whether of its own or another species. In fact, organisms go to great lengths to insure that they raise their offspring and not those of another organism, while certain parasitic species have evolved techniques to circumvent these mechanisms. The cowbird is one of the best-known examples of such a parasitic species.

Stealing can be productive in conceptual evolution because ownership is not incorporated in the structure of conceptual replicators. Instead it is brought about by scientists appending their names to their publications and other scientists citing them. Scientists can steal the work of other scientists merely by appending their own names to it. Scientists must use the contributions of other scientists, but why should they bother to give credit? Why is not stealing rampant in science? In the preceding chapters, I have argued that stealing in science is kept in check because scientists need each other's support in getting their views accepted by other scientists. Honorable scientists try to allocate credit where credit is due, but even those scientists who are less concerned with probity are still pressured to give credit so that they can gain support.

Perhaps from some Olympian view, one can see that certain scientists did not get the credit that they really deserved. Sometimes their work was used without ample credit. More often it was not used at all. This perspective is usually accompanied by a highly romantic and naive view of science and scientists. On this view, scientists who go out of their way to make sure that their fellow scientists fail to see the point of what they are doing are exemplars of proper scientific conduct. Anyone who pays attention to matters such as mode of presentation are condemned as opportunists, more interested in career advancement than the advancement of science. If truth does not prevail all on its own, it is not worth having. Examples of scientists who seemed bent on having their views ignored or who were dismissed as "crackpot" punctuate the narratives that I presented in the first half of this book, from Lamarck and Saint-Hilaire to Croizat.

Although there are places for naiveté and romanticism in life, the investigation of something as important as science is not one of them. When the romantic view of science is pushed to its limits, few are liable to espouse it, but certainly more than a hint of this attitude can be detected in the portrayals of science presented by scientists and students of science alike. Even so, it remains a fact about scientists from the inception of science to the present that they have been interested in obtaining credit for their contributions. Too much of their behavior is inexplicable on other hypotheses. Scientists are influenced both by the *results* of their labors in the field and laboratory and by the results of *their* labors in the field and laboratory. It is this interplay between these two concerns that gives science the peculiar character it has. Other people want credit for their contributions in other areas of human endeavor, and sometimes get it. Such things as reason, argument, and evidence matter in activities other than science. But nowhere else is the interplay between contributions and credit so finely tuned as it is in science.

In replication, structure is transmitted from entity to entity. If replication were the only process involved in selection, any sort of structure would do, but it is not. Some of the structure that is transmitted in replication sequences must have what it takes to count as information. Some of the structure of replicators must point outside itself. A systematic relationship must exist between it and something else. Not all stretches of the genetic material code for molecules that function in the construction of proteins, but some do. Depending on how successful the resulting proteins are at fulfilling the needs posed by the environment, replication is differential. Scientists do not spend all their time trying to test their views about the world by running experiments and making observations, but some of it is spent on these activities. In this chapter I discuss the interactions of scientists with that area of the empirical world they claim to be studying. As a result I am forced to discuss such traditional philosophical issues as the theory-ladenness of observation terms, operational definitions, and incommensurability. My views on these subjects are unusual in only two respects. First, I distinguish between the sort of piecemeal operationalizing that goes on when scientists attempt to test their ideas and the more global views of epistemologists. The former are extremely important in science; the latter are about as irrelevant as anything can get. Second, I contend along with Kitcher (1978) that meaning is best analyzed in terms of reference potential and that it in turn depends crucially on the existence of social groups. I differ from Kitcher in that I treat these groups as genuine social groups and not as defined by agreement on reference. In attempting to bring the natural world to bear on their theories, scientists are forced to engage in two sorts of "finagling," one in the messy process of "operationalizing" their concepts for the purpose of testing, the other in connection with broader issues of "meaning."

Observation and Theory in Science

If there is an area in which logical inference should play a crucial role, it is mathematics. Yet, in discussing the great advances that were made in mathematics during the eighteenth century, Kline (1972: 400) argues that progress in mathematics demanded almost a complete disregard of logical scruples in favor of intuitive and physical insights. "The conquest of new domains in mathematics proceeds somewhat as do military conquests. Bold dashes into enemy territory to capture strongholds. These incursions must then be followed up and supported by broader, more thorough and more cautious operations to secure what has been only tentatively and insecurely grasped."

Two different strategies have seemed attractive to scientists—bold, irresponsible incursions and cautious, careful explorations. The first strategy increases the magnitude of the achievement but at the cost of increased likelihood of error. Most scientists who produce the sort of bold conjectures recommended by Popper turn out to be mistaken. The second strategy decreases the likelihood of both error and major innovation. Which strategy has led to the greatest advances in science? If the histories of science written over the past few centuries are an accurate reflection, scientific change is largely punctuational. A very few scientists account for most change in science, and these changes are made within short periods of time. However, this impression may be more a function of what historians find interesting than of the actual facts of the matter. Histories of small improvements introduced by numerous unknown workers over protracted periods of time are, to say the least, difficult to write and even more boring to read. "Great Men" histories of science may have no more warrant than comparable histories of social change in human societies. However, the findings of sociologists of science support a view of science easily as elitist as Great Men histories presuppose.

The bold incursion strategy in science leads to frequent errors, errors which more careful investigators must later painfully expose. As a result, scientists have periodically urged greater caution. Instead of introducing views wholesale and then eliminating error later, scientists might be better advised to avoid the introduction of error in the first place. Error in science has two main sources—faulty observations and the misconstrual of observations because of related mistaken beliefs. For example, in the early days of genetics, chromosome counts were not easy to conduct. Mistakes were common. One of Goldschmidt's earliest controversies concerned the number of chromosomes in a particular species of trematode. As it turns out, all sides of the controversy had counted wrong. One cause of these errors was straightforwardly observational. The staining techniques and instruments were not good enough. Another source was the differing views held by the observers about what they *should* see. For example, Goldschmidt was strongly influenced by his observation of chromosome reduction of the sort predicted by Weismann (Stern 1980).

No one complains when observations "infect" a scientist's more general views, but periodic objections have arisen with respect to influences in the opposite direction. Observational errors of the more straightforward sort can be reduced by greater care and more powerful instruments and techniques. Errors introduced by the "biases" due to the views a scientist might have about issues and phenomena putatively related to those under investigation can be reduced by insulating observation from these other considerations. In fact, this source of error might be totally eliminated by insulating observation from everything else. If only scientists would restrict themselves to mindless look-see, the foundations of science would be much more secure. Since one of the chief sources of observational error in science is scientific theories, the goal of science should be the production of a reservoir of theory-free observation statements that all scientists can utilize, no matter what they might believe on other issues. If scientists disagree, they could then reconcile their differences simply by turning to this observational base.

The observational base might periodically be modified because of the discovery of an observational error. It of course would also be constantly supplemented by additional data, but no wholesale revisions would ever be necessary. Even though Aristotelian scientists were mistaken on almost every fundamental issue, if only they had presented their observations in a theory-free format, at least they would have produced observations useful to later scientists.

This view of science has proven to be very appealing, as several of the episodes discussed in the early pages of this book indicate. Scientists from Cuvier to Croizat have insisted that, in their work, they have stuck to the facts and nothing but the facts, while their opponents have speculated wildly. Such references can be explained in part as propaganda, but several scientific research programs incorporated a strongly inductive, operational construal of scientific method. Behavioral psychology is a paradigm example. In systematics, pheneticists and pattern cladists come close to advocating theory-free, totally operational classifications. According to the currently received view in philosophy of science, operationalism as a thesis about theoretical terms in science is bankrupt, and totally theory-free observation statements are a myth generated by erroneous epistemological theories.

I am of two minds with respect to the received view. If it is construed as a philosophical thesis about science, I agree. Theory-free observation languages and classifications are impossible. Even if they were possible, they would not be desirable. But all the attention that scientists pay to the operationalization of their concepts is not misplaced. Perhaps length is not equivalent to the various procedures scientists have devised to measure length, but scientists were not wasting their time when they devised these various methods of measurement. Similarly, the emphasis that pheneticists and cladists have placed on methods—explicit, carefully formulated, repeatable methods—have improved the science of systematics tremendously. The common response to these suggestions for improving classificatory methods was that they were not sophisticated enough—they left too much out. These responses were frequently accurate but nevertheless inappropriate. Operationalizations always leave things out. They could not function as operationalizations if they did not. I agree with the emphasis scientists place on operationalizing their concepts but draw back when they inflate this emphasis into a general philosophy of proper scientific method, especially when this method has an inherent temporal order and this order requires that scientists always begin their investigations with brute observations.

An excellent strategy in science is for a scientist temporarily to "bracket" a widely held scientific theory to see what the relevant phenomena look like without this presupposition. Such "bracketing" can never be complete, but even partial success can sometimes lead to startling new perspectives. This is precisely the strategy that Colin Patterson used in attempting to look at his own work in paleontology without presupposing evolution. In several places in this book, I have asked the reader to attempt similar willing suspensions of disbelief. For example, evolution looks very different if one does not take the levels of organization in the traditional organizational hierarchy as basic. Science also looks quite different if scientifically relevant social relations are taken as fundamental and conceptual similarity as derivative. However, I cannot even conceive of what it would be like to "bracket" all understanding and to view natural phenomena totally raw, the way that certain phenomenologists apparently recommend. Of greater importance, I have no idea what one might do with all these phenomenological reports once one had them. For over two generations, phenomenologists have worked earnestly at their philosophical research program. So, the story goes, the results of their labors should be the ideal starting point for genuine science. Thus far, the results of phenomenological analysis have proven to be the starting point for no successful scientific research programs whatsoever. The best candidates for such successful research programs occur in psychology, and they have been none too successful.

As my earlier discussions might have suggested, I think that similar observations hold for pheneticists and pattern cladists to the extent that they think that classifications can be and should be theory-free in the relevant sense. Since nothing answers to the name "theory-free observations," scientific classifications cannot begin with them. Philosophers of science commonly distinguish between theoretical and observation terms. Theoretical terms are those descriptive terms that appear in the most fundamental propositions of a theory, such terms as "mass," "acceleration," and "force." Because of the relative size, duration, and perceptual acuity of human beings, we have reasonably direct access to only a small band of middle-sized events and objects. Our windows to the world are both few and narrow. We can see litmus paper turn blue, a green fungus spread across the surface of a petri dish, and a spider ejaculate. Nothing precludes theoretical entities from falling within the narrow band of natural phenomena that we happen to be able to preceive, but few do. Most theoretical entities are too large, too small, or lack the sorts of properties that can impinge on our sense organs. Hence, their dependence on the theories in which they are postulated is clear. Those scientific terms that refer to entities and events we as sentient organisms can perceive might seem free of any such dependence.

For example, we can perceive many sorts of organisms. Many others exist at the lower limits of our perceptual acuity. Even so, "organism" counts as an observation term. It is also a theoretical term in biology. Organisms are those things that go through cycles of growth and reproduction, single-celled organisms presenting the limiting case. Because we can perceive organisms, we are tempted to assume that we can also perceive *that* something is an organism, but as the examples I enumerated earlier amply attest, we cannot. We can perceive a Portuguese man-of-war, but anyone who thinks that they can "simply see" that this colony is an organism is mistaken. The latter claim is a conclusion drawn on the basis of both observation and theory.

Some philosophers and not a few scientists are inclined to consider the accidents of the evolutionary development of our sense organs to be of fundamental importance to epistemology. The character of our sense organs in some significant sense "makes the world the way it is." It is certainly true that we cannot perceive what we cannot perceive, but the perceptual abilities of human beings vary extensively from person to person and within the lifespan of a single person, from time to time. Some people can hear tones much higher on the acoustical scale than can others. As time goes by, each of us loses our ability to hear higher tones, but I find it impossible to believe that our understanding of the science of acoustics is affected by such contingencies.

The human species might have evolved such that, in general, most of us were able to hear sounds much higher in the acoustical scale. Accidents of evolution are only one out of numerous sorts of contingencies that influence scientific development. If science is a selection process, then the particular paths that it takes are far from predetermined or predictable. As a result, the conceptual paths that physicists took in working out the physics of sound might well have been different. But in spite of such contingencies, scientists do seem to be able converge on remarkably similar conceptions of natural phenomena. Phenomenologically, red light and infrared light are quite different. Most people can see with one but not the other. Scientifically this difference is of no consequence. We know as much about one sort of light as the other.

Of course, human beings must have some sense organs or other if they are to learn about the world in which they live, but they need not have precisely the organs that they happen to have. Of course, human beings in learning about the world of their experience must begin somewhere or other, but it does not follow that they must all begin at precisely the same place or that the places at which they happen to begin have some sort of fundamental priority. If those who argue for the epistemological priority of perception mean more than this, I am unable to see what it is. A blind scientist who has never seen the color red can understand more about it than the millions of human beings who perceive it every day. I see no point in basing an entire branch of philosophy, not to mention all of science, on the happenstance of our evolutionary development.[1]

If the boundary between what human beings can perceive and cannot perceive were sharp, universal, and invariant, then there might be some point in attempting to distinguish sharply between observation terms and theoretical terms. Observation terms would refer to things that all human beings, or all normal human beings, or all well-trained normal human beings can perceive regardless of what they believe about the world. Observation terms would then be equally available to everyone. There would be no difference between seeing and seeing *that*. As the earlier discussions of the views of certain ideal morphologists, pheneticists, and cladists indicate, scientists themselves at times insist that they can produce a strictly descriptive, theory-neutral observation language. Part of this conviction stems from the fact that in the early stages of their development, human beings make observations even though they understand almost nothing. If babies can make theory-neutral observations, then such observations are possible, and all the opportunities for error that our beliefs about the world introduce into our descriptions of it might be avoidable. Perhaps the basic observational vocabulary of practicing scientists is not totally theory-neutral, but it is both possible and desirable to make it as theory-neutral as possible, or at least so the advocates of this bewitching position claim.

1. Through the years I periodically have heard about an early French physicist, Joseph Sauveur (1653–1716), who was a pioneer in the science of acoustics even though he was deaf and dumb since birth. If the story were true, Sauveur would be excellent evidence for the lack of any necessary connection between phenomenological awareness of a particular sort of natural phenomenon and scientific understanding of it. Unfortunately, this story appears to be a myth. According to the *Dictionary of Scientific Biography*, Sauveur could not speak until a relatively late age but was never deaf.

A second reason why scientists think that the goal of a theory-neutral observation language is obtainable is that they know so much that they forget how much even their simplest observation terms actually imply. For example, Antony van Leeuwenhoek (1632–1723), the Dutch lens maker who constructed several of the best early "microscopes," prided himself on describing what he saw without loading his descriptions with all sorts of theoretical baggage. As is usually the case, Leeuwenhoek's claims about sticking to the facts and nothing but the facts were more self-serving propaganda than anything else (Ruestrow 1984). However, one of his descriptions presents a good example of how theory-laden an apparently theory-neutral description can actually be. In a letter to the Royal Society of London, Leeuwenhoek described the little animalcules he saw swimming about in a drop of water taken from a gutter outside his window. That he called them "animalcules" and not "plantacules" reveals something about Leeuwenhoek's beliefs about plants and animals. If it swims about, it is an animal. But some things that can swim about also contain chlorophyll.

Of course, the commonsense distinction between plants and animals implicit in Leeuwenhoek's terminology is not very "theoretical" in the sense of being connected to a well-formulated scientific theory, but numerous examples of such close interdependence can be given. The astronomer who says that he can look up at the night sky and see the temperatures of the various stars is so embedded in his own theoretical perspective that he no longer realizes how complex and theory-laden inferences from the apparent color of a star to its temperature actually are. A common response to the early work by Michener and Sokal (1957) was that their success depended on the character of Michener's data sets, and Michener constructed his data sets in the context of a lifetime of knowledge of all aspects of the bees under investigation. The data that he and Sokal used in their program had already been extensively sanitized before they were fed into the computer. Well-structured data in, clear patterns out.

To avoid the dangers inherent in a scientist's knowledge "infecting" his observation reports, pheneticists suggested that ideally all data should be collected by intelligent ignoramuses. If the people collecting data know nothing about the phenomena under investigation, then their reports cannot be tainted by such considerations. It might be difficult to find data gatherers who are totally ignorant, e.g., they might hold some crude, everyday "theories" about the differences between plants and animals, but effects of these beliefs are likely to be less extensive and easier to discern than comparable effects of the sophisticated knowledge of professional scientists.

From Bacon to the present, the vision of armies of unlettered data gatherers has seemed attractive, but thus far it has never been realized. The crude inductivism implicit in this vision has been largely untried and when tried, spectacularly unsuccessful. The collection of data for the sake of data has proven to be an especially inefficient way of conducting scientific investigations. Rarely has the massive accumulation of data led to any significant improvement in our understanding of natural phenomena either on the part of the scientist collecting this mass of data or by others using it. As Mayr (1982: 647) has noted, Carl Friedrich von Gärtner (1772–1850) performed 10,000 separate breeding experiments among 700 species yielding 250 different hybrids. Nothing resulted from all this labor. Neither Gärtner nor later workers were ever able to use his results. They lie buried in the stacks of a few libraries around the world, as sterile as the day that they were conceived. Of course, Gärtner was neither an ignoramus nor a pure inductivist. He tried to produce hybrids between species because he thought that interspecific crosses were the key to unlock the mysteries of inheritance. He was mistaken. But his chances of success would not have been increased by his being more ignorant or proceeding more inductively.

Time and again, scientists would very much like to have used the data gathered by their more empirically oriented colleagues, but time and again they have been forced to run their own experiments and gather their own data because the data gathered by others are not good enough for their own needs. The major exception that I have been able to find to this generalization is the use Sewall Wright made of the data on guinea pigs collected by Walter J. Hall at the United States Department of Agriculture (Provine 1986). Even though Hall was not collecting data for any special purpose, these data were accurate enough and complete enough for Wright to use them for his peculiar purposes. One thing that Dobzhansky learned from his early work with Wright was that he should not collect data until Wright explained to him which sorts of data were needed to distinguish between the options under investigation (see also Schaeffer 1986 for the essential and pervasive role of theories in the ongoing process of science).

One source of the insistence of pheneticists that theoretical considerations should be excluded from the classificatory process in the early stages may well be their concern with the classificatory process itself. It one wants to understand how a classification was actually produced, then all the behind-the-scenes "sanitization" of data that goes on in systematics is subversive. For example, in studying a monograph on the Tettigoniidae (Katydids) of Australia produced by D. C. F. Rentz (1985), Colless (1985) suspected that Rentz had rejected some characters simply because they did not fit into his early tentative classification of these insects. He was surprised to discover a high degree of incompatibility among these presumably sanitized characters. He also discovered that, in spite of such extensive incompatibility, nearly all of Rentz's genera were extremely robust. Studies such as Colless's are extremely helpful in understanding the intuitive methods that traditional taxonomists use in the construction of their classifications. A few such studies are certainly called for, but requiring all taxonomists to include in their monographs all the "unsanitized" data that they utilized in early stages of the classificatory process would make the publication of such monographs, already expensive, financially prohibitive. I for one have not even considered publishing all the unsanitized data that I collected in my research over fifteen years, e.g., transcripts of my interviews with forty-six scientists that lasted from two to twenty hours each.

One would think that the pheneticists, who are committed to the production of general-purpose classifications equally useful to all scientists, would buttress their in-principle arguments about the superiority of phenetic classifications to all other classifications by listing all the general-purpose classifications they have produced and the scientists who had found these classifications especially useful. Thus far they have not. Conversely, the in-principle arguments that I have presented are unlikely to neutralize the conviction that theory-free classifications would be extremely useful if only systematists would produce them. However, continued failure might have some impact. As Rosenberg (1985: 186) has noted, the strongest argument against phenetic taxonomy is not philosophical but factual. In the past quarter of a century, the methods urged by the pheneticists "have simply not generated a workable taxonomy, one that will meet all or most of the needs for which biologists appeal to this discipline."

Pattern cladists are as opposed as pheneticists to letting anything save observations enter into the formation of their classifications. For pattern cladists, classifications are nothing more than synapomorphy schemes, the classifications implied by nested transformation sequences. Although Platnick (1985: 88) acknowledges that the "pheneticists of two decades ago tried to float the notion that their classifications were theory-free," he objects to the comparable claim made with respect to cladistics. To the contrary, "I know of no cladist so naive as to believe that there are any scientific statements whatsoever that are theory-free." But the only "theories" that Platnick accepts for cladistics are the assumption that natural groups exist, that classifications must correspond to data, and that only clustering by presence of characters will resolve natural taxa. These "theories" are a long way from the sorts of theories philosophers of science such as Popper insist permeate even the most observational terms. The theories of greatest concern to those philosophers who discuss the theory-ladenness of observation terms are such process theories as Newtonian theory and relativity theory (Popper 1962: 34), precisely the theories Platnick (1982: 283) rejects as being irrelevant to cladistic analysis:

> But one needs no causal theory to observe that of all the millions of species of organisms in the world, only about 35,000 of them have abdominal spinnerets. One also needs no causal theory to observe that of all the millions of species of organisms in the world, only about 35,000 of them have males with pedipalps modified for sperm transfer. One needs no causal theory to observe that the 35,000 species with abdominal spinnerets are precisely those 35,000 species with modified male pedipalps. One needs no causal theory to observe that of the thousands of (cladistically treated) characters that have been found to vary among those 35,000 species, not a single one has been shown to be unique to only some of those 35,000 species plus any species outside of those 35,000.

"Abdominal spinnerets" and "male pedipalps modified for sperm transfer" are hardly theory-free observational terms, and the theories that are involved are straightforward causal theories of all sorts including the very sort Platnick wants to exclude (Sober 1988). For example, one cannot merely look at germ cells and observe that one is male and the other female. In refining this distinction, evolutionary theory and the role of sex in the evolutionary process have played a central role. Regardless of what it might do to commonsense conceptions of the sexes, gametes are classed as male or female depending on their size and motility. Similarly, a present-day specialist in brachiopods can just look and see which shell is dorsal and which ventral without realizing how much theoretical labor went into deciding that brachiopod shells are dorsal and ventral rather than right and left or anterior and posterior (Dexter 1966). Something physiological and/or psychological is going on when a scientist watches a piece of litmus paper change color, a fungus spread across the surface of a petri dish, or a spider transfer sperm, but these statements refer to much more than these physiological and/or psychological occurrences.

When Platnick refers to characters being treated "cladistically," he is not referring to these sorts of considerations but merely to the results of seeing how various transformation sequences nest. Those characters that eventuate in perfectly nested transformation sequences are genuine characters, regardless of past practices; those that do not are not. For example, if the presence of abdominal spinnerets had not agreed in its distribution with other traditional arachnid characters, Platnick would have been forced to reject it as a genuine character. All sorts of causal theories, some of them no doubt mistaken, went into the sorts of arachnid characters Platnick learned to recognize when he first entered the field. Only when the effects of all these "theories" have been expunged from the work of arachnologists will cladistic classifications of spiders be theory-neutral in the relevant sense.

There are excellent reasons to want to eliminate the effects of mistaken theories. One way to do this is to eliminate all theories, but such total bracketing is impossible. Even if it were possible, it would have disastrous consequences for science. Science without process theories is not science at all. Once again the issue is not a particular scientific research program but the nature of science itself. One reason why scientists are partial to theory-neutral observation language is that they think that it would be worth having. It would be worth having because it would provide a common, reasonably safe point from which all scientists could start their investigations. It would also form a court of last appeal to decide subsequent disputes. If our theoretical beliefs influence how we present our observations, then choosing between conflicting theories becomes a very messy undertaking. Choosing between theories *is* a very messy activity, and I think that there is no way to make it much neater and more automatic than it currently is. Even if it were possible to produce a set of terms that referred to raw, uninterpreted observations, these terms would be scientifically useless.

As in the case of phenomenology and phenetics, the proof of the pudding is in the eating. Time and again, the goal of constructing classifications based on theory-neutral observation of patterns in nature has struck scientists as promising. Time and again, scientific research programs predicated on this perspective have failed. Only time will tell whether the theory-free patterns that pattern cladists propose to reflect in their classifications will prove useful to other scientists. If evolutionary biologists looking for patterns to test their theories about the evolutionary process find just what they need in the classifications produced by pattern cladists, in-principle arguments about how scientifically impoverished theory-neutral classifications are will prove as ineffectual in the evolution of science as is prayer in forestalling natural disasters. However, if theory-neutral classifications actually do turn out to be scientifically useful, the general philosophy that I have set out in this book will be seriously threatened because it in turn is predicated on the importance of a feedback between science and meta-science. Just as science proceeds by reciprocal illumination between theory and observation, philosophy of science proceeds by feedback between advances in science and advances in philosophy of science. The rejection of reciprocal illumination in science does not require a parallel rejection in philosophy of science, but it does increase the likelihood that it can be eliminated in philosophy of science as well.

Summa Contra Kuhn

Those scientists who accept the current received view on the impossibility of a theory-free observation language are caught in a bind. If the data that are to test a particular theory are structured by that theory, how in the world can they ever genuinely test it? An observation that apparently confirms a theory might confirm it merely because of the way that it was generated in the first place. Similarly, if each theory permeates the observation statements derivable from it, then one and the same observation statement cannot be derived from different theories ostensibly about the same natural phenomena. Hence, it should be impossible to choose between alternative theories on the basis of different observational implications. In fact, even calling these theories "alternatives" is not justified because different theories are about different phenomena. They cannot be brought into conflict because they are "incommensurable."

The trouble with the preceding arguments is that occurrences that should be impossible happen all the time. Sometimes observation statements generated within a particular research program, incorporating all the biases favorable to that program, nevertheless count against it. Sometimes two theories, as incommensurable as they might seem to be, do imply incompatible states of affairs. Even though the hare should never be able to catch the tortoise, sometimes it does. Whenever the impossible happens, one is forced to reexamine the conceptual system that generated this erroneous conclusion. In the case of the tortoise and the hare, the fault lay in how we conceived of space and time. In the case of incommensurability, the fault lies in our conceptions of scientific theories as they function in the scientific process.

In response to the obliteration of the observation/inference distinction, Fodor (1984) and his no-nonsense alter ego, Granny, object that organisms are strongly predisposed to certain observations in spite of past experience. For example, even after certain optical illusions have been pointed out and their sources explained, human beings still remain prone to them. If our perceptions are so strongly influenced by our background knowledge, we should come to see what we are supposed to see, but sometimes we do not. For example, we all know that the sun does not revolve around the earth, and with effort, we can force ourselves to see a sunset as the earth rotating away from the sun; but the second we relax, we flip right back to our old way of seeing. Perhaps the distinction between observation and inference is not sharp, but it is nevertheless very real. Although no observations may be totally theory-free, especially when the notion of "theory" is greatly expanded to include all sorts of half-articulated background knowledge, some are freer than others.

The paradoxes that have resulted from the theory-ladenness of observation stem from an unreal view of scientific theories and the relation of observation statements to them. The guiding metaphor in modern philosophy of science has been theories as conceptual nets mapping the world. According to the positivists (the all-purpose evil demons in philosophy), each node in the network must be connected to a corresponding atomic fact in nature. In his famous attack on the dogmas of "empiricism," Quine

(1951) reworked this metaphor so that only the nodes at the periphery of the net are actually tied down in some way to reality. Statements nearer the center of the conceptual network are more analytic, while those nearer the periphery become increasingly synthetic. Testing occurs at the periphery of conceptual networks, and any discrepancies discovered are accommodated, if possible, by adjusting relatively peripheral connections. If not, adjustments are made in the more central connections. On Quine's view, the analytic-synthetic distinction becomes a continuum. The more central a statement is, the more analytic it is. Nor are statements tested in isolation but only in the context of entire theories.

According to Fodor (1984), perception is "modular." Not all background knowledge is available all the time. Certain perceptual modules are more closely linked to certain areas of background knowledge than to others. As a result, certain perceptions and areas of background knowledge can be brought into contact quite easily, others only with great difficulty. At the conceptual level, this perspective implies that conceptual systems might be portrayed more appropriately as patchwork quilts than seamless webs. Each scientist usually works on a very limited area of the content of his or her science—differences between sexual and asexual reproduction, problems surrounding levels of selection, ways of reconstructing phylogenetic trees, and so on. In the course of research, a scientist will entertain slightly different versions of the problems he or she is trying to solve as well as different possible solutions. Each alternative formulation of problems and their solutions will bear somewhat differently on related areas. One solution might mesh neatly with the most promising solutions of related problems in one related area—that counts for it—and conflict with equally promising solutions in another related area. This conflict might force an investigator to shift his or her research to that area. For example, an evolutionary biologist who thinks that sexual and asexual organisms must function very differently in the evolutionary process might have to suspend research on this issue in order to straighten out the taxonomy of the organisms that he or she is studying. Given accepted classifications, the scientist's hypothesis is not working out, but then the classification might well be mistaken.

Although there are areas of systematicity in any scientist's conceptual system, few scientists succeed in forming a totally seamless worldview. Areas are partially dependent, partially independent. When one expands this picture to include other scientists, conceptual development in science becomes even patchier (see Campbell's 1969 fish-scale metaphor). As my earlier discussion indicates, no two scientists are ever in total agreement with each other even in their areas of most concentrated investigation. One will think that a particular innovation is most central to the more general system they are jointly developing, while his colleague will evaluate the situation somewhat differently. A generation or two after conceptual development in a particular area of science has ceased, all of this variation might well have disappeared, but it is an essential feature of science when scientists are having to make the decisions that eventually produce this consensus.

If one is interested in science as a process, conceptual systems cannot be viewed as seamless wholes. Science is a cooperative affair, each scientist contributing his or her special combination of abilities and areas of knowledge. A more appropriate picture is scientists casting their patchwork nets, one after the other, retrieving them, reworking them piecemeal on the basis of the most recent fit, and then casting them again. Conceiving of conceptual systems in this way has both its advantages and its disadvantages. As Simon (1962) has argued, causal localization is calculated to keep error from ramifying rapidly and indefinitely through a physical system. The price is, of course, no instant unity and global coherence (see Wimsatt 1974). Parallel observations hold for conceptual systems as well. One of the persistent goals of scientists is to present theories that are as universal and coherent as possible. Einstein and hosts of physicists have not labored to produce a unified field theory for nothing. Perhaps such a theory is impossible, but its production continues to be a consistent and important goal. The mistake is to treat such ideal goals as if they were currently realized, as if scientists can and should proceed as if they already possessed such idealizations. It is not a valid criticism of a theory that it is not currently totally global and coherent. Unlike some of my professional colleagues, I would hate to see all scientists give up this general goal.

According to the logical empiricist analysis of science, scientific theories are totally explicit, perfectly precise inferential systems. Either a statement is derivable from a particular theory or it is not. The conflict between two different theories must be needle sharp or nothing. Given the high level of precision assumed in this conception, slight differences in meaning can make a difference. But scientific theories as they function in science are much cruder. Perhaps a particular theory through the years might approach the philosophical ideal, but when scientists are attempting to test scientific theories, they are extremely far from this ideal state. Frequently it is very difficult to decide which observation statements follow from a particular theory without worrying about deriving a particular observation statement from one theory and its contradictory from another. Kelvin and Darwin produced very different theories about quite different phenomena, but these theories overlapped in one area—the age of the earth. Perhaps Kelvin and Darwin meant something subtly different by "age" and "earth," but the differences on this score between these two theories were so massive that any slight differences in meaning were irrelevant.

Scientists frequently *do* misunderstand each other, and sometimes these misunderstandings covary with adherence to different theories, research programs, worldviews, or what have you. For example, the term "cladogram" has caused massive confusion as its meaning has evolved. However, just as often, scientists who are in marked disagreement understand each other just fine. For example, one of the most bitter disputes in biology was between the early Mendelians and the biometricians. Opportunities for confusion were rife, and those involved took ample advantage of these opportunities, but they were able to resolve some of these confusions and selectively understand each other. As MacKenzie and Barnes (1979: 201) conclude, "Although the opposed communities did have occasional, temporary problems of communication, overall the evidence suggests that they were able to understand each other remarkably well."

If semantics has any implications for communication among people, then scientists working in different research programs should have a harder time communicating with each other than scientists working in the same research program. Of course, when the data turn out not to confirm such an operationalization, the advocates of a particular semantic theory can always retreat to the claim that semantic theories imply nothing about human communication. Nothing any person might or might not do can count either for or against a genuine semantic theory. If so, then incommensurability is one of those rarefied, in-principle problems that seem important to those working on general semantic theories but seldom cause any problems in actual practice. Do scientists working in different research programs understand each other perfectly? No, they do not, but then neither do scientists working in the same research program. For what it is worth, the scientists whom I interviewed claimed that difficulty in understanding did not especially covary with differences in overall theoretical outlook. If scientific theories actually were as pervasive and invisible as the air we breathe, then incommensurability might be a genuine problem, but scientists live in a much more polluted atmosphere. Like city dwellers, they are leery of breathing anything they cannot see.

Investigators who study science frequently express shocked disbelief with respect to actual scientific practice. According to popular conceptions, scientific theories are "scientific" because they are falsifiable, yet most of the research notes that see print confirm the hypothesis under investigation. Falsifications are rare. Sometimes an entire literature grows up on the basis of minimal data. As Raup (1986) notes, about the only hard data in the controversy over celestial causes for mass extinctions are the indications of increased iridium levels in the fossil record immediately preceding some mass extinctions. The only justification for postulating such things as Nemesis and the Oort Cloud is the very data they are designed to explain.

But, one might complain, such paucity of data characterizes science only in the earliest stages of an area of research. However, it is also characteristic of more settled areas. For example, from the beginning of the Modern Synthesis in biology to the present, the most common mechanism postulated by the balance school for the maintenance of genic variation is heterosis or overdominance. The usual example of this phenomenon is the sickling gene in human beings. Lewontin (1974: 37, 178) for one is embarrassed by the fact that it is the "only well-authenticated case of single-gene balanced polymorphism." Although we now have a few other reasonably good examples, the same examples are presented in exposition after exposition because these are close to the only examples that geneticists have. As Burchfield (1975: 202) notes, radioactive dating was widely accepted in 1931 even though only seven samples had been dated at the time with any degree of rigor. The universality of Kepler's laws was accepted on the basis of roughly a handful of observations on a single planet, while the universality of the genetic code was considered well established on the basis of information on an equally small number of species. Fortunately, these species were scattered throughout the phylogenetic tree, increasing the warrant of the conclusion. Parallel observations do not hold for the planets used to test Kepler's laws. They revolve around an obscure star at the edge of a single galaxy.

Physicists did proceed to investigate the paths of planets other than Mars, and geneticists did work out the genetic code of organisms other than the half dozen or so paradigm species. They were not especially surprised to discover what they expected to discover. But there are probably as many planets in the universe as there are species here on Earth. In testing, a point of diminishing returns is soon reached. Additional observations that bear directly on the particular phenomenon at issue are not worth the effort. The explanation why so little in the way of direct, hard data can produce such widespread acceptance so rapidly turns on the interconnectedness of scientific theories. Scientists are willing to accept certain problems as solved and proceed to new problem areas on the basis of admittedly minimal evidence because they are confident that error ramifies. If the hypotheses that they are accepting in order to attack new problems are mistaken, the results of related, though partially independent, research are likely to signal that something is wrong. The modular organization of conceptual systems allows scientists to localize their research. They do not have to study everything at once. To the extent that related areas of science are genuinely unified, errors can ramify rapidly. Any error found anywhere in the system calls the entire system into question. Modular organization slows down the diffusion of both apparent support and apparent conflict. A major change in one area might require only a minor adjustment in another.

Treating scientific method as a matter of reciprocal illumination poses problems for the methodologist who wants to provide a general analysis of scientific method (see, for example, Glymour's 1980 bootstrap-confirmation). Treating conceptual systems in terms of numerous versions related by descent presents a second dimension of complexity. Still another sort of complexity is added if each version is treated as being modular in organization. The end result may well prove to be intractable. For instance, when Simon (1983) himself turned to a general analysis of fitness requirements for scientific theories, he treated theories as being totally nonmodular. If one treats scientific theories in a highly idealized fashion, then one can make some reasonably simple general observations about such topics as confirmation. Even so, these analyses soon get extremely complicated. If, to the contrary, one accepts the complicated picture that I am recommending, then one cannot even begin with first-order approximations that are in any sense simple. The story is quite complicated right from the start.

Summa Contra Kripke

In the preceding discussion I have treated scientific theories and language in general as if they mirrored the world. Wittgenstein (1953) introduced a host of new problems when he convinced analytic philosophers that the primary function of languages is communication. Speakers can still describe the world but only in the context of community-based language games. Scientists can formulate scientific theories to describe certain sorts of natural phenomena, but the meanings of the constituent terms of these theories cannot be specified without reference to the scientists who formulated them. As a result, the very meaning of scientific terms has a social dimension.

Because of continuing difficulties with the notion of "meaning," contemporary analytic philosophers have attempted to find a way of handling general terms without any reference to it. The most influential recent treatment has been the causal theory of meaning sketched by Kripke (1972) and Putnam (1973). Proper names such as "Moses" and "Chicago" have been viewed traditionally as designating their referents rigidly. They denote what they denote, and that is it. At a particular time, one can describe the referent of a proper name so that others can pick it out. For example, right now Chicago is the largest city in Illinois, but such definite descriptions cannot count as definitions. In the absence of any spatiotemporal index, definite descriptions have to be constantly revised as the entities denoted change. For example, "the largest city in Illinois in 1837" does not denote Chicago. In order to find out which city it does designate, one must do a little historical research.

Kripke (1972) suggested that at least some general terms might be construed in much the same way. On his analysis, the reference of a speaker's term-tokens is fixed by means of an initiating event (possibly fictitious), a sort of "baptism," and then transmitted in a link-to-link reference-preserving chain. Hence, reference for such general terms as "water" and "gold" is fixed as rigidly as in the case of proper names. "Gold" is what was designated as gold by the first scientist to coin the term, regardless of any particular ideas the scientist might have had at the time about the stuff he was designating. As scientists rework their definitions, "gold" continues to denote gold because of the original baptismal act. If one has any doubts, one can trace the relevant reference-preserving chains back in time to the original naming ceremony to see what it was that was actually denoted. If it was gold as we conceive of it, then "gold" refers to gold. If not, then "gold" all along referred to something else, and we must reassess our terminology. Later, as our understanding of the world changes, this same exercise might lead to different results, but, at the very least, rigid designation provides some islands of stability in the vast sea of meaning-change.

To the extent that the preceding analysis is adequate, a whole raft of philosophical problems is solved or at least dissolved. Nearly all of the subsequent discussion in the philosophical literature has concerned the adequacy of rigid designation for general terms to solve such traditional problems as the existence of synthetic a priori statements. One aspect of the Kripke/Putnam proposal that has received surprisingly little attention is the link-to-link reference-preserving chains. According to Kripke and Putnam, a scientist designates a particular sample of stuff as gold and passes this usage on to other scientists, who continue to pass down this rigid connection in successive generations. Thus, on this analysis, reference is "causal" in two senses: the original designating act as well as the link-to-link chains are both equally causal. One need not ask all sorts of difficult questions about the meaning of "gold" if it denotes gold rigidly. All one has to do is to discover if it is part of one of the link-to-link chains emanating from the original naming ceremony.

To systematists, the preceding scenario should sound suspiciously familiar. It approximates how they name species. In the early years of systematics, nomenclature was chaotic. A single species might have different names in different languages, and even within the same language community it might vary from region to region, while different species were frequently called by the same name. Systematists began to limit this babel by replacing vernacular names with Latin appellations. Gradually they evolved a formal system of rules and regulations codified by successive international congresses to reduce ambiguity and synonomy. As these codes evolved, a very strange notion began to emerge and become central to taxonomic practice—the type specimen method. As the name indicates, initially systematists attempted to select a typical member of each species to serve as the type specimen of that species, preferably an unmutilated, vigorous, adult male. Under the pressure of increased data about variation in natural populations, they tended to replace single specimens with a type series designated to capture this variation. But throughout this period, systematists assumed that all the variation they perceived was due to deviations from the inherent constitution of the organisms under investigation. They studied variation but only in order to transcend it, to see through it to the essential features of the species.

On the assumption that each species can be characterized by a set of essential traits, later workers might decide that the organism initially designated as the type specimen for a particular species was not actually "typical" enough and substitute another specimen for it. The resulting confusion led systematists to rule that once a specimen is designated as the type specimen of a species, it can be replaced only in cases of duplication or accidental destruction. The sole function of the type specimen became to serve as the name bearer of its species. No matter in which species a type specimen might eventually be placed, the name must go with it. As a result, the functions of designating a species and of typifying it were disentangled and kept separate. For those species whose characters form unimodal distributions, the notion of a typical member makes some sense, but given the wide variety of character distributions exhibited by actual species, too often it does not. Instead, if one wants to display the various clines that characterize a species, several arrays of specimens are necessary. As Mayr (1969: 369) has put this point, "Species consist of variable populations, and no single specimen can represent this variability. No specimen can be typical in the Aristotelian sense." As Mayr (1983) has also pointed out, systematic practice with respect to type specimens is not entirely consistent, but I think a consistent theory underlies this practice.

Not all organisms belong to a species, but of those that do most belong to one species and one species only. Our ideas about character distributions, however, can be mistaken. A systematist may think that he or she has collected a variety of specimens from a single species, but later workers may decide that this "single" species actually consisted of two or more sibling species. Even in those cases in which the selection of a "typical" specimen is possible, the likelihood that increased knowledge will lead later systematists to change their minds about character distributions is sufficiently high to warrant excluding the name-bearing specimen from this activity. No matter how typical or aberrant systematists may take a type specimen to be at any one point in time, it can serve just as well as the name bearer of its species. As Mayr, Linsley, and Usinger (1953: 236) expressed this position in the first modern textbook on systematics:

> It is very difficult to characterize or to define a taxonomic entity solely by means of words. As a result, many of the Linnaean and early post-Linnaean species, particularly among the invertebrates, are unidentifiable on the basis of the descriptions alone. It is obvious that more secure "standards" are needed to tie scientific names unequivocally to objective taxonomic entities. These standards are the *types*, and the method using types to eliminate ambiguity is called the *type* method.

On a conservative reading, a type specimen in biology is much like the standard meter bar. The diagnoses published by systematists define the designated species. Though a type specimen may be somewhat aberrant, it must at least fall within the range of character variation established by the systematist. On a more radical reading, the reading that has come to prevail, diagnoses are at best definite descriptions. They describe a species and that is all. In the absence of a spatiotemporal index, these diagnoses must constantly be reformulated as the species changes. For example, in one generation, 80 percent of the plants might have yellow flowers, in the next generation 76 percent, and so on. The type specimen method works by tying down a name to one chunk of the genealogical nexus via a single node in that nexus. Like all specimens, the type specimen is merely one node in the genealogical nexus. No matter how the boundaries of a species are reassessed, the species that includes the type specimen node must be called by the name that the type specimen bears. The only exceptioin arises when it turns out that two specimens within the same nexus have been designated as types either because two systematists were ignorant of each other's work or they mistakenly thought they were dealing with two different species instead of one. In such cases, the first specimen designated as the type and its name are retained. As long as species are viewed as genealogical networks that split and occasionally merge, ancestor-descendant relations are primary, character distributions secondary. As important as character distributions are in the early stages of a taxonomic study, they do not *define* a species in the sense that "three-sided closed figure" defines "triangle."

The type specimen method in systematics looks amazingly similar to the system of rigid designation suggested by Kripke and Putnam, but there is one important difference. On my view species are individuals and their names are proper, while Kripke and Putnam introduced rigid designation to handle general terms such as "water" and "gold" as well. Of course, these philosophers also construe the names of species as general terms. Kripke mentions "tigers" and "cows." But if we are to take Putnam's (1975) division of linguistic labor seriously, scientific usage must take priority over ordinary usage. The reason that rigid designation works so well for species is that they themselves are historical entities. A species, like an organism, has a beginning, middle, and end. A name can be attached to it rigidly during any time-slice of its existence. No matter what changes take place during its existence, the same name applies. If one so chooses, one can subdivide a gradually evolving lineage into chronospecies, giving each a separate name, just as one can subdivide the life cycle of an organism into stages and change the name of the organism with each stage, but the logic of the situation remains unchanged. The lineage is basic; the characters are secondary.

Organisms and species are historical entities. Water and gold are not. They are substances, possibly even natural kinds. This difference in metaphysical category poses problems for the sort of naming procedure appropriate for each. For historical entities, issues of ambiguity can be resolved by historical investigations. Lewis Carroll was the same person as Charles Dodgson, and the evening star is the same heavenly body as the morning star. Similar observations hold for particular species even though the necessary historical investigations are more difficult to carry out. Two systematists exploring opposite coasts of an island might come across organisms belonging to the same species and give them different names. As these systematists trace the ranges of their species, they might well discover that they coincide and that the two species are actually one. Tracing species through time is a good deal more difficult, but the issues are the same—spatiotemporal continuity and organizational coherence.

Rigid designation works so well for historical entities because both the entities named and the link-to-link linguistic transmission of term-tokens form historical entities. A name is coined and affixed to a particular time-slice of a historical entity. Later both historical entities can be traced back in time to see if they intersect at the naming ceremony. If so, the name rigidly designates the entity. Although the same entity can be named repeatedly through time, redundancy can be discovered when it becomes apparent that a single entity extended through time has been baptized more than once. The continuity of the entity being named permits disambiguation. And all of this is accomplished without any reference to meaning or definitions.

However, substances such as water and gold can exist anywhere and at any time in the universe. Historical investigations are not sufficient for fixing reference. At one place and time, a sample of gold might receive one name intended to apply to all samples of gold regardless of space and time. At another place and time, it might receive another name. If these conceptual link-to-link lineages are traced back through time, they need not converge on each other. They can form distinct conceptual lineages. Since gold is not itself a historical entity, various samples of gold need not converge either. Why then treat these two names as being synonymous? The only answer available is sameness of structure. A physicist can discover that earlier investigators had affixed different names to two samples of the same substance because these two samples consisted predominantly of atoms with the same atomic number. Rigid designation might well work within each conceptual lineage but not between them.

In the preceding discussion, I have taken literally references to naming ceremonies and link-to-link transmission when the authors of rigid designation do not. For them fictitious baptismal acts will do, and the link-to-link transmission of rigid designators is decreed to be reference-preserving when frequently it is not. More recently, philosophers have suggested much needed improvements to the Kripke/Putnam analysis. For example, Kitcher (1978, 1982) has set out a theory of reference in which link-to-link reference chains can change their referents; even when they do not, the methods by which the reference of term-tokens is fixed can change. An example of the first sort of change is the term-type "planet." According to Kitcher, the extension of "planet" initially did not include Earth. After the acceptance of Copernicus's theory, it did. According to Kripke and Putnam, the term-type "planet" always referred to Earth regardless of what astronomers might have thought; they were just wrong. As an example of the way in which the mode of reference for a term-type can change, Kitcher (1982) traces the history of the term-type "gene." Early in the history of genetics, tokens of the term-type "gene" were applied only when Mendelian ratios were discerned. Later, the cis-trans test was added. Kitcher proposes to handle both sorts of change by means of something he terms "reference potential."

According to Kitcher (1982: 345), the reference potential of a term-type for an individual speaker is the "class of events which, given the speech dispositions of the speaker, can initiate productions of tokens of the type." The reference potential of a term-type for a community is, in turn, a "compendium of the ways in which the referents of tokens of the term are fixed for members of the community" (Kitcher 1982: 340). Hence, for Kitcher, the threads that intertwine to form conceptual continuity are furnished by individual actors and the communities they form. Although Kitcher rejects "mysterious" intensional entities, he does acknowledge a role for intentions of particular language users, a role he summarizes in three maxims—conformity, naturalness, and clarity. As Kitcher sees it, in most cases language users intend to conform to the usage of their community, although in crucial cases they may not. For example, for Darwin as well as for most of his contemporaries, the term "species" referred to such things as dogs, goats, and people, but most of Darwin's contemporaries were firmly convinced that species are immutable. If one putative species is shown to be connected by descent to another putative species, then automatically they must count as a single species. If Darwin was right and all present-day species arose from a half dozen common ancestors, then only a half dozen species actually exist. If all extant species arose from a single ancestral species, then all creatures that ever existed on this planet form a single species. Darwin decided to depart from common usage in this respect. When a species splits into two daughter species, it ceases to exist and the daughter species count as two new species. According to pre-evolutionary usage, anything that behaves in the way described by Darwin cannot count as species. Thus, on the immutabilist view, there are no species. If it can evolve, then it is not a species. According to later Darwinians, if it cannot evolve, then it is not a species.

The preceding example also illustrates Kitcher's second maxim. Sometimes people in general and scientists in particular intend to refer to natural kinds, even though the identifying description that they are using may be abandoned by later workers. On occasion, however, one might decide to stick to the description even if it means admitting that one had not been referring to a natural kind. Most of Darwin's contemporaries defined the term "species" so that particular species have the traditional characteristics of natural kinds. They are as eternal, immutable, and discrete as gold. On Darwin's view, particular species have none of these characteristics. *Dodo ineptus* is in no sense eternal. At one time it did not exist, later it evolved, and still later it became extinct. More importantly, once a species goes extinct, numerically that same species cannot re-evolve. Extinction is forever. For Darwin, particular species are children of time, contingent effects of natural forces, and not part of the framework of the universe. Nor are species immutable. Central to Darwin's theory was the view that new species arise out of preexisting species. Because Darwin thought that speciation was usually gradual, the boundaries between species are correspondingly fuzzy. (For the predisposition of people to conceptualize in terms of natural kinds, see Atran 1985, 1987a, 1987b; for general doubts about natural kinds, see Dupré 1981, 1983, 1986.)

Two alternatives are open at this point. The first is to deny that species are natural kinds. After all, they have none of the characteristics traditionally used to define "natural kind." The second alternative is to claim that the notion of natural kind has itself evolved. It has changed its reference. It used to refer to those things that are eternal, immutable, and discrete. Now it refers to entities that are temporary, modifiable, and amorphous. Stranger things have happened in the evolution of concepts. After all, cladism seems to have evolved into its opposite. However, as far as I can tell, the second alternative has not occurred. Although the philosophical notion of a natural kind has undergone considerable modification through the years, it has yet to evolve into its opposite. It might, but it has not. The only significant alteration that some philosophers are willing to allow is for natural-kind terms to have fuzzy boundaries in conceptual space. Cluster concepts can count as natural-kind terms. This leaves the first alternative. Generations of natural historians have intended to refer to things very much like natural kinds when they referred to such things as dogs, goats, and people, but they were mistaken. Kitcher (1982: 344) dramatizes these alternatives in terms of two different soliloquies:

> The first type consists in saying: "I intend to pick out a natural kind, and I hope that this description characterizes a kind: if it turns out that it doesn't, I'll abandon the description." The second consists in saying: "I intend to pick out whatever satisfies this description, and I hope it's a natural kind; if it turns out that it isn't, I'll stick with the description."

Kitcher's third maxim is that, on occasion, scientists intend to refer to that which they can specify. Philosophers have not had much patience with the emphasis that scientists place on operational definitions. After all, they are not definitions. But Kitcher sees an important role for the tests that scientists devise to help them apply their terms. For example, if a cis-trans test turns out in a particular way, a geneticist is likely to produce a token of the term-type "gene." Although there is much more to a theoretically significant term-type than the methods scientists use at any one time to apply tokens of it, I do not think that it seriously misrepresents the behavior of scientists to claim that they intend a referent in such circumstances to be whatever was causally linked to the production of this token or satisfied a particular description.

In general, I find Kitcher's analysis a significant improvement on the Kripke/Putnam analysis. Link-to-link sequences of term-tokens do change their referents through time. Ways of fixing reference also change. I also agree with the role that Kitcher assigns to individual language users and the communities they form. However, I think that Kitcher's analysis needs one important amendment concerning the way in which he delimits his linguistic communities. According to Kitcher (1982: 346), "With respect to a particular expression type, two speakers belong to the same linguistic community if they are disposed to count exactly the same events as initiating events for production of tokens of the same type." According to Kitcher, the only agreement necessary for scientists to belong to the same linguistic community is agreement over initiating events. As a result, he avoids all the variation due to differences in beliefs about the "meanings" of the terms at issue. In this connection, Kitcher (1982: 347) emphasizes two points:

> First, two members of a linguistic community may differ greatly in their beliefs. What is crucial is that they agree on the ways in which the referent of a term should be fixed. Second, not all community shared beliefs which use a particular term may be employed in fixing the reference of that term. It is quite possible that each member of a linguistic community should be prepared to assert that the things referred to by a particular term lack a particular property, and yet use that term to refer to entities which have that property. So long as the belief is not used to fix reference, a false belief may prevail throughout a community.

On Kitcher's analysis, members of the same linguistic community can disagree about all sorts of things, just so long as they agree about the way in which the reference of the term at issue is fixed as well as the initiating event that fixes its reference. Although each term-type delimits its own linguistic community, these communities might be coextensive for large families of terms. If nothing else, Kitcher's analysis explains why scientists are so interested in tracing back their usage to its introduction and thus declaring their usage "correct." For example, the literature on the proper use of such terms as "homology" and "monophyly" is extensive. Rather than scientists simply declaring that they are going to use these terms in a particular way, like it or lump it, they are intent on showing that these terms were introduced by some great authority in just the ways that they are using them. They are using "homology" correctly to refer to elements in ideal patterns because that is the way that Owen introduced this term into Victorian science in 1837 (Desmond 1982: 214), while their opponents are redefining it to presuppose descent. Similarly, all sides insist that they are using "monophyly" correctly because this is the way that Haeckel used this term when he introduced it in German, while their opponents are redefining it (see Ashlock 1971, 1972, Nelson 1971b, 1973b, Farris 1974, Mayr 1974, 1978, 1981, Platnick 1976, 1977a, 1977b).

However, Kitcher's linguistic communities do not coincide with scientific research groups as I have defined them. If research groups are defined in terms of cooperation, then scientists can belong to the same group even though they are not in total agreement on a variety of issues, *including* initiating events for the production of term-tokens of the same term-type. By and large, scientists who belong to the same research group will be in substantial agreement on a variety of issues including the initiating events for many of the term-types they use, but an occasional disagreement on this score does not automatically exclude a scientist from his or her research group. For Kitcher, linguistic communities are individuated solely for the purposes of fixing reference. As such, they are to some extent social fictions. I prefer to stick with groups defined in terms of the relevant social relations. A scientist can belong to several research groups of increasing inclusiveness, but restrictions on time and opportunity limit the number of collateral groups to which any one scientist can belong at the same time. For example, for a time, Michener, Sokal, Rohlf, and a few others belonged to the Kansas group of numerical/phenetic taxonomists. All three were also systematists. Michener worked with Sokal and Rohlf to improve the principles of systematics, but he also worked with other colleagues and students in studying bees. Because of the time that research takes, a scientist can participate in only a very few research groups, narrowly defined.

On the analysis I prefer, term-tokens must be organized into trees and networks solely on the basis of transmission. In these trees the structure of the tokens themselves can change; e.g., gradually tokens of "pangene" were transcribed as "gene" and "allelomorph" became "allele." Even so, these tokens can still be part of the same token trees. Initially, a particular term-token may well be introduced by a particular initiating event. However, later replicates may well be associated with their referents by quite different initiating events. For example, de Vries introduced "pangene" before publishing his mutation theory but then later utilized it in the context of his more general theory. Pangenes are the things that change to produce new species in the space of a single generation. His paradigms were changes in the evening primrose. If anyone questioned him about what he meant by "mutated pangenes," he would have pointed to the changes he observed taking place in this species of plant. His intent, however, was more general—"mutated pangene" denoted other changes in other species as well. Johannsen modified the term in the context of his distinction between genotype and phenotype and applied it to the "determiners" in the genotype that specify characters under certain conditions. Genes are the things that he selected to produce his pure lines in garden beans. Both the genotype-phenotype distinction and the gene concept have changed considerably since Johannsen introduced his terms.

Although term-tokens of "gene" continue to be transcribed as "gene," new events continue to be introduced to fix the reference of these term-tokens as geneticists have devised additional operational definitions. For example, genes are those things that exhibit cis-trans effects. As Kitcher (1982: 345) notes, scientific terms frequently have "reference potentials which include diverse initiating events." What binds these various reference-fixing events together is the continuity of conceptual replication. Although Kitcher does not emphasize it, one additional point must be made: these initiating events must be actual. Fictitious initiating events will not do.

Kitcher (1982: 339) would have us "describe the phenomena of conceptual change by charting the shifts in referential relations between words and the world." As difficult as it is in practice to trace such token-trees and to specify the various initiating events that scientists periodically use to connect a particular term-token to an instance of a particular natural phenomenon, the general outlines of the undertaking are clear. Term-tokens form trees, occasionally networks. Every once in a while, a particular term-token is connected to a particular event. The tokens may change their structure, the character of the reference-fixing events might even change, but in the face of such change, the replication sequences can be traced through time. Their continuity is the causal relation that makes the system work. However, a major problem remains: how to partition such conceptual trees into types. Isolated term-tokens are not sufficient for either communication or description. Even lineages composed of successive replicates of term-tokens are not good enough. Language requires term-types as well as term-tokens.

The same spectrum of alternatives is available for conceptual change as for biological change. Cladists insist that phylogenetic trees be subdivided into strictly monophyletic taxa. All replicates that eventuate from a particular node must be included in a single higher taxon, and all the nodes at a particular time-slice included in a particular higher taxon must be traceable to a single initiating node. Pheneticists disregard phylogeny entirely. They propose to produce taxa that are truly general. Both the characters they use and the taxa they delimit are intended to be spatiotemporally unrestricted. Phenetic taxa cannot function as theoretically significant *individuals* because they are spatiotemporally unrestricted. However, because they are genuine classes of the sort that can function in laws of nature, they might fulfill *this* function. They do not happen to.

Phylogenetic systematists insist that all taxa must be clades. As clades they are spatiotemporally continuous. Each lineage is also cohesive; entire clades are not. Hence, individual lineages are individuals of the sort that a natural kind might denote. Entire clades are not. More importantly, an important scientific theory refers crucially to lineages. Hence, "lineage" is a theoretically significant class, a natural kind. Evolutionary systematists opt for a hybrid form of classification. It is a combination of genealogy and considerations of process. Because reptiles occupy a reasonably discrete adaptive zone, Reptilia should be recognized as a taxon even though it is not strictly monophyletic. Although branch points do not strictly determine taxonomic boundaries for evolutionary systematists, taxa must have some degree of "phyly." As a result, they are spatiotemporally restricted.

The sort of classification one prefers depends on the purposes to which it is to be put. If one wants to distinguish the individual entities that function in evolutionary processes or result from them, then clades are preferable. If, to the contrary, one wants to distinguish classes of entities whose names function in putative biological laws, spatiotemporal considerations cannot arise. The distinction is, once again, between Carnivora and carnivores. Carnivora is restricted to a particular section of the terrestrial phylogenetic tree; carnivores are not. They are scattered through the terrestrial phylogenetic tree, and when at long last we are visited by genuine extraterrestrials, we might discover to our dismay that they too are carnivorous. They could not, however, belong to the monophyletic order Carnivora.

Many of the hybrid taxa recognized in evolutionary classification are quite robust and intellectually satisfying. They seem so natural. But, on the general account of science that I have adopted, they are not just hybrids; they are chimeras. They can function neither as individuals for laws to range over nor as classes whose names function in these laws. If they were genuinely ecological notions, then small birds and large moths might be placed in the same taxon, but evolutionary systematists are likely to reject such a grouping because it is not "phyletic" enough. Nor are they likely to exclude flightless birds from Aves.

The parallel question at the level of terms is what functions they are supposed to perform. Natural phenomena can be divided into those that can change and those that cannot. Living creatures on Earth have changed through the millennia; life itself, as a property of these creatures, cannot change anymore than gravity itself can. Bodies of different masses attract each other with varying degrees of strength, but the relation captured in the law of universal gravitation is supposed to be constant. Correspondingly, some terms refer to things that can change; others to things that cannot change. How about terms themselves? Term-tokens themselves change in replication sequences, e.g., sequences of allelomorph-allelomorph-allelomorph gave way to allele-allele-allele. This sort of transition is the subject matter of historical linguists. Term-tokens also change the means by which they are connected to their referents. If term-tokens are grouped into types genealogically, then they are not genuine "types." Hence, they cannot function in any generalizations one might propose to make about conceptual change. Hence, the pangene-gene tree might function as an instance over which a generalization about such things as technical terms in science might range, but this tree itself is not a term-type. It is a sequence of term-tokens. Any term-types that one might formulate are sure to gerrymander term-token trees. In fact the case for terms is much worse than the political analogue. Gerrymandered political districts may have extremely convoluted boundaries, but at least they are topologically contiguous. When term-types are mapped onto term-token trees, the results are topologically disjoint.

The preceding dilemma can be resolved by distinguishing between two different roles for terms when science is interpreted as a selection process. Within the context of a particular controversy in science, the causal connection of terms in replication sequences is crucial. Term-tokens are the things that are being differentially perpetuated. Anyone who wants to understand scientific change at the local level must order term-tokens into trees. However, those involved in such activities intend some of their terms to be general. A particular test situation is by necessity highly particularized. "Mendelian genetics" is never tested against "the world." Instead, some part of a conceptual system is tested by confronting a very limited implication of that system for some highly restricted and particularized section of the natural world; e.g., what happens when seed-coat color and texture in garden peas are traced simultaneously through a succession of generations? But even this characterization is too general. First, all sorts of boundary conditions, simplifying assumptions, and other particular circumstances must be specified. How precisely are the ova fertilized? Are the seeds being used the result of extensive inbreeding or drawn haphazardly from different populations of garden peas? Although a scientist can never anticipate all such questions in advance, think about them, and choose, the construction of the experiment will embody decisions, like it or not.

Second, each scientist will be testing only a particular version of a theory, law, or hypothesis. Of course, each scientist thinks that his or her version is typical of the cluster of versions afloat at the time. His or her preferred version incorporates the essence of the conceptual system under investigation. To the extent that this scientist is successful, his or her preferred version will proliferate and become standard, but it need not have been standard at the time. Scientists intend to transmit term-types, but all they actually transmit are term-tokens, which are immediately interpreted as types.

Thus, conceptual change in science consists in term-tokens being tested and transmitted locally but interpreted globally as types. Term-tokens are simultaneously part of spatiotemporally restricted term-trees and instances of spatiotemporally unrestricted term-types. Hence, conceptual change interpreted as a selection process incorporates a systematic ontological equivocation. Only tokens connected in replication sequences are differently perpetuated, but some of these term-tokens become exemplars for widely held term-types. All scientists intend for their conceptual systems to be generally applicable, but in each generation only a very small percentage of instances of these systems gets passed on, and the version of a particular conceptual system that eventually comes to prevail may well not be the one that early scientists intended. As a result, scientific theories composed of term-types are underdetermined by anything that might count as evidence.

The modular structure of conceptual systems in science is matched at the social level by the demic structure of science. Initially in the history of science, scientists worked largely in isolation, but as it proceeded they started to form increasingly more inclusive groups. Some of these groups have succeeded in getting descendants of their ideas adopted by later generations; e.g., the Darwinians, Weismannians, neo-Darwinians, Mendelians, advocates of the Modern Synthesis (Darwinians again), evolutionary taxonomists, the MacArthur group, numerical taxonomists in their numerical guise, cladists, and sociobiologists. Other groups have been influential only to the extent that the more successful groups played off against them; e.g., the Quinarians, phrenologists, the opposition to evolution by Kelvin and his fellow conspirators, numerical taxonomists as advocates of phenetic philosophy, and Marxist opponents of sociobiology. Isolated individuals also have played a role in science but a role that was greatly reduced because of their social isolation; e.g., such idealist opponents of Darwinism in Great Britain as Owen, Romanes in his opposition to neo-Darwinism, East in his opposition to Mendelian genetics, a whole series of neo-Lamarckians, Goldschmidt in his opposition to both the atomistic genes of Mendelian genetics and the gradual evolution of the Modern Synthesis, isolated critics of evolutionary systematics such as Blackwelder, and Croizat in his opposition to dispersalist biogeography, until the cladists took up his cause.

The formation of research groups affects conceptual change in science in two ways. One is generic—it increases the speed of conceptual change regardless of the specific content of the accompanying research program—but these allegiances and alliances can also have at least short-term effects on the content of science. For example, the animosities that developed between Huxley and Hooker on one side and Owen on the other generated the nucleus of the Darwinians even before Darwin published. Thus, these pre-evolutionary disputes influenced the content of the Darwinian research program when it did emerge. For instance, evolutionary theory as promulgated by the original, British Darwinians was much less progressive than the version adopted by other groups of Darwinians. The animosities that developed between Sokal and Farris strongly colored the subsequent development of cladistics and numerical taxonomy. Although such social relations are traditionally held not to be "internal" to science, this category might well be expanded to include them. If anything has characterized science throughout its existence, it has been this sort of cooperation and competition.

The effects of these group allegiances are local, but then so are the selection processes that serve as the motor of conceptual change in science, so local that theoretically relevant investigations of the ongoing process of science are all but impossible to conduct. Most sociologists and historians of science study science at such a distance that the most they can hope to discern are the distal effects of the causal factors operative in conceptual change in science. Some of these causal factors are internal to science—the effects of data, carefully run experiments, inferential connections to other laws and theories, and so on. However, the effects of these internal factors are very localized because they are tied to the cognitive resources of the small selection of scientists actually involved in particular disputes. For example, Huxley was drawn to Quinarian systematics because the relations he had stumbled upon in a couple of groups of invertebrates were duplicated in the grander Quinarian system. When Hooker took Strickland's place in Darwin's research group, one effect was a subtle shift to a more populational view of species. The content of numerical taxonomy was strongly influenced by developments in computer science, the special problems of bacterial systematics, and Sokal's inability to do intuitive taxonomy. In the development of science, historical contingency is crucial.

In the early pages of this book, I acknowledged that the descriptions of the social relations and conceptual changes in science that I relate here are anything but theory-neutral. No one can describe something like the course of science in a totally theory-free way. Assumptions about the nature of science are bound to "intrude." However, I also eschewed any justification for dictating proper scientific method on the basis of any epistemological first principles. The only necessity that I might eventually claim is the sort of nomic necessity that characterizes the more fundamental generalizations of science. Although the distinction between genuine laws of nature and accidentally true universal generalizations remains as problematic as ever, I think that it also remains central to our understanding of nature as well as the nature of science. The likelihood of finding any lawlike generalizations governing particular species or research programs is next to nil. However, general characteristics of selection processes as such might well belong near the nomic end of the spectrum. For example, the effects of the demic structure of biological species, scientific communities, and conceptual systems on the rate of change might well count as being more than just contingent.

Conclusion

If scientific change is to be understood as a social process, then one must distinguish between two sorts of historical entities—one social and one conceptual. Single scientists exist through time, sometimes engendering professional descendants, sometimes not. Scientists also form social groups integrated by a variety of professional relations. The most local of these groups tend to be quite ephemeral, lasting for only one or two decades. Even so, their membership does undergo some change, as new members join the group and others leave. Scientists are also members of more diffuse but longer-lived invisible colleges. The chief activity of scientists is the production of conceptual systems. At any one time, a scientist is usually entertaining slightly different versions of his or her conceptual system, playing off one against the other. As a result, this small family of conceptual versions changes through time. The relations that integrate these versions into a single conceptual historical narrative are partly inferential, but only partly. The fundamental relation is that of conceptual descent, and one proposition can give rise both to contraries and to its contradictory. When scientists are organized in cooperative groups, the multiplicity only increases, as certain versions are added and others eliminated.

The picture of scientific development that I have incorporated in the first half of this book is that of two sorts of historical entities changing through time, splitting, merging, and going extinct. Although the changes that occur at the social and conceptual levels are far from perfectly concordant, they are related often enough that one can move back and forth between the two, for example, following the Darwinians as they enlist new members while old ones die or abandon the program, then following Darwinism as certain views become more prevalent and others are played down or quietly abandoned. Because both sorts of historical entity are variable through time and internally heterogeneous at any one time, they cannot be individuated in terms of "essences," whether essential members or essential tenets. Even cluster analysis will not do, in part because sometimes the distributions that form are multimodal, in part because the methods of cluster analysis are not up to distinguishing between numerous alternative clusters even when the distributions are unimodal. Instead, a method akin to the type-specimen method in systematics must be utilized. If one wants to individuate a scientific group, all one has to do is to pick a member, any member, and trace out his or her professionally relevant social relations. If one wants to individuate a conceptual system, all one has to do is to select a particular token of a particular tenet and trace out its conceptual relations, both inferential and genealogical. When we trace out inferential relations, the only ones that count are those that were actually made.

The preceding recommendations are directed at those who are engaged in the study of science, not at scientists themselves. Scientists themselves, to some extent, seem to individuate their conceptual systems and social groups in this way, but scientists are not the audience I have in mind for the type-specimen method applied to science. Of course, one and the same person can both study science and participate in it, but the activities are distinguishable if not always distinct. Scientists, as they are engaged in the ongoing process of getting their views accepted as their views by their fellow scientists, are constantly negotiating the content of their conceptual systems. In doing so, they pick certain scientists and tenets to serve as exemplars, and they insist quite emphatically that both sorts of exemplars are truly exemplary. For example, they might insist that Darwin was a Darwinian but not Huxley or Lyell, or they might be just as adamant that gradualism was (or was not) essential to Darwinism.

As interesting as these activities are in their own right, those of us who are studying science have other goals. Scientists and those who study science have their agendas, but they are at least partially independent of each other. For example, it makes little difference to my own research program whether mass extinctions are periodic or not, or which explanations of this presumed periodicity come to prevail. Scientists involved in the disputes I discuss in the first half of this book are liable to become irritated that I do not opt often enough for one side or the other. After all, the term "monophyly" was introduced at some time or other by some biologist or other with a determinant meaning. If it was originally used the way that Hennig later insisted that it be used, then I should say so and award the prize to Hennig. If to the contrary Mayr's usage has historical precedence, then I should opt for his definition.

One role of the history of science, especially when it is written by scientists themselves, is to declare victors and vanquished. But one can chronicle scientific development without choosing on such issues. One can, but some issues debated by scientists are very intimately connected to those that divide students of science. For example, I am interested in portraying science as an evolutionary process in which selection plays an important role. Hence, debates over the role of selection in biological evolution are not irrelevant to my own activities. Thus, Ernst Mayr did not object to the selectionist character of my theory of scientific development, while Dan Brooks was taken aback. In both cases, because of our professional relationships, the criticisms offered by both men were supportive.

Similarly, a recurrent controversy in the philosophy of science has been the relation between scientific classifications and scientific theories. I am strongly committed to the view that the two are closely and necessarily intertwined. Hence, my discussions of the pheneticists and pattern cladists are hardly disinterested descriptive chronicles. If I thought that descriptions can be made totally theory-neutral, then I might try to present a totally theory-neutral description of the war between the pheneticists and cladists, but I do not, so I cannot.

When the type-specimen method was introduced, it applied to species as classes. Other organisms belonged in the same species as the type specimen because they were similar enough to it. As systematics became increasingly evolutionary, genealogical relations became increasingly prominent in the use of the type-specimen method. Because in my formulation of the type-specimen method as applied to science I have concentrated on genealogical relations, I am at least indirectly committed to some form of evolutionary or phylogenetic view of biological systematics. If it is good enough for those of us who are studying science, then possibly it is good enough for those who study biological evolution.

In the individuation of a research group, all one has to do is to pick a scientist to serve as one node in the social relations that integrate the group into a genuine social group. For example, to individuate the Kansas group of numerical taxonomists, I might pick Bob Sokal in 1963. In doing so, I am not committed to Sokal being the most important or influential member of this school. I could have picked Michener, Rohlf, Ehrlich, or any one else who actually belonged to the group at the time. Ehrlich ceased being an active member of this group in the late 1960s. What if I picked him to serve as the type specimen for the group? Biological type specimens rarely join or leave their species, but scientists can and frequently do join and leave their research groups. Hence, in order for the type-specimen method to work for scientific groups, these social type specimens must have a temporal index—Ehrlich in 1963.

The point of social type specimens is to fix the reference of such phrases as "the Kansas group of numerical taxonomists," "early Darwinians," and the like in a way that is independent of the beliefs that these scientists actually held. To be sure, membership in these groups was not totally independent of the beliefs these scientists held, but the two are not perfectly concordant either. Huxley differed with Darwin on several key issues, but if the two men had not agreed on enough issues, they were unlikely ever to have joined forces to push evolutionary theory. The same can be said for Darwin and Gray. However, the areas of agreement and disagreement between Darwin and Gray were different from those between Darwin and Huxley. The type-specimen method applied to social groups permits the individuation of such groups even though their membership might change, even though not all members at any one time might agree with each other totally even over fundamentals.

Parallel observations hold for conceptual systems. A particular version of a particular theory held by a particular scientist at a moment in time may be characterizable in terms of a few essential tenets. Later, once all the *Sturm und Drang* is over, the more global end-product might also be characterizable in these terms, but at the time that a conceptual system is under active development, it is not. Conceptual systems while they are evolving are internally heterogeneous, as they must be if they evolve at least in part as the result of a selection process. One way to individuate conceptual systems in the face of all this heterogeneity is by selecting a particular token of a particular tenet at a moment in time as a conceptual type specimen, e.g., Nelson's 1971 claim that the principle of dichotomy is essential to Hennig's phylogenetics. In doing so, nothing is implied about the importance of this tenet, let alone whether it is or was actually essential. It may turn out that all systematists throughout the world who considered themselves to be doing taxonomy the way that Hennig recommended accepted dichotomy, or that most did, or a few, or possibly only Nelson. Even if this tenet were widely accepted at the time, it may have dropped out since. It does not matter. Its only function is to serve as a point of entry to discern the outlines of this conceptual system, one node in a conceptual nexus. Just as "Nelson in 1971" can fix the reference for "the American Museum cladists," "Nelson's specification of dichotomy as essential in 1971" can be used to fix the reference for "American Museum cladism."

Although my discussion in this chapter has concerned science, I think that it can be used in the study of *any* historical entities (Hull 1975). However, conceptual historical entities differ from other sorts. Some of them, at least, refer. The interconnections between conceptual systems and the people who produce them are important, but conceptual systems can have a second interface as well. Some are supposed to be descriptive of the world. In the preceding discussion, however, I limited myself to conceptual systems in science. Scientific conceptual systems in their most global sense cannot be tested because they include numerous different versions of the same theory, frequently versions so different that they are clearly contradictory. Even if one restricts oneself to a single version of a theory, it cannot be tested in all its parts all at once. Instead, some limited aspect of this version is tested. Even this aspect is further particularized by the introduction of the various operationalizations necessary to bring it to bear on the empirical world. Depending on the results of the test, a method of operationalization can become more firmly established, modified, or brought into question, possibly abandoned. The same alternatives apply to the version of the theory being tested.

As a result, testing is a highly particularized activity. However, the intent of scientists in testing their views is highly general. The goal is to bring as many scientists as possible to accept the same general formulation of a particular view. At the very least, these scientists must be brought to give assent to the same terminological formulations. But even more than this, the goal is to rework the conceptual systems of individual scientists so that they will react similarly in similar situations, perform the same sorts of tests, react in the same ways given the results of these same tests, and so on—in short, to mean the same thing when they present similar tokens so that term-tokens in a term-token tree increasingly become tokens of the same term-type.

14

Conclusion

Writing about cooperation and solidarity means writing at the same time about rejection and mistrust. Solidarity involves individuals being ready to suffer on behalf of the larger group and their expecting other individual members to do as much for them. It is difficult to talk about these questions coolly. They touch on intimate feelings of loyalty and sacredness. Anyone who has accepted trust and demanded sacrifice or willingly given either knows the power of the social bond.

Mary Douglas, *How Institutions Think*

[Such categories as time, space, and casuality] represent the most general relations which exist between things; surpassing all other ideas in extension, they dominate all the details of our intellectual life. If men did not agree upon these essential ideas at any moment, if they did not have the same conceptions of time, cause, number, etc., all contact between minds would be impossible, and with that, all life together. Thus, society could not abandon these categories to the free choice of the individual without abandoning itself.

Emile Durkheim, *Les formes élémentaires de la vie religieuse*

THE POINT of the introductions to the early chapters of this book is to impress upon the reader that, contrary to common sense, neither particular social groups nor particular conceptual systems have essences. The Darwinians had no essential members. Even Darwin might have absented himself or been ejected from the group without the group ceasing to exist or losing its identity. Social groups per se might well have an essence, that is, anything that is to count as a social group might well have to fulfill certain requirements, without particular social groups exhibiting any peculiarities that distinguish them from all other social groups. Nor do conceptual systems need to have essences. Although each advocate of a particular conceptual system insists that it has an essence, to wit, the particular set of tenets that this advocate happens to hold, conceptual systems can be characterized by a particular set of principles only at the cost of stripping them of their power to grow. However, the notion of conceptual system per se might well have an essence even though particular conceptual systems do not.

These counterintuitive claims have two justifications—one evidential, the other theoretical. If one studies particular social groups and conceptual systems, one can see that the elements that make them up and the relations that integrate these elements are variable both at any one time as well as through time. Social groups and conceptual systems can be made to have essences only by means of distortion of the most extreme sort. One reason I have included such extensive investigations of various research groups and their programs in the first half of this book is to support this contention. But, so the inevitable objection goes, they *must* have essences. Unless all people are in fundamental agreement on at least fundamentals, "all contact between minds would be impossible." Although no evidence is ever brought in support of this conviction, its force remains all but irresistible.

The best antidote to this a priori belief is to point out that the parallel conviction about biological species is mistaken. More often than not, more variation exists within a species than between closely related species. Sometimes a single set of genes exists that all, or nearly all, the organisms belonging to a particular species possesses; sometimes not. It does not matter. Yet, if any two organisms are chosen at random from the same species, they will have identically the same alleles at nearly all their loci. This same sort of variation characterizes social groups and conceptual systems. One is tempted to argue that all the members of a particular language community must be in total agreement about certain aspects of their language or else communication would be impossible. But such universal agreement is typically absent, and people communicate just fine. Once again, the hare cannot catch the tortoise, but it does. Pick any two people from a language community and they are likely to use the vast majority of terms in very nearly the same variety of ways. However, it does not follow from this that there are certain basic terms about which nearly everyone agrees.

The theoretical justification for treating social groups and conceptual systems the way that I do is that they could not evolve via selection processes if the sort of variation that characterizes species did not also characterize them. Selection processes have essences—certain requirements must be met before selection can occur and one of them is the presence of not just variation but heritable variation. The point of my presenting historical narratives rather than unconnected vignettes is that selection processes require continuity through time. Change in systems that are evolving by means of selection is a function of the differential perpetuation of the elements that make up these systems. Generations and identity by descent are required. Of course, this justification is liable to carry weight only for those committed to viewing conceptual change in science as a selection process in the first place.

Selection processes are complicated because they are the result of the interplay through successive generations of two processes—replication and interaction. Conceptual change in science is even more complicated because it is the result of the interplay of at least a half dozen processes that intersect at the level of individual scientists. They are the most important vehicles for both replication and interaction. Recently a vocal school of commentators on science has argued that factors which traditional students of science have treated as being "external" to science are actually internal to it. One of the main messages of this book is that the more interesting of these factors have no general influence on science. Such things as social class and sexual preference have, at most, contingent, idiosyncratic effects on the conceptual development of science. However, one social force that might appear to some to be external to science *is* internal to science—the sort of cooperative competition that characterizes the social relations among scientists.

Scientists exchange credit for contributions. Credit comes in a variety of forms from prestigious prizes to citations. Of these, one sort of credit is most fundamental—the use that one scientist makes of the work of another. The "success" that is central to science is not career advancement but mutual use. Science has the cumulative character it has in part because of this sort of credit. Because scientists must use the work of other scientists, they are forced to "cooperate" in a metaphorical sense with even their closest competitors, i.e., use their work. Science can proceed and on occasion does proceed in the absence of cooperation of the more literal sort. However, many problems require an array of cognitive resources that no single scientist is liable to possess. As a result, scientists frequently join together for periods in their careers to develop a particular area of science or to investigate a particular set of problems. These research groups serve to enhance a scientist's conceptual inclusive fitness. Conceptual demes and larger invisible colleges are responsible for the "tribal" behavior of scientists. Scientists continue to cooperate and compete with each other but as members of particular research groups and conceptual demes. One would expect the presence of research groups and conceptual demes to have the same effects on conceptual evolution that population biologists have shown that kinship groups and biological demes have on biological evolution.

Scientists compete in a quite literal sense as they race to solve an easily identifiable, conspicuous problem. One danger in science is for career interests of the crassest sort to supersede the desire to produce work that other scientists can use. The two need not go together. However, in a sufficiently high percentage of cases, the best way for scientists to further their careers is to fulfill the loftier goals of their profession. Conflicts between these two goals are sufficiently rare that scientists are usually able to avoid admitting to themselves that they even have career interests. Competition can also become too intense. Because the scientist who is first to make a contribution public gets almost all the credit, scientists are sometimes led to publish too quickly or to adopt research strategies that are risky but, if successful, work quickly. The first sort of tendency is countered to some extent by the fear of getting the reputation of not producing work that other scientists can safely use. The second tendency does not always need countering. Those problems that are amenable to sequences of short-term, opportunistic forays are likely to be solved most rapidly, but not all problems in science are of this sort. Some require the adoption of a line of research that the scientists involved know is long-term and even then not guaranteed of success. Although the reward system in science does not especially encourage the adoption of such long-term strategies, it at least allows it.

Any system can be subverted, and in biological evolution the science-fiction monster of totally parasitic DNA is realized in junk DNA. All it does is replicate. In conceptual evolution, the great fear of scientists from the beginning has been "junk ideas," those views that are replicated but never tested, as epitomized somewhat unfairly by the occult qualities of the Scholastics. From the secure position that science enjoys in several present-day societies, commentators on science tend to disparage the great emphasis that so many scientists place on operational definitions and falsifiability, but these concerns are not misplaced. It is very easy for a particular research program to degenerate into a self-perpetuating ideology. At one time phrenology, Freudian psychology, and Marxian economics were all genuine scientific research programs. However, through the years, the protective belt that surrounds the most fundamental tenets of any scientific research program can become so protective that nothing can possibly penetrate it. The adherents of ideologies cease serious testing. Even though the same charge has been made repeatedly against evolutionary theory, it has remained open throughout its long history to successive challenges. It has avoided extinction for so long by the only method available under such conditions—by changing. Only if one thinks that scientific theories, to be legitimate, must have eternal, immutable essences do such changes become a fault.

Except in the rare case of mass extinctions due to some global modification of the environment, selection processes are always localized. Avatars of a particular species succeed or fail in a variety of ecological niches. Which avatars survive and proliferate depends on the sequential availability of the appropriate niches. Science is no less localized. People in general are overly influenced by their most recent experiences. Scientists are no different. No one knows in advance which experiences are going to turn out to be generalizable, but the "community" of scientists is extremely heterogeneous. Most scientists are done in by the idiosyncrasies of their conceptual upbringing, but every once in a while one happens to experience just the right antecedents to be able to piece them together in an interesting new way, possibly adding a new element now and again.

Even though the genesis of a new conceptual system is highly contingent, one fundamental goal of scientists has been and continues to be the construction of internally consistent, totally general scientific theories. To the extent that science is a selection process, such ideal entities cannot be tested. A scientific theory as a completely general conceptual entity cannot be made to confront the world in its entirety. Instead, particular scientists test parts of one version of a particular theory. Although the number of permutations of the parts of a theory are indefinitely large, only a few are ever tested. If the wrong ones get tested early on, the theory is likely to be rejected even though other combinations might well have survived had anyone entertained them. Although scientists conceptualize science as involving the selection of types, selection operates only on tokens. However, each scientist interprets his token as the ideal type. Thus, the proliferation of a particular token can be interpreted as the conquest of a single type—the correct theory.

In this book I have raised no objections to treating scientific theories and ideas in general as types. However, these types are not the things that function in selection processes. Just as in biological evolution, traits that "twinkle" are not genuine traits, in conceptual evolution themes or unit ideas that crop up now and again cannot count as the "same" in an evolutionary sense—they are not conceptual evolutionary homologies. Although both biological and conceptual evolution can be construed as historical entities wandering through preestablished ideal types, this way of viewing evolution is extremely strained and is likely to lead to misunderstandings.

An interplay between science and philosophy has also characterized science since its inception. Initially, the interplay was hardly noticeable because the two activities were not distinguished. Everyone engaged in intellectual activities was equally a "philosopher." However, as science developed, these activities became increasingly detached. Even so, their interconnections were never totally severed, recurrent calls for such severance notwithstanding. For example, both Woodger and Gilmour were attracted to the tenets of logical positivism, and through them this philosophical research program influenced systematics. The cladists in turn were attracted to the philosophy of Karl Popper. That the cladists simultaneously appealed to the authority of Popper and transformed his philosophical views caused some comment, but this practice is common in science, whether the authority is a philosopher such as Popper or a biologist such as Darwin or Hennig.

Because selection processes are so opportunistic, epistemological reconstructions of scientific method that portray the scientific process as a prescribed sequence of activities are exceedingly misleading. In the past, epistemologists searched for the epistemic givens upon which all knowledge must be based and the infallible method that would lead from these firm foundations to certain truth. Any scientific method that incorporates either of these premises is sure to be mistaken. Science by necessity must always be conducted in medias res. There is no one preferred place to begin, and no final resting place. Although this conclusion is widely held by those philosophers who study science, numerous scientists continue to be convinced that an order of epistemological priority is inherent in science and that this epistemological order should be translated into a temporal order in the conduct of scientific investigations.

If there are any lessons to be learned from recent controversies over systematics, one of them is surely that phylogeny reconstruction is not very closely linked to evolutionary theory. One need know very little about evolutionary theory in order to reconstruct phylogeny, and very little of this knowledge about the process is relevant to the reconstructions produced. Even so, the few connections that do exist are of central importance. Phylogenies are composed of spatiotemporally connected lineages, and the characters used to infer these connections must be evolutionary homologies. Another lesson to be learned is that hierarchical relations in nature can be reflected in a hierarchical classification only if great care is taken and then only with great difficulty. The apparent isomorphism between representations of phylogenetic trees and the subheadings of classifications is only apparent. The isomorphism between cladograms and cladistic classifications is not.

As mistaken as those scientists who want to segregate concept formation from theorizing may be, the attempt to disentangle the various elements in the scientific process is well worth it. Scientists themselves, while they are engaged in scientific investigations, need not always keep these distinctions in mind, but those who study science must. The meaning of theoretical terms cannot be totally exhausted by operational definitions, but the ways in which theoretical terms function in science cannot be understood in the absence of the ways in which they are operationalized. The basic units in biological classifications cannot be defined in total isolation from evolutionary theory. Nor can the preference for transformation series over nested clusters be justified without reference to phylogeny. However, it is important to notice exactly how minimal the theoretical input can be made.

Because contingency plays so central a role in selection processes, the particular paths that the resulting lineages take cannot in practice be predicted. If species were closed systems, their future states could be predicted; but if they were closed systems, there would be no future states to predict. Periodically scientists have tried their hand at predicting the future course of a particular area of science. In a provocative book, Stent (1969) argued that, with the advent of molecular biology, biology had reached its Golden Age. Soon all fundamental problems in biology would be solved, and nothing would be left for future biologists to do except tidy up a few loose ends. Although Stent may ultimately be proven right, almost twenty years later biology has still not fallen upon post-Cambrian times. Numerous fundamental problems are no closer to solution than when Stent wrote his book, and new ones have arisen.

Darwin was especially rankled when Owen predicted that his theory would be forgotten in ten years. Because we use a decimal system for counting, predictions of success or failure within ten years are common. In 1911 Wegener predicted that the old views about the stability of continents would not survive the decade (Hallam 1983: 128). In 1940 Julian Huxley thought that the time was ripe for rapid advances in systematics. He was mistaken. Two decades later, Cain once again perdicted a revolution in systematics predicated on a clear separation between phylogenetic weighting and the covariation of traits. Shortly thereafter, Ehrlich (1961a) presented a detailed list of the changes that numerical taxonomy was going to bring about by the year 1970. Although Ehrlich was a bit optimistic with respect to the time frame, some of his predictions have come true but not exactly as he had envisaged them. Electronic data-processing machines have become an important tool of the taxonomist, both for the more pedestrian task of keeping track of collections and for the more interesting task of discerning patterns in taxonomic data, but the programs that have been most widely used are exactly the sort that numerical taxonomists on the basis of their phenetic philosophy would disapprove. They are designed to reconstruct phylogenetic descent.

In 1977 Ed Wiley and Larry Martin were arguing about the virtues of phylogenetic versus numerical taxonomy. Out of exasperation each made his own prediction and sealed it in an envelope to be opened in five years. Martin predicted that "5 years from now the prevailing theoretical superstructure of systematics will differ as much from 'cladism' as 'cladism' now differs from numerical taxonomy," while Wiley thought that in "5 years phylogenetic systematics will be the prevailing theoretical superstructure." When they opened the envelope in 1983, both claimed to be right, and in a sense they both were right. Who could have predicted the emergence of pattern cladism at a time when cladists were reluctant to admit that something properly termed "cladism" actually existed? Although phylogenetic cladism differs markedly from numerical taxonomy, the gap between pattern cladism and numerical taxonomy is not nearly so great. Was phylogenetic systematics (in either sense) the prevailing theoretical superstructure in biological systematics in 1983?

For pheneticists, the ultimate arbiter in science is use. Systematists should construct a general-purpose classification equally useful for all scientists. In the account of science that I have presented, use is also central, not all-purpose usefulness, but the particular uses that scientists make of their own work and the work of others in pursuing their investigations. On this criterion, how have the various schools of systematics fared? In the 1940s, when all the hubbub about systematic methodology began to develop, most systematists just classified without thinking too much about it. One studied one's specimens until a pattern emerged and then constructed a classification in a way that apparently reflected that pattern. Another systematist with another subliminal program might recognize another pattern, but that could not be helped. If Buck (1986) is right, traditional taxonomists continue to classify in this way.

Figures for the actual use of various taxonomic methods are difficult to come by. When Mayr suggested in 1983 that the Society of Systematic Zoology poll its members to discover their taxonomic allegiance, the council of the society soundly rejected the idea. When Gould conducted an informal poll of twenty invertebrate paleontologists, in which he asked them to list five subjects relative to their research that had been most fruitful and five that had been most disappointing in the years since the Darwin centennial in 1959, he was surprised by the intensity of feeling roused with respect to taxonomic philosophy. Strong backing came for new techniques of preparing, storing, and manipulating data, including the use of computers, but just as strongly his respondents rejected numerical taxonomy. Gould concluded that old hostilities die slowly. Gould's respondents were mixed about the virtues of cladistics, and it is a sign of the intensity of feeling aroused that Gould preferred not to plunge into the "Hennigian maelstrom." In that same year, R. F. Johnston predicted a rapprochement between cladists and numerical taxonomists within five or six years. The six years are up, and these old hostilities have not been reduced, let alone eliminated.

From the point of view of encouraging the use of mathematical techniques throughout biology at large, the numerical taxonomy movement seems to have had significant impact. In fact, in the social sciences, the techniques of numerical taxonomy are taken to exemplify scientific classification in the biological sciences. But in biological systemat cs itself, numerical taxonomy has been much less successful in two senses. Biological systematists continue to retain their preference for "phylogenetic" classifications of some sort. Although they might use some of the quantitative techniques first developed by numerical taxonomists, their goal is to reflect phylogenetic development. The opinions of practicing systematists about the appropriate role for "a priori speculation" and "theorizing" in systematics are more difficult to gauge, especially since advocates of one branch of cladistic taxonomy voice similar reservations. To the extent that the appropriate distinctions are made, systematists remain mixed in their opinions. My best guess is that most think that reference to theories cannot be excluded from systematics, but they are not exactly sure about the precise way in which theoretical considerations should enter into the construction of classifications.

Within the confines of biological systematics, cladistics has been extremely successful, a state of affairs that its opponents put down to the fanaticism of its proponents and the gullibility of younger systematists. In the early stages of a budding scientific research program, its advocates must be at least enthusiastic. In the early years of numerical taxonomy, Sokal and Ehrlich could not understand how other systematists could misconstrue their enthusiasm for fanaticism. Later, numerical taxonomists reacted to the enthusiasm of the cladists with similar charges of fanaticism. Later still, cladists found the enthusiasm of Croizat's New Zealand defenders too fanatical. Attitudes that look like enthusiasm from the inside take on the appearance of fanaticism when viewed from the outside, but even fanaticism has its uses in science. Mendelians pushed an extremely simple explanatory device to its limits, in some cases beyond its limits. Nelson pursued the logic of branching diagrams to its logical conclusion. I have done the same for selection processes not only in biological evolution but also in conceptual change. None of these activities is especially "reasonable." Enthusiasts are necessary for science. Scientists have to believe that mysteries can be solved. The problem is how to make the most of enthusiasts without letting them have total sway.

The relative success of cladism (both phylogenetic and pattern) over numerical taxonomy (both phenetic and quantitative) seems to have stemmed from two sources. First, numerical taxonomists branched out too quickly. Before they had succeeded in establishing their methods for classification in biological systematics, they dissipated their energies in applying quantitative techniques in too many areas. They have had some influence in several areas of biology, but they have not come to predominate in any. As Jensen (1983:89) remarked in his report on the Sixteenth Numerical Taxonomy Conference, "Admittedly, much of what transpired here was not directly related to classification."

As far as methodology is concerned, cladists have largely confined themselves to biological systematics. They have continued to refine their methods and to push relentlessly for their adoption by other systematists. They too have branched out, trying to show how various areas of biology can be improved by the adoption of a cladistic perspective. For example, some cladists have shown how the principles of cladistic analysis can facilitate biogeography, a topic traditionally related to systematics. Others have shown how different the evolutionary process looks from a cladistic perspective. Although these activities have met with mixed results, the goal continues to be to show how one preferred systematic methodology produces classifications useful in biology at large.

Of greater importance, cladistics has appeared to be promising to systematists no matter the stage of their career. Here was a reasonably unitary method that they could learn relatively easily and apply to their own work. Few systematists care about mathematics for the sake of mathematics or tree diagrams for the sake of tree diagrams. They do not want to become mathematicians or topologists. They want to classify plants and animals. Numerical taxonomists offered systematists a plethora of techniques, each with its own strengths, each with its own weaknesses. The best advice that Rohlf (1970:80) could give "to someone who needs practical results from numerical taxonomy is to try a variety of techniques for summarizing data. Different methods will expose somewhat different aspects of the phenetic relationships." For mathematicians and mathematically inclined biologists, the literature produced by numerical taxonomists was exhilarating, but for the practicing systematist, it was immobilizing. The cladists presented them with a method—one method—and they could use it without becoming experts.

Of course other outcomes are still possible. In the past, advocates of "Darwinian" forms of evolution have proven to be as opportunistic as the version of the evolutionary process they prefer. T. H. Huxley usually gets the credit for an observation about science that can be traced back at least to Agassiz. Acording to this view, a new doctrine in science must go through three stages. "First, people say that it isn't true, then that it is against religion, and, in the third stage, that it has long been known" (Gould and Eldredge 1986: 143). Co-optation is a powerful technique in science. Stebbins and Ayala (1981) embrace both neutralism and punctualism in the all-engulfing pseudopods of Darwinism. From the beginning, Mayr had a similar fate in mind for challenges to evolutionary systematics. He proposed to incorportate the best aspects of all the suggested methodologies into his own synthetic approach, a spirit of ecumenism that would do the One True Church proud. If Mayr succeeds, phenetics and cladistics will be viewed in the future merely as preliminary steps on the path to finished synthetic classifications.

The history of attempts to produce an account of science that is structurally similar to biological evolution is long and far from reassuring. In this book I have attempted to drive an evolutionary-selectionist approach to science to its logical conclusion. If this is not what advocates of this philosophical research program have in mind, then what? If this approach is to be successful, it must suggest manageable problems for future research. I have portrayed science as a matter of curiosity, credit, and checking. Curiosity and the desire for credit are primitive in my theory. Both may have a genetic basis. In a wide variety of environments, people are liable to be curious and to desire status in the eyes of their social group. Science is so organized as to both encourage and channel these tendencies. Credit is tempered by the need for support, curiosity by the need to check one's speculations. Lying is punished so much more severely in science than stealing because lying harms everyone who uses these faulty contributions while stealing hurts only the victim.

Apparently, lying is much more prevalent in science than stealing, but exactly how great are the discrepancies? How prevalent can stealing become before the likelihood that one will get credit for one's contributions is so reduced that the system ceases to work? Is it true that scientists violate financial trust much less frequently than other professionals or are they simply much better at hiding their transgressions? Does science develop more quickly in areas characterized by competing factions than in areas where scientists work largely alone? What effect does the demic structure of science have on science? Social cohesion is necessary in maintaining research programs. How much cohesion is enough? How much is too much? Can group selection influence selection at the level of individual scientists? How much competition is too much competition? What happens when science becomes too competitive? Scientists aggressively seek credit. Can scientists be too agressive in their quest? What effects does extreme aggression have on the careers of individual scientists? On science itself?

Scientists want credit, but they also need support. What are the forces that lead scientists to conduct their quests for credit under the banner of another scientist? In this process the contributions of earlier workers are transformed by later scientists as they substitute their own views for those of their predecessors, but under what conditions do scientists claim as much originality as possible versus building on the solid foundation of earlier work? What influence do various sorts of "priority" have on science? How commonly do scientists publicly refute their own work? How often do they refute the work of others in their own research group or conceptual deme versus refuting the contributions of their competitors?

In evolutionary biology, two poles of reproductive strategy are distinguished, r and K selection. At the r end of this continuum, organisms produce vast numbers of offspring without any attempt to increase the likelihood that any one of them will survive to reproduce. At the K end, organisms produce only a few offspring but invest considerable effort in caring for them. According to population geneticists, the r strategy should prevail in density independent regimes when the population is well below carrying capacity, while in stable, crowded environments the K strategy is a more appropriate response. Scientists have two sorts of offspring: their contributions to the substantive content of science and, more literally, their scientific descendants. With respect to the first sort of offspring, some scientists generate numerous bright ideas and produce masses of publications, while others cherish every publication. With respect to the second sort of offspring, some scientists work very much in social isolation and discourage the advances made by other scientists to become their descendants, while others are extremely social, maintaining large numbers of graduate students and junior colleagues. The prevalence of one stategy over the other varies with respect to individual scientists, but it also varies with respect to fields of inquiry at particular times. In certain areas of science, the r strategy seems common; in others, the K strategy prevails. Is this variation happenstance, or does the mix of strategies that characterizes a field of inquiry vary in any sort of a systematic way?

Conceptual heterogeneity is necessary for the continued development of science if science is construed as a selection process. Some commentators on science praise "pluralism." Let a thousand theories bloom. Others emphasize the winnowing aspect of science. They value the rigorous pruning that goes on in science. How much heterogeneity is optimal for maximal change in science? Cole and Cole (1972: 372) note that most scientists produce very little during the course of their careers, and even fewer produce work that others actually use. Hence, perhaps the "number of scientists could be reduced without affecting the rate of advance." Perhaps, but selection processes, though they are highly effective, are also extremely inefficient. Selection requires waste, and if biological evolution is any sign, a great deal of waste. How efficient can science be made without decreasing its rate of conceptual growth?

In order to address the preceding questions productively, conceptual change in science must be addressed quantitatively. In this book I have made a start at gathering data to bear on several of these questions, in particular the effect of group allegiance on the refereeing process, the role of personality type and individual behavior in gaining supporters for new research programs, and the effect that the stage in one's career has on the alacrity with which one adopts a new view. For at least a generation, sociologists of science have also gathered data on such questions, data that by and large support the mechanism for scientific change I have urged in this book. More formal treatments of the sort to be found in Boyd and Richerson's (1985) "dual inheritance" model are needed. In their work, Boyd and Richerson emphasize the genetic contribution to the production of organisms capable of both genetic and conceptual inheritance. In my exposition I concentrate almost exclusively on conceptual inheritance. On my view, conceptual entities are replicators, not traits.

Kuhn concluded his highly influential *The Structure of Scientific Revolutions* by suggesting an evolutionary view of science. Later, in a postscript, he urged greater attention be paid both to exemplars in science and to the social structure of science. Later, Toulmin (1972) argued that a single analysis of selection processes could be produced that is equally applicable to social, conceptual, and biological evolution. One way to construe this book is the fulfillment of Kuhn's and Toulmin's suggestions. For the tastes of some, my account may well appear too extreme and exotic. I have pursued these goals too literally and at too great a length. To these objections, all I can reply is that further work needs to be done. The fantasies of Oz must be made as familiar as Kansas. There really is no place like home.

Appendices

Appendix A

Annual Meetings of the Society of Systematic Zoology, Numerical Taxonomy Conference, and Willi Hennig Society.

	Society of Systematic Zoology	Numerical Taxonomy Conferences	Hennig Society	Systematics Congresses
1967	New York	Lawrence		Ann Arbor
1968	Dallas	Boulder		
1969	Boston	Stony Brook		
1970	Chicago	Ann Arbor		
1971	Philadelphia	Toronto		
1972	Washington	Philadelphia		
1973	Boulder	Boulder		Boulder
1974	Tucson	Oreiras, Portugal		Oreiras, Portugal
1975	Corvallis	Montreal		
1976	New Orleans	Lawrence		
1977	Toronto	Madison		
1978	Richmond	Stony Brook		
1979	Tampa	Cambridge		
1980	Seattle	Norman	Lawrence	
1981	Dallas	Ann Arbor	Ann Arbor	Berkeley
1982	Louisville	Notre Dame	College Park	Bad Windheim, W. Ger.
1983	Grand Forks	Ottawa	Grand Forks	
1984	Denver	Cornell	London	
1985	Baltimore	Montreal	Miami	
1986	Nashville	Stony Brook	New York	

Appendix B

Presidents of the Society of Systematic Zoology.

1948–49	W. L. Schmitt	1960	W. I. Follett	1971	R. F. Inger
1950	A. Petrunkevitch	1961	R. E. Blackwelder	1972	N. D. Newell
1951	C. L. Hubbs	1962	C. W. Sabrosky	1973	H. H. Ross
1952	A. Romer	1963	G. G. Simpson	1974–75	T. Downs
1953	H. B. Hungerford	1964	R. W. Pennak	1976–77	D. Rosen
1954	H. W. Stunkard	1965	H. H. Smith	1978–79	R. F. Johnston
1955	L. M. Klauber	1966	E. Mayr	1980–81	E. C. Olson
1956	G. W. Wharton, Jr.	1967	R. L. Usinger	1982–83	J. Slater
1957	R. C. Moore	1968	P. L. Illg	1984–85	D. L. Hull
1958	A. E. Emerson	1969	C. D. Michener	1986–87	G. Nelson
1959	L. H. Hyman	1970	J. O. Corliss	1988–89	E. O. Wiley

Appendix C

Editors of *Systematic Zoology*

1952, no. 1	R. E. Blackwelder	1974–1976	G. J. Nelson
1952, no. 2–1958, no. 2	J. L. Brooks	1977–1979	R. T. Schuh
1958, no. 3–1963, no. 3	L. H. Hyman	1980–1982	J. D. Smith
1964, no. 4–1966	G. W. Byers	1983–1986	G. D. Schnell
1967–1970	R. F. Johnston	1987–	R. L. Schipp
1971–1973	A. J. Rowell		

Appendix D

Articles published in *Systematic Zoology* cited at least 50 times between 1961 and 1983 in other journals sampled. The last citation, however, appeared in 1976, indicating a six-year citation lag. The three most-cited papers were cited between 1955 and 1960 an additional 27, 26, and 9 times, respectively. The information was supplied by Eugene Garfield of the Institute for Scientific Information.

No. of Citations	Year	Article
213	1956	W. L. Brown and E. O. Wilson. Character displacement, 5:49–64.
156	1974	John Avise. Systematic value of electrophoretic data, 23:465–81.
142	1955	C. L. Hubbs. Natural hybridization in fishes, 4:1–20.
117	1970	J. S. Farris. Methods for computing Wagner trees, 19:83–92.
115	1969	A. G. Kluge and J. S. Farris. Quantitative phyletics and the evolution of Anurans, 18:1–32.
97	1971	W. E. Johnson and R. K. Selander. Protein variation and systematics in kangaroo rats (Genus Dipodomys), 20:377–405.
91	1966	R. C. Lewontin. On the measurement of relative variablity, 15:141–42.
89	1961	R. R. Sokal. Distance as a measure of taxonomic similarity, 10:70–79.
81	1971	W. M. Fitch. Toward defining the course of evolution: Minimum change for a specific tree topology, 20:406–16.
79	1969	V. M. Sarich. Pinniped origins and the rate of evolution of carnivore albumins, 18:286–95.
75	1964	G. G. Simpson. Species density of North American recent mammals, 13:57–73.
74	1973	J. -F. Pechere, J. -P. Capony, and J. Demaille. Evolutionary aspects of the structure of muscular parvalbumins, 22:533–48.
69	1953	C. L. Hubbs and C. Hubbs. An improved graphical analysis, 2:49–56.
69	1965	E. Mayr. Numerical phenetics and taxonomic theory, 14:73–97.
69	1967	L. Orloci. Data centering: A review and evaluation with reference to component analysis, 16:208–12.
68	1968	J. L. Brooks. The effects of body size: Introduction to a symposium, 12:272.
66	1974	L. Croizat, G. Nelson, and D. R. Rosen. Centers of origin and related topics, 25:265–87.
60	1976	W. R. Atchley, C. T. Gaskins, and D. Anderson. Statistical properties of ratios, 25:137–48.
58	1970	F. J. Rohlf. Adaptive hierarchical clustering schemes, 19:58–82.
58	1970	J. S. Farris, A. G. Kluge, and M. J. Eckardt. A numerical approach to phylogenetic systematics, 18:172–89.
56	1968	R. T. Holmes and F. A. Pitelka. Food overlap among coexisting sandpipers on northern Alaska tundra, 17:305–18.
55	1975	L. R. Maxson and A. C. Wilson. Albumin evolution and organismal evolution in tree frogs (Hylidae), 24:1–15.
54	1972	J. L. Patton, R. K. Selander, and M. H. Smith. Genic variation in hybridizing populations of gophers (Genus Thomomys), 21:263–70.
53	1974	J. Cracraft. Phylogenetic models and classification, 23:71–90.
52	1968	J. L. Brooks. The effects of prey size selection by lake planktivores, 17:273–91.
52	1968	N. P. Ashmole. Body size, prey size, and ecological segregation in five sympatric tropical terns (Aves: Laridae), 17:292–304.
51	1975	E. O. Wiley. Karl Popper, Systematics and classification: A reply to Walter Bock and other evolutionary taxonomists, 24:233–43.
50	1974	J. O. Corliss. The changing world of ciliate systematics, 23:91–137.

Appendix E

Founding Fellows of the Willi Hennig Society selected 18 August 1982 by James S. Farris, Charles Mitter, Mary F. Mickevich, and F. Chris Thompson.

Kare Bremer
Sadie Coats
Virginia Ferris
John M. Ferris
Bill Fink
Arnold Kluge
S. Løvtrup
Gary Nelson
Norman Platnick
Bill Presch
Mike Richardson
Toby Schuh
J. D. Smith

Tommy Allen
Peggy Bolick
Dan Brooks
Don Buth
Vicky Funk
Chris Humphries
Mike Novacek
Peter Stevens
Hans-Erik Wanntorp
Ed Wiley

Appendix F

The most frequently cited works in three general texts on cladistic analysis, including self-citation.

a) Eldredge, N., and J. Cracraft, 1980, *Phylogenetic patterns and the evolutionary process: Method and theory in comparative biology.*

	No. of Citations
1. Simpson, G. G., 1944, *Tempo and mode in evolution*	32
2. Mayr, E., 1969, *Principles of systematic zoology*	21
3. Simpson, G. G., 1961, *Principles of animal zoology*	18
4. Bock, W. J., 1977, in *Major patterns in vertebrate evolution*	14
5. Hennig, W., 1966, *Phylogenetic systematics*	13
6. Simpson, G. G., 1953, *The major features of evolution*	13
7. Bush, G. L., 1975, in *Annual Review of Ecology and Systematics*	11
8. Eldredge, N., and S. J. Gould, 1972, in *Models in paleontology*	10
9. Dobzhansky, T., 1937, *Genetics and the origin of species*	7
10. Mayr, E., 1963, *Animal species and evolution*	7
11. Patterson, C., and D. Rosen, 1977, in *Bulletin of the American Museum of Natural History*	7

b) Wiley, E. O., 1981, *Phylogenetics: The theory and practice of phylogenetic systematics.*

1. Hennig, W., 1966, *Phylogenetic systematics*	73
2. Wiley, E. O., 1979, in *Systematic Zoology*	51
3. Simpson, G. G., 1961, *Principles of animal taxonomy*	32
4. Mayr, E., 1969, *Principles of systematic zoology*	30
5. Mayr, E., 1963, *Animal species and evolution*	23
6. Mayr, E., 1974, in *Z. Zool. Syst. Evol.-forsch.*	23
7. Grant, V., 1971, *Plant speciation*	22
8. Patterson, C. D., and D. E. Rosen, 1977, in *Bull. of the Amer. Mus. of Nat. Hist.*	22
9. Rosen, D. E., 1979, in *Bulletin of the American Museum of Natural History*	20
10. Wiley, E. O., 1976, in *Misc. Publ. Mus. Nat. Hist. Univ. Kansas*	19
11. Wiley, E. O., 1978, in *Systematic Zoology*	19
12. Ashlock, P. H., 1971, in *Systematic Zoology*	17
13. Farris, S. J., 1977, in *Major patterns in vertebrate evolution*	16
14. Patterson, C., 1973, in *Interrelationships of fishes*	16

c) Nelson, G., and N. Platnick, 1981, *Systematics and biogeography: Cladistics and vicariance.*

1. de Candolle, A.-P., 1820, *Géographie botanique*	13
2. Darlington, P. J., Jr., 1957, *Zoogeography*	9
3. Hennig, W., 1966, *Phylogenetic systematics*	9
4. Wallace, A. R., 1855, in *Ann. Mag. Nat. Hist.*	7
5. Darwin, C., 1859, *On the origin of species*	6
6. Lyell, C., 1832, *Principles of geology*	6
7. Sneath, P. H. A., and R. R. Sokal, 1973, *Numerical taxonomy*	6
8. Wallace, A. R., 1876, *The geographic distribution of animals*	6
9. Hennig, W., 1950, *Grundzüge einer Theorie der phylogenetischen Systematik*	5
10. McKenna, M. C., 1975, in *Phylogeny of primates*	5

Cumulative Summary of Most-Cited Works, Excluding Self-Citation

Hennig (1966)	95	Mayr (1963)	30
Mayr (1969)	55	Patterson and Rosen (1977)	29
Simpson (1961)	50	Mayr (1974)	23
Simpson (1944)	33		

Most-Cited Authors

a) In Eldredge and Cracraft (1980)

Simpson	86
Eldredge	49
Mayr	40
Bock	36
Gould	27
Hennig	20
Nelson	20
Wiley	20
Platnick	16
Dobzhansky	14
Farris	12

b) In Wiley (1981)

Wiley	132
Mayr	109
Hennig	97
Nelson	65
Farris	61
Simpson	45
Patterson	44
Rosen	40
Ashlock	35
Platnick	29
Grant	26
Sneath	23
Løvtrup	22

c) In Nelson and Platnick (1981)

de Candolle	21
Linnaeus	19
Mayr	17
Simpson	17
Wallace	16
Hennig	15
Darlington	11
Croizat	8
Rosa	8
Sneath & Sokal	8

d) Cumulative Summary of the Most-Cited Authors, Excluding Self-Citation

Mayr	153
Simpson	148
Hennig	132
Nelson	85
Farris	73
Patterson	59
Bock	46
Rosen	44

Appendix G

Questionnaire used by Krantz and Wiggins (1982) to determine personality characteristics of four founders of behavioral psychology: Clark Hull, B. F. Skinner, Kenneth Spence, and E. C. Tolman. The results below are from administering this same questionnaire in 1981 to students and colleagues of George Ball, Robert R. Sokal, and Gareth Nelson. The choices were very accurate (VA), accurate (A), moderately accurate (MA), slightly accurate (SA), innaccurate (IA), and not applicable (NA).

	VA	*A*	*MA*	*SA*	*IA*	*NA*
1. The investigator was convinced of the value of his own systematic position.						
Ball	5	3	0	0	0	0
Sokal	6	2	0	0	0	0
Nelson	8	0	0	0	0	0
2. Allowed me autonomy in a choice of research problems.						
Ball	6	0	1	0	0	1
Sokal	0	2	1	3	0	2
Nelson	2	0	2	1	0	3
3. Took himself seriously.						
Ball	2	4	1	1	0	0
Sokal	6	2	0	0	0	0
Nelson	7	0	0	1	0	0
4. Expected strong commitment to his approach.						
Ball	3	1	1	0	3	0
Sokal	3	1	2	0	1	1
Nelson	5	0	1	1	0	1
5. Accepted criticism well.						
Ball	2	4	0	1	1	0
Sokal	1	3	1	1	2	0
Nelson	0	0	2	1	5	0
6. Inspiring.						
Ball	6	2	0	0	0	0
Sokal	3	2	2	0	1	0
Nelson	5	1	2	0	0	0
7. In teaching presented primarily his own position.						
Ball	0	0	2	0	5	1
Sokal	0	3	1	2	1	1
Nelson	1	2	1	1	1	2

	VA	*A*	*MA*	*SA*	*IA*	*NA*
8. Concerned about my professional future.						
Ball	4	2	1	0	1	0
Sokal	2	2	2	1	0	1
Nelson	2	1	1	1	2	1
9. Critical of others' systematic position.						
Ball	2	2	1	1	2	0
Sokal	3	3	0	1	0	1
Nelson	7	0	0	1	0	0
10. Authoritarian.						
Ball	1	1	1	2	3	0
Sokal	3	3	1	1	0	0
Nelson	4	0	1	1	2	0
11. Continually reformulated his systematic position.						
Ball	2	3	2	1	0	0
Sokal	1	1	2	3	0	1
Nelson	4	0	3	1	0	0
12. Set high standards for my work.						
Ball	5	2	0	0	0	1
Sokal	4	3	0	0	0	1
Nelson	3	2	0	1	1	1
13. Was supportive in his criticisms of my work.						
Ball	6	2	0	0	0	0
Sokal	3	3	1	0	0	1
Nelson	3	3	1	0	1	0
14. Allowed me to know him as a person.						
Ball	5	1	2	0	0	0
Sokal	0	3	2	1	1	0
Nelson	1	3	1	1	1	1
15. Broadly versed.						
Ball	5	0	1	1	1	0
Sokal	3	5	0	0	0	0
Nelson	4	3	1	0	0	0
16. Open-minded.						
Ball	3	2	2	1	0	0
Sokal	0	4	3	1	0	0
Nelson	1	2	4	1	0	0
17. Concerned about my position within the department.						
Ball	3	2	2	0	1	0
Sokal	1	3	1	1	0	1
Nelson	1	2	2	1	1	1

	VA	*A*	*MA*	*SA*	*IA*	*NA*
18. Helped me in obtaining support for my research.						
Ball	4	2	1	0	0	1
Sokal	1	3	0	2	0	2
Nelson	1	1	2	1	2	1
19. Humorous.						
Ball	4	3	1	0	0	0
Sokal	0	1	4	1	1	0
Nelson	1	2	2	1	1	1
20. Philosophically oriented.						
Ball	3	2	1	2	0	0
Sokal	1	4	1	0	1	1
Nelson	5	3	0	0	0	0
21. Overly influenced by others' views of his approach.						
Ball	0	1	0	3	4	0
Sokal	0	1	0	1	5	1
Nelson	0	0	1	3	4	0
22. Understanding of my needs.						
Ball	4	2	0	1	0	1
Sokal	0	2	3	1	0	2
Nelson	0	2	2	1	1	1
23. Aggressive.						
Ball	3	3	2	0	0	0
Sokal	3	2	1	1	0	0
Nelson	5	2	0	1	0	0
24. Creative.						
Ball	2	4	2	0	0	0
Sokal	4	1	1	0	0	2
Nelson	8	0	0	0	0	0
25. Outgoing.						
Ball	4	3	0	0	0	1
Sokal	0	0	4	2	0	2
Nelson	0	1	3	3	1	0
26. Helped me in publishing my research.						
Ball	3	2	0	1	1	1
Sokal	0	0	4	2	0	2
Nelson	0	1	3	3	1	0
27. Distant Interpersonally.						
Ball	0	0	1	2	5	0
Sokal	0	1	1	3	1	2
Nelson	1	3	0	1	3	0

	VA	*A*	*MA*	*SA*	*IA*	*NA*
28. Treated me as an intellectual equal.						
Ball	5	2	1	0	0	0
Sokal	1	3	2	1	0	1
Nelson	1	2	3	1	0	1
29. Variable in his moods and attitudes.						
Ball	2	3	1	1	1	0
Sokal	0	2	1	2	2	1
Nelson	3	4	1	0	0	0
30. Conferred appropriate credit for others' contributions to his systematic approach.						
Ball	5	3	0	0	0	0
Sokal	1	4	1	1	0	1
Nelson	5	3	0	0	0	0
31. Strong-willed.						
Ball	5	3	0	0	0	0
Sokal	4	3	0	0	0	1
Nelson	8	0	0	0	0	0
32. Receptive to my ideas about his work.						
Ball	3	2	3	0	0	0
Sokal	0	2	4	0	0	2
Nelson	1	0	4	1	0	2

Appendix H

Age in 1859 of British scientists who accepted the evolution of species within 10 years of the publication of Darwin's *Origin of Species,* compared to their age at acceptance and the age in 1869 of those who did not accept evolution; slightly modified from Hull, Tessner, and Diamond (1978).

Name and Dates	Age in 1859	Age at earliest evidence of acceptance	Age in 1869 of continued rejectors
Babington, C. C. (1808–95)	51	—	61
Balfour, J. H. (1808–84)	51	—	61
Bates, H. W. (1835–92)	34	32	—
Bell, T. (1792–1880)	67	—	77
Bennet, A. W. (1833–1902	26	36	—
Bentham, G. (1800–1884)	59	63	—
Busk, G. (1807–86)	52	54	—
Butler, A. G. (1831–1909)	28	38	—
Carpenter, W. B. (1813–85)	46	47	—
Dawkins, W. F. (1837–1929)	22	28	—
Dunkan, P. M. (1821–91)	38	44	—
Fawcett, H. (1833–84)	26	27	—
Flower, W. H. (1831–99)	28	29	—
Frankland, E. (1825–99)	34	39	—
Galton, F. (1822–1911)	37	38	—
Geikie, A. (1835–1924	24	24	—
Gosse, P. H. (1810–88)	49	—	59
Gray, J. F. (1800–75)	59	—	69
Grove, W. R. (1811–96)	48	55	—
Gunther, A. C. L. (1830–1914)	29	—	39
Haughton, S. (1821–93)	38	—	48
Herschel, J. F. W. (1792–1871)	67	69	—
Hirst, T. A. (1830–92)	29	34	—
Holland, H. (1788–1873)	71	72	—
Hooker, J. D. (1817–1911)	42	41	—
Humphrey, G. M. (1820–96)	39	46	—
Hunt, J. (1833–69)	26	—	36
Hutton, F. W. (1836–1905)	23	25	—
Huxley, T. H. (1825–95)	34	33	—
Jardine, W. (1800–74)	59	—	69
Jeffreys, J. (1809–85)	50	59	—
Jenkin, F. (1833–85)	26	—	36
Jenyns, L. (1800–93)	59	60	—
Jevons, W. S. (1835–82)	24	34	—
Jukes, J. B. (1811–69)	48	49	—
Kingsley, C. (1819–75)	40	44	—
Lankester, E. (1814–74)	45	55	—
Lewes, G. H. (1817–78)	42	51	—
Lubbock, J. (1834–1913)	25	26	—
Lyell, C. (1797–1875)	62	70	—
Mill, J. S. (1806–73)	53	—	63
M'Intosh, W. C. (1838–1931)	21	—	31

Name and Dates	Age in 1859	Age at earliest evidence of acceptance	Age in 1869 of continued rejectors
Mivart, J. (1827–1900)	32	33	—
Molesworth, W. N. (1816–96)	43	46	—
Morris, F. O. (1810–93)	49	59	—
Murchison, R. I. (1792–1871)	67	—	77
Murray, A. (1812–78)	47	56	—
Newton, A. (1829–1907)	30	29	—
Page, D. (1814–79)	45	50	—
Parker, W. K. (1823–90)	36	41	—
Ramsey, A. C. (1814–91)	45	46	—
Rolleston, G. (1829–81)	30	31	—
Sclater, P. L. (1829–1913)	30	31	—
Scott, J. G. (1838–80)	21	26	—
Sedgwick, A. (1785–1873)	74	—	84
Spottiswoode, W. (1825–83)	34	39	—
Stokes, G. G. (1814–83)	40	—	50
Tegetmeier, W. B. (1816–1912)	43	48	—
Thompson, A. (1809–84)	50	60	—
Thomson, C. W. (1830–82)	29	38	—
Thomson, W. (1824–1907)	35	45	—
Thwaites, G. H. K. (1811–82)	48	49	—
Tristam, H. B. (1822–1906)	37	37	—
Tyndall, J. (1820–93)	39	44	—
Watson, H. C. (1804–81)	55	56	—
Westwood, J. O. (1805–93)	54	—	64
Wood, S. V. (1798–1880)	61	62	—
Young, J. (1835–1902)	24	31	—

Appendix I

Ernst Mayr Student Award for best paper delivered at the annual meeting of the Society of Systematic Zoology between 1976 and 1986.

Diane Calabrese, 1976
Sharon Simpson, 1977
Sadie Coats, 1979
Kevin de Queiroz, 1981
Jonathan Coddington, 1982
Janine Caira, 1983
Richard O'Grady, 1983
Eldredge Bermingham, 1984
David Cannatella, 1985
Linda Dryden, 1986

Appendix J

Systematists and evolutionary biologists interviewed, in chronological order.

Tom Schopf, University of Chicago, Chicago, Ill., May 21, 1973
Joel Cracraft, Chicago, Ill., May 23, 1973
Leigh Van Valen, University of Chicago, Chicago, Ill., May 25, 1973
Richard Lewontin, University of Chicago, Chicago, Ill., June 6, 1973
Hymen Marx, Field Museum of Natural History, Chicago, Ill., June 12, 1973
Rolf Singer, Field Museum of Natural History, Chicago, Ill., June 12, 1973
Robert Inger, Field Museum of Natural History, Chicago, Ill., June 12, 1973
Donald Simpson, Field Museum of Natural History, Chicago, Ill., June 18, 1973
Sara Bretsky, State University of New York, Stony Brook, N.Y., July 9, 1973
James Rohlf, State University of New York, Stony Brook, N.Y., July 9, 1973
Gareth Nelson, American Museum of Natural History, New York, N.Y., July 10, 1973
Humphrey Greenwood, American Museum of Natural History, New York, N.Y., July 11, 1973
Donn Rosen, American Museum of Natural History, New York, N.Y., July 11, 1973
Bobb Schaeffer, Museum of Natural History, New York, N.Y., July 11, 1973
Robert R. Sokal, State University of New York, Stony Brook, N.Y., July 12, 1973
Eugene Gaffney, Museum of Natural History, New York, N.Y., July 13, 1973
A. E. Emerson, Hullett's Landing, Lake St. George, N.Y., July 14, 1973
James S. Farris, Boulder, Colo., August 8, 1973
G. C. D. Griffith, Boulder, Colo., August 8, 1973
Vernon Heywood, Boulder, Colo., August 8, 1973
David Rogers, University of Colorado, Boulder, Colo., August 8, 1973
Hildemar Scholz, Boulder, Colo., August 9, 1973
L. A. S. Johnson, Boulder, Colo., August 9, 1973
Lars Brundin, Cambridge, Mass., November 2, 1973
Stephen Jay Gould, Harvard University, Cambridge, Mass., November 2, 1973
Ernst Mayr, Harvard University, Cambridge, Mass., November 2, 1973
Peter Raven, Missouri Botanical Garden, St. Louis, Mo., November 29, 1973
Michael Ghiselin, Bodega Bay, Calif., December 12, 1973
Theodore Crovello, Notre Dame, Ind., November 2, 1974
Donald Colless, Chicago, Ill., November 29, 1974
Leon Croizat, Caracas, Venezuela, August 7, 1974
G. G. Simpson, Tucson, Ariz., July 19, 1978
Walter Bock, Chicago, Ill., January 1, 1979
E. O. Wiley, University of Kansas, Lawrence, Kans., February 7, 1979
Peter Ashlock, University of Kansas, Lawrence, Kans., February 8, 1979
George Byers, University of Kansas, Lawrence, Kans., February 9, 1979
Richard Johnston, University of Kansas, Lawrence, Kans., February 9, 1979
C. D. Michener, University of Kansas, Lawrence, Kans., February 8, 1979
Albert Rowell, University of Kansas, Lawrence, Kans., February 9, 1979
Roger Kaesler, University of Kansas, Lawrence, Kans., February 10, 1979
Ronald McGinley, University of Kansas, Lawrence, Kans., February 10, 1979
Soren Løvtrup, Lawrence, Kans., October 13, 1980
Dan Brooks, Berkeley, Calif., March 27, 1981
Walter Fitch, Berkeley, Calif., March 27, 1981
Paul Ehrlich, Stanford University, Stanford, Calif., March 28, 1981
Nicholas Jardine, Cambridge University, Cambridge, England, April 7, 1981
Colin Patterson, British Museum (Natural History), London, England, April 7, 1981

Roy Crowson, University of Glasgow, Glasgow, Scotland, April 17, 1981
George Estabrook, University of Michigan, Ann Arbor, Mich., October 5, 1981
Herbert Wagner, University of Michigan, Ann Arbor, Mich., October 5, 1981
Rainer Zangerl, Rockeville, Ind., October 12, 1981
Richard Blackwelder, Chicago, Ill., September 22, 1983

Questions asked of the preceding biologists

1. What are your major fields of interest?
What percentage is taxonomy proper? Which groups?
Do you think that your work in these taxonomic groups has influenced your general attitude toward taxonomy? How?

2. How did you get into science? Taxonomy?
Did you have any formal training in taxonomy? Was it worthwhile?
Which school of taxonomy attracted you most in your early years?
Who influenced you most? Which books?

3. In your early years, did you have easy access to the authorities in your field? To publication? Grants?
What journals do you read regularly? Have you found the refereeing practices of these journals what they should be? Fair? Thorough?
Have you ever had any difficulty in getting anything published? What are the purposes of citations? How have you been treated?

4. What are the most prestigious scientific societies?
Do you agree with the ways in which the membership of these societies is chosen?
What effects do such societies have on science?

5. When you started in taxonomy, were there any outstanding achievements which impressed you especially?
What were the outstanding problems in the field?
Do any of them remain today?
How do you select your research problems?
Do you think that scientific theories are fundamentally provisional? Data?
Do you feel that science proceeds via the competition of ideas?

6. If you had to label yourself according to one of the schools of taxonomy, which would it be?
What are the tenets of these schools of taxonomy—evolutionary systematics, phenetics, and cladistics?
Who are the strongest spokesmen for each of these schools?
What factors do you think gave rise to each of these schools? What contributions have they made to systematics? Chief weaknesses?

7. Have you been influenced very strongly by the literature arguing the pros and cons of these various schools? Who?
With whom do you communicate on scientific issues most frequently?
Of those with whom you disagree, whom do you admire most?
Do you find it impossible to communicate with members of rival schools over the issues that separate you?
Have you changed your mind on any fundamental issues since your early years? What?

8. What is the proper subject matter of taxonomy? Its purpose?
Should biological classifications be theoretically neutral?
Should the history of organisms enter into biological classifications?
Do you find taxonomy an especially atypical science? How?
Do you find taxonomists to be especially atypical scientists? How?

9. Do you think that the synthetic theory of evolution is fundamentally correct?
What are its chief weaknesses?
What is the role of speculation in science?
What do you view your task as a scientist to be?
Are there any tenets in science that are not open to challenge?

10. Is science a rational enterprise? Taxonomy?
How much of a role does tacit knowledge play in your work?
Do you think that there are important differences in the principles of botanical and zoological classification?
Of all the factors that influence taxonomy, which do you think are the most irrelevant?
Could evolutionary theory be false?

11. How would you define "homology"?
What are phenetic characters? Overall similarity?
What are sister groups? Matching coefficients?
What is the present status of each of the schools of taxonomy?
Do you think that scientists should compete with each other?

12. Who are the most prestigious taxonomists today?
Should scientists take sociological and psychological factors into account in setting out their views?
Should a scientist use whatever influence (power) he has in the scientific community to get his ideas accepted? Others rejected?
Have you felt the influence of sociological factors on yourself?
What are the operative factors in a scientist's adopting one view or another? An example from your own experience?

13. Can you think of a scientist whose work is consistently ignored?
Do you think that your older colleagues are less willing to listen to you than your younger colleagues? Less able to understand?
What would you think if a colleague published a book at his own expense?

14. Do you work largely alone or with others?
What do you consider your main contribution to science? How important is your work to your life as a whole?
Are you familiar with the views of R. E. Blackwelder? Leon Croizat?

Given the general orientation of the preceding questions, were there any which you felt should have been asked but were not?

Were there any questions which were asked in ways liable to bias the answers strongly?

References

Abele, L. 1982. Vacariants and the holy writ. *Paleobiology* 8:79–82.

Abir-am, P. G. 1982. How scientists view their heroes: some remarks on the mechanisms of myth construction. *Journal of the History of Biology* 15:281–315.

Adams, M. B. 1980. Sergei Chetverikov, the Kol'tsov Institute and the evolutionary synthesis. In *The evolutionary synthesis,* ed. E. Mayr and W. B. Provine, 242–78. Cambridge: Harvard University Press.

Adanson, M. 1763. *Families des plantes.* Paris: Vincent.

Agassiz, L. 1840. *Studies on glaciers.* Trans. A. V. Carozzi (1956), New York: Hafner.

Alberch, P. 1985. Problems with the interpretation of developmental sequences. *Systematic Zoology* 34:46–58.

Alexander, R. D. 1971. The search for an evolutionary philosophy of man. *Proceedings of the Royal Society of Victoria* 84:99–120.

———. 1974. The evolution of social behavior. *Annual Review of Ecology and Systematics* 5:325–83.

———. 1979. *Darwinism and human affairs.* Seattle: University of Washington Press.

Allen, D. E. 1976. *The naturalist in Britain.* London: Allen Lane.

Allen, E., et al. 1975. Against sociobiology. *New York Review of Books* 18:43–44.

Allen, E., et al. 1976. Sociobiology—another biological determinism. *BioScience* 26:182, 184–86.

Allen, G. E. 1979. Naturalists and experimentalists: the genotype and the phenotype. *Studies in the History of Biology* 3:179–209.

Andrews, F. M., ed. 1979. *Scientific productivity: the effectiveness of research groups in six countries.* New York: Cambridge University Press.

Anonymous. 1979. Vicariance, in "Talk of the Town." *The New Yorker,* September 10, pp. 37–9.

———. 1981a. Darwin's death in South Kensington. *Nature* 289:735.

———. 1981b. Loose ends in evolution. *Nature* 292:1–2.

Appel, T. A. 1987. *The Cuvier-Geoffroy debate: French biology in the decades before Darwin.* Oxford: Oxford University Press.

Armstrong, J. E., and B. A. Drummond. 1981. Macroevolution conference. *Science* 211:774.

Arnold, A. J., and K. Fristrup. 1982. The theory of evolution by natural selection: a hierarchical expansion. *Paleobiology* 8:113–29.

Ashlock, P. D. 1971. Monophyly and associated terms. *Systematic Zoology* 20:63–69.

———. 1972. Monophyly again. *Systematic Zoology* 21:430–37.

Atran, S. 1985. Pre-theoretical aspects of Aristotelian definition and classification of animals: the case for common sense. *Studies in the History and Philosophy of Science* 16:113–63.

———. 1987a. Origins of the species and genus concepts. *Journal of the History of Biology* 20:195–279.

———. 1987b. Ordinary constraints on the semantics of living kinds: a common sense alternative to recent treatments of natural-object terms. *Mind and Language* 2:27–63.

Austad, S. N. 1984. A classification of alternative reproductive behavior and methods for field testing ESS models. *American Zoologist* 24:309–19.

Ayala, F. J. 1974. Biological evolution: natural selection vs. random walk? *American Scientist* 62:692–701.

———. 1978. The mechanisms of evolution. *Scientific American* 239:56–68.

Babbage, C. 1830. *Reflections on the decline of science in England and on some of its causes.* London: Fellowes and Boothe.

Babchuck, N., and A. P. Bates. 1962. Professor or producer: the two faces of academic man. *Social Forces* 40:341–48.

Bacon, F. 1620. *The new organon and related writings.* Indianapolis: Bobbs-Merrill (1960).

Baker, J. R. 1981. Room for all. *Nature* 290:623.

Baldridge, W. S. 1984. The geological writings of Goethe. *American Scientist* 72:163–67.

Ball, H. W., et al. 1981. Darwin's survival. *Nature* 290:82.

Ball, I. R. 1981. The order of life—towards a comparative biology. *Nature* 294:675–76.

Balzer, W., and C. M. Dawe. 1986. Structure and comparison of gene theories: (1) classical genetics. *British Journal for the Philosophy of Science* 37:55–69.

Bannister, R. C. 1979. *Social Darwinism: science and myth in Anglo-American social thought.* Philadelphia: Temple University Press.

Barash, D. 1977. *Sociobiology and behavior.* Amsterdam: Elsevier.

———. 1979. *The whisperings within.* New York: Penguin Books.

Barber, B. 1961. Resistance by scientists to scientific discovery. *Science* 74:596–602.

Barlow, G. W., and J. Silverberg, eds. 1980. *Sociobiology: beyond nature/nurture?* AAAS selected symposium 35. Boulder: Westview Press.

Barnes, B. 1977. *Interests and the growth of knowledge.* London: Routledge & Kegan Paul.

———. 1981. On the 'hows' and 'whys' of cultural change (response to Woolgar). *Social Studies of Science* 11:481–98.

———. 1985a. Response to Roll-Hansen. *Social Studies of Science* 15:175.

———. 1985b. Ethnomethodology as science. *Social Studies of Science* 15:751–62.

Barnes, B., and D. Bloor. 1982. Relativism, rationalism and the sociology of knowledge. In *Rationality and Relativism,* ed. M. Hollis and S. Lukes, 21–47. Cambridge: MIT Press.

Bateson, W. 1894. *Materials for the study of variation, treated with especial regard to discontinuity in the origin of species.* London: Macmillan.

———. 1901. Problems of heredity as a subject for horticultural investigation. *Journal of the Royal Horticultural Society* 25:54–61.

———. 1909. Heredity and variation in modern lights. In *Darwin and modern science,* ed. A. C. Stewart, 85–101. Cambridge: At the University Press.

Bather, F. A. 1927. Biological classification: past and future. *Proceedings of the Geological Society* 83:63–104.

Beatty, J. 1982. Classes and cladists. *Systematic Zoology* 31:25–34.

Bell, T. 1860. Presidential address. *Journal of the Linnean Society* 4:viii–ix.

Bennett, A. W. 1870. The genesis of species. *Nature* 3:270–73.

Bernal, J. D. 1971. *Science in history.* 4 vols., 3d ed. Cambridge: MIT Press.

Bernstein, H., H. Byerly, F. Hopf, and R. E. Michod. 1985. Sex and the emergence of species. *Journal of Theoretical Biology* 117:665–90.

Bigelow, R. S. 1956. Monophyletic classification and evolution. *Systematic Zoology* 5:145–46.

———. 1958. Classification and Phylogeny. *Systematic Zoology* 7:49–59.

Birke, L. 1986. *Women, feminism and biology.* New York: Methuen.

Blackmore, J. 1978. Is Planck's 'principle' true? *British Journal for the Philosophy of Science* 29:347–61.

Blackwelder, R. E. 1959. The functions and limitations of classification. *Systematic Zoology* 8:202–11.

———. 1977. Twenty-five years of taxonomy. *Systematic Zoology* 26:107–37.

Blackwelder, R. E., and A. A. Boyden. 1952. The nature of systematics. *Systematic Zoology* 1:26–33.

Blau, J. R. 1978. Sociometric structures of a scientific discipline. *Research in Sociology of Knowledge, Sciences and Art* 1:91–206.

Bliss, M. 1982. *The discovery of insulin.* Chicago: University of Chicago Press.

Bloor, D. 1976. *Knowledge and social imagery.* London: Routledge and Kegan Paul.

———. 1982. Durkheim and Mauss revisited: classification and the sociology of knowledge. *Studies in the History and Philosophy of Science* 13:267–331.

Blute, M. 1979. Sociocultural evolutionism: an untried theory. *Behavioral Science* 24:46–59.

Bock, W. J. 1968. Phylogenetic systematics, cladistics, and evolution (review of Hennig 1966). *Evolution* 22:646–48.

———. 1973. Philosophical foundations of classical evolutionary classifications. *Systematic Zoology* 22:375–92.

———. 1977. Foundations and methods of evolutionary classification. In *Major patterns in vertebrate evolution.* ed. M. K. Hecht, P. C. Goody, and B. M. Hecht, 851–95. New York: Plenum.

Bock, W. J., and G. von Wahlert. 1963. Two evolutionary theories—a discussion. *British Journal for the Philosophy of Science* 14:140–46.

Boesiger, E. 1980. Evolutionary biology in France at the time of the evolutionary synthesis. In *The evolutionary synthesis,* ed. E. Mayr and W. B. Provine, 309–20. Cambridge: Harvard University Press.

Bondi, H. 1975. What is progress in science? In *Problems of scientific revolution.* ed. R. Harré, 1–10. Oxford: Clarendon Press.

Bonner, J. T. 1967. *The cellular slime molds.* Princeton: Princeton University Press.

———. 1974. *On development.* Cambridge: Harvard University Press.

Borgmeier, T. 1957. Basic questions of systematics. *Systematic Zoology* 6:53–69.

Bosk, C. L. 1979. *Forgive and remember, managing medical failure.* Chicago: University of Chicago Press.

Bowen, E. 1986. Radicals in conservative garb. *Time* 128:71–72.

Bowen, F. 1860. Darwin on the origin of species. *North American Review* 90:474–506.

Bowler, P. J. 1976. *Fossils and progress: paleontology and the idea of progressive evolution in the nineteenth century.* New York: Science History Publications.

———. 1983. *The Eclipse of Darwinism: anti-Darwinian evolution theories in the decades around 1900.* Baltimore: The Johns Hopkins University Press.

Box, J. F., ed. 1978. *R. A. Fisher: the life of a scientist.* New York: John Wiley.

Boyd, R., and P. J. Richerson. 1985. *Culture and the evolutionary process.* Chicago: University of Chicago Press.

Bradie, M. 1986. Assessing evolutionary epistemology. *Biology and Philosophy* 1:401–60.

Brady, R. 1979. Natural selection and the criteria by which a theory is judged. *Systematic Zoology* 28:600–621.

———. 1982a. Theoretical issues and "pattern cladists." *Systematic Zoology* 31:286–91.

———. 1982b. Dogma and doubt. *Biological Journal of the Linnean Society* 17:79–96.

———. 1985. On the independence of systematics. *Cladistics* 1:113–26.

———. 1987. Form and cause in Goethe's morphology. In *Goethe and the sciences,* ed. F. Amrine, F. J. Zucker, and H. Wheeler, 257–300. Dordrecht: Reidel.

Brandon, R. N. 1985. Adaptation explanations: Are adaptations for the good of replicators or interactors? In *Evolution at a crossroads,* ed. D. Depew and B. Wheeler, 81–96. Cambridge: MIT Press.

Brandon, R. N., and R. M. Burian, eds. 1984. *Genes, organisms, populations.* Cambridge: MIT Press.

Brandon, R. N., and N. Hornstein. 1986. From icons to symbols: some speculations on the origin of language. *Biology and Philosophy* 1:169–89.

Brannigan, A. 1979. The reification of Mendel. *Social Studies of Science* 9:423–54.
Branscomb, L. M. 1985. Integrity in science. *American Scientist* 73:421–23.
Bresler, J. B. 1968. Teaching effectiveness and government awards. *Science* 160:164–67.
Bridgman, P. W. 1959. *The Logic of Modern Physics* after thirty years. *Daedalus* 88:518–26.
Broad, W., and N. Wade. 1982. *Betrayers of the truth: fraud and deceit in the halls of science.* New York: Simon and Schuster.
Brock, W. H., and R. M. MacLeod. 1976. The scientists' declaration: reflexions on science and belief in the wake of *Essays and reviews,* 1864–5, *British Journal for the History of Science* 11:39–66.
Bromberger, S. 1962. An approach to explanation. In *Analytical philosophy,* ed. R. J. Butler, 72–105. Oxford: Basil Blackwell.
Brooks, D. R. 1981. Classifications as languages of empirical comparative biology. In *Advances in cladistics,* vol. 1, ed. V. A. Funk and D. R. Brooks, 61–70. New York: Columbia University Press.
Brooks, D. R., and E. O. Wiley. 1985. Theories and methods in different approaches to phylogenetic systematics. *Cladistics* 1:1–11.
———. 1986. *Evolution and entropy: toward a unified theory of biology.* Chicago: University of Chicago Press.
Brooks, J. L. 1984. *Just before the origin: Alfred Russel Wallace's theory of evolution.* New York: Columbia University Press.
Brown, C. H. 1985. Mode of subsistence and folk biological taxonomy. *Current Anthropology* 26:43–64.
Browne, J. 1980. Darwin's botanical arithmetic and the "principle of divergence," 1854–1858. *Journal of the History of Biology* 13:53–89.
Brundin, L. 1966. Transantarctic relationships and their significance, as evidenced by chironomid midges. *Kunglica svenska Vetensk apsakademicns Handlingar* 11:1–472.
———. 1968. Application of phylogenetic principles in systematic and evolutionary theory. *Current problems of lower vertebrate phylogeny,* ed. T. Ørvig, 471–95. New York: Wiley.
———. 1972. Evolution, causal biology, and classification. *Zoologica Scripta* 1:107–20.
Brush, S. G. 1974. Should the history of science be rated X? *Science* 183:1164–72.
Buck, W. R. 1986. Traditional methods in taxonomy; a personal approbation. *Taxon* 35:306–11.
Buican, D. 1982. Milestones in the philosophic and historical framework of the development of biology in France. *Scientia* 117:609–28.
———. 1984. *Histoire de la génétique et de l'évolutionnisme en France.* Paris: Presses Universitaires de France.
Burchfield, J. D. 1975. *Lord Kelvin and the age of the earth.* New York: Science History Publications.
Burkhardt, R. W., Jr. 1977. *The spirit of system: Lamarck and evolutionary biology.* Cambridge: Harvard University Press.
Burtt, B. L. 1966. Adanson and modern taxonomy. *Edinburgh Royal Botanic Gardens, Notes* 26:427–31.
Bush, G. L. 1975. Modes of animal speciation. *Annual Review of Ecology and Systematics* 6:339–64.
Bynum, W. F. 1984. Charles Lyell's *Antiquity of Man* and its critics. *Journal of the History of Biology* 17:153–88.
Cahn, R. 1984. Science policy à la française. *Nature* 308:563–64.
Cain, A. J. 1959a. Deductive and inductive methods in post-Linnaean taxonomy. *Proceedings of the Linnean Society of London,* 170th session, 185–217.
———. 1959b. The post-Linnaean development of taxonomy. *Proceedings of the Linnean Society of London,* 170th session, 234–44.
———. 1962. The evolution of taxonomic principles. In *Microbial classification,* ed. G. C. Ainsworth and P. H. A. Sneath, 1–13. New York: Cambridge University Press.
Cain, A. J., and G. A. Harrison. 1958. An analysis of the taxonomist's judgement of affinity. *Proceedings of the Zoological Society, London* 131:85–98.

———. 1960. Phyletic weighting. *Proceedings of the Zoological Society, London* 135:1–31.
Cambrosio, P., and P. Keating. 1985. Studying a biotechnology research centre: a note on local socio-political issues. *Social Studies of Science* 15:723–37.
Camin, J. H., and R. R. Sokal. 1965. A method for deducing branching sequences in phylogeny. *Evolution* 19:311–26.
Campbell, D. T. 1965. Variation and selective retention in socio-cultural evolution. In *Social change in developing areas,* ed. H. R. Barringer, G. I. Blanksten, and R. W. Mack, 19–49. Cambridge: Schenkman Press.
———. 1969. Ethnocentrism of disciplines and the fish-scale model of omniscience. In *Interdisciplinary relationships in the social sciences,* ed. M. Sherif and C. W. Sherif. 328–48. Chicago: Aldine.
———. 1970. Natural selection as an epistemological model. In *A handbook of methods in cultural anthropology,* ed. R. Naroll and R. Cohen, New York: Columbia University Press.
———. 1974. Evolutionary epistemology. In *The philosophy of Karl R. Popper,* ed. P. A. Schilpp, 413–63. LaSalle: Open Court.
———. 1979a. A tribal model of the social system vehicle carrying scientific knowledge. *Knowledge: creation, diffusion, utilization.* 1:181–201.
———. 1979b. Comments on the sociobiology of ethics and moralizing. *Behavioral Science* 24:37–45.
Campbell, M. 1976. Explanations of Mendel's results. *Centaurus* 20:159–74.
———. 1980. Did de Vries discover the law of segregation independently? *Annals of Science* 37:639–55.
Candolle, A. P., de. 1813. *Théorie élémentaire de la botanique.* Paris: Déterville.
———. 1820. Géographic botanique. In *Dictionnaire des sciences naturelles* 18:359–422.
Cannon, S. F. 1978. *Science in culture: the early Victorian period.* New York: Dawson and Science History Publication.
Cantor, G. N. 1975a. The Edinburgh phrenology debate: 1803–1828. *Annals of Science* 32:195–218.
———. 1975b. A critique of Shapin's social interpretation of the Edinburgh phrenology debate. *Annals of Science* 32:245–56.
Caplan, A. 1982. Stalking the wild culturgen. *Behavioral and Brain Sciences* 5:8–9.
———. 1984. Sociobiology as a strategy in science. *The Monist* 67:143–60.
Carlson, E. A. 1966. *The gene: a critical history.* Philadelphia: Saunders.
Carroll, L. 1898. *Sylvie and Bruno, concluded.* London: Macmillan.
Carr, N. G., and B. A. Whitton, eds. 1973. *The biology of blue-green algae.* Oxford: Blackwell Scientific Publications.
Carson, H. L. 1975. The genetics of speciation at the diploid level. *American Naturalist* 109:83–92.
———. 1981. Macroevolution conference. *Science* 211:773.
Carter, G. S. 1963. Two evolutionary theories by M. Grene: a further discussion. *British Journal for the Philosophy of Science* 14:345–48.
Cartwright, N. 1983. *How the laws of physics lie.* Oxford: Oxford University Press.
Cavalli-Sforza, and M. W. Feldman. 1981. *Cultural transmission and evolution.* Princeton: Princeton University Press.
Chambers, R. 1844. *Vestiges of the Natural History of Creation.* London: John Churchill; 10th ed. 1853.
———. 1845. *Explanation: a sequel to the Vestiges.* London: John Churchill.
Charig, A. J. 1982. Systematics in biology: a fundamental comparison of some major schools of thought. In *Problems of phylogenetic reconstruction,* ed. K. A. Joysey and A. E. Friday, 363–440. London: Academic Press.
Chetverikov, S. S. 1961. On certain aspects of the evolutionary process from the standpoint of modern genetics. *American Philosophical Society* 105:167–95; translation of 1926 paper in Russian.

Chomsky, N. 1976. The fallacy of Richard Herrnstein's IQ. In *The IQ controversy,* ed. N. J. Block and G. Dworkin, 285–98. New York: Pantheon.
Chubin, D. E., and S. Moitra, 1975. Content analysis of references: adjunct or alternative to citation counting? *Social Studies of Science* 5:423–41.
Churchill, F. B. 1980. The modern synthesis and the biogenetic law. In *The evolutionary synthesis,* ed. E. Mayr and W. B. Provine, 112–22. Cambridge: Harvard University Press.
Churchland, P. M., and C. A. Hooker. 1985. *Images of science.* Chicago: University of Chicago Press.
Clark, R. W. 1968. *J. B. S.: The life and works of J. B. S. Haldane.* New York: Coward-McCann.
Cock, A. G. 1973. William Bateson, Mendelism, and biometry. *Journal of the History of Biology* 6:1–36.
______. 1983. William Bateson's rejection and eventual acceptance of chromosome theory. *Annals of Science* 40:19–60.
Cohen, I. B. 1981. *The Newtonian revolution.* New York: Cambridge University Press.
Cohen, L. J. 1973. Is the progress of science evolutionary? *British Journal for the Philosophy of Science* 24:41–61.
Cole, J. R., and S. Cole. 1972. The Ortega hypothesis. *Science* 178:368–75.
______. 1973. *Social stratification in science.* Chicago: University of Chicago Press.
Cole, J. R., and H. Zuckerman. 1975. The emergence of a scientific specialty: the self-exemplifying case of the sociology of science. In *The idea of social structure,* ed. L. A. Coser, 139–74. New York: Harcourt, Brace, Jovanovich.
Cole, S. 1975. The growth of scientific knowledge. In *The idea of social structure,* ed. L. A. Coser, 175–220. New York: Harcourt, Brace, Jovanovich.
Cole, S., L. Rubin, and J. R. Cole. 1978. *Peer review in the National Science Foundation.* Washington, D.C.: National Academy of Science.
Coleman, W. 1970. Bateson and chromosomes: conservative thought in science. *Centaurus* 15:228–314.
Colless, D. H. 1967a. An examination of certain concepts in phenetic taxonomy. *Systematic Zoology* 16:6–27.
______. 1967b. The phylogenetic fallacy. *Systematic Zoology* 16:289–95.
______. 1969. The interpretation of Hennig's "Phylogenetic systematics." A reply to Dr. Schlee. *Systematic Zoology* 18:134–44.
______. 1980. Congruence between morphometrics and allozyme data for *Menidia* species: a reappraisal. *Systematic Zoology* 29:288–99.
______. 1981. Predictivity and stability in classification: some comments on recent studies. *Systematic Zoology* 30:325–30.
______. 1985. An appendix to *The Tettigoniinae,* vol. 1, 373–84. Australia: Commonwealth Scientific and Industrial Research Organization.
Collins, H. M. 1981a. Stages in the empirical programme of relativism. *Social Studies of Science* 11:3–10.
______. 1981b. Son of seven sexes: the social destruction of a physical phenomenon. *Social Studies of Science* 11:33–62.
______. 1981c. What is TRASP? The radical programme as a methodological imperative. *Philosophy of the Social Sciences* 11:215–24.
______. 1982a. Knowledge, norms and rules in the sociology of science. *Social Studies of Science* 12: 299–308.
______. 1982b. Special relativism—the natural attitude. *Social Studies of Science* 12:139–43.
______. 1985. Response to Roll-Hansen. *Social Studies of Science* 15:175.
Collins, H. M., T. J. Pinch, and S. Shapin. 1984. Authors' preface. *Social Studies of Science* 14:ii.
Colp, R. 1982. The myth of the Marx-Darwin letter. *History of political economy* 14:461–82.
Comte, A. 1830–42. *Cours de philosophie positive.* Paris: Chez Bachelier.
Connor, S. 1987. AIDS: science stands on trial. *New Scientist* 113:49–58.

Conry, Y. 1974. *L'introduction du darwinisme en France au xixe siècle.* Paris: J. Vrin.
Cook, R. E. 1980. Reproduction by duplication. *Natural History* 89:88–93.
Cooter, R. J. 1976. Phrenology; the provocation of progress. *History of Science* 14:211–34.
Correns, C. 1900. G. Mendel's law concerning the behavior of progeny of varietal hybrids. *Berichte der deutschen botanischen gesellschaft* 18:158–68.
Corsi, P., and P. J. Weindling. 1985. Darwinism in Germany, France, and Italy. In *The Darwinian heritage,* ed. D. Kohn, 683–730. Princeton: Princeton University Press.
Cracraft, J. 1973. Discussion of symposium papers on contemporary systematic philosophies. *Systematic Zoology* 22:393–400.
———. 1974. Phylogenetic models and classification. *Systematic Zoology* 23:71–90.
———. 1982. Geographic differentiation, cladistics, and vicariance biography; reconstructing the tempo and mode of evolution. *American Zoologist* 22:411–24.
———. 1985. Biological diversification and its causes. *Annals of the Missouri Botanical Gardens* 72:794–822.
Crane, D. 1972. *Invisible colleges: diffusion of knowledge in scientific communities.* Chicago: University of Chicago Press.
Craw, R. 1982. Phylogenetics, areas, geology and the biogeography of Croizat: a radical view. *Systematic Zoology* 31:304–16.
———. 1983. Panbiogeography and vicariance cladistics: are they truly different? *Systematic Zoology* 32:431–37.
———. 1984. Never a serious scientist: the life of Leon Croizat. *Tuatara* 27:5–7.
Craw, R., and P. Weston. 1984. Panbiogeography: a progressive research program? *Systematic Zoology* 33:1–13.
Crawford, M. 1987. NIMH finds a case of "serious misconduct." *Science* 235:1566–67.
Croizat, L. 1958. *Panbiography.* 3 vols. Caracas: published privately by the author.
———. 1964. *Space, time, form: the biological synthesis.* Caracas: published privately by the author.
———. 1976. Biogeografîa analitica y sintética ("panbiogeografia") de las Américas. Caracas: biblioteca de la academia de ciencias fisicas, matemáticas y naturales. Vols. 15–16.
———. 1978. Hennig (1966) entre Rosa (1981) y Løvtrup (1977): medio siglo de "systemática filogenética." *Boletin de la academia de ciencias fisicas matematicas y naturales* 38:59–147.
———. 1982. Vicariance/vicariism, panbiogeography, "vicariance biogeography," etc: a clarification. *Systematic Zoology* 31:291–303.
Croizat, L., G. Nelson, and D. E. Rosen. 1974. Centers of origin and related concepts. *Systematic Zoology* 23:265–87.
Crowson, R. A. 1970. *Classification and biology.* New York: Atherton Press.
Culliton, B. 1983. Coping with fraud: the Darsee case. *Science* 220:31–35.
———. 1986. Harvard researchers retract data in immunology paper. *Science* 234:1069.
———. 1987. Integrity of research papers questioned. *Science* 235:422–23.
Curd, M. V. 1984. Kuhn, scientific revolutions, and the Copernican revolution. *Nature and System* 6:1–14.
Cuvier, G. 1812. *Discours préliminaire sur les révolutions du globe.* Introduction to *Recherches sur les ossemens fossiles des quadrupédes.* 4 vols. Paris: Déterville.
Damuth, J. 1985. Selection among "species": a formulation in terms of natural functional units. *Evolution* 39:1132–46.
Damuth, J., and L. L. Heisler. Forthcoming. Alternative formulations of multilevel selection. *Biology and Philosophy.*
Darden, L. 1985. Hugo de Vries's lecture plates and the discovery of segregation. *Annals of Science* 40:233–42.
Darden, L., and J. A. Cain. 1988. Selection type theories. *Philosophy of Science.* Forthcoming.
Darlington, C. D. 1939. *The evolution of genetic systems.* Cambridge: Cambridge University Press.

Darlington, P. J., Jr. 1957. *Zoogeography: the geographical distribution of animals.* New York: John Wiley.

______. 1965. *Biogeography of the southern end of the world.* Cambridge: Harvard University Press.

______. 1970. A practical criticism of Hennig-Brundin "phylogenetic systematics" and antarctic biogeography. *Systematic Zoology* 19:1–18.

Darwin, C. 1839. *The voyage of the Beagle.* New York: Anchor Books (1962).

______. 1859. *On the origin of species.* A facsimile of the first edition (1859) with introduction by Ernst Mayr (1966). Cambridge: Harvard University Press.

______. 1872. *The expression of the emotions in man and animals.* London: Murray.

______. 1877. *The various contrivances by which orchids are fertilized by insects.* London: Murray.

______. 1881. *The formation of vegetable mould, through the action of worms, with observations on their habits.* London: Murray.

______. 1958. *The autobiography of Charles Darwin.* ed. N. Barlow. London: Collins.

Darwin, F. 1899. *The life and letters of Charles Darwin.* Vols. 1 and 2. New York: D. Appleton.

______. 1903. *More letters of Charles Darwin.* London: Murray.

Davies, G. L. 1969. *The earth in decay, 1578–1878.* New York: Elsevier.

Davis, B. P. 1986. *Storm over biology: essays on science, sentiment, and public policy.* New York: Prometheus Books.

Davis, D. D. 1949. Comparative anatomy and the evolution of vertebrates. In *Genetics, paleontology, and evolution,* ed. G. L. Jepsen, E. Mayr, G. G. Simpson, 64–89. New York: Atheneum.

Davis, P. H., and V. H. Heywood. 1963. *Principles of angiosperm taxonomy.* New York: D. Van Nostrand.

Dawkins, R. 1976. *The selfish gene.* New York: Oxford University Press.

______. 1978. Replication selection and the extended phenotype. *Zeitschrift für tierpsychologie* 47:61–76.

______. 1981. Selfish genes in race or politics. *Nature* 289:528.

______. 1982a. Replicators and vehicles. In *Current problems in sociobiology,* ed. King's College Sociobiology Group, 45–64. Cambridge: Cambridge University Press.

______. 1982b. *The extended phenotype.* San Francisco: Freeman.

______. 1986. *The Blind Watchmaker.* New York: W. W. Norton.

de Beer, G. R. 1958. Further unpublished letters of Charles Darwin. *Annals of Science* 14:82–115.

De Giustino, D. A. 1975. *Conquest of mind: phrenology and Victorian social thought.* London: Croom Helm.

Dendy, A. 1924. *Outlines of evolutionary biology.* London: Constable.

Dennett, D. 1983. Intentional systems in cognitive ethology: the "Panglossian paradigm" defended, with peer commentaries. *The Behavioral and Brain Sciences* 6:343–90.

Desmond, A. 1982. *Archetypes and ancestors: palaeontology in Victorian London, 1850–1875.* London: Blond and Briggs.

de Vries, H. 1889. *Intracellular pangenesis.* Jena: Gustav Fischer. English trans. 1910 by E. S. Gager. Chicago: Open Court.

______. 1900a. Sur la loi de disjonction des hybrides. *Comptes rendues de l'Académie des sciences* 130:845–47.

______. 1900b. Das spaltungsgesetz der bastarde. *Berichte der Deutschen botanischen Gesellschaft* 18:83–90.

______. 1900c. Sur les unités des caractères spécifiques et leur application a l'étude des hybrides. *Revue générale de botanique* 12:257–71.

______. 1901. *Die mutationstheorie.* Vol. 1. Leipzig: Veit & Co. English translation J. B. Farmer and A. D. Darbishire, 1909–1910. Chicago: Open Court. Vol. 2, 1903.

Dexter, R. W. 1966. Historical aspects of studies on the Brachiopoda by E. E. Morse. *Systematic Zoology* 15:241–43.

Diamond, A. M. 1980. Age and the acceptance of cliometrics. *Journal of Economic History* 40:838–41.

di Gregorio, M. A. 1984. *T. H. Huxley's place in natural science*. New Haven: Yale University Press.

Dobzhansky, T. 1937. *Genetics and the origin of species*. New York: Columbia University Press.

———. 1951. *Genetics and the origin of species*. 3d ed. New York: Columbia University Press.

———. 1970. *Genetics of the evolutionary process*. New York: Columbia University Press.

Dobzhansky, T., F. Ayala, G. Stebbins, and J. W. Valentine. 1977. *Evolution*. San Francisco: Freeman.

Dolby, R. G. A. 1971. Sociology of knowledge in natural sciences. *Science Studies* 1:3–21.

Doolittle, W. F., and C. Sapienza. 1980. Selfish genes, the phenotype paradigm and genome evolution. *Nature* 284:601–3.

Douglas, M. 1986. *How institutions think*. Syracuse, N.Y.: Syracuse University Press.

Dover, G. 1982. Molecular drive: a cohesive mode of species evolution. *Nature* 299:111–17.

Duffy, J. 1976. *The healers*. New York: McGraw-Hill.

Duncan, T., and T. F. Stuessy, eds. 1984. *Cladistics: perspectives on the reconstruction of evolutionary history*. New York: Columbia University Press.

Duncan, T., R. B. Phillips, and W. H. Wagner, Jr. 1980. A comparison of branching diagrams derived by various phenetic and cladistic methods. *Systematic Botany* 5:264–93.

Dupré, J. 1981. Natural kinds and biological taxa. *Philosophical Review* 90:66–90.

———. 1983. The disunity of science. *Mind* 92:321–46.

———. 1986. Sex, gender, and essence. *Midwest Studies in Philosophy* 11:441–57.

Dupuis, C. 1979. La "Systématique Phylogénétique" de W. Hennig. *Cahiers des Naturalistes* 34:1–69.

Durkheim, E. 1912.: *Les formes élémentaires de la vie religieuse: le système totémique en Australia*. Paris: Félix Alcan; 1915. Trans. J. W. Swain, London: Allen & Unwin; reprint 1961, New York: Collier Books.

Edge, D., and M. J. Mulkay. 1976. *Astronomy transformed: the emergence of radio astronomy in Britain*. New York: John Wiley.

Egerton, F. N. 1970. Reputation and conjecture: Darwin's response to Sedgwick's attack on Chambers. *Studies in the History and Philosophy of Science* 1:176–83.

Ehrlich, P. R. 1961a. Systematics in 1970: some unpopular predictions. *Systematic Zoology* 10:157–58.

———. 1961b. Has the biological species concept outlived its usefulness? *Systematic Zoology* 10:167–76.

Ehrlich, P. R., and A. H. Ehrlich. 1967. The phenetic relationships of the butterflies. *Systematic Zoology* 16:301–27.

Ehrlich, P. R., and D. D. Murphy. 1983. Butterfly nomenclature, stability, and the perils of obligatory categories. *Systematic Zoology* 32:451–53.

Einstein, A. 1949. *Albert Einstein: philosopher-scientist*, ed. P. A. Schilpp. LaSalle, Ill.: Open Court.

Eldredge, N. 1983. Phenomenological levels and evolutionary rates. *Systematic Zoology* 31:338–47.

———. 1985. *Unfinished synthesis*. New York. Oxford University Press.

———. 1986. Information, economics, and evolution. *Annual Review of Ecology and Systematics* 17:351–69.

Eldredge, N., and J. Cracraft. 1980. *Phylogenetic patterns and the evolutionary process: method and theory in comparative biology*. New York: Columbia University Press.

Eldredge, N., and S. J. Gould. 1972. Punctuated equilibria: an alternative to phyletic gradualism. In *Models in paleobiology*, ed. T. J. M. Schopf, 82–115. San Francisco: Freeman, Cooper.

Elliott, H. 1914. Introduction to Lamarck's *Zoological philosophy,* xvii–xcii. New York: Hafner.
Elster, J. 1979. *Ulysses and the sirens: studies in rationality and irrationality.* Cambridge: Cambridge University Press.
Endler, J. A. 1986. *Natural selection in the wild.* Princeton: Princeton University Press.
Engelberg, J., and L. L. Boyarsky. 1979. The noncybernetic nature of ecosystems. *American Naturalist* 114:317–24.
Fang, J. 1970. *Bourbaki.* New York: Paideia Press.
Farley, J. 1974. The initial reaction of French biologists to Darwin's *Origin of species. Journal of the History of Biology* 7:275–300.
Farris, J. S. 1966. Estimation of conservation of characters by constancy within biological populations. *Evolution* 20:587–91.
———. 1967a. The meaning of relationship and taxonomic procedure. *Systematic Zoology* 16:44–51.
———. 1967b. Comment on psychologism. *Systematic Zoology* 16:345–47.
———. 1969a. Informal discussion. In *Systematic Biology,* ed. C. G. Sibley, 64–66. Washington, D.C.: National Academy of Science Publications.
———. 1969b. On the cophenetic correlation coefficient. *Systematic Zoology* 18:279–85.
———. 1970. Methods for computing Wagner trees. *Systematic Zoology* 19:83–92.
———. 1971. The hypothesis of nonspecificity and taxonomic congruence. *Annual Review of Ecology and Systematics* 2:227–302.
———. 1974. Formal definitions of paraphyly and polyphyly. *Systematic Zoology* 23:548–54.
———. 1977a. Review of the *Proceedings of the eighth international conference on numerical taxonomy. Systematic Zoology* 26:228–30.
———. 1977b. On the phenetic approach to vertebrate classification. In *Major patterns in vertebrate evolution,* ed. M. K. Hecht, P. C. Goody, and B. M. Hecht, 823–50. New York: Plenum.
———. 1979. On the naturalness of phylogenetic classification. *Systematic Zoology* 28:200–214.
———. 1980. Naturalness, information, invariance, and the consequences of phenetic criteria. *Systematic Zoology* 29:360–81.
———. 1982. Simplicity and informativeness in systematics and phylogeny. *Systematic Zoology* 31:413–44.
———. 1985a. The pattern of cladistics. *Cladistics* 1:190–201.
———. 1985b. Review of *Numerical taxonomy,* ed. J. Felsenstein. *Cladistics* 1:97–102.
Felsenstein, J., ed. 1983. *Numerical taxonomy,* New York: Springer Verlag.
———. 1986. Waiting for post-neo-Darwinism. *Evolution* 40:883–89.
Ferris, V. R. 1980. A science in search of a paradigm? Review of the symposium, "Vicariance biogeography: a critique." *Systematic Zoology* 29:67–76.
Feyerabend, P. K. 1970. Consolations for the specialist. In *Criticism and the growth of knowledge,* ed. I. Lakatos and A. Musgrave, 197–230. Cambridge: Cambridge University Press.
———. 1975. *Against method: outline of an anarchist theory of knowledge.* New Left Books. London.
Fine, A. 1986a. *The shaky game: Einstein, realism, and the quantum theory.* Chicago: University of Chicago Press.
———. 1986b. Unnatural attitudes: realist and instrumentalist attachment to science. *Mind* 95:149–79.
Fink, S. 1982. Report on the second annual meeting of the Willi Hennig society. *Systematic Zoology* 31:180–96.
Fink, W. L. 1982. The conceptual relationship between ontogeny and phylogeny. *Paleobiology* 8:254–64.
Fisher, R. A. 1930. *The genetical theory of natural selection.* Oxford: Clarendon Press.
———. 1936. Has Mendel's work been rediscovered? *Annals of Science* 1:115–37.

Firth, R. 1981. Epistemic merit, intrinsic and instrumental. *Addresses of the American Philosophical Association* 55:5–23.
Flessa, K. W., and J. S. Levinton. 1975. Phanerozoic diversity patterns: tests for randomness. *Journal of Geology* 83:239–48.
Fodor, J. 1984. Observation reconsidered. *Philosophy of Science* 51:23–43.
Fogg, G. E., W. D. P. Stewart, P. Fay, and A. E. Walsby. 1973. *The blue-green algae*. New York: Academic Press.
Forbes, E. 1854. On the manifestation of polarity in the distribution of organized beings in time. *Royal Institution of Great Britain Proceedings* 1:428–33.
Ford, E. B. 1931. *Mendelism and evolution*. London: Methuen.
———. 1940. Polymorphism and taxonomy. In *The new systematics*, ed. J. Huxley, 493–513. Oxford: Oxford University Press.
———. 1980. Some recollections pertaining to the evolutionary synthesis. In *The evolutionary synthesis*, ed. E. Mayr and W. B. Provine, 334–42. Cambridge: Harvard University Press.
Foucault, M. 1966. *Les mots et les choses: une archéologie des sciences humaines*. Paris: Gallimard.
Fox, M. F. 1983. Publication productivity among scientists: a critical review. *Social Studies of Science* 13:285–305.
Friedland, W. H., and T. Kappel. 1979. *Production or perish*. Santa Cruz: University of California.
Friedman, S. M., S. Dunwoody, and C. L. Rogers. 1986. *Scientists and journalists: reporting science as news*. New York: The Free Press.
Funk, V. A., and D. R. Brooks. 1981a. National science foundation workshop on the theory and application of cladistic methodology, organized by T. Duncan and T. Stuessy, University of California, Berkeley, March 22–28, 1981. *Systematic Zoology* 30:491–97.
———. eds. 1981b. *Advances in cladistics*. Vol. 1. New York: Columbia University Press.
Futuyma, D. J., R. C. Lewontin, G. C. Mayer, J. Seger, and J. W. Stubblefield III. 1981. Macroevolution conference. *Science* 211:770.
Gage, A. T. 1938. *History of the Linnean society of London*. London: Linnean Society.
Gall, F. J., and J. G. Spurzheim. 1810. *Anatomie et physiologie au systéme nerveux en général et du cerveau en particulier*. Paris: Atlas.
Galton, F. 1871. Experiments in pangenesis, by breeding from rabbits of a pure variety, into whose circulation blood taken from other varieties had previously been largely transferred. *Proceedings of the Royal Society* 19:393–410.
———. 1874. *English men of science: their nature and nurture*. London: Macmillan.
Garvey, W. D. 1979. *Communication: the essence of science*. New York: Pergamon Press.
Gaston, J. 1978. *The reward system in British and American science*. New York: John Wiley and Son.
Geertz, C. 1973. *The interpretation of cultures*. New York: Basic Books.
Gellner, E. 1985. *Relativism and the social sciences*. New York: Cambridge University Press.
Ghent, W. J. 1902. *Our benevolent feudalism*. New York: Macmillan.
Ghiselin, M. T. 1967. Further remarks on logical errors in systematic theory. *Systematic Zoology* 16:347–48.
———. 1969a. *The triumph of the Darwinian method*. Berkeley: University of California Press.
———. 1969b. Non-phenetic evidence in phylogeny. *Systematic Zoology* 18:460–61.
———. 1974a. *The economy of nature and the evolution of sex*. Berkeley: University of California Press.
———. 1974b. A radical solution to the species problem. *Systematic Zoology* 23:536–44.
———. 1985. Mayr versus Darwin on paraphyletic taxes. *Systematic Zoology* 34: 460–62.
———. 1987. Species concepts, individuality, and objectivity. *Biology and Philosophy* 2:127–45.
Ghiselin, M. T., and L. Jaffe. 1973. Phylogenetic classification in Darwin's *Monograph on the sub-class Cirripedia*. *Systematic Zoology* 22:132–40.
Giere, R. 1973. History and philosophy of science: intimate relationship or marriage of convenience? *British Journal for the Philosophy of Science* 24:282–97.

———. 1984. *Understanding scientific reasoning.* 2d ed. New York: Holt, Rinehart, and Winston.
———. 1987. The cognitive study of science. In *The process of science,* ed. N. J. Nersessian, 139–59. Dordrecht: Martinus Nijhoff.
———. 1988. *Explaining science: a cognitive approach.* Chicago: University of Chicago Press.
Gieryn, T. F. 1982. Relativist/constructivist programmes in the sociology of science. *Social Studies of Science* 12:279–97.
———. 1983. Making the demarcation of science a sociological problem: Boundary-work by John Tyndall, Victorian scientist. *Working papers in science and technology,* ed. R. Laudan, 2:57–86. Virginia Technical Center for the Study of Science and Society.
Gilmour, J. S. L. 1940. Taxonomy and philosophy. In *The new systematics,* ed. J. Huxley, 401–74. Oxford: Oxford University Press.
Gilmour, J. S. L., and S. M. Walters. 1963. Philosophy and classification. In *Vistas in botany,* ed. W. B. Turrill, 4:1–22. London: Pergamon Press.
Gingerich, P. D. 1984. Punctuated equilibria—where is the evidence? *Systematic Zoology* 33: 335–38.
Glick, T. G., ed. 1974. *The comparative reception of Darwinism.* Austin: University of Texas Press.
Glymour, C. 1980. *Theory and evidence.* Princeton: Princeton University Press.
Goldschmidt, R. 1933. Some aspects of evolution. *Science* 78:539–47.
———. 1938. The theory of the gene. *Science Monthly* 46:268–73.
———. 1940. *The material basis of evolution.* New Haven: Yale University Press.
———. 1950. Fifty years of genetics. *American Naturalist* 84:313–40.
———. 1960. *In and out of the ivory tower.* Seattle: University of Washington Press.
Goodman, M., J. Czelusniak, G. W. Moore, A. E. Romero-Herrera, and G. Matsuda. 1979. Fitting the gene lineage into its species lineage, a parsimony strategy. *Systematic Zoology* 28:132–67.
Gordon, M. D. 1984. How authors select journals: a test of the reward maximization model of submission behavior. *Social Studies of Science* 14:27–43.
Goudge, T. A. 1961. *The ascent of life.* Toronto: University of Toronto Press.
Goudsmit, S. A. 1974. Citation analysis. *Science* 183:28.
Gould, S. J. 1973. Systematic pluralism and the uses of history. *Systematic Zoology* 22:322–23.
———. 1976. Darwin's untimely burial. *Natural History* 85:24–30.
———. 1977a. Eternal metaphors of palaeontology. In *Patterns of evolution as illustrated by the fossil record,* ed. A. Hallan, 1–26. New York: Elsevier.
———. 1977b. *Ontogeny and phylogeny.* Cambridge: Harvard University Press.
———. 1977c. The return of hopeful monsters. *Natural History* 86:22–30.
———. 1977d. Caring groups and selfish genes. *Natural History* 86:20–24.
———. 1979. On the importance of heterochrony for evolutionary biology. *Systematic Zoology* 28:224–26.
———. 1980. Is a new and general theory of evolution emerging? *Paleobiology* 6:119–30.
———. 1981. Museum debate. *Nature* 289:742.
———. 1982a. Introduction to Goldschmidt's *The material basis of evolution.* New Haven: Yale University Press.
———. 1982b. Darwinism and the expansion of evolutionary theory. *Science* 216:380–87.
———. 1982c. The meaning of punctuated equilibrium and its role in validating a hierarchical approach to macroevolution. In *Perspectives on evolution,* ed. R. Milkman, 83–104. Sunderland: Sinauer.
———. 1984. Balzan prize to Ernst Mayr. *Science* 223:255–57.
———. 1985. Recording marvels: the life and works of George Gaylord Simpson. *Evolution* 39:229–32.

Gould, S. J., and N. Eldredge. 1977. Punctuated equilibria: the tempo and mode of evolution reconsidered. *Paleobiology.* 3:115–51.
———. 1986. Punctuated equilibrium at the third stage. *Systematic Zoology* 35:143–48.
Gould, S. J., N. I. Gilinsky, and R. Z. German. 1987. Asymmetry of lineages and the direction of evolutionary time. *Science* 236:1437–41.
Gould, S. J., and R. C. Lewontin. 1979. The spandrels of San Marco and the Panglossian paradigm: a critique of the adaptational programme. *Proceedings of the Royal Society of London, B* 205:581–98.
Gould, S. J., and E. Vrba. 1982. Exaptation—a missing term in the science of form. *Paleobiology* 8:4–15.
Grafen, A. 1985. A geometric view of relatedness. In *Oxford surveys in evolutionary biology,* ed. R. Dawkins and M. Ridley, 2:28–89. Oxford: Oxford University Press.
Granger, H. 1985. The scala naturae and the continuity of kinds. *Phronesis* 30: 181–200.
———. 1987. Deformed kinds and the fixity of species. *Classical Quarterly* 37:110–16.
Grant, V. 1971. *Plant speciation.* New York: Columbia University Press.
———. 1985. *The evolutionary process.* New York: Columbia University Press.
Grayson, D. K. 1985. The first three editions of Charles Lyell's *The geological evidences of the antiquity of man. Archives of Natural History* 13:105–21.
Greenberg, D. S. 1967. *The practice of pure science.* New York: New American Library.
Greene, J. C. 1981. *Science, ideology, and world view.* Berkeley: University of California Press.
Gregg, J. R. 1950. Taxonomy, language, and reality. *American Naturalist* 84:421–33.
Gregor, T. 1979. Short people. *Natural History* (2) 88:14–23.
Gregory, M. S., A. Silvers, and D. Sutch, eds. 1978. *Sociobiology and human nature.* San Francisco: Jossey-Bass.
Grehan, J. 1984. Evolution by law: Croizat's "orthogeny" and Darwin's "laws of growth." *Tuatara* 27:14–19.
Grehan, J. R., and R. Ainsworth. 1985. Orthogenesis and evolution. *Systematic Zoology* 34:174–92.
Grene, M. 1958. Two evolutionary theories. *British Journal for the Philosophy of Science* 9:110–27 and 185–93.
Griffin, B. C., and N. Mullins. 1972. Coherent social groups in scientific change. *Science* 177: 959–64.
Grinnell, F. 1987. *The scientific attitude.* Boulder: Westview Press.
Groeben, C. V., ed. 1982. *Charles Darwin and Anton Dohrn: correspondence.* Naples: Macchiaroli.
Gunther, A. 1975. *A century of zoology.* New York: Science History Publications.
Haeckel, E. 1909. Darwin as an anthropologist. In *Darwin and Modern Science,* ed. A. C. Stewart, 137–51. Cambridge: At the University Press.
Hagstrom, W. O. 1965. *The scientific community.* New York: Basic Books.
Hahlweg, K. 1981. Progress through evolution? An inquiry into the thought of C. H. Waddington. *Acta Biotheoretica* 30:103–20.
———. 1986. Popper versus Lorenz: an exploration into the nature of evolutionary epistemology. In *PSA 1986,* ed. A. Fine and P. Machamer, 1:172–82. Ann Arbor: Philosophy of Science Association.
Haldane, J. B. S. 1932. *The causes of evolution.* London and New York: Harper.
Hall, M. B. 1981. Salomon's house emergent: the early royal society and cooperative research. In *The analytic spirit,* ed. H. Woolf, 177–94. New York: Cornell University Press.
———. 1984. *All scientists now: the Royal Society in the nineteenth century.* Cambridge: Cambridge University Press.
Hall, T. S. 1968. On biological analogs of Newtonian paradigms. *Philosophy of Science* 35:6–27.
Hallam, A. 1983. *Great geological controversies.* Oxford: Oxford University Press.

Hallpike, C. R. 1982. The "culturgen": science or science fiction? *Behavioral and Brain Science* 5:12–13.
Halstead, L. B. 1978. The cladistic revolution—can it make the grade? *Nature* 276:759–60.
______. 1980. Museum of errors. *Nature* 288:208.
______. 1981. Halstead replies. *Nature* 289:106–7.
Hamburger, V. 1980. Embryology and the modern synthesis in evolutionary theory. In *The evolutionary synthesis,* ed. E. Mayr and W. B. Provine, 112–22. Cambridge: Harvard University Press.
Hamilton, W. D. 1964. The genetical evolution of social behavior. *Journal of Theoretical Biology* 7:1–51.
______. 1975. Review of Ghiselin (1974a) and Williams (1975). *Quarterly Review of Biology* 50:175–79.
Hanson, N. R. 1958. *Patterns of discovery.* Cambridge: Cambridge University Press.
______. 1961. The Copernican disturbance and the Keplerian revolution. *Journal of the History of Ideas* 22:169–84.
Harding, S. 1986. *The science question in feminism.* Ithaca: Cornell University Press.
Hardy, G. H. 1908. Mendelian proportions in a mixed population. *Science* 28:49–50.
Harnad, S., ed. 1982. *Peer commentary on peer review: a case study in scientific quality control.* Cambridge: Cambridge University Press.
Harper, C. W., and N. I. Platnick. 1978. Phylogenetic and cladistic hypotheses: a debate. *Systematic Zoology* 27:354–61.
Harper, J. L. 1977. *Population biology of plants.* London: Academic Press.
Harré, R. 1979. *Social being.* Oxford: Blackwell.
Harris, H. 1966. Enzyme polymorphism in man. *Proceedings of the Royal Society, Series B* 164:298–310.
Harris, S. J. 1979. A corporation is hardly a "family" if it can sacrifice individuals. *Chicago Sun-Times.* April 4, 1979.
______. 1986. Honesty is policy of newspapermen. *Chicago Sun-Times.* November 4, 1986.
Harwood, J. 1979. Heredity, environment, and the legitimation of social policy. In *Natural order,* ed. B. Barnes and S. Shapin, 231–52. Beverly Hills: Sage Publications.
______. 1984. The reception of Morgan's chromosome theory in Germany: inter-war debate over cytoplasmic inheritance. *Medizin historisches Journal* 19:3–32.
______. 1985. Genetics and the evolutionary synthesis in inter-war Germany. *Annals of Science* 42:279–301.
Haughton, S. 1860. Bioyéveosis. *Natural History Review* 7:23–32.
Heads, M. J. 1984. *Principia botanica:* Croizat's contributions to botany. *Tuatara* 27:26–48.
______. 1985. On the nature of ancestors. *Systematic Zoology* 34:205–15.
Hecht, M. K., P. C. Goody, and B. M. Hecht, eds. 1977. *Major patterns in vertebrate evolution.* New York: Plenum.
Hennig, W. 1950. *Grundzüge einer theorie der phylogenetischen systematik,* Berlin: Deutscher zentralverlag.
______. 1965. Phylogenetic systematics. *Annual Review of Entomology* 10:97–116.
______. 1966. *Phylogenetic systematics.* Urbana: University of Illinois Press.
______. 1969. *Die stammesgeschichte der insekten.* Frankfurt am Main: Waldemar Kramer.
______. 1975. Cladistic analysis or cladistic classification? A reply to Ernst Mayr. *Systematic Zoology* 24:244–56.
Heywood, V. H. 1964. Introduction. *Phenetic and phylogenetic classification,* ed. V. H. Heywood and J. McNeill, 1–4. London: The Systematics Association.
______. 1973. Discussion of symposium papers on contemporary systematic philosophies. *Systematic Zoology* 22:393–400.
Hill, C. R. 1981. From the museum. *Nature* 290:540.

Hill, C. R., and J. M. Camus. 1986. Evolutionary cladistics of Marattialean ferns. *Bulletin of the British Museum (Natural History)*. Botany series. 14:219–300.

Ho, M. W., and P. T. Saunders. 1982. The epigenetic approach to the evolution of organisms—with notes on its relevance to social and cultural evolution. In *Learning, development, and culture: essays in evolutionary epistemology,* ed. H. C. Plotkin, 343–61. New York: Wiley.

Hodge, M. J. S. 1983. Darwin and the laws of the animate part of the terrestrial system (1835–1837): on the Lyllian origins of his zoonomical explanatary program. *Studies in the History of Biology* 6:1–106.

———. 1985. Darwin as a lifelong generation theorist. In *The Darwinian Heritage,* ed. D. Kohn, 207–43. Princeton: Princeton University Press.

Hofstadter, R. 1955. *Social Darwinism in American thought.* Boston: Beacon Press.

Holden, C. 1987a. NIH moves to debar cholesterol researcher. *Science* 237:718–19.

———. 1987b. Stanford psychiatrist resigns under a cloud. *Science* 237:479–80.

Hollis, M., and S. Lukes, eds. 1982. *Rationality and relativism.* Cambridge: MIT Press.

Holton, G. 1973. *Thematic origins of scientific thought: Kepler to Einstein.* Cambridge: Harvard University Press.

———. 1978. *The scientific imagination: case studies.* Cambridge: Cambridge University Press.

Hood, C. S., and J. D. Smith. 1982. Cladistical analysis of female reproductive histomorphology in Phyllostomatoid bats. *Systematic Zoology* 31:241–51.

Hopkins, W. 1860. Physical theories of the phenomena of life. *Fraser's Magazine* 61:739–52; 62:74–90.

Howden, H. F. 1972. Systematics and zoogeography: science or politics? *Systematic Zoology* 21:129–31.

Hsui, A. T. 1987. Borehole measurement of the Newtonian gravitational constant. *Science* 237:881–83.

Hubbard, R. 1979. Have only men evolved? In *Discovering reality,* ed. S. Harding and M. B. Hintikka, 45–69. Cambridge, Mass.: Schenkman.

Hubby, J. L., and R. C. Lewontin. 1966. A molecular approach to the study of genetic heterozygosity in natural populations. *Genetics* 54:577–94.

Hufbauer, K. 1979. A test of the Kuhnian theory. *Science* 204:744–45.

Hull, D. L. 1964. Consistency and monophyly. *Systematic Zoology* 13:1–11.

———. 1968. The operational imperative: sense and nonsense in operationism. *Systematic Zoology* 17:438–57.

———. 1969. The natural system and the species problem. In *Systematic biology,* ed. C. G. Sibley, 56–61. Washington, D.C.: National Academy of Science.

———. 1974. Darwinism and historiography. In *The comparative reception of Darwinism,* ed. T. F. Glick, 388–402. Austin: University of Texas Press.

———. 1975. Central subjects and historical narratives. *History and Theory* 14:253–74.

———. 1976. Are species really individuals? *Systematic Zoology* 25:174–91.

———. 1978. A matter of individuality. *Philosophy of Science* 45:335–60.

———. 1979a. The limits of cladism. *Systematic Zoology* 28:414–38.

———. 1979b. Sociobiology: scientific bandwagon or traveling medicine show? In *Sociobiology and Human Nature,* ed. M. S. Gregory, A. Silvers, and D. Sutch, 136–63. San Francisco: Jossey-Bass.

———. 1980. Individuality and selection. *Annual Review of Ecology and Systematics* 11:311–32.

———. 1981. Historical narratives and integrating explanations. *Pragmatism and purpose: essays presented to Thomas Goudge,* ed. L. W. Summer, J. G. Slater, and F. Wilson, 172–188. Toronto: University of Toronto Press.

———. 1982a. Biology and philosophy. *Contemporary philosophy.* Vol. 2, *Philosophy of science,* ed. Guttorm Fløistad, 281–316. The Hague: Martinus Nijhoff.

———. 1982b. The naked meme. In *Learning, development and culture,* ed. H. C. Plotkin, 273–327. New York: John Wiley & Sons.

———. 1983a. Thirty-one years of *Systematic Zoology. Systematic Zoology* 32:315–42.

———. 1983b. Darwin and the nature of science. In *Evolution from molecules to men,* ed. D. S. Bendall, 63–80. Cambridge: Cambridge University Press.

———. 1983c. Exemplars and scientific change. In *PSA 1982,* ed. P. D. Asquith and T. Nickles, 1:479–503. East Lansing: Philosophy of Science Association.

———. 1983d. Karl Popper and Plato's metaphor. In *Advances in cladistics,* vol. 2, ed. N. Platnick and V. A. Funk, 177–89. New York: Columbia University Press.

———. 1984a. Lamarck among the Anglos. Introduction to Lamarck's *Zoological Philosophy,* xi–lxvi. Chicago: University of Chicago Press.

———. 1984b. Historical entities and historical narratives. In *Minds, machines and evolution,* ed. C. Hookway, 17–42. Cambridge: Cambridge University Press.

———. 1985a. Bias and commitment in science: phenetics and cladistics. *Annals of Science* 42:319–38.

———. 1985b. Darwinism as a historical entity: a historiographic proposal. In *The Darwinian heritage,* ed. by D. Kohn, 773–812. Princeton: Princeton University Press.

———. 1985c. Openness and secrecy in science: their origins and limitations. *Science, Technology, & Human Values* 10:4–13.

———. 1985d. Linné as an Aristotelian. In *Contemporary perspectives on Linnaeus,* ed. J. Weinstock, 37–54. Lanham MD: University Press of America.

———. 1987. Genealogical actors in ecological plays. *Biology and Philosophy* 1:44–60.

Hull, D. L., P. Tessner, and A. Diamond. 1978. Planck's principle. *Science* 202:717–23.

Humboldt, A. 1818. *Personal narrative of travels to the equinoctial regions of the new continent during the years 1799 to 1804.* Trans. H. N. Williams. London: Longmans.

Humphries, C. J., and J. Camus. 1986. Contemporary issues in systematics. *Cladistics* 2:85–99.

Hutton, J. 1795. *Theory of the earth, with proofs and illustrations.* 2 vols. Edinburgh: Cadell and Davies; facsimile reprinted in 1959 by Wheldon and Wesley, Codicate, Herts.

Huxley, A. 1981. Anniversary address of the president. Supplement to *Royal Society News.* Nov. 1981, 12:i–vii.

Huxley, J., ed. 1940. *The new systematics.* Oxford: Oxford University Press.

———. 1942. *Evolution: the modern synthesis.* London: Allen and Unwin.

Huxley, L. 1901. *Life and letters of Thomas Henry Huxley.* 2 vols. New York: D. Appleton.

Huxley, T. H. 1852. Upon animal individuality. In *The scientific memoirs of Thomas Henry Huxley,* ed. M. Foster and E. R. Lankaster, 1:146–51. London: Macmillan.

———. 1853. On the morphology of the Cephalous Mollusca. *Philosophical Transactions of the Royal Society, London* 143: 29–66.

———. 1854. Review of *Vestiges of the natural history of creation.* 10th ed. *British and Foreign Medico-Chirurgical Review* 19:425–39.

———. 1863. *Evidence as to man's place in nature.* London: Williams & Norgate.

———. 1874. On the classification of the animal kingdom. *Nature* 11:101–2.

———. 1893. *Collected essays.* 9 vols. London: Macmillan.

Hyman, L. H. 1958. In regard to nema the thread. *Systematic Zoology* 7:133.

———. 1959. Further remarks on the word *Nema. Systematic Zoology* 8:57.

Inger, R. R. 1958. Comments on the definition of genera. *Evolution* 12:370–84.

Jablonski, D. 1986. Background and mass extinctions: the alternative of macroevolutionary regimes. *Science* 231:129.

———. 1987. Heritability at the species level: an analysis of geographic ranges of cretaceous mollusks. *Science* 238:360–63.

Jackson, C. I. 1986. *Honor In science.* New Haven: Sigma Xi, The Scientific Research Society.

———. 1986. *A new agenda for science.* New Haven: Sigma Xi, The Scientific Research Society.

Jackson, J. B. C. 1986. Distribution and ecology of clonal and aclonal benthic invertebrates. In *Population biology and evolution of clonal organisms,* ed. J. B. C. Jackson, L. W. Buss, and R. E. Cook, 297–356. New Haven: Yale University Press.

Jackson, J. B. C., L. W. Buss, and R. E. Cook, eds. 1986. *Population biology and evolution of clonal organisms.* New Haven: Yale University Press.

Janowitz, M. F. 1979. A note on phenetic and phylogenetic classification. *Systematic Zoology* 28:197–99.

———. 1980. Similarity measures on binary data. *Systematic Zoology* 29:342–59.

Janvier, P. 1984. Cladistics: theory, purpose, and evolutionary implications. In *Evolutionary Theory: Paths into the Future,* ed. J. W. Pollard, 39–75. New York: John Wiley and Sons.

Janzen, D. W. 1977. What are dandelion and aphids? *American Naturalist* 111:586–89.

Jardine, N. 1967. The concept of homology in biology. *British Journal for the Philosophy of Science* 18:125–39.

———. 1978. Science as problem-solving. *Science* 199:415–16.

Jardine, N., and R. Sibson, 1971. *Mathematical taxonomy.* New York: Wiley.

Jenkin, F. 1867. The origin of species. *North British Review* 46:149–71.

Jensen, R. S. 1983. Report on sixteenth international numerical taxonomy conference. *Systematic Zoology* 32:83–89.

Johannsen, W. 1896. *Arvelighed.* Copenhagen: Det Schubotheske Forlag.

———. *1909. Elemente der Exukten Erblichkeitslehre.* Jena: Gustav Fischer.

Johnson, M. S. 1975. Biochemical systematics of the atherinid genus *Menida. Copeia* 1975: 662–91.

Johnston, R. F. 1969. Character variation and adaptation in European sparrows. *Systematic Zoology* 18:206–31.

Joravsky, D. 1970. *The Lysenko affair.* Chicago: University of Chicago Press.

Joyce, J. 1928. *A portrait of the artist as a young man.* New York: Random House.

Jukes, T. H. 1981. Creationists persist. *Nature* 291:186.

Kawata, M. 1987. Units and passages: a view for evolutionary biology and ecology. *Biology and Philosophy* 2:415–34.

Kelly, A. 1981. *The descent of Darwin: the popularization of Darwinism in Germany, 1860–1914.* Chapel Hill: University of North Carolina Press.

Kevles, D. 1980. Genetics in the United States and Great Britain, 1890–1930. *Isis* 71:441–45.

Kierkegaard, S. 1841. *The concept of irony.* Trans. L. M. Capel (1965). New York: Harper and Row.

Kimura, M. 1968. Evolutionary rate at the molecular level. *Nature* 217:624–26.

———. 1985. Natural selection and natural evolution. In *What Darwin began,* ed. L. R. Godfrey, 73–93. Boston: Allyn and Bacon.

Kimura, Motoo, and T. Ohta. 1971. *Theoretical aspects of population genetics.* Princeton: Princeton University Press.

King, J. L., and T. H. Jukes. 1969. Non-Darwinian evolution. *Science* 164:788–98.

Kingsland, S. E. 1985. *Modeling nature: episodes in the history of population ecology.* Chicago: University of Chicago Press.

Kingsley, C. 1863. *The water-babies.* London: Macmillan.

———. 1890. *Charles Kingsley, his letters and memories of his life.* London: Macmillan.

Kiriakoff, S. G. 1959. Phylogenetic systematics versus typology. *Systematic Zoology* 8:117–18.

———. 1962. On the neo-Adansonian school. *Systematic Zoology* 11:180–85.

———. 1963. Comment on James' letter. *Systematic Zoology* 12:93–94.

———. 1965. Some remarks on Sokal and Sneath's *Principles of numerical taxonomy. Systematic Zoology* 14:61–64.

———. 1966. Cladism and phylogeny. *Systematic Zoology* 15:91–93.

Kirwan, R. 1794. Examination of the supposed origin of stony substances. *Transactions of the Royal Irish Academy,* pp. 51–81.

Kitcher, P. 1978. Theories, theorists, and theoretical change. *Philosophical Review* 87:519–47.

———. 1982. Genes. *British Journal for the Philosophy of Science* 33:337–59.

———. 1984. Against the monism of the moment: a reply to Elliott Sober. *Philosophy of Science* 51:616–30.

Kitts, D. B. 1977. Karl Popper, verifiability, and systematic zoology. *Systematic Zoology* 26:185–94.

———. 1980. Theories and other scientific statements: a reply to Settle. *Systematic Zoology* 29:190–92.

Klein, M. 1972. *Mathematical thought from ancient to modern times*. New York: Oxford University Press.

Kluge, A. G. 1983. Preface. In *Advances in cladistics*. Vol. 2, ed. N. Platnick and V. Funk, vii–viii. New York: Columbia University Press.

———. 1985. Ontogeny and phylogenetic systematics. *Cladistics* 1:13–28.

Kluge, A. G., and J. S. Farris. 1969. Quantitative phyletics and the evolution of Anurans. *Systematic Zoology* 18:1–52.

Knight, D. M. 1976. *The nature of science*. London: André Deutsch.

Kohn, A. 1986. *False prophets*. Oxford: Basil Blackwell.

Kohn, D., ed. 1985. *The Darwinian heritage*. Princeton: Princeton University Press.

Kolata, G. 1977. Social Anthropologists learn to be scientific. *Science* 195:770.

Koshland, D. E. 1987. Fraud in science. *Science* 235:141.

Kottler, M. J. 1978. Charles Darwin: Biological species concept and theory of geographic speciation: the transmutation notebooks. *Annals of Science* 35:275–98.

———. 1979. Hugo de Vries and the rediscovery of Mendel's laws. *Annals of Science* 36:517–38.

———. 1985. Charles Darwin and Alfred Russel Wallace: two decades of debates over natural selection. In *The Darwinian heritage*, ed. D. Kohn, 367–432. Princeton: Princeton University Press.

Krantz, D. L., and L. Wiggins. 1973. Personal and impersonal channels of recruitment in the growth of theory. *Human Development* 16:133–56.

Kripke, S. A. 1972. Naming and necessity. In *Semantics and natural language*, ed. D. Davidson and G. Harman, 253–355. Dordrecht: Reidel.

Kuhn, T. S. 1962. *The structure of scientific revolutions*. Chicago: University of Chicago Press.

———. 1970a. Postscript, 1969. *The structure of scientific revolutions*. 2d ed., 1970, 174–210. Chicago: University of Chicago press.

———. 1970b. Reflections on my critics. In *Criticism and the growth of knowledge*, ed. I. Lakatos and A. Musgrave, 31–278. Cambridge: Cambridge University Press.

———. 1977. Second thoughts on paradigms. In *The structure of scientific theories*, ed. F. Suppe, 459–82, 500–517. Urbana: University of Illinois Press.

Lack, D. 1966. *Population studies in birds*. London: Oxford University Press.

Ladd, E. C., and S. M. Lipset. 1977. Survey of 4,400 faculty members at 161 colleges and universities. *The Chronicle of Higher Education*, 21 November, p. 12; 28 November, p. 2.

Laderman, R. S. 1987. Scientific fraud and prosecution. *Science* 236:1613.

Lakatos, I. 1970. Falsification and the methodology of scientific research programmes. In *Criticism and the growth of knowledge*, ed. I. Lakatos and A. Musgrave, 91–196. Cambridge: Cambridge University Press.

———. 1971. History of science and its rational reconstruction. In *Boston studies in the philosophy of science*, ed. R. C. Buck and R. S. Cohen, 8:91–136. Dordrecht: Reidel.

Lakatos, I., and A. Musgrave, eds. 1970. *Criticism and the growth of knowledge*. Cambridge: Cambridge University Press.

Lamarck, J. B. 1809. *Philosophie zoologique*. Paris: Dentu. Reprinted 1984, trans. H. Elliot, Chicago: University of Chicago Press.

Lande, R. 1985. Expected time for random genetic drift of a population between stable phenotypic states. *Proceedings of the National Academy of Science, U.S.A.* 82:7641.

Laudan. L. 1977. *Progress and its problems: towards a theory of scientific growth.* Berkeley: University of California Press.
———. 1981a. The pseudo-science of science? *Philosophy of the social Sciences* 11:173–98.
———. 1981b. *Science and hypothesis.* Dordrecht: Reidel.
———. 1982. A note on Collins' blend of relativism and empiricism. *Social Studies of Science* 12:131–32.
———. 1984. *Science and values.* Berkeley: University of California Press.
———. 1986. Some problems facing intuitionist meta-methodologies. *Synthèse* 67:115–29.
Laudan, R. 1987. *From mineralogy to geology: the foundations of the earth sciences: 1650–1830.* Chicago: University of Chicago Press.
Lavoisier, A. L. 1777. Réflexions sur le phlogiston. In *Oeuvres de Lavoisier* (1862). Vol. 2. Paris: Imprimerie Impériale.
Lawrence, G. H. M., ed. 1963. *Adanson.* Vol. 1. Pittsburgh: The Hunt Botanical Library.
———. 1964. *Adanson.* Vol. 2. Pittsburgh: The Hunt Botanical Library.
Leeds, A. 1974. Darwinian and "Darwinian" evolutionism in the study of society and cultures. In *The comparative reception of Darwinism,* ed. T. Glick, 437–76. Austin: University Of Texas Press.
Leigh, E. G. 1977. How does selection reconcile individual advantage with the good of the group? *Proceedings of the National Academy of Science* 74:4542–46.
Lennox, J. G. 1980. Aristotle on generia, species, and "the more and the less." *Journal of the History of Biology* 13:321–46.
———. 1982. Teleology, chance, and Aristotle's theory of spontaneous generation. *Journal of the History of Philosophy* 20:219–38.
———. 1985. Are Aristotelian species eternal? In *Aristotle on nature and living things: philosophical and historical studies,* ed. A. Gotthelf, 67–94. Pittsburgh: Mathesis Publications.
Leplin, J. 1986. Methodological realism and scientific rationality. *Philosophy of Science* 53:31–51.
Lerner, M. M. 1950. *Population genetics and animal improvement.* Cambridge: Cambridge University Press.
Levins, R., and R. Lewontin. 1985. *The dialectical biologist.* Cambridge: Harvard University Press.
Levinton, J. 1986. Punctuated equilibrium. *Science* 231:1490.
Lewin, R. 1980. Evolutionary theory under fire. *Science* 210:883–87.
———. 1986. Punctuated equilibrium is now old hat. *Science* 231:672–73.
Lewontin, R. 1961. Evolution and the theory of games. *Journal of Theoretical Biology* 1:382–403.
———. 1970. The units of selection. *Annual Review of Ecology and Systematics* 1:1–18.
———. 1974. *The genetic basis of evolutionary change.* New York: Columbia University Press.
———. 1977. Introduction to *Biology as a social weapon,* pp. 1–5. Minneapolis: Burgess Publishing Company.
———. 1980. Theoretical population genetics in the evolutionary synthesis. In *The evolutionary synthesis,* ed. E. Mayr and W. B. Provine, 58–68. Cambridge: Harvard University Press.
———. 1981. Credit due to Nabi. *Nature* 291:608.
———. 1982. *Human diversity.* New York: Scientific American Books.
Lightfield, T. E. 1971. Output and recognition of sociologists. *The American Sociologist* 6:128–33.
Limoges, C. 1980. A second glance at evolutionary biology in France. In *The evolutionary synthesis,* ed. E. Mayr and W. B. Provine, 322–32. Cambridge: Harvard University Press.
Lindroth, S. 1983. The two faces of Linnaeus. In *Linnaeus, the man and his work,* ed. T. Frängsmyr, 1–62. Berkeley: University of California Press.
Lloyd, L. 1988. A structural approach to defining units of selection. *Philosophy of Science.* Forthcoming.
Loeb, J. 1912. *The mechanistic conception of life.* Chicago: University of Chicago Press.

Loewenberg, B. 1932. The reaction of American scientists to Darwinism. *American Historical Review* 38:687–701.

Longino, H., and R. Doell, 1983. Body, bias and behavior: a comparative analysis of reasoning in two areas of biology. *Signs* 9:206–27.

Losee, J. 1977. Limitations of an evolutionist philosophy of science. *Studies in History of Science* 8:349–52.

Lotka, A. J. 1926. The frequency of distribution of scientific productivity. *Journal of the Washington Academy of Science* 16:317–23.

Løvtrup, S. 1973. *Epigenetics.* New York: Wiley.

———. 1977. *The phylogeny of the Vertebrata.* New York: Wiley.

———. 1978. On von Baerian and Haeckelian recapitulation. *Systematic Zoology* 27:348–52.

———. 1983. Victims of ambition: comments on the Wiley and Brooks approach to evolution. *Systematic Zoology* 32:90–96.

Lugg, A. 1980. Theory choice and resistance to change. *Philosophy of Science* 47:227–43.

Lumsden, C. J., and E. O. Wilson. 1981. *Genes, mind, and culture: the coevolutionary process.* Cambridge: Harvard University Press.

———. 1982. Précis of *Genes, Mind, and Culture. The Behavioral and Brain Sciences* 5:1–37.

Lundberg, J. G. 1972. Wagner networks and ancestors. *Systematic Zoology* 21:398–413.

———. 1973. More on primitiveness, higher level phylogenies and ontogenetic transformations. *Systematic Zoology* 22:327–29.

Lyell, C. 1830, 1832, 1833. *Principles of geology.* 3 vols. London: Murray.

———. 1873. *The geological evidences of the antiquity of man.* 4th ed. London: Murray.

Lyell, K. M., ed. 1881. *Life, letters and journals of Sir Charles Lyell, Bart.* London: Murray.

McBeth, N. 1971. *Darwin retried.* New York: Dell.

McBride, N. K. 1981. Room for faith. *Nature* 292:96.

McCann, H. G. 1978. Chemistry transformed. Norwood, N.J.: Ablex.

McCartney, P. J. 1976. Charles Lyell and G. B. Brocchi: a study in comparative historiography. *British Journal for the History of Science* 11:175–89.

McCosh, J. 1881. On causation and development. *Princeton Review* 7:367–89.

———. 1890. *Religious aspects of evolution.* New York: Scribner.

McDowall, R. M. 1978. Geneological tracks and dispersal in biogeography. *Systematic Zoology* 27:88–104.

McGinn, C. 1976. On the necessity of origin. *The Journal of Philosophy* 73:127–35.

McGinty, L. 1981. Evolutionary lines. *New Scientist* 91:817–18.

Mach, E. 1910. Die leitgedanken meiner naturwissenschaftlichen erkenntislehre unde ihre aufnahme durch die zeitgenosssen. *Physikalische Zeitschrift* 11:599–606; reprinted as, The guiding principles of my scientific theory of knowledge and its reception by my contemporaries. In *Physical reality: philosophical essays on twentieth-century physics* (1970), ed. S. Toulmin, 28–43. New York: Harper and Row.

McKenna, M. C. 1975. Toward a phylogenetic classification of the Mammalia. In *Phylogeny of the primates,* ed. W. P. Luckett and F. S. Szalay, 21–46. New York: Plenum Press.

———. 1981. More museums. *Nature* 289:626–27.

MacKenzie, D., and B. Barnes. 1979. The biometry-Mendelism controversey. In *Natural order,* ed. B. Barnes and S. Shapin, 191–210. Beverly Hills: Sage Publishers.

MacKie, E. W. 1981. Too tolerant. *Nature* 292:403.

McMullin, E. 1976. The fertility of theory and the unit for appraisal in science. In *Essays in memory of Imre Lakatos,* ed. R. S. Cohen, P. K. Feyerabend, and M. W. Wartofsky, 395–432. Dordrecht: Reidel.

McNeill, J. 1982. Phylogenetic reconstructions and phenetic taxonomy. *Zoological Journal of the Linnean Society of London* 74:337–44.

MacRoberts, M. H. 1985. Was Mendel's paper on *Pisum* neglected or unknown? *Annals of Science* 42:339–45.

MacRoberts, M. H., and B. R. MacRoberts. 1984. The negational reference: or the art of dissembling. *Social Studies of Science* 14:91–93.

Mahoney, M. J. 1976. *Scientist as subject.* Cambridge: Ballinger.

———. 1979. Psychology of the scientists: an evaluative review. *Social Studies of Science* 9:349–75.

Malthus, T. R. 1798. *An essay on the principles of population as it affects the future improvement of society with remarks on the speculation of Mr. Godwin, M. Condorcet and other writers.* London: Macmillan.

Manier, E. 1978. *The young Darwin and his cultural circle.* Dordrecht: Reidel.

Manten, A. A. 1980. Publication of scientific information is not identical with communication. *Scientometrics* 2:303–8.

Margolis, H. 1982. *Selfishness, altruism and rationality.* Cambridge: At the University Press.

Marshall, E. 1986. San Diego's tough stand on research fraud. *Science* 234:534–35.

Martin, B. 1978. The selective usefulness of game theory. *Social Studies of Science* 8:85–110.

Masters, R. D. 1982. Is sociobiology reactionary? The political implications of inclusive-fitness theory. *Quarterly Review of Biology* 57:275–92.

Matthew, P. 1860. Nature's law of selection. *The Gardeners' Chronicle and Agricultural Gazette* 14:312–13.

Maxson, L. R., and A. C. Wilson. 1975. Albumin evolution and organismal evolution of tree frogs (Hylidae). *Systematic Zoology* 24:1–15.

May, R. M., and J. Seger. 1986. Ideas in ecology. *American Scientist* 74:256–67.

Maynard Smith, J. 1961. Evolution and history. In *Darwinism and the study of society,* ed. M.P. Banton, 83–93. London: Tavistock.

———. 1972. Game theory and the evolution of fighting. In *On evolution,* ed. J. Maynard Smith, 8–28. Edinburgh University Press.

———. 1974. The theory of games and the evolution of animal conflict. *Journal of Theoretical Biology* 47:209–21.

———. 1978. *The evolution of sex.* Cambridge: Cambridge University Press.

———. 1982. *Evolution and the theory of games.* Cambridge: Cambridge University Press.

———. 1983. The economics of sex. *Evolution* 37:872–73.

———. 1986. Natural selection of culture? *New York Review of Books* 17(33):11–12.

Mayo, D. G., and N. L. Gilinsky. 1987. Models of group selection. *Philosophy of Science* 54:515–38.

Mayr, E. 1942. *Systematics and the origin of species.* Cambridge: Cambridge University Press.

———. 1954. Change of genetic environment and evolution. In *Evolution as a process,* ed. J. Huxley, A. C. Hardy, and E. B. Ford, 157–80. London: Allen and Unwin.

———. 1961. Cause and effect in biology. *Science* 134:1501–6.

———. 1963. *Animal species and evolution.* Cambridge: Harvard University Press.

———. 1965. Numerical phenetics and taxonomic theory. *Systematic Zoology* 14:73–97.

———. 1969. *Principles of systematic zoology.* New York: McGraw-Hill.

———. 1973. The recent historiography of genetics. *Journal of the History of Biology* 6:125–54.

———. 1974. Cladistic analysis or cladistic classification? *Zeitschrift für Zoologische Systematik und Evolutionsforschung* 12:94–128.

———. 1975. The unity of the genotype. *Biologisches Zentrelblatt* 94:377–88.

———. 1976. *Evolution and the diversity of life.* Cambridge: Harvard University Press.

———. 1978a. Evolution. *Scientific American* 239:46–55.

———. 1978b. Origin and history of some terms in systematics and evolutionary biology. *Systematic Zoology* 27:83–88.

———. 1980a. Prologue: some thoughts on the history of the evolutionary synthesis. In *The evolutionary synthesis,* ed. E. Mayr and W. B. Provine, 1–50. Cambridge: Harvard University Press.

———. 1980b. How I became a Darwinian. In *The evolutionary synthesis*, ed. E. Mayr and W. B. Provine, 413–23. Cambridge: Harvard University Press.

———. 1980c. Problems of the classification of birds, a progress report. *Separatum ex actis xvii congressus internationalis ornithologici*, Berlin: *Deutsche ornithologen-gesellschaft*, 95–112.

———. 1981. Biological classification: toward a synthesis of opposing methodologies. *Science* 214:510–16.

———. 1982. *The growth of biological thought*. Cambridge: Harvard University Press.

———. 1983. Comments on David Hull's paper on exemplars and type specimens. In *PSA 1982*, ed. P. D. Asquith and T. Nickles, 1:504–11. Ann Arbor: Philosophy of Science Association.

———. 1985. Darwin and the definition of phylogeny. *Systematic Zoology* 34:97–98.

———. 1987. The ontological status of species. *Biology and Philosophy* 2:145–66.

———, and Amadon, D. 1951. A classification of recent birds. *American Museum Novitates*, no. 1496:1–42.

———, E. G. Linsley, R. Usinger. 1953. *Methods and principles of systematic zoology*. New York: McGraw-Hill Publications.

———, and W. Provine, eds. 1980. *The evolutionary synthesis*. Cambridge: Harvard University Press.

Mazur, A., and L. S. Robertson. 1972. *Biology and social behavior*. New York: Macmillan.

Medawar, P. B. 1972. *The hope of progress*. London: Methuen.

———. 1977. Unnatural science. *New York Review of Books*, February 3, pp. 13–18.

———. 1982. *Pluto's republic*. Oxford: Oxford University Press.

Meese, E. 1986. Dissent from Edwin Meese. *Time* 128:8.

Meijer, O. 1985. Hugo de Vries no Mendelian? *Annals of Science* 42:189–232.

Meltzer, L. 1942. Science and technology in a democratic order. *Journal of Legal and Political Sociology* 1:115–26.

———. 1956. Scientific productivity in organizational settings. *Journal of Social Issues* 12:32–40.

Menard, H. W. 1971. *Science: Growth and change*. Cambridge: Harvard University Press.

Merton, R. K. 1973. *The sociology of science*, ed. N. W. Storer. Chicago: University of Chicago Press.

Merton, R. K., and H. Zuckerman. 1973. Age, aging, and age structures in science. In *The sociology of science*, ed. N. W. Storer, 497–559. Chicago: University of Chicago Press.

Michener, C. D. 1963. Some future developments in taxonomy. *Systematic Zoology* 12:151–72.

———. 1978. Dr. Nelson on taxonomic methods. *Systematic Zoology* 27:112–18.

Michener, C. D., and R. R. Sokal. 1957. A quantitative approach to a problem in classification. *Evolution* 11:130–62.

Michener, C. D., and P. J. Brothers. 1974. Were workers of eusocial Hymenoptera initially altruistic or oppressed? *Proceeding of the National Academy of Science, U.S.A.* 71:671–74.

Michod, R. E. 1981. Positive heuristics in evolutionary biology. *British Journal for the Philosophy of Science* 32:1–36.

———. 1982. The theory of kin selection. *Annual Review of Ecology and Systematics* 13:23–55.

———. 1986. On fitness and adaptedness and their roles in evolutionary explanations. *Journal of the History of Biology* 19:289–302.

Mickevich, M. F. 1978. Taxonomic congruence. *Systematic Zoology* 27:143–58.

———. 1980. Taxonomic congruence: Rohlf and Sokal's misunderstanding. *Systematic Zoology* 29:162–76.

Mickevich, M. F., and J. S. Farris. 1981. The implications of congruence in *Menidia*. *Systematic Zoology* 30:351–70.

Mickevich, M. F., and M. S. Johnson. 1976. Congruence between morphological and allozyme data in evolutionary inference and character evolution. *Systematic Zoology* 25:260–70.

———. 1977. Author's corrections. *Systematic Zoology* 26:252.
———. 1979. Corrections. *Systematic Zoology* 28:255.
Mill, J. S. 1843. A system of logic. London: Longmans, Green.
Mishler, B. D. 1987. Sociology of science and the future of Hennigian phylogenetic systematics. *Cladistics* 3:55–60.
Mishler, B. D., and R. N. Brandon. 1987. Individuality, pluralism, and the phylogenetic species concept. *Biology and Philosophy* 2:397–414.
Mishler, B. D., and M. J. Donoghue. 1982. Species concepts: a case for pluralism. *Systematic Zoology* 31:491–503.
Mitchell, S. D. 1987. Competing units of selection? A case of symbiosis. *Philosophy of Science* 54:351–67.
Mitroff, I. 1974. Norms and counter-norms in a select group of the Apollo moon scientists: a case study of the ambivalence of scientists. *American Sociological Review* 39:579–95.
Mitroff, I. I., T. Jacob, and E. T. Moore. 1977. On the shoulders of the spouses of scientists. *Social Studies of Science* 7:303–27.
Mivart, St. G. J. 1892. Pearson's *Grammar of science. Nature* 46:269.
Moore, J. R. 1979. *The post-Darwinian controversies.* Cambridge: Cambridge University Press.
Moravcsik, M. J., and P. Murugesan. 1975. Some results on the function and quality of citations. *Social Studies of Science* 5:86–91.
Morgan, T. H. 1919. *The physical basis of heredity.* Philadelphia: J. B. Lippincott.
Morgan, T. H., A. H. Sturtevant, H. J. Muller, and C. B. Bridges. 1915. *The mechanism of Mendelian heredity.* New York: Holt.
Morrell, J., and A. Thackray. 1981. *Gentlemen of science.* Oxford: Oxford University Press.
Morris, D. 1967. *The naked ape.* New York: McGraw-Hill.
Morse, E. 1876. Address: on the contribution of American zoologists to the Darwinian theory of evolution. *American Association for the Advancement of Science Proceedings* 25:137–76; 36 (1887):1–43.
Moss, W. W. 1983. Taxa, taxonomists, and taxonomy. In *Numerical taxonomy,* ed. J. Felsenstein, 72–75. New York: Springer-Verlag.
Muller, H. J. 1950. Our load of mutations. *Journal of Human Genetics* 2:111–76.
Myers, G. 1985. Texts as knowledge claims: the social contribution of two biology articles. *Social Studies of Science* 15:593–630.
Nabi, I. 1980. Satyrical comment on the reductionist theory of Edward O. Wilson & Co. *Science and Nature,* November 3, 71–73.
Naef, A. 1919. *Idealistische morphologie und phylogenetik.* Jena: Gustav Fischer.
Nagel, T. 1980. *The limits of objectivity.* Salt Lake City: University of Utah Press.
Nanney, D. L. 1982. On winning in science. *Science* 32:171.
Nelson, G. J. 1969. The problem of historical biogeography. *Systematic Zoology* 18:243–46.
———. 1970. Outline of a theory of comparative biology. *Systematic Zoology* 19:373–84.
———. 1971a. "Cladism" as a philosophy of classification. *Systematic Zoology* 20:373–76.
———. 1971b. Paraphyly and polyphyly: redefinition. *Systematic Zoology* 20:471–72.
———. 1972a. Phylogenetic relationship and classification. *Systematic Zoology* 21:227–30.
———. 1972b. Science or politics? A reply to H. F. Howdin. *Systematic Zoology* 21:341–42.
———. 1972c. Review of *Die rekonstruktion der phylogenese mit Hennig's prinzip* by D. Schlee (1971). *Systematic Zoology* 21:350–52.
———. 1973a. The higher-level phylogeny of vertebrates. *Systematic Zoology* 22:87–91.
———. 1973b. Monophyly again? A reply to P. D. Ashloch. *Systematic Zoology* 22:310–12.
———. 1973c. Comments on Leon Croizat's biogeography. *Systematic Zoology* 22:312–19.
———. 1973d. Negative gains and positive losses: a reply to J. C. Lundberg. *Systematic Zoology* 22:330.
———. 1973e. Classification as an expression of phylogenetic relationships. *Systematic Zoology* 22:344–59.

———. 1974a. Darwin-Hennig classification: a reply to Ernst Mayr. *Systematic Zoology* 23:452–58.

———. 1974b. Historical biogeography: an alternative formalization. *Systematic Zoology* 23:555–58.

———. 1977. Review of *Biogeografia analitica y sintética* ("*Panbiogeografia*") *de las Américas* by L. Croizat (1976). *Systematic Zoology* 26:449–52.

———. 1978a. Professor Michener on phenetics—old and new. *Systematic Zoology* 27:104–12.

———. 1978b. Ontogeny, phylogeny, paleontology, and the biogenetic law. *Systematic Zoology* 27:324–45.

———. 1979. Cladistic analysis and synthesis: principles and definitions, with a historical note on Adanson's *Familles des Plantes* (1763–1764). *Systematic Zoology* 28:1–21.

———. 1981. More museums. *Nature* 289:627.

———. 1985. Outgroups and ontogeny. *Cladistics*. 1:29–45.

———. 1987. Review of *Evolution and classification* by M. Ridley (1986). *Cladistics* 3:72.

Nelson, G., and N. Platnick. 1981. *Systematics and biogeography: cladistics and vicariance.* New York: Columbia University Press.

Nelson, G., and D. E. Rosen, eds. 1981. Preface. *Vicariance biogeography: a critique,* xi–xii. New York: Columbia University Press.

Neyman, J. 1967. R. A. Fisher (1890–1962): an appreciation. *Science* 156:1456–60.

Nickles, T. 1977. Heuristics and justification in scientific research: comments on Shapere. In *The structure of scientific theories,* ed. F. Suppe, 571–89. Urbana, Ill.: University of Illinois Press.

Niles, H. E. 1923. The method of path coefficients: an answer to Wright. *Genetics* 8:256–60.

Nisbett, R., and L. Ross. 1980. *Human inference: strategies and shortcomings of social judgment.* Englewood Cliffs: Prentice-Hall.

Nitecki, M. H., J. L. Lemke, H. W. Pullman, and M. E. Johnson. 1978. Acceptance of plate tectonic theory. *Geology* 6:661–64.

Norman, C. 1983. Broad public support found for R & D. *Science* 222:1311.

———. 1984. Reduce fraud in seven easy steps. *Science* 224:581.

———. 1987. Prosecution urged in fraud case. *Science* 236:1057.

Olby, R. C. 1966. *Origins of Mendelism.* New York: Schocken.

———. 1974. *The path to the double helix.* New York: Macmillan.

———. 1979. Mendel no Mendelian? *History of Science* 17:53–72.

———. 1985. *Origins of Mendelism.* 2d ed. Chicago: University of Chicago Press.

Oldroyd, D. R. 1984. How did Darwin arrive at his theory? The secondary literature to 1982. *History of Science* 22:325–74.

Olson, E. 1981. Macroevolution conference. *Science* 211:773–74.

Orel, V. 1971. A reconstruction of Mendel's experiments and an attempt at an explanation of Mendel's way of presentation. *Folia Mendeliana* 6:45.

Orgel, L. E., and F. H. C. Crick. 1980. Selfish DNA: the ultimate parasite. *Nature* 284:604–7.

Ortega y Gasset, J. 1932. The revolt of the masses. New York: Norton.

Outram, D. 1984. *Georges Cuvier: vocation, science and authority in post-revolutionary France.* Manchester: Manchester University Press.

Owen, R. 1860. Palaeontology or a systematic summary of extinct animals and their geological relations. Edinburgh: Adam and Charles Block.

Pancaldi, G. 1983. *Darwin in Italia: Impresa scientifica e frontiere culturali.* Bologna: Il Mulino.

Panchen, A. L. 1982. The use of parsimony in testing phylogenetic hypotheses. *Zoological Journal of the Linnaen Society* 74:305–28.

Paradis, J. G. 1978. *T. H. Huxley: man's place in nature.* Lincoln: University of Nebraska Press.

Parssinen, T. M. 1974. Popular science and society: the phrenology movement in early Victorian Britain. *Journal of Social History* 8:1–20.

Passmore, J. 1983. Why philosophy of science? In *Science under scrutiny: the place of history and philosophy of science,* ed. R. W. Home, 5–29. Dordrecht: Reidel.
Pastore, N. 1949. *The nature-nurture controversy.* New York: Columbia University Press.
Patrusky, B. 1986–87. Mass extinction: the biological side. *Mosaic* 17:2–13.
Patterson, C. 1973. Interrelationships of holosteans. In *Interrelationships of fishes,* ed. H. Greenwood, R. S. Miles, and C. Patterson, 233–305. London: Academic Press.
———. 1980. Museum pieces. *Nature* 288:430.
———. 1981. Darwin's survival. *Nature* 290:82.
———. 1982. Classes and cladists or individuals and evolution. *Systematic Zoology* 31:284–86.
Patterson, C., and D. Rosen. 1977. Review of Ichthyodectiform and other mesozoic teleost fishes and the theory and practice of classifying fossils. *Bulletin of the American Museum of Natural History* 158:81–172.
Paul, H. W. 1974. Religion and Darwinism: varieties of Catholic reactions. In *Comparative reception of Darwinism,* ed. T. G. Glick, 403–36. Austin: University of Texas Press.
Pearson, K. 1892. *The grammar of science.* London: Adam & Charles Block.
Piternick, L. K. 1980. *Richard Goldschmidt: controversial geneticist and creative biologist.* Basal: Birkhäuser Verlag.
Planck, M. 1910. Zur Machschen theorie der physikalischen erkenntnis. *Physikalischen Zeitschrift* 11:1186–90; reprinted in *Physical reality: philosophical essays on twentieth-century physics* (1970), ed. S. Toulmin, 44–52. New York: Harper and Row.
———. 1950. *Scientific autobiography and other papers.* Trans. F. Gaynor. London: Williams & Norgate.
Platnick, N. I. 1976. Are monotypic genera possible? *Systematic Zoology* 25:198–99.
———. 1977a. Paraphyletic and polyphyletic groups. *Systematic Zoology* 26:195–200.
———. 1977b. Monotypy and the origin of higher taxa: a reply to E. O. Wiley. *Systematic Zoology* 26:355–57.
———. 1979. Philosophy and the transformation of cladistics. *Systematic Zoology* 28:537–46.
———. 1982. Defining characters and evolutionary groups. *Systematic Zoology* 31:282–84.
———. 1985. Philosophy and the transformation of cladistics revisited. *Cladistics* 1:87–94.
———. 1986. "Evolutionary cladistics" or evolutionary systematics? *Cladistics* 2:288–96.
Platnick, N. I., and V. A. Funk. 1983. *Advances in cladistics.* Vol. 2. New York: Columbia University Press.
Platnick, N. I., and E. S. Gaffney. 1977. Systematics: a Popperian perspective. *Systematic Zoology* 26:361–65.
———. 1978. Evolutionary biology: a Popperian perspective. *Systematic Zoology* 27:132–41.
Playfair, J. 1802. *Illustrations of the Huttonian theory of the earth.* Edinburgh: Cadell & Davies.
Plotkin, H. C. 1987. Evolutionary epistemology. *Biology and Philosophy* 2:295—313.
Plotkin, H. C., and F. J. Odling-Smee. 1984. Evolution: its levels and its units. *The Behavioral and Brain Sciences* 7:318–20.
Popper, K. R. 1934. *Logic der Forschung.* Vienna: Julius Springer Verlag.
———. 1957. *The poverty of historicism.* London: Routledge & Kegan Paul.
———. 1959. *The logic of scientific discovery.* New York: Basic Books.
———. 1962. *Conjectures and refutations.* New York: Basic Books.
———. 1972. *Objective knowledge: an evolutionary approach.* Oxford: Clarendon Press.
———. 1974. Darwinism as a metaphysical research programme. In *The philosophy of Karl Popper,* ed. P. A. Schilpp, 133–43. LaSalle: Open Court.
———. 1975. The rationality of scientific revolutions. In *Problems of scientific revolution,* ed. R. Harré, 72–101. Oxford: Clarendon Press.
———. 1976. *Unended quest: an intellectual autobiography.* LaSalle: Open Court.
———. 1978. Natural selection and the emergence of mind. *Dialectica* 32:339–55.
———. 1980. Evolution. *New Scientist* 87:611.

Poulton, E. B. 1908. *Essays in evolution.* Oxford: Clarendon Press.
Pratt, V. 1972. Numerical taxonomy. *Journal of Theoretical Biology* 36:581–92.
Price, D. de Solla. 1963. *Little science, big science.* New York: Columbia University Press.
———. 1965. Is technology historically independent of science? A study in statistical historiography. *Technology and Culture* 1:553–68.
Provine, W. B. 1971. *The origins of theoretical population genetics.* Chicago: University of Chicago Press.
———. 1973. Geneticists and the biology of race crossing. *Science* 182:790–96.
———. 1980. Epilogue. In *The evolutionary synthesis,* ed. E. Mayr and W. B. Provine, 399–411. Cambridge: Harvard University Press.
———. 1981. Origins of *The genetics of natural populations* series. In *Dobzhansky's genetics of natural populations,* ed. R. C. Lewontin, J. A. Moore, W. B. Provine, and B. Wallace, 5–83. New York: Columbia University Press.
———. 1985a. Adaptation and mechanisms of evolution after Darwin. In *The Darwinian heritage,* ed. D. Kohn, 825–66. Princeton: Princeton University Press.
———. 1985b. The R. A. Fisher–Sewall Wright controversy and its influence upon modern evolutionary biology. In *Oxford surveys in evolutionary biology,* ed. R. Dawkins and M. Ridley, 2:197–219. Oxford: Oxford University Press.
———. 1986. *Sewall Wright and evolutionary biology.* Chicago: University of Chicago Press.
Pusey, J. R. 1984. *China and Charles Darwin.* Cambridge: Harvard University Press.
Putnam, H. 1973. Meaning and reference. *Journal of Philosophy* 7:699–711.
———. 1975. The meaning of "meaning." *Minnesota Studies in the Philosophy of Science* 7:131–93.
———. 1982. Why reason can't be naturalized. *Synthesis* 52:3–23.
Quinn, P. L. 1984. The philosopher of science as expert witness. In *Science and reality: recent works in the philosophy of science,* ed. J. Cushing et al., 32–53. Notre Dame: University of Notre Dame Press.
Quine, W. v. O. 1951. Two dogmas of empiricism. *Philosophical Review* 60:20–43.
Quinton, A. 1982. *Thoughts and thinkers.* London: Duckworth.
Rachootin, S. P., and K. S. Thomson. 1981. Epigenetics, paleontology, and evolution. In *Evolution today,* ed. G. G. E. Scudder and J. L. Reveal, 181–94. Pittsburgh: Hurst Institute.
Rainger, R. 1985. Paleontology and philosophy: a critique. *Journal of the History of Biology* 18:267–88.
Raup, D. M. 1985. Magnetic reversals and mass extinctions. *Nature* 314:341–42.
———. 1986. *The nemesis affair.* New York: W. W. Norton.
Raup, D., S. J. Gould, T. J. M. Schopf, and D. S. Simberloff. 1973. Stochastic models of phylogeny and the evolution of diversity. *Journal of Geology* 81:525–42.
Raup, D. M., and J. J. Sepkoski, Jr. 1986. Periodic extinctions of families and genera. *Science* 231:833–36.
Rehbock, P. F. 1983. *The philosophical naturalists: theories in early nineteenth-century British biology.* Madison: University of Wisconsin Press.
Reif, W. -E. 1986. Evolutionary theory in German paleontology. In *Dimensions of Darwinism,* ed. M. Grene, 173–204. Cambridge: Cambridge University Press.
Reingold, N. 1980. Through paradigm-land to a normal history of science. *Social Studies of Science* 10:475–96.
Rensberger, B. 1975. The politics in a debate over sociobiology. *New York Times,* Nov. 9, vol. 125, sec. 4, p. 16.
Rensch, B. 1971. *Biophilosophy.* Trans. C. A. M. Sym. New York: Columbia University Press.
Rentz, D. C. F. 1985. *The Tettigoniinae.* Australian Commonwealth Scientific and Industrial Organization.
Rescher, N. 1977. *Methodological pragmatism.* Oxford: Blackwell.
Reskin, B. F. 1977. Scientific productivity and the reward structure of science. *American Sociological Review* 42:491–504.

Richards, E. 1987. A question of property rights: Richard Owen's evolutionism reassessed. *British journal for the history of science* 20:129–71.

Richards, R. J. 1981. The natural selection model and other models in the historiography of science. In *Knowing and validation in the social sciences*, ed. M. Brewer and B. Collins, 37–76. San Francisco: Jossey-Bass.

———. 1987. *Darwin and the emergence of evolutionary theories of mind and behavior.* Chicago: University of Chicago Press.

Richerson, P. J., and R. Boyd. 1985. *Culture and the evolutionary process.* Chicago: University of Chicago Press.

Richter, M. N., Jr. 1972. *Science as a cultural process.* Cambridge: Shenkman.

Ridley, M. 1983. *The explanation of organic diversity.* Cambridge: Clarendon University Press.

———. 1986. *Evolution and classification: the reformation of cladism.* New York: Longmans.

Roberts, F. 1922. Insulin. *British Medical Journal* December 16:1193–4.

Rodman, J. E. 1982. Life, death and taxes. *Evolution* 36:1327–29.

Roe, A. 1952. *The making of a scientist.* New York: Dodd, Mead.

Rogers, D. J., and T. T. Tanimoto. 1960. A computer program for classifying plants. *Science* 132:1115–18.

Rohlf, F. J. 1963. The congruence of larval and adult classification in *Aedes*. *Systematic Zoology* 12:97–117.

———. 1970. Adaptive hierarchical clustering schemes. *Systematic Zoology* 19:58–82.

———. 1981. Comparing numerical taxonomic studies. *Systematic Zoology* 30:459–90.

Rohlf, F. J., D. H. Colless, and G. Hart. 1983. Taxonomic congruence re-examined. *Systematic Zoology* 32:144–84.

Rohlf, F. J., and R. R. Sokal. 1980. Comments on taxonomic congruence. *Systematic Zoology* 29:97–101.

———. 1981. Comparing numerical taxonomic studies. *Systematic Zoology* 30:459–90.

Roll-Hansen, N. 1978. The genotype theory of Wilhelm Johannsen and its relation to plant breeding and the study of evolution. *Centaurus* 22:201–35.

———. 1980. The controversy between biometricians and Mendelians: a test case for the sociology of scientific knowledge. *Social Science Information* 19:501–17.

———. 1983. The death of spontaneous generation and the birth of the gene: two case studies of relations. *Social Studies of Science* 13:481–520.

———. 1984. E. S. Russell and J. H. Woodger: the failure of two 20th-century opponents of mechanistic biology. *Journal of the History of Biology* 17:399–428.

———. 1985. Reply to Barnes and Collins. *Social Studies of Science* 15:179–80.

Romer, A. S. 1956. *The vertebrate body.* Philadelphia: W. B. Saunders.

Root-Bernstein, R. S. 1983. Mendel and methodology. *History of Science* 21:275–95.

Rorty, R. 1980. Pragmatism, relativism, and irrationalism. *Proceedings and Addresses of the American Philosophical Association* 53:719–38.

Rosa, D. 1918. *Ologenesi nuova teoria dell' evoluzione e della distribtuzione dei Viventi.* Florence-Palermo: Bemporad.

Rose, S. 1981. Genes and race. *Nature* 289:335.

Rosen, B. R. 1982. Review of Nelson and Rosen (1981). *Paleontological Association Circular* 107:11.

Rosen, D. E. 1974. Cladism or gradism? A reply to Ernst Mayr. *Systematic Zoology* 23:446–51.

———. 1979. Fishes from the uplands and intermontane basins of Guatemala: revisionary studies and comparative biogeography. *Bulletin of the American Museum of Natural History* 162:262–376.

———. 1981. Introduction. In *Vicariance biogeography: a critique*, ed. G. Nelson and D. E. Rosen, 1–5. New York: Columbia University Press.

———. 1982. Do current theories of evolution satisfy the basic requirements of explanation? *Systematic Zoology* 31:76–84.

———, and P. G. Buth. 1980. Empirical evolutionary research versus neo-Darwinian speculation. *Systematic Zoology* 29:300–308.

Rosenberg, A. 1982. Are there culturgens? *The Behavioral and Brain Sciences* 5:22–24.

———. 1985. *The structure of biological science.* Cambridge: Cambridge University Press.

Rosenzweig, R. M. 1985. Research as intellectual property: influences within the university. *Science, Technology, & Human Values* 10:41–48.

Ross, H. H. 1964. Review of Sokal and Sneath (1963). *Systematic Zoology* 13:106–8.

Rottenberg, D. 1980. Tracking down medical misfits. *Parade,* May 25, p. 6.

Rudwick, M. J. S. 1972. *The meaning of fossils.* New York: Science History Publications.

———. 1982. Charles Darwin in London: the integration of public and private science. *Isis* 73:186–206.

———. 1985. *The great Devonian controversy: the making of scientific knowledge among gentlemanly specialists.* Chicago: University of Chicago Press.

Ruestrow, E. G. 1984. Leewenhoek and the campaign against spontaneous generation. *Journal of the History of Biology* 17:225–48.

Ruse, M. 1973. *Philosophy of science.* London: Hutchinson University Library.

———. 1975. Woodger on genetics. *Acta Biotheoretica* 24:1–13.

———. 1977. Karl Popper's philosophy of biology. *Philosophy of Science* 44:638–61.

———. 1979a. *The Darwinian revolution.* Chicago: University of Chicago Press.

———. 1979b. *Sociobiology: sense or nonsense?* Dordrecht: Reidel.

———. 1979c. Falsifiability, consilience, and systematics. *Systematic Zoology* 28:530–36.

———. 1982. Creation science is not science. *Science, Technology, & Human Values* 7:72–78.

Rynasiewicz, R. 1986. The universality of laws in space and time. In *PSA 1986,* ed. A. Fine and P. Machamer, 1:66–75. East Lansing: Philosophy of Science Association.

Sabine, E. 1864. President's address. *Proceedings of the Royal Society of London* 13:505–10.

Sahlins, M. 1976. *The use and abuse of biology.* Ann Arbor: University of Michigan Press.

Sahlins, M. D., and E. R. Service. 1960. *Evolution and culture.* Ann Arbor: University of Michigan Press.

Salthe, S. 1985. *Evolving hierarchical systems: their structure and representation.* New York: Columbia University Press.

Samuelson, P. 1966. The general theory. In *The collected scientific papers of Paul Samuelson,* ed. J. E. Stiglitz, 2:1517–33. Cambridge: MIT Press.

———. 1975. Social Darwinism. *Newsweek* July 7, 1975, p. 55.

Sandler, I. 1979. Some reflections on the protean nature of the scientific precursor. *History of Science* 17:170–90.

Schaefer, H. F., III. 1986. Methylene: a paradigm for computational quantum chemistry. *Science* 231:1100–7.

Schaeffer, B., M. K. Hecht, and N. Eldredge. 1972. Paleontology and phylogeny. *Evolutionary Biology* 6:31–46.

Schafersman, S. D. 1985. Anatomy of a controversy: Halstead vs. the British Museum (Natural History). In *What Darwin began,* ed. L. R. Godfrey, 186–220. Boston: Allyn and Bacon.

Scheffler, I. 1967. *Science and subjectivity.* New York: Bobs-Merrill.

Schindewolf, O. H. 1936. *Paläontologie, entwicklungslehre und genetik.* Berlin: Borntraeger.

Schlee, D. 1969. The interpretation of Hennig's systematics, an "intuitive, statistico-phenetic taxonomy"? *Systematic Zoology* 18:127–34.

Schleicher, A. 1863. *Die Darwinsche theorie und die sprachwissenschaft.* Weimer: Bölan.

Schmaus, W. 1981. Fraud and sloppiness in science. *Perspectives on the professions* 1:1–14.

———. 1983. Fraud and the norms of science. *Science, Technology, & Human Values* 8:12–22.

Schmid, R. 1986. Leon Croizat's standing among biologists. *Cladistics* 2:105–11.

Schopf, T. J. M. 1973. Ergonomics of polymorphism: its relation to the colony as the unit of natural selection in species of the phylum Ectoprocta. In *Animal colonies,* ed. R. S. Boardman, A. H. Cheetham, and W. A. Oliver, Jr., 247–94. Pennsylvania: Dowden, Hutchinson and Ross.

Schopf, J. W. 1978. The evolution of the earliest cells. *Scientific American* 239:110–37.

Schopf, J. W., B. N. Haugh, R. E. Molnar, and D. E. Satterthwait. 1973. On the development of metaphyta and metazoons. *Journal of Paleontology* 47:1–9.

Schrödinger, E. 1947. *What is life? The physical aspect of the living cell.* New York: Doubleday Anchor Book.

Schuh, R. T. 1981. William Hennig society: report of first annual meeting. *Systematic Zoology* 30:76–81.

Schuh, R. T., and J. S. Farris. 1981. Methods for investigating taxonomic congruence and their application. *Systematic Zoology* 33:331–50.

Schuh, R. T., and J. T. Polhemus. 1980. *Systematic Zoology* 29:1–26.

Segerstrale, U. 1986. Colleagues in conflict: an "in vivo" analysis of the sociobiology controversy. *Biology and Philosophy* 1:53–88.

Settle, T. 1979. Popper on "When is a science not a science?" *Systematic Zoology* 28:521–29.

Shapere, D. 1974. Scientific theories and their domains. In *The structure of scientific theories,* ed. F. Suppe, 518–65. Urbana: University of Illinois Press.

Shapin, S. 1975. Phrenological knowledge and the social structure of early 19th-century Edinburgh. *Annals of Science* 32:219–43.

———. 1979a. *Homo Phrenologicus:* anthropological perspective on an historical problem. In *Natural order,* ed. B. Barnes and S. Shapin, 41–72. Beverly Hills: Sage Publications.

———. 1979b. The politics of observation: cerebral anatomy and social interests in the Edinburgh phrenology disputes. *The Sociological Review Monographs* 27:39–178.

———. 1982. History of science and its sociological reconstruction. *History of Science* 20:157–211.

Shapin, S., and B. Barnes. 1977. Science, nature and control: interpreting mechanics' institutes. *Social Studies of Science* 7:31–74.

Shapley, D. 1978. Colorado professor fired over false accounts. *Science* 200:418, 1365.

Shields, W. M. 1982. *Philopatry, inbreeding, and the evolution of sex.* Albany: State University of New York Press.

Shimao, E. 1981. Darwin in Japan, 1877–1927. *Annals of Science* 38:93–102.

Shrader, D. 1980. The evolutionary development of science. *Review of Metaphysics* 34:273–96.

Sibley, C. G. 1969. Foreword to *Systematic Biology.* Washington, D.C.: Publication 1692, National Academy of Science.

Signor, P. W., III. 1985. Real and apparent trends in species richness through time. In *Phanerozoic diversity patterns,* ed. J. Valentine, 129–50. Princeton: Princeton University Press.

Simberloff, D. 1983. Competition theory, hypothesis-testing, and other community ecological buzzwords. *American Naturalist* 122:626–38.

Simon, H. A. 1962. *The sciences of the artificial.* Cambridge: MIT Press.

———. 1983. Fitness requirements for scientific theories. *British Journal for the Philosophy of Science* 34:355–65.

Simpson, G. G. 1944. *Tempo and mode in evolution.* New York: Columbia University Press.

———. 1953. *The major features of evolution.* New York: Columbia University Press.

———. 1961. *Principles of animal taxonomy.* New York: Columbia University Press.

———. 1964. Numerical taxonomy and biological classifications. *Science* 144:712–13.

———. 1965. Current issues in taxonomic theory. *Science* 148:1078.

———. 1978a. *Concessions to the improbable, an unconventional autobiography.* New Haven: Yale University Press.

———. 1978b. Variations and details in macroevolution. *Paleobiology* 4:217–21.

———. 1981. Exhibit dismay. *Nature* 290:286.
Simpson, G. G., and A. Roe. 1939. *Quantitative zoology.* New York: McGraw-Hill.
Simpson, G. G., A. Roe, and R. Lewontin. 1960. *Quantitative Zoology,* revised edition. New York: Harcourt, Brace.
Sloan, P. R. 1973. The idea of racial degeneracy in Buffon's *Histoire Naturelle. Studies in Eighteenth-Century Culture* 3:293–321.
———. 1987. From logical universals to historical individuals: Buffon's idea of biological species. In *Histoire du concept d'espèce dans les science de la vie,* ed. J. Roger and Jean-Louis Fischer, 101–40. Paris: éditions de la fondation Singer-Polignai.
Smart, J. J. C. 1963. *Philosophy and scientific realism.* London: Routledge & Kegan Paul.
———. 1968. *Between science and philosophy.* New York: Random House.
Smart, R. 1964. The importance of negative results in psychological research. *Canadian Psychologist* 5:225–32.
Smith, C. A. B. 1984. Fisher from inside. *Nature* 307:299.
Smith, R. J. 1980. Drug-making topples eminent anthropologist. *Science* 210:296–300,993.
Sneath, P. H. A. 1957a. Some thoughts on bacterial classification. *Journal of General Microbiology* 17:184–200.
———. 1957b. The application of computers to taxonomy. *Journal of General Microbiology* 17:201–26.
———. 1968. International conferences on numerical taxonomy, 1967. *Systematic Zoology* 17:88–92.
———. 1983. Philosophy and method in biological classification. In *Numerical Taxonomy,* ed. J. Felsenstein, 22–37. New York: Springer-Verlag.
Sneath, P. H. A., and R. R. Sokal. 1973. *Numerical taxonomy.* San Francisco: W. H. Freeman.
Sober, E. 1980. Evolution, population thinking, essentialism. *Philosophy of Science* 47:350–83.
———. 1984a. *The nature of selection.* Cambridge: MIT Press.
———. 1984b. Discussion: sets, species, and evolution: comments on Philip Kitcher's "species." *Philosophy of Science* 51:334–41.
———. 1988. *Reconstructing the past: parsimony, evolution, inference.* Cambridge: MIT Press.
Sober, E., and R. C. Lewontin. 1982. Artifact, causes, and genic selection. *Philosophy of Science* 49:147–76.
Sokal, R. R. 1959. Comments on quantitative systematics. *Evolution* 13:420–23.
———. 1962. Typology and empiricism in taxonomy. *Journal of Theoretical Biology* 3:230–67.
———. 1967. Principles of taxonomy (review of Hennig 1966). *Science* 156:1356.
———. 1975. Mayr on cladism—and his critics. *Systematic Zoology* 24:257–62.
———. 1983a. A phylogenetic analysis of the Caminalcules I. The data base. *Systematic Zoology* 32:159–84.
———. 1983b. A phylogenetic analysis of the Caminalcules II. Estimating the true cladogram. *Systematic Zoology* 32:185–201.
———. 1983c. A phylogenetic analysis of the Caminalcules III. Fossils and classification. *Systematic Zoology* 32:248–58.
———. 1983d. A phylogenetic analysis of the Caminalcules IV. Congruence and character stability. *Systematic Zoology* 32:259–75.
———. 1985. The principles of numerical taxonomy: twenty-five years later. *Computer-assisted bacterial systematics,* ed. M. Goodfellow, 1–20. Orlando: Academic Press.
Sokal, R. R., J. H. Camin, F. J. Rohlf, and P. H. A. Sneath. 1965. Numerical taxonomy: some points of view. *Systematic Zoology* 14:237–43.
Sokal, R. R., and T. Crovello. 1970. The biological species concept: a critical evaluation. *American Naturalist* 104:127–53.
Sokal, R. R., and F. J. Rohlf. 1981. Taxonomic congruence in the Leptopodomorphia reexamined. *Systematic Zoology* 30:309–24.

Sokal, R. R., and C. D. Michener. 1958. A statistical method for evaluating systematic relationships. *University of Kansas Science Bulletin* 38:1409–38.

Sokal, R. R., and P. H. A. Sneath. 1963. *The principles of numerical taxonomy.* San Francisco: W. H. Freeman.

Solzhenitsyn, A. 1969. *Cancer ward.* New York: Bantam Books.

Spencer, H. 1852. The development hypothesis. *The Saturday Analyst and Leader* pp. 280–81.

———. 1857. Progress: its law and cause. *Westminster Review* 11:445–85.

Sprat, T. 1667. *History of the royal society.* London: Routledge & Kegan Paul.

Stafleu, F. A. 1963. Adanson and his "Familles des plantes." In *Adanson,* ed. G. H. M. Lawrence, 1:123–64. Pittsburgh: The Hunt Botanical Library.

Stanley, S. M. 1975. Clades versus clones in evolution: why we have sex. *Science* 190:382–83.

———. 1979. *Macroevolution, pattern and process.* San Francisco: W. H. Freeman.

Stebbins, R. E. 1974. France. In *The comparative reception of Darwinism.* ed. T. Glick, 117–63. Austin: University of Texas.

Stebbins, G. L., and F. Ayala. 1981. Is a new evolutionary synthesis necessary? *Science* 213:967–71.

Steele, E. J. 1981. *Somatic selection and adaptive evolution.* Chicago: University of Chicago Press.

Stent, G. S. 1969. *The coming of the golden age: a view of the end of progress.* Garden City: Natural History Press.

———. 1977. You can take the ethics out of altruism but you can't take the altruism out of ethics. *Hastings Center Report* 7:33–36.

Stern, C. 1980. Richard Benedict Goldschmidt (1878–1958): a biographical memoir. In *Richard Goldschmidt: controversial geneticist and creative biologist,* ed. L. K. Piternick, 68–99. Basel: Birkhäuser.

Stevens, P. F. 1983. Report of the third annual Willi Hennig society meeting. *Systematic Zoology* 32:285–90.

Stevenson, L. 1974. *Seven theories of human nature.* Oxford: Clarendon Press.

Stewart, J. A. 1986. Drifting continents and colliding interests: a quantitative application of the interests perspective. *Social Studies of Science* 16:261–79.

Stewart, W. W., and N. Feder. 1987. The integrity of the scientific literature. *Nature* 325:207–16.

Steyskal, G. C. 1953. Trans. of Hennig 1950. *Systematic Zoology* 2:41.

Stroud, C. P. 1953. Factor analysis in termite systematics. *Systematic Zoology* 2:76–92.

Sulloway, F. J. 1982. Darwin and his finches: the evolution of the legend. *Journal of the History of Biology* 15:1–53.

Sumner, F. 1932. Genetic distributional and evolutionary studies of the subspecies of deer-mice (Peromyscus). *Bibliotheca genetica* 9:1–106.

Suppe, F. 1977. Exemplars, theories, and disciplinary matrixes. In *The structure of scientific theories,* ed. F. Suppe, 483–99. Urbana: University of Illinois Press.

Swainson, W. 1835. *A treatise on the geography and classification of animals.* London: D. Lardner.

Tait, P. 1869. Geological time. *North British Review* 50:215–33.

Tassy, P. 1981. Le crâne de *Moeritherium* (Proboscidea, Mammalia) de l'Eocène de Dor el Talha (Lybie) et le problème de la classification phylogénétique du genre les Tethytheria McKenna, 1975. *Bulletin du Museum national d'Histoire naturelle Paris* 3:87–147.

Tattersall, I., and N. Eldredge. 1977. Fact, theory, and fantasy in human paleontology. *American Scientist* 65:204–11.

Taubes, G. 1987. *Nobel dreams: power, deceit, and the ultimate experience.* New York: Random House.

Templeton, A. R., and L. V. Giddings. 1981. Macroevolution conference. *Science* 211:770–73.

Thagard, P. R. 1978. Why astrology is a pseudoscience. *PSA 1978*, ed. P. D. Asquith and I. Hacking, 1:223–34. East Lansing: Philosophy of Science Association.

______. 1980. Against evolutionary epistemology. *PSA 1980*, ed. P. D. Asquith and R. N. Giere, 1:187–96. East Lansing: Philosophy of Science Association.

Thoday, J. M. 1966. Mendel's work as an introduction to genetics. *Advancement of Science* 23:120–34.

Thorson, T. L. 1970. *Biopolitics*. New York: Holt, Rinehart & Winston.

Thomson, K. S. 1985. Essay review: the relationship between development and evolution. In *Oxford series in evolutionary biology*, ed. R. Dawkins and M. Ridley, 2:220–33. Oxford: Oxford University Press.

Toulmin, S. 1972. *Human understanding*. Princeton: Princeton University Press.

Trivers, R. L. 1971. The evolution of reciprocal altruism. *Quarterly Review of Biology* 40:35–57.

______. 1974. Parent-offspring conflict. *American Zoologist* 14:249–64.

Tschermak-Seysenegg, E. von. 1951. Historischer rückblick auf die wiederentdeckung der Gregor Mendelschen. *Verhandlungen der Zoologisch-Botanischen Geselschaft* 92:30–1.

Turner, J. R. G. 1983. The hypothesis that explains mimetic resemblances explains evolution: the gradualist-saltationist schism. In *Dimensions of Darwinism*, ed. M. Grene, 129–69. Cambridge: Cambridge University Press.

Turrill, W. B. 1963. *Joseph Dalton Hooker*. London: Thomas Nelson.

Turton, W. 1802. Trans. of Linnaeus (1735), *A general system of nature, through the three grand kingdoms of animals, vegetables, and minerals*. London: Lackington Allen.

van Fraassen, B. C. 1980. *The scientific image*. Oxford: Clarendon Press.

Van Valen, L. 1981. Nabi—a life. *Nature* 293:422.

______. 1984. A resetting of phanerozoic community evolution. *Nature* 307:50–52.

______. 1963. On evolutionary theories. *British Journal for the Philosophy of Science*. 14:146–51.

Venn, J. 1889. *The principles of empirical or inductive logic*. London: Macmillan.

Vetta, A. 1981. Natural selection and sociobiology. *Behavioral and Brain Sciences* 4:255.

Vining, D. R. 1986. Social versus reproductive success: the central theoretical problems of human sociobiology. *The Behavioral and Brain Sciences* 9:167–216.

Voorzanger, B., and W. J. van der Steen. 1982. New perspectives on the biogenetic law? *Systematic Zoology* 31:202–5.

Vrba, E. S. 1984. Why species selection? *Systematic Zoology* 33:318–28.

Vrba, E. S., and N. Eldredge. 1984. Individuals, hierarchies, and processes: towards a more complete evolutionary theory. *Paleobiology* 10:146–71.

Waddington, C. H. 1961. The human evolutionary system. In *Darwinism and the study of society*, ed. M. Banton, 63–81. London: Tavistock Publications.

______. 1975. Mindless societies. *The New York Review of Books* 22:30–32.

Wade, M. J. 1978. A critical review of the models of group selection. *Quarterly Review of Biology* 53:101–14.

Wade, N. 1976. Sociobiology: troubled birth of a new discipline. *Science* 191:1151–55.

______. 1981. *The Nobel duel*. Garden City: Anchor Press.

Wagner, M. 1868. *Die Darwin'she theorie und das migrationsgesetz der organismen*. Leipzig: Duncker und Humblot.

Wagner, W. H. 1961. Problems in the classification of ferns. In *Recent advances in botany*, ed. D. L. Bailey, 1:841–44. Toronto: University of Toronto Press.

Wake, D. 1980. A view of evolution. *Science* 210:1239–40.

Wallace, A. R. 1855. On the law which has regulated the introduction of new species. *Annals and Magazine of Natural History* 16:184–96.

______. 1857. Letters to S. Stevens, August 21, 1856. *Zoologist* 5414–15.

______. 1864. The origin of human races and the antiquity of man deduced from the theory of natural selection. *Journal of the Anthropological Society of London* 2:clvii–clxxvii.

———. 1870. *Contributions to the theory of natural selection.* New York: Macmillan.
———. 1876. *The geographic distribution of animals.* London: Macmillan.
Wallis, C. 1986. Weeding out the incompetents. *Time* 127:57–58.
Waters, C. K. 1986. Taking analogical inference seriously: Darwin's argument from artificial selection. In *PSA 1986,* ed. A. Fine and P. Machamer, 1:502–13. East Lansing: Philosophy of Science Association.
Watson, J. D., and F. H. C. Crick. 1983. Molecular studies of nucleic acid: a structure for deoxyribose nucleic acid. *Nature* 171:737–38.
Webster, N. 1831. *The American spelling book.* Middleton: Wm. H. Niles.
Weiling, F. 1966. Hat J. G. Mendel bei seinen Versuchen "zu genau" gearbeitet? *Der Züchter* 36:359–65.
———. 1986. What about R. A. Fisher's statement of the "too good" data of J. G. Mendel's Pisum paper? *The Journal of Heredity* 77:281–83.
Weimer, D. B. 1985. The roots of "Michurinism": transformist biology and acclimatization as currents in the Russian life sciences. *Annals of Science* 42:243–60.
Weinberg, S. 1977. *The first three seconds.* New York: Basic Books.
Weinberg, W. 1908. On the demonstration of inheritance in man. *Jahresheft des Vereins für Vaterländische Naturkunde in Württemberg, Stuttgart.* 64:368–82.
Weinstein, A. 1980. Morgan and the theory of natural selection. In *The evolutionary synthesis,* ed. E. Mayr and W. B. Provine, 432–45. Cambridge: Harvard University Press.
Weismann, A. 1883. *Über die Vererbung.* Jena: Gustav Fischer.
———. 1885. *Die Kontinnität des Keimplasmas als Gründlage einer Theorie der Vererbung.* Jena: Gustav Fischer.
———. 1892. *Das keimplasma: eine theorie der vererbung.* Jena: Gustav Fischer.
Weldon, W. F. R. 1894. The study of animal variation. *Nature* 50:25–26.
Westfall, R. S. 1972. Newton and the fudge factor. *Science* 179:751–58.
Wetzel, W. 1985. Goethe and Linné. In *Contemporary Perspectives on Linnaeus,* ed. J. Weinstock, 135–51. Lanham, Md.: University Press of America.
Whewell, W. 1834. Review of *On the Connexion of the Physical Sciences* by Mary Somerville (1834). *Quarterly Review* 51:54–68.
———. 1847. *The philosophy of the inductive sciences, founded upon their history.* London: John W. Parker & Sons.
———. 1851. On the transformation of hypotheses in the history of sciences. *Transactions of the Cambridge Philosophical Society* 9:134–47.
———. 1853. *Of the plurality of worlds.* London: John W. Parker & Sons.
White, R. R. 1987. Accuracy and truth. *Science* 235:1447.
Wiley, E. O. 1975. Karl Popper, systematics, and classification. *Systematic Zoology* 24:233–43.
———. 1976. The phylogeny and biogeography of fossil and recent gars (Actinopterygii: Lepisosteidae). *Miscellaneous Publications of the Museum of Natural History, University of Kansas* 64:1–111.
———. 1977. Are monotypic genera paraphyletic? A response to Norman Platnick. *Systematic Zoology* 26:352–54.
———. 1978. The evolutionary species concept reconsidered. *Systematic Zoology* 27:17–26.
———. 1979. The annotated Linnaean hierarchy, with comments on natural taxa and competing systems. *Systematic Zoology* 28:308–37.
———. 1981a. *Phylogenetics, the theory and practice of phylogenetic systematics.* New York: John Wiley & Sons.
———. 1981b. Evolution's Waterloo. *Nature* 290:730.
Wiley, E. O., and D. R. Brooks. 1982. Victims of history—a nonequilibrium approach to evolution. *Systematic Zoology* 31:1–24.
Williams, G. C. 1966. *Adaptation and natural selection.* Princeton: Princeton University Press.
———. 1971. *Group selection.* New York: Aldine-Atherton.
———. 1975. *Sex and evolution.* Princeton: Princeton University Press.

______. 1985. A defense of reductionism in evolutionary biology. In *Oxford Surveys in Evolutionary Biology,* ed. R. Dawkins and M. Ridley, 1:1–27. Oxford: Oxford University Press.
Williams, M. B. 1970. Deducing the consequences of evolution: a mathematical model. *Journal of Theoretical Biology* 29:343–85.
Wilson, D. L. 1980. *The natural selection of populations and communities.* Menlo Park: Benjamin Cummings.
______. 1983. The group selection controversy: history and current status. *Annual Review of Ecology and Systematics* 14:159–89.
Wilson, E. O. 1971. *The insect societies.* Cambridge: Harvard University Press.
______. 1975a. *Sociobiology: the new synthesis.* Cambridge: Harvard University Press.
______. 1975b. Human decency in animals. *New York Times Magazine,* pp. 38–50.
______. 1975c. For sociobiology. *The New York Review of Books* 22:60–61.
______. 1976. Academic vigilantism and the political significance of sociobiology. *Biological Science* 26:183, 187–90.
______. 1981. Who is Nabi? *Nature* 290:623.
Wilson, R., ed. 1967. *Darwinism and the American intellectual.* Homewood, Ill.: Dorsey Press.
Wimsatt, W. 1974. Complexity and organization. In *PSA 1972,* ed. K. Schaffner and R. S. Cohen, 67–86. Dordrecht: Reidel.
______. 1980. Reductionistic research strategies and their basis in the units of selection controversy. In *Scientific discovery,* ed. T. Nickles, 2:213–59. Dordrecht: Reidel.
______. 1981. Units of selection and the structure of multi-level genomes. In *PSA 1980,* ed. P. D. Asquith and R. N. Giere, vol. 2, 122–83. East Lansing: Philosophy of Science Association.
Winsor, M. P. 1976. *Starfish, jellyfish and the order of life.* New Haven: Yale University Press.
______. 1985. Review of P. R. Rehbock, *The philosophical naturalists* (1983). *Isis* 76:252–53.
Wisdom, J. O. 1974. The nature of normal science. In *The philosophy of Karl Popper,* ed. P. A. Schilpp, Vol. 2. LaSalle: Open Court.
Wittgenstein, L. 1922. *Tractatus logico-philosophicus.* London: Routledge & Kegan Paul.
______. 1953. *Philosophical investigations.* New York: Macmillan.
Wolgart, E. 1984. The invisible paw. *The Monist* 67:230–50.
Wolin, L. 1962. Responsibility for raw data. *American Psychologist* 17:657–58.
Woodger, S. 1937. *The axiomatic method in biology.* Cambridge: Cambridge University Press.
______. 1945. On biological transformations. In *Essays on growth and form,* ed. W. E. LeGros Clark and P. B. Medawar, 95–120. Oxford: Clarendon Press.
______. 1952. From biology to mathematics. *The British Journal for the Philosophy of Science* 3:1–21.
Woolgar, S. 1981. Interests and explanation in the social study of science. *Social Studies of Science* 11:365–94.
Wright, S. 1921. Correlation and causation. *Journal of Agricultural Research* 20:557–85.
______. 1929. Fisher's theory of dominance. *American Naturalist* 63:274–79.
______. 1931a. Statistical theory of evolution. *Journal of the American Statistical Association* 26:201–8.
______. 1931b. Evolution in Mendelian populations. *Genetics* 16:97–100, 155–59.
______. 1968. *Evolution and the genetics of populations.* Chicago: University of Chicago Press.
Wynne-Edwards, V. C. 1962. *Animal dispersion in relation to social behavior.* Edinburgh: Oliver & Boyd.
______. 1963. Intergroup selection in the evolution of social systems. *Nature* 200:623–26.
Yeo, R. 1984. Robert Chambers and *Vestiges of the Natural History of Creation. Victorian Studies* 28:5–31.
Yoels, W. C. 1973. On "publishing or perishing": fact or fable? *The American Sociologist* 8:128–34.
Young, R. M. 1968. The functions of the brain, Gall to Ferris. *Isis* 59:251–68.

———. 1970. *Mind, brain and adaptation in the nineteenth century: cerebral localization and its biological context from Gall to Ferris*. Oxford: Oxford University Press.

Yule, G. U. 1902. Mendel's laws and their probable relations to intra-race heredity. *New Phytologist* 1:194–238.

Zangerl, R. 1948. The methods of comparative anatomy and its contribution to the study of evolution. *Evolution* 4:351–74.

Ziman, J. M. 1968. *Public knowledge: an essay concerning the social dimension of science*. Cambridge: Cambridge University Press.

Zimmerman, W. F. 1978. Sociobiology and human evolution. *Berkshire Review* 13–29.

Zuckerman, H. 1977. Deviant behavior and social control in science. In *Deviance and social change*, ed. E. Sagarin, 87–138. Beverly Hills: Sage Publications.

———. 1984. Norms and deviant behavior in science. *Science, Technology, & Human Values* 9:7–13.

Name Index

Subject Index

Lightning Source UK Ltd.
Milton Keynes UK
UKHW05f1806130318
319305UK00007B/532/P